Dynamic Nuclear Magnetic Resonance Spectroscopy

Contributors

R. D. Adams
F. A. L. Anet
Ragini Anet
Gerhard Binsch
F. A. Cotton
Ray Freeman
Ernest Grunwald
H. S. Gutowsky
H. D. W. Hill
R. H. Holm

L. M. Jackman
J. P. Jesson
W. G. Klemperer
E. L. Muetterties
Earle K. Ralph
L. W. Reeves
M. Saunders
S. Sternhell
L. A. Telkowski
K. Vrieze

Dynamic Nuclear Magnetic Resonance Spectroscopy

EDITED BY

Lloyd M. Jackman

Department of Chemistry
The Pennsylvania State University
University Park, Pennsylvania

F. A. Cotton

Department of Chemistry
Texas A & M University
College Station, Texas

Academic Press   *New York   San Francisco   London*   *1975*
A Subsidiary of Harcourt Brace Jovanovich, Publishers

COPYRIGHT © 1975, BY ACADEMIC PRESS, INC.
ALL RIGHTS RESERVED.
NO PART OF THIS PUBLICATION MAY BE REPRODUCED OR
TRANSMITTED IN ANY FORM OR BY ANY MEANS, ELECTRONIC
OR MECHANICAL, INCLUDING PHOTOCOPY, RECORDING, OR ANY
INFORMATION STORAGE AND RETRIEVAL SYSTEM, WITHOUT
PERMISSION IN WRITING FROM THE PUBLISHER.

ACADEMIC PRESS, INC.
111 Fifth Avenue, New York, New York 10003

United Kingdom Edition published by
ACADEMIC PRESS, INC. (LONDON) LTD.
24/28 Oval Road, London NW1

Library of Congress Cataloging in Publication Data

Jackman, Lloyd Miles
    Dynamic nuclear magnetic resonance spectroscopy.

    Includes bibliographies.
    1.   Nuclear magnetic resonance spectroscopy.
I. Cotton, Frank Albert, Date.   joint author.
II. Title.
QC762.J32   538'.3   74-1643
ISBN 0-12-378850-1

PRINTED IN THE UNITED STATES OF AMERICA

Contents

List of Contributors	xi
Preface	xiii

1 Time-Dependent Magnetic Perturbations
H. S. Gutowsky

I. Background	1
II. Dipole–Dipole Interactions	3
III. Chemical Exchange	8
IV. Coherent Time-Dependent Fields	15
V. Accomplishments and Prospects	16
References	20

2 Delineation of Nuclear Exchange Processes
W. G. Klemperer

I. Introduction	23
II. Dynamic Stereochemistry of Symmetric Molecules	24
III. NMR Differentiable Reactions	30
IV. Examples	35
References	44

3 Band-Shape Analysis
Gerhard Binsch

I. Introduction	45
II. Classical Theory	47

	III. Quantum-Mechanical Theory	53
	IV. Practical Band-Shape Analysis	70
	V. Treatment of Dynamic Data	75
	VI. Conclusion	78
	References	78

4 Application of Nonselective Pulsed NMR Experiments— Diffusion and Chemical Exchange
L. W. Reeves

	I. Introduction	83
	II. Diffusion and Flow	84
	III. Chemical Exchange Rates from Spin-Echo Studies	105
	IV. Chemical Exchange Rates from Rotating Frame Relaxation	125
	References	128

5 Determination of Spin–Spin Relaxation Times in High-Resolution NMR
Ray Freeman and H. D. W. Hill

	I. Introduction	131
	II. Definitions	132
	III. Experimental Methods	138
	IV. Concluding Remarks	159
	References	161

6 Rotation about Single Bonds in Organic Molecules
S. Sternhell

	I. Introduction	163
	II. Rotation about sp^3–sp^3 Carbon–Carbon Single Bonds	164
	III. Rotation about sp^3–sp^2 Carbon–Carbon Single Bonds	170
	IV. Rotation about sp^2–sp^2 Carbon–Carbon Single Bonds	178
	V. Rotation about Carbon–Nitrogen Single Bonds	185
	VI. Rotation about Miscellaneous Single Bonds	191
	References	196

7 Rotation about Partial Double Bonds in Organic Molecules
L. M. Jackman

	I. Introduction	203
	II. Amides and Related Systems (R^1R^2N—COX)	204

Contents   vii

    III. Amidines, Thionamides, Selenamides, and Related Compounds [R—C(=X)—N<; X = N—R′, S, Sc]   215
    IV. Enamines   218
    V. Enolates, Enol Ethers and Thio Enol Ethers, Acylides, Diazoketones, and Aminoboranes   233
    VI. Potentially Aromatic Systems   237
    VII. Cumulenes   243
    VIII. Inversion versus Rotation—Imines and Related Systems   244
    References   250

8 Dynamic Molecular Processes in Inorganic and Organometallic Compounds
J. P. Jesson and E. L. Muetterties

    I. Introduction   253
    II. Recent Advances in Instrumentation and Technique   258
    III. Applications   261
    IV. Conclusion   313
    References   313

9 Stereochemically Nonrigid Metal Chelate Complexes
R. H. Holm

    List of Abbreviations   317
    I. Introduction   318
    II. Four-Coordinate Systems   320
    III. Six-Coordinate Systems   330
    IV. Systems of Higher Coordination Number   369
    V. Summary   372
    References   374

10 Stereochemical Nonrigidity in Organometallic Compounds
F. A. Cotton

    Introduction   377
    I. Cyclopolyenylmetal Systems   378
    II. Cyclopolyenemetal Systems   403
    III. Systems with Two Independent Metal Atoms on One Ring   417
    IV. Scrambling of Differently Bonded Rings   419
    V. Fluxionality in the Solid State   425
    VI. Miscellaneous Systems   427
    VII. Some Nonfluxional Molecules, or the Dog in the Night-Time   435
    References   437

11 Fluxional Allyl Complexes
K. Vrieze

I. Introduction	441
II. Bonding and Movements of the Allyl Group in Metal–Allyl Compounds	442
III. Survey of Reactions Occurring for Metal–π-Allyl Compounds	457
IV. Fluxional Metal–σ-Allyl and Ionic Metal–Allyl Compounds	480
V. Concluding Remarks	483
References	483

12 Stereochemical Nonrigidity in Metal Carbonyl Compounds
R. D. Adams and F. A. Cotton

I. Introduction	489
II. Permutational and Geometrical Isomerizations	491
III. Bridge-Terminal Exchanges and Scrambling of CO Groups	500
Note Added in Proof	519
References	520

13 Dynamic NMR Studies of Carbonium Ion Rearrangements
L. A. Telkowski and M. Saunders

I. Introduction	523
II. Generation of Stable Carbonium Ion Solutions	524
III. Types of Carbonium Ion Rearrangements Studied	526
References	539

14 Conformational Processes in Rings
F. A. L. Anet and Ragini Anet

I. Introduction	543
II. Definitions of Conformational Processes	544
III. Conformational Energy Calculations	554
IV. Experimental Data and General Discussion	574
References	613

15 Proton Transfer Processes
Ernest Grunwald and Earle K. Ralph

I. Introduction	621
II. Rate of Proton Exchange	622

III.	Kinetic Analysis	624
IV.	Comparison with Relaxation Methods	630
V.	Representative Results and Conclusions	632
VI.	Concluding Remarks	646
	References	646

Subject Index   649

List of Contributors

Numbers in parentheses indicate the pages on which the authors' contributions begin.

R. D. Adams* (489), Department of Chemistry, Texas A & M University, College Station, Texas

F. A. L. Anet (543), Department of Chemistry, University of California, Los Angeles, California

Ragini Anet (543), Department of Chemistry, University of California, Los Angeles, California

Gerhard Binsch (45), Institute of Organic Chemistry, University of Munich, Munich, Germany

F. A. Cotton (377, 489), Department of Chemistry, Texas A & M University, College Station, Texas

Ray Freeman† (131), Varian Associates Instrument Division, Palo Alto, California

Ernest Grunwald (621), Department of Chemistry, Brandeis University, Waltham, Massachusetts

H. S. Gutowsky (1), Department of Chemistry, University of Illinois, Urbana, Illinois

* *Present address:* Department of Chemistry, State University of New York at Buffalo, Buffalo, New York.

† *Present address:* Physical Chemistry Laboratory, South Parks Road, Oxford, England.

H. D. W. Hill (131), Varian Associates Instrument Division, Palo Alto, California

R. H. Holm (317), Department of Chemistry, Massachusetts Institute of Technology, Cambridge, Massachusetts

L. M. Jackman (203), Department of Chemistry, The Pennsylvania State University, University Park, Pennsylvania

J. P. Jesson (253), Central Research Department, E. I. Du Pont de Nemours and Company, Experimental Station, Wilmington, Delaware

W. G. Klemperer (23), Department of Chemistry, Columbia University, New York, New York

E. L. Muetterties (253), Department of Chemistry, Cornell University, Ithaca, New York

Earle K. Ralph (621), Department of Chemistry, Memorial University of Newfoundland, St. John's, Newfoundland

L. W. Reeves (83), Department of Chemistry, University of Waterloo, Waterloo, Ontario, Canada

M. Saunders (523), Department of Chemistry, Yale University, New Haven, Connecticut

S. Sternhell (163), Department of Organic Chemistry, University of Sydney, Sydney, New South Wales, Australia

L. A. Telkowski* (523), Department of Chemistry, Yale University, New Haven, Connecticut

K. Vrieze (441), Anorganisch Chemisch Laboratorium, The University of Amsterdam, Amsterdam, The Netherlands

---

* *Present address:* School of Chemistry, University of New South Wales, Kensington, New South Wales, Australia.

Preface

Basic research is an activity which consists of asking questions of Nature and of finding ways to answer them. The two phases are closely related since there is no point in asking questions, however deep or pertinent, unless there is a way of answering them. Many times the thrust and direction of fundamental research in chemistry have been profoundly altered by the advent of a new experimental technique which brought a new category of questions into the realm of the answerable. This book documents such an occurrence.

Over approximately the past eight years, both the experimental capability and the methods of data analysis and interpretation required to observe by nmr a wide variety of nuclear exchange processes, at temperatures from $-170°$ to well above room temperature, have become effective, practical tools. Chemists can now explore all manner of phenomena involving internal rotations, fast conformational flips, fluxionality, polytopal rearrangements, "ring whizzing," proton transfers, etc., which occur on the "nmr time scale," typically 10^3 sec^{-1}, at some temperature within the experimentally accessible range. Because such phenomena *can* be investigated, they have been. Many extremely subtle and important details of molecular dynamics are now being revealed where previously it would have been idle to enquire. A large body of results has already accumulated.

Each of the editors independently concluded that the subject of dynamic nuclear magnetic resonance (dnmr) spectroscopy, as we propose to call it, had reached the point where a comprehensive progress report would be useful. We decided to pool our efforts in order to bring together as many aspects of the subject as possible. It was also immediately apparent that broad and authoritative coverage could only be achieved by calling upon many of the leading workers to join us in the effort.

The early chapters describe the theoretical basis and practical techniques which have or will be used for extracting kinetic data from dnmr spectra. The subsequent chapters provide reviews of the many areas in which dnmr spectroscopy has been applied. We hope that this volume will provide a literature guide, source book, and progress report which will be helpful to all those who will continue or will begin work in this field.

We are particularly pleased that Professor H. S. Gutowsky, who has played a pioneering part in virtually every chemically useful application of nmr spectroscopy, including the interpretation of line shapes for systems undergoing site exchange, agreed to write Chapter 1.

Lloyd M. Jackman
F. A. Cotton

Dynamic Nuclear Magnetic Resonance Spectroscopy

1

Time-Dependent Magnetic Perturbations

H. S. GUTOWSKY

I.	Background..	1
II.	Dipole–Dipole Interactions.................................	3
III.	Chemical Exchange...	8
	A. Discovery of Chemical Exchange Effects...............	8
	B. Bloch Equations Modified for Chemical Exchange......	9
	C. Early Applications......................................	11
IV.	Coherent Time-Dependent Fields............................	15
V.	Accomplishments and Prospects............................	16
	References..	20

I. Background

The various chapters comprising this work describe a multiplicity of methods applied to a truly remarkable number and variety of systems, ranging from the exotic to the common garden variety. The breadth of subject is far too wide and the variety of detail much too great for me to be sanguine about developing a useful preview, synopsis, or initial evaluation of it. Instead, as one who had the good fortune to be in the right place at the right time and who therefore entered nmr research 25 years ago, I will seek to place the present work in some sort of historical and general perspective.

The possibility that magnetic resonance might be used to measure chemical exchange rates was suggested explicitly in 1953 (Gutowsky and Saika, 1953), and it was reduced to practice in 1956 (Arnold, 1956; Gutowsky and Holm, 1956). Twenty years later, over 500 papers are being published annually that deal in some way or another with the study of chemical exchange by means of magnetic resonance. In the intervening period, such studies have given us a vividly detailed picture of the conformational flexibility displayed by molecules as simple as SF_4 and PF_5 and by others as complex as cumulenes and organometallic complexes of cyclooctatetraene. Especially significant is the sensitivity of the magnetic resonance phenomena to exchange processes in which reactants and products are indistinguishable by other means. Also, the rates readily measurable by magnetic resonance techniques extend from about 1 sec^{-1} to 10^6 sec^{-1}, a substantial and important range that is too fast for more conventional approaches. These two aspects of magnetic resonance are the *raison d'être* of this book, and they are the basis of its impressive diversity and length.

In retrospect, the present activity in this field may be less remarkable in some ways than the extended period required for its development. Usually, scientific discovery looks easier after the fact than before. But even if one corrects for this with the hindsight coefficient V_{20-20}, it can be argued that the dynamic aspects of magnetic resonance were slow to be appreciated. Thirty years ago, time-dependent perturbation theory certainly was well understood. However, it was used primarily to calculate transition probabilities for the interactions of atoms and molecules with electromagnetic radiation. The importance in magnetic resonance of time dependent magnetic interactions arising from lattice motions in the broad sense really was not anticipated—neither the spin–lattice relaxation nor the line–shape phenomena.

This lack and poor fortune seem to be the most likely reasons that the first attempt to observe magnetic resonance in a bulk sample was unsuccessful. Gorter and Broer (1942) used a sample of LiF, which we now know to have a rigid lattice and therefore a broad, dipolar-broadened resonance with a long spin–lattice relaxation time T_1. Both factors can greatly decrease the peak signal-to-noise (S/N) in a resonance experiment, and undoubtedly both contributed to the failure of the attempt. On the other hand, in the independent first successful experiments a few years later, Bloch, Hansen, and Packard (1946) used a water sample doped with a paramagnetic salt, and Purcell, Torrey, and Pound (1946) used a sample of paraffin, both experiments with the sample at room temperature. Bloch's intent was to enhance the proton spin–lattice relaxation by the thousandfold ratio of electronic to nuclear *g*-values; Purcell's was to have a high concentration of protons. In both cases, the dipolar broadening of the proton resonances was unwittingly narrowed by molecular rotation and self-diffusion, which also made T_1 relatively short; and this is probably why their attempts were successful.

Another factor that may have been important in these three early experiments is the instrumental broadening of the resonance absorption by inhomogeneities in H_0, the static magnetic field, which also serve to degrade the peak S/N. Certainly such broadening delayed for several years the discovery in liquids of the chemical shift (Knight, 1949; Proctor and Yu, 1950; Dickinson, 1950) and the indirect spin–spin splitting (Gutowsky and Hoffman, 1950; Hahn, 1950), the sources of the local magnetic fields ordinarily involved in chemical exchange studies.

II. Dipole–Dipole Interactions

In any case, when nuclear magnetic resonance was first observed for protons in bulk samples in 1946, the instrumental broadening of the resonances was several kHz, which is one to two orders of magnitude smaller than the dipolar broadening in rigid lattice solids and larger than the chemical shifts and scalar coupling constants by a similar proportion. Therefore, attention was focussed initially upon the local magnetic fields produced by neighboring nuclear (or electronic) magnetic moments and upon the consequences of the various types of lattice motions, such as vibrations and molecular or ionic reorientation and translation, which can cause the local dipolar fields to be time dependent. The first analysis of this problem was the classic work of BPP (Bloembergen et al., 1948), which dealt primarily with liquids but takes a rigid-lattice solid as the point of departure.

The BPP paper provided a basic treatment of two very important aspects of such time dependent magnetic interactions, namely their contribution to the spin–lattice relaxation and the circumstances under which time averaging of local magnetic fields becomes significant. In its simplest form, the problem consists of a pair of like nuclear spins, with magnetogyric ratio γ and spin $I = \frac{1}{2}$, separated by a fixed distance r, reorienting randomly at a rate described by a single correlation time τ_c. The Hamiltonian of this system is

$$\mathcal{H} = \gamma\hbar(\mathbf{I}_1 + \mathbf{I}_2) \cdot \mathbf{H}_0 + \gamma\hbar[\mathbf{I}_1 \cdot \mathbf{I}_2(1/r^3) - 3\mathbf{I}_1 \cdot \mathbf{r}\mathbf{I}_2 \cdot \mathbf{r}] \qquad (1)$$

the second term of which is the dipole–dipole perturbation of the nuclear Zeeman energy. For the rigid lattice, i.e., when $\tau_c \to \infty$, the dipolar interaction may be described semiclassically (Pake, 1948) as a local magnetic field, which adds to H_0, taken to be in the z-direction. The magnitude of this local field is given by

$$H_{lz} = \pm \tfrac{3}{2}(\mu/r^3)(1 - 3\cos^2\theta) \qquad (2)$$

where μ is the magnetic moment γ/I, and θ is the angle between the internuclear vector $\mathbf{r}$ and the static field $\mathbf{H}_0$. The $\pm$ sign describes the spin state of the nucleus producing H_{lz} at its partner. Therefore, the magnetic resonance of a single crystal of isolated pairs of nuclei, all with the same orientation θ with respect to H_0, would be a doublet with a splitting of $2H_{lz}$, as given by Eq. (2).

In principle, the easiest way in which to make the dipolar perturbation be time dependent is to have each nuclear pair reorient at random about a single axis, which makes the same angle with all $\mathbf{r}$'s and with $\mathbf{H}_0$. However, at the time of the BPP work, such cases of limited molecular or group reorientational mobility had yet to be observed by magnetic resonance methods (Gutowsky and Pake, 1950). Perhaps because of this, as well as because of their emphasis upon liquids, BPP chose to consider the more complex case of completely isotropic reorientations, in which θ itself is a random function of time. For this case, Fourier analysis of the complete time-dependent dipole–dipole interaction of Eq. (1) gave them the result

$$\frac{1}{T_1} = \frac{3}{10} \frac{\gamma^4 \hbar^2}{r^6} \left\{ \frac{\tau_c}{1 + \omega_0^2 \tau_c^2} + \frac{2\tau_c}{1 + 4\omega_0^2 \tau_c^2} \right\} \quad (3)$$

in which ω_0 is the Larmor frequency. The nature of Eq. (3) is shown graphically in Fig. 1. It is seen that there is a minimum in T_1 separating two straight line portions. The high frequency (short τ_c) portion, which occurs when $\omega_0 \tau_c \ll 1$, is often referred to as the extreme narrowing limit. The correlation frequency or rate of the reorientations is, of course, $1/\tau_c$, and they are most

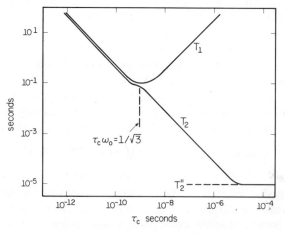

Fig. 1 The dependence of T_1 and T_2 upon the correlation time τ_c for random, isotropic reorientations of a pair of spin-½ nuclei. (Adapted from Bloembergen et al., 1948.)

effective in inducing transitions in the nuclear spin states when the reorientations occur at the Larmor frequency; hence the minimum in T_1 is at $\omega_0 \tau_c = 3^{-1/2}$.

The general approach of BPP was very quickly applied to many other more complex systems and types of motion, giving expressions for $1/T_1$ that are similar in form to Eq. (3). Such theoretical analyses have been widely used in measuring the rates of the various dynamic processes that cause fluctuations in the dipole–dipole interactions and thereby govern T_1 in solids and gases as well as in liquids (see, e.g., Abragam, 1961, or Slichter, 1963). If τ_c describes a thermally activated process, one may write

$$\tau_c = \tau_c^0 \exp(E_a/RT) \quad (4)$$

and therefore, measurements of $1/T_1$ as a function of temperature will reflect the changes in τ_c. Accordingly, a semilog plot of $1/T_1$ versus reciprocal temperature, $1/T$, is similar to Fig. 1. The slopes of the straight line portions of such plots enable one to determine the activation energy E_a for the process. However, conversion of the relaxation time data into absolute values of τ_c requires either the observation of the T_1 minimum or a reliable value for the coefficient(s) of the terms in τ_c in Eq. (3).

The analysis of the line-narrowing aspects of molecular reorientations and other lattice motions is more complex than that for the spin–lattice relaxation, and the BPP approach was less successful. The chief complication is the continuous character of the dipole–dipole interaction. The local field falls off as the inverse cube of the distance from a magnetic dipole, as stated in Eq. (2). However, the number of nuclei at a given distance r from the dipole increases in proportion to r^2. As a consequence, in a homogeneous phase the dipolar field from neighboring nuclei decreases only as $1/r$. Furthermore, even in rigid lattice solids, one deals most commonly with crystal powders, for which the $(1 - 3\cos^2\theta)$ angular dependence spreads the effective dipolar splitting of a given pair of nuclei over a threefold range. For these reasons, the quantitative calculation of dipolar broadened line shapes is a formidable problem which has been done for only a few cases of magnetically isolated, simple sets of spin-$\frac{1}{2}$ nuclei with high symmetry. These include the two-spin set of protons in $CaSO_4 \cdot 2H_2O$ (Pake, 1948), the equilateral three-spin set in CH_3CCl_3 (Andrew and Bersohn, 1950), and the four-spin tetrahedral set in NH_4Cl (Gutowsky et al., 1954).

Because of such difficulties, BPP approximated the rigid-lattice dipolar broadening with a single, inverse linewidth parameter T_2''. For our isolated two-spin example, T_2'' is related to the root mean square angular frequency broadening as follows:

$$[\langle \Delta\omega^2 \rangle_{av}]^{1/2} \equiv (1/T_2'') = \tfrac{3}{2}\gamma^2 \hbar [(1 - 3\cos^2\theta)^2 r^{-6}]^{1/2} \tag{5}$$

If the line shape is Gaussian, the quantity $1/(2\pi)^{1/2}T_2''$ is the full width of the line between its half-maximum points, on a frequency scale. As in the case of T_1, BPP made a Fourier analysis of the effects upon T_2'' of random reorientations of the two-spin sets. They found that the reorientations and the resultant time-dependence of the dipolar interaction begins to average out the rigid-lattice broadening when the reorientation rate becomes fast enough that $\tau_c \to T_2''$. In this event, the inverse linewidth parameter for the observed dipolar broadening T_2', defined in the same units as T_2'', becomes

$$\left(\frac{1}{T_2'}\right)^2 = \frac{9}{10\pi} \frac{\gamma^4 \hbar^2}{r^6} \tan^{-1} \frac{2\tau_c}{T_2'} \tag{6}$$

As τ_c becomes shorter than T_2'', the dipolar broadening becomes proportionately smaller, with

$$\frac{1}{T_2'} \simeq \frac{9\pi}{5} \frac{\gamma^4 \hbar^2}{r^6} \tau_c \qquad (7)$$

The spin–lattice relaxation also contributes to the total inverse linewidth parameter T_2, the latter being given by

$$(1/T_2) = (2\pi)^{1/2}(1/T_2') + \tfrac{1}{2}T_1 \qquad (8)$$

In Fig. 1, the quantity T_2 is given as a function of τ_c, on a log–log plot. It is seen that for $\omega_0 \tau_c > 1$, T_2 is dominated by the dipolar broadening described by T_2', becoming T_2'' for the rigid lattice. On the other hand, for $\omega_0 \tau_c < 1$, T_2 is dominated by T_1, and the residual dipolar broadening is only about 1% of the T_1 width. Probably the most significant aspect of the BPP analysis of the motional narrowing is the point that the narrowing occurs as the rate constant $1/\tau_c$ becomes comparable with and larger than the frequency width $1/T_2''$ of the perturbation which is time dependent. This relationship defines the "nmr time scale." Dipolar line widths in rigid-lattice solids are roughly in the 10–100 kHz range; hence, the line narrowing is sensitive to lattice motions having correlation times of the order of 10^{-4} to 10^{-5} sec, as is evident in Fig. 1.

Efforts to generalize and apply Eq. (6) as a means of obtaining quantitative values of reorientation rates from dipolar determined linewidths have not been nearly as successful as the use of T_1 via Eq. (3) and related expressions. Van Vleck (1948) developed a very useful general method for calculating the second and higher moments of a magnetic resonance absorption line from the nuclear properties and structural parameters of the rigid crystal lattice. Equation (5) is a simplified form of Van Vleck's expression for the second moment, which consists of such terms summed over all spin–spin pairs. Although the method gives neither the apparent second moment or the lineshape for partially narrowed lines, it was extended at an early date by Gutowsky and Pake (1950) to cases in which the motion is fast, but limited in nature. A typical example that they studied is the reorientation of the CH_3 group in solid CH_3CCl_3, which occurs only about the C—C bond of an otherwise "rigid" lattice.

The effects of such motions, in the limit of extreme (but limited) narrowing, can be calculated by taking the time average over $\theta(t)$, and if necessary also over $r(t)$, of the appropriate expressions for the dipole–dipole interactions, such as in Eq. (2). This approach has been widely employed to approximate the degree to which the observed width and second moment of an absorption line would be narrowed by particular types of limited motion, and in some degree also to approximate the corresponding line shapes for simple sets of spin-$\frac{1}{2}$ nuclei. It is now common practice to use second moment and lineshape considerations to identify the nature of the dynamic processes that narrow the dipolar broadening, and to determine the rates of the processes by measuring T_1 or $T_{1\rho}$, the latter being the spin–lattice relaxation time in the rotating frame (Abragam, 1961). Some early examples of this approach are described by Stejskal *et al.* (1959) and in papers cited therein.

Having devoted several pages to a historical synopsis of some of the commonest aspects of time-dependent dipole–dipole interactions, I wish now to point out that they are not the main subject of this treatise on dynamic nmr spectroscopy. Instead, the book deals primarily with rearrangements or other changes in molecular conformation and with the exchange of nuclei between different molecular species. These chemical exchange processes usually have little or no effect upon the dipole–dipole interactions. They are either too slow to be effective or, if not, overall molecular motions are faster and produce motional narrowing and spin–lattice relaxation before the chemical exchange effects become important. Therefore, chemical exchange studies by magnetic resonance methods are limited ordinarily to liquids and gases, with sensitivity considerations a strong deterrent to gas phase studies.

There are some circumstances, however, in which the motional narrowing or spin–lattice relaxation associated with dipole–dipole interactions can be employed to measure the rate in a solid at which one part of a molecule rotates with respect to the rest of the molecule. A simple case is provided by methylchloroform, CH_3CCl_3, in which the dipole–dipole interactions are dominated by those within each CH_3 group. Proton lineshape studies (Gutowsky and Pake, 1950; Powles and Gutowsky, 1955) have shown that there is a substantial temperature range in the solid for which the motional narrowing corresponds to restricted reorientations of the CH_3 group about its threefold axis. The temperature dependence of the proton T_1, when fitted by Eq. (4) and the appropriate form of Eq. (3), gives an activation energy E_a for the reorientations of 4.1 kcal/mole (Stejskal *et al.*, 1959).

These data alone do not establish whether the CH_3 group is rotating with respect to the CCl_3 group or the two groups are locked together and the molecule is rotating as a whole. However, in this case, other experimental evidence indicates that the CCl_3 groups remain "fixed" in the lattice at temperatures well above those at which τ_c for CH_3 group reorientations is 10^{-7} seconds or shorter. This may be demonstrated in some cases by observing two different species of magnetic nuclei located in different parts of a molecule. For example, in CH_3–CF_3 or C_6H_5–CF_3, the ^{19}F line shapes and T_1 values can be observed as well as those for the protons. If, at a given temperature, one species indicates a rigid lattice and the other a reorientating group, there must be internal rotation within the molecule. Examples of this nature in solids are not uncommon. However, more subtle conformational changes are likely to be obscured by intermolecular dipole–dipole interactions. Nonetheless, there have been some recent studies of "fluxionality in the solid state," summarized by Cotton in Section V of his chapter on organometallic compounds.

III. Chemical Exchange

A. Discovery of Chemical Exchange Effects

For the reasons cited in the preceding paragraphs, the discovery of magnetic resonance methods for studying chemical exchange awaited the characterization of interactions—other than the dipole–dipole variety described by Eq. (1)—that can lead to time-dependent magnetic perturbations. These are, of course, the chemical shift and the scalar or indirect coupling of nuclear spins. The electron–nuclear hyperfine interaction is also of this nature and it is being used quite widely in studies of electron exchange as well as chemical exchange; however, it is a rather different concern, the applications of which are not included in this book.

By 1949–1950, the magnets in use for magnetic resonance experiments had enough stability and magnetic field homogeneity for chemical shifts to be observed in liquids (Knight, 1949; Proctor and Yu, 1950; Dickinson, 1950) and for the fine structure due to the scalar coupling to be resolved (Gutowsky and Hoffman, 1950; Hahn, 1950). A historical account has been given elsewhere of some of those exciting days, including a few carefully selected personal reminiscences (Slichter, 1974; Gutowsky, 1974). It was in the course of our characterization during 1951–1952 of the scalar coupling that we turned up some anomalies which led us to recognize the importance of chemical exchange in determining whether such coupling produces observable splittings (Gutowsky et al., 1953).

In liquids, the correlation times for molecular reorientation and diffusion are of the order of 10^{-8} to 10^{-12} seconds, well within the extreme narrowing range for the dipolar broadening. At the time of our early experiments, the linewidths we observed for liquids were instrumentally determined at about 0.01 G. Mutual splittings of several tenths of a gauss were observed in the proton, fluorine, and phosphorus spectra of compounds such as PF_3 and PH_3, and were attributed to the scalar coupling of the nuclei involved. But the proton resonances in HF, HBF_4, and HPF_6 were found to be single, as were the fluorine resonances in HF and HBF_4, even though the similarity of the various compounds led us to believe that there should be detectable scalar coupling between H and F in HF and between B and F in HBF_4. After much discussion, we attributed the apparent absence to rapid chemical exchange of fluorines between different molecules, in the case of HF by processes such as

$$HF \rightleftharpoons H^+ + F^-$$
$$F^- + HF \rightleftharpoons (FHF)^- \rightleftharpoons FH + F^-$$

The situation is not unlike the motional narrowing of the dipole–dipole splitting. In fact, it is much simpler to handle theoretically, because it involves a small number of discrete local fields (narrow lines) and simple interchanges among them, rather than a continuous distribution of local fields (a single broad line) and a complex distribution of exchanges among them. For example, consider liquid HF for which the proton and fluorine resonances in the absence of chemical exchange would consist of symmetrical doublets with a splitting J of about 600 Hz (Solomon and Bloembergen, 1956). The net effect of the chemical exchange in HF is that a proton (or fluorine nucleus) moves from the local field of one fluorine nucleus (or proton) to that of another. The spin state of the new nuclear partner may be the same as that of the old, or it may be the opposite. If it is the opposite, the sign of the local field is reversed, so the exchange process consists of the nucleus whose resonance is being observed, hopping from one line to the other in a random manner described by a correlation or exchange time τ_{ex}, or simply τ (or 2τ) in the more customary notations. By reference to the BPP expression for motional narrowing, Eq. (6), coalescence of the 600 Hz doublets in HF should occur when $(1/\tau) \gtrsim J \cong 600$ Hz.

B. Bloch Equations Modified for Chemical Exchange

In any event, the absence of such splittings led Slichter (Gutowsky *et al.*, 1953) to develop a theoretical analysis of the effect, based upon the Bloch equations (1946). For a single set of nuclei in the presence of an rf field $\omega_1 = \gamma H_1$ rotating at a frequency ω in the plane perpendicular to H_0, the bulk magnetic moment of the sample may be written as

$$(du/dt) + \Delta\omega v = -u/T_2$$
$$(dv/dt) - \Delta\omega u = -(v/T_2) - \omega_1 M_z$$
$$(dM_z/dt) - \omega_1 v = -(M_z - M_0)/T_1 \qquad (9)$$

The quantities u, v, and M_z are the three components of the nuclear magnetization along a set of axes rotating about H_0 at the adjustable angular frequency ω, with H_1 taken along u; $\Delta\omega$ is $(\gamma H_0 - \omega)$; and M_0 is the value of M_z at thermal equilibrium. Solution of Eq. (9) under the usual, steady-state resonance absorption conditions $(\gamma H_1 \to 0)$ gives a Lorentzian lineshape with a full width at half-maximum of $(1/\pi T_2)$ sec^{-1}. Two sets of nonexchanging nuclei, displaced equally from ω_0 by amounts $\pm \delta\omega/2$, may be described by two sets of Bloch equations in which $\Delta\omega$ is replaced by

$$\Delta\omega_a = \Delta\omega + (\delta\omega/2) \quad \text{and by} \quad \Delta\omega_b = \Delta\omega - (\delta\omega/2) \tag{10}$$

Chemical exchange between the two sets of nuclei couples the two sets of equations. The initial treatment by Gutowsky et al. (1953) of the exchange effects was based upon the steady-state solution of the Bloch equations for a single set of nonexchanging nuclei, namely

$$\begin{aligned}G_j &\equiv u_j + iv_j \\ &= (M_{0j}T_2/2)(\Delta\omega_j T_2 - iT_2)/(1 + \Delta\omega_j^2 T_2^2)\end{aligned} \tag{11}$$

For two equally populated sets of nuclei, the total nuclear magnetization in the u–v plane is the sum for the two sets, each averaged over the exchange; that is,

$$\langle G \rangle_{ex} = \tfrac{1}{2}\langle G_a \rangle_{ex} + \tfrac{1}{2}\langle G_b \rangle_{ex} \tag{12}$$

The calculation of $\langle G \rangle_{ex}$ is relatively straightforward, but somewhat complex. If one assumes that T_1 and T_2 are the same for both sets of nuclei, the end result is

$$\langle G \rangle_{ex} = \frac{-iM_0\{2 + [(1/T_2) - i\Delta\omega]\tau\}}{(1 + \alpha_a \tau)(1 + \alpha_b \tau) - 1} \tag{13}$$

in which

$$\alpha_j \equiv (1/T_2) - i\Delta\omega_j \tag{14}$$

with τ defined as the mean lifetime of a nucleus at a given local field (site). For $\tau \to \infty$, Eq. (13) gives a doublet absorption line with a total splitting of $\delta\omega$, while for $\tau \to 0$, it gives a single line at the midpoint of the doublet and with twice the peak intensity. Also, the transition from doublet to a single line occurs when $\tau\delta\omega \simeq 1$, analogous to Eq. (6) for motional narrowing of the dipole–dipole interaction.

A somewhat simpler approach is to add exchange terms directly to the Bloch equations and solve the resulting coupled set. For the equally populated, two-site case, the modified Bloch equations are of the form

$$(du_a/dt) + \Delta\omega_a v_a = -(u_a/T_2) - (u_a/\tau_a) + (u_b/\tau_b)$$
$$(dv_a/dt) - \Delta\omega_a u_a = -(v_a/T_2) - (v_a/\tau_a) + (v_b/\tau_b) - \omega_1 M_{za}$$
$$(dM_{za}/dt) - \omega_1 v_a = -(M_{za} - M_{0a})(1/T_1) - (M_{za}/\tau_a) + (M_{zb}/\tau_b) \tag{15}$$

with an equivalent set in which the indices a and b are interchanged. Direct solution of these equations gives, of course, the same result as Eq. (13); it is obtained by setting dG_j/dt equal to zero and solving the resulting linear equations for the imaginary part of $G = \Sigma_j G_j$. Hahn and Maxwell (1952) employed this formulation to derive an expression equivalent to Eq. (13) to account in spin-echo experiments for the disappearance of the doublet splitting of the CH_3 group resonance in CH_3OH that contains H_2O. It was also apparently used by Arnold (1956) in his work on ethanol and was later rediscovered in steady-state terms by McConnell (1958).

C. Early Applications

Application of these potentialities in any systematic quantitative way was relatively slow in developing. As early as 1951, Liddel and Ramsey (1951) suggested that the temperature-dependent shift of the OH line in ethanol, observed by Arnold and Packard (1951), was attributable to rapid exchange among low-lying, hydrogen-bonded states of the molecule. Also, Hahn and Maxwell (1952) approximated the proton exchange rate in a particular solution of methanol from their spin-echo data for it. However, the proton exchange is catalyzed by traces of water, acid, or base and is not very easy to study quantitatively. It was simpler to observe and analyze the concentration dependence of proton shifts in the fast-exchange limit (Gutowsky and Saika. 1953). For example, in aqueous acetic acid, the fast exchange between the CO_2H and H_2O protons results in a single line, the position of which may be calculated via Eqs. (10)–(15) and is given by

$$\delta = p_1 \delta_{HAc} + p_2 \delta_{H_2O} \tag{16}$$

where p_1 and p_2 are the fractions of protons in the acid and in the water; and δ_{HAc} and δ_{H_2O} are the proton chemical shifts of HAc and H_2O, respectively. In strong acids such as HCl and HNO_3, the dissociation introduces H_3O^+ as a third important participant in the exchange

$$HA + H_2O \rightleftharpoons H_3O^+ + A^-$$

and the concentration dependence of the averaged shift provided evidence for the incomplete dissociation of the acid HA.

In addition, we calculated a set of lineshapes in the intermediate region between fast and slow exchange for the equally populated two-site case. Typical results are given in Fig. 2 as a function of the exchange lifetime τ. These lineshapes led to the suggestion that rather short chemical lifetimes in the range of 10^{-2} to 10^{-4} second could be measured by such an approach. I was even bold enough to propose that, "A combination of temperature and magnetic field changes in the transition region would permit evaluation of the activation energy for the exchange." Reduction of this to practice was delayed by the difficulty of finding suitable systems with exchange rates in the right region. At that time there just were no other methods for measuring reactions

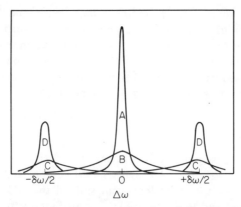

Fig. 2 High-resolution nmr line shapes as a function of the exchange lifetime τ between two equally populated absorption lines separated by an amount $\delta\omega$ in the absence of exchange. $2\tau\delta\omega$: A, 10^{-2}; B, 1; C, 10; D, 100. (Gutowsky and Saika, 1953.)

with half-lives in the millisecond range, so no well-known suitable examples were available for us to study via the proposed new method. However, when Phillips (1955) reported that the N-methyl proton resonance was a doublet in N,N-dimethylformamide, we became interested in it as a candidate for line-shape rate studies.

The structure of such amides may be described by the two canonical forms shown below.

One of the methyls is cis to the C—O group at the other end of the molecule, while the second methyl is trans, so they have chemically shifted lines. Furthermore, the partial double-bond character of the central N–C bond hinders internal rotation about that bond and prevents exchange narrowing of the shift at room temperature. The internal rotation is acid and base catalyzed; also, it is a thermally activated process. A successful study was made (Gutowsky and Holm, 1956) of its temperature dependence through the transition from slow to fast exchange. The spectra and their analysis are crude by today's standards, but they did give reasonable values for the exchange rates and enabled the activation energy to be determined. A more recent set of proton lineshapes observed as a function of temperature in N,N-dimethyltrichloroacetamide (Allerhand and Gutowsky, 1964) is reproduced in Fig. 3.

The dimethylamide case is a very simple example of the type of conformational change that may be profitably studied by dynamic nmr methods.

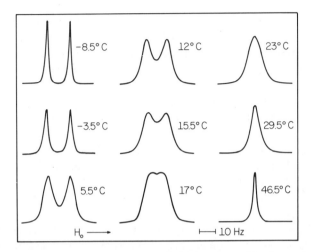

Fig. 3 The proton resonance of N,N-dimethyltrichloroacetamide observed as a function of temperature at 60 MHz. (Allerhand and Gutowsky, 1964.)

Several aspects merit special attention. The initial and final states of the molecule are indistinguishable except for the chemical shift and nuclear spin labeling of the methyl groups. Therefore, the exchange rate is not measurable by any of the more traditional relaxation methods such as T-jump. Furthermore, the exchange is stationary in time so its rate is described by a pseudo first-order rate constant, $k \equiv 1/\tau$ sec^{-1}, even though the mechanism may involve binary collisions. Mention should also be made of the assumption implicit in Eq. (12) that the time spent in the transition state is negligible compared with the average time τ spent in each of the two sites. Finally, it should be noted that the dipole–dipole interactions among the protons will be completely in the extreme narrowing limit because of the fast, overall molecular reorientations and diffusion in the liquid; the slow internal rotations serve only to modulate the overall rotation about the N—C bond, with negligible effects upon the dipolar interactions.

The exchange of nuclei among different molecules was first studied quantitatively by Arnold (1956) and by Solomon and Bloembergen (1956). Arnold investigated the proton exchange in ethanol, which is acid and base catalyzed. A possible mechanism for the former may be written as

$$\text{EtOH}^* + \text{H}^+ \rightleftharpoons \text{EtOH}^*\text{H}^+ \rightleftharpoons \text{EtOH} + \text{H}^{*+}$$

In this case, the pseudo first-order rate constant describing the exchange at a given temperature is concentration dependent, and one would expect to find

$$k \equiv (1/\tau) = k'[\text{H}^+] \tag{17}$$

Arnold (1956) estimated τ from the lineshape for the OH group protons at several concentrations of H^+ and OH^-. However, the scatter in the data was too great to draw any conclusions about the mechanism, probably because of the difficulties in establishing and maintaining a known concentration of H^+.

Solomon and Bloembergen (1956) studied in liquid HF the proton and fluorine exchange, which is catalyzed by traces of moisture, and obtained numerical estimates of its rate in several samples at two or three temperatures. The determinations were based upon the internuclear Overhauser effect which they observed in double resonance experiments on the compound (see Section IV). The intramolecular dipole–dipole interaction gives a positive contribution and the scalar coupling a negative one in the compound. The latter dominates when the chemical exchange is slow and gives a net negative effect. However, when the exchange is fast enough to average out the scalar coupling, the net effect becomes positive. Thereby, the two interactions can be separated and the exchange rate determined. The system is an excellent one for studying the basic phenomenological aspects of magnetic resonance in a heteronuclear two-spin system; however, such ideal cases are rare and the method is too specialized for general use.

Mention should also be made of a phenomenon described by Hahn (1950), which can be used to study diffusion rates in fluids. If H_0 is inhomogeneous, an absorption line will be broadened by the distribution of local magnetic fields over the sample. However, diffusion will move a molecule from one local field to another, and if fast enough and far enough, such exchange will average out the inhomogeneities. Therefore, by imposing a known adjustable gradient in H_0 over the sample and observing the extent to which its effects are reduced by the diffusion, the rate can be determined. Spin–echo methods are ordinarily used for such measurements, which have been employed quite broadly, as described in the chapter by Reeves.

The chemical exchange processes described in this section involve time-dependent perturbations with an amplitude (γH_l) of 10^1 to 10^3 $\sec^{-1}$. The dynamic processes considered in Section III on dipole–dipole interactions are generally larger, say 10^3 to 10^5 $\sec^{-1}$. There is a corresponding difference in the range of rates measurable by means of these perturbations. Moreover, via the theory of absolute reaction rates for thermally activated processes, different ranges of activation energies are selected by the different time-dependent perturbations. The dipole–dipole interactions are affected most commonly by processes with relatively low activation energies, in the 1–10 kcal/mole range, while a range of larger values, about 7–20 kcal/mole, is sampled by the chemical shifts and scalar coupling. These are approximate ranges only, inasmuch as they depend upon the temperature range employed in the experiments, as well as upon the exchange rates that can be measured.

IV. Coherent Time-Dependent Fields

This introductory commentary on time-dependent magnetic perturbations would be incomplete without some reference to coherent perturbations. As used here, coherent perturbation refers to local fields that are repetitive in time with a fixed frequency for the whole sample. There are two classes of these. One class consists of radiofrequency fields at or near the Larmor frequency, applied in the plane perpendicular to H_0, which are employed in observing and/or modifying magnetic resonance phenomena. The other includes coherent fluctuations in H_0. The simplest of these to treat is the case of a single rf field of amplitude $\omega_1 = \gamma H_1$, at an adjustable frequency $\omega \sim \omega_0$ for a given set of nuclei. Its effects are included in the Bloch equations, Eq. (9), which are widely used to analyze spin–echo and transient effects, as well as to obtain the steady state solutions mentioned in Section III,B. If a second rf field, $\omega_2 = \gamma' H_2$, is applied at a different frequency ω' to a different set of nuclei in the same sample, it will perturb the resonance of the first set by an amount and in a manner depending upon the extent and nature of any interaction between the two sets of nuclei. The two sets in question may be different nuclear species or chemically shifted sets of the same species in the same or different molecules.

Consider for example the ^{19}F resonance in the aqueous PO_3F^{2-} ion. In the ordinary, single-resonance experiment, it is a 1:1 doublet with a splitting of about 860 Hz due to the scalar coupling, $2\pi J \mathbf{I}_F \cdot \mathbf{I}_P$, between the fluorine and phosphorus nuclei in each molecule. However, when a strong second rf field $\omega_2 = \gamma_P H_2$ is applied at the central Larmor frequency of the ^{31}P doublet, it causes the phosphorus nuclei to change coherently between their two spin states at a rate that increases monotonically with ω_2. Except for its coherence, this "stirring" of the ^{31}P nuclei is equivalent to chemical exchange of the ^{31}P nuclei among its two spin states and, when it is fast enough ($2\omega_2 \gtrsim \pi J$), the ^{19}F resonance becomes a single line with a peak intensity twice that of the original doublet (Bloom and Shoolery, 1955). At intermediate stages the appearance of the ^{19}F resonance depends upon the position as well as the magnitude of ω_2 and on whether the spectrum is observed in a field- or frequency-sweep mode. The main effect of the coherence in the spin states of the ^{31}P nuclei is that the intermediate lineshapes consist of discrete lines rather than a continuous distribution as in the case of chemical exchange, the latter reflecting the continuous spectral density function for the exchange. Such double (or multiple) resonance experiments have proved to be very useful in simplifying and analyzing high-resolution nmr spectra.

In addition to its coherent modulation of any spin-dependent perturbations, the application of a second rf field will also tend to equalize the populations of the energy levels between which it induces transitions. If these levels are also involved in the transitions observed by the weak rf field, the intensities found by the latter will be modified. The importance of such effects was first suggested by Overhauser (1953) who predicted that the nuclear–electron hyperfine interaction would enable the intensity of a nuclear resonance to be enhanced by as much as the ratio of electronic to nuclear g-values, $\gamma_e/\gamma_n \sim 10^3$, via saturation of the electron resonance. This expectation was verified shortly thereafter as were similar internuclear Overhauser effects (see, e.g., Slichter, 1963; Abragam, 1961; Solomon, 1955).

Coherent fluctuations in H_0 may be produced by modulation coils through which a fluctuating current is passed or by mechanical rotation of the sample in an inhomogeneous H_0. The basic theory applicable in these instances was developed at an early date by Karplus (1948) in treating the effects of electric field, i.e., Stark modulation upon the microwave spectra of gases. The overall effects of sample spinning and field modulation are very similar (Williams and Gutowsky, 1956). Sideband resonances appear symmetrically about each otherwise single line, displaced by integral multiples ($n = 0, 1, 2, \ldots$) of the modulation or spinning frequency ω_m and ω_s. The relative intensities of the fundamental and of the several side bands are proportional to the products of the Bessel functions of the first kind, $J_n(k)$. The argument of the functions is of the general form $k = \gamma H_1^m/\omega_m$ (or ω_s), where H_1^m is a constant proportional to the amplitude of the fluctuation in the local field, depending upon its wave form and spatial distribution. As $k \to 0$ (ω_m and ω_s become large), the side bands move out and decrease in intensity, giving in both cases a single narrow line in the "fast-exchange" limit. It is interesting to compare the results of such coherent fluctuations in the local field with motional averaging by random processes such as chemical exchange or molecular reorientations.

V. Accomplishments and Prospects

So far, this account has touched upon those aspects of nmr which were discovered during its first decade and which led to the demonstration that conformational and intermolecular exchange processes could be investigated by lineshape changes. This occupied the first decade of nmr, 1946–1956. The next decade, 1956–1966, was one of exploration, development, and refinement of theory, experimental methods, data analysis, and systems suitable for and important to study. Reviews are available for much of this period (Pople *et al.*, 1959; Loewenstein and Connor, 1963; Johnson, 1965; Allerhand *et al.*, 1966). A brief listing of the major directions involved may be helpful in providing entries to the literature for those who wish details and background beyond those provided in the subsequent chapters.

The general theory of the lineshape for exchange among a large number of local fields was treated as early as 1954 by Anderson and by Kubo (Anderson, 1954; Anderson and Weiss, 1953; Kubo, 1954) and subsequently by Sack (1958). The application of the modified Bloch equations to complex spectra with second-order corrections to the splittings and intensities presents some difficulties. In order to handle such complications more readily, density matrix formulations have been evolved (Kaplan, 1958a,b; Alexander, 1962; Johnson, 1965; Gutowsky et al., 1965). The most widely used experimental method for observing chemical exchange effects continues to be the steady-state, high-resolution approach, which was first employed for the purpose as described in Section III. In addition, fast passage (McConnell and Thompson, 1957, 1959; Meiboom, 1961) and double resonance methods (Forsén and Hoffman, 1963) have been proposed and used to some degree in measuring relatively slow exchange rates, those in the 0.1–1 sec^{-1} range.

Also, considerable effort has been spent on the application of spin–echo experiments to the study of chemical exchange (Allerhand et al., 1966, and papers cited therein). A primary objective of such work has been to extend the range of measurable rates at both the slow and fast ends of the scale. The rationale for this is apparent from the corresponding spectra in Fig. 3, for which

$$(1/T_2)_{\text{total}} = (1/T_2^0) + (1/T_2)_{\text{instr}} + (1/T_2)_{\text{ex}} \qquad (18)$$

where $(1/T_2^0)$ is the natural linewidth in the absence of exchange, $(1/T_2)_{\text{instr}}$ represents the inhomogeneity and other instrumental broadening, and $(1/T_2)_{\text{ex}}$ describes the exchange broadening. The limiting rates observable in the steady state experiments occur as the exchange broadening becomes comparable with and smaller than the sum of $(1/T_2^0)$ and $(1/T_2)_{\text{instr}}$. However, the instrumental broadening can be an order of magnitude larger than the natural linewidth in the absence of exchange, thereby limiting the accessible range. The spin–echo method of Hahn (1950; Hahn and Maxwell, 1952) provides a way of eliminating a large part of the instrumental effects and thereby it can be used to extend substantially the range of observable rates (Allerhand and Gutowsky, 1964). The current state of the art, as applied to chemical exchange and also to the measurement of diffusion rates, is described in the chapter by one of its early protagonists, Reeves.

Along with the development of alternate experimental methods, the procedures for analyzing the experimental data have become increasingly sophisticated and more accurate. An extensive, comparative analysis of the various experimental and data reduction methods was given by Allerhand et al. (1966). In the early lineshape work, single-parameter approximations were often used, which introduced systematic errors in the apparent rates and quite large errors in activation parameters. Study of the set of spectra in Fig. 3 discloses the several options: (a) the exchange broadening in the limits of slow and fast exchange; (b) the separation of partially coalesced peaks compared with that in the absence of exchange; (c) the peak-to-central intensity ratio of partially coalesced lines; (d) the width of the single line at rates above coalescence. The inaccuracies in such approximations and the broad availability of computers led to the development and general application of complete lineshape fitting procedures (Jonas et al., 1965). The best fit of the digitized experimental spectra is obtained by an iterative, computerized process in which theoretical line shapes are calculated without crippling approximations and compared with those observed. The present status of lineshape methods and their recent development are reviewed in the chapter by G. Binsch (Chapter 3).

We are now approaching the close of the third decade of nmr, 1966–1976, which is the second decade of dynamic nmr spectroscopy. The achievements of this decade are the subject of this book. After a 20-year induction period, the methods have become widely applied to an increasingly rich variety and large number of particular systems. Greatly improved, commercially available nmr instrumentation has been essential for such applications, especially cryostats that have extended the range of temperatures and of systems available for study. Furthermore, magnet technology has evolved to the point that instrumental broadening is no longer the major limiting factor it was for chemical exchange studies at the outset of the decade. The increased miniaturization and sophistication of computers and the continued rapid drop in cost per computation have also made possible many of the results surveyed in this book.

Prospects for the continued development and utilization of dynamic nmr spectroscopy appear to be bright. Prediction is chancy at best, but it seems safe to say, "Toujours gay. Toujours gay. There's life in the old girl yet!" My own interests have focussed on the phenomenological aspects and the methodology, so I won't comment on the new types of systems or dynamic processes that may be studied or on the significance of the results. However, there are some trends discernible that will no doubt have an impact upon the field during the next decade, and a few comments on them seem desirable.

Probably the most important trend in nmr at present is the rapid conversion from steady-state to the more sensitive Fourier transform spectrometers as the instrument of choice for general-purpose use. The increased sensitivity makes it possible to observe ^{13}C spectra at natural abundance, and this in turn is enabling the conformational changes of the carbon skeletons to be studied directly rather than by looking at protons or fluorine nuclei attached to the carbons. The range of ^{13}C chemical shifts is more than tenfold larger than that of protons, and this will in principle permit one to study a wider range of rates and of systems. In practice, one seldom gets something for nothing, and if the shifts being exchange averaged are too large, the peak S/N in the coalescence region will be too small for convenient observation. Nonetheless, a broader range of nuclei and of problems will become accessible for study. Also, the increased sensitivity will enable proton studies to be made in more dilute solutions or in the gas phase, and larger molecules will be studied. However, as a word of caution, past experience with other methods suggests that the use of Fourier transform spectra to study chemical exchange will have its own interpretational pitfalls.

Another significant instrumental trend is the use of superconducting solenoids to attain higher and still higher magnetic fields. A major reason for this is the desire to "spread out" the proton spectra of large molecules such as enzymes to study molecular structure, complex formation, and other problems. Proton spectrometers operating at 200–300 MHz are not uncommon now, and further increases up to 500 MHz seem likely to be accomplished during the coming decade. These developments will extend the range of observable systems and rates. Also, they are finally making feasible the reduction to practice of a much earlier prediction (Gutowsky and Saika, 1953) that "magnetic field changes in the transition region would permit evaluation of the activation energy for the exchange." The point involved is that at the coalescence point the line shape is most sensitive to changes in the exchange rate. Observation of the coalescence temperature for a thermally activated process or the coalescence concentration for exchange between two or more species as a function of the Larmor frequency is quite simple as well as relatively accurate. The approach has been treated recently by H. N. Cheng (Gutowsky, 1974), who observed the coalescence temperatures for several amides at 60, 100, and 220 MHz and obtained thereby the activation parameters with a high degree of internal consistency in the determinations. The results appear promising.

There will no doubt be other trends of more specialized interest. For example, consider the recent development of chemical shift reagents. These are an interesting example in themselves of a dynamic nmr process. For solutions of them in polar solvents, a short electron T_1 exchange averages the electron–nuclear hyperfine interaction to give contact shifts characteristic of the hyperfine constant for each set of solvent nuclei, depending upon the temperature and the applied field. These shifts are adjustable via the concentration or nature of the shift reagent and can be used to develop or extend the chemical shift changes associated with a chemical exchange process. The effect depends upon the formation of a weak complex between the shift reagent and the molecule under study, and this can affect the dynamics in question. However, a preliminary study (Cheng and Gutowsky, 1972) of our perennial victim, the dimethylamides, indicates some promising though limited applications of the shift reagents to chemical exchange processes. Another possibility that warrants attention is the use of clathrates or zeolites as a way of isolating molecules so as to study low-barrier conformational changes at low temperatures in what is effectively a rigid but "open" matrix where the molecules would have sufficient reorientational mobility to average out the intramolecular dipole–dipole interactions.

In spite of all the instrumental improvements and theoretical advances during nearly three decades of nmr, there still remains a major technical obstacle sorely limiting the accuracy of determining chemical exchange rates. This is, of course, the inadequacy of the variable temperature nmr probes commonly used. Too many compromises have been made in the design of spectrometers for them to be very well suited for rate studies. Traditionally, reaction rates have been measured in large constant-temperature baths, well stirred and of high heat capacity and stability. It is not uncommon to have reactants at temperatures constant to a millidegree or so and known to an accuracy of 5 millidegrees or better. Contrast this with the usual nmr probe, the sample in a long, narrow glass tube, spun by an air blast in a magnet gap of 2–3 cm. Temperature gradients along the sample of several degrees are not uncommon. Moreover, insofar as I know, no really good way has yet been found for measuring the temperature of the sample while it is spinning, which means that the temperature is seldom known to an accuracy as good as 500 millidegrees, much less 50. Ten years ago, such inaccuracies in temperature control and measurement were comparable with the accuracy of the methods used to extract exchange rates from the nmr data. However, this is no longer the case. The potentialities of dynamic nmr spectroscopy for quantitative studies during the next decade can be fulfilled only by substantial, generally applicable improvements in sample temperature control and measurement. I hope someone, somewhere is rising to or has already met the challenge.

ACKNOWLEDGMENT

Financial support of the U.S. Office of Naval Research and of the National Science Foundation is gratefully acknowledged.

REFERENCES

Abragam, A. (1961). "The Principles of Nuclear Magnetism." Oxford Univ. Press, London and New York.
Alexander, S. (1962). *J. Chem. Phys.* **37**, 967 and 974.
Allerhand, A., and Gutowsky, H. S. (1964). *J. Chem. Phys.* **41**, 2115.
Allerhand, A., Gutowsky, H. S., Jonas, J., and Meinzer, R. A. (1966). *J. Amer. Chem. Soc.* **88**, 3185.
Anderson, P. W. (1954). *J. Phys. Soc. Jap.* **9**, 935.
Anderson, P. W., and Weiss, P. K. (1953). *Rev. Mod. Phys.* **25**, 269.
Andrew, E. R., and Bersohn, R. (1950). *J. Chem. Phys.* **18**, 159.

Arnold, J. T. (1956). *Phys. Rev.* **102**, 136.
Arnold, J. T., and Packard, M. E. (1951). *J. Chem. Phys.* **19**, 1608.
Bloch, F. (1946). *Phys. Rev.* **70**, 460.
Bloch, F., Hansen, W. W., and Packard, M. (1946). *Phys. Rev.* **69**, 127.
Bloembergen, N., Purcell, E. M., and Pound, R. V. (1948). *Phys. Rev.* **73**, 679.
Bloom, A. L., and Shoolery, J. N. (1955). *Phys. Rev.* **97**, 1261.
Cheng, H. N., and Gutowsky, H. S. (1972). *J. Amer. Chem. Soc.* **94**, 5505.
Dickinson, W. C. (1950). *Phys. Rev.* **77**, 736.
Forsén, S., and Hoffman, R. A. (1963). *J. Chem. Phys.* **39**, 2892.
Gorter, C. J., and Broer, L. J. F. (1942). *Physica (Utrecht)* **9**, 591.
Gutowsky, H. S. (1974). *Proc. Int. Symp. Magn. Resonance, 6th, 1974* (in press).
Gutowsky, H. S., and Hoffman, C. J. (1950). *Phys. Rev.* **80**, 110.
Gutowsky, H. S., and Holm, C. H. (1956). *J. Chem. Phys.* **25**, 1228.
Gutowsky, H. S., and Pake, G. E. (1950). *J. Chem. Phys.* **18**, 162.
Gutowsky, H. S., and Saika, A. (1953). *J. Chem. Phys.* **21**, 1688.
Gutowsky, H. S., McCall, D. W., and Slichter, C. P. (1953). *J. Chem. Phys.* **21**, 279.
Gutowsky, H. S., Pake, G. E., and Bersohn, R. (1954). *J. Chem. Phys.* **22**, 643.
Gutowsky, H. S., Vold, R. L., and Wells, E. J. (1965). *J. Chem. Phys.* **43**, 4107.
Hahn, E. L. (1950). *Phys. Rev.* **80**, 550.
Hahn, E. L., and Maxwell, D. E. (1952). *Phys. Rev.* **88**, 1070.
Johnson, C. S., Jr. (1965). *Advan. Magn. Resonance* **1**, 33.
Jonas, J., Allerhand, A., and Gutowsky, H. S. (1965). *J. Chem. Phys.* **42**, 3396.
Kaplan, J. (1958a). *J. Chem. Phys.* **28**, 278.
Kaplan, J. (1958b). *J. Chem. Phys.* **29**, 492.
Karplus, R. (1948). *Phys. Rev.* **73**, 1027.
Knight, W. D. (1949). *Phys. Rev.* **76**, 1259.
Kubo, R. (1954). *J. Phys. Soc. Jap.* **9**, 935.
Liddel, U., and Ramsey, N. F. (1951). *J. Chem. Phys.* **19**, 1608.
Loewenstein, A., and Connor, T. M. (1963). *Ber. Bunsenges. Phys. Chem.* **67**, 280.
McConnell, H. M. (1958). *J. Chem. Phys.* **28**, 430.
McConnell, H. M., and Thompson, D. D. (1957). *J. Chem. Phys.* **26**, 958.
McConnell, H. M., and Thompson, D. D. (1959). *J. Chem. Phys.* **31**, 85.
Meiboom, S. (1961). *J. Chem. Phys.* **34**, 375.
Overhauser, A. W. (1953). *Phys. Rev.* **92**, 212.
Pake, G. E. (1948). *J. Chem. Phys.* **16**, 327.
Phillips, W. D. (1955). *J. Chem. Phys.* **23**, 1363.
Pople, J. A., Schneider, W. G., and Bernstein, H. J. (1959). "High-Resolution Nuclear Magnetic Resonance." McGraw-Hill, New York.
Powles, J. G., and Gutowsky, H. S. (1955). *J. Chem. Phys.* **23**, 1692.
Proctor, W. G., and Yu, F. C. (1950). *Phys. Rev.* **77**, 717.
Purcell, E. M., Torrey, E. C., and Pound, R. V. (1946). *Phys. Rev.* **69**, 37.
Sack, R. A. (1958). *Mol. Phys.* **1**, 163.
Slichter, C. P. (1963). "Principles of Magnetic Resonance." Harper, New York.
Slichter, C. P. (1974). *Proc. Int. Symp. Magn. Resonance, 6th, 1974* (in press).
Solomon, I. (1955). *Phys. Rev.* **99**, 559.
Solomon, I., and Bloembergen, N. (1956). *J. Chem. Phys.* **25**, 261.
Stejskal, E. O., Woessner, D. E., Farrar, T. C., and Gutowsky, H. S. (1959). *J. Chem. Phys.* **31**, 55.
Van Vleck, J. H. (1948). *Phys. Rev.* **74**, 1168.
Williams, G. A., and Gutowsky, H. S. (1956). *Phys. Rev.* **104**, 278.

2

Delineation of Nuclear Exchange Processes

W. G. KLEMPERER

I. Introduction... 23
II. Dynamic Stereochemistry of Symmetric Molecules............... 24
    A. Configurations and Reactions............................. 24
    B. Configurational Symmetry Groups 26
    C. Symmetry Equivalent Reactions........................... 27
    D. Mechanism and Steric Course 29
III. NMR Differentiable Reactions................................ 30
    A. The Effective NMR Symmetry Group 30
    B. Definition of NMR Nondifferentiable Reactions 31
    C. Complete Sets of NMR Differentiable Reactions 33
    D. Enumeration of NMR Differentiable Reactions 34
IV. Examples... 35
    A. 3-Hydroxy-2,4-dimethylcyclobutenone 35
    B. The Cation 1,2-Bis(diphenylphosphino)ethane(2-methylallyl)-
       palladium(II).. 38
    C. Dihydridotetrakis(diethoxyphenylphosphine)ruthenium(II)...... 39
    D. Sulfur Tetrafluoride 40
    References.. 44

I. Introduction

Dynamic nuclear magnetic resonance spectroscopy has been used extensively for the investigation of chemical reaction mechanisms. The usual approach has been to postulate several possible reaction mechanisms, determine the nuclear site exchange scheme implied by each mechanism, and then simulate rate-dependent spectra based on each of these site exchange schemes. Postulated mechanisms are considered to be consistent with experiment if the simulated spectra implied by these mechanisms are similar to the experimental spectra. Using this approach, however, the possibility always exists that the true mechanism was overlooked when postulating mechanisms, and that simulated spectra implied by the true mechanism match the experimental spectra better than those implied by the mechanisms postulated. Clearly, there is a need for a systematic and general approach that guarantees that all possible site exchange schemes that might imply different rate-dependent nmr line shapes are considered. Such an approach is formulated in this chapter.

II. Dynamic Stereochemistry of Symmetric Molecules

Before turning to the interpretation of dynamic nmr spectra, a brief introduction to the dynamic stereochemistry of symmetric molecules is in order. The material reviewed in this section has been presented in detail elsewhere (Klemperer, 1973a,b).

A. Configurations and Reactions

A configuration is a set of labeled ligands (chemical substituents) that have been assigned to labeled sites called skeletal positions. Assume that a set of n ligands contains n_1 ligands of a given chemical identity, n_2 ligands of another chemical identity, and n_3 ligands of a third chemical identity. Then $n = n_1 + n_2 + n_3$. These ligands are assigned labels from the set $\mathbf{L} = \{A_1, A_2, \ldots, A_{n_1}, B_{n_1+1}, B_{n_1+2}, \ldots, B_{n_1+n_2}, C_{n_1+n_2+1}, C_{n_1+n_2+2}, \ldots, C_n\}$ such that the letter indicates the chemical identity of each ligand and the integral subscript provides a unique index for each ligand. The skeletal positions of a configuration are assigned labels from the set $\mathbf{S}^W = \{s_1^W, s_2^W, \ldots, s_n^W\}$ such that ligands having chemical identities A, B, and C occupy skeletal positions labeled by $\{s_1^W, s_2^W, \ldots, s_{n_1}^W\}$, $\{s_{n_1+1}^W, s_{n_1+2}^W, \ldots, s_{n_1+n_2}^W\}$, and $\{s_{n_1+n_2+1}^W, s_{n_1+n_2+2}^W, \ldots, s_n^W\}$, respectively. Here, the superscript W serves to identify the geometry of the configuration, while the integral subscript provides a unique index for each skeletal position.

Chemically speaking, a configuration represents a set of labeled, rigid molecules, each having a definite orientation in space. In other words, a configuration may consist of only one, or of several molecules. Mathematically, a configuration is conveniently described by a $2 \times n$ matrix

$$\begin{pmatrix} l \\ s \end{pmatrix}_i^W = \begin{pmatrix} 1 & 2 & 3 & \cdots & n \\ j & k & p & \cdots & q \end{pmatrix}^W$$

where ligand indices are listed in the top row, and below each ligand index is written the index of the skeletal position which that ligand occupies. The superscript W identifies the geometry of the configuration, and the subscript

2. Delineation of Nuclear Exchange Processes

i is an integral label for each matrix. The reference configuration having geometry W is defined by

$$\begin{pmatrix} l \\ s \end{pmatrix}_e^W \equiv \begin{pmatrix} 1 & 2 & 3 & \cdots & n \\ 1 & 2 & 3 & \cdots & n \end{pmatrix}^W$$

Examples of this nomenclature are shown in Fig. 1a–c. Although discussion will proceed in terms of configurations that contain ligands having only three different chemical identities, extension to the general case should be clear.

A reaction is a transformation that converts one configuration into another configuration in such a way that measurable stereochemical change occurs. Mathematically, this transformation is defined by an operation h_i^{VW} that

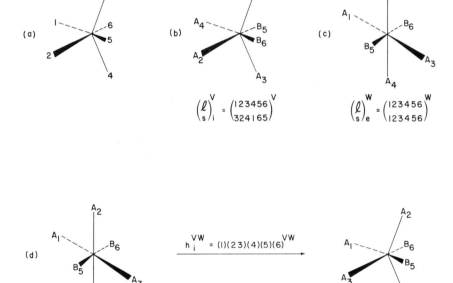

Fig. 1 Some configurations and an isomerization reaction of $RuH_2[P(C_6H_5)(OC_2H_5)_2]_4$. The skeletal positions of the cis isomer are indexed in (a). In (b) is shown a configuration of the cis isomer. The reference configuration for the trans isomer is given in (c). By designating a reference configuration, the indexing of skeletal positions is defined. Illustration (d) shows a trans–cis isomerization reaction.

transforms any configuration having geometry W, a reactant configuration $\begin{pmatrix} l \\ s \end{pmatrix}_j^W$, into a configuration having geometry V, a product configuration, $\begin{pmatrix} l \\ s \end{pmatrix}_k^V$, by letting the permutation operation h_i act on the indices of the skeletal positions listed in the bottom row of $\begin{pmatrix} l \\ s \end{pmatrix}_j^W$, and replacing the superscript W in $\begin{pmatrix} l \\ s \end{pmatrix}_j^W$ by V. The operation indicated in Fig. 1d describes the reaction illustrated since the bottom row of the matrix defining the product configuration is generated when $h_i = (1)(23)(4)(5)(6)$ operates on the bottom row of the matrix defining the reactant configuration. We write

$$(1)(23)(4)(5)(6)^{VW}\begin{pmatrix} 1 & 2 & 3 & 4 & 5 & 6 \\ 1 & 2 & 3 & 4 & 5 & 6 \end{pmatrix}^W = \begin{pmatrix} 1 & 2 & 3 & 4 & 5 & 6 \\ 1 & 3 & 2 & 4 & 5 & 6 \end{pmatrix}^V$$

Reactant and product configurations may have the same geometry, i.e., a reaction h_i^{WW} is a special case of particular importance. The only general restriction is that reactant and product configurations must have the same ligand set.

Given a set of n ligands labeled as above, we define a permutation group H, the group of allowed permutations, which acts on the set of numbers $\{1, 2, \ldots, n\}$. Operations in H permute numbers within the sets $\{1, 2, \ldots, n_1\}$, $(n_1 + 1, n_1 + 2, \ldots, n_1 + n_2\}$, and $\{n_1 + n_2 + 1, n_1 + n_2 + 2, \ldots, n\}$ among themselves in all possible ways. The group H therefore contains $n_1! \, n_2! \, n_3!$ operations. All possible reactions that convert configurations having geometry W into configurations having geometry V are contained in the set H^{VW}. The product of $h_j^{XV} \in H^{XV}$, i.e., an operation h_j^{XV} which is an element in the set H^{XV}, and $h_i^{VW} \in H^{VW}$ is defined by

$$h_j^{XV} \cdot h_i^{VW} \equiv (h_j \cdot h_i)^{XW}$$

and indicates the net result of h_i^{VW} followed by h_j^{XV}. Of course products like $h_j^{XV} \cdot h_j^{XV}$ are undefined. The inverse of h_i^{VW}, called the reverse reaction of h_i^{VW}, is defined by

$$(h_i^{VW})^{-1} \equiv (h_i^{-1})^{WV}$$

B. Configurational Symmetry Groups

Certain operations in the group H^{WW} do not represent reactions, since they do not represent measurable stereochemical change. Specifically, operations that represent molecular rotations, molecular translations, and/or internal motions that are immeasurably rapid on the observational time scale, and in any case rapid relative to the lifetime of the configuration, do not represent measurable stereochemical change. These operations form a group W^{ww}, the proper configurational symmetry group. Operations in W^{ww} are said to represent rotation of configuration. If all the molecules in a configuration have achiral skeletal frameworks, we say the configuration has an achiral skeletal framework. Then a transformation that inverts one or more molecules in a configuration having geometry W can be expressed by an operation in H^{ww}. A group $\overline{W}^{ww}$, the full configurational symmetry group, is formed by all operations in H^{ww} which (a) represent inversion of one or more molecules in the configuration or (b) are operations in W^{ww}. An operation in $\overline{W}^{ww}$, which represents inversion of all the molecules in a configuration, is said to represent inversion of configuration. If one or more molecules in a configuration having geometry W have chiral skeletal frameworks, we say the configuration has a chiral skeletal framework. In such a case we define a reference configuration $\binom{l}{s}_e^{w*}$, which is generated by inverting all (labeled) molecules in the reference configuration $\binom{l}{s}_e^{w}$. Then the proper configurational symmetry groups W^{ww} and W^{w*w*} are both representations of the same permutation group W. Reactions in the set $W^{ww}e^{ww*}W^{w*w*} = W^{ww*}$ and their reverse reactions represent inversion of configuration, where e is the identity permutation.

C. Symmetry Equivalent Reactions

If a chemical transformation is characterized solely by the stereochemical relationship between reactant and product molecules, then certain reactions as defined here characterize the same chemical transformation. This is because a reaction is defined by the relationship between configurations, and configurations represent sets of rigid molecules having definite orientations in space. If we assume that molecules are free to rotate and translate rapidly in space and possibly undergo rapid internal motions, then certain configurations become equivalent, i.e., physically indistinguishable, and therefore certain reactions become equivalent. Such equivalent reactions are defined to be nondifferentiable in a chiral environment. Formally, two reactions h_i^{vw} and h_j^{vw} in H^{vw} are nondifferentiable in a chiral environment if Eq. (1) holds for some

$$h_i = v_k \cdot h_j \cdot w_l \tag{1}$$

$v_k^{VV} \in V^{VV}$ and some $w_l^{WW} \in W^{WW}$ (Klemperer, 1973a). If Eq. (1) does not hold, then h_i^{VW} and h_j^{VW} are differentiable in a chiral environment. Equation (1) is an equivalency relation which may be used to partition H^{VW} into equivalency classes. If one reaction is selected from each class of equivalent reactions, a complete set of reactions in H^{VW} differentiable in a chiral environment is formed.

When reactions occur in an achiral environment, certain reactions that are not nondifferentiable in a chiral environment must occur with equal probability. Such reactions, enantiomeric reactions, are formally defined as follows. A "mirror image" h_j^{VW} of a reaction h_i^{VW} is defined by Eq. (2), where $\bar{v}_k^{VV} \in \bar{V}^{VV}$ and $\bar{w}_l^{WW} \in \bar{W}^{WW}$ both represent

$$h_j = \bar{v}_k \cdot h_i \cdot \bar{w}_l \tag{2}$$

inversion of configuration. If h_i^{VW} and h_j^{VW} are in addition differentiable in a chiral environment, they are defined to be enantiomeric reactions and each reaction is said to be a chiral reaction. If a reaction h_i^{VW} and its mirror images are nondifferentiable in a chiral environment, then h_i^{VW} is an achiral reaction. If two reactions are differentiable in a chiral environment and are not mirror images, they are defined to be diastereomeric. The set of reactions in H^{VW} can be partitioned into equivalency classes such that two reactions are in the same class if they are nondifferentiable in a chiral environment or enantiomeric. Choosing one reaction from each class, a complete set of diastereomeric reactions in H^{VW} is formed.

Equation (2) is of use only when both reactant and product configurations have achiral skeletal frameworks. If configurations having geometry V and/or configurations having geometry W have chiral skeletal frameworks, then all reactions in H^{VW} and H^{WV} are defined to be chiral reactions. When configurations having geometry V have chiral skeletal frameworks but configurations having geometry W have achiral skeletal frameworks, two reactions h_i^{VW} and h_j^{V*W} are enantiomeric if

$$h_i = v_k \cdot h_j \cdot \bar{w}_l$$

for some $v_k^{VV} \in V^{VV}$ and some $\bar{w}_l^{WW} \in \bar{W}^{WW}$ which represents inversion of configuration. Similarly, h_i^{WV} and h_j^{WV*} are enantiomeric if

$$h_i = \bar{w}_l \cdot h_j \cdot v_k$$

for some $\bar{w}_l^{WW} \in \bar{W}^{WW}$ which represents inversion of configuration and some $v_k^{VV} \in V^{VV}$. When configurations having geometry V and configurations having geometry W have chiral skeletal frameworks, two reactions h_i^{VW} and h_j^{V*W*} are enantiomeric reactions if h_i^{VW} and h_j^{VW} are nondifferentiable in a chiral environment [see Eq. (1)].

When dealing with equilibrium situations, the principle of microscopic reversibility implies that if a reaction h_i^{YW} occurs, then its reverse reaction $(h_i^{-1})^{WV}$ must also occur. When $V = W$, h_i^{VV} and $(h^1)_i^{-VV}$ must occur with equal probability.

D. Mechanism and Steric Course

A chemical reaction mechanism may be defined in terms of a potential energy surface in multidimensional space where reactant, intermediate, and product configurations are represented by points on the surface and mechanistic pathways are represented by paths on the surface which interconnect these points. Muetterties (1969) has defined the connectivity of a configuration to be the number of these paths which meet at the point which represents that configuration. If a single mechanism interconverts configurations having geometry W into configurations having geometry V, and all intermediate configurations have connectivities equal to two, then all reactions which characterize these interconversions are formally nondifferentiable from one reaction h_i^{YW}, its reverse reaction, or possibly its enantiomer or the reverse reaction of its enantiomer, in a chiral environment.

If intermediate configurations have connectivities greater than two, it may be necessary to use many reactions differentiable in a chiral environment to define the stereochemical changes implied by a given reaction mechanism.

$$\psi^{XV}(h_q^{XW}; h_p^{WV}) = \sum_{i=1}^{D_{XV}} a_i h_i^{XV} \tag{3}$$

Equation (3) defines an expression that indicates the stereochemical changes implied by the reaction sequence h_p^{WV} followed by h_q^{XW}. The reactant configurations have geometry V, the product configurations have geometry X, and the only intermediate configurations having connectivities greater than two have geometry W. In Eq. (3), D_{XV} is the number of reactions in a complete set of reactions in H^{XV} differentiable in a chiral environment, the set of h_i^{XV}'s appearing in the expression comprise a complete set of reactions in H^{XV} differentiable in a chiral environment, and a_i is the relative probability of a reaction nondifferentiable from h_i^{XV} in a chiral environment occurring each time a configuration having geometry V is converted into a configuration having geometry $\dot{X}$ via an intermediate configuration having geometry W. When $X = V$, an additional term $a_0 e^{WW}$ is added to the sum in Eq. (3), which indicates the relative probability of no net reaction occurring when h_p^{WV} is followed by $(h_p^{-1})^{VW}$. If memory effects are ruled out, $\psi^{XV}(h_q^{XW}; h_p^{WV})$ may be calculated (Klemperer, 1973a). The form of Eq. (3) is modified slightly when any of the configurations involved have chiral skeletal frameworks (Klemperer, 1973b).

Frequently, certain reactions in H^{VW} do not represent chemically allowed processes in that they would necessitate surmounting energy barriers that are effectively insuperable under the experimental conditions at hand. It is then often convenient to consider only reactions in H'^{VW}, a subset of H^{VW}, where H' is a subgroup of H called the restricted group of allowed permutations. Reactions in H^{VW} that are not in H'^{VW} are not "chemically allowed" reactions. A maximal set of reactions in H'^{VW} differentiable in a chiral environment may be formally derived from a complete set of reactions in H^{VW} differentiable in a chiral environment by eliminating those reactions which are not operations in H'^{VW}.

III. NMR Differentiable Reactions

A complete set of reactions in H^{VW} differentiable in a chiral environment defines all possible steric courses of reactions which convert configurations having geometry W into configurations having geometry V. Dynamic nuclear magnetic resonance spectroscopy, however, cannot distinguish between certain reactions even if they are differentiable in a chiral environment, since certain reactions imply site-exchange schemes that must generate identical rate-dependent line shapes. Such reactions are called nmr nondifferentiable reactions. In this section, a formal equivalency relation is derived that defines the relationship between two nmr nondifferentiable reactions.

A. The Effective NMR Symmetry Group

The formalisms presented in the preceding section may be used to define nuclear configurations if the word "ligand" is taken to mean nucleus or group of magnetically equivalent nuclei treated as a composite particle (e.g., the three ^{1}H nuclei in a freely rotating methyl group or three ^{19}F nuclei in PF_3), and the expression "skeletal position" is taken to mean nuclear site. Accordingly, ligand indices are nuclear labels, while the indices of the skeletal positions are nuclear coordinates.

The chemical shift of the nucleus (or magnetically equivalent nuclei) occupying skeletal position s_i^W is denoted v_i^W, while the coupling constant between the nuclei occupying skeletal positions s_i^W and s_j^W is denoted J_{ij}^W. The sets of chemical shifts and coupling constants for nuclei in configurations having geometry W are denoted $\{v_i^W\}$ and $\{J_{ij}^W\}$, respectively. If the nucleus occupying skeletal position s_i^W has no nuclear spin, then $v_i^W \equiv 0$ and $J_{ij}^W \equiv 0$ for all j. Since couplings within sets of magnetically equivalent nuclei do not affect nmr spectra (Gutowsky et al., 1953), J_{ij}^W is set equal to zero if nuclei occupying skeletal positions s_i^W and s_j^W are magnetically equivalent. Also, if

2. Delineation of Nuclear Exchange Processes

the resonances of certain nuclei are not of interest and these nuclei do not interact or exchange with those nuclei whose resonances are of interest, then the chemical shifts assigned to the sites of these nuclei are set equal to zero. It will be assumed throughout this chapter that all nuclei having a given chemical identity have the same nuclear spin.

For configurations having geometry W, we define a group $\hat{W}^{WW}$, the effective nmr symmetry group (Woodman, 1970), which contains all operations in H^{WW} that leave elements in the sets $\{v_i^W\}$ and $\{J_{ij}^W\}$ fixed when operating on the indices of the skeletal positions indicated as subscripts. The group $\hat{W}^{WW}$ is therefore a subgroup of H^{WW}, and the groups $\overline{W}^{WW}$ and W^{WW} are subgroups of $\hat{W}^{WW}$. Note that the definition of the effective nmr symmetry group given by Woodman (1970) has been modified only in that (a) $\hat{W}^{WW}$ may act on nonmagnetic nuclei, (b) $\hat{W}^{WW}$ acts on nuclear coordinates, not nuclear labels, (c) a group of magnetically equivalent nuclei may be treated as a composite particle and assigned one label and one coordinate, and (d) the configuration with which $\hat{W}^{WW}$ is associated may contain more than one molecule.

B. Definition of NMR Nondifferentiable Reactions

When analyzing dynamic nmr spectra, the assumption is customarily made that site exchange occurs so rapidly that although the Hamiltonian of the spin system may change, the individual spin wave functions do not (Kaplan and Fraenkel, 1972). For each nuclear configuration, a Hamiltonian can be written, and a reaction detectable by dynamic nmr spectroscopy implies a switch from one Hamiltonian to another (Johnson, 1964, 1965). If there is a one-to-one correspondence between configurations and Hamiltonians, then different reactions imply different switches of Hamiltonians. In many cases, however, there is a many-to-one correspondence between configurations and Hamiltonians, and therefore certain reactions, nmr nondifferentiable reactions, imply the same switch of Hamiltonians. The problem of formally defining nmr nondifferentiable reactions is that of defining precisely which reactions imply the same switch of Hamiltonians.

The appropriate Hamiltonian for the reference configuration having geometry W is given in Eq. (4) (Binsch, 1968). This Hamiltonian assumes a frame

$$\mathcal{H}_e^W = -\sum_i (v_i^W - v)I_i^z + \sum_{i<j} J_{ij}^W \mathbf{I}_i \cdot \mathbf{I}_j + \gamma H_1 \sum_i I_i^x \qquad (4)$$

rotating with angular frequency $\omega = 2\pi v$, v_i^W and J_{ij}^W were defined above, H_1 is the strength of the rotating rf field, and the subscripts on the spin operators are nuclear labels. For each configuration in the set $H^{WW} \binom{l}{s}_e^W$, a Hamiltonian may be written. The Hamiltonian $\mathcal{H}_j^W$ for the configuration $h_i^{WW}\begin{pmatrix} l \\ s \end{pmatrix}_e^W$ is generated by letting h_i act on the nuclear coordinates which index the chemical shifts and coupling constants in $\mathcal{H}_e^W$. We write

$$\mathcal{H}_j^W = h_i^{WW}(\mathcal{H}_e^W)$$

Certain operations in H^{WW}, operations in the effective nmr symmetry group $\hat{W}^{WW}$, leave the Hamiltonians for configurations having geometry W fixed,* i.e.,

$$\mathcal{H}_j^W = \hat{w}_i^{WW}(\mathcal{H}_j^W)$$

for any $\hat{w}_i^{WW} \in \hat{W}^{WW}$. Therefore two configurations $h_i^{WW}\begin{pmatrix} l \\ s \end{pmatrix}_e^W$ and $\hat{w}_i^{WW} \cdot h_i^{WW}\begin{pmatrix} l \\ s \end{pmatrix}_e^W$ have the same Hamiltonian. Formally, there is a one-to-one correspondence between right cosets of $\hat{W}^{WW}$ in H^{WW} and Hamiltonians for configurations having geometry W. The set of all $|H^{WW}|/|\hat{W}^{WW}|$ different Hamiltonians for configurations having geometry W is denoted $\{\mathcal{H}^W\}$.

A reaction $h_i^{VW} \in H^{VW}$ defines a mapping of $\{\mathcal{H}^W\}$ onto $\{\mathcal{H}^V\}$. If h_i^{VW} maps $\mathcal{H}_l^W$ to $\mathcal{H}_k^V$, we write

$$\mathcal{H}_k^V = h_i^{VW}(\mathcal{H}_l^W) \tag{5}$$

The operation indicated in Eq. (5) is performed by letting h_i permute the nuclear coordinates that index the chemical shifts and coupling constants in $\mathcal{H}_l^W$ and then replacing W by V wherever it appears in $\mathcal{H}_l^W$. If two reactions in H^{VW} imply the same mapping of $\{\mathcal{H}^W\}$ onto $\{\mathcal{H}^V\}$, they are nmr nondifferentiable reactions. Formally, h_i^{VW} and h_j^{VW} are nmr nondifferentiable if

$$h_i = \hat{v}_m \cdot h_j \cdot \hat{w}_n \tag{6}$$

for some $\hat{v}_m^{VV} \in \hat{V}^{VV}$ and some $\hat{w}_n^{WW} \in \hat{W}^{WW}$. If Eq. (6) does not hold, h_i^{VW} and h_j^{VW} are nmr differentiable reactions. Equation (6) is proved as follows. Assume that Eq. (5) holds for an arbitrary $\mathcal{H}_l^W \in \{\mathcal{H}^W\}$. We must first show that if Eqs. (5) and (6) hold, then Eq. (7) holds.

$$\mathcal{H}_k^V = h_j^{VW}(\mathcal{H}_l^W) \tag{7}$$

---

* It is important to note that operations in H^{WW} are coordinate transformations, not operators. Formally, the operators in the set $\{\mathcal{H}^W\}$ are invariant under any transformation $\hat{w}_i^{WW} \in \hat{W}^{WW}$, but it is meaningless to define commutation relations unless the operations in the effective nmr symmetry group are represented by Dirac permutation operators.

2. Delineation of Nuclear Exchange Processes

Substituting Eq. (6) into Eq. (5), Eq. (8) is obtained.

$$\mathcal{H}_k{}^V = (\hat{v}_m \cdot h_j \cdot \hat{w}_n)^{VW}(\mathcal{H}_l^W) = \hat{v}_m^{VV} \cdot h_j^{VW} \cdot \hat{w}_n^{WW}(\mathcal{H}_l^W) \tag{8}$$

Left multiplying both sides of Eq. (8) by $(\hat{v}_m^{-1})^{VV}$ yields Eq. (9).

$$(\hat{v}_m^{-1})^{VV}(\mathcal{H}_k^V) = h_j^{VW} \cdot \hat{w}_n^{WW}(\mathcal{H}_l^W) \tag{9}$$

Since $(\hat{v}_m^{-1})^{VV}(\mathcal{H}_k^V) = \mathcal{H}_k^V$ and $\hat{w}_n^{WW}(\mathcal{H}_l^W) = \mathcal{H}_l^W$, Eq. (7) follows. To complete the proof, Eq. (6) is rearranged to give Eq. (10).

$$h_j = \hat{v}_m^{-1} \cdot h_i \cdot \hat{w}_n^{-1} \tag{10}$$

Following the procedure used in the first half of the proof, Eq. (10) may be substituted into Eq. (7) to yield Eq. (5).

When dealing with chiral configurations, further classes of nmr nondifferentiable reactions may be determined. Inversion of a chiral configuration does not alter the environments of any nuclei, and therefore the Hamiltonian for that configuration is invariant under the inversion operation. If configurations having geometry V have chiral skeletal frameworks, the Hamiltonians $\mathcal{H}_i^V$ and $\mathcal{H}_i^{V*} = e^{V*V}(\mathcal{H}_i^V)$ are identical, and we let $\{\mathcal{H}^V\}$ be the set of Hamiltonians for configurations having geometry V^*. Consequently, if all configurations having geometry V and W have chiral skeletal frameworks, the reactions h_i^{VW}, $h_i^{VW*} = h_i^{VW} \cdot e^{WW*}$, $h_i^{V*W} = e^{V*V} \cdot h_i^{VW}$, and $h_i^{V*W*} = e^{V*V} \cdot h_i^{VW} \cdot e^{WW*}$ are nmr nondifferentiable. Also, if configurations having geometry V have chiral skeletal frameworks and configurations having geometry W have achiral skeletal frameworks, then h_i^{VW} and $h_i^{V*W} = e^{V*V} \cdot h_i^{VW}$ as well as h_j^{WV} and $h_j^{WV*} = h_j^{WV} \cdot e^{VV*}$ are nmr nondifferentiable.

For the sake of completeness, we note that since $\overline{W}^{WW}$ is a subgroup of $\hat{W}^{WW}$, enantiomeric reactions must always be nmr nondifferentiable. However, there is no guarantee that a reaction h_i^{VV} and its reverse reaction $(h_i^{-1})^{VV}$ are nmr nondifferentiable.

C. Complete Sets of NMR Differentiable Reactions

The equivalency relation defined by Eq. (6) may be used to partition H^{VW} into equivalency classes. The number of these equivalency classes, $D_{\hat{v}\hat{w}}$, is called the number of nmr differentiable reactions* in H^{VW}. By selecting one reaction from each equivalency class, a complete set of nmr differentiable reactions in H^{VW} is formed. It must be emphasized that there is in general no guarantee that two nmr differentiable reactions generate dependences of line

---

* If $V = W$ and $\hat{V}^{VV} = V^{VV}$, then one of the $D_{\hat{v}\hat{v}}$ equivalency classes in H^{VV} generated by Eq. (6) contains no reactions, and therefore the number of nmr differentiable reactions in H^{VV} is $D'_{\hat{v}\hat{v}} = D_{\hat{v}\hat{v}} - 1$.

shape on rate that are distinguishable. The important point is that nmr nondifferentiable reactions must generate identical dependences of line shape on rate, and therefore any reaction in H^{VW} will generate line shapes identical to those generated by some reaction in the complete set of nmr differentiable reactions in H^{VW}. Accordingly, an expression of the form

$$\hat{\psi}^{VW} = \sum_{i=1}^{D\hat{v}\hat{w}} a_i h_i^{VW} \tag{11}$$

may be used to characterize the site exchange scheme implied by any process that converts configurations having geometry W into configurations having geometry V. In Eq. (11), the h_i^{VW}'s appearing in the summation form a complete set of nmr differentiable reactions in H^{VW}, and a_i is the relative probability of reactions occurring that are nmr nondifferentiable from h_i^{VW}.

Since W^{WW} is a subgroup of $\hat{W}^{WW}$, if two reactions are nondifferentiable in a chiral environment, then they must be nmr nondifferentiable. Thus if the steric course of a reaction sequence [see Eq. (3)] is characterized by Eq. (12),

$$\psi^{XV}(h_q^{XW}; h_q^{WV}) = a_1 h_1^{XV} + a_2 h_2^{XV} + a_3 h_3^{XV} \tag{12}$$

and h_1^{XV} and h_2^{XV} are nmr nondifferentiable but h_1^{XV} and h_3^{XV} are nmr differentiable, then Eq. (13) must hold.

$$\hat{\psi}^{XV}(h_q^{XW}; h_p^{WV}) = (a_1 + a_2)h_1^{XV} + a_3 h_3^{XV} \tag{13}$$

Using arguments presented elsewhere (Klemperer, 1973a,b), the reader may verify that if $\{h_1^{VW}, h_2^{VW}, \ldots, h_{D\hat{v}\hat{w}}^{VW}\}$ is a complete set of nmr differentiable reactions* in H^{VW}, then

(a) $\{(h_1^{-1})^{WV}, (h_2^{-1})^{WV}, \ldots, (h_{D\hat{v}\hat{w}}^{-1})^{WV}\}$ is a complete set of nmr differentiable reactions in H^{WV},

(b) $\{h_1^{V^*W}, h_2^{V^*W}, \ldots, h_{D\hat{v}\hat{w}}^{V^*W}\}$ is a complete set of nmr differentiable reactions in H^{V^*W} if configurations having geometry V have chiral skeletal frameworks,

(c) $\{h_1^{VW^*}, h_2^{VW^*}, \ldots, h_{D\hat{v}\hat{w}}^{VW^*}\}$ is a complete set of nmr differentiable reactions in H^{VW^*} if configurations having geometry W have chiral skeletal frameworks.

D. *Enumeration of NMR Differentiable Reactions*

When determining a complete set of nmr differentiable reactions in H^{VW}, it is often of great help to know in advance the number of reactions in that set, $D_{\hat{v}\hat{w}}$. A formula that enumerates the set of nmr differentiable reactions in H^{VW} is provided here.

* See footnote on p. 33.

2. Delineation of Nuclear Exchange Processes

First, however, we define the generalized cyclic type, $(j_1, j_2, \ldots, j_{n_1}; k_1, k_2, \ldots, k_{n_2}; l_1, l_2, \ldots, l_{n_3})$, of a permutation $h_i \in H$. Recall that operations in H permute numbers in the sets $\{1, 2, \ldots, n_1\}$, $\{n_1 + 1, n_1 + 2, \ldots, n_1 + n_2\}$, and $\{n_1 + n_2 + 1, n_1 + n_2 + 2, \ldots, n\}$ among themselves. The array $(j_1, j_2, \ldots, j_{n_1}; k_1, k_2, \ldots, k_{n_2}; l_1, l_2, \ldots, l_{n_3})$ indicates that the permutation h_i contains j_p cycles of length p which permute the numbers $\{1, 2, \ldots, n_1\}$ among themselves, k_q cycles of length q which permute the numbers $\{n_1 + 1, n_1 + 2, \ldots, n_1 + n_2\}$ among themselves, and l_r cycles of length r which permute the numbers $\{n_1 + n_2 + 1, n_1 + n_2 + 2, \ldots, n\}$ among themselves. For example, the generalized cyclic type of the permutation h_i, which is indicated in Fig. 1d, is (2, 1, 0, 0; 2, 0).

Equation (14) is used to count the number of nmr differentiable reactions* in H^{VW}. This formula is a trivial modification of a theorem by de Bruijn (1964)

$$D_{\hat{V}\hat{W}} = \frac{1}{|\hat{V}^{VV}||\hat{W}^{WW}|} \sum_{\hat{V}^{VV}, \hat{W}^{WW}} (h^{\hat{V}}_{j_1 \cdots j_{n_1}, k_1 k_2 \cdots k_{n_1}, l_1 l_2 \cdots l_{n_3}})$$

$$\times (h^{\hat{W}}_{j_1 j_2 \cdots j_{n_1}, k_1 k_2 \cdots k_{n_2}, l_1 l_2 \cdots l_{n_3}}) \prod_{p=1}^{n_1} (j_p! \, p^{j_p}) \prod_{q=1}^{n_2} (k_q! \, q^{k_q}) \prod_{r=1}^{n_3} (l_r! \, r^{l_r}) \quad (14)$$

and may be derived following procedures used by Klemperer (1972a,b) in another context. In Eq. (14), $|\hat{V}^{VV}|$ and $|\hat{W}^{WW}|$ are the numbers of operations in $\hat{V}^{VV}$ and $\hat{W}^{WW}$, respectively; the summation extends over all generalized cyclic types of permutations in $\hat{V}$ and $\hat{W}$; and $h^{\hat{V}}_{j_1 j_2 \cdots j_{n_1}, k_1 k_2 \cdots k_{n_2}, l_1 l_2 \cdots l_{n_3}}$ and $h^{\hat{W}}_{j_1 j_2 \cdots j_{n_1}, k_1 k_2 \cdots k_{n_2}, l_1 l_2 \cdots l_{n_3}}$ are the number of operations in $\hat{V}$ and $\hat{W}$, respectively, having generalized cyclic type $(j_1, j_2, \ldots, j_{n_1}; k_1, k_2, \ldots, k_{n_2}; l_1, l_2, \ldots, l_{n_3})$.

Note that the use of Eq. (14) does not demand explicit knowledge of the permutations in $\hat{V}$ and $\hat{W}$, but only the generalized cyclic types of permutations in these groups. When dealing with large groups, $\hat{V}$ and $\hat{W}$ may often be expressed in terms of simpler groups exactly as the groups V and W may be expressed in terms of simpler groups (Klemperer, 1973a,b). Then the cyclic index (Harary, 1969) of the simple groups may be calculated and used to generate the generalized cyclic types of permutations in $\hat{V}$ and $\hat{W}$.

IV. Examples

A. 3-Hydroxy-2,4-dimethylcyclobutenone

Chickos et al. (1972) have studied the methine proton exchange of 3-hydroxy-2,4-dimethylcyclobutenone by dynamic nuclear magnetic resonance

---
* See footnote on p. 33.

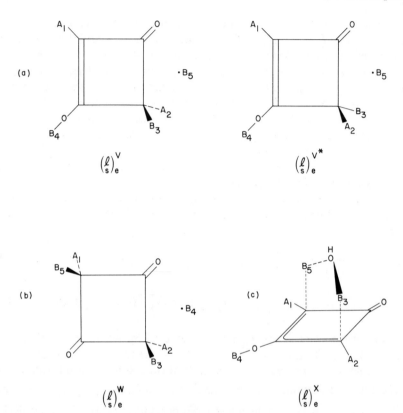

Fig. 2 Configurations representing 3-hydroxy-2,4-dimethylcyclobutenone in aqueous solution.

spectroscopy. The reference configurations of interest are shown in Fig. 2a. Here, skeletal positions s_1^V, s_2^V, s_1^{V*}, and s_2^{V*} are occupied by methyl groups; s_3^V and s_3^{V*} are occupied by methine protons; s_4^V and s_4^{V*} are occupied by hydroxyl protons; and s_5^V and s_5^{V*} are occupied by protons in solution. The group of allowed permutations contains $2! \cdot 3! = 12$ operations. If we assume that the hydroxyl proton exchanges with protons in solution at an immeasurably rapid rate on the nmr time scale, then $\hat{V}^{VV}$ and $\hat{V}^{V*V*}$ representations of the group $\hat{V}$ which contains two operations, $\hat{v}_1 = (1)(2)(3)(4)(5)$ and $\hat{v}_2 = (1)(2)(3)(45)$.

Using Eq. (14),

$$D_{\hat{V}\hat{V}} = \frac{1}{2 \cdot 2} \{1 \cdot 1 \cdot 2! \cdot 1^2 \cdot 3! \cdot 1^3 + 1 \cdot 1 \cdot 2! \cdot 1^2 \cdot 1! \cdot 1^1 \cdot 1! \cdot 2^1\} = 4$$

2. Delineation of Nuclear Exchange Processes

The following four reactions form a complete set of nmr differentiable reactions in H^{VV}.

$$h_1^{VV} = (1)(2)(3)(45)^{VV}$$
$$h_2^{VV} = (12)(3)(4)(5)^{VV}$$
$$h_3^{VV} = (1)(2)(35)(4)^{VV}$$
$$h_4^{VV} = (12)(35)(4)^{VV}$$

Consequently, the sets of reactions $\{h_1^{V \cdot V}, h_2^{V \cdot V}, h_3^{V \cdot V}, h_4^{V \cdot V}\}$, $\{h_1^{VV^*}, h_2^{VV^*}, h_3^{VV^*}, h_4^{VV^*}\}$, and $\{h_1^{V \cdot V^*}, h_1^{V \cdot V^*}, h_3^{V \cdot V^*}, h_4^{V \cdot V^*}\}$ form complete sets of nmr differentiable reactions in $H^{V \cdot V}$, H^{VV^*}, and $H^{V \cdot V^*}$, respectively, but for a given $i = 1$–4, the reactions h_i^{VV}, $h_i^{V \cdot V}$, $h_i^{VV^*}$, and $h_i^{V \cdot V^*}$, are nmr nondifferentiable. Since $h_1^{VV} \in \hat{V}^{VV}$, this reaction implies no change in nmr line shapes in the exchange region of interest. Therefore, all possible site exchange schemes may be generated using combinations of the reactions h_2^{VV}, h_3^{VV}, and h_4^{VV}.

An intramolecular, suprafacial [1,3] hydride shift implies $h_2^{V \cdot V}$ and its enantiomer $h_2^{VV^*}$. This mechanism involves no intermediate configurations having connectivities greater than two. For certain intermolecular mechanisms, connectivities of intermediate configurations may be greater than two. Consider, for example, a mechanism involving an intermediate configuration having the geometry shown in Fig. 2b. The reaction $h_5^{WV} = (1)(2)(3)(4)(5)^{WV}$ describes the transformation of $\begin{pmatrix} l \\ s \end{pmatrix}_e^V$ into $\begin{pmatrix} l \\ s \end{pmatrix}_e^W$. Note that the groups W^{WW} and $\overline{W}^{WW}$ contain but two operations,

$$w_1^{WW} = \overline{w}_1^{WW} = (1)(2)(3)(4)(5)^{WW} = h_5^{WW}$$
$$w_2^{WW} = \overline{w}_2^{WW} = (12)(35)(4)^{WW} = h_4^{WW}$$

and both operations represent rotation of configuration as well as inversion of configuration. Therefore, h_5^{WV} and its enantiomer, $h_5^{WV^*} = h_5^{WW} \cdot h_5^{WV} \cdot e^{VV^*}$, must occur with equal probability in an achiral environment. The net reactions implied by this mechanism are contained in the sets $h_5^{VW} W^{WW} h_5^{WV} = W^{VV}$, $h_5^{V \cdot V} W^{WW} h_5^{WV} = W^{V \cdot V}$, $h_5^{WV} W^{WW} h_5^{WV^*} = W^{VV^*}$, and $h_5^{V \cdot W} W^{WW} h_5^{WV^*} = W^{V \cdot V^*}$. The steric course of this mechanism is therefore defined by Eq. (15),

$$\psi^{V \cdot V, VV}(h_5^{VW}, h_5^{V \cdot W}; h_5^{WV}) = e^{V \cdot V} + h_4^{V \cdot V} + e^{VV} + h_4^{VV} \qquad (15)$$

and Eq. (16) characterizes the site exchange scheme implied by the mechanism.

$$\hat{\psi}^{VV}(h_5^{VW}; h_5^{WV}) = h_1^{VV} + 2h_4^{VV} \qquad (16)$$

Turning to a mechanism involving an intermediate configuration having the geometry shown in Fig. 2c, we note

$$\psi^{V \cdot V, VV}(h_4^{V \cdot X}, h_5^{VX}; h_5^{XV}) = h_4^{V \cdot V} + e^{VV} \qquad (17)$$

and

$$\hat{\psi}^{VV}(h_5^{YX}; h_5^{XV}) = h_4^{VV} \tag{18}$$

Comparing Eqs. (16) and (18), it is evident that these two mechanisms imply identical dependency of the nmr line shapes on rate. However, comparison of Eqs. (15) and (17) reveals that the two mechanisms have different steric courses and therefore suitable experiments could distinguish between these two mechanisms. Replacing the methyl groups with isopropyl groups might allow such a choice to be made.

B. *The Cation 1,2-Bis(diphenylphosphino)ethane(2-methylallyl)palladium(II)*

The pyridine catalyzed permutational isomerization of the cation 1,2-bis(diphenylphosphino)ethane(2-methylallyl)palladium(II) has been studied by Tibbets and Brown (1970) using dynamic nuclear magnetic resonance techniques. The configuration shown in Fig. 3a represents this species, and the groups W^{WW}, $\overline{W}^{WW}$, and $\hat{W}^{WW}$ contain these operations:

$$w_1^{WW} = \overline{w}_1^{WW} = \hat{w}_1^{WW} = (1)(2)(3)(4)(5)(6)(7)^{WW}$$
$$\overline{w}_2^{WW} = \hat{w}_2^{WW} = (14)(23)(56)(7)^{WW}$$

To simplify discussion, it is assumed that breaking of carbon–carbon and carbon–hydrogen bonds is not allowed. Therefore only 16 of the $4! \cdot 2! \cdot 1! = 48$ operations in H^{WW} are "chemically allowed." These 16 operations form a group, H'^{WW}, which is a representation of the restricted group of allowed permutations, H'. Only 15 of the 16 operations in H'^{WW} are reactions:

$$h_2'^{WW} = (14)(23)(56)(7)^{WW}$$
$$\{h_3'^{WW} = (1)(2)(3)(4)(56)(7)^{WW}$$
$$\,h_4'^{WW} = (14)(23)(5)(6)(7)^{WW}$$
$$\{h_5'^{WW} = (12)(34)(5)(6)(7)^{WW}$$
$$\,h_6'^{WW} = (13)(24)(56)(7)^{WW}$$
$$\{h_7'^{WW} = (12)(34)(56)(7)^{WW}$$
$$\,h_8'^{WW} = (13)(24)(5)(6)(7)^{WW}$$
$$\{h_9'^{WW} = (12)(3)(4)(5)(6)(7)^{WW} \quad h_{10}'^{WW} = (1)(2)(34)(5)(6)(7)^{WW}$$
$$\,h_{11}'^{WW} = (1423)(56)(7)^{WW} \quad h_{12}'^{WW} = (1324)(56)(7)^{WW}$$
$$\{h_{13}'^{WW} = (12)(3)(4)(56)(7)^{WW} \quad h_{14}'^{WW} = (1)(2)(34)(56)(7)^{WW}$$
$$\,h_{15}'^{WW} = (1423)(5)(6)(7)^{WW} \quad h_{16}'^{WW} = (1324)(5)(6)(7)^{WW}$$

All fifteen reactions are differentiable in a chiral environment, enantiomeric reactions have been written on the same line, and brackets enclose nmr nondifferentiable reactions. The set $\{h_2'^{WW}, h_3'^{WW}, h_5'^{WW}, h_7'^{WW}, h_9'^{WW}, h_{13}'^{WW}\}$ forms a maximal set of nmr differentiable reactions in H'^{WW}.

In the absence of a catalyst, nmr line shapes show no temperature de-

2. Delineation of Nuclear Exchange Processes

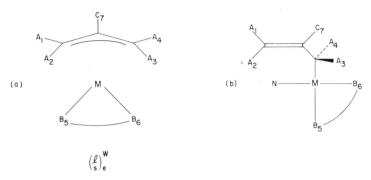

Fig. 3 The configuration shown in (a) represents the cation 1,2(diphenylphosphino)-ethane(2-methylallyl)palladium. In illustration (b) is shown the geometry of a postulated intermediate for the pyridine catalyzed permutational isomerization of this cation. The pyridine ligand is represented by N.

pendence. This observation implies that either permutational isomerization is not occurring or a reaction nmr nondifferentiable from $h_2'^{WW}$ is occurring. Five rearrangement mechanisms have been proposed for intramolecular permutational isomerization reactions of allyl complexes: (I) planar allyl flip, (II) π-rotation, (III) allyl flip with syn–anti exchange, (IV) rotation about an allyl carbon–carbon bond, and (V) π–σ equilibrium. As shown by Klemperer (1973b), each mechanism implies a site exchange scheme characterized by the indicated combination of reactions: (I) $h_2'^{WW}$, (II) $h_3'^{WW}$, (III) $h_5'^{WW}$, (IV) $h_9'^{WW}$, (V) $h_3'^{WW} + h_9'^{WW} + h_{13}'^{WW}$. Thus, only one of these mechanisms is consistent with experiment.

In the presence of increasing concentrations of pyridine, nmr line broadening and coalescence is observed. Line shape simulations indicate that syn–anti proton exchange but no net rotation of the allyl ligand relative to the diphos ligand occurs. Such an exchange scheme is implied by $h_5'^{WW}$ and $h_9'^{WW}$, but not by $h_2'^{WW}$, $h_3'^{WW}$, $h_7'^{WW}$, or $h_{13}'^{WW}$. Tibbetts and Brown realized that more than one mechanism is consistent with this exchange scheme, and they chose a mechanism involving the intermediate configuration shown in Fig. 3b on the basis of ancillary evidence. The reader may verify that the steric course of this mechanism is defined by $h_{11}'^{WW} + h_{12}'^{WW}$.

C. Dihydridotetrakis(diethoxyphenylphosphine)ruthenium(II)

The complex $RuH_2[P(C_6H_5)(OC_2H_5)_2]_4$ exists in solution as cis and trans isomers (see Fig. 1) and the interconversion of these isomers is detectable by dynamic nuclear magnetic resonance spectroscopy (Meaking et al., 1973). The slow exchange limit is reached at 21°C, where resonances for both isomers are observed. We ask the question here whether temperature dependent line shape analysis might provide any information regarding the mechanism of cis–trans interconversion.

Using the labeling schemes shown in Fig. 1, $\hat{V}^{VV} = C_{2v}^{VV}$ and $\hat{W}^{WW} = H^{WW}$, since both protons as well as all phosphorus nuclei in the trans isomer are magnetically equivalent. Using Eq. (14), we find $D_{\hat{W}\hat{V}} = 1$, i.e., all reactions in H^{WV} are nmr nondifferentiable. As a result, temperature-dependent lineshape analysis can yield no mechanistic information.

D. Sulfur Tetrafluoride

In the preceding examples, relatively simple systems have been examined in order to show how the mathematical formalisms developed in Sections II and III rigorously provide information which might have been obtained by less formal, intuitive reasoning. When dealing with more complex systems, however, the formal approach becomes mandatory. In this final example, we shall examine the somewhat more complex fluorine exchange reactions of sulfur tetrafluoride.

Sulfur tetrafluoride has C_{2v} symmetry and in the low temperature limit an A_2B_2 ^{19}F spectrum is observed (Bacon et al., 1963). As the temperature is raised, these lines coalesce and to a single peak (Cotton et al., 1958). Muetterties and Phillips (1959) found that the exchange rate is dependent on SF_4 concentration, and also ruled out wall reaction and homolytic dissociation as mechanistic possibilities. In spite of subsequent spectroscopic (Redington and Berney, 1965, 1967; Frey et al., 1971) and chemical (Muetterties and Phillips, 1967) studies, it is unclear whether the fluorine exchange process is intermolecular or intramolecular, or whether the exchange is catalyzed by fluoride impurities. Of interest, therefore, is the question of how much mechanistic information might be obtained by dynamic nmr studies.

1. Exchange Involving Fluoride Impurities

Exchange reactions involving fluoride impurities interconvert configurations having the geometry shown in Fig. 4a. Here, the ligand occupying skeletal position s_5^W may be any fluoride impurity or simply the fluoride anion. The effective nmr symmetry group $\hat{W}^{WW}$ contains the following operations:

$$\hat{w}_1^{WW} = (1)(2)(3)(4)(5)^{WW}$$
$$\hat{w}_2^{WW} = (13)(24)(5)^{WW}$$
$$\hat{w}_3^{WW} = (1)(24)(3)(5)^{WW}$$
$$\hat{w}_4^{WW} = (13)(2)(4)(5)^{WW}$$

2. Delineation of Nuclear Exchange Processes

The set of $5! - 2 = 118$ reactions in H^{WW} contains $D_{\hat{W}\hat{W}} = 11$ nmr differentiable reactions. A complete set of nmr differentiable reactions in H^{WW} is formed by the following reactions:

$$h_2^{WW} = (1)(24)(3)(5)^{WW}$$
$$h_3^{WW} = (12)(3)(4)(5)^{WW}$$
$$h_4^{WW} = (12)(34)(5)^{WW}$$
$$h_5^{WW} = (15)(2)(3)(4)^{WW}$$
$$h_6^{WW} = (15)(2)(34)^{WW}$$
$$h_7^{WW} = (1)(25)(3)(4)^{WW}$$
$$h_8^{WW} = (1)(25)(34)^{WW}$$
$$h_9^{WW} = (1)(2)(345)^{WW}$$
$$h_{10}^{WW} = (1)(2)(354)^{WW}$$
$$h_{11}^{WW} = (12)(345)^{WW}$$
$$h_{12}^{WW} = (12)(354)^{WW}$$

Reactions h_2^{WW}–h_4^{WW} are permutational isomerization reactions, while the remaining eight reactions represent intermolecular exchange processes. Note that h_9^{WW} and h_{10}^{WW} as well as h_{11}^{WW} and h_{12}^{WW} must occur with equal probability, since reactant and product configurations have identical free energies and thus a reaction and its reverse reaction must occur with equal probability. Consequently, only nine of the eleven processes in fact imply distinct site exchange schemes.

a. *Associative Mechanisms.* A reasonable C_{4v} intermediate for an associative mechanism is shown in Fig. 4b. The set of reactions in H^{XW} contains sixteen reactions differentiable in a chiral environment and eleven diastereomeric reactions. The following reactions form a complete set of diastereomeric reactions in H^{XW}:

$$h_{13}^{XW} = (14352)^{XW}$$
$$h_{14}^{XW} = (142)(35)^{XW}$$
$$h_{15}^{XW} = (1452)(3)^{XW}$$
$$h_{16}^{XW} = (1432)(5)^{XW}$$
$$h_{17}^{XW} = (1)(2435)^{XW}$$
$$h_{18}^{XW} = (1)(245)(3)^{XW}$$
$$h_{19}^{XW} = (1)(2453)^{XW}$$
$$h_{20}^{XW} = (1)(24)(3)(5)^{XW}$$
$$h_{21}^{XW} = (1)(254)(3)^{XW}$$
$$h_{22}^{XW} = (1)(2)(354)^{XW}$$
$$h_{23}^{XW} = (1)(2)(3)(45)^{XW}$$

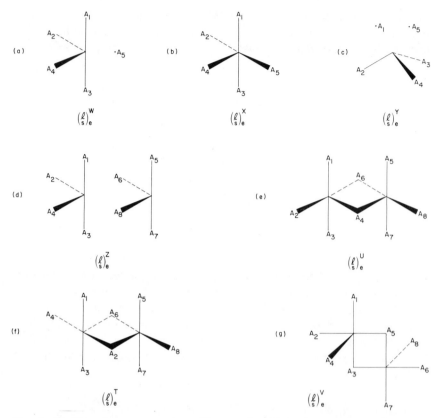

Fig. 4 Configurations representing sulfur tetrafluoride and postulated intermediates for fluorine exchange mechanisms.

Here, h_{14}^{XW}, h_{15}^{XW}, h_{17}^{XW}, h_{19}^{XW}, and h_{22}^{XW} are chiral, while the remaining six reactions are achiral. The reader may verify that

$$\hat{\psi}^{WW}(h_{13}^{-1WX}; h_{13}^{XW}) = \hat{\psi}^{WW}(h_{16}^{-1WX}; h_{16}^{XW}) = 2h_8^{WW} + h_5^{WW}$$
$$\hat{\psi}^{WW}(h_{14}^{-1WX}; h_{14}^{XW}) = \hat{\psi}^{WW}(h_{15}^{-1WX}; h_{15}^{XW})$$
$$= 2h_3^{WW} + 2h_5^{WW} + 2h_7^{WW} + 4h_8^{WW} + 2h_9^{WW} + 2h_{10}^{WW}$$
$$\hat{\psi}^{WW}(h_{17}^{-1WX}; h_{17}^{XW}) = \hat{\psi}^{WW}(h_{19}^{-1WX}; h_{19}^{XW})$$
$$= 2h_3^{WW} + 2h_5^{WW} + 4h_6^{WW} + 2h_7^{WW} + 2h_9^{WW} + 2h_{10}^{WW}$$
$$\hat{\psi}^{WW}(h_{18}^{-1WX}; h_{18}^{XW}) = \hat{\psi}^{WW}(h_{20}^{WX}; h_{20}^{XW}) = 2h_6^{WW} + h_7^{WW}$$
$$\hat{\psi}^{WW}(h_{21}^{-1WX}; h_{21}^{XW}) = \hat{\psi}^{WW}(h_{23}^{-1WX}; h_{23}^{WX}) = 2h_4^{WW}$$
$$\hat{\psi}^{WW}(h_{22}^{-1WX}; h_{22}^{XW}) = 2h_2^{WW} + 8h_3^{WW} + 4h_4^{WW}$$

This result implies that only six different classes of associative mechanisms involving a C_{4v} intermediate might be differentiated by dynamic nmr spectroscopy. We have assumed, of course, that the lifetime of the metastable C_{4v} intermediate is very short on the nmr time scale. Should the intermediate be detectable by nmr, less mechanistic information may be obtained, since $D_{\hat{x}\hat{w}} = 3$ and $\{h_{13}^{XW}, h_{17}^{XW}, h_{21}^{XW}\}$ forms a complete set of nmr differentiable reactions in H^{XW}.

b. *Dissociative Mechanisms.* A reasonable cationic C_{3v} intermediate for a dissociative mechanism is shown in Fig. 4c. The set of reactions in H^{YW} contains only four dissociation reactions differentiable in a chiral environment; each of which is achiral:

$$h_1^{YW} = (1)(2)(3)(4)(5)^{YW}$$
$$h_2^{YW} = (1)(24)(3)(5)^{YW}$$
$$h_3^{YW} = (12)(3)(4)(5)^{YW}$$
$$h_{24}^{YW} = (142)(3)(5)^{YW}$$

The reader may further verify that

$$\hat{\psi}^{WW}(h_1^{WY}; h_1^{YW}) = \hat{\psi}^{WW}(h_2^{WY}; h_2^{YW}) = h_2^{WW} + 2h_3^{WW} + 2h_6^{WW} + h_5^{WW}$$
$$\hat{\psi}^{WW}(h_3^{WY}; h_3^{YW}) = \hat{\psi}^{WW}(h_4^{-1WY}; h_4^{YW}) = h_2^{WW} + 2h_3^{WW} + 2h_8^{WW} + h_7^{WW}$$

Note, however, that if a truly dissociative mechanism is operative, then association of a given SF_3^+ cation with the same fluoride anion that originally dissociated from SF_4 to form that cation will occur with negligible probability. Therefore, the reactions h_2^{WW} and h_3^{WW} will occur with negligible probability.

2. *Bimolecular Reactions of* SF_4

Muetterties and Phillips (1959, 1967) and Frey *et al.* (1971) have proposed different bimolecular mechanisms for fluorine exchange in SF_4. The pertinent configurations are shown in Fig. 4d–g. We shall examine some bimolecular mechanisms that involve metastable intermediate configurations having geometry U, T, and V in order to determine whether they imply nmr differentiable reactions in H^{ZZ}.

The implications of the following association reactions will be investigated:

$$h_1^{UZ} = (1)(2)(3)(4)(5)(6)(7)(8)^{UZ}$$
$$h_2^{UZ} = (16432)(5)(7)(8)^{UZ}$$
$$h_3^{UZ} = (1432)(5)(6)(7)(8)^{UZ}$$
$$h_4^{UZ} = (16785432)^{UZ}$$
$$h_1^{TZ} = (1)(2)(3)(4)(5)(6)(7)(8)^{TZ}$$
$$h_5^{TZ} = (1)(2)(3)(4)(5678)^{TZ}$$
$$h_6^{TZ} = (1432)(5678)^{TZ}$$
$$h_1^{VZ} = (1)(2)(3)(4)(5)(6)(7)(8)^{VZ}$$

All of these reactions are achiral. The reader may verify that

$$\hat{\psi}^{ZZ}(h_1^{ZU}; h_1^{UZ}) = (1)(2)(3)(46)(5)(7)(8)^{ZZ}$$
$$\hat{\psi}^{ZZ}(h_2^{-1ZU}; h_2^{UZ}) = \hat{\psi}^{ZZ}(h_3^{-1ZU}; h_3^{UZ}) = (15876234)^{ZZ}$$
$$\hat{\psi}^{ZZ}(h_4^{-1ZU}; h_4^{UZ}) = (15)(2)(3)(4)(6)(7)(8)^{ZZ}$$
$$\psi^{ZZ}(h_1^{ZT}; h_1^{TZ}) = \psi^{ZZ}(h_6^{-1ZT}; h_6^{TZ}) = (1)(2)(3)(4)(5)(6)(7)(8)^{ZZ}$$
$$\hat{\psi}^{ZZ}(h_5^{-1ZT}; h_5^{TZ}) = (1234)(5876)^{ZZ}$$
$$\hat{\psi}^{ZZ}(h_1^{ZV}; h_1^{VZ}) = (156732)(4)(8)^{ZZ}$$

All of the reactions in H^{ZZ} listed here are nmr differentiable. Some of the association reactions may be ruled out immediately as the sole process responsible for fluorine exchange, since they do not imply axial–equatorial exchange.

ACKNOWLEDGMENT

Acknowledgment is made to the National Science Foundation for a predoctoral fellowship. The author is also grateful to Professor F. A. Cotton for his encouragement.

REFERENCES

Bacon, J., Gillespie, R. J., and Quail, J. W. (1963). *Can. J. Chem.* **41**, 1016.
Binsch, G. (1968). *Top. Stereochem.* **3**, 97–192.
Chickos, J. S., Larsen D. W., and Legler, L. E. (1972). *J. Amer. Chem. Soc.* **94**, 4266.
Cotton, F. A., George, J. W., and Wauch, J. S. (1958). *J. Chem. Phys.* **28**, 994.
de Bruijn, N. G. (1964. *In* "Applied Combinatorial Mathematics" (E. F. Beckenbach, ed.), pp. 144–184. Wiley, New York.
Frey, R. A., Redington, R. L., and Khidir Aljibury, A. L. (1971). *J. Chem. Phys.* **54**, 344.
Gutowsky, H. S., McCall, D. W., and Slichter, C. P. (1953). *J. Chem. Phys.* **21**, 279.
Harary, F. (1969). "Graph Theory," p. 184. Addison-Wesley, Reading, Massachusetts.
Johnson, C. S., Jr. (1964). *J. Chem. Phys.* **41**, 3277.
Johnson, C. S., Jr. (1965). *Advan. Magn. Resonance* **1**, 33.
Kaplan, J. I., and Fraenkel, G. (1972). *J. Amer. Chem. Soc.* **94**, 2907.
Klemperer, W. G. (1972a). *J. Chem. Phys.* **56**, 5478.
Klemperer, W. G. (1972b). *Inorg. Chem.* **11**, 2668.
Klemperer, W. G. (1973a). *J. Amer. Chem. Soc.* **95**, 380.
Klemperer, W. G. (1973b). *J. Amer. Chem. Soc.* **95**, 2105.
Meakin, P., Muetterties, E. L., and Jesson, J. P. (1973). *J. Amer. Chem. Soc.* **95**, 75.
Muetterties, E. L. (1969). *J. Amer. Chem. Soc.* **91**, 1636.
Muetterties, E. L., and Phillips, W. D. (1959). *J. Amer. Chem. Soc.* **81**, 1084.
Muetterties, E. L., and Phillips, W. D. (1967). *J. Chem. Phys.* **46**, 1967.
Redington, R. L., and Berney, C. V. (1965). *J. Chem. Phys.* **43**, 2020.
Redington, R. L., and Berney, C. V. (1967). *J. Chem. Phys.* **46**, 2862.
Tibbetts, D. L., and Brown, T. L. (1970). *J. Amer. Chem. Soc.* **92**, 3031.
Woodman, C. M. (1970). *Mol. Phys.* **19**, 753.

3

Band-Shape Analysis

GERHARD BINSCH

```
 I. Introduction.................................................. 45
II. Classical Theory ............................................. 47
    A. The Bloch Equations ..................................... 47
    B. Exchange between Two Sites ............................. 49
    C. Exchange between Many Sites ............................ 51
III. Quantum-Mechanical Theory .................................. 53
    A. Density Matrices and the Liouville Representation ............ 53
    B. Relaxation and Exchange Superoperators .................... 56
    C. Unsaturated Steady-State Single Resonance ................. 60
    D. Double Resonance in the Linear Approximation .............. 64
    E. Nonlinear Response....................................... 66
    F. Transient Response ...................................... 69
IV. Practical Band-Shape Analysis ................................ 70
    A. Experimental Procedures.................................. 70
    B. Computer Programs....................................... 72
    C. Determination of Static Parameters......................... 73
    D. Determination of Dynamic Parameters...................... 74
 V. Treatment of Dynamic Data................................... 75
    A. Calculation of Activation Parameters ....................... 75
    B. Error Analysis........................................... 76
VI. Conclusion .................................................. 78
    References ................................................. 78
```

I. Introduction

The primary objective of spectroscopy is the spectrum, commonly recorded as a photograph, an oscilloscope tracing, a curve drawn by a pen on a piece of chart paper, or a series of data points stored in analog or digital form in a computer. The spectral information may in each case be represented by a plot of the signal intensity S versus some judiciously chosen variable. In this chapter we shall be concerned with somewhat idealized nmr experiments in which the only variable is the angular frequency ω of the observing channel, all other instrumental parameters being regarded as either constant in time or periodically varying with constant frequency and amplitude, and we shall also ignore certain instrumental artifacts such as beat patterns with a second radiofrequency. The plot of $S(\omega)$ versus ω will then be called the band shape. With these restrictions, the formulas for $S(\omega)$ apply equally well to field-sweep single resonance experiments, but not to field-sweep multiple resonance spectra. Except for some borderline cases briefly discussed in Section III,F, we shall in addition confine ourselves to steady-state conditions. Transient phenomena and the great variety of modern pulse techniques will receive coverage in the succeeding two chapters.

The ultimate objective of spectroscopy is to interpret a given spectrum in terms of macroscopic and microscopic properties of the sample being investigated. In many applications of high-resolution nmr spectroscopy, one obtains a well-resolved line spectrum in which the shapes of the lines are only of secondary interest. By replacing them with delta functions the band shape is reduced to a stick diagram, which may then be analyzed in terms of a static or quasistatic model, i.e., by solving an eigenvalue problem for the nuclear spins. For extracting information about the detailed nuclear spin dynamics from steady-state spectra, however, it will in general be necessary to analyze the total band shape.

Dynamic effects on high-resolution nmr line shapes are usually attributed to two different phenomena, relaxation and exchange, but it should be realized that this distinction is largely a matter of convenience. Relaxation and exchange processes may in fact be so closely interlinked that a separate treatment cannot be justified. In the past, nuclear spin relaxation has mostly been studied by nonequilibrium techniques, whereas the bulk of information about the rates of nuclear exchange processes stems from steady-state band-shape work. In this chapter, the emphasis will be on exchange, but in view of the growing interest in detailed relaxation information obtainable from band-shape analyses, it is essential to devote some space also to the discussion of relaxation. Since this general topic forms the subject of another chapter in this book, however, we can here be satisfied with a rather cursory treatment. In particular, no attempt will be made to present a systematic account of the theory of nuclear spin relaxation or of the multitude of possible relaxation mechanisms.

The considerable complexity of the relevant material necessitates further restrictions for an article of prescribed length. Since it is believed that this book will be consulted mainly by those research workers interested in practical applications of dynamic nmr spectroscopy, we shall make an effort to concentrate on the principal line of reasoning in the theoretical discussion and to pay scant attention to many of the interesting contributions concerned with the possibility of simplifying the general formalism under a variety of special circumstances. These contributions have greatly added to our understanding of dynamic effects in nmr spectroscopy, but because of their frequently rather intricate mathematics are likely to be somewhat distracting to the uninitiated. We have adopted the view that a basic appreciation of the general features of the theory in compact form provides an adequate preparation for an intelligent application of the methods developed by the specialists. Readers left dissatisfied with such an approach have recourse to a large body of authoritative literature. Besides the imposing treatise of Abragam (1961) and the excellent introduction in Slichter's (1963) book, we particularly recommend the articles by Lynden-Bell (1967) on density matrix theory, by Hoffman (1970) on single-resonance, and by Nageswara Rao (1970) on double-resonance line shapes. As an entry to the extensive literature on exchange effects we mention the recent review by Sutherland (1971).

II. Classical Theory

A. The Bloch Equations

If we are dealing with an ensemble of identical nuclear spins in identical chemical environments subject to a constant magnetic field H_0 in the z direction and a linearly oscillating field of strength $2H_1$ and frequency ω in the x direction, a classical equation of motion for the macroscopic magnetization $\mathbf{M}$ provides an accurate description of the resonance phenomenon; it is always valid for spins with spin quantum number $I = \frac{1}{2}$ and valid under certain restrictions also for higher spin quantum numbers (Wangsness and Bloch, 1953). The linear differential Eq. (1), first proposed by Bloch (1946),

$$d\mathbf{M}/dt = \gamma \mathbf{M} \times \mathbf{H} - (\mathbf{i}M_x + \mathbf{j}M_y)/T_2 - \mathbf{k}(M_z - M_0)/T_1 \qquad (1)$$

has been discussed at great length numerous times in the textbook and review literature, so we here only need to summarize the results and point out some of the consequences. In Eq. (1), γ is the magnetogyric ratio and $\mathbf{i}$, $\mathbf{j}$, and $\mathbf{k}$ are the unit vectors in the direction of the coordinate axes. Owing to the cylindrical symmetry of the system in the magnetic field H_0 it is necessary to allow for two different phenomenological time constants, the longitudinal relaxation time T_1 and the transverse relaxation time T_2, which are inversely related to the first-order rate constants of the processes by which the components of $\mathbf{M}$ would return to the thermal equilibrium value $\mathbf{M}_{eq} = \mathbf{k}M_0$ after the perturbing radio-frequency field H_1 has been switched off.

The components of Eq. (1) are most conveniently analyzed in a coordinate system rotating about the z axis with the angular frequency $-\omega \mathbf{k}$. We then obtain

$$dM_x/dt = (\omega_0 - \omega)M_y - M_x/T_2 \qquad (2)$$

$$dM_y/dt = -(\omega_0 - \omega)M_x + \gamma H_1 M_z - M_y/T_2 \qquad (3)$$

$$dM_z/dt = -\gamma H_1 M_y - (M_z - M_0)/T_1 \qquad (4)$$

where the Larmor frequency $\omega_0 = \gamma H_0$ ($\boldsymbol{\omega}_0 = -\omega_0 \mathbf{k}$) corresponds to the resonance frequency. By defining the complex transverse magnetization $M_+ = M_x + iM_y$ in the rotating frame, Eqs. (2) and (3) may also be written as

$$dM_+/dt = -i(\omega_0 - \omega)M_+ + i\gamma H_1 M_z - M_+/T_2 \qquad (5)$$

In band-shape studies, we are interested in the steady-state solution for the transverse components

$$M_x = \gamma H_1 M_0 \frac{(\omega_0 - \omega)T_2^2}{1 + (\omega_0 - \omega)^2 T_2^2 + \gamma^2 H_1^2 T_1 T_2} \qquad (6)$$

$$M_y = \gamma H_1 M_0 \frac{T_2}{1 + (\omega_0 - \omega)^2 T_2^2 + \gamma^2 H_1^2 T_1 T_2} \qquad (7)$$

corresponding to the dispersion and absorption modes, respectively. For low H_1 fields, the third term in the denominator of Eqs. (6) and (7) may be neglected, so that we obtain for the unsaturated steady-state absorption line shape the simple formula

$$S(\omega) = CT_2/[1 + (\omega_0 - \omega)^2 T_2^2] \qquad (8)$$

which corresponds to a Lorentzian absorption line whose full natural width W (in hertz) at half height is related to T_2 (in seconds) by

$$T_2 = (\pi W)^{-1} \qquad (9)$$

The constant C contains nuclear and instrumental parameters and may be treated as a scaling factor. In practical applications, Eq. (9) is rarely useful because the actual line width $W^{\text{eff}} = W + \Delta W$ is in most cases entirely dominated by the field-inhomogeneity term ΔW, so that one can only calculate an effective transverse relaxation time T_2^{eff}. For this reason, relaxation time measurements on single lines must usually be performed with pulse techniques.

In favorable cases of relatively short T_2, it is, however, possible to obtain both T_2 and T_1 from band-shape studies by exploiting Eq. (7), as recently demonstrated by Harris and Worvill (1973a). The absorption line remains Lorentzian even under saturation conditions, and one may therefore extract an apparent T_{2s} from the partially saturated band shape. It can be shown that

3. Band-Shape Analysis

this quantity satisfies the equation

$$\frac{1}{T_{2s}^2} = \frac{1}{T_2^2} + \frac{\gamma^2 H_1^2 T_1}{T_2} \qquad (10)$$

so that a plot of $1/T_{2s}^2$ versus H_1^2 yields the relaxation times from the intercept and slope. Although it is unlikely that this technique will become popular for single lines, we shall later demonstrate that relaxation studies by band-shape analysis have certain advantages over pulse methods for more complicated spin systems.

B. Exchange between Two Sites

The phenomenological Bloch equations also apply to an ensemble of nuclear spins that differ in magnetogyric ratio or chemical environment, provided that each distinct species gives rise to a single resonance whose separation from other lines is large compared to the natural line width. For identical spins in environments characterized by shielding constant σ_j and chemical shifts $\omega_j = \gamma H_0(1 - \sigma_j)$, an additional relaxation mechanism may be provided by nuclear exchange processes between the various sites. It is generally assumed that such exchange processes can be adequately described by a stochastic model. For two sites r and s with populations p_r and p_s, we can then write the pair of coupled differential equations (Hahn and Maxwell, 1952; McConnell, 1958)

$$dM_{+r}/dt = -[i(\omega_r - \omega) + 1/T_{2r}]M_{+r} + i\gamma H_1 M_{zr} - k_{rs}M_{+r} + k_{sr}M_{+s} \qquad (11a)$$

$$dM_{+s}/dt = -[i(\omega_s - \omega) + 1/T_{2s}]M_{+s} + i\gamma H_1 M_{zs} - k_{sr}M_{+s} + k_{rs}M_{+r} \qquad (11b)$$

where k_{rs} is the macroscopic first-order (for intramolecular exchange) or pseudo-first-order (for intermolecular exchange) rate constant for transfers from site r to site s. Since detailed balancing demands that

$$p_r k_{rs} = p_s k_{sr} \qquad (12)$$

it is convenient to introduce the single variable τ defined as

$$\tau = \frac{p_r}{k_{sr}} = \frac{p_s}{k_{rs}} \quad \text{or} \quad \frac{1}{\tau} = k_{rs} + k_{sr} \qquad (13)$$

The unsaturated steady-state solution for the absorption band shape is found to be

$$S(\omega) = \text{Im}\{-iC\tau[2p_r p_s - \tau(p_r \alpha_s + p_s \alpha_r)]/(p_r p_s - \tau^2 \alpha_r \alpha_s)\} \qquad (14)$$

where Im stands for the imaginary part and where

$$\alpha_r = -[i(\omega_r - \omega) + 1/T_{2r} + p_s/\tau] \tag{15}$$

In practical applications of the band-shape formula (14), one may add another term to Eq. (15) to account for field-inhomogeneity effects, provided they can be assumed to have a Lorentzian distribution. If such an assumption is not justified, the whole band shape should be convoluted with an empirical broadening function.

If the line widths in the absence of exchange are dominated by Lorentzian field-inhomogeneity broadening, rough estimates of the exchange rate constants may be calculated from the observed line widths W (in hertz) in the slow-exchange and fast-exchange regions by using the approximate formulas (16) and (17)

$$k_{rs} = \pi W_r - 1/T_2^{\text{eff}} \qquad \text{(slow exchange)} \tag{16}$$

$$k_{rs} = 4\pi p_r p_s^2 \Delta_{rs}^2/(W - 1/\pi T_2^{\text{eff}}) \qquad \text{(fast exchange)} \tag{17}$$

For the frequently occurring case of equal populations, it is usually possible to identify the point at which the two separate lines have just merged into a single broad line. At this point the rate constant $k = k_{rs} = k_{sr} = 1/2\tau$ is given approximately by

$$k = \pi \Delta_{rs}/2^{1/2} \qquad \text{(coalescence)} \tag{18}$$

For unequal populations, a graphic method may be employed (Shanan-Atidi and Bar-Eli, 1970). In Eqs. (17) and (18), Δ_{rs} stands for the chemical shift difference expressed in hertz.

Interesting new effects arise when relaxation contributes significantly to one of the line widths, for it is then possible to have partial transfer of relaxation broadening to the other line mediated by exchange. Of great practical importance, especially for studies on biomolecules (Stengle and Baldeschwieler, 1967; Cohn and Reuben, 1971), are those examples where intermolecular ligand exchange takes place between free and complexed sites and where the sensor nuclei either have a quadrupole moment or are rapidly relaxed by paramagnetic ions in the complexed site. It is usually adequate to use Eq. (14) for a full band-shape analysis, provided that the correlation time τ_c of the relaxation process satisfies the extreme narrowing condition $\omega\tau_c \ll 1$ in the case of quadrupolar nuclei of half-integral spin, and a variety of simple line-width formulas have been derived (Swift and Connick, 1962; Leigh, 1971) that apply under certain limiting conditions. For very fast exchange, for instance, one obtains

$$W = p_r W_r + p_s W_s \tag{19}$$

i.e., the line width W of the collapsed line is simply the weighted average of

3. Band-Shape Analysis

the individual widths in the absence of exchange. However, as recently pointed out by a number of authors (Anderson and Fryer, 1969; Marshall, 1970; Wennerström, 1972), modified Bloch equations cannot in general be assumed to be valid if the exchange rate becomes comparable to or larger than $1/\tau_c$, for two reasons. First, it is then no longer feasible to treat relaxation and exchange as distinct mechanisms. Rather, one must consider the joint probability for changes in site and in the principal-axis direction of the relaxation interaction tensor. Second, one finds that exchange between identical sites may also contribute to the band shape. Clearly it is impossible to incorporate corresponding rate terms in Eq. (11), for they would simply cancel out. Further complications ensue if the extreme narrowing condition cannot be assumed to hold (Bull, 1972). The more sophisticated methods necessary in such cases will be discussed in Section III.

C. Exchange between Many Sites

The extension of the modified Bloch Eq. (11) to n sites is accomplished most economically with matrix methods (Sack, 1958). The unsaturated steady-state absorption band shape may be written as

$$S(\omega) = \text{Im}(-iC\mathbf{1}^\dagger \mathbf{A}^{-1}\mathbf{P}) \tag{20}$$

where $\mathbf{1}^\dagger$ is an n-dimensional row vector whose components are all equal to 1, and the column vector $\mathbf{P}$ contains the populations p_r of the n sites r. The matrix $\mathbf{A}$ is composed of

$$\mathbf{A} = -i(\mathbf{\Omega} - \omega\mathbf{I}) + \mathbf{R} + \mathbf{X} \tag{21}$$

where the diagonal matrices $\mathbf{\Omega}$ and $\mathbf{R}$ contain the site frequencies ω_r and transverse relaxation rates $-1/T_{2r}$, respectively. If all line widths are dominated by field-inhomogeneity broadening in the absence of exchange, the relaxation matrix may be represented by a multiple of the n-dimensional unit matrix $\mathbf{I}$, i.e.,

$$\mathbf{R} = -\mathbf{I}/T_2^{\text{eff}} \tag{22}$$

The exchange matrix $\mathbf{X}$ has elements

$$X_{rr} = -\sum_{s(\neq r)} k_{rs} \tag{23a}$$

$$X_{rs} = k_{sr} \quad (r \neq s) \tag{23b}$$

and because of the balancing conditions, Eq. (12), $\mathbf{X}$ may contain at most $n(n-1)/2$ independent rate constants. The chemical mechanisms underlying the exchange processes frequently place additional constraints on the number of independent rate constants and may in fact demand that all of them are interrelated. In the latter case, the exchange matrix simplifies to

$$\mathbf{X} = k\mathbf{Q} \tag{24}$$

where k is the macroscopic rate constant of the elementary process responsible for the changes in the band shape and $\mathbf{Q}$ is a statistical matrix in which each off-diagonal element Q_{rs} specifies the probability that a transfer out of site s terminates in site r; the diagonal elements Q_{rr} are then easily calculated from the relations (23a). In practice, one is often faced with the opposite problem, namely to enumerate all possible permutational schemes that are in principle distinguishable by nmr and to identify the one (or several, or possibly a linear combination) that is compatible with the observed band shapes. Chemical intuition is helpful in carrying out such a program, but to avoid mistakes, it is advisable to resort to the systematic procedure discussed in Chapter 2 of this book. Although a unique permutational scheme does not necessarily also define a unique chemical mechanism, it is often a sufficient criterion for rejecting a large number of incompatible mechanisms. Multisite exchange studies by dynamic nmr spectroscopy are thus seen to be an important tool in mechanistic chemistry, and the many interesting applications discussed in the remainder of this book amply testify to the great power of this technique.

Although modified Bloch equations and hence Eq. (20) do not rigorously apply to coupled spin systems (Section III), an approximate classical treatment is often accurate enough in the presence of first-order multiplet structure. Here the diagonal frequency matrix $\mathbf{\Omega}$ contains the resonance frequencies of all individual lines in the absence of exchange and the dimensions of the other matrices and vectors must be enlarged correspondingly. The entries in the population vector $\mathbf{P}$ have to be split up in such a way as to take proper account of the relative intensities within a group of multiplet lines. For setting up the exchange matrix $\mathbf{X}$, it is now essential to distinguish between intramolecular and intermolecular exchange. In the former every line can be assigned to a transition of a certain nucleus i with all other nuclei j being in a definite spin state, and the switching process can only connect those lines of i that correspond to an identical spin–state combination for nuclei j. Degeneracies must be accounted for by suitable statistical factors. As a consequence, the $\mathbf{A}$ matrix splits up into unconnected blocks, of which each may be treated separately in the band-shape computation. On the other hand, an additional randomization obtains in the case of intermolecular exchange, since a spin arriving from another molecule may be in any of its possible spin states with equal probability. In other words, the spin states of a certain molecule right after the switch are not completely determined by the spin states of the same molecule just before the switch.

3. Band-Shape Analysis

The solution of Eq. (20) requires the inversion of the complex matrix **A**. In the early computer programs written for classical multisite exchange, which are still being used today, such a matrix inversion has to be performed for each value of the variable frequency ω. As we shall see in Section III, much more efficient computational algorithms are now known for tackling this problem.

III. Quantum-Mechanical Theory

Besides the limitations already alluded to, a classical approach cannot be expected to yield reasonable results for strongly coupled spins. Here it is meaningless to speak of transitions of individual nuclei and one must therefore consider each system of coupled spins as a single entity. A quantum-mechanical treatment of the problem is most consistently formulated in the framework of density matrix theory.

A. Density Matrices and the Liouville Representation

The linear superposition principle of quantum mechanics implies that the state function ψ of a microscopic system can be represented by the expansion

$$|\psi\rangle = \sum_{\mu=1}^{n} c_\mu |\phi_\mu\rangle \quad \text{or} \quad \langle\psi| = \sum_{\mu=1}^{n} c_\mu^* \langle\phi_\mu| \tag{25}$$

where the complete set of orthonormal eigenkets $\{|\phi\rangle\}$ of some suitable linear Hermitian operator $\hat{O}$ span a Hilbert space of dimension n and where the expansion coefficients are given by

$$c_\mu = \langle\phi_\mu|\psi\rangle \quad \text{or} \quad c_\mu^* = \langle\psi|\phi_\mu\rangle \tag{26}$$

The state function only serves as a mathematical device for calculating observables $\bar{P}$ as expectation values of corresponding operators $\hat{P}$

$$\bar{P} = \langle\psi|\hat{P}|\psi\rangle = \sum_{\mu\nu} c_\nu c_\mu^* \langle\phi_\mu|\hat{P}|\phi_\nu\rangle = \sum_{\mu\nu} \rho_{\nu\mu} P_{\mu\nu} = \text{Tr}(\rho P) = \text{Tr}(P\rho) \tag{27}$$

and may be eliminated in favor of the density matrix ρ, which may itself be regarded as the representation of a density operator $\hat{\rho}$ defined by the identities

$$\rho_{\nu\mu} = c_\nu c_\mu^* = \langle\phi_\nu|\psi\rangle\langle\psi|\phi_\mu\rangle = \langle\phi_\nu|\hat{\rho}|\phi_\mu\rangle \tag{28}$$

The density matrix for a microscopic system in a "pure" state, i.e., a state describable by a state function, is thus seen to have the properties of an

Hermitian projection operator

$$\rho = \rho^\dagger \tag{29}$$

$$\rho^2 = \rho \tag{30}$$

$$\mathrm{Tr}(\rho) = 1 \tag{31}$$

and each diagonal element $\rho_{\mu\mu}$ specifies the probability that a measurement of the observable $\bar{O}$ will yield the eigenvalue $\langle \phi_\mu | \hat{O} | \phi_\mu \rangle$.

If the knowledge about the system of interest can only be gained from measurements on a macroscopic sample consisting of a large number of subsystems, the full information contained in all the individual state functions is no longer accessible, but a description in terms of a density matrix remains realistic, provided one redefines ρ as

$$\rho_{\nu\mu} = \overline{c_\nu c_\mu^*} \tag{32}$$

where the bar denotes the ensemble average. Of the previously cited properties, Eq. (30) ceases to be valid and the criterion $\mathrm{Tr}(\rho^2) < 1$ indicates that one is dealing with a "mixed" state.

The total system evolves in time under the action of its Hamiltonian according to the Liouville–von Neumann equation

$$d\rho/dt = -i[\mathscr{H}, \rho] \tag{33}$$

from which Planck's constant $\hbar$ has been omitted; a convention adopted throughout this chapter. It is convenient to split the Hamiltonian into three parts

$$\mathscr{H} = \mathscr{H}^\mathrm{s} + \mathscr{H}^\mathrm{b} + \mathscr{H}^\mathrm{sb} \tag{34}$$

where $\mathscr{H}^\mathrm{s}$ acts on the spin system(s) of interest and depends only on spin variables, $\mathscr{H}^\mathrm{b}$ is the Hamiltonian for the remainder of the macroscopic sample, summarily called the "bath," and $\mathscr{H}^\mathrm{sb}$ describes the interaction between system(s) and bath. We will also assume that at some time $t = 0$ it is permissible to write the total density matrix as the direct product

$$\rho = \sigma \otimes \rho^\mathrm{b} \tag{35}$$

of a spin density matrix σ and a bath density matrix ρ^b.

Instead of arranging the elements of ρ in a square matrix it is advantageous to interpret them as components of a state vector $\boldsymbol{\rho}$ in an extended vector space of dimension n^2. The last part of Eq. (27) may then be written as

$$\bar{P} = \mathbf{P}^\dagger \boldsymbol{\rho} \tag{36}$$

where the row vector $\mathbf{P}^\dagger$ is formed from the matrix elements of P, and Eq. (33) becomes

$$d\boldsymbol{\rho}/dt = -i\mathbf{L}\boldsymbol{\rho} \tag{37}$$

which describes a rotation of the density vector ρ in Liouville space. The Liouville representation of quantum mechanics, as developed systematically especially by Fano (1964), has only recently been used explicitly in dynamic nmr (Binsch, 1968b, 1969; Blume, 1968; Hoffman, 1970; Kleier and Binsch, 1970a; Anderson, 1971; Ben-Reuven, 1971; Meakin *et al.*, 1971; Pyper, 1971a–e; Vold and Chan, 1971; Ayant *et al.*, 1972), but is implicit in much concurrent and earlier work in this field (Gutowsky *et al.*, 1965; Allerhand and Thiele, 1966; Alger *et al.*, 1967) including the famous Wangsness–Bloch–Redfield (WBR) theory of relaxation (Wangsness and Bloch, 1953; Bloch, 1956, 1957; Redfield, 1957, 1965). It recommends itself not only because of its notational compactness and because the equations can be written in operator form independent of a specific representation, but particularly because certain invariance properties not present in ordinary Hilbert space become apparent only in Liouville space (Kleier and Binsch, 1970a; Pyper, 1971b,e). Since the Liouville space is spanned by an operator basis, all operators in this space are called superoperators, in extension of an earlier and slightly more restrictive use of this term (Banwell and Primas, 1963). The Liouville superoperator **L** occurring in Eq. (37) is a particular example of a derivation superoperator and may be generated from the Hamiltonian by the prescription

$$\mathbf{L} = \mathscr{H} \otimes E - E \otimes \mathscr{H}^* \tag{38}$$

where E is the identity operator in ordinary Hilbert space. Because of the trace relationship of Eq. (31), it is sometimes advisable to work with the special Liouville representation (Fano, 1964), which corresponds to the space of dimension $n^2 - 1$ that is orthogonal to the unity basis operator. The matrix elements of superoperators may be characterized by two indices when referred to an operator basis or by four (or more) when indirectly referred to a function basis (or function bases) in ordinary Hilbert space(s). The latter convention is usually easier to handle in practical applications.

In the subsequent discussion, we will have occasion to make use of three different spin Liouville representations (Binsch, 1969):

(a) A primitive Liouville representation in a space spanned by the ensemble-averaged spin density components of a uniquely labeled nuclear spin system in a single molecule l. Unless stated otherwise, we will henceforth assume that we are only dealing with identical spin-$\frac{1}{2}$ nuclei. The dimension of the primitive spin Liouville space for n_l coupled spins is then 2^{2n_l}.

(b) A composite spin Liouville space of dimension $\sum_{l=1}^{m} (2^{2n_l})$ simultaneously describing m different types of molecules l. The total set of all distinctly labeled nuclear spin systems in all m different molecules will be called the "nuclear configurations." The composite density vector is constructed as the direct sum of the renormalized primitive density vectors, the renormalization factors f_l being

$$f_l = n_l \lambda_l \left(\sum_{l=1}^{m} n_l \lambda_l \right)^{-1} \tag{39}$$

where the λ_l represent the relative abundances of the m molecules l. If all molecules contain the same number of spin-$\frac{1}{2}$ nuclei, the renormalization factors are simply equal to the populations p_l.

(c) An interaction spin Liouville space of dimension $\prod_l 2^{2n_l}$, representing the direct product space of the constituent primitive Liouville spaces.

B. Relaxation and Exchange Superoperators

1. Relaxation

In the WBR theory of relaxation one proceeds from the Hamiltonian of Eq. (34), whose first part is decomposed into

$$\tilde{\mathscr{H}}^s = \tilde{\mathscr{H}}_0 + \sum_k \tilde{\mathscr{H}}_k(t) \tag{40}$$

for each nuclear configuration. The time-independent term $\tilde{\mathscr{H}}_0$ consists of the nuclear Zeeman and spin-coupling Hamiltonians

$$\tilde{\mathscr{H}}_0 = -\sum_i \omega_i I_{zi} + 2\pi \sum_{i>j} J_{ij} \mathbf{I}_i \cdot \mathbf{I}_j \tag{41}$$

involving the coupling constants J (in hertz) and the spin operators $\mathbf{I}$, and the time-dependent terms

$$\tilde{\mathscr{H}}_k(t) = -2 H_k \gamma \cos \omega(k) t \sum_i I_{xi} \tag{42}$$

arise from the k radio-frequency fields of frequencies $\omega(k)$ and strengths $2H_k$ in the x direction. If we are dealing with a single radio-frequency field of frequency $\omega(1) = \omega$, the time dependence of $\tilde{\mathscr{H}}_1(t)$ can be eliminated by transforming all equations to a rotating frame by means of the exponential operator

$$U = \exp(-i\omega t F_z) \tag{43}$$

where

$$F_u = \sum_i I_{ui} \tag{44}$$

in which case Eq. (40) becomes

$$\mathscr{H}^s = \mathscr{H}_0 + \mathscr{H}_1 = -\sum (\omega_i - \omega) I_{zi} + 2\pi \sum_{i>j} J_{ij} \mathbf{I}_i \cdot \mathbf{I}_j - \gamma H_1 F_x \tag{45}$$

3. Band-Shape Analysis

By further transforming Eq. (37) to an interaction representation, by expanding the density vector $\rho(t)$ in a time-dependent perturbation series about $\rho(0)$, and by postulating that the bath stays at thermal equilibrium even under the influence of the mutual interaction one finally derives the WBR master Eq. (46) for the spin density vector in primitive spin Liouville space

$$d\sigma/dt = -i(\mathbf{L}_0 + \mathbf{L}_1)\sigma + \mathbf{R}(\sigma - \sigma_0) \tag{46}$$

Corresponding equations can also be derived in composite and interaction spin Liouville space. It can be shown that the sole effect of treating the bath quantum-mechanically is to introduce the thermal equilibrium spin density vector σ_0, whose nonvanishing components consist of diagonal density matrix elements in the eigenrepresentation of $\mathcal{H}_0$ and also, to a very good approximation, in the eigenrepresentation of the F_z operator, i.e.,

$$\sigma_0 = \frac{\exp(-\hbar\mathcal{H}_0/kT)}{\text{Tr}\{\exp(-\hbar\mathcal{H}_0/kT)\}} \cong \frac{E + \hbar\omega_0 F_z/kT}{N} \tag{47}$$

where N is the dimension of the spin Hilbert space and T the absolute temperature. It is therefore permissible to treat the bath classically and to insert σ_0 afterward, by writing

$$\mathcal{H}^{\text{sb}} = \sum_{pq}(-1)^q K_{pq} Y_{p-q}(t) \tag{48}$$

where K are spin operators and $Y(t)$ stationary random functions of the classical bath variables, and where both K and $Y(t)$ are expressed in irreducible spherical tensor form; the indices q and p label the tensor components and relaxation mechanisms, respectively.

The perturbation expansion leading to Eq. (46) may be expected to converge if the elements of the relaxation superoperator $\mathbf{R}$ remain small with respect to the characteristic frequency of the molecular motions responsible for the relaxation, usually expressed as the inverse of a correlation time τ_c, and if $\gamma H_1 \tau_c \ll 1$. We then obtain

$$R_{\mu\nu,\kappa\lambda} = J_{\mu\kappa\lambda\nu}(\omega_{\lambda\nu}) + J_{\lambda\nu\mu\kappa}(\omega_{\mu\kappa}) - \delta_{\nu\lambda}\sum_{\eta}J_{\mu\eta\eta\kappa}(\omega_{\eta\kappa}) - \delta_{\mu\kappa}\sum_{\eta}J_{\eta\nu\lambda\eta}(\omega_{\lambda\eta}) \tag{49}$$

where the matrix elements of $\mathbf{R}$ in primitive Liouville space are indirectly referred to a function basis in Hilbert space. The spectral densities J are calculated from the (one-sided) Fourier transforms of the correlation functions

$$J_{uvwx}(\omega_{yz}) = \sum_{pp'}\sum_{qq'}(-1)^{q+q'}\langle u|K_{pq}|v\rangle\langle w|K_{p'q'}|x\rangle$$

$$\times \int_0^\infty G_{pp'qq'}(\tau)\exp(i\omega_{yz}\tau)\,d\tau \tag{50}$$

where

$$G_{pp'qq'}(\tau) = \overline{Y_{p-q}(t)Y_{p'-q'}(t+\tau)} \tag{51}$$

and become frequency-independent in the extreme narrowing limit, $\omega_0\tau_c \ll 1$. Even though the matrix elements of the relaxation superoperator **R** must satisfy a variety of symmetry relationships and sum rules and the evaluation of the correlation functions themselves can be simplified by exploiting the transformation properties of irreducible spherical tensors (Hubbard, 1969; Pyper, 1971b,e), the computation of the **R** matrix for complex spin systems remains a formidable problem. Progress in this area would no doubt be greatly facilitated if sufficiently versatile computer programs were developed for performing such calculations automatically from a minimum of input information.

A more detailed discussion of the many intricacies of relaxation is beyond the scope of this article, but we wish to mention one particularly important property of the **R** matrix. If the matrix elements of superoperators in primitive Liouville space are indirectly referred to a function eigenbasis of F_z and if the density components are grouped according to the eigenvalues of the derivation superoperator $\mathbf{F}_z^D$

$$\mathbf{F}_z^D = F_z \otimes E - E \otimes F_z^* \tag{52}$$

the matrix elements of **R** connecting different groups become strongly nonsecular, since the corresponding eigenvalues of $\tilde{\mathbf{L}}_0$ differ approximately by integral multiples of ω_0, and may thus be neglected. In other words, the **R** matrix blocks out into submatrices, whose dimensions are equal to the degeneracies of the eigenvalues of $\mathbf{F}_z^D$.

2. Intramolecular Mutual Exchange

Just as for the classical Bloch equations, the WBR master Eq. (46) is now easily generalized to include exchange, by writing

$$d\boldsymbol{\sigma}/dt = -i(\mathbf{L}_0 + \mathbf{L}_1)\boldsymbol{\sigma} + \mathbf{R}(\boldsymbol{\sigma} - \boldsymbol{\sigma}_0) + \mathbf{X}\boldsymbol{\sigma} \tag{53}$$

provided the exchange rates are small compared to $1/\tau_c$, which is the case normally of interest in practice. If the spin Hamiltonians of the nuclear configurations are identical except for the labeling of the particles, i.e., if we are dealing with intramolecular mutual exchange, Eq. (53) can be formulated in the primitive spin Liouville space of one particular nuclear reference configuration $r = 1$. The exchange superoperator is then given by the formula (Kleier and Binsch, 1970a)

$$\mathbf{X}_\lambda = k_\lambda \sum_{r=2}^{s_\lambda} (\mathbf{P}_{1r}^\lambda - \mathbf{I}) \tag{54}$$

3. Band-Shape Analysis

for each permutationally nonequivalent (cf. Chapter 2) exchange mechanism λ with rate constant $k_\lambda = k_{rs}^\lambda = k_{sr}^\lambda$ and degeneracy s_λ, where $\mathbf{I}$ is the identity superoperator in primitive Liouville space and

$$\mathbf{P}_{1r}^\lambda = P_{1r}^\lambda \otimes P_{1r}^\lambda \tag{55}$$

The matrix elements of the exchange operator P in Hilbert space are most easily calculated in a simple spin product function basis (Alexander, 1962a), but P and hence $\mathbf{X}$ can of course be transformed to any other basis if so desired. If more than one permutationally nonequivalent exchange mechanism is operative, Eq. (54) must be summed over all λ.

3. Intramolecular Nonmutual Exchange

Here the master equation must be formulated in composite Liouville space (Johnson, 1964; Binsch, 1969) and the reference labeling must be maintained for all nuclear configurations. The composite Liouville superoperators are the direct sums of the constituent primitive Liouville superoperators and a corresponding statement also holds for the composite relaxation superoperator. The individual subspaces are connected through the composite exchange superoperator, whose elements, indirectly referred to a function basis, may be calculated by means of the formula (Binsch, 1969)

$$X_{\mu_r \nu_r, \kappa_s \lambda_s} = \delta_{\mu\kappa} \delta_{\nu\lambda} [-\delta_{rs} \sum_{t(\neq r)} k_{rt} + (1 - \delta_{rs}) k_{sr}] \tag{56}$$

4. Intermolecular Exchange

This case was first treated in the framework of density matrix theory by Kaplan (1958a,b) and Alexander (1962b). Here the exchange takes place in collision complexes, and the theory must therefore be formulated in the corresponding interaction spin Liouville spaces. It is furthermore advantageous to decompose the primitive spin Liouville spaces of each type of molecule into direct products of Liouville subspaces referring to the exchanging and nonexchanging spins; the dimensions of the former must be identical for all molecules, but those of the latter may of course be different. For simplicity, we will assume that exchange always takes place during a collision of only two molecules and that the exchange is uniquely defined by a permutation of certain nuclei within the collision complex. If the chemistry of the molecules permits more than one such possibility, all collision complexes have to be treated as distinct. With these assumptions, it is possible to describe the individual exchanges in all distinct dual-interaction spin Liouville spaces.

The form of the exchange superoperator in the interaction-free representation is then found after projection into composite Liouville space. The result is (Binsch, 1969)

$$X_{\mu_r u_r, \nu_r v_r; \kappa_s x_s, \lambda_s y_s} = [d^{-1}\delta_{\mu\nu}\delta_{\kappa\lambda} - \delta_{\mu\kappa}\delta_{ux}\delta_{\nu\lambda}\delta_{vy}]\delta_{rs}\sum_{t(\neq r)} k_{rt}$$
$$+ n_s^{-1}\delta_{uv}\delta_{xy}(1 - \delta_{rs})k_{rs} \tag{57}$$

where the Greek indices refer to the basis functions of the exchanging group of nuclei in molecule $r(s)$ and the Latin indices to the nonexchanging group in molecule $r(s)$; d denotes the dimension of the Hilbert subsubspace of the exchanging group, and n_t is the dimension of the Hilbert subsubspaces of the nonexchanging group of nuclei in molecule t. If the intermolecular exchange takes place between identical molecules, it is possible to simplify the equations further by projecting into primitive spin Liouville space. An alternative method of handling intermolecular exchange, called the permutation of indices (PI) method, has recently been proposed by Kaplan and Fraenkel (1972).

C. Unsaturated Steady-State Single Resonance

The steady-state condition implies a vanishing time dependence of the density vector. By introducing the deviation density vector $\chi = \sigma - \sigma_0$ and by observing the equalities $L_0\sigma_0 = X\sigma_0 = 0$, Eq. (53) may be written as

$$[-i(L_0 + L_1) + R + X]\chi = iL_1\sigma_0 \tag{58}$$

For small H_1 fields (no saturation), L_1 may be neglected on the left-hand side of Eq. (58) (linear response). The band-shape equation may then be written as

$$S(\omega) = \text{Im}(F_-^{\dagger}\chi) = \text{Im}(iF_-^{\dagger}M_0^{-1}L_1\sigma_0) \tag{59}$$

provided the inverse of the complex non-Hermitian superoperator

$$M_0 = -iL_0 + R + X \tag{60}$$

exists (see Section III, D).

1. Trivial Relaxation

By "trivial relaxation" we mean that the widths of the lines in the absence of exchange are completely dominated by field inhomogeneity; hence, differential relaxation effects become unobservable for unsaturated single-resonance band shapes. By exploiting various commutation properties with a number of superoperators derived from F_z, it can be shown (Kleier and Binsch, 1970a) that under these conditions the band-shape Eq. (59) can be

3. Band-Shape Analysis

projected into a smaller spin Liouville subspace without loss of information, whence Eq. (59) becomes

$$S(\omega) = \text{Im}(i\mathbf{F}_-^\dagger \mathbf{M}_0^{-1} \mathbf{L}_+ \boldsymbol{\sigma}_0) \tag{61}$$

where

$$\mathbf{L}_+ = \mathscr{H}_+ \otimes E - E \otimes \mathscr{H}_+^* \tag{62}$$

$$\mathscr{H}_+ = -\gamma H_1 F_+/2 \tag{63}$$

and the relaxation superoperator in this contracted space is simply given by

$$\mathbf{R} = -\mathbf{I}/T_2^{\text{eff}} \tag{64}$$

It is convenient to introduce a scaling factor C and to define a vector $\mathbf{P}$ by

$$\mathbf{L}_+ \boldsymbol{\sigma}_0 = C\mathbf{P} \tag{65}$$

so that Eq. (61) can also be written as

$$S(\omega) = -C\,\text{Re}(\mathbf{F}_-^\dagger \mathbf{M}_0^{-1}\mathbf{P}) \tag{66}$$

which brings out the close analogy to the classical Eq. (20). In fact, for intramolecular exchange, the elements of $\mathbf{P}$ either become zero by virtue of Eq. (65) or may be taken as equal to the populations of the nuclear configurations, all other constants being absorbed in the scaling factor C.

One finds that in the contracted Liouville subspace the $\mathbf{M}_0$ matrix can be decomposed as

$$\mathbf{M}_0 = \mathbf{B} - i\omega \mathbf{I} \tag{67}$$

where $\mathbf{B}$ is independent of the radio frequency. This feature can be exploited for deriving an efficient algorithm for solving Eq. (66) (Binsch, 1968b, 1969; Gordon and McGinnis, 1968; Schirmer et al., 1969; Reeves and Shaw, 1970; Meakin et al., 1971). If one diagonalizes $\mathbf{B}$ by a similarity transformation

$$\mathbf{U}^{-1}\mathbf{B}\mathbf{U} = \boldsymbol{\Lambda} \tag{68}$$

the inverse of $\mathbf{M}_0$ may be written as

$$\mathbf{M}_0^{-1} = \mathbf{U}(\boldsymbol{\Lambda} - i\omega\mathbf{I})^{-1}\mathbf{U}^{-1} \tag{69}$$

and Eq. (66) becomes

$$S(\omega) = -C\,\text{Re}[\mathbf{F}_-^\dagger \mathbf{U}(\boldsymbol{\Lambda} - i\omega\mathbf{I})^{-1}\mathbf{U}^{-1}\mathbf{P}] \tag{70}$$

which may be written more conveniently as

$$S(\omega) = -C\,\text{Re}(\mathbf{V}^T \mathbf{W}) \tag{71}$$

where the elements of the complex vectors $\mathbf{V}$ and $\mathbf{W}$ are given by the formulas

$$V_\mu = (\Lambda_\mu - i\omega)^{-1} \tag{72}$$

$$W_\mu = (\mathbf{F}_-^\dagger \mathbf{U})_\mu (\mathbf{U}^{-1}\mathbf{P})_\mu \tag{73}$$

It is therefore only necessary to diagonalize a complex non-Hermitian matrix once per spectrum, whereas a brute-force solution of Eq. (66) would require a matrix inversion for each value of the variable ω. It is clear that the same type of algorithm can also be employed for the solution of the classical Eq. (20) (Reeves and Shaw, 1970). In addition to this advantage, it proves of great value in practical applications that there is a simple physical interpretation of the eigenvalues Λ_μ and the elements of $\mathbf{W}$. In the absence of exchange, we obtain Lorentzian absorption lines whose positions and widths are given by the imaginary and real components of Λ, respectively, and whose intensities are determined by the components of the real vector $\mathbf{W}$. As soon as exchange begins to contribute significantly to the band shape, the imaginary components of $\mathbf{W}$ become finite, and their magnitudes may be taken as a measure of the deviations from Lorentzian absorption shape. Alternatively, one may say that the "absorption" band shape now contains dispersion-like terms. This can be shown as follows (Reeves and Shaw, 1970; Vold, 1972). Defining

$$\Lambda_\mu = -\lambda_\mu + i\omega_\mu \tag{74}$$

and

$$W_\mu = w_\mu \exp(-i\phi_\mu) \tag{75}$$

and inserting in Eq. (71) yields

$$S(\omega) = C \sum_\mu \{w_\mu \lambda_\mu \cos \phi_\mu / [\lambda_\mu{}^2 + (\omega_\mu - \omega)^2] + w_\mu(\omega_\mu - \omega) \sin \phi_\mu / [\lambda_\mu{}^2 + (\omega_\mu - \omega)^2]\} \tag{76}$$

from which it is seen that each line μ consists of absorption- and dispersion-mode components with relative weights $\cos \phi_\mu$ and $-\sin \phi_\mu$, respectively.

We mention in passing that the band-shape Eq. (61) or (66) in the contracted Liouville subspace breaks up into a series of independent equations, of which each may be treated separately in practical computations, and that a further factorization ensues in the presence of symmetry and/or magnetic equivalence (Kleier and Binsch, 1970a; Meakin et al., 1971).

For very simple exchange problems, it is of course feasible to solve the equations algebraically. The frequently occurring case of mutual AB exchange was originally solved by Alexander (1962a). One obtains the formula

$$S(\omega) = C \, \mathrm{Im} \left\{ \frac{R_+ + \omega}{(A_+ + i\omega)(B_+ + i\omega) - Q_+} + \frac{R_- + \omega}{(A_- + i\omega)(B_- + i\omega) - Q_-} \right\} \tag{77}$$

with

$$\omega_0 = (\omega_A + \omega_B)/2; \quad A_\pm = -i(\omega_A \pm \pi J) - 1/T_2^{\mathrm{eff}} - k;$$
$$B_\pm = -i(\omega_B \pm \pi J) - 1/T_2^{\mathrm{eff}} - k; \quad Q_\pm = (\pm i\pi J + k)^2;$$
$$R_\pm = -\omega_0 \mp 2\pi J + 2ik + i/T_2^{\mathrm{eff}} \tag{78}$$

A rough estimate of the rate constant at the coalescence point can be obtained (Kurland et al., 1964) from

$$k = \pi[(\Delta_{AB}^2 + 6J^2)/2]^{1/2} \qquad (79)$$

2. Quadrupolar Effects

If relaxation contributes significantly to the line widths, a rigorous band-shape analysis requires the full machinery of the WBR formalism, but except for some of the additional factorizations feasible in the presence of trivial relaxation, the methods of the preceding section apply without change. Since the calculation of relaxation matrices falls outside the purview of the present chapter, nothing more needs to be said about those cases involving only identical nuclei of spin $\frac{1}{2}$.

In spin systems containing nuclei with spins $> \frac{1}{2}$, quadrupolar relaxation usually provides the most effective mechanism. The spin operators and spin Hamiltonians must be modified appropriately, using well-established procedures (Corio, 1966), and the large frequency differences caused by unequal magnetogyric ratios render additional elements in the relaxation matrix strongly nonsecular. Of particular interest in band-shape work are the subspectra of spin-$\frac{1}{2}$ nuclei that are spin-coupled to nuclei with spins $> \frac{1}{2}$, the latter undergoing rapid relaxation by the quadrupolar mechanism (for some recent references, see Cunliffe and Harris, 1968; Kintzinger et al., 1969; Bacon et al., 1970; Harris et al., 1970; Pyper, 1970, 1971a,c,d; Harris and Pyper, 1971; Brévard et al., 1972a,b).

The simplest case of a spin-$\frac{1}{2}$ nucleus A coupled through a coupling constant J_{AX} to a spin-1 nucleus X has first been treated by Pople (1958) on the basis of a stochastic model. One assumes that the A band shape can be described by a classical equation of the type (20), in which Ω contains the frequencies of the A multiplet lines and the relaxation matrix $\mathbf{R}$ is given by $1/T_{1X}$ times a statistical matrix, whose general properties are analogous to those of $\mathbf{Q}$ occurring in Eq. (24) and whose off-diagonal elements represent the transition probabilities among the corresponding spin states of X, the latter being calculated from the spectral densities of the quadrupole relaxation Hamiltonian. Pople derived a closed expression for the A band shape

$$S(\omega) = \frac{45 + \eta^2(5x^2 + 1)}{225x^2 + \eta^2(34x^4 - 2x^2 + 4) + \eta^4(x^6 - 2x^4 + x^2)} \qquad (80)$$

with

$$\eta = 10\pi T_{1X} J_{AX} \qquad (81)$$

and

$$x = (\omega_0 - \omega)/2\pi J_{AX} \qquad (82)$$

but for the general case of an X nucleus with spin > 1, it is more economical to solve Eq. (20) directly by numerical methods.

It has recently been shown by Pyper (1970) that the stochastic approach can be rigorously justified on the basis of WBR theory if the spin system contains just one single spin-1 nucleus, but that the stochastic equations become invalid (Pyper, 1971d) in the presence of several quadrupolar nuclei. In the latter case, one has no choice but to resort to a full relaxation matrix treatment.

D. Double Resonance in the Linear Approximation

We have seen that unsaturated single-resonance band shapes often do not depend on the details of relaxation and that this feature constitutes a distinct advantage if one is only interested in exchange phenomena. A variety of sophisticated pulse techniques has recently been developed (Chapter 5) for investigating relaxation processes in complex spin systems, but their application is limited to well-resolved spectra and the interpretation of the results is complicated by the fact that the magnetization decay of the individual lines is not necessarily governed by a single exponential. Steady-state double-resonance band-shape studies, which also yield detailed information on relaxation (the earlier literature is reviewed by Nageswara Rao, 1970; for some recent references, see Sinivee, 1969; Lippmaa and Rodmar, 1969; Kumar *et al.*, 1970; Krishna and Nageswara Rao, 1971a,b, 1972; Gestblom and Hartmann, 1971; Kumar and Gordon, 1971; Krishna *et al.*, 1973), do not suffer from these drawbacks but are confronted with difficulties of their own. The analysis is considerably more involved than for single resonance, and although the incorporation of an exchange superoperator into the equations is easy enough (Anderson, 1971; Yang and Gordon, 1971), exchange and relaxation effects are now intimately interrelated, so that the former can no longer be studied in isolation from the latter. Furthermore, in contrast to single resonance, field inhomogeneity is known (Freeman and Anderson, 1962; Freeman and Gestblom, 1967, 1968; Sinivee, 1969; Lippmaa and Rodmar, 1969) to affect double-resonance line shapes in a complicated fashion. Finally, whereas the strength of the radio-frequency field only contributes through a scaling factor to single-resonance band shapes, the band shapes in double resonance are a sensitive function of the absolute value of the strength of the second radio-frequency field H_2, which places stringent requirements on the accuracy of its measurement and the time-stability of certain spectrometer variables.

It is customary to eliminate the strong time dependence introduced by the double-irradiation frequency $\omega(2)$ by transforming the equations to a frame

3. Band-Shape Analysis

rotating with $-\omega(2)\mathbf{k}$. The spin Hamiltonian may then be represented by the equations

$$\mathcal{H}^s(t) = \mathcal{H}_0 + \mathcal{H}_2 + \mathcal{H}(t) \tag{83}$$

$$\mathcal{H}_0 = -\sum_i [\omega_i - \omega(2)]I_{zi} + 2\pi \sum_{i<j} J_{ij}\mathbf{I}_i \cdot \mathbf{I}_j \tag{84}$$

$$\mathcal{H}_2 = -\gamma H_2 F_x \tag{85}$$

$$\mathcal{H}_1(t) = -\gamma H_1[F_+ \exp(i\omega't) + F_- \exp(-i\omega't)]/2 \tag{86}$$

$$\omega' = \omega(1) - \omega(2) \tag{87}$$

In the domain of linear response one must require the observing field H_1 to be weak enough to avoid saturation and $|\omega'| \ll \gamma H_2$. It is then permissible to separate the effects of H_2 and $H_1(t)$ by writing

$$\mathbf{\sigma}(t) = \mathbf{\sigma}_0 + \mathbf{\chi} + \mathbf{\eta}(t) \tag{88}$$

whence the equation of motion of the (quasi) steady-state spin density vector

$$d\mathbf{\sigma}(t)/dt = [-i\mathbf{L}^s(t) + \mathbf{R} + \mathbf{X}]\mathbf{\sigma}(t) \tag{89}$$

breaks up into the two coupled equations

$$[-i(\mathbf{L}_0 + \mathbf{L}_2) + \mathbf{R} + \mathbf{X}]\mathbf{\chi} = \mathbf{M}\mathbf{\chi} = i\mathbf{L}_2\mathbf{\sigma}_0 \tag{90}$$

$$d\mathbf{\eta}(t)/dt = [-i(\mathbf{L}_0 + \mathbf{L}_2) + \mathbf{R} + \mathbf{X}]\mathbf{\eta}(t) - i\mathbf{L}_1(t)[\mathbf{\sigma}_0 + \mathbf{\chi}] \tag{91}$$

which can be combined by writing

$$d\mathbf{\eta}(t)/dt = [-i(\mathbf{L}_0 + \mathbf{L}_2) + \mathbf{R} + \mathbf{X}]\mathbf{\eta}(t) - i\mathbf{L}_1(t)[\mathbf{\sigma}_0 + i\tilde{\mathbf{M}}^{-1}\mathbf{L}_2\mathbf{\sigma}_0] \tag{92}$$

The matrix $\mathbf{M}$ is singular owing to the trace relationship

$$\text{Tr}(\chi) = 0 \tag{93}$$

and can therefore not be directly inverted, but a nontrivial solution for χ can nevertheless be shown to exist by formulating Eq. (90) in the special Liouville representation (Section III,A). Hoffman (1970) has described a projection operator technique for circumventing this problem, but in practice it is probably easiest to first diagonalize $\mathbf{M}$ and then delete the eigenvalue zero together with the corresponding rows and columns of the transforming matrices before calculating $\tilde{\mathbf{M}}^{-1}$ from the reverse transformation. A stationary solution of Eq. (92) is now obtained by developing $\eta(t)$ into a Fourier series and picking out the terms oscillating at ω'. It can be shown that the resulting equation need only be formulated in a contracted Liouville subspace in which the corresponding matrix $\mathbf{M}'$ multiplying the stationary vector $\mathbf{\eta}$ is nonsingular. Since the dimensions of the equations tend to become uncomfortably large with the increase in the number of spins, it is of utmost importance in practical applications to exploit all simplifications feasible in specific cases. Yang and Gordon (1971) have recently described a general algorithm for accomplishing this task, but a discussion of these technical matters would lead us too far afield.

Finally we need to comment on the incorporation of field inhomogeneity effects into the theory. A rigorous treatment would require a separate solution of the band-shape equation for each subensemble, over which the field H_0 may be considered to be sufficiently homogeneous. These solutions would then have to be summed (or integrated) with weighting factors extracted from an empirical broadening function. To avoid solving such a cumbersome problem in the general case, one might try to subtract a suitable constant from each "pure" linewidth term $R_{\mu\nu,\mu\nu}$ in the relaxation matrix, but there is as yet not enough experience available to be certain whether such a procedure can really be regarded as satisfactory.

In view of the considerable complexity of a theoretical and practical nature, it is not surprising to find that double-resonance band-shape work has so far been actively pursued only by a handful of specialists, but the technique seems to be ripe for more widespread applications in chemistry. The foregoing, admittedly rather terse, exposition may perhaps give a glimpse of the potential power of this method.

E. Nonlinear Response

1. Saturation

The problem of overcoming the leveling effect of field inhomogeneity broadening on detailed relaxation information may also be solved by investigating saturation phenomena in single resonance. We have already mentioned the case of a single-line amenable to a classical Bloch treatment in Section II, A, but there a band-shape study is definitely inferior to nonequilibrium techniques. This situation is reversed for strongly coupled spin systems with closely spaced lines. The relevant experimental work is of rather recent origin (Hoffman, 1970; Gestblom and Hartmann, 1972; Harris and Worvill, 1972, 1973b) and has been limited to AB spin systems, but the area holds considerable promise for the future. In comparison with the double-resonance band-shape method of the preceding section, the single-resonance saturation technique has the advantage that one does not have to worry about the differential effects of field inhomogeneity.

The nonlinear effects of the radio-frequency field on the density vector are

3. Band-Shape Analysis

accounted for by employing the full Liouville superoperator in the rotating frame on the left-hand side of Eq. (58). We thus obtain the bandshape formula

$$S(\omega) = -C \operatorname{Re}(\mathbf{F}_-{}^\dagger \tilde{\mathbf{M}}^{-1} \mathbf{L}_1 \sigma_0) \tag{94}$$

where

$$\mathbf{M} = -i(\mathbf{L}_0 + \mathbf{L}_1) + \mathbf{R} + \mathbf{X} \tag{95}$$

Since Eq. (95) refers to the full Liouville spin space, the matrix $\mathbf{M}$ is singular and the calculation of the inverse $\tilde{\mathbf{M}}^{-1}$ in the special Liouville representation must be approached with one of the methods mentioned in Section III,D. Harris and Worvill (1973b) have listed algebraic equations for an AB spin system in terms of the real and imaginary components of the density vector.

2. Nonlinear Effects in Multiple Resonance

Nonlinear effects become important for double-resonance band shapes either when H_1 gets large or when $\omega(1)$ gets too close to $\omega(2)$ (we are not referring to the well-known beat patterns in double resonance, which are of purely instrumental origin). The problem could be handled by carrying more terms in a Fourier expansion, but for strongly nonlinear behavior and for more than two radio-frequency fields, this procedure becomes rather unwieldy and convergence is not easily judged. In such extreme situations, the phenomenon begins to acquire the characteristics of a multiphoton inelastic scattering process on the spin system, which suggests that a quantized field formulation should be more appropriate. Such a quantized field formalism has recently been developed by a group of Italian workers (Di Giacomo and Santucci, 1969; Bucci et al., 1970a,b,c, 1972). Although the examples considered in these papers are perhaps a little too exotic (one case involving up to 27 photons) to be of direct relevance in the context of the present paper, we briefly mention some of the salient ideas, since it is conceivable that they might also be applied with advantage to the more pedestrian problems that are normally treated by rotating frame methods.

The first essential point is to get rid of *all* time dependencies from the very beginning by including the quantized electromagnetic field in the unperturbed Hamiltonian. Equation (40) may then be written as

$$\mathcal{H}^s = \mathcal{H}_0 + \mathcal{H}_R + \mathcal{H}_I \tag{96}$$

where $\mathcal{H}_0$ is given by Eq. (41) (there denoted as $\tilde{\mathcal{H}}_0$), $\mathcal{H}_R$ is the Hamiltonian of the radiation field

$$\mathcal{H}_R = \sum_k \omega(k) a_k{}^\dagger a_k \tag{97}$$

and $\mathcal{H}_\mathrm{I}$ represents the interaction between the spin system and the photon field contained in the volume V

$$\mathcal{H}_\mathrm{I} = -[\gamma/(2V)^{1/2}]F_x \sum_k [\omega(k)]^{1/2}(a_k^\dagger + a_k) \tag{98}$$

The creation and annihilation operators $a^\dagger$ and a are defined by their action on the n-photon states

$$a^\dagger|n\rangle = (n+1)^{1/2}|n+1\rangle \quad \text{and} \quad a|n\rangle = n^{1/2}|n-1\rangle \tag{99}$$

The density matrix ζ for spin system plus photon field is written as

$$\zeta = \sigma \otimes \rho^\mathrm{R} \tag{100}$$

since one may assume that the field contains so many photons that ρ^R remains unaffected by the interaction. All equations are now expressed in a representation in which $\mathcal{H}_0 + \mathcal{H}_\mathrm{R}$ is diagonal, i.e., the basis states are chosen from the infinite manifold

$$|m\rangle \prod_k |n_k \pm p_k\rangle \tag{101}$$

where m labels the eigenstates of $\mathcal{H}_0$ and p_k are the number of photons of frequency $\omega(k)$ that are to be added to or subtracted from the n_k-photon states.

The second important step in the development consists in the realization that the major part of the interaction can be accounted for if the equations are projected into a small subspace spanned by an approximately degenerate submanifold of the infinite set (101). In single resonance, such a submanifold might, for instance, look like follows

$$\begin{aligned} &|m; n\rangle \\ &|m-1; n+p\rangle \\ &\quad\vdots \\ &|-m; n+2mp\rangle \end{aligned} \tag{102}$$

in which the eigenstates of $\mathcal{H}_0$ (m is here understood to be a fixed label) differ approximately by the energies of p photons of frequency ω. Another simple example is provided by a progressively connected three-level system of $\mathcal{H}_0$ in which the energy differences between $|m_1\rangle$ and $|m_2\rangle$ and between $|m_2\rangle$ and $|m_3\rangle$ are approximately matched by single-photon frequencies $\omega(1)$ and $\omega(2)$, respectively. Here one would take the submanifold

$$\begin{aligned} &|m_1; n_1+1, n_2\rangle \\ &|m_2; n_1, n_2\rangle \\ &|m_3; n_1, n_2-1\rangle \end{aligned} \tag{103}$$

3. Band-Shape Analysis

The interaction Hamiltonian $\mathcal{H}_I$ is now expanded in terms of its resolvent

$$\overline{\mathcal{H}_I} = P\mathcal{H}_I P + P\mathcal{H}_I Q(E - \mathcal{H}_0 - \mathcal{H}_R - Q\mathcal{H}_I Q)^{-1} Q\mathcal{H}_I P \quad (104)$$

where P is the projection operator on the approximately degenerate submanifold, $Q = 1 - P$ the projection operator into the rest of Hilbert space and E the energy of the submanifold. The first part $\overline{\mathcal{H}_I} = P\mathcal{H}_I P$ is then treated exactly and the rest by time-independent perturbation theory.

Without going into any further details, it is clear that this formulation gets at the heart of the two principal obstacles of current band-shape theories: (1) The dimensionality of the problem in the conventional approach increases so steeply with the number of spins that a fairly rigorous treatment soon becomes impossible in practice. (2) As a consequence, it is mandatory to make drastic simplifications, but more often than not these approximations are arrived at by physical intuition. Although physical intuition is not to be despised, it seems not unreasonable to expect that a division of the problem into a major and minor part from the very outset should be of help in developing more objective criteria.

The power of the second-quantization formalism has already been demonstrated in a number of special situations. Whether substantial advantages can be realized for chemically interesting studies on multispin systems by reformulating some of the theories presently in use must at this point of time remain an open question.

F. Transient Response

A general treatment of transient phenomena will be given in Chapters 4 and 5, but there are two papers dealing with time-dependent effects that, in a certain sense, belong to the realm of band-shape analysis as defined in the present chapter. Gore and Gutowsky (1968) have considered the case where a spin system is prepared in a special way at time $t = 0$ and its "steady-state" band shape examined with a weak stimulating field H_1 at some later instant of time; in other words, one makes the somewhat artificial assumption that no time is needed to sweep the spectrum while still maintaining "steady-state" conditions. To solve this problem, one starts from the time-dependent Eq. (53), which is then integrated subject to the initial conditions at time $t = 0$. Gore and Gutowsky have calculated band shapes for an AB spin system in the two situations where $H_1 = 0$ or $H_1 = \infty$ at $t = 0$. More recently, Goodwin and Wallace (1972) have published calculated spectra obtained by numerical integration based on a fourth-order Runge–Kutta–Gill algorithm for the general case, involving one or more radio-frequency fields and including the effects of finite sweep rates.

IV. Practical Band-Shape Analysis

A. Experimental Procedures

The great majority of nmr band-shape studies described in the chemical literature were motivated by the desire to obtain quantitative information about the rates of nuclear exchange processes at a number of fixed temperatures. To simplify the theoretical analysis, one normally makes an attempt to satisfy the unsaturated steady-state conditions and to choose spin systems for which the band shapes are not complicated by nontrivial relaxation effects. This requires that one pays attention to the following points:

1. The sweep rates employed in recording the spectra must be sufficiently slow to avoid transient phenomena. This requirement becomes particularly stringent in those regions of the spectrum where one encounters relatively sharp or closely spaced lines. On the other hand, excessively long recording times are undesirable, since there always exists the danger that certain other parameters, which have to be treated as constant in the theoretical analysis, will experience a significant drift; a danger especially pronounced for the temperature in the sample. A reasonable compromise can usually be achieved by selecting appropriate sweep rates for each part of the spectrum as judged on the basis of a preliminary spectrum recorded with a fairly rapid sweep rate.

2. The formulas of Sections II,B,C and III,C only apply in the absence of saturation and demand "sufficiently small" H_1 fields. Although it is easy enough to specify the necessary conditions for negligible saturation theoretically, it is often rather difficult in practice to be certain that saturation effects are completely absent, especially in the case of very broad lines where the requirement of low H_1 fields is in conflict with the practical necessity to increase the overall intensity of the spectrum and to enhance the signal-to-noise ratio. Unfortunately, there is virtually no unambiguous information to be found in the pertinent literature as to how serious such saturation errors may become for strongly exchange-broadened band shapes in particular instances. In general, it is probably preferable to tolerate a relatively noisy spectrum rather than to risk saturation. By the same token, one should avoid excessive filtering, which is also known to cause band-shape distortions.

3. The homogeneity of the external field should be carefully adjusted. Experience has shown that in highly homogeneous fields the lines not broadened by exchange are practically indistinguishable from true Lorentzian shape except possibly far out in the wings, which are, however, usually buried in noise anyway; hence, the residual inhomogeneity effects can be

3. Band-Shape Analysis

accounted for by using an effective transverse relaxation time in the computations. In principle, it is also possible to analyze band shapes that are strongly distorted by field inhomogeneity by convoluting the theoretical curves or deconvoluting the experimental curves with an empirical broadening function, obtainable, for instance, from a reference signal. But such spectra are difficult to reproduce in practice. Before resorting to these computational devices it appears advisable, therefore, first to go back and try to improve on the experiments.

4. Nontrivial relaxation is frequently caused by coupling of the exchanging spins to quadrupolar nuclei such as deuterium or nitrogen-14 and can conveniently be removed by recording the spectra of the spin system of interest while simultaneously decoupling the quadrupolar nuclei.

In many applications, it is important to determine the exchange rate constants at a series of fixed and accurately measured temperatures. The variable-temperature accessories supplied with current commercial spectrometers are usually adequate for maintaining a sufficiently constant temperature in the sample, but they should not be relied upon for accurate temperature measurements, since the readings are taken in the gas stream on the outside of the sample tube. Two superior methods are presently in common use. In the substitution method, an nmr tube partially filled with a suitable solvent and containing a thermocouple or thermistor is exchanged for the sample tube immediately before and immediately after recording a spectrum, and readings are taken after sufficient time has elapsed for complete equilibration with the surroundings in the sample cavity of the nmr probe. The maximum accuracy achievable with this procedure is probably of the order of $\pm 1°$ between $-50°C$ and $+100°C$, but is likely to deteriorate outside these limits. Better results can be obtained by using an "nmr thermometer liquid" contained in a sealed capillary within the sample tube. Any substance giving rise to sharp nmr lines whose chemical shift difference is a sensitive function of the temperature is in principle suitable for this purpose, provided it contains the same species of nuclei as those of interest in the sample. This chemical shift difference can then be recorded simultaneously with the spectrum of the sample. Popular thermometer liquids for proton nmr are methanol and ethylene glycol (containing a trace of hydrochloric acid), for which carefully measured calibration curves are available (van Geet, 1968, 1970).

For analyzing double-resonance or partially saturated band shapes, one also needs to know the strength of the radio-frequency fields (H_2 in the former, H_1 in the latter). It appears that the most reliable procedure is to employ the transient nutation method originally described by Torrey (1949) and recently developed further by Leigh (1968) and Harris and Worvill (1973a).

B. Computer Programs

1. Simulation Programs

Although it is sometimes possible to extract dynamic information from band shapes with the help of simple formulas, such as Eq. (9) for calculating effective transverse relaxation times from the widths of single lines or Eq. (18) and (79) for obtaining exchange rate constants at coalescence temperatures, the band-shape formulas are in general too complicated to be amenable to an algebraic analysis and one must therefore resort to numerical calculations. Even in the case of single lines, a complete band-shape analysis is more accurate than single-parameter methods (Allerhand, 1970; Harris and Worvill, 1973a), since it tends to compensate for residual errors in phase, base-line definition and sweep nonlinearity and allows a more objective analysis of noisy spectra.

The computational core common to all dynamic nmr computer programs is the simulation of a theoretical band shape from static and dynamic input parameters. Such programs have been developed by numerous research workers and copies are in most cases available through private requests to the authors. In the following we limit ourselves to mentioning a few programs that are either well documented in the official or semiofficial literature, or seem to be in widespread use, or are in some way unique.

The most frequently investigated cases are the two-site classical exchange and mutual AB exchange, for which we have given closed expressions in Eq. (14) and Eq. (77). These formulas are so simple that most workers in the area have written their own programs. Source program listings are reproduced in a review article (Binsch, 1968a). One of the earliest programs for classical multisite exchange was written by M. Saunders at Yale University (see Saunders, 1963) and seems to have been widely distributed. A similar program frequently cited in the literature was developed by G. M. Whitesides at MIT (see Whitesides and Fleming, 1967). Source program listings for the inversion of complex matrices are reproduced in a paper by Yandle and Maher (1969), but it is known that the diagonalization procedure mentioned in Section III,C provides a much more efficient algorithm; it has in fact been incorporated into more recent programs for classical multisite exchange (Chan and Reeves, 1973). General simulation programs for intramolecular exchange involving strongly coupled spin systems are available from the Quantum Chemistry Program Exchange (Binsch and Kleier, 1969; Kleier and Binsch, 1970b). Among the more sophisticated programs for treating quadrupolar effects we mention those described by Bacon *et al.* (1970), Harris and Pyper (1971), and Pyper (1971d). The most versatile double-resonance band-shape simulation program seems to be that of Yang and Gordon (1971). Computer

3. Band-Shape Analysis

programs for dealing with nonlinear and transient response were developed by the authors cited in Sections III,E and F.

2. Iterative Programs

Further progress in practical band-shape analysis has become possible by the development of computer programs that automatically iterate to the best (in a least-squares sense) fit of a simulated band shape on an experimental spectrum. Many nmr spectrometers are now interfaced with dedicated computers and can therefore easily produce the experimental spectrum in digital form suitable for further computer manipulation. A least-squares analysis eliminates some of the subjective features inevitably associated with visual comparisons of theoretical and experimental band shapes, gives more reliable information on errors, and allows a more extensive and systematic variation of the static and dynamic parameters. A great number of such programs of various degrees of sophistication are presently in use for relatively simple spin systems (e.g., Jonas et al., 1965; Zumdahl and Drago, 1967; Inglefield et al., 1968; Rabinovitz and Pines, 1969; Drakenberg, et al., 1970; Cox et al., 1970; Reeves and Shaw, 1971; Reeves et al., 1971; Harris and Pyper, 1971; Spaargaren et al., 1971; Harris and Worvill, 1973b). A general iterative program for intramolecular exchange involving strongly coupled spin systems is available from J. Heinzer at the Swiss Federal Institute of Technology, Zurich (unpublished).

C. Determination of Static Parameters

Before one can begin to carry out a dynamic band-shape analysis, one must know the chemical shifts and coupling constants and for nonmutual exchange problems also the populations of the species at equilibrium, or one must at least have reasonably good first approximations to these static parameters. For double-resonance or saturation studies of relaxation, this information can usually be obtained by first subjecting the unsaturated single-resonance spectra to an iterative static analysis using standard procedures. Strongly exchange-broadened spectra are not amenable to such a direct static analysis. It is therefore necessary first to freeze out the exchange by lowering the temperature. In practice, this requires that the barrier of the exchange process be not substantially smaller than about 7 kcal/mole. A static analysis of the frozen spectra carried out at a number of different temperatures yields the standard enthalpy and entropy differences of the species from the thermodynamic equations

$$\Delta G_{ij}^0 = -RT \ln (p_j/p_i) \qquad (105)$$

$$\Delta G_{ij}^0 = \Delta H_{ij}^0 - T \Delta S_{ij}^0 \qquad (106)$$

The populations in the exchange-broadened region can then be calculated by back-substitution into Eqs. (106) and (105). In a similar fashion, one may fit the temperature dependencies of the chemical shifts and coupling constants as extracted from the frozen spectra to simple (frequently linear) polynomials in T or $1/T$ and then extrapolate to the exchange-broadened band shapes. The fast-exchange spectra at higher temperatures serve as a convenient check of the adequacy of such extrapolation procedures, since these spectra must be reproduced by the corresponding time-averaged static parameters. The final refinement of the static parameters extrapolated to the broad spectra can in principle be accomplished by a full least-squares analysis of the total band shape, in which the static as well as the dynamic parameters are treated as independent variables, but practical experience as to the reliability of such a general procedure is still very limited.

D. Determination of Dynamic Parameters

1. Relaxation

The determination of detailed relaxation information from steady-state band shapes of complex spin systems has not yet progressed beyond a rather primitive level. The molecules investigated in some detail were mostly chosen for methodological demonstration purposes and not for their intrinsic chemical interest; the spin systems were kept deliberately simple and the dynamic information content was sometimes artificially swamped by the addition of paramagnetic impurities in order to ensure the dominance of the isotropic random field relaxation mechanism. An exception is the extraction of relaxation information from band shapes broadened by coupling to quadrupolar nuclei, which has yielded results of genuine chemical interest; for details we refer to the original papers cited in Section III,C,2.

To set up the WBR relaxation matrices is quite straightforward in principle for any spin system and any relaxation mechanism, but turns out to be exceedingly tedious in practice for all but the most primitive cases. The lack of sufficiently general computer programs for accomplishing such tasks is still a major obstacle to progress in this area. From the few published hand calculations, it is already clear, however, that relaxation matrices depend only on a very small number of dynamic variables, so that their determination by least-squares band-shape analysis should be feasible without too many preconceived notions regarding the relative importance of various relaxation mechanisms. A great wealth of detailed and chemically highly interesting information should become accessible by such studies, but the extent to which this inherent potential can be exploited in practice remains to be demonstrated.

3. Band-Shape Analysis

For the practical analysis of exchange-broadened band shapes in the presence of trivial relaxation, one needs to know the effective transverse relaxation time as a function of temperature. If one makes an effort to maintain good field homogeneity at all temperatures, the difference between T_2^{eff} at the slow- and fast-exchange limits is likely to be very small; it is then permissible to use linearly interpolated values for the intermediate temperatures. Alternatively, one may extract T_2^{eff} from the width of an internal reference signal that is not participating in the exchange, or one may treat T_2^{eff} as a free variable in a least-squares computation. In favorable cases, it is even possible to extract T_2^{eff} from certain lines of the exchanging spin system directly throughout the entire temperature range (Steigel et al., 1972).

2. Exchange

In typical situations, the nmr band shapes are a sensitive function of nuclear exchange processes whose rate constants cover the range from about 10^{-1} to 10^3 sec^{-1}. A few trial simulations encompassing this range, in combination with simple approximate formulas, are usually adequate to provide rough first estimates of the rate constants, which can then be refined by further trial simulations or by iterative computations. A more economical procedure is to find the best rate constant(s) for one spectrum of a series taken at different temperatures and then to calculate estimates of the rate constants at the other temperatures by assuming a temperature-independent free energy of activation. Experience has shown that the values obtained with this approximation, which is equivalent to postulating a small entropy of activation, are normally close enough to the true rate constants to serve as good starting points for an iterative calculation.

For multisite exchange processes with chemical constraints, one is faced with the problem of setting up the correct exchange matrix or, equivalently, of choosing the correct set of nuclear configurations interconnected by the exchange. A general procedure for finding such sets is presented in Chapter 2, and numerous experimental examples are discussed in Chapters 8 and 9.

V. Treatment of Dynamic Data

A. Calculation of Activation Parameters

The measurement of relaxation times or exchange rate constants as a function of temperature offers the possibility of calculating activation parameters for the underlying physical or chemical processes. "Barriers" to molecular reorientation in liquids have often been obtained in this fashion from relaxation data, but the physical meaning of such numbers is somewhat obscure (Hertz, 1967), except in those cases where they can be related to the barriers of segmental motions within molecules. The great practical value of relaxation information lies elsewhere, as will be discussed in Chapter 5.

Nuclear exchange processes lend themselves to a clearcut chemical interpretation in most instances, and it is therefore standard practice to convert the corresponding rates to activation parameters. A free energy of activation may be calculated from a rate constant at a single temperature by means of the Eyring equation

$$k = \kappa(k_B T/h) \exp(-\Delta G^\ddagger/RT) \tag{107}$$

in which the symbols have their usual meaning. Rate constants measured over a range of temperatures may be converted to Arrhenius activation energies and frequency factors or to enthalpies and entropies of activation with the help of the formulas

$$k = A \exp(-E_a/RT) \tag{108}$$

$$k = \kappa(k_B T/h) \exp(-\Delta H^\ddagger/RT) \exp(\Delta S^\ddagger/R) \tag{109}$$

A popular method of analysis consists in fitting a straight line to the logarithmic variants of Eqs. (108) and (109) graphically and to extract the parameters of interest from the slopes and intercepts, but for reasons to be explained presently, it is much preferable to do a least-squares calculation on a computer.

B. Error Analysis

Dynamic parameters extracted from nmr band shapes are subject to a plethora of possible sources of errors of a systematic as well as a statistical nature. In the wake of the drastic discrepancies in the results reported by different groups of authors in the earlier literature, especially regarding Arrhenius parameters and enthalpies and entropies of activation, there has been a steadily growing awareness of the factors that are likely to give rise to serious systematic errors (see, in particular, Allerhand et al., 1966; Shoup et al., 1972). We have alluded to the most important ones at the appropriate points in the preceding discussion. It is well known to workers in the area that the free energy of activation is rather insensitive to errors, in striking contrast to the parameters occurring in Eqs. (108) and (109), and the patent incongruities of the latter are commonly blamed on systematic errors. This may well be correct in many instances, but the frequently drawn corollary that random errors in E_a and A or $\Delta H^\ddagger$ and $\Delta S^\ddagger$ are incapable of giving a realistic representation of the error situation is largely based on a myth

3. Band-Shape Analysis

attributable to a faulty statistical analysis of the data. It may thus be worthwhile to clarify matters by discussing a few illustrative examples.

An analysis of Eq. (107) yields the following formula for the linearized relative statistical error in $\Delta G^{\ddagger}$

$$(\sigma_{\Delta G^{\ddagger}}/\Delta G^{\ddagger})^2 = [\ln(k_B T/hk)]^{-2}(\sigma_k/k)^2 + [1 + \{\ln(k_B T/hk)\}^{-1}]^2(\sigma_T/T)^2 \quad (110)$$

or approximately

$$(\sigma_{\Delta G^{\ddagger}}/\Delta G^{\ddagger})^2 \cong [\ln(k_B T/hk)]^{-2}(\sigma_k/k)^2 + (\sigma_T/T)^2 \quad (111)$$

For a typical case with $T = 300°K$ and $k = 100 \text{ sec}^{-1}$, a relative statistical error of 100% in the rate constant introduces a relative error of 4% in $\Delta G^{\ddagger}$, and a temperature wrong by 6°K an error of 2% in $\Delta G^{\ddagger}$, which adds up to a total standard deviation of 4.5%, or, for a typical $\Delta G^{\ddagger}$ of 15 kcal/mole, to ± 0.7 kcal/mole. Similarly, errors in k and T of 25% and 3°K, respectively, yield a $\Delta G^{\ddagger}$ with a precision of ± 0.2 kcal/mole. Contrast the latter case with the error formula for the Arrhenius activation energy

$$(\sigma_{E_a}/E_a)^2 \cong [2T^2/(\Delta T)^2](\sigma_T/T)^2 + 2[\Delta(\ln k)]^{-2}(\sigma_k/k)^2 \quad (112)$$

where for a temperature range of $\Delta T = 20°K$ an error of 25% in k causes an error of 26% in the activation energy, and an error of 3°K in T an error of 21% in E_a, for a combined error of 34% (± 5 kcal/mole for $E_a = 15$ kcal/mole). For $\Delta T = 60°K$, the corresponding numbers are 8.5%, 7%, 11%, ± 1.7 kcal/mole.

It can thus be seen that statistical errors amply suffice to explain the absurdities in the reported activation parameters determined by dynamic nmr spectroscopy. By a similar line of argument, based on statistical principles alone, it is also easy to show that the discrepancies in the values of the pairs (E_a, A) and $(\Delta H^{\ddagger}, \Delta S^{\ddagger})$ have a tendency to be mutually compensating; for example, an activation enthalpy too small tends to be associated with an activation entropy too negative, and an activation enthalpy too large with an activation entropy too positive, etc. The reason why the statistical errors reported in the pertinent literature have in most cases been vastly underestimated is due to the fact that all data points were treated on an equal footing on a logarithmic scale. This is inevitable if one uses graphical methods or computer programs that only calculate the precision of the fit of the data points to a straight line. Such a procedure would be incorrect even if the variances of the rate constants and temperatures happened to be the same throughout the temperature range, for then it necessarily follows that they cannot be the same for plots of $\ln k$ versus $1/T$, for instance. In reality, the situation is much worse, especially for very simple exchanging spin systems, where the precision of the rate constants decreases steeply with distance from the coalescence region. The data points toward the temperature extremes, which would be most valuable for tying down an Arrhenius or Eyring line, thus become essentially worthless statistically. For carrying out a sensible error analysis, it is therefore indispensable to use a computer program capable of doing an error propagation calculation. Basic programs of this type are in plentiful supply at all modern computing centers. The required input variances are most objectively obtained from a least-squares band-shape computation.

The discussion following Eq. (112) points to the importance of a large temperature range, but at the same time it is clear that a large temperature range is not in itself a sufficient requirement for obtaining high-precision results. It is in addition necessary that the spectra remain sufficiently sensitive to changes in the rate constants at the temperature extremes. This condition is hard to satisfy with simple spin systems, but the difficulties can be overcome by deliberately choosing to study more complex spectra. Several experimental demonstrations of this conclusion have been published recently (e.g., Dahlqvist and Forsén, 1969, 1970; Drakenberg and Forsén, 1970; Kleier et al., 1970; Steigel et al., 1972).

VI. Conclusion

The desire to understand the nuclear spin dynamics has been a primary concern from the very inception of nmr spectroscopy, as documented by the fundamental work carried out by the two groups of physicists who discovered the phenomenon in bulk matter. When the method began to make its impact on chemistry about 20 years ago, however, there was a pronounced shift of emphasis toward the static aspects of this branch of spectroscopy. Band-shape analyses have been mostly confined to the study of nuclear exchange processes in simple spin systems, but the wealth of valuable chemical information obtained by such applications is nevertheless very impressive. There can be no doubt that the techniques discussed in this chapter hold far greater potential for applications than has hitherto been realized. The investigations of multisite exchange processes and of exchange in complicated spin systems published in the last 5 years already provide a demonstration of the great power of more advanced band-shape methods. It will be particularly interesting to see what can be learned from band-shape work of relaxation in complex spin systems in the near future.

REFERENCES

Abragam, A. (1961). "The Principles of Nuclear Magnetism." Oxford Univ. Press (Clarendon), London and New York.
Alexander, S. (1962a). *J. Chem. Phys.* **37**, 967.

Alexander, S. (1962b). *J. Chem. Phys.* **37**, 974.
Alger, T. D., Gutowsky, H. S., and Vold, R. L. (1967). *J. Chem. Phys.* **47**, 3130.
Allerhand, A. (1970). *J. Chem. Phys.* **52**, 3596.
Allerhand, A., and Thiele, E. (1966), *J. Chem. Phys.* **45**, 902.
Allerhand, A., Gutowsky, H. S., Jonas, J., and Meinzer, R. A. (1966) *J. Amer. Chem. Soc.* **88**, 3185.
Anderson, J. E., and Fryer, P. A. (1969). *J. Chem. Phys.* **50**, 3784.
Anderson, J. M. (1971). *J. Magn. Resonance* **4**, 184.
Ayant, Y., Belorizky, E., Gagniere, D., and Taieb, M. (1972). *J. Chem. Phys.* **56**, 5399.
Bacon, J., Gillespie, R. J., Hartman, J. S., and Rao, U. R. K. (1970). *Mol. Phys.* **18**, 561.
Banwell, C. N., and Primas, H. (1963). *Mol. Phys.* **6**, 225.
Ben-Reuven, A. (1971). *Phys. Rev. A* **4**, 2115.
Binsch, G. (1968a). *Top. Stereochem.* **3**, 97.
Binsch, G. (1968b). *Mol. Phys.* **15**, 469.
Binsch, G. (1969), *J. Amer. Chem. Soc.* **91**, 1304.
Binsch, G., and Kleier, D. A. (1969). "The Computation of Complex Exchange-Broadened NMR-Spectra," Program 140. Quantum Chemistry Program Exchange, Indiana University, Bloomington.
Bloch, F. (1946). *Phys. Rev.* **70**, 460.
Bloch, F. (1956). *Phys. Rev.* **102**, 104.
Bloch, F. (1957). *Phys. Rev.* **105**, 1206.
Blume, M. (1968). *Phys. Rev.* **174**, 351.
Brévard, C., Kintzinger, J. P., and Lehn, J. M. (1972a). *Tetrahedron* **28**, 2429.
Brévard, C., Kintzinger, J. P., and Lehn, J. M. (1972b). *Tetrahedron* **28**, 2447.
Bucci, P., Serra, A. M., Cavaliere, P., and Santucci, S. (1970a). *Chem. Phys. Lett.* **5**, 605.
Bucci, P., Cavaliere, P., and Santucci, S. (1970b). *J. Chem. Phys.* **52**, 4041.
Bucci, P., Martinelli, M., and Santucci, S. (1970c). *J. Chem. Phys.* **53**, 4524.
Bucci, P., Martinelli, M., Santucci, S., and Serra, A. M. (1972). *J. Magn. Resonance* **6**, 281.
Bull, T. E. (1972). *J. Magn. Resonance* **8**, 344.
Chan, S. O., and Reeves, L. W. (1973). *J. Amer. Chem. Soc.* **95**, 673.
Cohn, M., and Reuben, J. (1971). *Accounts Chem. Res.* **4**, 214.
Corio, P. L. (1966). "Structure of High-Resolution NMR Spectra." Academic Press, New York.
Cox, B. G., Riddell, F. G., and Williams, D. A. R. (1970). *J. Chem. Soc., B* p. 859.
Cunliffe, A. V., and Harris, R. K. (1968). *Mol. Phys.* **15**, 413.
Dahlqvist, K. I., and Forsén, S. (1969). *J. Phys. Chem.* **73**, 4124.
Dahlqvist, K. I., and Forsén, S. (1970). *J. Magn. Resonance* **2**, 61.
Di Giacomo, A., and Santucci, S. (1969). *Nuovo Cimento B* **63**, 407.
Drakenberg, T., and Forsén, S. (1970). *J. Phys. Chem.* **74**, 1.
Drakenberg, T., Dahlqvist, K. I., and Forsén, S. (1970). *Acta Chem. Scand.* **24**, 694.
Fano, U. (1964). *In* "Lectures on the Many-Body Problem" (E. R. Caianiello, ed.), Vol. 2, p. 217. Academic Press, New York.
Freeman, R., and Anderson, W. A. (1962). *J. Chem. Phys.* **37**, 2053.
Freeman, R., and Gestblom, B. (1967). *J. Chem. Phys.* **47**, 2744.
Freeman, R., and Gestblom, B. (1968). *J. Chem. Phys.* **48**, 5008.
Gestblom, B., and Hartmann, O. (1971). *J. Magn. Resonance* **4**, 322.
Gestblom, B., and Hartmann, O. (1972). *J. Magn. Resonance* **8**, 230.
Goodwin, B. W., and Wallace, R. (1972). *J. Magn. Resonance* **8**, 41.
Gordon, R. G., and McGinnis, R. P. (1968). *J. Chem. Phys.* **49**, 2455.
Gore, E. S., and Gutowsky, H. S. (1968). *J. Chem. Phys.* **48**, 3260.

Gutowsky, H. S., Vold, R. L., and Wells, E. J. (1965). *J. Chem. Phys.* **43**, 4107.
Hahn, E. L., and Maxwell, D. E. (1952). *Phys. Rev.* **88**, 1070.
Harris, R. K., and Pyper, N. C. (1971). *Mol. Phys.* **20**, 467.
Harris, R. K., and Worvill, K. M. (1972). *Chem. Phys. Lett.* **14**, 598.
Harris, R. K., and Worvill, K. M. (1973a). *J. Magn. Resonance* **9**, 383.
Harris, R. K., and Worvill, K. M. (1973b). *J. Magn. Resonance* **9**, 394.
Harris, R. K., Pyper, N. C., Richards, R. E., and Schulz, G. W. (1970). *Mol. Phys.* **19**, 145.
Hertz, H. G. (1967). *Progr. NMR Spectrosc.* **3**, 159.
Hoffman, R. A. (1970). *Advan. Magn. Resonance* **4**, 87.
Hubbard, P. S. (1969). *Phys. Rev.* **180**, 319.
Inglefield, P. T., Krakower, E., Reeves, L. W., and Stewart, R. (1968). *Mol. Phys.* **15**, 65.
Johnson, C. S. (1964). *J. Chem. Phys.* **41**, 3277.
Jonas, J., Allerhand, A., and Gutowsky, H. S. (1965). *J. Chem. Phys.* **42**, 3396.
Kaplan, J. I. (1958a). *J. Chem. Phys.* **28**, 278.
Kaplan, J. I. (1958b). *J. Chem. Phys.* **29**, 462.
Kaplan, J. I., and Fraenkel, G. (1972). *J. Amer. Chem. Soc.* **94**, 2907.
Kintzinger, J. P., Lehn, J. M., and Williams, R. L. (1969). *Mol. Phys.* **17**, 135.
Kleier, D. A., and Binsch, G. (1970a). *J. Magn. Resonance* **3**, 146.
Kleier, D. A., and Binsch, G. (1970b). "DNMR3: A Computer Program for the Calculation of Complex Exchange-Broadened NMR Spectra. Modified Version for Spin Systems Exhibiting Magnetic Equivalence or Symmetry," Program 165. Quantum Chemistry Program Exchange, Indiana University, Bloomington.
Kleier, D. A., Binsch, G., Steigel, A., and Sauer, J. (1970). *J. Amer. Chem. Soc.* **92**, 3787.
Krishna, N. R., and Nageswara Rao, B. D. (1971a). *Mol. Phys.* **20**, 981.
Krishna, N. R., and Nageswara Rao, B. D. (1971b). *Mol. Phys.* **22**, 937.
Krishna, N. R., and Nageswara Rao, B. D. (1972). *Mol. Phys.* **23**, 1013.
Krishna, N. R., Yang, P. P., and Gordon, S. L. (1973). *J. Chem. Phys.* **58**, 2906.
Kumar, A., and Gordon, S. L. (1971). *J. Chem. Phys.* **54**, 3207.
Kumar, A., Krishna, N. R., and Nageswara Rao, B. D. (1970). *Mol. Phys.* **18**, 11.
Kurland, R. J., Rubin, M. B., and Wise, W. B. (1964). *J. Chem. Phys.* **40**, 2426.
Leigh, J. S. (1968). *Rev. Sci. Instrum.* **39**, 1594.
Leigh, J. S. (1971). *J. Magn. Resonance* **4**, 308.
Lippmaa, E., and Rodmar, S. (1969). *J. Chem. Phys.* **50**, 2764.
Lynden-Bell, R. M. (1967). *Progr. NMR Spectrosc.* **2**, 163.
McConnell, H. M. (1958). *J. Chem. Phys.* **28**, 430.
Marshall, A. G. (1970). *J. Chem. Phys.* **52**, 2527.
Meakin, P., Muetterties, E. L., Tebbe, F. N., and Jesson, J. P. (1971). *J. Amer. Chem. Soc.* **93**, 4701.
Nageswara Rao, B. D. (1970). *Advan. Magn. Resonance* **4**, 271.
Pople, J. A. (1958). *Mol. Phys.* **1**, 168.
Pyper, N. C. (1970). *Mol. Phys.* **19**, 161.
Pyper, N. C. (1971a). *Mol. Phys.* **20**, 449.
Pyper, N. C. (1971b). *Mol. Phys.* **21**, 1.
Pyper, N. C. (1971c). *Mol. Phys.* **21**, 961.
Pyper, N. C. (1971d). *Mol. Phys.* **21**, 977.
Pyper, N. C. (1971e). *Mol. Phys.* **22**, 433.
Rabinovitz, M., and Pines, A. (1969). *J. Amer. Chem. Soc.* **91**, 1585.
Redfield, A. G. (1957). *IBM J. Res. Develop.* **1**, 19.
Redfield, A. G. (1965). *Advan. Magn. Resonance* **1**, 1.

Reeves, L. W., and Shaw, K. N. (1970). *Can. J. Chem.* **48**, 3641.
Reeves, L. W., and Shaw, K. N. (1971). *Can. J. Chem.* **49**, 3671.
Reeves, L. W., Shaddick, R. C., and Shaw, K. N. (1971). *Can. J. Chem.* **49**, 3683.
Sack, R. A. (1958). *Mol. Phys.* **1**, 163.
Saunders, M. (1963). *Tetrahedron Lett.* p. 1699.
Schirmer, R. E., Noggle, J. H., and Gaines, D. F. (1969). *J. Amer. Chem. Soc.* **91**, 6240.
Shanan-Atidi, H., and Bar-Eli, K. H. (1970). *J. Phys. Chem.* **74**, 961.
Shoup, R. R., Becker, E. D., and McNeel, M. L. (1972). *J. Phys. Chem.* **76**, 71.
Sinivee, V. (1969). *Mol. Phys.* **17**, 41.
Slichter, C. P. (1963). "Principles of Magnetic Resonance." Harper, New York.
Spaargaren, K., Korver, P. K., van der Haak, P. J., and de Boer, T. J. (1971). *Org. Magn. Resonance* **3**, 605.
Steigel, A., Sauer, J., Kleier, D. A., and Binsch, G. (1972). *J. Amer. Chem. Soc.* **94**, 2770.
Stengle, T. R., and Baldeschwieler, J. D. (1967). *J. Amer. Chem. Soc.* **89**, 3045.
Sutherland, I. O. (1971). *Annu. Rep. NMR (Nucl. Magn. Resonance) Spectrosc.* **4**, 71.
Swift, T. J., and Connick, R. E. (1962). *J. Chem. Phys.* **37**, 307; erratum, **41**, 2553 (1964).
Torrey, H. C. (1949). *Phys. Rev.* **76**, 1059.
van Geet, A. L. (1968). *Anal. Chem.* **40**, 2227.
van Geet, A. L. (1970). *Anal. Chem.* **42**, 679.
Vold, R. L. (1972). *J. Chem. Phys.* **56**, 3210.
Vold, R. L., and Chan, S. O. (1971). *J. Magn. Resonance* **4**, 208.
Wangsness, R. K., and Bloch, F. (1953). *Phys. Rev.* **89**, 728.
Wennerström, H. (1972). *Mol. Phys.* **24**, 69.
Whitesides, G. M., and Fleming, J. S. (1967). *J. Amer. Chem. Soc.* **89**, 2855.
Yandle, J. R., and Maher, J. P. (1969). *J. Chem. Soc., A* p. 1549.
Yang, P. P., and Gordon, S. L. (1971). *J. Chem. Phys.* **54**, 1779.
Zumdahl, S. S., and Drago, R. S. (1967). *J. Amer. Chem. Soc.* **89**, 4319.

4

Application of Nonselective Pulsed NMR Experiments — Diffusion and Chemical Exchange

L. W. REEVES

I. Introduction.. 83
II. Diffusion and Flow... 84
III. Chemical Exchange Rates from Spin-Echo Studies................ 105
IV. Chemical Exchange Rates from Rotating Frame Relaxation......... 125
References.. 128

I. Introduction

The discovery of spin echoes by Hahn reported in 1950 included a treatment of the effect of diffusion and laid a sound basis for the further application of pulsed nuclear magnetic resonance (nmr) to diffusion studies. The recognition that chemical exchange, which can be regarded also as modulation of the Larmor frequency, would produce related effects on the amplitude of spin echoes was first reported by Woessner (1961b). This chapter will discuss how these two applications developed.

The subject of measurements of rates of chemical exchange by nuclear magnetic resonance has now been thoroughly investigated and a much more complete picture can be given. It is interesting to compare the present state of knowledge with that reported in the last reviews on the subject (Johnson, 1965; Reeves, 1965). Much of the basic understanding of the observation of nmr signals using pulsed radio-frequency sources will already be treated in this book by Freeman and Hill (Chapter 5).

II. Diffusion and Flow

The important concept in a spin-echo experiment is the variation with time of the distribution of Larmor frequencies, from an arbitrary initial state, of an assembly of nuclear spins. The variation of Larmor frequency with time due to spin diffusion in an inhomogeneous magnetic field alters the form of an imposed distribution. Sampling of the distribution at the time of the echo, or at any other point, where observable signal is available gives us information about the loss of transverse magnetization due to changes in frequency or phase of a spin isochromat. Spin isochromats are considered for the purpose of the diffusion experiment to be macroscopic subspin assemblies which have the same precession frequency. Natural transverse relaxation processes can be treated phenomenologically according to Bloch (1946) and either carried through the whole calculation or simply added at the conclusion. We assume that:

$$\frac{dM_x}{dt} = \frac{-M_x}{T_2}; \qquad \frac{dM_y}{dt} = \frac{-M_y}{T_2} \tag{1}$$

Where M_x, M_y are transverse magnetizations, T_2 is the spin–spin or transverse relaxation time, and derivatives are with respect to time. Since the modulations of Larmor frequencies by spin diffusion under all reasonable conditions are of low frequency, we need only consider the effects on transverse magnetization (Farrar and Becker, 1971). The Bloch equations are then:

$$\dot{u} + [\Delta\omega + \delta(t)]v = -u/T_2 \tag{2}$$

$$\dot{v} - [\Delta\omega + \delta(t)]u - \omega_1 M_z = -v/T_2 \tag{3}$$

where $\dot{u}$ and $\dot{v}$ are in phase and out of phase time derivatives of magnetization in the rotating frame at a frequency ω radius sec^{-1}; u and v correspond to dispersion and absorption mode, respectively; $\Delta\omega$ is the difference in frequency $(\omega_0 - \omega)$, where ω_0 is a frequency, usually the most probable one, of an arbitrary distribution $g(\Delta\omega)$ and ω is the angular frequency of the rotating frame (the radio frequency, rf).

The term $\delta(t)$ describes the time dependence of the Larmor frequencies due to diffusion processes and is appropriately included in a linear fashion as modifying $\Delta\omega$ with time. The movement of a liquid molecule, which is the vehicle for the precessing spins, is basically a random walk phenomena, and $\delta(t)$ will depend on the form of the inhomogeneity of the magnetic field, which is theoretically assumed to be linear.

Within good precision this condition can be achieved experimentally. The usual conditions pointed out by Hahn (1950) for pulse experiments must be met. These are:

(a) Absence of relaxation of any type during the period t_w of an rf pulse, i.e., $t_w \ll T_1, T_2$.

4. Diffusion and Chemical Exchange

(b) Sufficient amplitude of the rf pulse to ensure all nuclei in the distribution in resonance, i.e., $\gamma H_1 = \omega_1 \gg (\Delta\omega)^{1/2}$. Where $(\Delta\omega)^{1/2}$ is the half width usually of an assumed Lorentzian or Gaussian distribution of Larmor frequencies that is imposed on the system by the basic inhomogeneity of the magnet.

(c) Pulse intervals must always be very large compared to t_w. Any relaxation of this condition requires a more complex analysis, but in cases where it has been carried out numerically, it has not been usual to find large differences in the result.

If a single pulse is applied to the spin system that satisfies the above conditions and produces significant nutation of the total magnetization from the z direction, then on removal of the pulse at a time $t = t_i'$, i.e., when $\omega_1 = \gamma H_1 = 0$, the xy plane magnetization is given by:

$$F(t) = F(t_i') \exp[-(t - t_i')/T_2^*] \exp i[\Delta\omega(t - t_i') + \int_{t_i'}^{t} \delta(t'') \, dt''] \quad (4)$$

$F(t_i')$ is the initial xy magnetization at the time t_i' immediately after the pulse is removed and depends on the preparation of the system of spins. In future argument $F(0)$ will indicate t_i' to be zero time. The first exponential, a decay, is merely the normal transverse magnetization in an inhomogeneous magnetic field. This is composed of the natural component T_{2_0} and the imposed component T_2' from a magnetic field that is assumed to impose an overall Lorentzian distribution in Larmor frequencies.

$$\frac{1}{T_2^*} = \left(\frac{1}{T_{2_0}} + \frac{1}{T_2'}\right) \quad (5)$$

The second exponential is the oscillatory component derived from the difference between the Larmor frequency and the imposed rf frequency, while the third is a term derived from the diffusion in an inhomogeneous field. Note that $F = (u + iv)$.

The changes in phase of spin isochromats derived from the third exponential is calculated assuming a linear field gradient:

$$G = (dH_0/dz)_{\text{av}} \quad (6)$$

$(dH_0/dz)_{\text{av}}$ is the variation with length z of the large field H_0. The H_1 or rf field is assumed to be highly homogeneous. The diffusing spins will gain or lose phase with respect to their original phase in the distribution and this can be expressed:

$$\int_{t_i'=0}^{t} \delta(t'') dt'' = \Phi(t) - \Phi(0) \quad (7)$$

$\Phi(t) - \Phi(0)$ is the total phase shift accumulated in the period $(t - t_i')$. This must be made an average of all possible phase shifts for all possible initial phases $\Phi(0)$ in the distribution and all possible pathways for the given phase shift with time. The result of this calculation has been derived by several different methods (Hahn, 1950; Carr and Purcell, 1954; Muller and Bloom 1960).

$$\delta(t'') = \gamma G z(t'') \tag{8}$$

where γ is the magnetogyric ratio of the nucleus being observed. Equation (8) shows that the phase changes are dependent on the variation of the distance diffused z as a function of time $z(t'')$. The averaging in Eq. (4) can be executed using a probability function $P(\Phi, t)$ in the integral:

$$\int_{-\infty}^{\infty} \{\exp i[\Phi(t) - \Phi(t')]\} P(\Phi, t) \, d\Phi \tag{9}$$

The influence of diffusion on the phase difference is more clearly seen using a random walk approach of Carr and Purcell (1954).

With a linear z field gradient $(dH/dz) = G$ gauss per centimeter, we assume the molecule in which nuclei are embedded remains fixed for a time τ_c, then jumps randomly in the z direction to a new position ξa_i, where $a_i = \pm 1$ and ξ is a fixed distance. This fixed distance is an average root mean square distance over a large number of displacements. The magnetic field changes with z position and thus also the precession frequency of the nucleus because of the field gradient. If at time $t = 0$ the z field is $H_z(0)$, then some time later $t = j\tau_c$, where j is an integer, the field becomes $H_z(j\tau)$ given by:

$$H_z(j\tau_c) = H_z(0) + G\xi \sum_{i=1}^{j} a_i \tag{10}$$

If a large number N steps have been made after a time $t = N\tau_c$, the phase difference Φ_D accumulated because of a diffusion process can be expressed:

$$\Phi_D = \Phi(t) - \Phi(0)$$
$$= \sum_{j=1}^{N} \gamma \tau_c [H_z(j\tau_c) - H_z(0)] \tag{11}$$

Substituting from Eq. (10),

$$\Phi_D = G\xi\gamma\tau_c \sum_{j=1}^{N} \sum_{i=1}^{j} a_i$$
$$= G\xi\gamma\tau_c \sum_{j=1}^{N} (N + 1 - j)a_j \tag{12}$$

where the sum of random values of ± 1 has been evaluated. The value $(N + 1 - j)$ is integral and may be written as j, a number to be summed. Thus,

$$\Phi_D = G\xi\gamma\tau_c \sum_{j=1}^{N} ja_j \tag{13}$$

4. Diffusion and Chemical Exchange

In the limit of large N, the distribution of the mean square value of Φ_D will be Gaussian and

$$\langle \Phi_D \rangle_{av} = G^2 \xi^2 \gamma^2 \tau_c^2 \sum_{j=1}^{N} j^2$$
$$= \tfrac{1}{6} G^2 \xi^2 \gamma^2 \tau_c^2 N(N+1)(2N+1) \quad (14)$$

The dominant term, N being very large, is $N^3/3$, thus we can write:

$$\langle \Phi_D^2 \rangle_{av} = \tfrac{1}{3} G^2 \xi^2 \gamma^2 \tau_c^2 N^3 \quad (15)$$

Using the well-known mean square probability of displacement which agrees with Ficks second law of diffusion (Moore, 1960) in the above notation, we obtain for the displacement ξ^2

$$\xi^2 = 2\tau_c D \quad (16)$$

Substituting into Eq. (15)

$$\langle \Phi_D^2 \rangle_{av} = \tfrac{2}{3} G^2 \gamma^2 \tau_D^2 D N^3$$
$$= \tfrac{2}{3} \gamma^2 G^2 D t^3 \quad (17)$$

The probability distribution in phase after a time t becomes:

$$P(\Phi_D) = (\tfrac{4}{3}\pi \gamma^2 G^2 D t)^{-1/2} \exp(-3\Phi_D^2 / 4 G^2 \gamma^2 D t^3) \quad (18)$$

This result must be extended to cover the reversal of phases during a second pulse. For convenience of description, the first pulse is considered to be 90° and a second, τ seconds later, 180°. The extension to n pulses in a Carr–Purcell sequence (1957) $\pi/2, \tau, \pi, \tau$, echo, τ, π, τ, echo, $\tau, \pi, \ldots$ can be equally easily treated and then the final result taken for the specific case ($\pi/2, \tau, \pi, \tau$, echo) of a two-pulse experiment. After a 180° or π pulse, the phases of the distribution are all reversed. Thus, Φ_D after a series of π pulses is written as:

$$\Phi_D = \Phi - \Phi_0 = G \xi \gamma \tau_c \sum_{j=1}^{N} b_j \sum_{i=1}^{j} a_i \quad (19)$$

If $t = N\tau_c$ at the time of the nth echo and n 180° pulses have been applied with the first at a time $\tau = N\tau_c/2n$, we have even and odd periods between pulses where the coefficient b_j is ± 1 depending on the sign of phase accumulation: $b_j = 1$ for $0 \leq j \leq N/2n$; $3N/2n \leq j \leq 5N/2n$; $b_j = -1$ for $N/2n < j \leq 3N/2n$; $5N/2n \leq j \leq 7N/2n$. Equation (19) represents two sums with $b_j = +1$ and $b_j = -1$. Evaluating these sums separately and combining the mean square deviation in phase angle from the initial phase the result becomes:

$$\langle \Phi_D^2 \rangle_{av} = G^2 \xi^2 \tau_c^2 \gamma^2 [(N^3/12n^2) - (N/6)] \quad (20)$$

or with N very large and using Eq. (16) and $n \neq 0$,

$$\langle \Phi_D^2 \rangle_{av} = G^2 \gamma^2 D t^3 / 6n^2 \qquad (21)$$

Where t is the time at the nth echo $t^2/4n^2 = \tau^2$, where τ is the time interval between the $\pi/2$ and π pulses, then

$$\langle \Phi_D^2 \rangle_{av} = (\tfrac{2}{3} G^2 \gamma^2 D \tau^2) t \qquad (22)$$

If $n = 1$, then the result of the two-pulse ($\pi/2$, π, echo) experiment becomes:

$$\langle \Phi_D^2 \rangle_{av} = (\tfrac{2}{3} G^2 \gamma^2 D) \tau^3 \qquad (23)$$

The nmr instrument used is usually sensitive to the phase of the signal, either using a mechanical leakage with paddles between receiver and transmitter (Bloch *et al.*, 1946a,b) or by having electronic leakage (Meiboom and Gill, 1958) to compare phases. The component of the transverse magnetization detected along the v direction of the rotating frame ($F = u + iv$) can be written:

$$v(t) = M_0 \int_{-\infty}^{+\infty} \cos \Phi_D \, P(\Phi_D) \, d\Phi_D \qquad (24)$$

where M_0 is the equilibrium magnetization and t is the time of the nth echo. With the natural transverse relaxation time T_{2_0} included, we have the height of the echo at time t given by:

$$v(t) = M_0 \exp[(-T_{2_0}^{-1} - \tfrac{1}{3} \gamma^2 G^2 D \tau^2) t] \qquad (25)$$

As τ is made smaller and smaller, the second damping term can be rendered negligible, and in the limit of extremely fast pulse repetition rates the decay has a time constant $R_1 = T_{2_0}^{-1}$.

For a two pulse experiment, $t = 2\tau$ and $n = 1$, so that:

$$v(t) = M_0 \exp[-(2\tau/T_{2_0}) - \tfrac{2}{3} \gamma^2 G^2 D \tau^3] \qquad (26)$$

The results expressed in Eqs. (25) and (26) have been derived by other methods than the random walk approach (Torrey, 1956; Douglass and McCall, 1958; McCall *et al.*, 1963a). Muller and Bloom (1960) investigated the dependence on arbitrary pulse widths. Some disagreement in the original treatments was cleared up in this paper (Muller and Bloom, 1960). Hahn (1950) with a specialized treatment for angles of nutation $\theta_1 = \pi/2$, $\theta_2 = \pi/2$, or any $\theta_1 = \theta_2$ derived a factor b multiplying $\gamma^2 G^2 D \tau^3$ in the second exponential of Eq. (26) as $\tfrac{8}{3}$. Das and Saha (1954) recorded a factor $b = \tfrac{5}{3}$ for the same case while Carr and Purcell for $\theta_1 = \pi/2$ and $\theta_2 = \pi$ derived $b = \tfrac{2}{3}$. Muller and Bloom show that for all cases, θ_1 and θ_2 arbitrary, the factor $b = \tfrac{2}{3}$. The experiment now in general use is $\theta_1 = \pi/2$, $\theta_2 = \pi$. The effect of a third rf pulse at a time T with respect to the first pulse where $2\tau < T < T_2$

4. Diffusion and Chemical Exchange

gives rise to additional secondary echoes and these were observed and treated by Hahn (1950). Echoes occur at $T + \tau$, $2T - 2\tau$, $2T - \tau$, and $2T$. This scheme of echoes appears in Hahn's paper and is shown diagrammatically in the work of Muller and Bloom (1960). The echo at $2T$ arises because of the first and last pulses of the three-pulse sequence and the echo at $2T - \tau$ because of the second and third pulses of the three-pulse array. Each of these are really primary echoes. The echo at $2T - 2\tau$ can be derived by considering the first primary echo at 2τ as effectively a pulse, which with the third pulse of the sequence gives a further echo. The echo at $(T + \tau)$ is called the "stimulated echo" of the three-pulse sequence. If τ is short compared to T and T is smaller than T_1, Hahn (1950) showed that the stimulated echo amplitude could be used to measure T_1 the spin–lattice relaxation time. All secondary echoes are affected by diffusion effects. Of these, three "primary" type echoes have amplitudes which decay with exponents $\tfrac{1}{3}k\tau^3$ where $k = 2\gamma^2 G^2 D$ and might be used in fact in conjunction with the original primary echo to determine D when this part of the decay is dominant. The stimulated echo decay has a different time dependence for the effect of diffusion. The equations of Hahn and Herzog (see Solomon, 1954 and Muller and Bloom, 1960) agree for the stimulated echo which again has the factor b independent of the pulse lengths:

$$v(\tau + T) = \tfrac{1}{2} M_0 \sin \theta_1 \sin \theta_2 \sin \theta_3 \exp[-\tfrac{1}{3}\gamma^2 G^2 D(3\tau^2 T - \tau^3)] \qquad (27)$$

θ_1, θ_2, and θ_3 are the nutation angles in successive pulses. The above result is modified in the presence of spin–spin and spin–lattice relaxation. In the absence of diffusion and with the condition $k\tau^2 \ll T_1/T$ maintained by choosing τ very small, the stimulated echo can be used to measure T_1. It has not survived as a common method, since a 180°/90° sequence is simpler and generally preferred (Farrar and Becker, 1971). Woessner (1961a) has computed the spin-echo attenuation by molecular diffusion for three- and four-pulse sequences including the stimulated echo. He lists the amplitude of twelve echoes present for the most complex pulse sequence and evaluates the effect of diffusion on each one. Independently of Muller and Bloom (1960), the echo attenuation is shown to be the same form whatever the nutation angles θ of the rf pulses. Diffusion is potentially an active experimental parameter in sophisticated pulse schemes (Solomon, 1954; Freeman and Wittekoek, 1969; Wells and Abramson, 1969) that have been suggested in more recent studies, but it has long been regarded as a nuisance, and experimental steps are taken to eliminate it. The measurements of diffusion constants per se is most generally achieved with the simplest pulse scheme. The measurement of diffusion constants was greatly advanced by the idea suggested by McCall et al. (1963a) and first used by Stejskal and Tanner (1965), of pulsing the field gradient.

Stejskal and Tanner (1965) use as a starting point the Bloch equations as formulated by Abragam (1961):

$$\frac{\partial M(\mathbf{r}, t)}{\partial t} = \gamma[\mathbf{M} \times H(\mathbf{r}, t)] - \frac{(M_x \mathbf{i} + M_y \mathbf{j})}{T_2} - \frac{(M_z - M_0)}{T_1}\mathbf{k} + D\nabla^2 M \quad (28)$$

This differs from the simple vectorial Bloch equations (6) in making both the nuclear magnetization $M(\mathbf{r}, t)$ and the magnetic field $H(\mathbf{r}, t)$ functions of the position vector $\mathbf{r}$ of the spin and time and adding a term $D\nabla^2 M$. This last term is the contribution of diffusion to the rate of change of magnetization The static magnetic field H_z is given by:

$$H_z = H_0 + (\mathbf{r} \cdot \mathbf{G}) \quad (29)$$

where $\mathbf{G}$ is the constant field gradient vector. In the absence of a pulse Eq. (28) becomes:

$$\left(\frac{\partial F}{\partial t}\right) = i\omega_0 F - \frac{F}{T_2} - i\gamma(\mathbf{r} \cdot \mathbf{G})F - D\nabla^2 F \quad (30)$$

where only the transverse magnetization $F = u + iv$ is considered. Substituting:

$$F = \psi \exp(i\omega_0 t - (t/T_2)) \quad (31)$$

to separate the precession and transverse relaxation a function $\psi(\mathbf{r}, t)$ is introduced and we obtain:

$$\partial \psi/\partial t = -i\gamma(\mathbf{r} \cdot \mathbf{G})\psi + D\nabla^2 \psi \quad (32)$$

The static field H_z is conventionally in the z direction of a Cartesian system and the field gradient $\mathbf{G}(t)$ becomes, in the pulsed gradient experiment, a function of time. The field gradient is given by $H_z = H_0 + (\mathbf{r} \cdot \mathbf{G})$ as in Eq. (29). In the absence of diffusion and with the boundary condition $\psi = A$ at the time of the first pulse the solutions of Eq. (32) are

$$\psi = A \exp(-i\gamma \mathbf{r} \cdot \mathbf{F}) \quad (33)$$

where

$$\mathbf{F}(t) = \int_0^t \mathbf{G}(t')\, dt' \quad (34)$$

and after the second pulse, at time τ:

$$\psi = A \exp[-i\gamma \mathbf{r} \cdot (\mathbf{F} - 2\mathbf{f}) + i\phi] \quad (35)$$

where $\mathbf{f} = \mathbf{F}(\tau)$. The phase angle ϕ can be made zero between the first 90° and second 180° pulse. A more general expression for ψ in all time intervals is given by:

$$\psi = A \exp\{-i\gamma \mathbf{r} \cdot [\mathbf{F} + (\xi - 1)\mathbf{f}] \quad (36)$$

$$\xi = +1 \quad \text{for} \quad 0 < t < \tau; \quad \xi = -1 \quad \text{for} \quad t > \tau$$

4. Diffusion and Chemical Exchange

The effect of diffusion is to make A a function of t the time. Thus, by assuming $A = A(t)$ in Eq. (36), we can investigate the solution of this form by using Eq. (32), whence:

$$dA/dt = -\gamma^2 D[\mathbf{F} + (\xi - 1)\mathbf{f}]^2 A \tag{37}$$

integrating this between $t = 0$ and $t = \tau'$ gives:

$$\ln[A(\tau')/A(0)] = -\gamma^2 D[\int_0^{\tau'} F^2 \, dt - 4\mathbf{f} \cdot \int_\tau^{\tau'} \mathbf{F} \, dt + 4f^2(\tau' - \tau)] \tag{38}$$

Since natural relaxation effects have been removed, then $A(\tau')/A(0)$ represents the amplitude change at the time of the echo due to diffusion. This ratio depends on $\mathbf{G}(t)$ and τ and can be solved for various cases. The usual experimental case is to apply two equal field gradient pulses an equal time each side of the 180° rf pulse of short duration. The $\mathbf{G}(t)$ then becomes:

$$\begin{array}{llll} \mathbf{g}_0 & \text{when} \quad 0 < t < t_1; & \mathbf{g}_0 + \mathbf{g} & \text{when} \quad t_1 < t < (t_1 + \delta); \\ & \mathbf{g}_0 & \text{when} \quad (t_1 + \delta) < t < (t_1 + \Delta) \end{array}$$

where $(t_1 + \Delta) > t$, i.e., second field gradient pulse occurs after the 180° rf pulse.

$$\begin{array}{ll} \mathbf{g}_0 + \mathbf{g} & \text{when} \quad (t_1 + \Delta) < t < (t_1 + \Delta + \delta); \\ \mathbf{g}_0 & \text{when} \quad (t_1 + \Delta + \delta) < t \end{array}$$

The gradient $\mathbf{g}_0$ is that present due to the inherent inhomogeneity of the magnet used in the experiment. This value of $\mathbf{g}_0$ can be rendered negligibly small with the magnets used for high-resolution nmr work. The echo appears at a time 2τ with the above choice of field gradient pulses. Evaluating Eq. (38) for the echo amplitude decrease due to the diffusion term, it becomes:

$$\ln[A(2\tau)/A(0)] = -\gamma^2 D\{\tfrac{2}{3}\tau^2 g_0^2 + \delta^2(\Delta - \tfrac{1}{3}\delta)g^2 - \delta[(t_1^2 + t_2^2) \\ + \delta(t_1 + t_2) + \tfrac{2}{3}\delta^2 + 2\tau^2]\mathbf{g}\cdot\mathbf{g}_0\} \tag{39}$$

In the case where $\mathbf{g}$ is negligibly small only the term in g_0^2 remains and the result is the same as that for the conventional two-pulse spin-echo experiment. When $\mathbf{g}_0$ tends to zero in a homogeneous magnet, only the g^2 term is significant and Eq. (39) is greatly simplified

$$\ln[A(2\tau)/A(0)] = -\gamma^2 D\delta^2(\Delta - \tfrac{1}{3}\delta)g^2 \tag{40}$$

If the interval δ during which the field gradient is applied is reduced such that $\delta/3 \ll \Delta$ and yet $g\delta$ is kept sufficiently large at the same time, then:

$$\ln[A(2\tau)/A(0)] = -\gamma^2 D\delta^2 \Delta g^2 \tag{41}$$

A sinusoidal variation of field gradient can be visualized by making:

$$G(t) = \mathbf{g}_1 + \mathbf{g}_2[1 - \cos(2\pi t/\tau)] \tag{42}$$

In this case the echo amplitude due to diffusion becomes:

$$\ln\left[\frac{A(2\tau)}{A(0)}\right] = -\gamma^2 D \tau^3 \left[\frac{2}{3} g_1{}^2 + \left(\frac{2}{3} + \frac{5}{4\pi^2}\right) g_2{}^2 + \left(\frac{4}{3} + \frac{1}{\pi^2}\right) \mathbf{g}_1 \cdot \mathbf{g}_2 \right] \tag{43}$$

The advantages of the pulsed field gradient experiment are several: (a) The requirement that $H_1 \gg (\Delta\omega)^{1/2}$ is much easier to satisfy during rf pulses. (b) The diffusion constant range available can be extended by at least two orders of magnitude in the direction of slower diffusion, since g can now be much larger than any steady field gradient. (c) The time during which diffusion affects the amplitude of the echo is much better defined. (d) The echo becomes broader, being affected only by the magnitude of g_0. This has considerable practical advantages in lowering requirements for high rates of digitization of intensity data points necessary for very sharp echoes. It also has one advantage not yet utilized in practice, namely that in a multicomponent mixture of liquids the Fourier transform of the echo in a steady homogeneous field ($\mathbf{g}_0$ small) will result in measurement of decreased echo or peak amplitudes for several components at the same time. Thus, in principle, diffusion constants for all components in a mixture can be measured in one experiment.

The advantages of spin-echo methods in general in the measurement of diffusion constants are many. Unlike classic methods, there is no perturbation of the system (no isotope effects), and the method requires no radioactive label. The experiment can be completed in a very short time and is easily adaptable to varied temperature and pressure. The precision of the spin-echo technique can be questionable if the measurements are not made very carefully. With some care, a relaxation time can be measured to $\pm 5\%$, but exceptional care is needed to remove all systematic and random errors to give a value within $\pm 1\%$. The measurement of echo amplitudes in a properly experimentally defined system is somewhat more difficult to achieve with measurements of diffusion. The field gradient appears in the exponent as the square and time intervals (which can be measured much more precisely) appear as the square or the cube, depending on the experiment. The Carr–Purcell sequence is notoriously difficult to achieve with high precision and probably is the least attractive as a method to measure diffusion constants. The experimental challenge can be met, but at the moment the literature indicates many divergent values for water diffusion at 25°C (Douglass and McCall 1959; McCall et al., 1959a; Fratiello and Douglass, 1963; Valiev and Emel'yanov, 1964; Valiev et al., 1964; McCall and Douglass, 1965; Trappeniers et al., 1965; Dass and Vanshneya, 1969; Weiss and Nothnagel, 1971).

The transport of molecules in material is more complex in an anisotropic or colloidal system. In general, the diffusion coefficient becomes a tensor and in addition diffusive movement may be restricted by solid walls, membranes

4. Diffusion and Chemical Exchange

or other interfaces. This requires time and space dependence and an anisotropy to be included in the diffusion coefficient. While classic diffusion measurements are unable to cope with these complications, the spin-echo method has already achieved limited success in this area (Woessner, 1963; Boss and Stejskal, 1965; Stejskal, 1965; Tanner and Stejskal, 1968).

Following Bloch (1946) and Torrey (1956), Stejskal (1965) reformulates Eq. (32) of this text to include liquid flow and a diffusion constant tensor.

$$\partial\psi/\partial t = -i\gamma(\mathbf{r}\cdot\mathbf{G})\psi - \nabla\cdot\mathbf{v}\psi + \nabla\cdot\mathbf{D}\cdot\nabla\psi \tag{44}$$

where $\mathbf{D}$ is now a diffusion tensor and $\mathbf{v}(t)$ is the velocity of flow of spins due to flow of the medium itself in one direction. The time-dependent field gradient $\mathbf{G}(t)$ is the same form as before, namely along the z axis of the static magnetic field.

The time dependence of $\mathbf{D}$ and $\mathbf{v}$ can be considered separately for a small volume of the sample, and the spatial dependence can later be attempted by integrating ψ over the whole volume of the system. Without diffusion or flow, solutions of (44) are equivalent to Eq. (36)

$$\psi = A \exp[-i\gamma\,\mathbf{r}\cdot(\mathbf{F} - 2\xi\mathbf{f}) + i\xi(2\phi - \pi)] \tag{45}$$

where
$$\xi = 0 \quad \text{when} \quad 0 < t < \tau; \quad \xi = 1 \quad \text{when} \quad t > \tau$$

and $F(t)$ and $\mathbf{f} = \mathbf{F}(\tau)$ are given before. $\psi = A$ at end of 90° pulse, i.e., $t = 0$ and $\psi(\mathbf{r}, \tau)$ becomes $\psi^*(\mathbf{r}, \tau) \exp[i(2\phi - \pi)]$ at the 180° pulse, which is ϕ out of phase with the rf of the 90° pulse.

The dependence of $A(t)$ in Eq. (45) on the velocity of flow $\mathbf{v}(t)$ is accommodated by the equation:

$$A(t) = B \exp[i\Phi(t)] \tag{46}$$

Introducing this expression for $A(t)$ in Eq. (45) and using Eq. (44), the equation for Φ in the exponent of Eq. (46) becomes:

$$d\Phi/dt = \gamma\mathbf{v}\cdot(\mathbf{F} - 2\xi\mathbf{f}) \tag{47}$$

The solution for Φ can be written:

$$\Phi = \gamma[K - 2\xi\mathbf{f}\cdot(\mathbf{S} - \mathbf{s}) - 2\xi k]$$

or
$$\Phi = \gamma[\mathbf{S}\cdot(\mathbf{F} - 2\xi\mathbf{f}) - H + 2\xi h] \tag{48}$$

where $\mathbf{S}(t) = \int_0^t \mathbf{v}\, dt'$, $\mathbf{s} = \mathbf{S}(\tau)$, $K(t) = \int_0^t \mathbf{v}\cdot\mathbf{F}\, dt'$, $k = K(\tau)$, $H(t) = \int_0^t \mathbf{S}\cdot\mathbf{G}\, dt'$, and $h = H(\tau)$. The effect of diffusion $\mathbf{D}(t)$ can be treated now in the time dependence of $B(t)$ from Eq. (46).

$$d(\ln B)/dt = -\gamma^2(\mathbf{F} - 2\xi\mathbf{f})\cdot\mathbf{D}\cdot(\mathbf{F} - 2\xi\mathbf{f}) \tag{49}$$

Integration gives

$$\ln[B(t)/B(0)] = -\gamma^2 \left[\int_0^t \mathbf{F} \cdot \mathbf{D} \cdot \mathbf{F} \, dt' - 4\xi \int_\tau^t \mathbf{F} \cdot \mathbf{D} \cdot \mathbf{f} \, dt' + 4\xi \int_\tau^t \mathbf{f} \cdot \mathbf{D} \cdot \mathbf{f} \, dt' \right] \quad (50)$$

Since f is not a function of time, the above expression simplifies if $\mathbf{D} \neq \mathbf{D}(t)$. Combined solutions for flow and diffusion with no spatial dependence in a small volume of liquid gives:

$$\psi = B(t) \exp\{-i\gamma[(\mathbf{r} - \mathbf{S}) \cdot (\mathbf{F} - 2\xi \mathbf{f}) + H - 2\xi h] + i\xi(2\phi - \pi)\} \quad (51)$$

where $B(t)$ is obtained from Eq. (50). Spatial dependence may be included in principle by integrating over the volume of the sample. The echo defined by:

$$\mathbf{F}(\tau') - 2\xi \mathbf{f} = 0 \quad (52)$$

when $t = \tau'$ makes ψ independent of $\mathbf{r}$ and there is no volume integration needed.

$$\psi(\tau') = B(\tau') \exp\{-i\gamma[H(\tau') - 2\xi h] + i\xi(2\phi - \pi)\} \quad (53)$$

The diffusion attenuates the principal echo but flow only changes its phase. If the field gradient $\mathbf{G}(t)$ is pulsed with conditions as set out earlier in the limit $\mathbf{g}_0 \to 0$, $\delta \to 0$. $\delta\mathbf{g}$ finite are observed, the echo at $\tau' = 2\tau$ is given by:

$$\psi(2\tau) = B(0) \exp[-\gamma^2 \delta^2 \mathbf{g} \cdot \mathbf{D} \cdot \mathbf{g} \Delta - i\gamma \delta \mathbf{g} \cdot \mathbf{S}(\Delta) + i(2\phi - \pi)] \quad (54)$$

for a constant $\mathbf{D}$. $\mathbf{S}(\Delta)$ is given by: $[\mathbf{S}(t_1 + \Delta) - \mathbf{S}(t_1)]$, the displacement of the whole system during the time Δ. Equilibrium macroscopic z magnetizations are assumed for all parts of the sample. The time interval Δ may be of the order of 1 msec, and restricted diffusion on this time scale can affect the results giving values of $\mathbf{D}$ dependent on Δ.

The equation of McCall *et al.* (1963a) modified for a pulsed field gradient:

$$\psi(2\tau) = B(0) \exp[i(2\phi - \pi)] \int_{v_0} P_0(\mathbf{r}_0) \int_v P(\mathbf{r}_0/\mathbf{r}, \Delta) \\ \times \exp[-i\gamma\delta(\mathbf{r} - \mathbf{r}_0) \cdot \mathbf{g}] \, dV \, dV_0 \quad (55)$$

can be considered as giving $\psi = \psi(\mathbf{g}, \Delta)$ where $P(\mathbf{r}_0|\mathbf{r}, \Delta)$ is the probability that a nucleus with initial position $\mathbf{r}_0$ when the first gradient pulse is applied has a position $\mathbf{r}$ a time Δ later, $P(\mathbf{r}_0|\mathbf{r}, \Delta)$ satisfies:

$$\partial P/\partial t = -\nabla \cdot \mathbf{v} P + \nabla \cdot \mathbf{D} \cdot \nabla P \quad (56)$$

The problem must account for different starting points $\mathbf{r}_0$ in a distribution if necessary. For a uniform constant diffusion constant tensor, the solution of

4. Diffusion and Chemical Exchange

Eq. (56) gives:

$$P(\mathbf{r}_0|\mathbf{r}, t) = (64\pi^3 D_{xx}D_{yy}D_{zz}t^3)^{-1/2} \qquad (57)$$
$$\times \exp\left[-\frac{(x - x_0 - S_x)^2}{4D_{xx}t} - \frac{(y - y_0 - S_y)^2}{4D_{yy}t} - \frac{(z - z_0 - S_z)^2}{4D_{zz}t}\right]$$

with the coordinate system oriented along the principal axes of the diffusion ellipsoid.

Diffusion in a lamellar system is a common situation in lamellar solids such as vermiculite neat soap phases or smectic liquid crystals. The layers in which diffusion is possible vary in thickness and transport in the plane of the layers is unrestricted while perpendicular to the layers it is restricted by bounding surfaces. The condition $\Delta \gg a^2/2D$ is such that the bounding surfaces are restricting transport in a time Δ. In a time $\Delta \ll a^2/2D$, the diffusion is isotropic and in all directions given by D.

$$P(\mathbf{r}_0|\mathbf{r}, \Delta) = (4\pi a D\Delta)^{-1} \exp[-(\mathbf{r} - \mathbf{r}_0)_\parallel^2/4D\Delta] \qquad (58)$$

inside the layer

$$P(\mathbf{r}_0|\mathbf{r}, \Delta) = 0$$

outside the layer

$$\psi(\mathbf{g}, \Delta) = B(0)\left[\frac{2(1 - \cos \gamma\delta g_\perp a)}{\gamma^2\delta^2 g_\perp^2 a^2}\right]\exp[-\gamma^2\delta^2 g_\parallel^2 D\Delta + i(2\phi - \pi)] \qquad (59)$$

This case of diffusion in layers is difficult in theory to distinguish from true anisotropic diffusion, where the value of **D** is a tensor irrespective of the value of Δ.

$$\psi(g, \Delta) = B(0) \exp[-\gamma^2\delta^2 g_\parallel^2 D_\parallel\Delta - \gamma^2\delta^2 g_\perp^2 D\Delta + i(2\phi - \pi)] \qquad (60)$$

Distinction between the two cases will sometimes be possible on the grounds of the physical system studied, but in any case variation of ψ as a function of both Δ and **g** (magnitude and direction) will separate truly restricted diffusion from just anisotropic diffusion. The application of restricted diffusion theory to the spin-echo experiment has ambiguities, since restrictions on transport can take many forms in physical, especially living, systems.

Some experimental tests of the affect of restricted diffusion on the echo amplitude as a function of Δ have been made by Tanner and Stejskal (1968). For the lamellar system, the factor R by which diffusion attenuates the echo amplitude is modified for varying Δ and given by:

$$R = \exp(-\gamma^2\delta^2 g_\parallel^2 \Delta D)\left[\frac{2(1 - \cos(\gamma\delta g_\perp a))}{(\gamma\delta g_\perp a)^2} + 4(\gamma\delta g_\perp a)^2\right.$$
$$\left.\times \sum_{n=1}^{\infty} \exp\left(\frac{-n^2\pi^2 D\Delta}{a^2}\right)\frac{1 - (-1)^n \cos(\gamma\delta g_\perp a)}{[(\gamma\delta g_\perp a)^2 - (n\pi)^2]}\right] \qquad (61)$$

the second term is important when the condition $\Delta \not> \frac{1}{3}\delta$ and includes the effect of restriction of transport by bounding surfaces. The first term is merely the expression for unrestricted diffusion in a pulsed gradient experiment. As Δ increases without bound, R reaches an asymptote when essentially all diffusing particles encounter the bounding surfaces in a time Δ. The asymptote can be defined:

$$R_\infty = \lim_{\Delta \to \infty} R = \frac{2[1 - \cos(\delta \gamma g a)]}{(\gamma \delta g a)^2} \tag{62}$$

and

$$\ln R_\infty^0 \equiv \lim_{\gamma \delta g \to 0} (\ln R_\infty) = -(\gamma \delta g a)^2/12 \tag{63}$$

This affords in an ideal physical system a means of estimating a, the thickness of the layer. Experiments on several systems including a model system of water layers $1.27 \pm 0.1 \times 10^{-3}$ cm in thickness trapped between mica sheets were made. The results qualitatively give the right dependence of R on $(\Delta - \frac{1}{3}\delta)$ with a clear symptotic value for large gradient spacings, but the computed value of a for the spacing of layers was somewhat too high even after a correction factor was applied. The diffusion constants in the unrestricted system were within experimental error the same both parallel and perpendicular to the layers. Experiments in much more heterogeneous systems, such as yeast cells and apple and tobacco pith, clearly indicate restricted diffusion and estimates of cell sizes give the correct magnitude when compared with microscopic studies. In general, the pulsed gradient experiment shows good promise in studying bounded diffusion and will undoubtedly contribute in the future to our studies of living cells.

Bounded diffusion has also been treated for the steady field gradient experiment (Robertson, 1966; Wayne and Cotts, 1966). Robertson (1966) uses the same Torrey (1956) modifications of the Bloch equations as a starting point, including the diffusion terms. An effective diffusion constant $D'(t)$ is measured as a function of t, the time from the 90° pulse to the echo, in terms of the ratio R of the amplitudes of the echo with and without a steady field gradient G assumed linear as in the Hahn paper (1950).

$$D'(t) = -12 \ln R/\gamma^2 G^2 t^3 \tag{64}$$

The value of $D'(t)$ decreases with t due to bounding of the transport process. A universal curve for the echo attenuation can be constructed using D'/D, the ratio of the restricted to the free diffusion constant, when the denominator in Eq. (64) becomes $\alpha = (a^3 \gamma G/D)$ defining t' as equal to (Dt/a^2), the mean time for diffusing a distance a. The value of R can be computed from theory and experiment for any value of a and agreement is found (Wayne and Cotts, 1966). The Carr–Purcell sequence has a decay given by Eq. (25) for the echo amplitudes. At a constant field gradient G, the exponential decay constant in the presence of restricted diffusion is not linear with τ^2, the pulse interval squared, but falls off more rapidly as τ gets longer due to a decrease in the effective diffusion constant. The restricted diffusion experiments of Wayne and Cotts (1966) were carried out on a model system of methane gas under pressure confined between Teflon separators at calibrated distances. Packer (1973) has concluded that restricted diffusion through locally inhomogeneous magnetic fields is not a significant factor in transverse spin relaxation, whereas Cooke and Wein (1971), who studied solution of F actin and G actin, suggest the reverse. The striated muscle tissue investigated by Packer (1973) is known to have smaller values of the dimensions of barriers restricting diffusion. Corresponding studies have been made in phospholipid/water (Chan et al. 1971; Finer et al., 1972) and lyotropic mesophases (Hansen and Lawson, 1970; Tiddy, 1971). The essential difference between thermotropic smectic phases (Murphy and Doane, 1971) and lyotropic lamellar neat phases is that in the former diffusion is genuinely anisotropic, while in the latter, at least the water diffusion is bounded. The anisotropic diffusion constants of tetramethylsilane in nematic p-methoxybenzilidene-p'-n-butylaniline are $D_\| = 3.2 \times 10^{-5}$ cm^2 sec^{-1} and $D_\perp = 0.8 \times 10^{-5}$ cm^2 sec^{-1} as determined by a steady field gradient experiment (Murphy and Doane, 1971).

In all diffusion measurements the existence of convection currents or bulk liquid flow remain a potential source of error in affecting echo amplitudes. Carr and Purcell (1954) treat the case of convection for their multipulse sequence with a steady field gradient and show that the echoes are changed in phase and thus apparent amplitude, assuming an apparatus is used which is phase sensitive to the spin isochromats. The differential phase accumulation due to bulk flow corresponding to h units of field increase per unit time along the field gradient gives:

$$\phi_c = \gamma h \tau_1^2 \tag{65}$$

during the time τ_1 between the first pulse and the first echo while at the second echo and

$$\phi_c = \gamma h \tau_2^2 - 2\gamma h \tau_1 \tau_2 \tag{66}$$

τ_2 being the time of the second echo. The phase accumulations depend linearly on h and may lead to a more intense second echo than the first.

Tanner (1970, 1972) has extended measurements of isotropic diffusion into the region of solid plastic crystals by the pulsed field gradient method for stimulated echoes. Only Reich (1963) and Garwin and Reich (1959) in studies of the diffusion of helium-3 in solid helium had previously studied solids by the spin-echo technique. Normal mobile liquids have diffusion constants of the order of $D \sim 10^{-5}$ cm² sec⁻¹ but viscous fluids may have diffusive rate constants two orders of magnitude smaller. The diffusion constants measured by Tanner for solid plastic crystals of cyclohexane, neopentane, hydrogen sulfide, and hydrogen chloride below the melting point are in the region 3×10^{-9} cm² sec⁻¹. The stimulated echo was used to take advantage of a situation in which $T_1 \gg T_2$, and field gradient pulses were applied with gradients of the order of 500 G cm⁻¹.

Burnett and Harmon (1972) also measured diffusion by the technique of relaxation in the rotating frame T_{1_ρ} and claim an ultimate limit of $\sim 5 \times 10^{-13}$ cm² sec⁻¹ for D. Their method is really an extension of the indirect one, namely that of the contribution of translational diffusion to nuclear spin relaxation. We are concerned in this chapter with direct application of pulse methods to diffusion measurements through transverse relaxation in field gradients externally applied. In the region where T_{1_ρ} and pulse field gradients are both applicable, $10^{-7} > D > (5 \times 15^8)$, the direct determinations of Gross and Kosfield (1969) for glycerol agree with those determined using T_{1_ρ} (Burnett and Harmon, 1972). The change of relaxation rate R_{1_ρ} in the rotating frame with Larmor frequency ω_1 in that frame is related to the diffusion constant (Burnett and Harmon, 1972) by:

$$\frac{d(R_{1_\rho})}{d(\omega_1^{1/2})} = \frac{-\sqrt{2}\gamma^4\hbar^2\pi n}{20 D^{3/2}} \tag{67}$$

where n is the number density of resonant nuclei. Within some experimental error, the plot of R_{1_ρ} versus $\omega_1^{1/2}$ is linear for glycerol (Burnett and Harmon, 1972).

Douglas and McCall (1958) measured diffusivities of normal paraffins C_5, C_6, C_7, C_8, C_9, C_{10}, C_{18}, and C_{32} over a considerable range of temperatures using the Carr–Purcell pulse sequence and noted an increase in activation energy with molecular weight. Diffusion constants of all the paraffins were correlated with the reduced temperature parameter T_c/T. Douglass, McCall, and Anderson measured diffusion constants for nearly spherical molecules and neopentane and tetramethylsilane (1961) as a function of both pressure and temperature, also by the Carr–Purcell method (1954). Several other liquids of varied polarity, including water, were studied by the same authors, also at varied temperature and pressure (Douglass and McCall, 1958; McCall et al., 1959a,b). Extension of the diffusion studies to linear siloxanes of varied chain length to derive a constant-volume activation energy was made in an article by McCall et al. (1961).

Diffusion in gaseous helium-3 has been measured by a three-pulse sequence using the two echoes separated by $2(T - \tau)$ mentioned earlier (Luszczynski et al., 1962). The measurements were made at varied gas densities. Gavin et al. (1963) used a two-pulse steady-field gradient method to measure diffusive transport in liquid methane between the melting point and the normal boiling point. Low-molecular-weight polyisobutylene has a self diffusion obeying the Arrhenius equation with an activation energy of 5.5 kcal mole^{-1} (McCall et al., 1963b), while in solutions the activation energy is lower. Infinite dilution values of the polymer diffusion constant were determined in carbon disulfide and carbon tetrachloride. Constant linear field gradients were used to measure the self-diffusion constants of liquid sulfur hexafluoride between 221° and 315°K (Hackleman and Hubbard, 1963). Hausser et al. (1966) review the spin-echo methods for determining diffusion constants and use a steady field gradient technique to determine the self-diffusion constants of benzene and water in the liquid state up to the critical points.

Measurements of diffusion in binary solutions by deuterating one component have been reported by McCall and Douglass (1967) with corrections for the isotope effect. Echoes were recalculated for the undeuterated binary mixtures to correct for isotope effects on the diffusion in the isotopically labeled mixtures. Agreement with previous studies by tracer methods where these overlapped were satisfactory. All measurements were made using the steady field gradient technique at 25 ± 1°C. The self diffusion in liquid lithium metal has been studied using the nmr signal of lithium-7 (Murday and Cotts, 1968) by the pulsed field gradient technique. Since the metal was present in a sample of finely divided particles, the background inhomogeneity was rather large (0.7 G cm^{-1}). A larger-pulsed field gradient was therefore required (10 G cm^{-1}) to minimize the error of background gradient, but the echo width still corresponded to the gradient of 0.7 G cm^{-1} and could be satisfactorily measured without an excessive bandwidth in the detector system. Requirements on pulsed-field gradient separations were also met with the small particles so that bounded diffusion constants could be avoided.

Woessner et al. (1969) chose the Carr–Purcell (1954) sequence to measure self and mutual diffusion in high pressure mixtures of methane and propane gas in a cell of 1 ml capacity. A careful study of self diffusion in methane at varied pressures and temperatures was completed by Oosting and Trappeniers by using the constant field gradient technique (1971) with a two pulse sequence. Self diffusion in liquid NH_3 and ND_3 has been investigated by O'Reilly et al. (1973) by steady field gradient technique between 200° and 298°K.

Some discussion of the experimental aspects of diffusion measurements by spin-echo techniques is appropriate. The spin-echo experiment itself (Hahn, 1950; Carr and Purcell, 1954; Meiboom and Gill, 1958; Farrar and Becker, 1971) will be adequately treated in other chapters in this book (Freeman and Hill, this volume, Chapter 4). The application, measurement and pulsing of external field gradients is peculiar to the diffusion experiment. Variable temperature and pressure apparatus is also a feature of most other nmr work and are somewhat easier to achieve with the nonspinning sample used in diffusion studies.

The theories of experiments with field gradients assume a linear field gradient in the z direction, that is, in the direction of the large static magnetic field. Several arrangements have been used for the auxiliary coils. If the spin-echo method is not used as an absolute method, knowledge of the field gradient is not necessary, since a calibrating liquid can be used whose diffusion constant is well known. If relaxation times T_{20} are known for the two comparison liquids and are preferably about the same value, then using Eq. (26), for instance, for two liquids gives the diffusion constant of one in terms of the other.

$$\frac{-2\tau/(T_2)_1 - \tfrac{2}{3}\gamma^2 D_1 G^2 \tau^3}{-2\tau/(T_2)_2 - \tfrac{2}{3}\gamma^2 D_2 G \tau^3} = \frac{A_1 (A_2)_0}{(A_1)_0 A_2} \qquad (68)$$

Subscripts 1 and 2 refer to the two liquids, and A_1 and $(A_1)_0$ refer to amplitudes with and without field gradients. Usually the spin-echo method is applied as an absolute method of determining the diffusion constant.

Auxiliary coils described by McCall et al. (1963a) are based on two parallel straight conductors parallel to the axis of the cylindrical sample and equidistant from it. In order to make currents more reasonable, the conductors are made of bundles of insulated wires with the current direction such as to cause opposing fields at the sample. Two circular coils each side of the sample are a more generally used arrangement (Carr and Purcell, 1954; Woessner, 1960). The field gradient is more linear if the Helmholtz coil dimensions are used but with magnetic fields from each coil opposed at the sample. Tanner (1965) discusses the design of such coils, taking into account image currents in the magnet pole faces. Gerritsma and Trappeniers (1971) also discuss the image currents with Helmholtz coils and utilized tapered coils wound on formers which have an outside diameter 75 mm and a diameter closest to the sample of 18 mm with a thickness of 2 mm (Berger and Butterweck, 1956). Gradients of 12.5 ± 0.02 G cm^{-1} A^{-1} were achieved with a variation of **G** of less than 1% over a sample volume of 1 cm^3. The maximum gradient generated was 75 G cm^{-1}.

The use of pole caps with a varying gap at the sample has also been used to produce linear field gradients (Pfeiffer and Weiss, 1961), while Hausser et al. (1966), working in a restricted magnet space, used steel bars placed parallel to the sample.

All absolute diffusion measurements require an accurate knowledge of the linear field gradient **G** defined earlier. Several methods have been in use for this measurement: (a) theoretical calculations from gradient coil geometry without measurement (Trappeniers et al., 1965); (b) measurement of resonant frequency as a function of field with a sample of small dimensions (Murday and Cotts, 1971); (c) the shape of the free induction decay (FID) or spin echo (Carr and Purcell, 1954); this shape may show both positive and negative going signals if the detector is phase sensitive or only positive going signals with a diode detector.

The use of a theoretical calculation alone from coil geometry, however accurate, leaves uncertain the error from systematic and random sources, while moving a capillary sample accurately over distances (usually less than 1 cm) in a magnetic field and measuring beats to a reference frequency or a signal from a second fixed sample is difficult to achieve in practice. The third alternative (c) is the most frequently used method. Following Carr and Purcell, the sample is assumed to be a cylinder in shape with the axis perpendicular to the field gradient direction z. If the detector measures the transverse magnetization along the x direction, then the FID has the form:

$$M(t) = M(0)[2J_1(\gamma GRt)/(\gamma GRt)] \cos(\omega_0 - \gamma H)$$
$$= M(0)[2J_1(\alpha)/\alpha] \cos(\omega_0 - \gamma H) \tag{69}$$

H is the applied field, ω_0 the precession frequency of center of distribution, R is the cylinder radius, G is the gradient in the z direction. J_1 is the first order Bessel Function. Ideally, the whole FID, or half an echo, must be digitized and fitted with a minimum error analysis to the function given by Eq. (69), and this is the best procedure. In almost all cases the value of G is derived from the zero values of the function (69).

Errors can creep into this measurement, unfortunately many of them systematic (Murday, 1973). These are nonorthogonality of gradient direction and sample cylinder axis, time constants in electronic circuits which distort FID, digitization rates of multichannel analyzer being too slow, baseline errors, sample meniscus and rounded bottom on sample tube, inherent nonlinear field gradients in the magnet, and inhomogeneities in the applied gradient. Since the diffusion measurement depends in a sensitive fashion on G, then the disparity of literature values for the same liquid from different laboratories is not so surprising. Murday (1973) has treated some of these systematic errors in some detail in a recent publication.

It is usual to avoid meniscus effects by using a tight-fitting cylinder in the sample that squares off the meniscus by contact with the liquid. The bottom of the sample can be squared off by good glassblowing to a flat-bottomed form. Baseline errors are best avoided in the gradient measurement by using diode detection in which case the minima or zero in the function (70)

$$M(t) = M(0)[2J_1(\gamma GRt)/\gamma GRt]$$
$$= M(0)[J(\alpha)/\alpha] \tag{70}$$

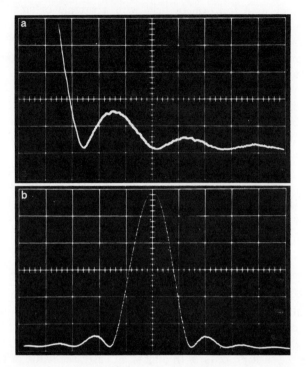

Fig. 1 Calibration of the field gradient. (a) Free induction decay (90°) in the presence of a linear field gradient for a cylindrical sample with diode detection. The current in the gradient coils was 0.025 A and radius of the sample 3.25 mm. The gradient can be calculated from the first order Bessel function described in the text. The gradient coils used gave a linear relation between gradient and current [see Eq. (72)]. Each oscilloscope time division is 0.2 msec. (b) The first spin-echo after a $\pi/2$, π pulse sequence in the presence of a field gradient. Gradients obtained at different currents are given by Eq. (72) and are represented in Fig. 2. The current was 0.021 A, sample radius 3.25 mm. Oscilloscope trace is calibrated with 0.5 msec per division.

can easily be identified. The zeros occur at $J_1(\alpha)/\alpha = 0$ for $x_1 = 3.8317$, $x_1 = 7.0156$, and $x_3 = 10.1735$, and to x_3 they are usually measurable with good precision. In diode detection, these appear as minima and the functional minima are inverted to a positive going signal and become maxima. In Fig. 1, the typical modulation of the free induction decay and echo in diode detection is shown with a field gradient applied (Ertl, 1973). In the figure, the gradient has been determined using time intervals Δt_1 and Δt_0 where

$$G = \frac{7.664}{\gamma R \Delta t_0} = \frac{3.184}{\gamma R \Delta t_1} \tag{71}$$

4. Diffusion and Chemical Exchange

With most gradient coil arrangements the value of G is proportional to the current passing. A typical calibration curve is shown in Fig. 2, where

$$G = 0.02698I \tag{72}$$

between 5 and 120 mA, with a random error of $\pm 0.9\%$ by least squares fitting (Ertl, 1973). The coil design was of the anti-Helmholtz type (Trappeniers et al., 1965).

In steady field gradients, the two-pulse $\pi/2-\pi$ sequence may be used in two ways. The pulse separation can be kept constant and **G** varied or **G** kept constant and the pulse separation varied. Figure 3a shows the effect of varying pulse interval. The FID is seen to the left of the figure and the broad echoes in the absence of the field gradient appears on the right with successive displacement of the time interval τ between the 90° and 180° pulses. The sharp echoes with an applied constant field gradient of 0.673 G cm^{-1} are visible at their maximum amplitude and are attenuated as the pulse interval τ is increased by the diffusion effect. In Fig. 3b, the pulse interval is maintained constant and the field gradient increased between 0.1 and 1.15 G cm^{-1}. The oscilloscope time base is adjusted so as to show clearly the decrease in amplitude of the echo with increasing field gradient as the echo signals move to the right side of the figure.

In Fig. 4, the linearity of the logarithm of the echo amplitude ratios with G^2, the field gradient at constant pulse interval and with the pulse interval (Δt in the figure) at constant field gradient is illustrated. It is important to

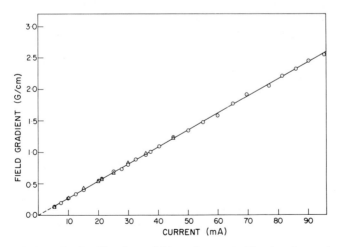

Fig. 2 Typical result of calibrating a field gradient as in Fig. 1 and equations of the first order Bessel function in the text. Coils were of anti-Helmholtz type with currents between 5 and 95 mA. $G = 0.0270I$ G/cm; $\bigcirc = \Delta t_0$ (spin-echo), $\triangle = \Delta t_1$ (90° decay).

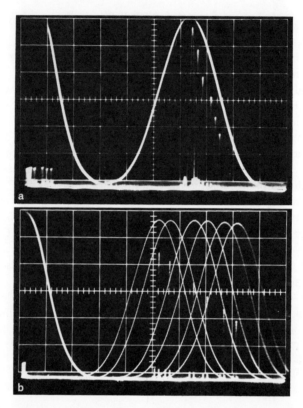

Fig. 3 Two methods of measuring the diffusion constant using a steady field gradient. (a) Free induction decay and first echo for n-heptane 62.6°C. The field gradient is varied between 0.25 G cm^{-1} to 1.2 G cm^{-1}. The single broad echo corresponds to the system without a field gradient and a period 2τ up to the echo of 65 msec. Currents in the gradient coils, whose calibrations are shown in Fig. 2, are 10, 18, 25, 30, 35, and 40 mA. The time base of the oscilloscope is adjusted to the right slightly after each increase of gradient. The echo amplitude is seen to decrease as the gradient increases and the echo shape becomes successively sharper in time. Each oscilloscope time division is 10 msec. (b) The free induction decay and first echo for water at 65.8°C. A fixed field gradient of 0.673 G cm^{-1} is applied and the pulse interval between the $\pi/2$ and π pulses is varied. The broad echoes are obtained in the absence of a field gradient and the sharp echoes decrease in amplitude as the period 2τ increases. In this case with constant field gradient, the echoes are all equally sharp in time. Oscilloscope trace is calibrated with 10 msec per division.

4. Diffusion and Chemical Exchange

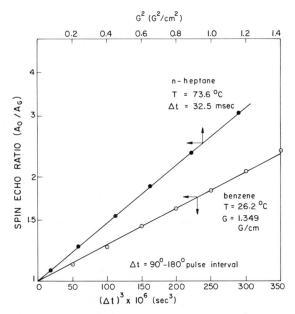

Fig. 4 The typical plots of logarithms of ratios of spin-echo heights against the field gradient squared at constant pulse interval τ (see Fig. 3a) and against the time of the first echo cubed at constant field gradient (Fig. 3b). Both methods give satisfactory linear plots from which the diffusion constant can be calculated.

remember that diode detectors are nonlinear at low signal inputs, and this leads to systematic errors in the estimation of true echo heights. The gain of the receiving circuit can, however, be calibrated using a signal generator and the echo amplitudes corrected. If no signal generator is available, the receiving circuit can be calibrated by using the height of the FID or echo with samples of fixed geometry and position in the field but varying the proton concentration in the sample over a wide range. If phase-sensitive detection is used, the problem is avoided but field drift then becomes important. In measuring the field gradient from the echo shape, the reference frequency must be maintained sufficiently different from the Larmor frequency so as not to confuse the overall Bessell function of the envelope.

III. Chemical Exchange Rates from Spin-Echo Studies

In the study of diffusion by spin-echo methods, the modulations of the Larmor frequency are artificially imposed upon the system by creating an external magnetic field gradient. Similar effects can be observed if the nuclear frequencies are modulated by chemical exchange or spin–spin coupling. The experimenter is a victim of his chemical system and does not have the useful experimental variable of a field gradient. The modulation of the nmr frequency itself is not in fact necessary, since an exchange process that transfers a nucleus between two sites of different relaxation rates reveals the exchange through a modulation of the relaxation rate. The effects on spin echoes of coherent modulations arising from indirect spin–spin coupling, for instance, will not be treated here (Hahn and Maxwell, 1952) and remarks will be confined to chemical exchange.

The spin transfer between sites of differing relaxation times T_1 or T_2 was treated by Zimmerman and Brittin (1957) for the general case of n sites being finite. In the fast exchange limit; $k_i \gg R_i$ for all sites (k_i being pseudo first order rate constants for the transfer out of the ith site and R_i being a nuclear spin relaxation rate in this site) the observed relaxation rate will be the weighted average of R for all sites

$$R_{\mathrm{av}} = \sum_{i=1}^{n} (P_i/T_i) = \sum_{i=1}^{n} P_i R_i \qquad (73)$$

where $\sum_i P_i = 1$. In the slow exchange limit, the condition $k_i \ll R_i$ applies. The overall decay of magnetization after polarization is not exponential but a weighted sum of exponentials

$$M(t) = M(0) \sum_{i=1}^{n} P_i \exp(-R_i t_i) \qquad (74)$$

The intermediate exchange rate region ($k_i \sim R_i$) was solved by Zimmerman and Brittin (1957) for two sites in terms of the eigenvalues M_1 and M_2 of a relaxation rate matrix:

$$\mathbf{F} = \begin{bmatrix} k_1 + R_1 & -k_1 \\ -k_2 & k_2 + R_2 \end{bmatrix} \qquad (75)$$

The magnetization $M(t)$ can be written in terms of these eigenvalues.

$$M(t) = a_1 \exp(-\mu_1 t) + a_2 \exp(-\mu_2 t) \qquad (76)$$

where coefficients a_1 and a_2 are functions of the eigenvalues μ_1 and μ_2 and the averaged relaxation rate. The same expressions apply for modulation of both R_{i_1} and R_{i_2}, the longitudinal and transverse relaxation rates.

Woessner (1961b) was the first to consider the effects of nuclear spin-transfer effects on pulse experiments. He was able to show that in a two-pulse experiment the transverse dephasing from chemical exchange between two sites of different Larmor frequency could be partly recuperated at the time of the first echo. A Bloch equation approach was used similar to that of McConnell (1958). The equations were solved for the effect of pulses and

4. Diffusion and Chemical Exchange

intervening periods of free precession. The FID following a 90° pulse is merely the Fourier transform of the well-known steady-state spectra, including the effects of chemical exchange (McConnell, 1958). For the special case of two sites of equal population with equal relaxation times T_2 in each site the result was obtained in a rather complicated form which was simplified in later theories. A graphical representation of the Woessner (1961b) predictions are shown in a series of figures computed by L. W. Reeves and E. J. Wells (unpublished work). In Fig. 5, the FID for the equally populated two-site case is plotted for diode detection (Fig. 5a and b) and for phase-sensitive detection along the out of phase y' direction of the rotating frame (Fig. 5c and d). The reference frequency is placed midway between the two sites ω_a and ω_b. The value $\rho = k/\frac{1}{2}(\omega_a - \omega_b) = k/\omega$ is the reduced rate constant in terms of the frequency difference between the two sites. $\Delta\omega = (\omega_a - \omega_b)$. The time axes are also reduced, therefore, in terms of the frequency differences. The decays are exponential in the fast exchange limit, where the steady state spectrum is ideally Lorentzian, with no modulation, but the decay becomes non-Lorentzian in the intermediate exchange region. In the slow exchange region modulation with a frequency $\omega = \frac{1}{2}(\omega_A - \omega_B)$ is observed for the phase detected signal and with frequency $\Delta\omega = (\omega_a - \omega_b)$ in the

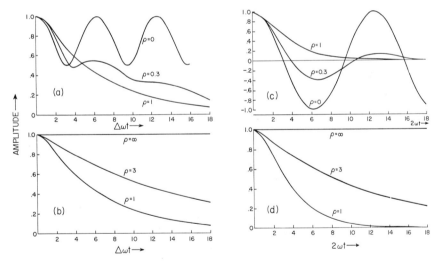

Fig. 5 Theoretical plots of the free induction decay for a two-site equally populated exchange system. The relaxation function $G(t)$ [Eq. (79)] has been omitted so that the curves represent the contribution due to exchange only. Parts (a) and (b) correspond to diode detection and (c) and (d) to phase sensitive detection along a direction perpendicular to the radio frequency H_1 $\rho = k/\omega = 2k/(\omega_a - \omega_b)$; $\Delta\omega = (\omega_a - \omega_b)$. The time axes are reduced in terms of the variables $\Delta\omega t$ [(a) and (b)] and $2\omega t$ [(c) and (d)].

diode detected signal. The overall decay, not shown for the nonexchanging system $\rho = 0$, is exponential in the slow exchange limit. The natural relaxation rate in each site R_{2_a} and R_{2_b} are assumed to be zero for the purpose of the figure, since these are merely additive effects in the decay constant. It is attractive in these days when the Fourier transform technique is established, to consider making quantitative chemical exchange studies by using the FID.

In Fig. 6, the first echo amplitude in a 90°/180° sequence is plotted as a function of the reduced time ($2\omega\tau$), where τ is the pulse time interval. ρ is the reduced exchange rate. In the intermediate exchange region the echo amplitude is modulated provided $\rho < 1$. The qualitative features of the FID in the presence of chemical exchange between two approximately equally populated sites was confirmed experimentally by Reeves and Wells (1962). The value of R_2^*, the transverse relaxation rate imposed by the magnet, must, of course, be considerably less than $(\omega_a - \omega_b)$ to resolve the modulations in the slow exchange limit. A good pulse experiment, in the expectation of the chemist, is much more difficult to perform than a continuous wave experi-

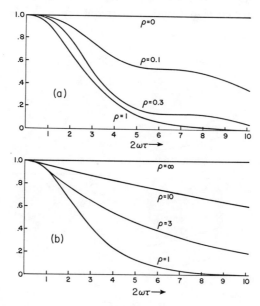

Fig. 6 The first echo amplitude in a 90°/180° two-pulse sequence as a function of the relative exchange rate $\rho = k/\omega$. The signal is assumed to be phase detected in a direction perpendicular to H_1 so as to give a positive going signal. Natural transverse relaxation of the magnetization is neglected, so that with $\rho = 0$ no exchange, the echo does not decay with increased pulse intervals τ. The echo amplitudes are modulated for $\rho < 1$ slow exchange.

4. Diffusion and Chemical Exchange

ment. Biological situations and many chemical exchange studies require line shapes of spins in low concentration, often in the presence of large solvent peaks. The relatively high-power ($H_1 > 1000$ Hz) pulse studies described here are not as suitable as the steady-state method or the more sophisticated low-power selective experiments now available (Freeman and Hill, this volume, Chapter 5). In a sense, the motivation for doing pulse experiments in the first place was to have an independent check on the steady-state method; in simple cases, in order to track down the systematic errors that mar chemical exchange studies in general (Reeves and Wells, 1962).

The Carr–Purcell pulse sequence (1954) with the Meiboom–Gill phase shift modification (1958) is also suitable for studying chemical exchange, especially in simple cases, provided all individual frequencies in the spectrum are subject to modulation by the exchange process. Luz and Meiboom (1963) reported a study of proton transfer between methylamine and the methylammonium ion in dilute aqueous solution using the Carr–Purcell sequence in the fast exchange limit only. In this particular event, the natural line width approaches that arising from the natural relaxation rate in the single exchange averaged peak. The pulse method eliminates inhomogeneity broadening due to an imperfect magnet and provides more reliable estimates of the exchange contribution to line width.

In this limit, Luz and Meiboom show that the envelope of the echo amplitude decay is exponential and obeys the equation:

$$R_{2_{\text{obs}}} = R_{2_0} + [1 - (2n/k)\tanh(k/2n)]k \sum_i p_i \delta_i^2 \tag{77}$$

with

$$\sum_i P_i \delta_i = 0 \tag{78}$$

P_i are site populations; $k \gg \delta_i$ mean inverse lifetime is greater than the frequency difference δ_i from the average position; n is the number of pulses per second; k = rate constant; R_{2_0} is the averaged natural relaxation rate for all sites. The pulse repetition rate n is assumed to be large.

The above first-order approximation has been extended both by numerical analysis (Allerhand and Gutowsky, 1964) and analytical derivation of closed equations for all pulse repetition rates in the two site case (Allerhand and Gutowsky, 1965a; Bloom et al., 1965). The Woessner (1961a) solutions were confined to two sites of arbitrary populations P_a and P_b with a maximum of two pulses while Luz and Meiboom (1963) treated the Carr–Purcell sequence for more sites than two but in the fast exchange limit only.

Allerhand and Gutowsky (1964) first followed the Bloch equation approach similar to Woessner (1961a) but solved the Bloch equations numerically for many pulses. The decay of the transverse magnetization in the Carr–Purcell sequence was shown to be a function of relaxation rates R_{2_a}, R_{2_b} in two sites, the chemical shift difference $\Delta\omega$, the site populations P_a and P_b, the pulse intervals in the 180° train 2τ, and time t. A study of the hindered rotation barrier about the —CN— bond in N,N-dimethyltrichloroacetamide (DMTCA) and dimethylcarbamyl chloride (DMCC) lead to frequency factors of 3.5×10^{12} and 7.1×10^{10}, respectively. The expected frequency factors of $\sim 10^{13}$ which give $\Delta S^* \sim 0$ are an indicator of systematic errors (Inglefield et al., 1968) in the rate measurements. Besides making the range of rate constants much more extensive, this first complete spin-echo study of rate constants provided much more reliable frequency factors. The random error is large, but the systematic errors seemed less serious in the pulse method. The chemical shift was determined from the numerical analysis of the experimental results. The chemical shift variation with temperature for DMTCA, except in the extreme exchange limits, was a monotonic function of temperature, but for DMCC, fitted chemical shifts varied somewhat randomly with temperature between 6.6 and 8.9 Hz in a rather unsatisfactory manner. The apparent relaxation times as a function of pulse repetition rate were measured to $\pm 10\%$. In the most sophisticated line shape studies at the present time, which take account of all exchanging sites and variations of line position with temperature, the rate constants can be determined with much better precision and in much more complex cases, but it must be remembered that the pulse equipment used at this time was much more elementary and field stability was not good (Allerhand and Gutowsky, 1965a; Reeves and Shaw, 1971; Reeves et al., 1971; Allan et al., 1972; Chan and Reeves, 1973). The advantage of the spin-echo method in extending to faster exchange rates still remains.

Allerhand and Gutowsky (1965) and Bloom et al. (1965) independently derived closed formulas for the echo amplitudes in the two-site case for equal population, the latter by use of probability theory in the transverse phase angles as a result of chemical exchange. Hahn and Maxwell (1952), in a paper which describes the codiscovery with Gutowsky et al. (1953) of scalar spin–spin coupling, were able to formulate and estimate the proton–hydroxyl life time on a methanol molecule in the water–methanol system. The stochastic approach (Bloom et al., 1965) will be treated here, as it brings out the exchange process rather well. The starting point is to compute a phase angle distribution $P(\phi, t)$ in the transverse plane of the rotating frame assuming first-order rate theory for exchange between two sites. The chemical exchange process does not affect the natural relaxation times so that these can be added at the end of the calculation. In the absence of exchange,

$$\mathbf{M}(t) = [(P_a \cos \omega_a t + P_b \cos \omega_b t)\mathbf{i} + (P_a \sin \omega_a t + P_b \sin \omega_b t)\mathbf{j}]G(t) \quad (79)$$

$G(t)$ is a relaxation function independent of site. The precession frequency of a nucleus becomes time dependent due to exchange $\omega(t)$ and the phase

4. Diffusion and Chemical Exchange

accumulated by a spin is

$$\phi(t) = \int_0^t \omega(t') \, dt' \tag{80}$$

In the absence of exchange, the phase accumulation is exactly $\omega_a t$ or $\omega_b t$, where $\omega_a = (\omega_a' - \omega_0)$, $\omega_b = \omega_b' - \omega_0$. ω_0 is chosen in the equal population case to be midway between ω_a' and ω_b', the two separate Larmor frequencies.

A computation of $\phi(t)$ over the exchange process leads to $\mathbf{M}(t)$ using Eq. (80). At the time of a π pulse $\phi(\tau)$ is instantaneously transformed into $-\phi(\tau)$ and echoes occur at $2n\tau$, where n is integral. Figure 7 depicts diagrammatically the fate of arbitrary spins under conditions of no exchange, few exchanges during 2τ, and many exchanges during 2τ. Natural relaxation is assumed to be absent or very slow compared to exchange rates.

The complex nuclear spin magnetizations at a time t in the two sites are $M_a(t)$ and $M_b(t)$ and are given by:

$$\begin{aligned} M_a(t) &= M_{a_x}(t) + iM_{a_y}(t) \\ &= f_a e^{i\phi_a} \langle \exp i\phi(t) \rangle_{aa} + f_b e^{i\phi_b} \langle \exp i\phi(t) \rangle_{ba} \end{aligned} \tag{81}$$

$$\begin{aligned} M_b(t) &= M_{b_x}(t) + iM_{b_y}(t) \\ &= f_b e^{i\phi_b} \langle \exp i\phi(t) \rangle_{bb} + f_a e^{i\phi_a} \langle \exp i\phi(t) \rangle_{ab} \end{aligned} \tag{82}$$

f_a and f_b are the *a priori* lengths of magnetization vectors in sites a and b, ϕ_a and ϕ_b are arbitrary initial phase angles in these sites and angular brackets with subscripts aa, ba, bb, and ab signify averages of the included exponential over spins which start in site a when $t = 0$ and end in site a at time t, start in site b and end in site a, respectively, etc.

The averaging is computed from the integral:

$$\langle \exp i\phi(t) \rangle_{ij} = \int_{-1}^{+1} \exp[i\phi(x, t)] P^{ij}(x, t) \, dx \tag{83}$$

The phase angle accumulation variable is changed to

$$x = (2\phi - \omega_a t - \omega_b t)/\Delta \omega t \tag{84}$$

where $\Delta\omega = \omega_a - \omega_b$, noting $\omega_b t < \phi < \omega_a t$ and $-1 < x < +1$ and P^{ij} denotes a probability for a spin starting in site i at $t = 0$ and being in site j at time t.

The functions $P^{ij}(x, t)$ are Bessel functions (Allerhand and Gutowsky, 1964). Differentiation of Eqs. (81) and (82) give the Hahn–Maxwell–McConnell equations including chemical exchange (Hahn and Maxwell, 1952; L. W. Reeves and E. J. Wells, unpublished work).

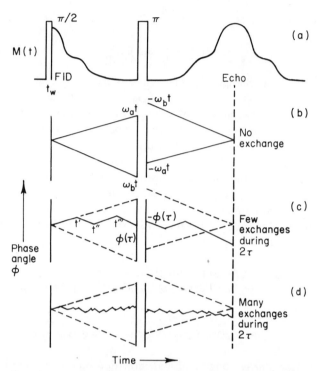

Fig. 7 (a) The transverse magnetization $M(t)$ according to Eq. (79) during a $\pi/2$, π pulse sequence. π pulse occurs at a time τ and the echo at a time 2τ. Pulse widths are neglected in calculations t_w, $2t_w \ll \tau$. (b) The rotating frame has some reference frequency $\omega_b < \omega_0 < \omega_a$. The phase angle ϕ of a single spin that does not relax or exchange accumulates at relative rates ω_a and ω_b having at the time of the π pulse $\omega_b t$ and $\omega_a t$ phase angles ϕ for sites a and b of different frequency. The phase angles are reversed at the end of the π pulse and decrease at rates $-\omega_a$ and $-\omega_b$ to refocus at 2τ. (c) A spin undergoing an exchange process at times t', t'', and t''' alternately accumulates phase at rates ω_a in the period $0 < t < t'$, ω_b in the period $t' < t < t''$, ω_a in the period $t' < t < t'''$, and ω_a in the period $t''' < t < \tau$. The particular spin depicted starts in the a site at $t = 0$, ends in the a site at the π pulse and at the echo time. All possible changes in the residence of the spin are allowed during the period 2τ. Note the exchange causes incomplete refocussing at the time 2τ. (d) A similar diagram to part (c) with many more changes in site during 2τ.

In calculating spin echoes by applying a pulse at a time τ, the value $\phi(\tau)$ is changed to $-\phi(\tau)$ at the pulse. Then:

$$M_a(t) = f_a e^{-i\phi_a}[\langle \exp -i\phi(\tau)\rangle_{aa}\langle \exp +i\phi(t-\tau)\rangle_{aa}$$
$$+ \langle \exp -i\phi(\tau)\rangle_{ab}\langle \exp +i\phi(t-\tau)\rangle_{ba}]$$
$$+ f_b e^{-i\phi_b}[\langle \exp -i\phi(\tau)\rangle_{bb}\langle \exp +i\phi(t-\tau)\rangle_{ba}$$
$$+ \langle \exp -i\phi(\tau)\rangle_{ba}\langle \exp +i\phi(t-\tau)\rangle_{aa}] \qquad (85)$$

4. Diffusion and Chemical Exchange

by interchanging labels b and a, the expression for $M_b(t)$ is obtained. The value of $M(2\tau)$ is important for computing the echo maximum. By defining matrix elements:

$$E_{ij} = \sum_k \langle \exp -i\phi(\tau)\rangle_{ik} \langle \exp +i\phi(\tau)\rangle_{kj} \tag{86}$$

the exchange process can be included in a matrix form:

$$\begin{pmatrix} M_a(2\tau) \\ M_b(2\tau) \end{pmatrix} = \begin{pmatrix} E_{aa} & E_{ab} \\ E_{ba} & E_{bb} \end{pmatrix} \begin{pmatrix} M_a^*(0) \\ M_b^*(0) \end{pmatrix} \tag{87}$$

In a Carr–Purcell sequence:

$$\mathbf{M}(4n\tau) = (\mathbf{EE}^*)^n \mathbf{M}(0) \tag{88}$$

The echo envelope is a superposition of two exponentials, one of which is experimentally negligible. Summarizing the results for the symmetric two-site case, the two exponential decays may be written:

$$M(4n\tau) = A_1 \exp(-r_1 4n\tau) + A_2 \exp(-r_2 4n\tau) \tag{89}$$

A_1 and A_2 depend on initial conditions with r_1, the dominant decay for the symmetric case

$$r_1 = k - (1/2\tau) \sinh^{-1} F \tag{90}$$

where for $\omega > k$

$$F = [k/(\omega^2 - k^2)^{1/2}] \sin 2v \tag{91}$$

and for $\omega < k$

$$F = [k/(k^2 - \omega^2)^{1/2}] \sinh 2u \tag{92}$$

with

$$v = (\omega^2 - k^2)^{1/2}\tau \quad \text{and} \quad u = (k^2 - \omega^2)^{1/2}\tau \tag{93}$$

If by chance $\omega = k$,

$$F = 2k\tau \tag{94}$$

k is the first-order rate constant for the exchange. In practice, the variation of the apparent T_2 has a strong variation on pulse repetition rate in the intermediate exchange region and a smaller and smaller variation as the fast and slow exchange limits are approached.

Some interesting approximate limits for the two exponents r_i in Eq. (89) are found.

$$\omega > k \quad \text{slow exchange} \quad \frac{2k \sin 2v}{(\omega^2 - k^2)^{1/2}} \ll 1$$

$$r_i \cong k \mp \frac{k \sin 2v}{2\tau(\omega^2 - k^2)^{1/2}} \pm \frac{k^3 \sin^3 2v}{12\tau(\omega^2 - k^2)^{3/2}} \tag{95}$$

$$\omega < k \quad \text{fast exchange} \quad \frac{2k \sinh 2u}{(k^2 - \omega^2)^{1/2}} \ll 1$$

$$r_i \cong k \mp \frac{k \sinh 2u}{2\tau(k^2 - \omega^2)^{1/2}} \pm \frac{k^3 \sinh^3 2u}{12\tau(k^2 - \omega^2)^{3/2}} \tag{96}$$

Fast-pulsing limit $k\tau \ll 1$, $\omega\tau \ll 1$ valid for $\omega > k$, $\omega < k$,

$$r_i \cong k \mp k \pm \tfrac{2}{3}k\omega^2\tau^2 \tag{97}$$

Slow-pulsing limit (sinh $2u \gg 1$) and fast exchange $\omega \ll k$,

$$r_i \cong k \mp k \pm (\omega^2/2k) \tag{98}$$

Equations (90–98) have all been experimentally tested and within the limitations of the equipment available have proved to be satisfactory for evaluating k, the rate constant.

A random walk approach in phase angle $\phi(t)$ gives a satisfactory result only for the limit of many exchanges during the period 2τ, depicted as situation (d) in Figure 7. In the limit of few random steps the appropriate averaging used in random-walk problems fails.

The complete Carr–Purcell decay for the symmetric two site case can be written:

$$1/T_{2_{CP}} = 1/T_{20} + r_1 \tag{99}$$

where $T_{2_{CP}}$ is the observed decay time and T_{20} is the natural transverse relaxation time.

In the fast-pulsing limit at all exchange rates, the longest decay corresponds to the natural relaxation time as in the diffusion experiment. A representative set of results is shown in Fig. 8 for the hindered rotation rates in N,N-dimethylnitrosamine (Abramson et al., 1966). Equation (97) is illustrated in Fig. 9 for the same molecule. The dependence of $T_{2_{CP}}$ on pulse repetition rates is shown for all exchange rate regions in Fig. 10 for the molecule N,N-dimethylformamide-d_1. Equation (90) with F taking values in Eqs. (91), (92), and (94), is used to fit the dependence of apparent relaxation time on pulse repetition rate in the Carr–Purcell train.

The echo amplitude envelope becomes more complex in analytical form even for the unsymmetric two-site case (Bloom et al., 1965), but with curve fitting procedures, the dependence of apparent relaxation rate on chemical shift, population difference, and exchange rate is possible. Gutowsky and co-workers using the Carr–Purcell method for the symmetric case, reported rate constants for cyclohexane and cyclohexane-d_{11} (Allerhand et al., 1965).

The J coupling in homonuclear spin-echo experiments provides a coherent modulation of the echo amplitudes (Hahn and Maxwell, 1952; Powles and Strange, 1962) in both two pulse and multipulse sequences. If a nucleus is J

4. Diffusion and Chemical Exchange

Pulse interval (2τ), (msec)		$T_{2 (C.P.)}$ (s)	Time base (sec/division)
1.30		10.1	2.0
2.05		9.21	2.0
2.58		6.30	2.0
8.11	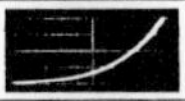	2.68	1.0
10.24		2.03	1.0
13.02		1.44	0.5
20.48		0.72	0.2
40.72		0.20	0.1

Fig. 8 Variation of the apparent relaxation rate measured in a Carr–Purcell experiment as a function of pulse interval 180°/180° of 2τ. The hindered rotation in N,N-dimethylnitrosamine is a two-site equal population exchange case to a high degree of approximation. The apparent relaxation rate at 135°C, the temperature of the experiment, varies between 0.2 seconds and 10.1 seconds for pulse intervals 2τ in the range 40.72 msec to 1.3 msec. [After Abramson et al. (1966).] Note that the time increases from right to left in the oscilloscope photographs.

coupled but not in resonance, the experiment fails to detect coupling unless there is chemical exchange. The variation of the modulation of echo amplitudes with pulse repetition rates was also studied by Wells and Gutowsky (1965). The modulation of indirect spin–spin coupling by chemical exchange is not generally accommodated by the classic theories of the spin-echo experiment (Woessner, 1961b; Luz and Meiboom, 1963; Allerhand and Gutowsky, 1965a; Allerhand et al., 1965; Bloom et al., 1965). The effect on the apparent J measured in a spin-echo experiment during chemical exchange of the hydroxyl proton in methanol was observed by Powles and Strange (1962) and by Hahn and Maxwell (1952) much earlier.

The problem of chemical exchange in more complex systems with spin coupling was formulated in terms of the density matrix by Kaplan (1958a,b). Alexander generalized the approach for both inter and intramolecular processes (1962). A theory applicable to spin-echo-type experiments was formulated in detail by Gutowsky et al. (1965).

The relaxation terms can be included in a total relaxation equation in matrix form.

$$d\mathbf{M}/dt = \mathbf{A} \cdot \mathbf{M} \tag{100}$$

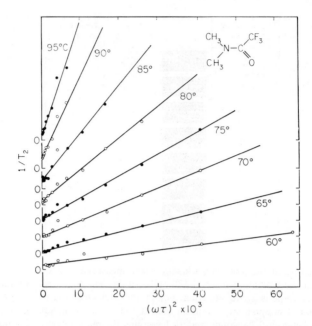

Fig. 9 Illustration of the experimental results which confirm the applicability of Eq. (97) in the text. For the fast pulsing limit, $\omega > k$ or $\omega < k$, but $\omega\tau \ll 1$ the apparent relaxation rate $R_2 = 1/T_2$ is linear with $(\omega\tau)^2$ the chemical shift times the pulse interval 90°/180° all squared. The molecule was N,N-dimethyltrifluoroacetamide, which is approximately a two-site equal population case. [After Abramson et al. (1966).]

M is a column vector containing the density matrix elements of the equation:

$$d\sigma(\alpha, \beta)/dt = i\omega_{\alpha\beta}\sigma(\alpha, \beta) + \sum_{\gamma,\delta} k_{\gamma\delta}\sigma(\gamma, \delta) \tag{101}$$

α, β, γ, and δ refer to eigen vectors of H_0 (time independent), $\omega_{\alpha\beta}$ is the frequency for the transition between states α and β, while $k_{\gamma\delta}$ is the probability per unit time that the spin assembly will change from state γ to state δ by a chemical exchange process. Each transition in Eq. (101) contributes a component to the vector **M**. σ is the spin density matrix. The classic approach uses site magnetizations (Reeves and Shaw, 1970), while the density matrix approach prescribes "spectral magnetizations" in the language of Gutowsky et al. (1965). Both are written as a column vector **M**.

The equation for the time dependence of the density matrix is the approximate equation suggested by Redfield (1957).

$$d\sigma/dt = i[\sigma, H_0] + \sum_i \tfrac{1}{2}k_i[(X_i, \sigma), X_i^\dagger] \tag{102}$$

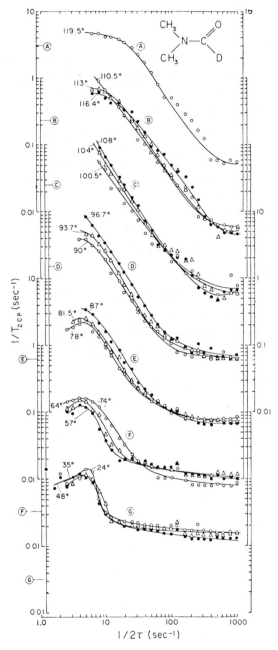

Fig. 10 Logarithmic plots of the apparent relaxation rate $R_2 = T_2^{-1}$ versus inverse pulse interval $(2\tau)^{-1}$ for all exchange regions using a Carr–Purcell sequence. The molecule was N,N-dimethylformamide-d_1, which is to a high degree of approximation an equally populated two-site case of chemical exchange. The solid lines are the theoretical best fit to Eqs. (90) and (92) from which the chemical shift and rate constant k can be derived. The natural relaxation rate is determined from the fast pulsing limit. Temperatures are indicated for each curve and the proton frequency was 40 MHz.

k_i is the first-order rate constant for the ith exchange process, the dagger signifying the adjoint of the exchange operators X_i. Uncorrelated exchange events are assumed. $\mathcal{H}_0$ is the time-independent Hamiltonian in the absence of exchange. The **A** matrix is obtained from Eq. (101) with the Hamiltonian $\mathcal{H}_0$ in the absence of pulses given by:

$$\mathcal{H}_0 = \sum_i (\omega_0 - \omega) I_{zi} + \sum_i \Omega_i I_{zi} + \sum_{k<l} A_{kl} \mathbf{I}_k \cdot \mathbf{I}_l \tag{103}$$

ω is the radio frequency and A_{kl} the spin–spin coupling constant in radians per second. Ω_i is the chemical shift, ω_0 is some average Larmor frequency for the purpose of reference.

In a manner somewhat similar to that of Bloom et al. (1965), the evolution of $\mathbf{M}(t)$ after a series of pulses and echoes are given by:

$$\mathbf{M}(t) = \mathbf{M}(0) \exp(\mathbf{A}t) \tag{104}$$

after the first pulse

$$\mathbf{M}(2\tau) = \exp(\mathbf{A}\tau) \exp(\mathbf{A}^\ddagger \tau) \mathbf{M}(0) \tag{105}$$

at a time 2τ of the first echo for nonselective pulses. $\mathbf{A}^\ddagger$ is an **A** matrix modified so that with H_1 along the x axis, the signs of the z and y components of spin operators are changed but the sign of the scalar interactions in H_0 are not modified.

Finally, defining an echo operator as:

$$\mathbf{E}^2 = \exp(\mathbf{A}\tau) \exp(2\mathbf{A}^\ddagger \tau) \exp(\mathbf{A}\tau) \tag{106}$$

the odd and even echoes in a Carr–Purcell sequence can be written:

$$\mathbf{M}(4n\tau) = \mathbf{E}^{2n} \mathbf{M}(0) \tag{107}$$

even echoes at time $(2n)(2\tau)$, and

$$\mathbf{M}[(4n+2)\tau] = \mathbf{E}^{2n} \mathbf{M}(2\tau) \tag{108}$$

odd echoes at time $(2n+1)(2\tau)$. If

$$\mathbf{M}(2\tau) = \mathbf{E}\mathbf{M}(0) \tag{109}$$

then odd and even echoes follow the same progression, but this is not necessarily true in general.

The computations are greatly simplified by diagonalizing $\mathbf{E}^2$

$$\boldsymbol{\Gamma} = \mathbf{V}^{-1} \mathbf{E}^2 \mathbf{V} \tag{110}$$

when the contributing decay constants to the echo amplitudes become:

$$\boldsymbol{\Delta} = -(1/4\tau) \ln \boldsymbol{\Gamma} \tag{111}$$

The classic case of two uncoupled sites treated previously (Allerhand and

4. Diffusion and Chemical Exchange

Gutowsky, 1965a; Bloom *et al.*, 1965), is easily verified using this formulation. The advance in understanding comes from the treatment of two coupled sites. In an intramolecular exchange process, the **A** matrix differs from the intermolecular exchange case. The intramolecular case will be treated first. In general the **A** matrix is a sum of three submatrices:

$$\mathbf{A} = i\omega + \mathbf{K} + \mathbf{R}_2 \tag{112}$$

ω is a diagonal matrix of the frequencies of spectral magnetizations. **K** is the kinetic transfer matrix (Reeves and Shaw, 1970) and $\mathbf{R}_2$ is a diagonal natural transverse relaxation matrix.

For two distinct sites A and B with subscript labels a and b, the **A** matrix is:

$$\mathbf{A} = \begin{bmatrix} i(\Omega - \tfrac{1}{2}A) - k - R_{2a} & k - \tfrac{1}{2}iA & 0 & 0 \\ k - \tfrac{1}{2}iA & i(-\Omega + \tfrac{1}{2}A) - k - R_{2b} & 0 & 0 \\ 0 & 0 & i(\Omega - \tfrac{1}{2}A) - k - R_{2b} & k + \tfrac{1}{2}iA \\ 0 & 0 & k + \tfrac{1}{2}iA & -i(\Omega + \tfrac{1}{2}A) - k - R_{2b} \end{bmatrix} \tag{113}$$

Partitioning the **A** matrix

$$\mathbf{A} = \mathrm{diag}(\mathbf{A}_1, \mathbf{A}_2) \tag{114}$$

where

$$\mathbf{A}_1 = (\tfrac{1}{2}iA - k - R)\mathbf{I} + \mathbf{A}_1' \tag{115}$$

$$\mathbf{A}_2 = (-\tfrac{1}{2}iA - k - R)\mathbf{I} + \mathbf{A}_2' \tag{116}$$

with equal relaxation times $R_{2a} = R_{2b}$.

If both nuclei are in resonance, homonuclear coupling case, the pulse changes the sign of Ω in **A** but does not alter the sign of the coupling constant A. The matrix **P** which transforms **A** to $\mathbf{A}^{\ddagger}$ is now of simple form:

$$\mathbf{A}^{\ddagger} = \mathbf{PAP} \tag{117}$$

where **P** is partitioned

$$\mathbf{P} = \mathrm{diag}(\mathbf{P}_1, \mathbf{P}_2) \tag{118}$$

with

$$\mathbf{P}_1 = \mathbf{P}_2 = \begin{bmatrix} 0 & 1 \\ 1 & 0 \end{bmatrix} \tag{119}$$

The echo operator matrix is also partitioned into two two-by-two matrices;

$$\mathbf{E} = \mathrm{diag}(E_1, E_2) \tag{120}$$

where
$$\mathbf{E}_j = \exp(\mathbf{A}_j \tau) \mathbf{P}_j \exp(\mathbf{A}_j \tau) \tag{121}$$

Four decay constants for the echo amplitudes occur:

$$D_{1_\pm} = k + R - \tfrac{1}{2}iA - \frac{1}{2\tau}\ln[F_1 \pm (F_1^2 + 1)^{1/2}] \tag{122}$$

$$D_{2_\pm} = k + R + \tfrac{1}{2}iA - \frac{1}{2\tau}\ln[F_2 \pm (F_2^2 + 1)^{1/2}] \tag{123}$$

where
$$F_j = \sin 2\phi_j \sin 2\lambda_j \tau \tag{124}$$

with

$$\begin{aligned}
\sin 2\phi_1 &= (k - \tfrac{1}{2}iA)/\lambda_1 \\
\sin 2\phi_2 &= (k + \tfrac{1}{2}iA)/\lambda_2 \\
\cos 2\phi_1 &= i\Omega/\lambda_1 \\
\cos 2\phi_2 &= i\Omega/\lambda_2 \\
\lambda_1 &= +(k^2 + \tfrac{1}{4}\Delta^2 - iAk)^{1/2} \\
\lambda_2 &= +(k^2 - \tfrac{1}{4}\Delta^2 + iAk)^{1/2} \\
\Delta &= (4\Omega^2 + A^2)^{1/2}
\end{aligned} \tag{125}$$

The form of the results is rather similar, except for the inclusion of A, the coupling constant, to the uncoupled case. In the slow and fast exchange limits, appropriate limiting cases can be developed.

The intermolecular exchange of proton B coupled to proton A has a different solution. In the exchange the chemical situation determines whether B becomes A in the second molecule or remains a B nucleus. The **A** matrix remains a 4×4 and cannot be partitioned so that numerical solutions have to be developed. In certain limits of fast and slow exchange useful approximation can be developed. Selective pulsing of transitions are not appropriate to this chapter and though mentioned by Gutowsky *et al.* (1965) are not included here.

For intramolecular cases, partitioning of the **A** matrix can be extended to the ABX and ABX_q system of spins and analytical solutions can be developed. The analysis is also useful in the A_nX system, where nuclei A are observed and X has $I > \tfrac{1}{2}$. In this case, the spin–spin coupling is modulated by the interruptions due to the shorter lifetimes of the quadrupolar nucleus relaxing in the fluctuating electric field gradients. Determination of the coupling constant from the Carr–Purcell sequence can be achieved as well as information about the relaxation rate R_1 of the X nucleus. Numerical methods are in general necessary.

Providing the equipment is satisfactory, the extension of the Carr–Purcell sequence for studying more complex cases seems possible but cumbersome in

4. Diffusion and Chemical Exchange

the numerical analysis. Nonselective pulse experiments always have the serious drawback of nonexchanging nuclei in resonance. Most chemical systems do not contrive to avoid this problem. The measurement of small coupling constants beyond the resolving capacity of man-made magnets is certainly an advantage that steady-state experiments cannot match, but this is beyond the scope of this chapter.

Allerhand and Gutowsky (1965b) made a theoretical and experimental study of the rates of chair to chair conversion in 1,1-difluorocyclohexane by observations of the ^{19}F echo amplitudes in a Carr–Purcell sequence. The density matrix formalism was analyzed numerically rather than using the closed forms solution of Gutowsky et al. (1965), which appeared shortly afterward. Recursion formulas were used to generate the amplitude of successive echoes. The experiments showed that the amplitudes were dependent on A, $\delta\omega$ the chemical shift, k the rate of exchange, and pulse separations τ. Coupling is revealed as a modulation of the amplitudes providing $k > 3A$. The spectrum was treated as an AB case where $\delta\omega$ was found to be 2482 radians sec^{-1} and $A = 1477$ radians sec^{-1}. The analysis ignores modulation of coupling to the 2,2 and 6,6 protons, which although not in resonance do lead to extra spectral magnetizations which are modulated by the exchange. Gutowsky et al. (1965) cite this as a reason for the discrepancy between the spin-echo (Allerhand and Gutowsky, 1965b) and steady-state results (Jonas et al., 1965) for this molecule. With the equipment used at this time, the experimental reproduction of rather complex modulations of echo amplitudes in the slow exchange limit were difficult to reproduce. In the fast exchange limit, modulation due to coupling disappears and the analysis of simple exponential decays is much easier. With improvements in instrumentation, such problems would not now be so serious.

A satisfactory improvement of the energy barrier for chair to chair conversion in perfluorocyclohexane (Tiers, 1960) by a spin-echo study was achieved by Gutowsky and Chen (1965). The fast exchange approximation was adopted with a chemical shift derived from low-temperature high resolution studies.

The degenerate Cope rearrangement of Bullvalene was discovered by Doering and Roth (1963) and studied by Saunders (1963) as a function of temperature by steady state nmr methods.

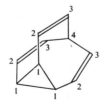

The following exchanges take place in the reaction:

H-3 ⇌ H-1 (twice)
H-3 ⇌ H-2
H-4 ⇌ H-1
H-2 ⇌ H-2

In the fast exchange limit, H-2 and H-3 have a chemical shift $\tau \sim 4.4$ (tau scale) and H-1 and H-4 tend to collapse with $\tau \sim 7.8\ \tau$. Thus, the four independent sites collapse to two sites. For a spin-echo study, the case is relatively complex, having more than two sites and spin–spin coupling. A limited range of pulse repetition rates applied in a fast exchange approximation was successful in obtaining rate constants essentially in agreement with the complete lineshape fit of the steady-state spectrum (Allerhand and Gutowsky, 1965c).

Critical reviews comparing the capability of the spin-echo and steady-state methods for determining rate constants based on a thorough analysis of errors were published (Allerhand et al., 1966; Inglefield et al., 1968). It is probable that the improvements in commercial pulse spectrometers for the chemist have eliminated the errors identified only a few years ago, and care in the use of field-frequency-locked spectrometers for high-resolution lineshapes make both methods satisfactory. The complete lineshape fit of steady-state spectra has much more flexibility but lacks precision in the slow and fast exchange regions. The spin-echo method, on the other hand, is really useful in eliminating the effect of inhomogeneity in the magnet in the limits of rate measurements, but the nonselective pulse schemes discussed here are very inflexible and in most systems inapplicable. Each case of exchange must be treated as a problem in itself and the following factors are important to take into account: (a) dependence of chemical shift or coupling constant on temperature; (b) inclusion of all sites or spectral magnetizations that affect the exchange process in the analysis; (c) the recognition of species contributing to the exchange that are below noise level (the system can be treated as a truncated case if this occurs); (d) proper classification of inter- and intra-molecular processes; (e) line broadening due to distinct relaxation rates in different sites or long range unresolved coupling; (f) interrupted scalar coupling to nuclei with quadrupole moments; (g) careful temperature control and measurement; and (h) magnetic field stability and the standard high grade performance of pulse gear. The 180° pulses should not droop as the pulse repetition rate increases. Errors due to diffusion and radiation damping.

In test comparison cases Allerhand et al. (1966) made thorough theoretical analyses of errors, while Inglefield et al. (1968) compared classic barrier studies using both the high-resolution method and the spin-echo method in

4. Diffusion and Chemical Exchange

the same laboratory. These workers found the spin-echo method was less satisfactory than good lineshape fit measurements on the same molecule. The barrier for hindered internal rotation about the C—N bond in N,N-dimethylformamide was determined as 20.8 ± 0.6 kcal mole^{-1} with $\Delta S^{\ddagger} = 0 \pm 1$ eu by high-resolution methods using a four-site formulation that included spin-spin coupling long range to the aldehyde proton. The spin-echo study of the same molecule gave an activation energy 21.6 ± 2.7 kcal mole^{-1}, but deuteration of the aldehyde proton was necessary to remove it from resonance. The standard error was much larger for the spin-echo study. The long-range couplings to the deuterium were ignored in the analysis, since this error is much smaller than random errors encountered in the measurement.

Allerhand and Thiele (1966) in an extended analysis of the Carr–Purcell method for chemical exchange show that in multisite cases for the fast exchange limit, one single exponential decay dominates in the echo amplitudes and in general the decay is a sum of r terms of the form:

$$M(t) = \sum_{j=1}^{r} [A_j + (-1)^n B_j] \exp(-n\tau R_j) \qquad (126)$$

where R is a positive function of τ, the apparent relaxation rate in the rotating frame.

For a multisite process for r sites or spectral magnetizations, there are, $\frac{1}{2}r(r-1)$ independent rates, $(r-1)$ relative chemical shifts or frequency differences, r natural transverse relaxation rates R_{20}, and $(r-1)$ fractional populations for the sites. Allerhand and Thiele (1966) show that in the fast exchange limit when one single Lorentzian line is observed in the steady state spectrum, the spin-echo method can give more information than the lineshape method. The analysis promises to be of real use in biologically interesting situations where a solvent such as water in great excess participates in exchange processes or in the case of nmr of ions which are mostly free in the solvent but also locate in active sites of biochemical systems. Such systems are very unfavorable for high-resolution analysis, but are certainly not a general case.

Vold and Gutowsky (1967) made the first study by the spin-echo method of a multisite case with spin–spin coupling in a region where echo amplitudes were modulated by coupling and damped by chemical exchange. The hindered rotation in 1,1-difluoro-1,2-dibromodichloroethane is a two-spin system with two spins in resonance, and echo decays depend on sixteen parameters as well as the usual pulse intervals. The steady state method was previously used by Newmark and Sederholm (1965) to study this threefold barrier, which has in the slow exchange limit two equally stable gauche and one distinct trans form. The interconversion rate constants can be written:

$$k_{ij} = (kT/h) \exp[-(G_{ij}^{\ddagger} - G_i)/RT] \qquad (127)$$

$i, j = 1, 2$, or 3 depending on the conformational (chemical) site for t, g_1, and g_2 rotamers. k_{ij} is the rate constant for the $i \rightarrow j$ process. $G_{ij}^{\ddagger}$ and G_i are Gibbs free energies of activation for eclipsed transition state and ground staggered states, respectively.

The echo trains were analyzed numerically in the slow exchange limit, but because of the number of parameters, spectral magnetization frequencies were derived from steady state spectra. The fast exchange limit gave exponential decays of echo amplitude. In Fig. 11 the complex modulation and decay pattern for the exchange process is illustrated for this molecule. Agreement with the steady-state method was satisfactorily achieved. Relative populations and peak positions are more easily and accurately estimated from steady-state spectra.

Hindered internal rotation in 1-fluoro-1,1,2,2-tetrachloroethane is associated with two gauche and one trans isomer (Alger et al., 1967). A compre-

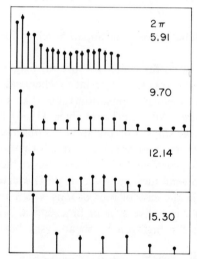

Fig. 11 Spin-echo amplitudes in a Carr–Purcell-type experiment with chemical exchange derived from hindered internal rotation in $CF_2Br-CCl_2Br$ at $-68°C$ and a fluorine frequency of 25.27 MHz. The echo amplitudes were computed numerically. There are two equivalent gauche forms and one trans form. Vertical lines represent experimental amplitudes, and filled circles are computed using chemical shifts, rate constants, natural relaxation time, and spin–spin coupling constants. In the steady-state spectrum in the slow exchange limit there are five observable transitions. A single resonance for the transform and two overlapping AB spectra for the gauche forms. [After Vold and Gutowsky (1967).]

hensive study by Alger, Gutowsky, and Vold of the rates of rotation about the C—C bond encompassed both fluorine and proton steady state lineshapes as well as Carr–Purcell spin-echo studies of the fluorine resonance at 25.27 MHz and the proton resonance at 26.8 and 17.7 MHz. An unusually large range of temperature was possible using both methods and also several internal checks on the consistency of data. The theory of Gutowsky, Vold, and Wells (1965) is entirely appropriate for all methods used. The **A** matrix is 12 × 12 and partitions into four diagonal submatrices, each of which contain gauche–gauche, and two gauche–trans exchange rates. It is possible to contract this to a two-site gauche–trans exchange with unequal populations. The **A** matrix for protons becomes:

$$\mathbf{A} = \mathrm{diag}(\mathbf{A}_+, \mathbf{A}_-) \qquad (128)$$

where

$$\mathbf{A}_\pm = \begin{bmatrix} R_2^0 - k_{gt} + i(\Omega_g \pm \tfrac{1}{2}A_g) & k_{tg} \\ k_{gt} & -R_2^0 - k_{tg} + i(\Omega_t \pm \tfrac{1}{2}A_t) \end{bmatrix} \qquad (129)$$

where subscripts t and g refer to trans and gauche forms and relaxation rates are equal in each site. A_g and A_t are proton–fluorine coupling constants in appropriate forms. The population of the d_1 and l gauche rotamers is P_g while that of the trans isomers is $(1 - P_g)$. A corresponding **A** matrix is written for the fluorine spectral magnetizations. The fluorine chemical shifts between trans and gauche forms are much larger than for the protons and the range of rates accessible also much larger. In Fig. 12, the Arrhenius plot for the activation energy is a composite of four independent measurements, and agreement is impressive. Some interdependence of the methods, such as chemical shifts of the fluorine resonances being derived solely from high-resolution low-temperature spectra, are not serious in comparisons of the precision of high resolution and spin-echo techniques. One of the errors of the spin-echo method not identified initially is the effect of field drift on amplitudes in a Carr–Purcell train. This has been mentioned by Inglefield et al. (1968) and treated theoretically by A. Allerhand (private communication; Allerhand and Moll, 1969).

IV. Chemical Exchange Rates from Rotating Frame Relaxation

The relaxation effects in a rotating frame are measured in the H_1 field by causing the macroscopic magnetization to orient in the xy plane of the rotating frame, then phase shifting the radio frequency/field relationship by some means so as to lock the transverse magnetization in the H_1 field direction. The spin locking can be achieved by pulse techniques (Look and Lowe,

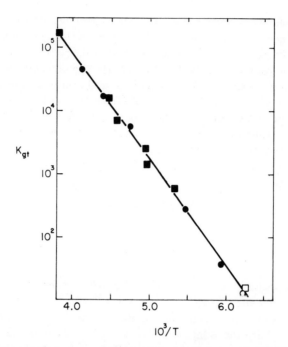

Fig. 12 Arrhenius plot for gauche–trans exchange rates in $CFCl_2$–$CHCl_2$. The experimental points are derived from steady-state lineshapes of the proton and fluorine resonance and corresponding studies using the Carr–Purcell method for both protons and fluorine. ● = ^{19}F line shape, ■ = ^{19}F spin echo, ○ = ^{1}H line shape, □ = spin echo. [After Alger et al. (1967).]

1966) or by adiabatic half passage methods (Meiboom, 1961). The H_1 field may be increased to the point of dielectric breakdown in specially designed probes, which seems to occur in the neighborhood of 80 G. The fluctuating transverse magnetizations caused by chemical exchange may be studied from very low frequencies in low H_1 fields to rates of exchange exceeding the fastest measurable by Carr–Purcell or steady state methods in fields of 80 G. Deverell et al. (1970) published a theory of the experiment and determined exchange for the classic chair-to-chair conversion in cyclohexane.

If $\delta\omega\tau_e \ll 1$ and $\delta\omega \ll \omega_1 \ll \omega_0$, where $\delta\omega$ is the chemical shift between two uncoupled sites and τ_e is the correlation time produced at the nucleus by a fluctuating magnetic field in the z direction due to chemical exchange, then in the density matrix formalism:

$$\frac{\partial \langle \mathbf{I} \rangle}{\partial t} = \gamma \{\langle \mathbf{I} \rangle \times \mathbf{H}_{eff}\} - \tfrac{1}{2} \int_{-\infty}^{+\infty} \langle [R^{-1}(\tau)\mathcal{H}_1(t-\tau)R(\tau), [\mathcal{H}_1(t), \mathbf{I}]] \rangle \, d\tau \quad (130)$$

4. Diffusion and Chemical Exchange

Equation (130) represents the time variation of the expectation value of **I**

$$\gamma\hbar\langle \mathbf{I}\rangle = \gamma\hbar \, \text{trace}[\sigma, I] \tag{131}$$

where σ is the density matrix in the frame of reference rotating with H_1 at ω_0. In the rotating frame $\mathbf{H}_{eff} = \mathbf{H}_1$. $\mathscr{H}_1(t)$ is the relaxation Hamiltonian in the rotating frame

$$\mathscr{H}_1(t) = \exp(i\omega_0 I_z t)\mathscr{H}_1(t)\exp(-i\omega_0 I_z t) \tag{132}$$

$\mathscr{H}_1(t) = \mathscr{H}_1(t)$ because of the commutation. $R(t) = \exp(i\omega_1 I_x t)$ for H_1 along the x axis.

The rotating axes are redefined so that the I_x axis becomes I_z'; I_z becomes $-I_x'$; I_y'; then:

$$R^{-1}(\tau)\mathscr{H}_1(t-\tau)R(\tau) = -\tfrac{1}{2}\delta\omega\,(t-\tau)(I_+\exp(-i\omega_1\tau) + I_-\exp(+i\omega_1\tau)) \tag{133}$$

Evaluating the commutations in Eq. (130), the time dependence of $\langle I_z'\rangle$ becomes:

$$\frac{\partial \langle I_z'\rangle}{\partial t} = \frac{-\langle I_z'\rangle}{T_{1\rho}} \tag{134}$$

where

$$1/T_{1\rho} = \tfrac{1}{2}\int_{-\infty}^{+\infty} \overline{\delta\omega(t)\,\delta\omega(t-\tau)} \cos\omega_1\tau \, d\tau \tag{135}$$

For a simple uncoupled two-site case, the correlation function becomes:

$$\overline{\delta\omega(t)\,\delta\omega(t-\tau)} = \tfrac{1}{4}\delta\omega^2 \exp(-\tau/\tau_e) \tag{136}$$

$\tfrac{1}{2}\tau_e = K$, the rate constant for the two site exchange. Thus:

$$1/T_{1\rho} = \tfrac{1}{4}(\delta\omega)^2[\tau_e/(1 + \omega_1^2\tau_e^2)] \tag{137}$$

Thus, $T_{1\rho}$ will be linear with ω_1^2 on a slope $[4\tau_e/(\delta\omega)^2]$ with intercept $\omega_1 \to 0$ of $T_{1\rho} = [4/(\delta\to)^2\tau_e]$. Both τ_e and $\delta\omega$ can thus be measured. As in the Carr–Purcell, sequence measurements other relaxation mechanisms must be subtracted off:

$$\frac{1}{T_{1\rho(ex)}} = \left(\frac{1}{T_{1\rho(obs)}} - \frac{1}{T_1}\right) = \frac{(\delta\omega)^2}{4}\frac{\tau_e}{1 + \omega_1^2\tau_e^2} \tag{138}$$

cf. Eq. (99) earlier.

The scope and precision of this new method has yet to be fully investigated. The rate measurements for cyclohexane inversion had considerable scatter but agreed with the best previously measured rates from high-resolution and spin-echo studies.

REFERENCES

Abragam, A. (1961). "The Principles of Nuclear Magnetism," Oxford Univ. Press, London and New York.
Abramson, K. H., Inglefield, P. T., Krakower, E., and Reeves, L. W. (1966). *Can. J. Chem.* **44**, 1685.
Alexander, S. (1962). *J. Chem. Phys.* **37**, 967 and 974.
Alger, T. D., Gutowsky, H. S., and Vold, R. L. (1967). *J. Chem. Phys.* **47**, 3130.
Allan, E. A., Hobson, R. F., Reeves, L. W., and Shaw, K. N. (1972). *J. Amer. Chem. Soc.* **94**, 6604.
Allerhand, A., and Gutowsky, H. S. (1964). *J. Chem. Phys.* **41**, 2115.
Allerhand, A., and Gutowsky, H. S. (1965a). *J. Chem. Phys.* **42**, 1587.
Allerhand, A., and Gutowsky, H. S. (1965b). *J. Chem. Phys.* **42**, 4203.
Allerhand, A., and Gutowsky, H. S. (1965c). *J. Amer. Chem. Soc.* **87**, 4092.
Allerhand, A., and Moll, R. E. (1969). *J. Magn. Resonance* **1**, 488.
Allerhand, A., and Thiele, E. (1966). *J. Chem. Phys.* **45**, 902.
Allerhand, A., Chen, F.-M., and Gutowsky, H. S. (1965). *J. Chem. Phys.* **42**, 3040.
Allerhand, A., Gutowsky, H. S., Jonas, J., and Meinzer, R. A. (1966). *J. Amer. Chem. Soc.* **88**, 3185.
Berger, W., and Butterweck, H. J. (1956). *Arch. Elektrotech.* (*Berlin*) **42**, 216.
Bloch, F. (1946). *Phys. Rev.* **70**, 460.
Bloch, F., Hansen, W. W., and Packard, M. (1946a). *Phys. Rev.* **69**, 127.
Bloch, F., Hansen, W. W., and Packard, M. (1946b). *Phys. Rev.* **70**, 474.
Bloom, M., Reeves, L. W., and Wells, E. J. (1965). *J. Chem. Phys.* **42**, 1615.
Boss, B. D., and Stejskal, E. O. (1965). *J. Chem. Phys.* **43**, 1068.
Burnett, L. J., and Harmon, J. F. (1972). *J. Chem. Phys.* **57**, 1293.
Carr, H. Y., and Purcell, E. M. (1954). *Phys. Rev.* **94**, 630.
Chan, S. I., Feigenson, G. W., and Seiter, C. M. A. (1971). *Nature* (*London*) **231**, 110.
Chan, S. O., and Reeves, L. W. (1973). *J. Amer. Chem. Soc.* **95**, 670 and 673.
Cooke, R., and Wein, R. (1971). *Biophys. J.* **11**, 1002.
Das, T. P., and Saha, A. K. (1954). *Phys. Rev.* **93**, 749.
Dass, N., and Vanshneya, N. C. (1969). *J. Phys. Soc. Jap.* **26**, 873.
Deverell, C., Morgan, R. E., and Strange, J. H. (1970). *Mol. Phys.* **18**, 553.
Doering, W. von E., and Roth, W. R. (1963). *Tetrahedron* **19**, 715.
Douglass, D. C., and McCall, D. W. (1958). *J. Phys. Chem.* **62**, 1102.
Douglass, D. C., and McCall, D. W. (1959). *J. Chem. Phys.* **31**, 569.
Douglass, D. C., McCall, D. W., and Anderson, E. W. (1961). *J. Chem. Phys.* **34**, 152.
Ertl, H. (1973). Ph.D. Thesis, University of Waterloo.
Farrar, T., and Becker, E. D. (1971). "Pulse and Fourier Transform NMR." Academic Press, New York.
Finer, E. G., Flook, A. G., and Hauser, H. (1972). *Biochim. Biophys. Acta* **260**, 59.
Fratiello, A., and Douglass, D. C. (1963). *J. Mol. Spectrosc.* **11**, 465.
Freeman, R., and Wittekoek, S. (1969). *J. Magn. Resonance* **1**, 238.
Garwin, R. L., and Reich, H. A. (1959). *Phys. Rev.* **115**, 1478.
Gavin, J. V., Waugh, J. S., and Stockmayer, W. H. (1963). *J. Chem. Phys.* **38**, 287.
Gerritsma, C. J., and Trappeniers, N. J. (1971). *Physica* (*Utrecht*) **51**, 365.
Gross, B., and Kosfeld, R. (1969). *Messtechnik* **77**, 171.
Gutowsky, H. S., and Chen, F. M. (1965). *J. Phys. Chem.* **69**, 3216.
Gutowsky, H. S., McCall, D. W., and Slichter, C. P. (1953). *J. Chem. Phys.* **21**, 279.
Gutowsky, H. S., Vold, R. L., and Wells, E. J. (1965). *J. Chem. Phys.* **43**, 4107.

Hackleman, W. R., and Hubbard, P. S. (1963). *J. Chem. Phys.* **39**, 2688.
Hahn, E. L. (1950). *Phys. Rev.* **80**, 580.
Hahn, E. L., and Maxwell, D. E. (1952). *Phys. Rev.* **88**, 1070.
Hansen, J. R., and Lawson, K. D. (1970). *Nature (London)* **225**, 542.
Hausser R., Maier, G., and Noack, F. (1966). *Z. Naturforsch. A* **21**, 1410.
Inglefield, P. T., Krakower, E., Reeves, L. W., and Stewart, R. (1968). *Mol. Phys.* **15**, 65.
Johnson, C. S. (1965). *Advan. Mag. Resonance* **1**, 33.
Jonas, J., Allerhand, A., and Gutowsky, H. S. (1965). *J. Chem. Phys.* **42**, 3396.
Kaplan, J. (1958a). *J. Chem. Phys.* **28**, 278.
Kaplan, J. (1958b). *J. Chem. Phys.* **29**, 462.
Look, D. C., and Lowe, I. J. (1966). *J. Chem. Phys.* **44**, 2995.
Luszczynski, K., Norberg, R. E., and Opfer, J. E. (1962). *Phys. Rev.* **128**, 186.
Luz, Z., Meiboom, S. (1963). *J. Chem. Phys.* **39**, 366.
McCall, D. W., and Douglass, D. C. (1965). *J. Phys. Chem.* **69**, 2001.
McCall, D. W., and Douglass, D. C. (1967). *J. Phys. Chem.* **71**, 987.
McCall, D. W., Douglass, D. C., and Anderson, E. W. (1959a). *J. Chem. Phys.* **31**, 1555.
McCall, D. W., Douglass, D. C., and Anderson, E. W. (1959b). *Phys. Fluids* **2**, 87.
McCall, D. W., Anderson, E. W., and Huggins, C. M. (1961). *J. Chem. Phys.* **34**, 804.
McCall, D. W., Douglass, D. C., and Anderson, E. W. (1963a). *Ber. Bunsenges. Phys. Chem.* **67**, 336.
McCall, D. W., Douglass, D. C., and Anderson, E. W. (1963b). *J. Polym. Sci., Part A* **1**, 1709.
McConnell, H. M. (1958). *J. Chem. Phys.* **28**, 430.
Meiboom, S. (1961). *J. Chem. Phys.* **34**, 375.
Meiboom, S., and Gill, D. (1958). *Rev. Sci. Instrum.* **29**, 688.
Moore, W. J. (1960). "Physical Chemistry," 2nd ed., pp. 448–449. Prentice-Hall, Englewood Cliffs, New Jersey.
Muller, B., and Bloom, M. (1960). *Can. J. Phys.* **38**, 1318.
Murday, J. S. (1973). *J. Magn. Resonance* **10**, 111.
Murday, J. S., and Cotts, R. M. (1968). *J. Chem. Phys.* **48**, 4938.
Murday, J. S., and Cotts, R. M. (1971). *Z. Naturforsch. A* **26**, 85.
Murphy, J. A., and Doane, J. W. (1971). *Mol. Cryst. Liquid Cryst.* **13**, 93.
Newmark, R. A., and Sederholm, C. H. (1965). *J. Chem. Phys.* **43**, 602.
Oosting, P. H., and Trappeniers, N. J. (1971). *Physica (Utrecht)* **51**, 418.
O'Reilly, D. E., Peterson, E. M., and Scheie, C. E. (1973). *J. Chem. Phys.* **58**, 4072.
Packer, K. J. (1973). *J. Magn. Resonance* **9**, 438.
Pfeiffer, H., and Weiss, K. H. p. 40. Akademie-Verlag, Berlin, 1961.
Powles, J. G., and Strange, J. H. (1962). *Discuss. Faraday Soc.* **34**, 30.
Redfield, A. G. (1957). *IBM J. Res. Develop.* **1**, 19.
Reeves, L. W., (1965). *Advan. Phys. Org. Chem.* **3**, 187.
Reeves, L. W., and Shaw, K. N. (1970). *Can. J. Chem.* **48**, 3641.
Reeves, L. W., and Shaw, K. N. (1971). *Can. J. Chem.* **49**, 3671.
Reeves, L. W., and Wells, E. J. (1962). *Discuss. Faraday Soc.* **34**, 177.
Reeves, L. W., Shaddick, R. C., and Shaw, K. N. (1971). *Can. J. Chem.* **49**, 3683.
Reich, H. A. (1963). *Phys. Rev.* **129**, 630.
Robertson, B. (1966). *Phys. Rev.* **151**, 273.
Saunders, M. (1963). *Tetrahedron Lett.* p. 1699.
Solomon, I. (1954). *Phys. Rev. Lett.* **2**, 301.
Stejskal, E. O. (1965). *J. Chem. Phys.* **43**, 3597.

Stejskal, E. O., and Tanner, J. E. (1965). *J. Chem. Phys.* **42**, 288.
Tanner, J. E. (1965). *Rev. Sci. Instrum.* **36**, 1086.
Tanner, J. E. (1970). *J. Chem. Phys.* **52**, 2523.
Tanner, J. E. (1972). *J. Chem. Phys.* **56**, 3850.
Tanner, J. E., and Stejskal, E. O. (1968). *J. Chem. Phys.* **49**, 1768.
Tiddy, G. J. T. (1971). *Nature (London), Phys. Sci.* **230**, 136.
Tiers, G. V. D. (1960). *Proc. Chem. Soc., London* p. 389.
Torrey, H. C. (1956). *Phys. Rev.* **104**, 563.
Trappeniers, N. J., Gerritsma, C. J., and Oosting, P. H. (1965). *Phys. Lett.* **18**, 256.
Valiev, K. A., and Emel'yanov, M. I. (1964). *Zh. Struct. Khim.* **5**, 7.
Valiev, K. A., Emel'yanov, M. I., and Samigullin, F. M. (1964). *Zh. Struct. Khim.* **5**, 371.
Vold, R. L., and Gutowsky, H. S. (1967). *J. Chem. Phys.* **47**, 2495.
Wayne, R. C., and Cotts, R. M. (1966). *Phys. Rev.* **151**, 264.
Weiss, A., and Nothnagel, K. H. (1971). *Ber. Bunsenges. Phys. Chem.* **75**, 216.
Wells, E. J., and Abramson, K. H. (1969). *J. Magn. Resonance* **1**, 378.
Wells, E. J., and Gutowsky, H. S. (1965). *J. Chem. Phys.* **43**, 3414.
Woessner, D. E. (1960). *Rev. Sci. Instrum.* **31**, 1146.
Woessner, D. E. (1961a). *J. Chem. Phys.* **34**, 2057.
Woessner, D. E. (1961b). *J. Chem. Phys.* **35**, 41.
Woessner, D. E. (1963). *J. Phys. Chem.* **67**, 1365.
Woessner, D. E., Snowden, B. S., George, R. A., and Melrose, J. C. (1969). *Ind. Eng. Chem., Fundam.* **8**, 779.
Zimmerman, J. R., and Brittin, W. E. (1957). *J. Phys. Chem.* **61**, 1328.

5

Determination of Spin–Spin Relaxation Times in High-Resolution NMR

RAY FREEMAN and H. D. W. HILL

I. Introduction... 131
II. Definitions.. 132
    A. The Rotating Reference Frame 132
    B. Transverse Relaxation 132
    C. Spin Echoes.. 133
    D. Selective and Nonselective Pulses......................... 134
    E. Echo Modulation 135
III. Experimental Methods....................................... 138
    A. Precise Refocussing of Echoes 138
    B. Absolute-Value Display................................. 142
    C. Selective Pulses 144
    D. High Pulse Repetition Rates............................. 145
    E. J-Spectra.. 149
    F. Heteronuclear Decoupling............................... 153
IV. Concluding Remarks 159
    References... 161

I. Introduction

Nuclear magnetic resonance investigations of chemical exchange have usually been restricted to two techniques, the measurement of spin–spin relaxation times in systems so simple that essentially only one resonance response is detected, or line width analysis of more complicated high-resolution spectra. These limitations may be lifted in recently developed techniques for determining transverse relaxation times in spectra with many component lines (Vold *et al.*, 1968), permitting much wider application of pulse methods to chemical exchange problems and extending the accessible time scale at the very slow and very fast exchange limits. The aim of this chapter is to review these new techniques with the hope that they may soon be applied to chemical exchange problems.

II. Definitions

A. The Rotating Reference Frame

Pulsed nmr experiments are most readily visualized in terms of the macroscopic nuclear magnetization in the rotating frame of reference. This is the resultant of all the individual nuclear spin magnetizations integrated over the effective volume of the sample, and has been shown by Bloch (1946) to obey classic mechanics. For nuclear spins at thermal equilibrium with their surroundings and not subject to radio-frequency fields, this magnetization may be represented by a vector of magnitude M_0 directed along the z axis of the reference frame. The x and y axes of this coordinate system are considered to be rotating about the z axis at an angular rate equal to the radio frequency of the spectrometer and in the direction that corresponds to the sense of the nuclear precession. Thus, any applied radio-frequency field is represented by a static vector H_1 (normally directed along the x axis) in this rotating frame, while any transverse components of nuclear magnetization that are static in this frame will induce signals at the radio frequency in a receiver coil fixed in the laboratory frame of reference. The y component represents absorption, the x component the dispersion mode.

B. Transverse Relaxation

In this picture a radio-frequency pulse of width τ seconds causes the nuclear magnetization to precess about the x axis through an angle α where

$$\alpha = \gamma H_1 \tau \qquad (1)$$

Clearly for nuclei initially at equilibrium a flip angle $\alpha = 90°$ produces the maximum transverse magnetization. In a perfectly homogeneous magnet, this transverse component would decay solely through spin–spin interactions with the environment, decreasing exponentially with a time constant defined as the spin–spin relaxation time T_2. As a corollary of this, the line shape measured in a continuous-wave slow-passage experiment in the absence of saturation is Lorentzian in form with a width at half-height $(\pi T_2)^{-1}$ Hz. In practical magnets, this "natural" line width is very often obscured by the more severe broadening by the inhomogeneity of the field over the effective volume of the sample even when sample spinning is employed. Similarly, the decay of transverse magnetization has a much shorter time constant than T_2, since magnetization "isochromats" from different sample regions mutually interfere and their initial phase coherence is destroyed. This "instrumental" time constant is usually designated T_2^*.

Spin–spin relaxation times can only be determined from slow passage line widths if it is known for certain that T_2 is markedly shorter than T_2^*; for marginal cases it may be necessary to correct the observed line width for small contributions from field inhomogeneity. However, for a majority of high-resolution spectra of small molecules recorded in present-day magnets, the natural line widths are completely obscured by instrumental effects, which sometimes include significant contributions from field/frequency instability as well as field nonuniformity.

C. Spin Echoes

Hahn (1950) was the first to realize that the decay of transverse magnetizaion due to field inhomogeneity is a reversible process and devised an experiment whereby the dephasing of spin isochromats from different sample regions could all be brought back into phase again or "refocussed." Hahn accomplished this by applying a second 90° pulse τ seconds after an initial 90° pulse had excited the transverse magnetization. Optimum refocussing occurred an equal time τ after the second pulse, and the reappearance of a transverse nmr signal at this time was called a spin echo. The process of echo formation is somewhat easier to visualize if the second pulse represents a 180° rotation, as suggested by Carr and Purcell (1954). The net magnetization is represented as the vector summation of many smaller isochromatic vectors in the xy plane of the rotating frame. These are all aligned along the y direction immediately after the initial 90° pulse, but fan out in the xy plane, since they all represent sample regions of slightly different static field intensity. The 180° pulse about the x axis at time τ translates all these vectors into mirror-image orientations with respect to the xz plane, and in this looking-glass world, the fast isochromats are strung out behind the slow isochromats in exactly the right disposition to bring them all into alignment with the $-y$ direction at time 2τ. This configuration thus resembles the start of the experiment, and further refocussing pulses at times 3τ, 5τ, etc. will generate further echoes at times 4τ, 6τ, etc., a Carr–Purcell train of spin echoes. The amplitude of these echoes is now independent of field inhomogeneity, provided that a given nuclear spin does not diffuse into an appreciably different static field in the interval 2τ, and the decay of the echo amplitude can then be attributed to spin–spin relaxation effects. It should follow an exponential curve of time constant T_2.

The Carr–Purcell sequence is now the accepted method for measuring transverse relaxation times. However, there are other experiments that generate spin echoes. A regular sequence of equal radio-frequency pulses repeated at a rate faster than T_2 but slower than T_2^* will entrain a steady-state regime where a negative half-echo appears just before each radio-frequency pulse, with a positive half-echo just after each pulse (Bradford et al., 1951; Bloom, 1955; Freeman and Hill, 1971a; Waldstein and Wallace, 1971). The interruption of the pulse sequence then produces a decaying sequence of spin echoes and under the right experimental conditions T_2 may be extracted from the decay rate (Kaiser et al., 1974). A quite different refocussing effect has been described by Solomon (1959a) and called "rotary echoes." A strong continuous radio-frequency field is applied to the nuclear spin system causing a continued precession of the magnetization vector about H_1 through many cycles in the yz plane of the rotating frame. The nonuniformity of the static field H_0 can be neglected while H_1 is on, but inhomogeneities of the radio-frequency field H_1 cause a loss of phase coherence in the yz plane. Sudden reversal of the sense of H_1 allows faster isochromats to catch up with slower components and a rotary spin echo is thus generated. In the absence of spin diffusion, the decay along a repeated train of rotary echoes is determined by both spin–spin and spin–lattice relaxation times according to the equation

$$\frac{1}{T} = \frac{1}{2}\left(\frac{1}{T_1} + \frac{1}{T_2}\right) \tag{2}$$

D. Selective and Nonselective Pulses

When spin-echo methods are applied to high-resolution nmr spectra with many component lines, some thought must be given to the method of separating several different transverse relaxation rates. For well-resolved first-order spectra, one simple approach is to excite only one single line (or group of closely-spaced lines) at a time, while avoiding any significant perturbation of the rest of the spectrum. This can be achieved (Freeman and Wittekoek, 1969a) with very weak radio-frequency pulses such that

$$|\nu_A - \nu_X| \gg \gamma H_1/2\pi \gg \Delta\nu \tag{3}$$

where $\nu_A - \nu_X$ is the separation of the chosen resonance from its nearest neighbor and $\Delta\nu$ is the width of the chosen resonance (or group of closely spaced lines). In order to satisfy Eq. (1) for 180° pulses, this entails widths of the order of 0.01–1.0 seconds for radio-frequency fields of the order of 0.1–10 mG (for protons), so the method is only really applicable to long spin–spin relaxation times where the relaxation during the pulse can be

5. Spin–Spin Relaxation

safely neglected. Moreover, unless the excitation of off-resonance spins is carefully avoided, problems can arise if the near-neighbor lines are part of the same coupled spin system. Nevertheless, the selective pulse technique has one marked advantage for measuring spin–spin relaxation times in coupled spin systems, as will be seen below.

The idea of employing nonselective radio-frequency pulses for spin–spin relaxation measurements stems from the realization that the slow-passage high-resolution spectrum can be derived by Fourier transformation of a spin echo, essentially a "free induction" signal (Vold et al., 1968). Since the amplitude of each component that makes up an echo will decay with its own characteristic spin–spin relaxation time, a set of spectra derived from progressively later echoes in the Carr–Purcell train will exhibit intensities decaying with the transverse relaxation times of the individual lines. In principle, the entire set of spectra could be derived from the successive echoes of a single Carr–Purcell train. In practice, this may severely restrict the attainable resolution by limiting the time for which each echo may be sampled (set by the interval between refocussing pulses). In order to accommodate high-pulse repetition rates, the pulse sequence is allowed to proceed without sampling and then interrupted to allow the last half-echo to decay completely. This is sampled and Fourier transformed. The time development of the spectrum is then followed by allowing the spins to return to thermal equilibrium and then repeating the Carr–Purcell experiment with a sequence of different length.

A common procedure is to record a set of several such spectra at regularly increasing values of the elapsed time t after the initiating 90° pulse. If these spectra are suitably stacked together, the impression of a three-dimensional diagram can be created showing intensity as a function of both frequency and time. Each component line may then be seen to decay exponentially with its own characteristic time constant T_2.

E. Echo Modulation

In fact, the result described above is only achieved if the high-resolution spectrum exhibits only chemical shifts and spin–spin coupling to non-resonant nuclei (e.g., heteronuclear coupling). Coupling to a resonant nucleus (homonuclear coupling) gives rise to a modulation of the echoes (Hahn and Maxwell, 1952; Crawford and Foster, 1956; Emshwiller et al., 1960; Abragam, 1961; Powles and Hartland, 1961; Powles and Strange, 1964). Consider the case of two coupled protons (AX), giving rise to a high resolution spectrum of two doublets. After a 90° pulse, the response from the A nuclei (Fig. 1b) will precess away from the y direction of the rotating frame (due to its chemical shift) and split into two components, one fast and one

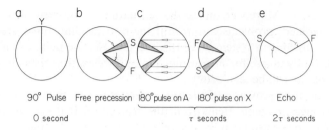

Fig. 1 Vector diagram illustrating the phase modulation of spin echoes from an AX spin system when both A and X are influenced by the 180° pulses. The magnetization vectors are visualized in projection on the xy plane of the rotating frame, and the 180° pulse is applied along the y axis.

slow (due to spin–spin splitting), while each component will become diffuse (field inhomogeneity). Application of a 180° pulse then has two distinct effects—the flipping of all isochromats into a mirror-image configuration (Fig. 1c) and the interchanging of the spin-states of the X nucleus, which interchanges the slow and fast components of the A doublet (Fig. 1d). As a result of this latter reversal, the slow and fast components continue to diverge after the 180° pulse, and although chemical shift and field inhomogeneity effects are refocussed at time 2τ, two echo components occur, each dephased from the "expected" direction, along the y axis of the rotating frame (Fig. 1e).

The phase deviation becomes progressively larger at each later echo of a Carr–Purcell train. For the simple first-order AX system, the phase angle is given by

$$\varphi = \pm 2\pi J n \tau \quad \text{radians} \tag{4}$$

where n is the number of the echo and $2n\tau$ the elapsed time from the initial 90° pulse. A spin multiplet with an odd number of components (for example, the A triplet of an AX_2 spectrum) shows an unmodulated central component flanked by outer lines with phase-modulation at twice the angular frequency of Eq. (4). Figure 2 illustrates the echo modulation exhibited by the AX_2 nmr spectra of 1,1,2-trichloroethane obtained by Fourier transformation of the last half-echo of Carr–Purcell trains of different lengths. Note that while the doublet components rotate their phase angle through one complete revolution, the outer triplet components achieve two complete revolutions.

In general, a spin multiplet arising from coupling to a group of equivalent X nuclei will exhibit phase-modulation on its component lines at angular frequencies determined by m_x, the total z spin component of the X group,

$$\varphi' = \pm 4\pi m_x J n \tau \quad \text{radians} \tag{5}$$

5. Spin–Spin Relaxation

This reflects the symmetry of the phase-modulation and illustrates that when X contains an equal number of spin-$\frac{1}{2}$ nuclei, there is a central unmodulated A component for which $m_X = 0$. In general, there will be more than two groups spin-coupled together, and the J-modulation is then correspondingly more complicated. The foregoing simple treatment only applies for the case of first-order coupling (Allerhand, 1966; Tokuhiro and Fraenkel, 1968), exact 180° refocussing pulses, and low-pulse repetition rates. However, even when these simplifying criteria are fulfilled, the general case of a molecule with several groups of coupled spins exhibits very complex echo modulation, making the determination of the individual spin–spin relaxation times quite

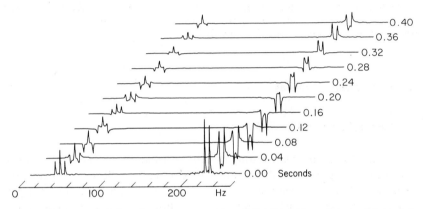

Fig. 2 Phase modulation in proton spectra of 1,1,2-trichloroethane obtained by Fourier transformation of the last half-echo of a Carr–Purcell train, with $2\tau = 20$ milliseconds. The time scale represents the duration of the Carr–Purcell sequence.

difficult. In contrast to almost all other aspects of high-resolution nmr, where weak unresolved spin–spin couplings can be safely neglected, echo modulation brings to light any small couplings that exceed the natural line widths. In the cases where these couplings are obscured by the instrumental line width, the echo modulation appears as an amplitude modulation, since the x components of magnetization cancel for reasons of symmetry. This is illustrated for the AX spectrum of 2,4,5-trichloronitrobenzene (Fig. 3), where the small para coupling (0.4 Hz) is not resolved in the spectrum, but a cosine-wave amplitude modulation at 0.2 Hz is quite evident when the spectra are followed as a function of the length of the echo train. J-Modulation is thus the major barrier to the measurement of spin–spin relaxation times of individual resonances of nmr spectra with several coupled groups. In the following sections, four methods of attacking this problem are considered.

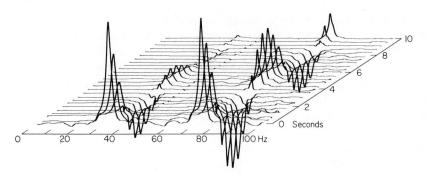

Fig. 3 Amplitude modulation of spectra obtained by Fourier transformation of the last half-echo of a Carr–Purcell train when there is a small unresolved homonuclear spin–spin coupling constant. The spectra are from the protons in 2,4,5-trichloronitrobenzene which has a para coupling of 0.4 Hz.

III. Experimental Methods

A. Precise Refocussing of Echoes

In the foregoing description of echo formation in the Carr–Purcell sequence, it has been tacitly assumed that a spin isochromat remains unchanged in frequency between refocussing pulses. In practice, self-diffusion within the liquid and field/frequency instabilities in the spectrometer both induce slight changes in this precession frequency, and thus a phase error is accumulated in the period 2τ between refocussing pulses. Fortunately, high-resolution spectrometers are designed to minimize field inhomogeneity and field/frequency instability, so the combined effects of diffusion and instability are usually slight, provided that the interval 2τ is kept reasonably short, in practice of the order of 100 milliseconds or less. For long spin–spin relaxation times, this may mean hundreds or thousands of refocussing pulses. Pulse imperfections that lead to cumulative errors must therefore be avoided at all costs. The modification suggested by Meiboom and Gill (1958) has the effect of compensating for a missetting of the pulse flip angle and for spatial inhomogeneity in the radio-frequency field. Figure 4 illustrates how, by applying the refocussing pulses along the y axis of the rotating frame, the Meiboom modification compensates pulse flip angle errors to first order on the even numbered echoes, while without this modification the errors are cumulative. A similar compensation is possible by alternating the sense of the refocussing pulses (along the $\pm x$ directions of the rotating frame) (Freeman and Wittekoek, 1969a).

Radio-frequency field homogeneity is expected to be better in a crossed-coil probe than in a single coil probe, since the effective volume of the receiver coil is normally smaller than the enclosed volume of the transmitter coil. Vold *et al.* (1973) have emphasized the importance of maintaining good orthogonality in the crossed-coil system to prevent degradation of the radio-frequency homogeneity at the sample. An electrostatic (Faraday) shield helps in this respect. Radio-frequency field homogeneity may be checked by comparing the nuclear spin responses to 90° and 270° pulses, or, in a more sensitive way, by following the response to a train of closely-spaced 90° pulses (Vaughan *et al.*, 1972). The adjustment of the flip angle of the refocussing pulses is best carried out in an experiment where a missetting has a cumulative effect, for example the Carr–Purcell experiment without the Meiboom modification. It may then be preferable to reintroduce the Meiboom modification as a 90° phase shift on the initial 90° pulse rather than on the repeated 180° pulses, in case the phase shifting network should perturb the radio-frequency amplitude.

The symmetry introduced into the refocussing process by the Meiboom modification also compensates to a certain extent for "off-resonance" effects which may arise because of the finite intensity of the radio-frequency field H_1. The usual criterion is that the field intensity H_1 should be at least ten times the displacement of the most distant nuclear precession frequency from resonance ($H_1 \geqslant 10 \, \Delta H$). Even then the effective field in the rotating frame

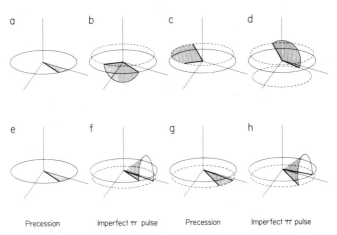

Fig. 4 The compensating action of the Meiboom modification in the Carr–Purcell sequence. The vector diagrams (a)–(d) show how a refocussing pulse that is too long has a cumulative effect in the unmodified Carr–Purcell sequence, driving the magnetization further and further from the *xy* plane. With the Meiboom phase shift (e)–(h), these errors are corrected on even-numbered echoes.

is slightly stronger than H_1 and is tilted slightly with respect to the xy plane. A slight modification of the vector diagram in Fig. 4 indicates that small non-idealities of this kind will be compensated to first order on the even-numbered echoes.

There are other kinds of pulse imperfection that cannot be corrected in this way. The effects of instabilities in the pulse flip angle or in the timing of the refocussing pulses have been analyzed by Weiss (1967); fortunately crystal-controlled digital timing methods are normally quite capable of ensuring adequate stability. In principle, unless the pulse envelope is made synchronous in phase with respect to the radio-frequency carrier, small fluctuations of the amplitude of the pulses can occur, depending on the radio-frequency phase at the leading edge of the pulse. However Vaughan et al. (1972) have shown that such effects are very small even for the very narrow pulses used for solid-state line-narrowing experiments, so no such phase synchronization appears to be necessary for spin-echo experiments on liquid samples. Finite rise and fall times of radio-frequency pulses can be shown to change the effective direction of the radio-frequency field in the rotating frame (Ellet et al., 1971; Mehring and Waugh, 1972), although this problem is only likely to be serious for very narrow radio-frequency pulses (with widths of the order of 10^{-6} seconds).

A general analysis of the impact of some of these pulse imperfections on the measurement of spin–spin relaxation times (in systems where homonuclear coupling is absent) has been presented by Vold et al. (1973). This is based on a numerical solution of the Bloch equations and indicates that in general the echo amplitude will decay with a time constant T_a not necessarily equal to T_2, will exhibit a damped oscillatory term that alternates in sign from echo to echo, and will eventually reach a finite steady-state level, which may consist of positive or negative echoes. The authors emphasize the importance of correcting the measurements for this finite steady-state echo response. Pulse imperfections, by carrying the precessing magnetization out of the xy plane, involve spin–lattice relaxation effects, and if T_1 is longer than T_2, the apparent spin–spin relaxation time T_a may be longer than T_2. If $T_1 = T_2$ then $T_a = T_2$.

Most of the classic spin-echo experiments on single resonance lines had no reason to consider spinning the sample, whereas for high-resolution nmr spectra, spinning is routine for reasons of resolution. Some thought must therefore be given to this new factor when spin–spin relaxation times are to be measured in high-resolution spectra with many component lines, in order to improve resolution and the effectiveness of any internal field/frequency regulation scheme. Since it does not seem practical to start the spinner *suddenly* at the end of an echo train in order to acquire a free induction signal with better resolution, the interaction of spinning on the refocussing process

5. Spin–Spin Relaxation

must be examined. At high pulse repetition rates the spinner motion moves any given nucleus through only a small arc in the time interval τ. The effect is thus similar to that of a slow linear drift of the magnetic field, and this can be shown to be compensated on even-numbered echoes. More serious effects are to be expected when the pulse repetition rate approaches the spinner rate (or a small multiple or submultiple of it). A recent analysis of the effects of sinusoidal field modulation on spin echoes (Allerhand, 1970) could clearly be adapted to this case. Spinning may also affect pulse flip angles and thus influence echo formation. During the pulse, H_1 is so intense that any static field inhomogeneity effects can be safely neglected; however, the H_1 intensity is modulated by spinning in an inhomogeneous transmitter field. This may well be the most serious spinner problem. Some of the experiments described below do not involve Fourier transformation of a free induction signal, but rather follow the time development of echo amplitudes; these experiments would thus be carried out with a stationary sample.

The compensatory effect of the Meiboom modification (or of the phase alternation method) depends on symmetry properties that are destroyed when there is J-modulation of the echoes. Figure 5 illustrates this effect by comparing the motion of two vectors split by heteronuclear coupling (upper diagrams) and homonuclear coupling (lower diagrams) for the case that the Meiboom modification is employed. Two vectors, **F** (fast) and **S** (slow), subjected to a refocussing pulse significantly less than 180° (Fig. 5a) precess to approximately mirror-image positions between the pulses (Fig. 5b) where an

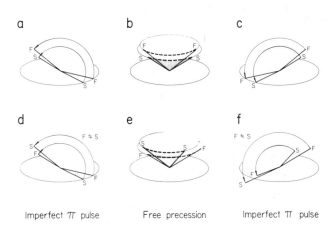

Fig. 5 Breakdown of the compensatory action of the Meiboom modification when the echoes are phase modulated by a homonuclear spin–spin coupling constant. A short refocussing pulse is compensated (c) if the coupling is heteronuclear, but for homonuclear coupling the interchange of fast and slow components destroys the symmetry (e) and inhibits the compensation (f).

equal refocussing pulse returns them to the xy plane with the fast component so disposed behind the slow component that the second echo shows negligible amplitude error (Fig. 5c). In contrast, when the dephasing of the two vectors results from homonuclear spin coupling, the refocussing pulse has the additional effect of interchanging the fast and slow components (Fig. 5d) and the precession paths lose their symmetry (Fig. 5e) so that the next refocussing pulse cannot return them to the xy plane (Fig. 5f). Spin echo measurements on systems with homonuclear coupling may therefore require much more careful attention to the pulse imperfections normally compensated by the Meiboom modification (Meiboom and Gill, 1958; Vold and Gutowsky, 1967).

B. Absolute-Value Display

Phase-modulation of spin echoes due to homonuclear spin–spin coupling is most simply avoided by displaying the transformed spectra in the absolute value or $(u^2 + v^2)^{1/2}$ mode rather than the conventional absorption or v-mode display. This mode is available in many Fourier transform programs, and is occasionally used if there are severe frequency-dependent phase distortions that the usual correcting procedures cannot handle. The disadvantages are well known (Bremser et al., 1971). First, the signals have very long tails on each side since the u-mode component extends much further out from resonance, as is readily seen from the steady-state solutions of the Bloch equations (for negligible saturation):

$$u = \frac{\gamma H_1 M_0}{\Delta\omega} \quad (6)$$

$$v = \frac{\gamma H_1 M_0}{\Delta\omega^2 T_2} \quad (7)$$

for offsets from resonance sufficiently large that $\Delta\omega T_2 \gg 1$. Furthermore, in contrast to the conventional absorption-mode display, the signal is not additive in the region of overlap of two lines, since the two u-mode signals are in opposite sense and thus interfere. This causes abnormally low minima between component lines of a spin multiplet and large "pedestals" on the outer edges of a spin multiplet where all components reinforce.

Overlap effects may be safely disregarded if the absorption-mode component at the center of one line,

$$v_0 = \gamma H_1 M_0 T_2 \quad (8)$$

5. Spin–Spin Relaxation

is much more intense than the tail of the dispersion-mode signal from any neighboring line (neglecting the much weaker absorption-mode tail).

$$v_0/u = \Delta\omega T_2 \qquad (9)$$

The imposition of instrumental broadening reduces the absorption-mode signal v_0 through the replacement of T_2 by T_2^*. The resolution criterion for negligible overlap is therefore that all lines be separated by several times the instrumental line width,

$$\Delta\omega \gg (T_2^*)^{-1} \qquad (10)$$

For spectra that conform to this resolution requirement, the absolute-value display mode may be used to follow spin–spin relaxation without the complications attendant upon phase-modulated echoes. [This idea has been suggested independently by McLaughlin et al. (1973).] Figure 6 compares the phase-modulated absorption-mode spectra obtained by Fourier transformation of echoes with the unmodulated absolute value spectra derived from the same experimental data.

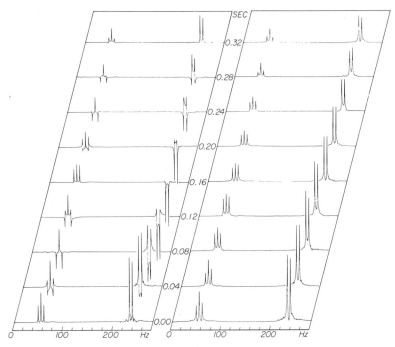

Fig. 6 A comparison of proton spectra of 1,1,2-trichloroethane from phase-modulated spin echoes in the absorption-mode display (left) and the absolute value display (right) as a function of the duration of the spin-echo train.

C. Selective Pulses

It is only the coupling to resonant spins that produces J-modulation of spin echoes. The problems of complex modulation in multicomponent spectra can therefore be circumvented if the radio-frequency pulses are made so weak as to affect only one spin multiplet, all coupled spins being so far away from resonance that they are not flipped by the 180° pulses. In this sense, they behave just like heteronuclear spins. The spin–spin relaxation of each chemically shifted group must then be examined in a separate experiment.

The selective pulse method presupposes first-order spectra and reasonably large chemical shifts in order to satisfy Eq. (3). Consider, for example, the proton nmr spectrum of 1,1,2-trichloroethane where $J_{AX} = 6$ Hz and $\delta_{AX} = 181$ Hz at 23 kG. In order to excite the doublet resonance as uniformly as possible, the radio-frequency field (expressed in frequency units) should be large compared with 3 Hz (since it will be centered between the two doublet components). An earlier treatment (Freeman and Wittekoek, 1969b) of the effects of field inhomogeneity broadening on spin echo experiments with weak pulses, indicated that the echoes could be seriously degraded as off-resonance components were flipped by an effective field that was tilted with respect to the xy plane of the rotating frame. The echoes lost their symmetrical shape, and since the off-resonance effects were cumulative, decayed more rapidly than the natural spin–spin relaxation rate. Similar effects were observed when the radio-frequency field was deliberately set off resonance. However the modification of Meiboom and Gill (1958) or the phase alternation method was found to compensate these errors quite effectively, as illustrated by the experimental observations of the apparent spin–spin relaxation time of chloroform subjected to weak off-resonance pulses (Fig. 7). When selective pulses are applied to spin multiplets with the intention of flipping all multiplet components, the Meiboom modification, or its equivalent, is therefore very important.

For the example of 1,1,2-trichloroethane, the second requirement, that the radio-frequency level (expressed in hertz) be weak with respect to the distance from the triplet resonance (181 Hz), results in a compromise level $\gamma H_1/2\pi = 25$ Hz. This entails a 180° pulse width of 20 milliseconds, and in turn limits the maximum permissible pulse repetition rate, since the duty cycle must be kept small enough that relaxation during the pulse can be reasonably neglected, because this involves spin–lattice interactions. In practice, the minimum pulse interval used was $2\tau = 200$ milliseconds. This leaves only a very narrow range of permissible pulse intervals, since a long interval increases the importance of self-diffusion effects.

Figure 8 illustrates the spin echo responses excited from the doublet resonance of 1,1,2-trichloroethane by a selective Carr–Purcell–Meiboom sequence.

5. Spin–Spin Relaxation

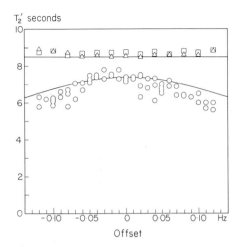

Fig. 7 The results of measuring the apparent spin-spin relaxation time T_2' of protons in chloroform with weak selective radio-frequency pulses ($\gamma H_1/2\pi = 1$ Hz) as a function of the distance from exact resonance. The Meiboom modification or the phase alternation technique are seen to compensate quite well for this kind of pulse imperfection. ○ = Carr–Purcell sequence; □ = Meiboom modification; △ = phase alternation. [Freeman and Wittekoek (1969b).]

An internal deuterium reference sample was employed for field/frequency regulation and the sample spinner was not used. A low-frequency beat appears on each echo representing the difference frequency between the doublet components (this is best seen in the inset of Fig. 8c). The echo decay is shown for three different pulse repetition rates, indicating a detectable acceleration at the lower repetition rates as self-diffusion becomes appreciable. A plot of the apparent spin–spin relaxation rate as a function of τ^2 illustrates how the true relaxation rate can be obtained by extrapolation and confirms that at $\tau = 100$ milliseconds the influence of diffusion is negligible in a magnetic field of this homogeneity (Fig. 9). Note that Fourier transformation is not involved in this experiment.

D. High Pulse Repetition Rates

Echo modulation arises when two groups of coupled nuclei experience the repeated train of 180° pulses. The Fourier spectrum of these radio-frequency pulses consists of the carrier frequency and a series of modulation sidebands separated by the pulse repetition rate and covering a wide frequency band. If the repetition rate is increased until it is much larger than the chemical shift

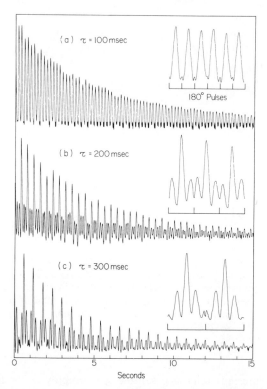

Fig. 8 Spin echoes observed by applying selective pulses to the CH_2 doublet of 1,1,2-trichloroethane with $\gamma H_1/2\pi = 25$ Hz. The modulation of individual echoes, visible in the lowest inset, is the beat frequency between the doublet components. The time scale on the insets is expanded four times.

between the two groups of nuclei, then only the carrier is close enough to resonance to have an appreciable influence, and all the sidebands are ineffective. The net effect on the nuclear spin system is then rather similar to that of a continuous radio-frequency field. Figure 10 illustrates this point schematically for the Carr–Purcell–Meiboom sequence. At slow repetition rates (top diagram), there is appreciable fanning out of individual nuclear spin vectors between refocussing pulses. At faster repetition rates (middle diagram), the loss of phase coherence between different components is greatly reduced, and in the limit where the repetition rate is fast compared with the range of nuclear precession frequencies, the fanning out is negligible, showing a close resemblance to the "spin-locking" experiment (Redfield, 1955; Solomon, 1959b) where a continuous radio-frequency field is applied along the y axis of the rotating frame. In this limit, the pulse modulation is not

5. Spin–Spin Relaxation

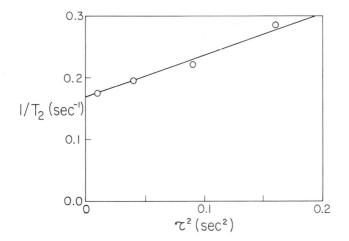

Fig. 9 A plot of the decay time constants observed in Fig. 8 as a function of the square of the pulse interval τ, indicating that self-diffusion effects can be neglected for τ values below 100 milliseconds.

effective in flipping spin states of the coupled neighbor nuclei, so that echo modulation disappears. This effect was demonstrated experimentally by Wells and Gutowsky (1965) for AB and AX type spectra. Theoretical treatments (Abragam, 1961; Powles and Hartland, 1961; Powles and Strange, 1962, 1964; Allerhand, 1966, 1969; Tokuhiro and Fraenkel, 1968, 1969; Gutowsky et al., 1965) predict the details of the disappearance of the echo modulation. For example, if the spectrum is first-order, or if the pulse repetition rate is large compared with the chemical shift δ Hz, then the amplitude of the nth echo can be shown to be

$$S = \left\{\cos(2\pi Jn\tau)\left[1 - \left(\frac{\sin 2\pi\delta\tau}{2\pi\delta t}\right)\right]\right\} \exp\left(-\frac{2n\tau}{T_2}\right) \qquad (11)$$

Here the phase modulation of the echo is reflected as an amplitude modulation since the components are not resolved. Experimental results (Wells and Gutowsky, 1965; Vold and Shoup, 1972) confirm this $(\sin x)/x$ dependence of the modulation frequency on pulse interval.

Figure 10 also suggests that the spin-locking experiment could be used to measure transverse relaxation rates in coupled spin systems without the complication of modulation due to scalar coupling. The spectra would be studied as a function of the duration of the spin-locking field H_1. When this field is extinguished, a free induction signal is observed that may be Fourier-transformed to give the slow-passage spectrum. The decay rate of individual

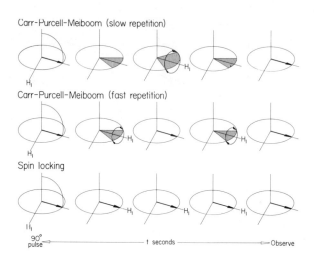

Fig. 10. Schematic representation of the relationship between the Carr–Purcell–Meiboom refocussing sequence and the spin-locking or forced transitory precession experiment. In the limit of repetition rates fast compared with the relevant chemical shifts, the maximum divergence of magnetization vectors in the spin-echo experiment is extremely small and approximates the motion in a spin-locking experiment.

lines is then the "spin–lattice relaxation time in the rotating frame" $T_{1\rho}$, usually a good approximation to T_2 for liquids, unless the correlation spectrum has anomalies at low frequencies. This technique has been used to measure transverse relaxation times of the carbon-13 nuclei in *o*-dichlorobenzene (Freeman and Hill, 1971d). One practical drawback is the problem of generating a sufficiently intense radio-frequency field applied for durations of several seconds without unduly heating the sample.

A more practical general solution to the problem is the Carr–Purcell sequence with high pulse repetition rate. Figure 11 illustrates spectra of a degassed sample of 1,1,2-trichloroethane observed as a function of the length of the Carr–Purcell sequence, with pulse interval $\tau = 1$ millisecond. Although this is not fast enough to satisfy the requirement $2\pi\delta\tau \ll 1$, no *J*-modulation is evident on the decays, which appear to fit exponential curves. It is interesting to note that one component of the central lines of the CH triplet decays much more slowly than the outer components (which decay at the same rate as the CH_2 doublet). The early work of Anderson (1956) on this compound reported a slight line width difference between the two components of the central line of the triplet, which were just resolved at the lower field he used (7 kG), and suggested that these line widths were affected by the lifetimes of the states of the other group (Bloch, 1956) and that the singlet (antisymmetri-

5. Spin–Spin Relaxation

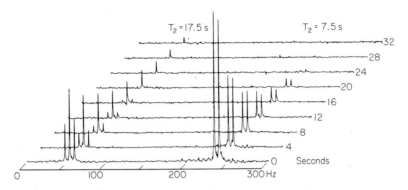

Fig. 11 Spin–spin relaxation traces obtained by Fourier transformation of the last half-echo of a Carr–Purcell sequence on the protons of a degassed sample of 1,1,2-trichloroethane. The pulse interval $2\tau = 2$ milliseconds, which is apparently short enough to suppress any J-modulation. Note the slow relaxation of the central line of the triplet.

cal) state was longer-lived than the three triplet (symmetrical) states, since it was insensitive to symmetrical relaxation perturbations. A similar effect is apparent in the spectrum of ethanol (Arnold, 1956).

The disadvantage of deriving spin–spin relaxation rates from spin echo trains obtained at high pulse repetition rates is that some other interesting contributions to transverse relaxation can be "pulsed out". In particular chemical exchange effects, the principal concern of investigators of dynamic nmr spectroscopy, are well known to be dependent on pulse repetition rate (Luz and Meiboom, 1963), and experimental studies of exchange usually involve a study of decay rate as a function of pulse repetition rate. This would seem to be incompatible with the condition for disappearance of the echo modulation.

E. J-Spectra

The fundamental problem of measuring spin–spin relaxation times in coupled spin systems is that of separating several different frequency components, each associated with its individual decay-time constant. Some simplification of this problem can be achieved by filtration of the time-domain spin echo train, suppressing all components except those in a narrow frequency band corresponding to a chosen chemical shift value. The echo modulation components are thereby reduced to those associated with spin couplings to that group only, and in multispin systems this represents a considerable simplification. Filtration may be accomplished by conventional analog electronics; the technique has been used for example to filter free-induction signals from a 180°–90° pulse sequence in order to follow the spin–lattice relaxation of individual lines one at a time (Freeman and Hill, 1969). Stempfle and Hoffmann (1970) have extended the idea of filtration to spin-echo trains but have used a digital filtration technique based on double Fourier transformation. A conceptually simpler experiment would be to filter a single echo and then follow its intensity as a function of time after the initiation pulse.

Clearly, the general problem of separating out many modulation components is most simply handled by Fourier transformation of the complex modulation envelope of the echoes. This should be clearly distinguished from Fourier transformation of a single echo (see Fig. 12). One sample is taken at the peak of each echo, and sampling is continued as far down the echo train as feasible. The general result of this transformation is known as a spin-echo spectrum. It covers quite a narrow frequency range from zero to typically several hertz and consists of narrow resonance responses corresponding to the modulation components of the spin-echo train, usually arising from scalar spin–spin coupling (Freeman and Hill, 1971b) or anisotropically averaged dipole–dipole interactions in a nematic liquid crystal phase (Vold and Chan, 1970). Ideally, the line widths of these resonance responses are not determined by field inhomogeneity (provided that self-diffusion of the liquid can be neglected) but represent the spin–spin relaxation rates of the individual

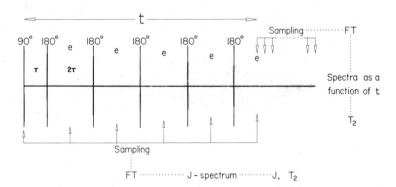

Fig. 12 There are two distinct techniques for deriving individual spin–spin relaxation times from a Carr–Purcell spin-echo train. The last half-echo may be sampled and Fourier transformed to give a high-resolution spectrum of several lines; the intensity of a chosen line is then followed as a function of t. Alternatively, the center of each echo may be sampled and the resulting complex modulation wave Fourier transformed to give a J-spectrum, where the individual spin–spin relaxation times are related to the corresponding line widths.

5. Spin–Spin Relaxation

responses. Because of the requirements of the sampling theorem, the echoes should be repeated at a rate at least twice that of the highest modulation component, since one-half the sampling rate defines a critical frequency beyond which any higher frequency responses appear folded back into the spin-echo spectrum.

It is convenient to define a particular class of spin-echo spectra, called "J-spectra," satisfying the following criteria:

(a) The spin echoes are recorded in a sufficiently intense magnetic field that the first-order approximation for spin–spin coupling is valid, $|\nu_A - \nu_X| \gg |J_{AX}|$ for all coupled resonant spins.

(b) The repetition rate for the Carr–Purcell refocussing pulses is sufficiently low in comparison with all chemical shift differences between coupled resonant spins that $2\pi|\nu_A - \nu_X|\tau \gg 1$. In this approximation, the frequencies of the responses are not appreciably affected by the pulse repetition rate.

(c) The refocussing pulses are accurately adjusted to the 180° condition; otherwise new responses appear in the transformed spectrum at frequencies involving chemical shifts.

(d) Echo modulation effects other than spin–spin coupling are absent (e.g., dipolar coupling in anisotropic liquid phases).

(e) The magnetic field is sufficiently homogeneous that self-diffusion effects can safely be neglected.

When these criteria are satisfied, the resonance responses appear at frequencies given by the absolute value of one-half the sums or differences of all the spin–spin coupling constants to a particular group A,

$$\nu = \tfrac{1}{2} | \pm J_{AK} \pm J_{AL} \pm J_{AM} \pm \cdots | \qquad (12)$$

with a similar set of frequencies for each chemically shifted group. For N nonequivalent spins, with $N(N-1)/2$ finite spin–spin coupling constants, the corresponding J-spectrum contains $N \cdot 2^{N-2}$ responses. For the special case $N = 2$, the two responses are degenerate.

When groups of chemically and magnetically equivalent spins are involved, the J-spectrum reflects the spin multiplet patterns observed in the conventional high-resolution spectrum. Suppose such a group contained N equivalent spin-$\tfrac{1}{2}$ nuclei, the J-spectrum would include a set of $(N+1)$ responses at frequencies given by Eq. (5), symmetrically disposed about zero frequency and with the usual binomial distribution of intensities.

The J-spectrum may thus be thought of as a collapsed high-resolution spectrum (Powles and Hartland, 1961) where the chemical shift and field inhomogeneity effects have been eliminated. In principle, the J-spectrum is perfectly symmetrical about zero frequency, but in practice, many Fourier transform programs will display only the positive frequency portion of the spectrum, which anyway contains all the useful information.

The J-spectrum of 1,1,2-trichloroethane serves as an example. The spin–spin coupling easily satisfies the first-order criterion with $\delta/J \approx 30$. The slow repetition rate requirement is satisfied with a pulse interval $2\tau = 40$ milliseconds, which corresponds to a sampling rate of 25 Hz and a spectral width of 12.5 Hz. Figure 13 shows the J-spectrum. It contains lines at zero frequency and 5.94 Hz attributable to the CH proton and a line at 2.97 Hz assigned to the CH_2 protons. These responses are much narrower than the lines of the conventional high-resolution spectrum, since the field inhomogeneity effects have been to a large extent eliminated. However, there remains the possibility of residual contributions, since even very slight departures from the ideal conditions may have cumulative effects that are not properly compensated by the Meiboom modification. The width of the line at 2.97 Hz seemed particularly susceptible to instrumental effects for reasons that are not understood. Improvements in experimental technique are expected to permit the recording of J-spectra without appreciable instrumental broadening of the natural line widths.

Figure 14 gives an illustration of a partial J-spectrum where the resolution is almost three times higher. It represents the Fourier transform of the echo envelope of the protons of 3-bromothiophene-2-aldehyde, where a narrow-band filter has been used to suppress all contributions to the echoes from protons other than the aldehyde proton, so that only two spin–spin coupling constants are involved. The pulse repetition rate was 2.92 Hz and 150 spin

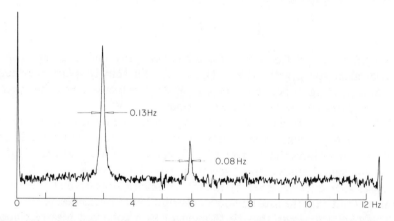

Fig. 13 J-spectrum obtained by Fourier transformation of the spin-echo envelope from the protons of 1,1,2-trichloroethane (vacuum degassed). The interval between echoes is $2\tau = 40$ milliseconds and echoes have been sampled for 30 seconds. Responses are observed at 0 Hz, 2.97 Hz, and 5.94 Hz.

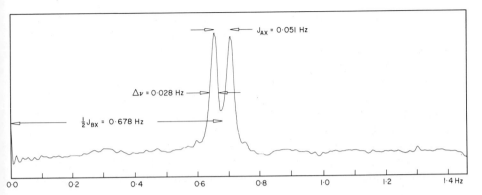

Fig. 14 *J*-spectrum obtained by Fourier transformation of the spin-echo envelope from a degassed sample of 3-bromothiophene-2-aldehyde. The spin-echo train has been passed through a narrow-band filter which retains only signals from the aldehyde proton. This experiment permits accurate determination of the long-range couplings into the thiophene ring. [Freeman and Hill (1971b).]

echoes were sampled covering a period of about 51 seconds, that is about 4.5 times the spin–spin relaxation time. The long-range coupling, $J_{AX} = 0.051$ Hz, is seen to be well resolved in this partial *J*-spectrum. The technique thus seems to be well suited to the study of exchange phenomena in the slow exchange limit.

F. Heteronuclear Decoupling

It has become the accepted practice to record carbon-13 spectra with simultaneous irradiation of protons with a broadband radio-frequency generator, the so-called noise decoupler (Ernst, 1966). This procedure results in a significant simplification of the carbon-13 spectrum by eliminating proton–carbon splittings, while at the same time improving sensitivity through the coalescence of multiplets and the nuclear Overhauser enhancement. Unfortunately, this incoherent irradiation adversely affects the formation of spin echoes in the carbon-13 spectrum. Ernst (1966) has given a theoretical treatment of noise decoupling and has shown that the observed spectra closely resemble those calculated for chemical exchange. Provided that the protons are weakly coupled and have only two spin states available, any random perturbation of the protons will induce the same behavior in the carbon-13 spectrum, whether it be incoherent irradiation or random chemical exchange. In particular, at intermediate levels of proton irradiation, where the power spectral density of the irradiation field is comparable with the proton–carbon spin-coupling constant, the carbon multiplets coalesce to a broad response. As the irradiation level is further increased, this line width decreases, and for conventional carbon-13 spectroscopy it is sufficient to reduce this contribution below the instrumental line width, often determined by field inhomogeneity.

For spin-echo work, this requirement must be much stricter so that incomplete decoupling does not obscure the natural line width. This artificial line broadening may be thought of as a superposition of many different residual splittings (Anderson and Freeman, 1962) randomly modulated in time, which inhibit the refocussing process essential to spin-echo formation. At sufficiently intense levels of proton irradiation, these residual splittings should be reduced below the point where echo formation is affected. Figure 15

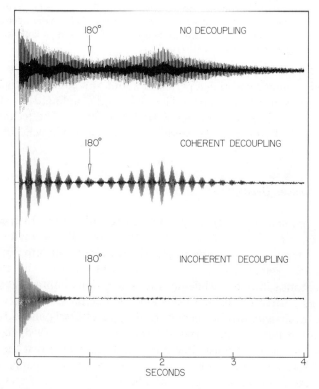

Fig. 15 The effect of incomplete decoupling on the formation of a single spin echo from carbon-13 in methyl iodide. Good echoes are observed at $t = 2$ seconds with no proton irradiation, or with off-resonance coherent proton irradiation. When the proton decoupler is made incoherent but other parameters kept unchanged, the echo is essentially eliminated, while the free induction signal is more rapidly damped. [Freeman and Hill (1971c).]

5. Spin–Spin Relaxation

illustrates the behavior of a single spin echo from carbon-13 in methyl iodide excited by a 90°–180° pulse sequence. Good echoes are observed either with no proton irradiation (top trace) or with coherent proton irradiation (center trace), but with incoherent irradiation (bottom trace), echo formation is almost completely inhibited, while the decay rate of the free induction signal is perceptibly increased. The decoupled traces represent averages of 20 measurements each. (Since the noise decoupler was known to have a small coherent component at the carrier frequency, proton irradiation was conducted several hundred hertz off resonance, which accounts for the low-frequency beat pattern on the center trace.) This represents a case where the irradiation field intensity is insufficient for complete decoupling. Clearly this could be a serious (but not insurmountable) problem for the determination of carbon-13 spin–spin relaxation times.

There appears to be a quite distinct mechanism inhibiting spin-echo formation under double resonance conditions. The origin is more subtle, involving resonance in the rotating reference frame (Hartmann and Hahn, 1960, 1962). It is suggested by a study of the decay of carbon-13 Carr–Purcell spin-echo trains as a function of the intensity of the proton irradiation field ($\gamma H_2/2\pi$) for coherent irradiation (Freeman and Hill, 1971c). The "expected" value for the spin–spin relaxation time of carbon-13 in methyl iodide was obtained only for certain settings of the proton irradiation intensity; other settings gave much faster and apparently nonexponential decays. Figure 16

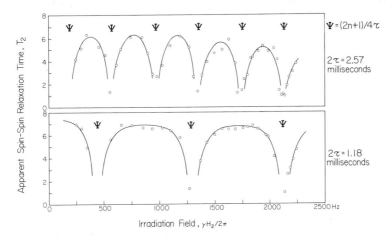

Fig. 16 Rotating frame resonance effects on the experimental decay rate of carbon-13 spin echoes from methyl iodide studied as a function of the intensity of the coherent proton irradiation field H_2. Note the dependence on the repetition rate of the refocussing pulses.

shows plots of the apparent relaxation time against $\gamma_H H_2/2\pi$, and it is evident that some kind of resonance effect is occurring and that it is a function of the interval 2τ between refocussing pulses of the Carr–Purcell sequence. The "resonances" appear at frequencies

$$\Psi = (2n + 1)/4\tau \tag{13}$$

where n is zero or a positive integer. It is believed that the response for $n = 0$ represents the fundamental frequency at which resonance in the rotating frame occurs, the responses for $n > 0$ corresponding to odd harmonics. As a consequence of the repeated 180° pulses, the carbon-13 nuclei execute a frequency-modulated precession about H_1 in the rotating frame at a fundamental frequency $1/(4\tau)$ Hz, the modulation giving the odd harmonics.

The rotating frame resonance is more easily seen in a spin-locking experiment on this same compound, since the continuous nature of the H_1 field in this experiment ensures that only the fundamental frequency $\gamma_C H_1/2\pi$ shows a resonance. The carbon-13 response is the Fourier transform of the free-induction signal after spin locking for 10 milliseconds along the strong radio-frequency field H_1. Studied as a function of the intensity of the proton irradiation field H_2, it shows a marked decrease (Fig. 17) near the rotating frame resonance condition,

$$\gamma_C H_1/2\pi = \gamma_H H_2/2\pi \tag{14}$$

When the precession frequency of carbon-13 spins about H_1 in the 25 MHz rotating frame coincides with the precession frequency of protons about H_2

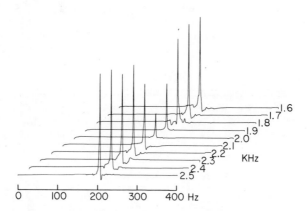

Fig. 17 Rotating frame resonance effects on the intensity of the carbon-13 signal of methyl iodide obtained by Fourier transformation of the free induction signal after spin-locking for 10 milliseconds along a strong radio-frequency field H_1. The spectra are shown as a function of the intensity of the coherent proton irradiation field, and rotating frame resonance occurs when $\gamma_H H_2/2\pi = \gamma_C H_1/2\pi \approx 2$ kHz.

5. Spin–Spin Relaxation

in the 100 MHz frame, energy can be transferred between the two spin reservoirs, the saturated protons "heating" the carbon-13 spins. Since in isotropic liquids intramolecular dipole–dipole interactions are rapidly modulated by molecular reorientation and time-averaged to zero, it is believed that the interaction responsible for energy transfer in this case is the scalar spin–spin coupling, J_{CH}.

Rotating frame resonance effects can be avoided by choosing H_1 and H_2 levels that do not satisfy Eq. (14), but it is suggested that with incoherent proton irradiation, the resonance condition becomes extremely broad so that energy exchange cannot be prevented. This appears to be the crux of the difficulty of T_2 measurements in the presence of a noise decoupler. The problem can only be circumvented. Since the undesirable interaction occurs during echo formation (or while spin locking), incoherent proton irradiation must be avoided during this period, but it can be applied for a suitably long period before initiation of the pulse sequence in order to establish a nuclear Overhauser enhancement and reapplied for the second half of the last echo (or for the free induction signal after spin locking), so that Fourier transformation will yield a spectrum that is free of carbon–proton splittings and thus simpler to interpret. Note that it is the z component of carbon-13 magnetization just before the 90° pulse that determines the nuclear Overhauser enhancement; any loss of proton saturation after this time is unimportant. In practical experiments, it has been found that a steady sample temperature is more easily maintained if coherent proton irradiation is used during the critical refocussing period, arranged to have the same average power, and a level that avoids the rotating frame resonance condition.

The utility of this mode-switching is illustrated in Fig. 18. These sets of

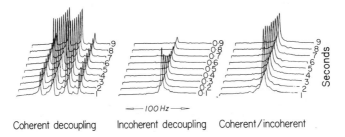

Coherent decoupling   Incoherent decoupling   Coherent/incoherent

Fig. 18 Carbon-13 signals of methyl iodide studied as a function of the duration of spin-locking along a strong radio-frequency field H_1. With coherent off-resonance proton decoupling, transverse relaxation is normal but a residual splitting remains. Incoherent decoupling greatly accelerates the decay of transverse magnetization (note the expanded time scale). Mode-switching of the proton decoupler restores normal relaxation and eliminates the residual splitting.

carbon-13 spectra of methyl iodide have been obtained by Fourier transformation of the free induction signal after variable periods of spin locking and show the decay due to transverse relaxation. With coherent (off-resonance) proton irradiation (left hand traces), the expected slow spin–spin relaxation is observed, whereas with incoherent irradiation the residual splitting is removed but the decay rate is considerably faster (note the shorter time scale in the center traces). When the decoupler mode is suitably switched from coherent to incoherent under computer control (right hand traces), normal spin–spin relaxation is restored and the spectra are free of residual splittings.

Figure 19 illustrates an application of such spin–spin relaxation measurements to carbon-13 decoupled from protons. The spectra have been obtained by Fourier transformation of the transient signal after a spin-locking experiment and show the three carbon-13 resonances of o-dichlorobenzene. For purposes to comparison, the spin–lattice relaxation spectra of this same sample are presented in the upper diagram. The low-field line, from carbon with chlorine directly attached, shows a very long spin–lattice relaxation time, since there is no close proton dipole. Its much shorter spin–spin relaxation time is ascribed to coupling to the fast-relaxing quadrupolar nuclei ^{35}Cl or ^{37}Cl.

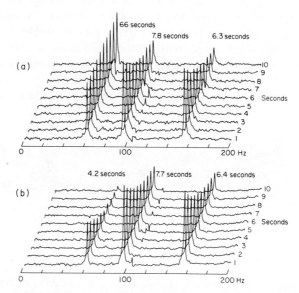

Fig. 19 Spin–lattice (a) and spin–spin relaxation (b) of the carbon-13 nuclei of o-dichlorobenzene. The low-field resonance, assigned to carbons with chlorine directly attached, has a long T_1 because there are no close protons, and a short T_2 because of scalar spin coupling to chlorine nuclei. The spectra in (b) were obtained by spin locking, using a mode-switched proton decoupler. [Freeman and Hill (1971d).]

IV. Concluding Remarks

Spin–spin relaxation measurements by the methods outlined in this chapter have not as yet been widely applied to high-resolution nmr problems, and no chemical exchange studies by Fourier transformation of spin echoes appear to have been published. Nevertheless, there are good reasons to believe that these techniques can extend conventional line-shape studies of exchange to significantly slower exchange rates or lift the usual restriction that spin-echo methods must be limited to spectra consisting of only a single line.

The principal complication of the spin-echo method would seem to be the echo modulation that appears whenever there is a homonuclear spin–spin coupling constant larger than the natural line width. This appears as a progressive phase modulation of the echoes, and one simple solution for well-resolved spectra is to compute the absolute-value mode of display, $(u^2 + v^2)^{1/2}$, instead of the usual v-mode or absorption signal. This alternative display mode is available in many of the standard Fourier transform computer routines.

Echo modulation arises because the spin states of both coupled nuclei are flipped by the 180° pulses; if the pulses can be made sufficiently selective in the frequency domain that only one spin multiplet is appreciably affected, the modulation disappears. This can be achieved through the use of extremely weak radio-frequency pulses (in the range 0.1–10 mG) of relatively long duration. In order to satisfy the selectivity criteria, this presupposes first-order spectra. If care is taken to minimize relaxation during the pulses and self-diffusion effects between pulses, reliable values for the spin–spin relaxation times can be obtained.

It is well known that echo modulation also disappears with strong nonselective radio-frequency pulses if the pulse repetition rate is high in comparison with the relevant chemical shift difference. This may be visualized as moving the sidebands of the Fourier spectrum of the pulses far from the nuclear precession frequencies. Then the motion of the nuclear spins resembles that in a spin-locking experiment where a continuous radio-frequency field is applied for a relatively long duration. Unfortunately, this does not seem to be an acceptable solution for the study of chemical exchange effects, since these are also dependent on pulse repetition rate. However, in the fast-exchange limit, where spin–spin splitting is eliminated by the exchange process, the pulse-rate dependence of the spin–spin relaxation time could be followed without any complications from echo modulation.

The separation of several different modulation components from the complex envelope of spin echoes is most satisfactorily achieved by Fourier transformation. The resulting frequency-domain spectrum has been called in general a "spin-echo spectrum," while under certain approximations (first-order spin coupling, low pulse repetition rate, and exact 180° refocussing pulses), the responses are at frequencies determined solely by spin–spin coupling constants, and the result is then known as a J-spectrum. Where self-diffusion effects and pulse imperfections can be eliminated, the line widths of individual components of a J-spectrum should reflect the corresponding spin–spin relaxation rates unaffected by static field inhomogeneity. This mode of display, where line widths as low as 0.03 Hz have already been achieved, emphasizes the potential for extending chemical exchange measurements to much slower exchange rates than are accessible through line-shape studies of conventional high-resolution spectra.

Spurious values of spin–spin relaxation times can be obtained when heteronuclear noise decoupling is employed, for example when carbon-13 spin echoes are recorded with incoherent proton irradiation. The simplest solution would be to examine the relaxation without any proton irradiation, but this sacrifices the simplicity of the noise decoupled carbon-13 spectrum, which may not be acceptable in complicated molecules. The fundamental problem appears to be the transfer of energy between protons and carbon when the precession frequency of protons about H_2 in the 100 MHz rotating frame equals the precession frequency of carbon-13 about H_1 in the 25 MHz frame. For incoherent proton irradiation, it does not seem possible to avoid this rotating frame resonance condition.

Signal enhancement through the nuclear Overhauser effect can be retained by allowing the coupled spin system to come to thermal equilibrium in the presence of incoherent proton irradiation, but from the time of the initial 90° pulse and throughout the refocussing sequence, the proton noise decoupler should be switched off. (In practice a steady sample temperature is more easily maintained in this period by coherent proton irradiation at an equivalent power level, avoiding the rotating frame resonance condition for the intensities of H_1 and H_2.) Once the refocussing cycle is complete, the proton noise decoupler can be reapplied, modifying the last echo so that it transforms to a noise-decoupled spectrum.

ACKNOWLEDGMENT

The authors are pleased to acknowledge several stimulating discussions with Drs. Robert and Regitze Vold, who also provided a copy of their paper prior to publication. Dr. A. C. McLaughlin was also kind enough to provide a copy of his paper prior to publication.

REFERENCES

Abragam, A. (1961). "The Principles of Nuclear Magnetism." Oxford Univ. Press London and New York.
Allerhand, A. (1966). *J. Chem. Phys.* **44**, 1.
Allerhand, A. (1969). *J. Chem. Phys.* **50**, 5429.
Allerhand, A. (1970). *Rev. Sci. Instrum.* **41**, 269.
Anderson, W. A. (1956). *Phys. Rev.* **102**, 151.
Anderson, W. A., and Freeman, R. (1962). *J. Chem. Phys.* **37**, 85.
Arnold, J. T. (1956). *Phys. Rev.* **102**, 136.
Bloch, F. (1946). *Phys. Rev.* **70**, 460.
Bloch, F. (1956). *Phys. Rev.* **102**, 104.
Bloom, A. (1955). *Phys. Rev.* **98**, 1105.
Bradford, R., Clay, C., and Strick, E. (1951). *Phys. Rev.* **84**, 157.
Bremser, W., Hill, H. D. W., and Freeman, R. (1971). *Messtechnik* **79**, 14.
Carr, H. Y., and Purcell, E. M. (1954). *Phys. Rev.* **94**, 630.
Crawford, G. J. B., and Foster, J. S. (1956). *Can. J. Phys.* **34**, 653.
Ellet, J. D., Gibby, M. G., Haeberlen, U., Huber, L. M., Mehring, M., Pines, A., and Waugh, J. S. (1971). *Advan. Magn. Resonance* **5**, 117.
Emshwiller, M., Hahn, E. L., and Kaplan, D. (1960). *Phys. Rev.* **118**, 414.
Ernst, R. R. (1966). *J. Chem. Phys.* **45**, 3845.
Freeman, R., and Hill, H. D. W. (1969). *J. Chem. Phys.* **51**, 3140
Freeman, R., and Hill, H. D. W. (1971a). *J. Magn. Resonance* **4**, 366.
Freeman, R., and Hill, H. D. W. (1971b). *J. Chem. Phys.* **54**, 301.
Freeman, R., and Hill, H. D. W. (1971c). *J. Chem. Phys.* **54**, 3367.
Freeman, R., and Hill, H. D. W. (1971d). *J. Chem. Phys.* **55**, 1985.
Freeman, R., and Wittekoek, S. (1969a). *Proc. Colloq. AMPERE (At. Mol. Etud. Radio Elec.)* **15**, 205.
Freeman, R., and Wittekoek, S. (1969b). *J. Magn. Resonance* **1**, 238.
Gutowsky, H. S., Vold, R. L., and Wells, E. J. (1965). *J. Chem. Phys.* **43**, 4107.
Hahn, E. L. (1950). *Phys. Rev.* **80**, 580.
Hahn, E. L., and Maxwell, D. E. (1952). *Phys. Rev.* **88**, 1070.
Hartmann, S. R., and Hahn, E. L. (1960). *Bull. Amer. Phys. Soc.* [2] **5**, 498.
Hartmann, S. R., and Hahn, E. L. (1962). *Phys. Rev.* **128**, 2042.
Kaiser, R., Bartholdi, E., and Ernst, R. R. (1974). *J. Chem. Phys.* **60**, 2966.
Luz, Z., and Meiboom, S. (1963). *J. Chem. Phys.* **39**, 366.
McLaughlin, A. C., McDonald, G. G., and Leigh, J. S. (1973). *J. Magn. Resonance* **11**, 107.
Mehring, M., and Waugh, J. S. (1972). *Rev. Sci. Instrum.* **43**, 649.
Meiboom, S., and Gill, D. (1958). *Rev. Sci. Instrum.* **29**, 688.
Powles, J. G., and Hartland, A. (1961). *Proc. Phys. Soc., London* **77**, 273.
Powles, J. G., and Strange, J. H. (1962). *Discuss. Faraday Soc.* **34**, 30.
Powles, J. G., and Strange, J. H. (1964). *Mol. Phys.* **8**, 169.
Redfield, A. G. (1955). *Phys. Rev.* **98**, 1787.
Solomon, I. (1959a). *Phys. Rev. Lett.* **2**, 301.
Solomon, I. (1959b). *C. R. Acad. Sci.* **248**, 92; **249**, 1631.
Stempfle, W., and Hoffmann, E. G. (1970). *Z. Naturforsch. B* **25**, 2000.
Tokuhiro, T., and Fraenkel, G. (1968). *J. Chem. Phys.* **49**, 3998.
Tokuhiro, T., and Fraenkel, G. (1969). *J. Chem. Phys.* **51**, 2769.

Vaughan, R. W., Elleman, D. D., Stacey, L. M., Rhim, W-K., and Lee, J. W. (1972). *Rev. Sci. Instrum.* **43**, 1356.
Vold, R. L., and Chan, S. O. (1970). *J. Chem. Phys.* **53**, 449.
Vold, R. L., and Gutowsky, H. S. (1967). *J. Chem. Phys.* **47**, 2495.
Vold, R. L., and Shoup, R. R. (1972). *J. Chem. Phys.* **56**, 4787.
Vold, R. L., Waugh, J. S., Klein, M. P., and Phelps, D. E. (1968). *J. Chem. Phys.* **48**, 3831.
Vold, R. L., Vold, R. R., and Simon, H. E. (1973). *J. Magn. Resonance* **11**, 283.
Waldstein, P., and Wallace, W. E. (1971). *Rev. Sci. Instrum.* **42**, 437.
Weiss, K. H. (1967). *Z. Instrumentenk.* **75**, 1.
Wells, E. J., and Gutowsky, H. S. (1965). *J. Chem. Phys.* **43**, 3414.

6

Rotation about Single Bonds in Organic Molecules

S. STERNHELL

I. Introduction... 163
II. Rotation about sp^3–sp^3 Carbon–Carbon Single Bonds............. 164
III. Rotation about sp^3–sp^2 Carbon–Carbon Single Bonds............. 170
IV. Rotation about sp^2–sp^2 Carbon–Carbon Single Bonds............. 178
V. Rotation about Carbon–Nitrogen Single Bonds 185
VI. Rotation about Miscellaneous Single Bonds..................... 191
References... 196

I. Introduction

In absence of large steric interactions, or some double bond character, rotation about single bonds is associated with small barriers and is therefore difficult to study by dnmr techniques. For this reason, the majority of examples found in the literature concern molecules with certain well-defined steric and electronic features. The aim of this review, in addition to summarizing the existing data, is to enable the reader to identify the molecular environments where barriers to rotation about single bonds are likely to be susceptible to investigation by dnmr. For this reason, some interesting examples where only qualitative observations were made are included.

Earlier work in this field has been summarized by Kessler (1970b) and Sutherland (1971) and an attempt was made here to cover the literature up to the end of 1972. The data are arranged according to the type of single bond whose rotational barrier was investigated, the most important categories being sp^3–sp^3, sp^3–sp^2, and sp^2–sp^2 carbon–carbon single bonds and carbon–nitrogen single bonds. Rotation about metal–carbon and metal-to-metal bonds has also been studied and the results are reviewed in Chapter 8. The term "partial double bond" is qualitative, and hence the subject matter discussed in this and the following chapter has been arbitrarily subdivided using principally the criterion of degree of *p–p* overlap. It is therefore suggested that the two chapters should be read in conjunction.

In many papers quoted in this review, the experimental conditions used did not permit the derivation of all kinetic parameters and, only the free energies of activation for the rotational process ($\Delta G^{\ddagger}$) are given. In other cases, although the complete sets of parameters have been derived, the uncertainties in the enthalpies and entropies of activation are usually too large to permit meaningful comparison with work from other laboratories. For this reason, generally only the free energies are taken into consideration in this review.

II. Rotation about sp^3–sp^3 Carbon–Carbon Single Bonds

Rotational barriers in halogenated alkanes have been the subject of many dnmr investigations, and fluorinated compounds were particularly suitable for this purpose, because, while the rotational barriers in such derivatives are generally small, the large range of chemical shifts of ^{19}F made it possible to obtain exchange-broadened spectra at experimentally accessible temperatures. The data summarized in Table I show that the barriers are dependent on the steric bulk of the halogens (e.g., entries 2 and 5 or entries 11 and 12), but it was noted early (Newmark and Sederholm, 1965) that the relationship between the bulk of the halogen and the magnitude of the rotational barrier is not simple and that it is not possible to derive useful values for the magnitudes of pairwise nonbonded interactions from such data. In a further study of eight heavily halogenated ethanes, Weigert et al. (1970) have shown that in fact the replacement of chlorine by bromine may not lead to an increase in the barrier height, but that the total number of halogen atoms (other than fluorine) is very significant. Thus, the rotational barriers for pentahalogenated ethanes fall into the range of 13.2–14.8 kcal/mole while those for tetrahalogenated ethanes fall between 9.0 and 10.2 kcal/mole.

Several of the authors considered polar interactions, but the situation is too complicated for definite conclusions. It may be pertinent, however, that the rotational barrier in $CHBr_2$–$CFBr_2$ (Govil and Bernstein, 1968a) varies significantly with the polarity of the solvent; the value quoted in entry 8 in Table I (10.0 kcal/mole) refers to a solution in $CFCl_3$ ($\varepsilon = 2.88$) and changes to 10.80 in acetone ($\varepsilon = 32.5$).

The special case of trifluoromethyl derivatives has been the object of a separate study (Weigert and Roberts, 1968; Weigert and Mahler, 1972). It can be seen (entries 13–15, Table I) that the progressive replacement of chlorine with bromine and iodine is associated with the expected increase in the height of the barrier, but that fluorine and the trifluoromethyl group (entries 16 and 17) are associated with similar barriers.

TABLE I
HALOGENATED ETHANE DERIVATIVES

Entry	Compound $R_1R_2R_3C-CR_4R_5R_6$						$\Delta G^\ddagger$ (kcal/mole)	(°C)[a]	Reference
	R_1	R_2	R_3	R_4	R_5	R_6			
1	F	F	Br	F	Br	Br	9.9 ± 0.2[b]		Newmark and Sederholm (1965)
2	F	Cl	Cl	F	Cl	Cl	9.65 ± 0.1[b]		Newmark and Sederholm (1965); Newmark and Graves (1968)
3	F	F	Br	Cl	Cl	Br	$\geq 12.0 \pm 0.5$[b,c]		Newmark and Sederholm (1965)
4	F	F	Br	F	Br	Cl	≥ 8.8[b]		Newmark and Sederholm (1965)
5	F	Cl	Br	F	Br	Br	≥ 12.0[b]		Newmark and Sederholm (1965)
6	F	Cl	Br	H	Cl	Cl	~ 8.2[b]		Newmark and Sederholm (1965)
7	F	Cl	Cl	H	Br	Cl	9.1[d]		Alger et al. (1967)
8	F	Br	Br	H	Br	Br	10.0 ± 0.18	(−93 and −60)	Govil and Bernstein (1968a)
9	Cl	Cl	Me	H	Cl	Cl	10.2 ± 1	(−101)	Heatley et al. (1972)
10	Br	Br	Me	Br	Me	Me	12.61 ± 0.01	(−30)	Hawkins et al. (1971)
11	Cl	Cl	Me	Cl	Me	Me	13.47 ± 0.01	(−31)	Hawkins et al. (1971)
12	Br	Br	Me	Br	Br	Me	15.96 ± 0.01	(+18)	Hawkins et al. (1971)
13	F	F	F	Cl	Cl	CF_3	6.1	(−140)	Weigert and Mahler (1972)
14	F	F	F	Br	Br	CF_3	6.7	(−120)	Weigert and Mahler (1972)
15	F	F	F	I	I	CF_3	7.2	(−107)	Weigert and Mahler (1972)
16	F	F	F	I	CF_3	F	5.5	(−150)	Weigert and Mahler (1972)
17	F	F	F	I	Cl	F	5.2	(−150)	Weigert and Mahler (1972)
18	F	F	F	Cl	Cl	I	8.5	(−72)	Weigert and Mahler (1972)
19	F	F	F	Cl	Cl	CCl_3	5.6	(−139)	Weigert and Mahler (1972)

[a] Free energy of activation for the rotational barrier about the central bond at temperatures indicated in parentheses. For cases where more than one energy barrier is involved, the largest barrier is quoted.
[b] The data are derived from spectra taken at low temperatures, but the method of analysis used makes the quotation of a single temperature meaningless. Free energies in this series are independent of temperature within limits used.
[c] Vold and Gutowsky (1967) obtained the value of 12.0 ± 0.2 for this barrier ($\Delta G^\ddagger$ gauche–gauche) using spin-echo techniques. They also obtained the value of 11.0 ± 0.05 for $\Delta G^\ddagger$ trans–gauche, in excellent agreement with 10.8 ± 0.1 obtained by Newmark and Sederholm (1965).
[d] Spin echo and line shape results for ^1H and ^{19}F resonances give an excellent Arrhenius plot; the ^{19}F data span four orders of magnitude of the rate constant.

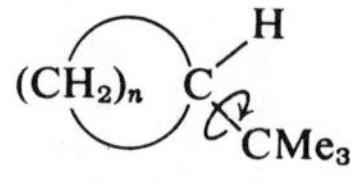

Hindered rotation about the bond joining a tertiary butyl group to an sp^3 hybridized carbon atom has been investigated by dnmr in numerous compounds (Table II). A systematic dependence of the magnitude of the rotational barrier on the bulk of the substituents has been noted by several authors (Anderson and Pearson, 1971a,b, 1972a,b; Hawkins et al., 1971; Bushweller and Anderson, 1972) for acyclic derivatives (entries 1–28, Table II), and Anderson and Pearson (1971b, 1972b) proposed that these barriers can be used as a basis for comparing the magnitudes of nonbonded interactions (P-values). The P-values for a number of groups were determined and found to correspond only approximately with the well-known A-values, derived from conformational equilibria in the cyclohexane series (Eliel, 1965; Hirsch, 1967), and this was explained on the basis of "anisotropy of the van der Waal's surface." The trends in the magnitudes of the rotational barriers in a series of t-butylcycloalkanes (Anet et al., 1968) (entries 29–34, Table II), and

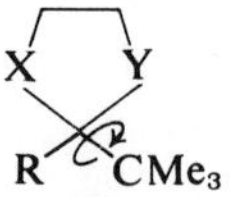

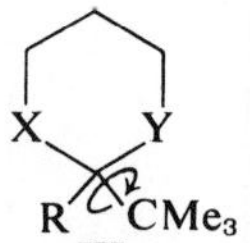

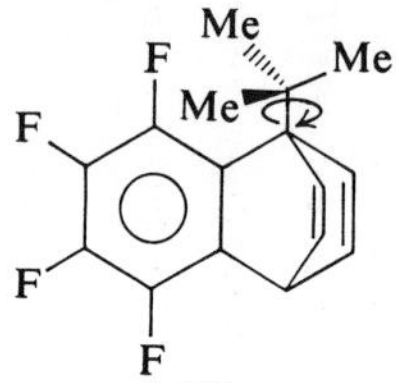

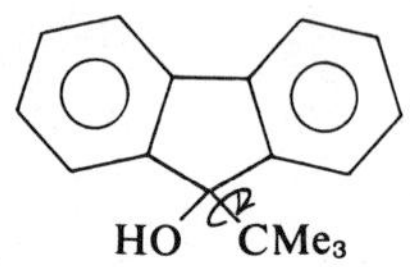

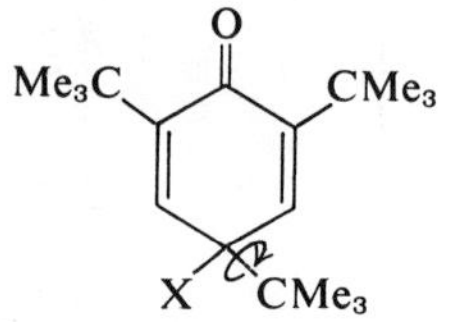

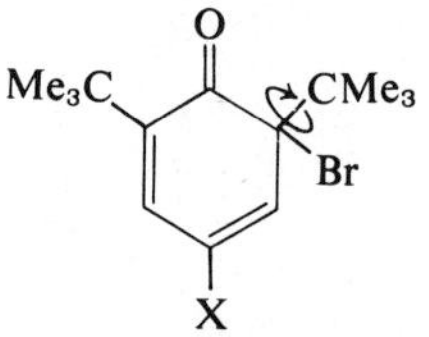

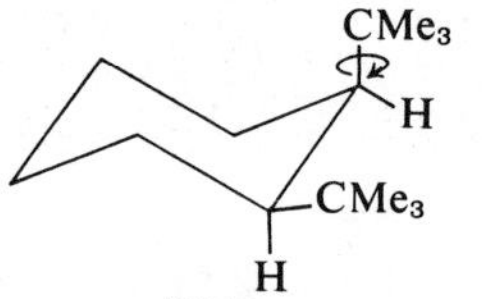

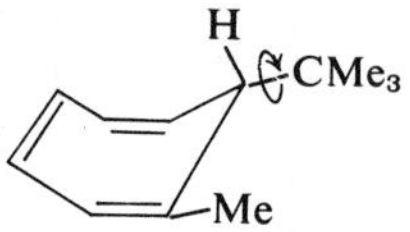

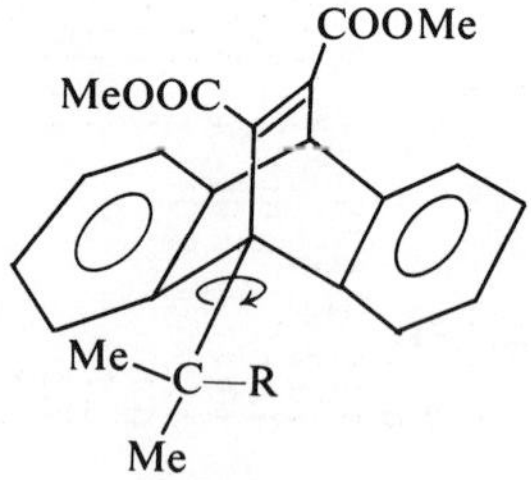

6. Rotation about Single Bonds

ketals and thioketals (Bushweller et al., 1972a) (entries 35–41, Table II) can be rationalized on steric grounds, but, especially in the latter series, other effects also appear to operate. The larger effective size of the sulfur atom is clearly demonstrated, e.g., in the sets of data shown in entries 35, 37, 39, 40, and 41.

Miscellaneous examples of relatively high rotational barriers about sp^3–sp^3 carbon–carbon single bonds are shown in a number of 9,9'-bifluoronyl derivatives (Bartle et al., 1970): **X**, R = H ($\Delta G^{\ddagger}$ = 17.7 kcal/mole at 64°C) (Ōki and Suda, 1971); **XI**, R = CH$_2$Cl or H ($\Delta G^{\ddagger}$ = 16 ± 2 kcal/mole at −23 °C) (Sergeyev et al., 1972); and in a number of highly alkylated ethanes (Bushweller and Anderson, 1972).

The first example of restricted rotation of a methyl group giving rise to a clearly identifiable dnmr phenomenon was reported (Nakanishi et al., 1973) for the 1-t-butyl-1,4-dihydronaphthalene 1,4-oxides (**XII**, X = Y = Cl and X = OMe, Y = H). The tertiary butyl groups in both compounds give rise to singlets at room temperature and to three singlets at −60°C, as expected. However, further lowering of temperature results in very significant broadening of two of the three methyl signals, which must be due to slow rotation of the methyl groups themselves.

Besides the work cited above, there exists a relatively large number of investigations concerned with the rotation about sp^3–sp^3 carbon–carbon single bonds that are not usually considered as involving dnmr because the data consist only of the variation in averaged (fast exchange) spectra with temperature or solvent. Good descriptions of the methods used to extract thermodynamic parameters for the conformational equilibria and the nmr parameters for the individual rotamers from such data are given by, inter alia, Abraham and Bernstein (1961) and by Gutowsky (1963). While the method is valuable because it is applicable to processes that cannot be slowed down sufficiently for normal dnmr studies, it suffers from two general limitations: first, the amount of information that can be extracted from the numerical matching procedures varies with the nature of the plot of the nmr

TABLE II
TERTIARY BUTYL DERIVATIVES

Entry	Compound $R_1R_2R_3C-CMe_3$			$\Delta G^\ddagger$ (kcal/mole)	(°C)[a]	Reference
	R_1	R_2	R_3			
1	Cl	Me	Me	9.82 ± 0.01	(−81)	Hawkins et al. (1971)
2	Br	Me	Me	10.80 ± 0.02	(−68)	Hawkins et al. (1971)
3	Cl	Cl	Me	10.81 ± 0.01[b]	(−61)	Hawkins et al. (1971)
4	Br	Br	Me	12.27 ± 0.01	(−38)	Hawkins et al. (1971)
5	Cl	Br	Me	11.90 ± 0.01	(−45)	Hawkins et al. (1971)
6	Cl	CD_2CD_3	Me	10.87 ± 0.01	(−71)	Hawkins et al. (1971)
7	Cl	CMe_3	Me	11.42 ± 0.02	(−58)	Hawkins et al. (1971)
8	Cl	Me	Me	10.05	(−87.9)	Anderson and Pearson (1971a)
9	Cl	Me	—CH_2Ph	10.55	(−87.6)	Anderson and Pearson (1971a)
10	Cl	Me	H	8.32 ± 0.2	(−115.8)	Anderson and Pearson (1971b)
11	Cl	Me	Me	10.43 ± 0.1	(−65.0)	Anderson and Pearson (1971b)
12	Cl	Me	Et	10.82 ± 0.2	(−69.6)	Anderson and Pearson (1971b)
13	Cl	Me	—CH_2Ph	10.85 ± 0.1	(−63.9)	Anderson and Pearson (1971b)
14	Cl	Me	—CMe_3	11.43 ± 0.1	(−60.0)	Anderson and Pearson (1971b)
15	Cl	Me	Ph	10.19 ± 0.1	(−62.5)	Anderson and Pearson (1971b)
16	Cl	Me	—CH_2—CMe_3	11.80		Anderson and Pearson (1972a)
17	Cl	Cl	Me	11.1		Anderson and Pearson (1972a)
18	Cl	Cl	H	8.87		Anderson and Pearson (1972a)
19	H	Me	Me	6.97	(−129.6)	Anderson and Pearson (1972b)
20	F	Me	Me	8.04	(−123.2)	Anderson and Pearson (1972b)
21	Cl	Me	Me	10.43	(−61)	Anderson and Pearson (1972b)
22	Br	Me	Me	10.73	(−58.8)	Anderson and Pearson (1972b)
23	I	Me	Me	11.14	(−60.7)	Anderson and Pearson (1972b)
24	H	H	Me	4.9 ± 0.5	(−181)	Bushweller and Anderson (1972)
25	H	Me	Me	6.9 ± 0.3	(−134)	Bushweller and Anderson (1972)

#	R1	R2	R3	Compound	ΔG‡	T (°C)	Reference
26	H	Me	Et		6.3 ± 0.5	(−146)	Bushweller and Anderson (1972)
27	H	Me	—CHMe$_2$		~6		Bushweller and Anderson (1972)
28	Me	Me	Et		8.3 ± 0.5	(−102)	Bushweller and Anderson (1972)
29				I, n = 4	6.0	(−155)	Anet et al. (1968)
30				I, n = 5	6.3	(−150)	Anet et al. (1968)
31				I, n = 6	7.4	(−126)	Anet et al. (1968)
32				I, n = 7	7.8	(−118)	Anet et al. (1968)
33				I, n = 8	7.3	(−130)	Anet et al. (1968)
34				I, n = 9	7.3	(−130)	Anet et al. (1968)
35				II, X = Y = S, R = Me	10.6 ± 0.1	(−70.2)	Bushweller et al. (1972a)
36				II, X = Y = S, R = H	7.5 ± 0.2	(−133.2)	Bushweller et al. (1972a)
37				II, X = O, Y = S, R = Me	9.8 ± 0.1	(−101.2)	Bushweller et al. (1972a)
38				II, X = O, Y = S, R = H	7.0 ± 0.2	(−139.6)	Bushweller et al. (1972a)
39				II, X = Y = O, R = Me	7.5 ± 0.1	(−124.7)	Bushweller et al. (1972a)
40				III, X = Y = S, R = Me	9.6 ± 0.1	(−83.5)	Bushweller et al. (1972a)
41				III, X = Y = O, R = Me	8.7 ± 0.1	(−109.8)	Bushweller et al. (1972a)
42				IV	—	c	Brewer et al. (1967, 1968)
43				V	9.4	(−85)	Rieker and Kessler (1969)
44				VI, X = Br	9.0	(−90)	Rieker et al. (1968)
45				VI, X = OMe	8.9	(−89)	Rieker et al. (1968)
46				VII, X = CN or NO$_2$	—	c	Rieker et al. (1968)
47				VII, X = —OC$_6$Cl$_5$	9		Kessler (1970b)
48				VIII	10.1		Kessler et al. (1968b); Kessler (1970b)
49				IX	10.3	(25)	Heyd and Cupas (1971)
50				X, R = Me	>25		Ōki and Suda (1971)

[a] Free energy of activation for rotation about the C—C bond in R$_1$R$_2$R$_3$C—CMe$_3$ or about the bond indicated by an arrow on the appropriate structural formula at temperatures indicated in parentheses.

[b] Rieker and Kessler (1969) give 11.1 for the same compound in the same solvent (CS$_2$).

[c] Not determined quantitatively, but variable temperature spectra indicate that the barrier is in a region that can be conveniently investigated by dnmr.

parameter against temperature, and second, an assumption must always be made regarding the inherent invariability of the nmr parameters of the individual rotamers with temperature. The latter condition is certainly not always met (Govil and Bernstein, 1967, 1968b). Studies of this type are well exemplified by investigations of Abraham *et al.* (1966), Whitesides *et al.* (1967), and Woller and Garbisch (1972a).

III. Rotation about sp^3–sp^2 Carbon–Carbon Single Bonds

The first recorded instance of rotation about an sp^3–sp^2 carbon–carbon single bond giving rise to a dnmr phenomenon is due to Dix *et al.* (1966), who determined the rotational barrier of the neopentyl groups in (**XIII**) (Entry 1, Table III) by observing the collapse of the anisochronicity of the methylene protons with increase in temperature. Since then, a large number of examples has been reported (Table III) and these include systematic studies in series of related compounds as well as a number of interesting isolated observations.

Unlike with ethane derivatives, where the low energy state is invariably assumed to be a staggered form (or a slightly distorted staggered form) while the high energy state giving rise to the barrier is an eclipsed form, the energy profile for the rotation about an sp^3–sp^2 carbon–carbon single bond may vary qualitatively with the nature of substituents. In fact, dnmr studies have often given information, not only about the heights of barriers to rotation, but also about the nature of the preferred conformations that may be deduced from the symmetry of the nmr spectra. For this reason, structural formulas (**XIII–XXXVIII**) have been drawn to show the preferred stereochemistry in the ground state wherever this had been obtained. In cases where two lower energy states exist, the stereochemistry of one of them is shown and the average $\Delta G^\ddagger$ value is quoted in Table III.

High rotational barriers in 9-fluorene derivatives (**XIV**) were observed qualitatively by means of variable temperature nmr spectra by Chandross and

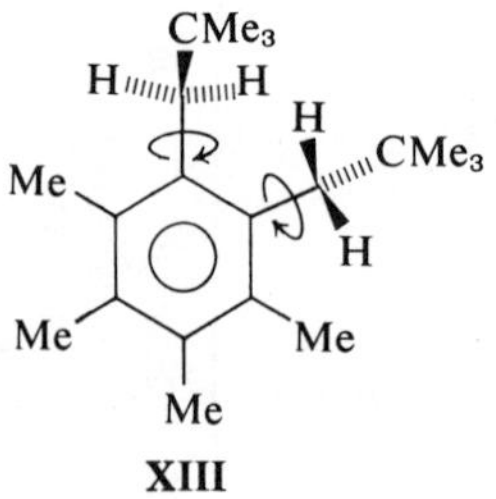

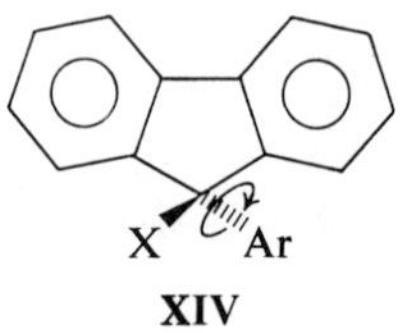

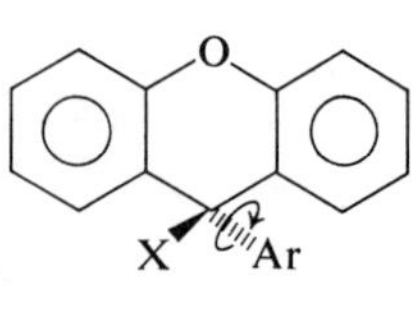

6. Rotation about Single Bonds

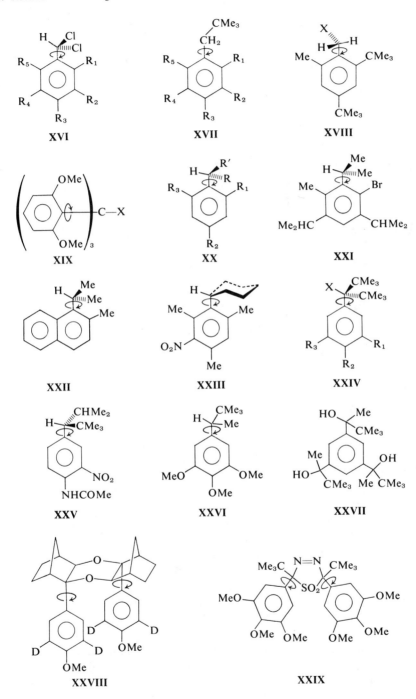

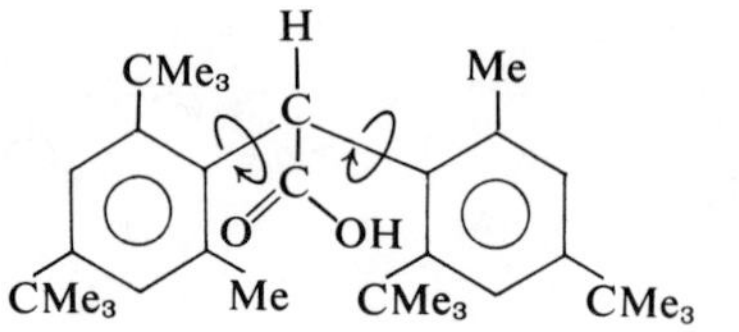

XXX 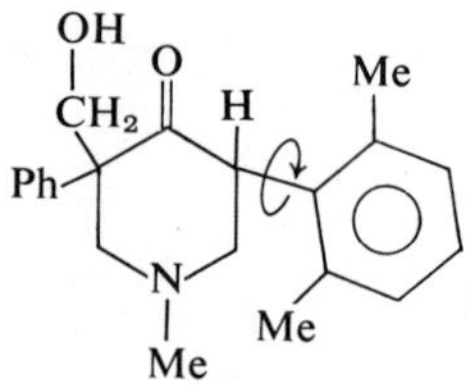 **XXXI** 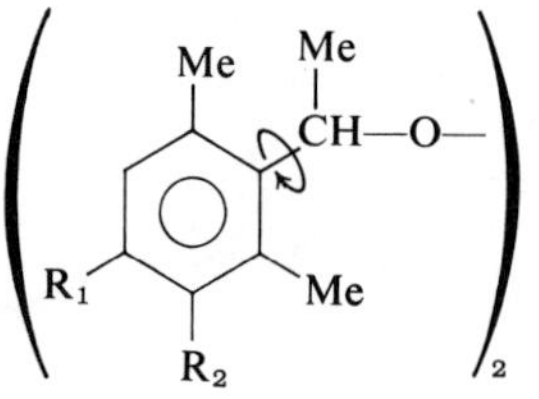 **XXXII** 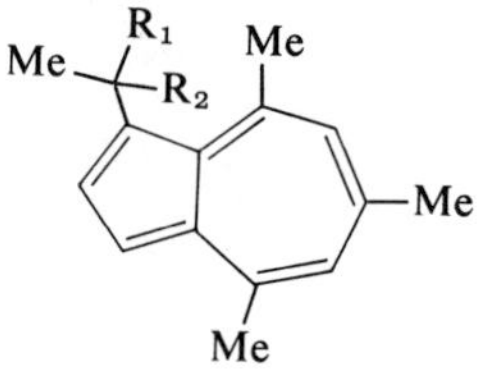 **XXXIII** 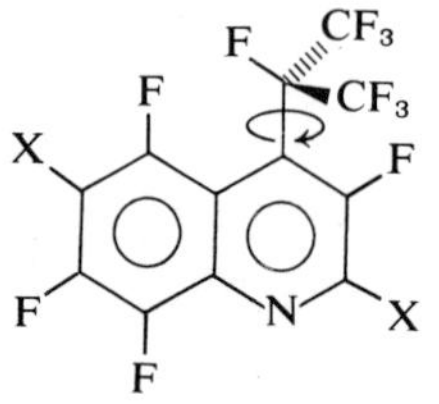 **XXXIV** 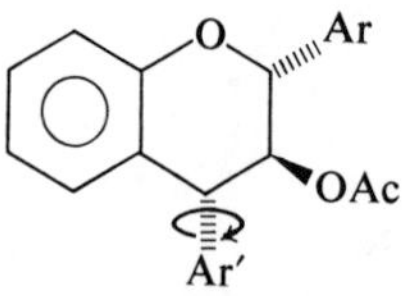 **XXXV** 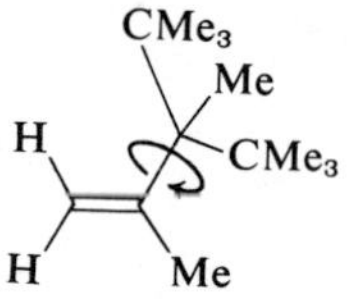 **XXXVI** 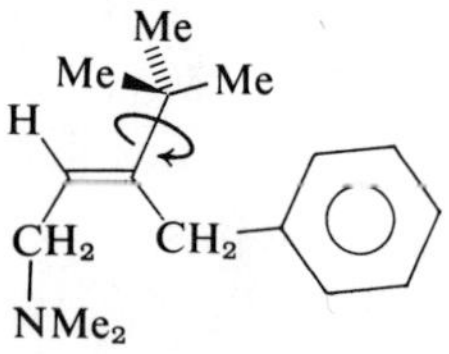 **XXXVII** 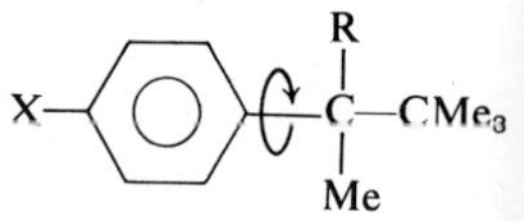 **XXXVIII**

Sheley (1968), and many quantitative data have now appeared in the literature (Table III, entries 2–16). It has been pointed out (Rieker and Kessler, 1969; Siddall and Stewart, 1969b; McKinley et al., 1970) that the regular decrease of the barrier height with the increasing size of the substituent at C-9 in **XIV** (cf., e.g., entries 2–4, 5–7, and 8–10 in Table III) is due to the destabilization of the ground state. The slow (on nmr time scale) process in **XIV** (X = Cl, Ar = 2,4,6-trimethylphenyl) was originally attributed (Chandross and Sheley, 1968) to reversible ionization of chlorine, but is almost certainly due to restricted rotation about the C-9—aryl bond (Rieker and Kessler, 1969; McKinley et al., 1970). Interestingly, the corresponding barriers in xanthyl derivatives (**XV**) are lower (cf., e.g., entries 2 and 17, 6 and 18, 9 and 19 in Table III), and this has been ascribed (McKinley et al., 1970) to the conformational flexibility of the six-membered heterocyclic ring. It was also recognized that the transition state involves a simultaneous interaction between the 2 and 6 substituents in the benzene nucleus at C-9 with the hydrogens at C-4 and C-5 (cf., e.g., entries 9 and 10, 17 and 19 in Table III) (McKinley et al., 1970).

6. Rotation about Single Bonds

Many systematic data have been collected for the rotational barriers of the dichloromethyl group attached to the benzene nucleus (entries 20–26 in Table III). An obvious example of bulk effect may be seen by comparing entries 20 and 21 in Table III; buttressing effects vary from insignificant (Barber and Schaefer, 1971) to quite severe (Mark and Pattison, 1971), and several examples (Stewart *et al.*, 1971; Peeling *et al.*, 1971) of substituent effects on rotamer distribution have been observed. Interestingly, the cis and trans conformers of **XVI** ($R_1 = R_2 = R_4 = R_5 = Cl$, $R_3 = CHCl_2$) are equally populated (Barber and Schaefer, 1971).

The series of trineopentyl benzenes (entries 27–31, Table III) exhibit concerted motion of the three bulky groups, which is simultaneously influenced by the nature of substituents around any of them, unsymmetrical substitution leading to lower barriers to rotation (Nilsson *et al.*, 1972). In the series of neopentyl benzenes with an ortho carbonyl substituent (entries 32–35, Table III), the rotational barrier is determined not only by the nature of the carbonyl derivative, but also by an unusual type of buttressing. The introduction of a methyl group ortho to the carbonyl lowers the rotational barrier of the neopentyl group, presumably by turning the carbonyl group at right angles to the benzene ring and thus making it effectively smaller (Reuvers *et al.*, 1971).

In a series of related halogen derivatives (entries 36–38, Table III), the barrier varies with the size of the halogen atom and corresponds to the passing of the halogen over the methyl group (Cupas *et al.*, 1968).

The rotation of alkyl or aryl–alkyl groups attached to a benzene ring and flanked by two substituents has been the subject of considerable study (Rieker and Kessler, 1969; Mannschreck and Ernst, 1971). In particular, Mannschreck and Ernst (1971) showed that the experimentally obtained barriers (entries 44–59, Table III) can in many cases be successfully reproduced by empirical calculations of strain energies.

When sufficiently bulky alkyl groups are attached to the benzylic carbon, high rotational barriers may be observed even in the absence of ortho substituents in the benzene ring. The first example of this type (entry 60, Table III) was observed early in these laboratories (Newsoroff and Sternhell, 1967) and numerous data (entries 61–71, Table III) have been accumulated since. It is interesting to note that in compound **XXIV** the replacement of the benzylic hydrogen by a hydroxyl group leads to a consistent (cf., e.g., entries 61 and 66, Table III) decrease in the height of the barrier as a result of destabilization of the ground state. The decrease in the size of one of the benzylic alkyl groups leads to a drastic decrease in the size of the rotational barrier (entries 70 and 71, Table III) as expected from the postulated (Gall *et al.*, 1972) transition state, which involves the eclipsing of the tertiary butyl group in **XXIV** by one of the ortho hydrogens. For this reason, the very high

TABLE III
BARRIERS TO ROTATION ABOUT sp^3–sp^2 CARBON–CARBON SINGLE BONDS

Entry	Compound	$\Delta G^{\ddagger}$ (kcal/mole)	(°C)[a]	Reference
1	XIII	$E_A = 11.5 \pm 0.6$		Dix et al. (1966)
2	XIV, X = H, Ar = 2,4,6-trimethylphenyl	>25[b]	(>190)	Rieker and Kessler (1969)
3	XIV, X = OH, Ar = 2,4,6-trimethylphenyl	20.2	(145)	Rieker and Kessler (1969)
4	XIV, X = Cl, Ar = 2,4,6-trimethylphenyl	16.2	(66)	Rieker and Kessler (1969)
5	XIV, X = H, Ar = 2,6-dimethoxyphenyl	20.6	(145)	Rieker and Kessler (1969)
6	XIV, X = OH, Ar = 2,6-dimethoxyphenyl	14.4	(24)	Rieker and Kessler (1969)
7	XIV, X = Cl, Ar = 2,6-dimethoxyphenyl	9.2	(−81)	Rieker and Kessler (1969)
8	XIV, X = OH, Ar = 2,6-dimethylphenyl	21.6	(150)	Siddall and Stewart (1968a, 1969b)
9	XIV, X = H, Ar = 2-methylphenyl	16.4	(6)	Siddall and Stewart (1969b)
10	XIV, X = H, Ar = 2,6-dimethylphenyl	>26	(>200)	Siddall and Stewart (1969b)
11	XIV, X = H, Ar = 3-methylphenyl	<9	(−85)	Siddall and Stewart (1969b)
12	XIV, X = 1-naphthyl	18.0	(60)	Siddall and Stewart (1969b); Bartle et al. (1970)
13	XIV, X = 2-methyl-1-naphthyl	29.2	(116)	Siddall and Stewart (1969b)
14	XIV, X = 2-methylphenyl, Ar = 4-amino-2,5-dimethylphenyl	13 ± 1	(−30)	Siddall and Stewart (1968b)
15	XIV, X = 2-methylphenyl, Ar = 4-amino-2-isopropyl-5-methylphenyl	>20		Siddall and Stewart (1968b)
16	XIV, X = 1-naphthyl, Ar = 4-amino-2-isopropyl-5-methylphenyl	>20		Siddall and Stewart (1968b)
17	XV, X = H, R = 2,4,6-trimethylphenyl	17.6	(25)	McKinley et al. (1970)
18	XV, X = OH, Ar = 2,6-dimethylphenyl	10.9	(25)	McKinley et al. (1970)
19	XV, X = H, Ar = 2-methylphenyl	9.4	(25)	McKinley et al. (1970)
20	XVI, $R_1 = R_3 = R_5 = Cl$, $R_2 = R_4 = H$	15 ± 0.1	(31)	Fuhr et al. (1970); Gyulai et al. (1970)
21	XVI, $R_1 = R_3 = R_5 = Br$, $R_2 = R_4 = H$	17.5 ± 0.1	(31)	Peeling et al. (1970)
22	XVI, $R_1 = R_2 = R_4 = R_5 = Cl$, $R_3 = CHCl_2$	15.4 ± 0.1	(15)	Barber and Schaefer (1971)
23	XVI, $R_1 = R_2 = R_5 = Cl$, $R_3 = R_4 = H$	15.4 ± 0.1	(25)	Stewart et al. (1971)

6. Rotation about Single Bonds

#	Structure	Value	(T)	Reference
24	XVI, $R_1 = R_3 = R_4 = R_5 = Cl$, $R_2 = CHCl_2$	14.9 ± 0.05	(20)	Peeling et al. (1971)
25	XVI, $R_1 = CHCl_2$, $R_2 = R_3 = R_4 = R_5 = Cl$	17.7	(83)	Mark and Pattison (1971)
26	XVI, $R_1 = CHCl_2$, $R_2 = R_5 = Cl$, $R_3 = R_4 = H$	17.4	(77)	Mark and Pattison (1971)
27	XVII, $R_1 = R_5 = Cl$, $R_2 = R_4 = -CH-CMe_3$, $R_3 = H$	14.3 ± 0.2	(25)	Carter et al. (1970)
28	XVII, $R_1 = R_5 = Br$, $R_2 = R_4 = -CH_2-CMe_3$, $R_3 = H$	16.2 ± 0.1	(25)	Carter et al. (1970)
29	XVII, $R_1 = Cl$, $R_2 = R_4 = -CH_2-CMe_3$, $R_3 = H$, $R_5 = I$	14.6 ± 0.3	(25)	Nilsson et al. (1972)
30	XVII, $R_1 = Br$, $R_2 = R_4 = -CH_2-CMe_3$, $R_3 = H$, $R_5 = I$	16.6 ± 0.3	(25)	Nilsson et al. (1972)
31	XVII $R_1 = Cl$, $R_2 = R_4 = -CH_2-CMe_3$, $R_3 = H$, $R_5 = I$	14.7 ± 0.5	(25)	Nilsson et al. (1972)
32	XVII, $R_1 = COOMe$, $R_2 = R_3 = R_4 = H$, $R_5 = Me$	14.6 ± 0.1	(25)	Reuvers et al. (1971)
33	XVII, $R_1 = COCl$, $R_2 = R_3 = R_4 = H$, $R_5 = Me$	14.6 ± 0.1	(25)	Reuvers et al. (1971)
34	XVII, $R_1 = CHO$, $R_2 = R_3 = R_4 = H$, $R_5 = Me$	13.2 ± 0.1	(25)	Reuvers et al. (1971)
35	XVIII, $R_1 = COOMe$, $R_2 = R_3 = R_4 = R_5 = Me$	13.8 ± 0.1	(25)	Reuvers et al. (1971)
36	XVIII, $X = Cl$	$E_A = 11.3 \pm 0.5$		Cupas et al. (1968)
37	XVIII, $X = Br$	$E_A = 12.5 \pm 0.5$		Cupas et al. (1968)
38	XVIII, $X = I$	$E_A = 15.9 \pm 0.7$		Cupas et al. (1968)
39	XIX, $X = H$	11.1	(-57)	Rieker and Kessler (1969); Kessler et al. (1969)
40	XIX, $X = OH$ or Cl	< 8.2		Rieker and Kessler (1969); Kessler et al. (1969)
41	XX, $R = R' = Ph$, $R_1 = R_2 = R_3 = Me$	8.8	(-90)	Rieker and Kessler (1969); Kessler et al. (1969)
42	XX, $R = R' = R_1 = R_2 = R_3 = Me$	12.8	(-35)	Rieker and Kessler (1969); Mannschreck and Ernst (1971)

(continued)

TABLE III—(continued)

Entry	Compound	$\Delta G^\ddagger$ (kcal/mole)	(°C)^a	Reference
43	XX, R = R' = Me, R$_1$ = R$_3$ = OMe, R$_2$ = H	9.6	(−93)	Rieker and Kessler (1969); Mannschreck and Ernst (1971)
44	XX, R and R' = (CH$_2$)$_5$, R$_1$ = R$_2$ = R$_3$ = Me	14.4	(−1.5)	Mannschreck and Ernst (1971)
45	XX, R and R' = (CH$_2$)$_4$, R$_1$ = R$_2$ = R$_3$ = Me	<7.5	(< −120)	Mannschreck and Ernst (1971)
46	XX, R and R' = (CH$_2$)$_2$, R$_1$ = R$_2$ = R$_3$ = Me	<7.5	(< −120)	Mannschreck and Ernst (1971)
47	XX, R = R' = Cl, R$_1$ = R$_2$ = R$_3$ = Me	13.9	(−1)	Mannschreck and Ernst (1971)
48	XX, R = OMe, R' = Ph, R$_1$ = R$_2$ = R$_3$ = Me	9.2	(−83)	Mannschreck and Ernst (1971)
49	XX, R = OH, R' = Ph, R$_1$ = R$_2$ = R$_3$ = Me	9.5	(−79.5)	Mannschreck and Ernst (1971)
50	XX, R = SH, R' = Ph, R$_1$ = R$_2$ = R$_3$ = Me	11.4	(−42)	Mannschreck and Ernst (1971)
51	XX, R = CN, R' = Ph, R$_1$ = R$_2$ = R$_3$ = Me	10.4	(−60)	Mannschreck and Ernst (1971)
52	XX, R = Me, R' = Ph, R$_1$ = R$_2$ = R$_3$ = Me	11.2	(−44.5)	Mannschreck and Ernst (1971)
53	XX, R = Cl, R' = Ph, R$_1$ = R$_2$ = R$_3$ = Me	11.4	(−43.5)	Mannschreck and Ernst (1971)
54	XX, R = Br, R' = Ph, R$_1$ = R$_2$ = R$_3$ = Me	12.7	(−19)	Mannschreck and Ernst (1971)
55	XX, R = R' = Me, R$_1$ = R$_2$ = Me, R$_3$ = Cl	13.7	(−28)	Mannschreck and Ernst (1971)
56	XX, R = R' = Me, R$_1$ = Me, R$_2$ = NO$_2$, R$_3$ = Br	14.5	(−9)	Mannschreck and Ernst (1971)
57	XXI	14.6	(−13)	Mannschreck and Ernst (1971)
58	XXII	12.7	(−45)	Mannschreck and Ernst (1971)
59	XXIII	14.8	(−7)	Mannschreck and Ernst (1971)
60	XXIV, X = OH, R$_1$ = R$_3$ = H, R$_2$ = OMe	—^{c}		Newsoroff and Sternhell (1967)
61	XXIV, X = OH, R$_1$ = R$_2$ = R$_3$ = OMe	19.7 ± 0.2	(25)	Gall et al. (1972)
62	XXIV, X = OH, R$_1$ = R$_2$ = OMe, R$_3$ = Br	21.9 ± 0.6	(25)	Gall et al. (1972)
63	XXIV, X = OH, R$_1$ = R$_3$ = OMe, R$_2$ = CMe$_3$	20.7 ± 0.2	(25)	Gall et al. (1972)
64	XXIV, X = OH, R$_1$ = OMe, R$_2$ = OMe or OCH$_2$Ph or OH, R$_3$ = H or Cl or NO$_2$	—^{c}		Gall et al. (1972); G. C. Brophy, R. E. Gall, D. Landman, and S. Sternhell (unpublished data, 1973)

6. Rotation about Single Bonds

#	Compound	ΔG‡	(T)	Reference
65	XXIV, X = OH, R_1 = NO_2, R_2 = NHCOMe, R_3 = H	21.3	(116)	Baas (1972a)
66	XXIV, X = H, R_1 = R_2 = OMe, R_3 = H	22.3	(136)	G. C. Brophy, R. B. Gall, D. Landman, and S. Sternhell (unpublished data, 1973)
67	XXIV, X = H, R_1 = R_2 = R_3 = OMe	22.3	(128)	G. C. Brophy, R. B. Gall, D. Landman, and S. Sternhell (unpublished data, 1973)
68	XXIV, X = H, R_1 = OMe, R_2 = OMe or OCH_2Ph or OH, R_3 = Br or H	—[c]		G. C. Brophy, R. B. Gall, D. Landman, and S. Sternhell (unpublished data, 1973)
69	XXIV, X = H, R_1 = NO_2, R_2 = NHCOMe, R_3 = H	22.2	(158)	Baas (1972)
70	XXV	11.8	(−42)	Baas (1972)
71	XXVI	9.5 ± 0.4	(−77)	Baas (private communication, 1972)
72	XXVIII	19.1 ± 0.1	(122)	Gerteisen et al. (1971)
73	XXIX	18.2 ± 0.1	(81)	G. C. Brophy, D. J. Collins, J. J. Hobbs, and S. Sternhell (unpublished data, 1973)
74	XXX	$\Delta H^\ddagger$ = 21.8 ± 0.7; $\Delta S^\ddagger$ = −3.3 ± 1.7		Akkerman (1970)
75	XXXI	—[c]		Misiti et al. (1970)
76	XXXII, R_1 = Me or OMe, R_2 = H or Br	—[c]		Kamikawa et al. (1968)
77	XXXVIII, R = Cl, X = NO_2	7.8	(−114.0)	Anderson et al. (1973)
78	XXXVIII, R = Cl, X = OMe	7.4	(−118.4)	Anderson et al. (1973)
79	XXXVIII, R = H, X = NO_2	9.01	(−101.0)	Anderson et al. (1973)
80	XXXVIII, R = H, X = OMe	8.79	(−101.0)	Anderson et al. (1973)

[a] Free energy of activation to rotation about the bond(s) indicated by an arrow on the appropriate structural formula at temperatures indicated in parentheses.

[b] Siddall and Stewart (1969a) report $\Delta G^\ddagger$ > 26 kcal/mole for this compound.

[c] Not determined quantitatively, but variable temperature spectra indicate that the barrier is in a region that can be conveniently investigated by dnmr.

rotational barrier ($\Delta G^{\ddagger} > 25$ kcal/mole) reported for **XXVII** (Martinson, 1969) appeared abnormal, and it has, in fact, been conclusively demonstrated (J. M. A. Baas, private communication, 1972) that the additional multiplicity in the nmr spectrum of **XXVII**, which persisted even at high temperatures, is due to the presence of several enantiomers.

In the heavily substituted diarylacetic acid (**XXX**), the enthalpies and entropies for the rotational process determined by Akkerman (1970) (entry 74, Table III) are in excellent agreement with those obtained from the rates of racemization ($\Delta H^{\ddagger} = 21.3 \pm 0.1$, $\Delta S^{\ddagger} = -5.2 \pm 0.2$). This is in contrast with several examples of rate processes where the kinetic parameters measured by dnmr are in poor agreement with those obtained from rates of racemization (Carter and Berntsson, 1968), suggesting that different processes are being observed by the two methods.

Examples of substantial, but not quantitatively evaluated, barriers to rotation about sp^3–sp^2 bonds in azulene derivatives (**XXXIII**, $R_1 = R_2 = $ Me and $R_1 = $ OH, $R_2 = $ H) (Ōki and Nakamura, 1971), perfluoro heteroaromatic systems [e.g., **XXXIV**, X = F or $CF(CF_3)_2$] (Chambers et al., 1972) and natural or synthetic biflavanoids of type **XXXV** (Weinges et al., 1970; du Preez et al., 1971; Thompson et al., 1972) have been reported, but only two examples—**XXXVI** (Bartlett and Tidwell, 1968) and **XXXVII**, R = H,Cl (Casy and Ison, 1969)—are known where the process does not involve the hindered rotation of an aromatic nucleus.

In all examples discussed so far, the rotational barriers were clearly dominated by bulk effects. However, a recent study (Anderson et al., 1973) of four *p*-substituted 2-phenyl-3,3-dimethylbutanes (**XXXVIII**) demonstrated not only another example of the ground state effect (entries 77 and 79, 78 and 80 in Table III), but a significant electronic effect (entries 77 and 78, 79 and 80) due to the remote substituent.

In analogy with work cited in Section II, a number of workers investigated the temperature dependence of averaged spectra of compounds where conformational motion about sp^3–sp^2 carbon–carbon single bonds is possible, the most notable examples being the work on propene derivatives (Bothner-By et al., 1966; Woller and Garbisch, 1972b). Nuclear magnetic resonance studies in nematic phase (Diehl et al., 1971a,b; Burnell and Diehl, 1972), as well as pulsed nmr studies in solid state (Allen and Cowking, 1968) and studies of relaxation times (Alger et al., 1972), give otherwise inaccessible information about the average conformation and small ($E_a \cong 1$–3 kcal/mole) rotational barriers of methyl groups attached to aromatic nuclei.

IV. Rotation about sp^2–sp^2 Carbon–Carbon Single Bonds

The potential energy profiles corresponding to this type of rotation may be dominated by either steric factors, which favor the configuration with the

6. Rotation about Single Bonds

p-orbitals at right angles, or electronic factors, which favor the coplanar configuration. Examples of both types have been studied by dnmr.

Hindered rotation in a number of biphenyl derivatives (Table IV) takes place at a rate suitable for dnmr investigation, which complement and partially overlap (Theilacker and Böhm, 1967) the range accessible to study by measuring the rates of racemization. Most of the studies involved biphenyl derivatives with substituents which can be used to sense the presence of a chiral center, e.g., —CH_2—X and —$CHMe_2$.

XXXIX **XL** **XLI**

No systematic work on the influence of the bulk of substituents at C-2 and C-2′ on the size of the barrier to rotation in simple biphenyls has so far been reported, but the influence of through-conjugation, as well as of bond-bending effects (Ōki and Yamamoto, 1971), can be seen in the results for 2-isopropyl-2′-methoxybiphenyls (entries 5–13, Table IV). In particular, there is a significant decrease in the barrier height in 4,4′-nitromethoxy derivatives as compared with the unsubstituted compound (entries 5, 10, and 12 in Table IV) as expected from the stabilization of the high-energy coplanar state by resonance. Another systematic study (Ōki et al., 1971) deals with the influence of OH–π bonding on the rotational barriers in a number of 2-isopropyl-2′-hydroxybiphenyls (entries 14–18, Table IV).

In aromatic aldehydes and derivatives of acetophenone (Table V), it can be clearly seen that the introduction of electron-donating para substituents causes an increase in the height of the barrier by stabilizing the coplanar ground state through resonance effects. Similarly, protonation with SbF_5– FSO_3H (Jost et al., 1971a,b) or complexation with boron trifluoride (Greenvald and Rabinovitz, 1969) causes an increase in the height of the barrier, while the substitution of a methyl group for the aldehydic proton causes a marked decrease in the height of the barrier (Klinck et al., 1967), presumably by destabilizing the coplanar ground state. In the series of para-substituted α-halogenoacetophenones (entries 12–19, Table V), the above trends are

TABLE IV
BARRIERS TO ROTATION IN BIPHENYL DERIVATIVES

Entry	Biphenyl	$\Delta G^{\ddagger}$ (kcal/mole)	(°C)[a]	Reference
1	2,2'-di-CH$_2$—O—C(=O)—CH$_3$	20.2	(127)	Meyer and Meyer (1963); Kessler (1970b)
2	2,2'-di-CHMe$_2$	>26.1	(>200)	Kessler (1970b); Ōki and Yamamoto (1971)
3	2,2'-di-C(OH)Me$_2$	>24.5	(>190)	Kessler (1970b); Ōki and Yamamoto (1971)
4	2-CHMe$_2$-2'-COOMe	—[b,c]		Cable et al. (1972)
5	2-CHMe$_2$-2'-OMe	19.2 ± 0.3	(86)	Ōki and Yamamoto (1971)
6	2-CHMe$_2$-2'-OMe-4'-OMe	18.7 ± 0.3	(86)	Ōki and Yamamoto (1971)
7	2-CHMe$_2$-2'-OMe-4'-NO$_2$	19.2 ± 0.3	(86)	Ōki and Yamamoto (1971)
8	2-CHMe$_2$-2'-OMe-4-OMe	18.3 ± 0.3	(86)	Ōki and Yamamoto (1971)
9	2-CHMe$_2$-2'-OMe-4,4'-di-OMe	18.0 ± 0.3	(86)	Ōki and Yamamoto (1971)
10	2-CHMe$_2$-2'-OMe-4-OMe-4'-NO$_2$	18.2 ± 0.3	(86)	Ōki and Yamamoto (1971)
11	2-CHMe$_2$-2'-OMe-4-NO$_2$	18.9 ± 0.3	(86)	Ōki and Yamamoto (1971)
12	2-CHMe$_2$-2'-OMe-4-NO$_2$-4'-OMe	18.0 ± 0.3	(86)	Ōki and Yamamoto (1971)
13	2-CHMe$_2$-2'-OMe-4,4'-di-NO$_2$	19.1 ± 0.3	(86)	Ōki and Yamamoto (1971)
14	2-CHMe$_2$-2'-OH	19.2	(86)	Ōki et al. (1971)
15	2-CHMe$_2$-2'-OH-4'-NO$_2$	19.2	(86)	Ōki et al. (1971)
16	2-CHMe$_2$-2'-OH-5'-NO$_2$	18.7	(86)	Ōki et al. (1971)
17	2-CHMe$_2$-2'-OH-4-NO$_2$	18.9	(86)	Ōki et al. (1971)
18	2-CHMe$_2$-2'-OH-5-NO$_2$	18.4	(86)	Ōki et al. (1971)
19	XXXIX[d]	16	(25)	House et al. (1970)
20	XL, R = —COMe	18.6 ± 0.3	(46)	Colebrook and Jahnke (1968)
21	XL, R = —Me	14.2 ± 0.2	(−11)	Colebrook and Jahnke (1968)
22	XLI	—[c,e]		Bentley et al. (1969)

[a] Free energy of activation to rotation about the bond indicated by an arrow on the appropriate structural formula at temperatures indicated in parentheses. Where two ground states are unequally populated, the average value of $\Delta G^{\ddagger}$ is quoted.
[b] Not determined quantitatively, but variable temperature spectra indicate that the barrier is in region that can be conveniently investigated by dnmr.
[c] The temperature of coalescence is increased by > 50°C on addition of Eu(dpm)$_3$.
[d] Variation of nmr spectra with temperature was also reported for several other 1,8-diarylnaphthalenes, but no quantitative evaluations could be made.
[e] This compound, a degradation product of thebaine, can be obtained in an optically active form.

TABLE V
Barriers to Rotation in Aromatic Aldehydes and Aryl Methyl Ketones

Entry	Compound	$\Delta G^{\ddagger}$ (kcal/mole)	(°C)[a]	Reference
1	XLII, R = X = H	7.9	(−123)	Anet and Ahmad (1964)
2	XLII, R = H, X = NMe$_2$	10.8	(−71)	Anet and Ahmad (1964)
3	XLII, R = H, X = NMe$_2$[b]	10.5	(−56)	Klinck et al. (1967)
4	XLII, R = H, X = OMe	9.2	(−99)	Anet and Ahmad (1964)
5	XLII, R = H, X = OMe[b]	9.4	(−75)	Klinck et al. (1967)
6	XLII, R = H, X = OMe, BF$_3$ complex	14.2	(−17)	Greenvald and Rabinovitz (1969)
7	XLII, R = H, X = Me[c]	16.3	(40)	Jost et al. (1971a)
8	XLII, R = H, X = Cl[c]	14.8		Jost et al. (1971a)
9	XLII, R = H, X = Br[c]	14.8		Jost et al. (1971a)
10	XLII, R = Me, X = NMe$_2$	8.5	(−92)	Klinck et al. (1967)
11	XLII, R = Me, X = OMe	≤7.3	(−116)	Klinck et al. (1967)
12	XLII, R = CH$_2$F, X = Me	12.45 ± 0.5[d]		Jost et al. (1971b)
13	XLII, R = —CH$_2$Cl, X = Me	12.42 ± 0.5[d]		Jost et al. (1971b)
14	XLII, R = —CH$_2$Br, X = Me	12.21 ± 0.5[d]		Jost et al. (1971b)
15	XLII, R = —CH$_2$I, X = Me	11.95 ± 0.5[d]		Jost et al. (1971b)
16	XLII, R = —CH$_2$F, X = Cl	11.28 ± 0.5[d]		Jost et al. (1971b)
17	XLII, R = —CH$_2$Cl, X = Cl	11.20 ± 0.5[d]		Jost et al. (1971b)
18	XLII, R = —CH$_2$Br, X = Cl	11.19 ± 0.5[d]		Jost et al. (1971b)
19	XLII, R = —CH$_2$I, X = Cl	11.10 ± 0.5[d]		Jost et al. (1971b)
20	XLIII, X = CH	10.9 ± 0.2	(−23)	Dunlop et al. (1971)
21	Furan-2-aldehyde	10.7 ± 0.3[e]	(−57 to −81)	Dahlquist and Forsén (1965)[f]
22	N-Methylpyrrole-2-aldehyde	10.7 − 12.5[e]	(−57)	Arlinger et al. (1970b)[f]
23	Methyl 2-furanyl ketone	9.1 ± 0.6[e]	(−70 to −90)	Arlinger et al. (1970a)[f]

[a] Free energy of activation to rotation about the bond indicated by an arrow on the appropriate structural formula at temperatures indicated in parentheses.
[b] 3,5-Dideutero derivative.
[c] Protonated form.
[d] At temperature of coalescence.
[e] Averaged value for unequally populated planar rotamers.
[f] Closely related systems also exhibit a significant variation in the magnitude of the nuclear Overhauser effect between the aldehydic proton and H-3 with temperature (Roques et al., 1971; Combrisson et al., 1971) that can be used to obtain an independent measure of the dynamic processes involved.

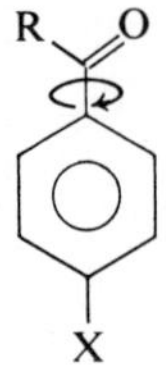

XLII 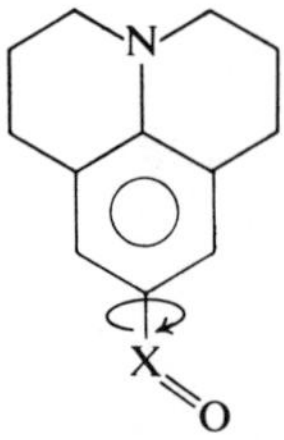 XLIII

further exemplified, the increase in the bulk of the halogen and the decrease in the electron-donating power of the para substituent both being associated with a lowering of the barrier to rotation.

By contrast, in the case of more heavily substituted aromatic ketones, the ground state corresponds to a perpendicular arrangement of the benzene ring and the carbonyl system and the barriers (Table VI) are dominated by steric effects.

It can be seen that the barriers to rotation in this series are very appreciable and, in fact, compounds such as **XLIV** ($R = CMe_3$, $R_1 = R_3 = R_5 =$ Me, $R_2 = H$, $R_4 = COOH$) can be resolved (Pinkus et al., 1968). Certain readily recognizable trends emerge from the series of aryl isopropyl and aryl

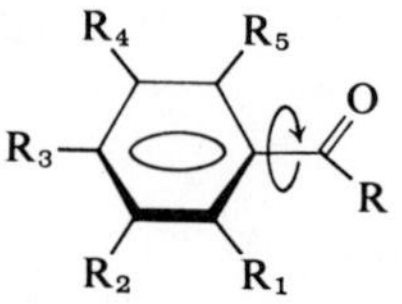

XLIV

benzyl ketones (Table VI, entries 5–10). The height of the barrier depends (Nakamura and Ōki, 1972) on the bulk of the ortho substituents (cf., e.g., entries 7 and 8, Table VI) and on the bulk of the alkyl groups (cf., e.g., entries 8 and 10, Table VI) and is appreciably increased by buttressing effects (cf., e.g., entries 5 and 7, Table VI). On the other hand, the decrease in the barrier height associated with decreased ortho substitution noted in entries 1 and 3, Table VI, is unexpected and was attributed (Lauer and Staab, 1969) to crowding in the ground state in the first case. The preferred conformation of 2,4,6,2′,4′,6′-hexa-*t*-butylbenzil (Lauer and Staab, 1969) (entry 10, Table VI) cannot be inferred in a straightforward manner from its nmr spectrum.

Numerous examples of appreciable barriers to rotation about the aryl–carbonyl bond have been detected by dnmr in amides, and the available data have been summarized by Stewart and Siddall (1970). Thus, e.g., the methoxy groups in **XLV** (Siddall and Garner, 1966) give rise to two pairs of signals due

TABLE VI
Barriers to Rotation in Heavily Substituted Ketones

Entry	Compound	$\Delta G^{\ddagger}$ (kcal/mole) (°C)[a]		Reference
1	XLIV, R = 2,4,6-tri-t-butylphenyl, $R_1 = R_3 = R_5 = $ H, $R_2 = R_4 = $ Me	17.7 ± 0.2	(91)	Lauer and Staab (1969)
2	XLIV, R = 2,4,6-tri-t-butylphenyl, $R_1 = R_3 = R_5 = $ OMe, $R_2 = R_4 = $ H	18.5 ± 0.2	(91)	Lauer and Staab (1969)
3	XLIV, R = 2,4,6-tri-t-butylphenyl, $R_1 = R_3 = R_5 = $ Me, $R_2 = R_4 = $ H	17.3 ± 0.4	(81)	Lauer and Staab (1969)
4	XLIV, R = —CO-2,4,6-tri-t-butylphenyl, $R_1 = R_3 = R_5 = $ t-butyl, $R_2 = R_4 = $ H	16.7 ± 0.2	(62)	Lauer and Staab (1969)
5	XLIV, R = CHMe$_2$, $R_1 = R_4 = $ Br, $R_2 = R_5 = $ Me, $R_3 = $ H	14.5 ± 0.2	(−3)	Nakamura and Ōki (1972)
6	XLIV, R = CHMe$_2$, $R_1 = R_4 = $ Cl, $R_2 = R_5 = $ Me, $R_3 = $ H	12.5 ± 0.2	(−40)	Nakamura and Ōki (1972)
7	XLIV, R = CHMe$_2$, $R_1 = R_3 = $ Me, $R_2 = R_4 = $ H, $R_5 = $ Br	10.1	(−77)	Nakamura and Ōki (1972)
8	XLIV, R = CHMe$_2$, $R_1 = R_3 = $ Me, $R_2 = R_4 = $ H, $R_5 = $ OMe	9.0	(−95)	Nakamura and Ōki (1972)
9	XLIV, R = CH$_2$Ph, $R_1 = R_4 = $ Br, $R_2 = R_5 = $ Me, $R_3 = $ H	12.4 ± 0.2	(−27)	Nakamura and Ōki (1972)
10	XLIV, R = CH$_2$Ph, $R_1 = R_3 = $ Me, $R_2 = R_4 = $ H, $R_5 = $ OMe	8.8	(−95)	Nakamura and Ōki (1972)

[a] Free energy of activation to rotation about the bond indicated by arrow on the appropriate structural formula at temperatures indicated in parentheses.

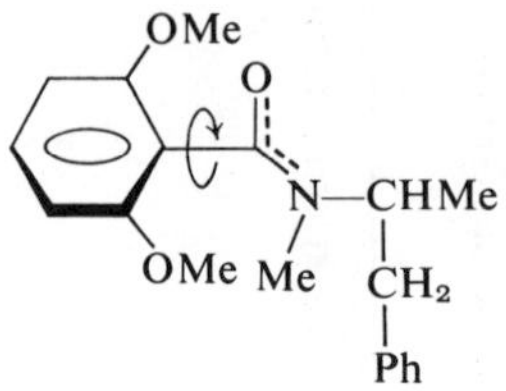

to the presence of cis–trans isomerism about the peptide bond and to restricted rotation about the aryl–carbonyl bond, where the barrier is estimated at more than 20 kcal/mole. The signals coalesce to a doublet at 130°C and finally to a singlet at 165°C.

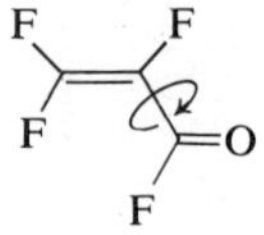

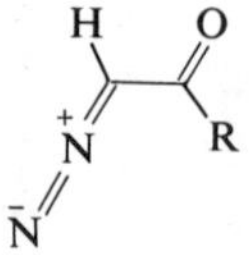

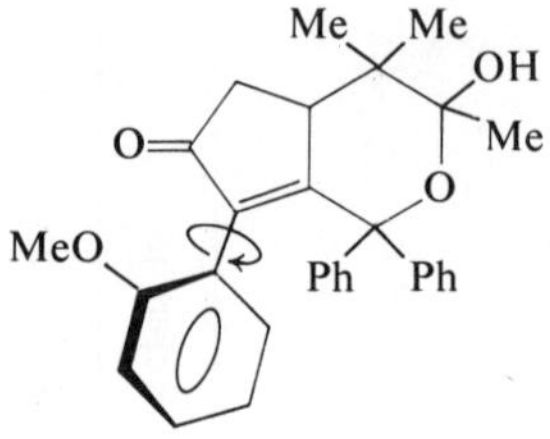

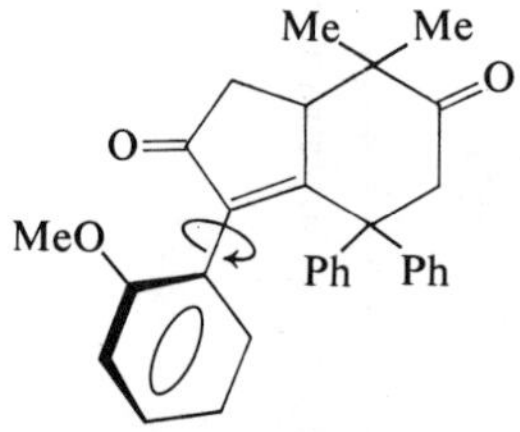

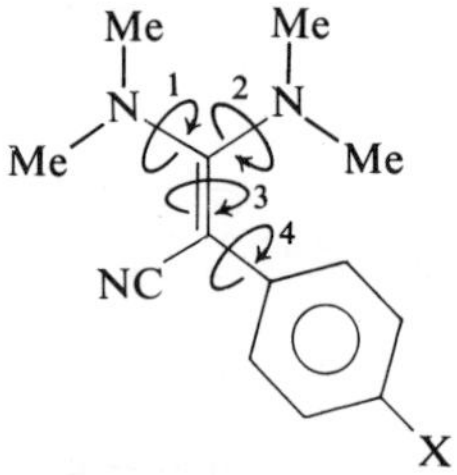

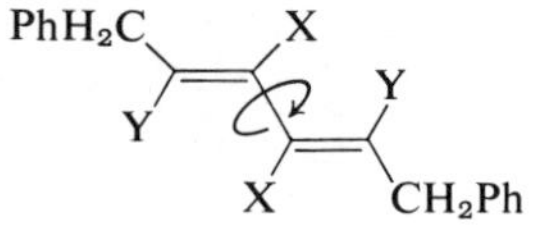

Other carbonyl derivatives where rotation about single bonds is slow on the nmr time scale are **XLVI** (Brey and Ramey, 1963) and a series of diazoketones (**XLVII**) (Kaplan and Meloy, 1966). In the latter case the barrier is clearly of electronic origin and is closely analogous to restricted rotation about the peptide bond (see the following chapter). The height of the barrier is virtually unaffected by the nature of the alkyl group ($\Delta G^{\ddagger}_{25°C}$ = 15.3 – 15.4 kcal/mole for R = Me, Et, or CH_2Ph) but is significantly lower for the diazo esters ($\Delta G^{\ddagger}_{25°C}$ = 12.8 and 13.3 kcal/mole for R = OMe and OEt, respectively), presumably due to the competitive electron back-donation by the oxygen.

The only examples of high rotational barriers of steric origin about a single bond joining a benzene ring to a nonaromatic sp^2 hybridized carbon atom are found in the interesting rearrangement products (Kleinfelter et al., 1969) **XLVIII** and **XLIX**, which have been investigated in these laboratories (G. C. Brophy, D. C. Kleinfelter, and S. Sternhell, unpublished data, 1973). The free energies of activation are quite high and slightly, but significantly, different (**XLVIII**, $\Delta G^{\ddagger}_{142°C}$ = 21.6 kcal/mole; **XLIX**, $\Delta G^{\ddagger}_{102°C}$ = 20.6 kcal/mole), and in each case the populations of the two conformers are not identical, although only remote effects operate.

The temperature dependence of the nmr spectra of a series of 1,1-bis-

(dimethylamino)-2-aryl-2-cyanoethylenes (**L**) investigated by Kessler (1970a) showed that four different rotational processes took place, among them the rotation about the =C—aryl bond (process 4, $\Delta G^{\ddagger}$ = 10–11 kcal/mole) which increases with the increase of the electron-withdrawing power of the group X. Clearly, the barrier is of electronic origin and can be rationalized in simple resonance terms as the above increase is accompanied by a parallel decrease in the barrier to rotation about the formal double bond (process 3).

Some highly substituted benzyl-1,3-butadienes (Köbrich *et al.*, 1972) assume ground-state conformations with the two double bonds perpendicular, thus giving rise to chiral systems and hence, at low rates of rotation about the central bond, to nonequivalence within the pairs of benzylic methylene protons. Within the series **LI**, X = Cl, Br, and I, Y = —CH$_2$Ph, the barrier increases with the size of the halogen as the high-energy planar state involves eclipsing of the halogen and a benzyl group ($\Delta G^{\ddagger}$ is about 21 kcal/mole and > 24 kcal/mole for Y = CH$_2$Ph and X = Cl and I, respectively). The replacement of the eclipsing benzyl group by the smaller chlorine leads to a lowering of the barrier: $\Delta G^{\ddagger}_{75°C}$ for **LI** (X = Y = Cl) is 17.6 ± 0.3 kcal/mole.

V. Rotation about Carbon–Nitrogen Single Bonds

Dynamic nuclear magnetic resonance studies of processes of this type are complicated by the fact that in many cases the slowing down of inversion about the pyramidal nitrogen atom and of rotation about the bond joining this atom to a carbon atom may have the same qualitative effect, e.g., the appearance of a chiral center at low rates. Further complications may arise in the case of protonated amines, amides, or related compounds, where additional dynamic processes (proton exchange and cis–trans isomerization about the peptide bond, respectively) often occur at a rate that is reflected in nmr spectra. The identification of individual processes in such cases is most reliably based on symmetry grounds, but may have to be made on the basis of arguments involving the actual magnitude of the barriers.

In many examples (see below), the barrier to rotation about a formal C—N single bond is clearly due to considerable double bond character, and in at least one case (**LII**) (Mannschreck and Köelle, 1969), the rotation about

LII

TABLE VII
BARRIERS TO ROTATION IN AMINES

Entry	Compound	$\Delta G^{\ddagger}$ (kcal/mole)	(°C)[a]	Reference
1	LIII, R_1 = Me, R_2 = CH_2Ph	6.2 ± 0.2[b]	(−138)	Bushweller et al. (1971a)
2	LIII, R_1 = H, R_2 = CMe_3	6.0	(−153)	Bushweller et al. (1971b)
3	LIII, R_1 = R_2 = Cl	9.4 ± 0.2	(−92)	Bushweller et al. (1973)
4	LIII, R_1 = Cl, R_2 = Me	8.6 ± 0.2	(−114)	Bushweller et al. (1973)
5	LIII, R_1 = Br, R_2 = Me	8.3 ± 0.2	(−121)	Bushweller et al. (1973)
6	LIII, R_1 = Cl, R_2 = CH_2CD_3	8.6 ± 0.2	(−109)	Bushweller et al. (1973)
7	LIII, R_1 = Cl, R_2 = CH_2Ph	8.3 ± 0.2[c]	(−122)	Bushweller et al. (1973); Bushweller and O'Neil (1971)
8	LIII, R_1 = Cl, R_2 = $CH(CD_3)_2$	7.0 ± 0.4	(−143)	Bushweller et al. (1973)
9	LIII, R_1 = R_2 = Me	6.0 ± 0.1	(−153)	Bushweller et al. (1973)
10	LIV	10.0 ± 0.1	(−79)	Bushweller et al. (1971c)
11	LV	6.1	(−146)	Brownstein et al. (1970)
12	LVI, R = $CHMe_2$	<9.6	(0)	Price et al. (1967); Brownstein et al. (1970)
13	LVII, R = CMe_3	15.6	(0)	Price et al. (1967); Brownstein et al. (1970)
14	LVII, R = $CHMe_2$	$\Delta H^{\ddagger} = 10.8 \pm 0.6$		Brownstein et al. (1969)
15	LVII, R = CMe_3	$\Delta H^{\ddagger} = 16.9 \pm 0.3$		Brownstein et al. (1969)
16	LVIII, Ar = o-tolyl	24.3	(25)	Bentz et al. (1970)
17	LVIII, Ar = 1-naphthyl	23.1	(141)	Colebrook et al. (1972)

6. Rotation about Single Bonds

#	Compound	Substitution	E_A	Reference
18	LIX		—[d]	Khetan and George (1969)
19	LX, $R_1 = D$, $R_2 = Me$, $R_3 = R_4 = R_5 = NO_2$		$E_A = 14.5 \pm 0.3$	Heidberg et al. (1964)
20	LX, $R_1 = H$, $R_2 = NPh_2$, $R_3 = R_4 = R_5 = NO_2$		$E_A = 12.5 \pm 0.2$	Heidberg et al. (1964)
21	LX, $R_1 = H$, $R_2 = n\text{-Pr}$, $n\text{-Bu}$, $n\text{-hexyl}$, or $n\text{-octyl}$, $R_3 = R_4 = R_5 = NO_2$		$10.7 - 11.1$ (25)	Jouanne and Heidberg (1971)
22	LX, $R_1 = H$, $R_2 = Me$, $R_3 = R_4 = R_5 = NO_2$		10.5 (25)	Jouanne and Heidberg (1973)
23	LX, $R_1 = H$, $R_2 = CHMe_2$, $R_3 = R_4 = R_5 = NO_2$		9.2 (25)	Jouanne and Heidberg (1973)
24	LX, $R_1 = R_2 = H$, $R_3 = R_4 = H$, $R_5 = COMe$		10.6 ± 0.5 (−40)	Wasylishen and Schaefer (1971)
25	LXI		14.9 (10)	Shoup et al. (1972a,b); Martin and Reese (1967)
26	LXII		15.6 (24)	Almog et al. (1972)
27	LXIII		17.2 (70)	Almog et al. (1972)
28	LXIV		12.1 (−28)	Almog et al. (1972)
29	LX, $R_1 = R_3 = Me$, $R_2 = CH_2Ph$, $R_4 = Cl$		15.1 ± 0.2 (30)	Mannschreck and Muensch (1968)
30	LX, $R_1 = R_3 = Me$, $R_2 = CH_2Ph$, $R_4 = Br$		15.8 ± 0.2 (43)	Mannschreck and Muensch (1968)
31	LX, $R_1 = R_3 = R_4 = Me$, $R_2 = CHMePh$		17.6 ± 0.2 (63.5)	Mannschreck and Muensch (1968)

[a] Free energy of activation to rotation about the bond indicated by an arrow on the appropriate structural formula at temperatures indicated in parentheses.
[b] There is a common transition state for inversion and rotation.
[c] The barrier to inversion is somewhat larger ($\Delta G^\ddagger = 9.0$ at $-84°C$).
[d] Not determined quantitatively, but the process is slow on the nmr time scale.

a formal =C—N bond (process 1) is more hindered than the rotation about the conjugated formal carbon–carbon double bond (process 2), the free energies of activation being 21.3 and 19.2 kcal/mole, respectively.

Rotational barriers in aliphatic amines have been extensively investigated (entries 1–11, Table VII) in spite of the generally low energies of activation and the complicating presence of inversion processes. A particularly extensive study concerns the rotation of the *t*-butyl group in (**LIII**). It can be seen that in several examples the increased bulk of the substituent at the nitrogen (entries 4, 6, and 8 in Table VII) may be associated with a decrease in the barrier height, presumably due to the destabilization of the ground state.

LIII

LIV

$(PhCH_2)_2NCMe_3$

LV

LVI

LVII

LVIII

LIX

LX

LXI

LXII

LXIII

LXIV

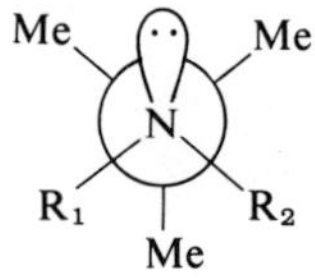

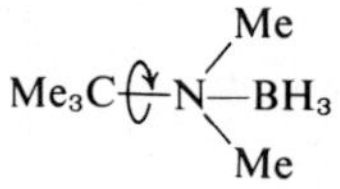

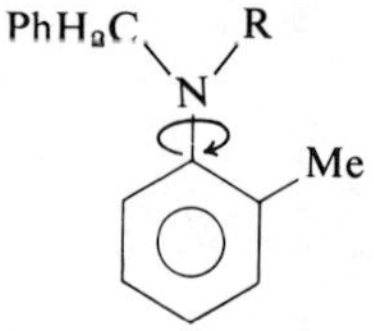

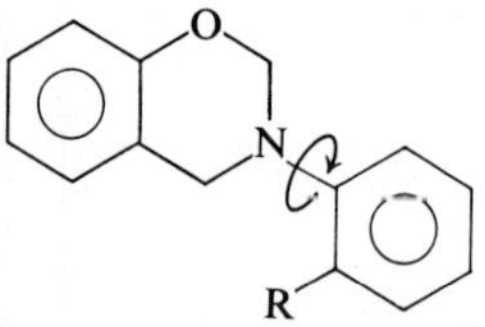

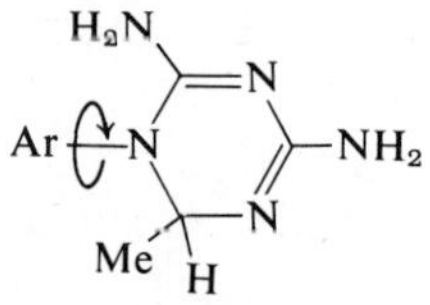

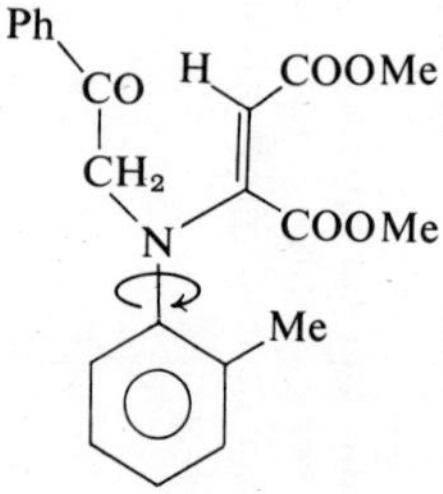

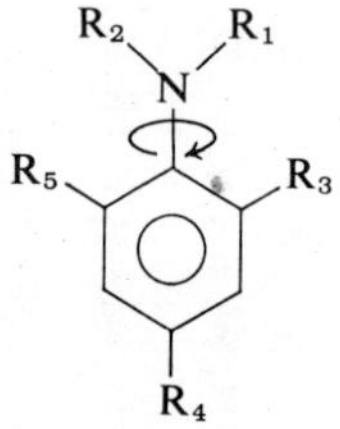

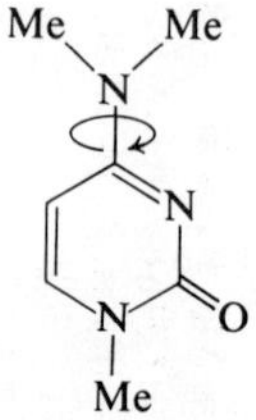

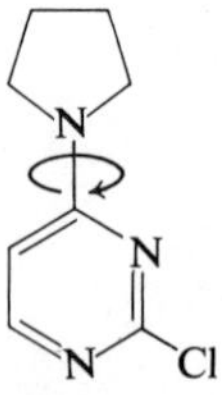

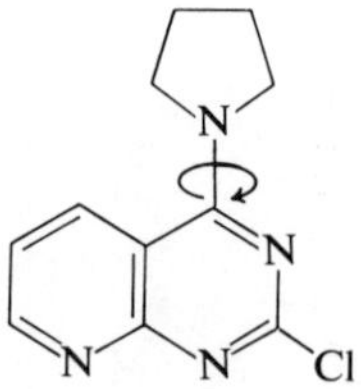

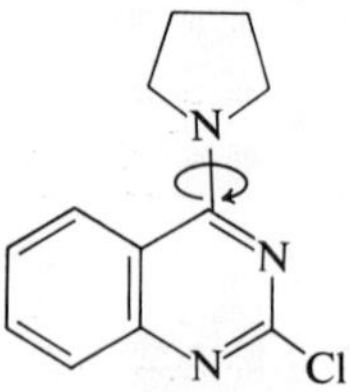

6. Rotation about Single Bonds

The rotation about the aryl–nitrogen bond in aromatic amines may be dominated by steric factors (cf., e.g., entries 12–17 and 29–31 in Table VII), by intramolecular hydrogen bonding (entries 19–24, Table VII), or by back-donation of the nitrogen lone pair into a heterocyclic ring (entries 25–28, Table VII). The latter effect has been the subject of some theoretical calculations (Almog et al., 1972).

Evidence for restricted rotation in protonated amines (cf., e.g., Reynolds and Schaefer, 1964: Vigevani et al., 1971) and anilines (Mannschreck et al., 1970) has been obtained from nmr data, but few quantitative results are available. In the case of protonated anilines (Mannschreck and Muensch, 1968) the barrier to rotation about the nitrogen–aryl bond is larger than in the corresponding unprotonated amines. The preferred conformation in **LXV** and the barrier to rotation ($\Delta G^{\ddagger}_{26°C}$ = 15.9 ± 0.3 kcal/mole) have been established by dnmr (Mannschreck and Ernst, 1968).

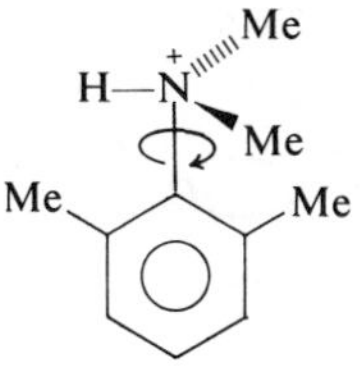

LXV

Many examples of high barriers to rotation about the nitrogen–aryl bond in anilides and related compounds have been observed by dnmr techniques (Siddall and Prohaska, 1966; Price et al., 1967; Shvo et al., 1967; Seymour and Jones, 1967; Kessler, 1968; Siddall and Stewart, 1969a,c; Bentz et al., 1970; Colebrook et al., 1972), and the field has been thoroughly reviewed by Stewart and Siddall (1970). The process is dominated by steric factors and the transition state is generally planar and as shown in **LXVI**. The energy barriers may be high, e.g., **LXVI**, R = H, $\Delta G^{\ddagger}_{124°C}$ = 20.5 kcal/mole; **LXVI**, R = Me, $\Delta G^{\ddagger}_{-53°C}$ = 23.0 kcal/mole (Siddall and Stewart, 1969c), thus exhibiting a typical buttressing effect.

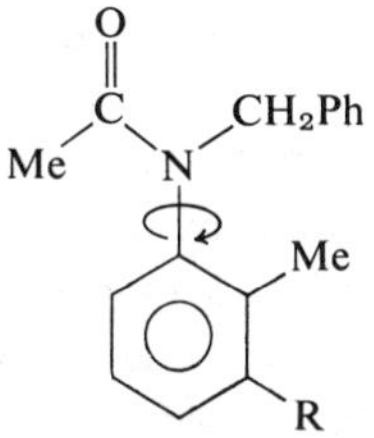

LXVI 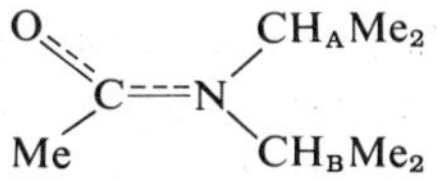   **LXVII**

An unusual type of temperature dependence of nmr spectra has been reported for N,N-diisopropylacetamide (**LXVII**) (Siddall and Stewart, 1968c); the resonances assigned to H_A and H_B are sharp at 0°C. The downfield resonance assigned to H_A remains sharp at $-37°C$, while that assigned to H_B is selectively broadened. At $-60°C$, both resonances are equally sharp and very close to their original positions without any additional signals appearing. Analogous changes were also observed in the nmr spectra of related amides, and while a complete interpretation was not possible, the rotation about the N—CH_B bond is apparently slow on the nmr time scale.

High rotational barriers in conjugated enamines (e.g., **LII**) (Mannschreck and Köelle, 1969) have already been alluded to and are still substantial in less conjugated enamines. Thus, the barrier in **LXVIII** (X = CH) has $\Delta G^\ddagger_{16.5°C} = 13.4$ kcal/mole (Mannschreck and Köelle, 1967), while the

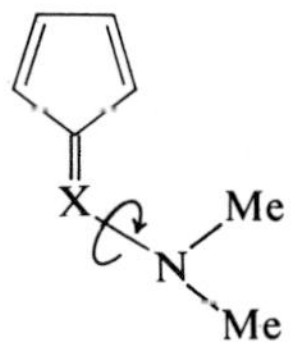

LXVIII 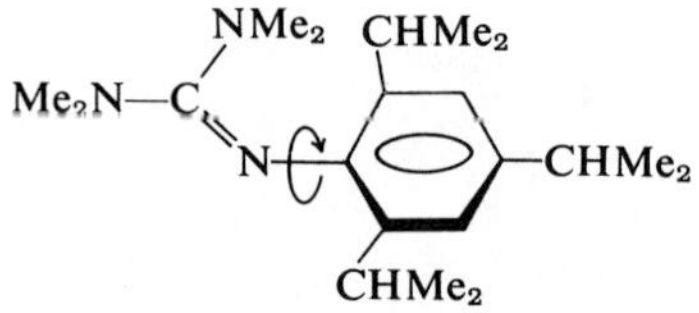   **LXIX** 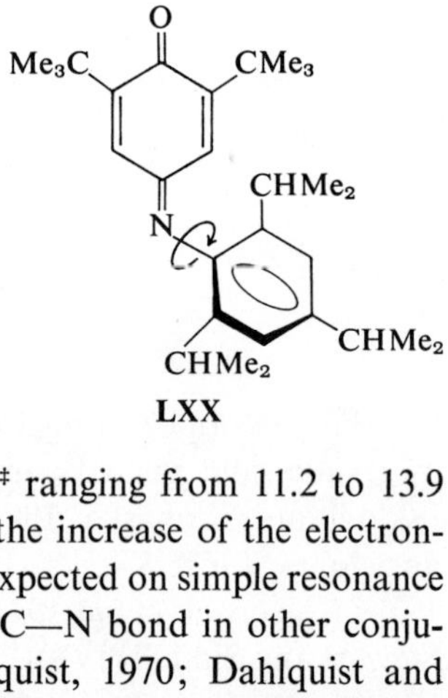   **LXX**

barriers to the processes 1 and 2 in **L** have $\Delta G^\ddagger$ ranging from 11.2 to 13.9 kcal/mole (Kessler, 1970a) and increasing with the increase of the electron-withdrawing power of the substituent X in **L**, as expected on simple resonance considerations. Restricted rotation about the =C—N bond in other conjugated enamines has also been reported (Dahlquist, 1970; Dahlquist and Forsén, 1970; Shvo and Shanan-Atidi, 1969; Dabrowski and Kamienksa-Trela, 1972), and in each case, the process is clearly dominated by electronic factors. In fact, in most of these examples, the compounds can be considered as vinylogous amides or closely related species. By contrast, the barriers to rotation in nitrogen–aryl imines ($\Delta G^\ddagger$ for **LXIX** = 11.6 kcal/mole; $\Delta G^\ddagger_{62°C}$ for **LXX** = 18.8 kcal/mole) appear to be dominated by steric factors (Kessler, 1968; Kessler and Leibfritz, 1970, 1971; Rieker and Kessler, 1967).

The restricted rotation about the Ar—NO bond in p-nitroso-N,N-dimethylaniline has been extensively investigated by dnmr (MacNicol et al., 1965; Korver et al., 1966; MacNicol, 1969; Mackenzie and MacNicol, 1970) and found to be associated with a free energy of activation of about 12 kcal/mole. The process is clearly dominated by electronic factors and the barrier is substantially increased ($\Delta G^\ddagger_{33°C} = 16.8$ kcal/mole) by protonation, which takes place on the oxime oxygen (MacNicol et al., 1966). In fact, the barrier height to this process has been found to correlate well with the Hammett constants of the para substituent in a series of para-substituted nitrosobenzenes (Calder and Garratt, 1969). Interestingly, the barrier to rotation of the nitroso group in **XLIII** (X = N) is significantly higher ($\Delta G^{\ddagger}_{23°C}$ = 14.2 ± 0.2 kcal/mole, Dunlop et al., 1971) than in p-nitroso-N,N-dimethylaniline, suggesting more efficient back-donation of the lone pair on the nitrogen atom due to steric constraint. Nuclear magnetic resonance data of some relevance to the problem of restricted rotation in aromatic nitroso derivatives are also given by Calder et al. (1967), Norris and Sternhell (1969), and Matsubayashi et al. (1970).

A more detailed study (MacNicol, 1969; Mackenzie and MacNicol, 1970) revealed that a second dynamic process can be studied by dnmr in p-nitroso-N,N-dimethylaniline, viz., the rotation about the aryl–nitrogen bond ($\Delta G^{\ddagger}_{76°C}$ = 10.6 ± 0.2 kcal/mole). Somewhat lower barriers to rotation about the same bond were found in p-N,N-dimethylaminobenzaldehyde ($\Delta G^{\ddagger}_{140°C}$ = 8.6 kcal/mole) and p-N,N-dimethylaminoacetophenone ($\Delta G^{\ddagger}_{140°C}$ = 7.6 kcai/mole) (Mackenzie and MacNicol, 1970). The comparison of these values with the corresponding values in heteroaromatic systems (cf. entries 25–28, Table VII) affords an interesting insight into the relative electron-withdrawing powers of heteroaromatic rings and benzene rings with − R substituents in the para positions.

VI. Rotation about Miscellaneous Single Bonds

As the exact electronic configuration of a carbon atom bearing a formal positive or negative charge is not always known, rotation about C—C single bonds in carbanions and carbonium ions was not considered in Sections II–IV above.

A number of authors (Schuster et al., 1968; Breslow et al., 1968; Rakshys et al., 1971) have reported temperature dependence of both ^{1}H and ^{19}F nmr spectra of various triarylmethylcarbonium ions and concluded that the dynamic process involved rotation about the aryl—C$^+$ bonds. Several related processes can actually be postulated, e.g., concerted and individual rotation of the three aryl groups, and it has been possible to decide between them on symmetry grounds in some cases. Few quantitative data are available, but the three-ring flips in some fluorinated tritylcarbonium ions have enthalpies of activation in the range of 8.6–9.1 kcal/mole (Schuster et al., 1968).

Slow rotation about carbon–carbon bonds has been observed in isomeric pentadienyllithium derivatives (Bates et al., 1967), diphenyl methide (Kloosterziel, 1968), and the phenylallyl salts of lithium, sodium, and potassium

(**LXXI**), where the very substantial barrier to rotation ($\Delta G^\ddagger = 15.7$–20.1 kcal/mole, depending on solvent and counterion) indicates extensive charge delocalization (Sandel *et al.*, 1968).

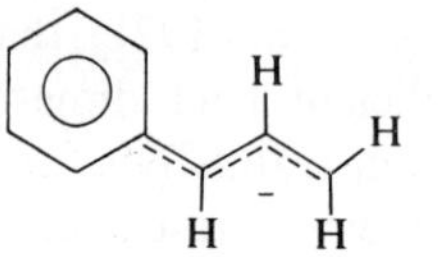

LXXI

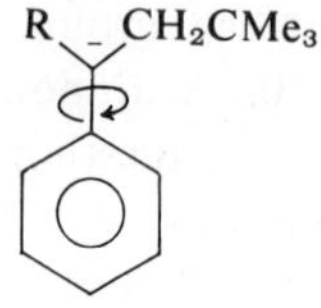

LXXII

Steric factors also influence the heights of the barriers to rotation about formal single bonds joining phenyl groups to the anionic center. Thus, in the styryl anions (**LXXII**), the barrier increases from 11.9 ± 0.8 kcal/mole for R = H to 14 ± 1 kcal/mole for R = Me (Brownstein and Worsfold, 1972). However, the barriers are clearly connected with the delocalization of the negative charge into the benzene rings, because they are substantially higher than those in comparable hydrocarbons (cf. entries 70 and 71, Table III).

Slow rotation about aryl–oxygen bonds has been studied by dnmr in several classes of compounds. In the esters (**LXXIII**), the barrier increases from about 11 kcal/mole for R = Me to larger values for R = $CHMe_2$

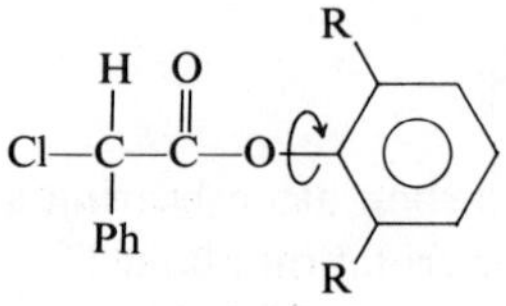

LXXIII

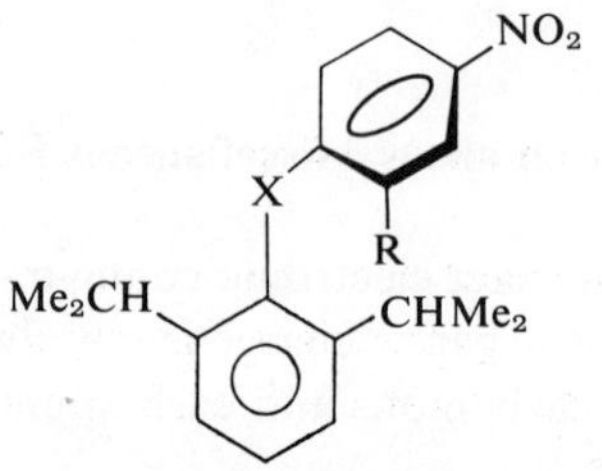

LXXIV

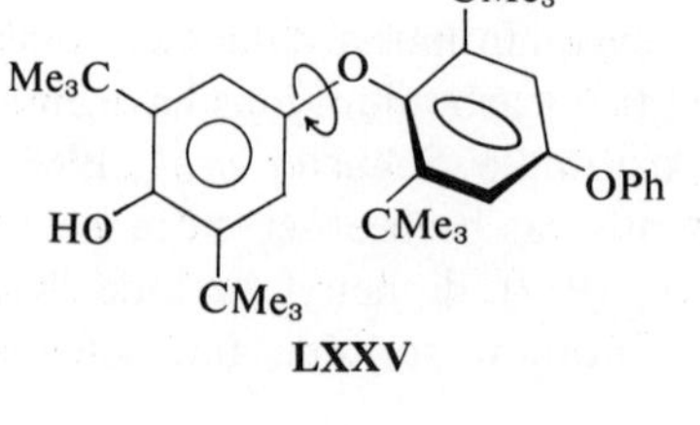

LXXV

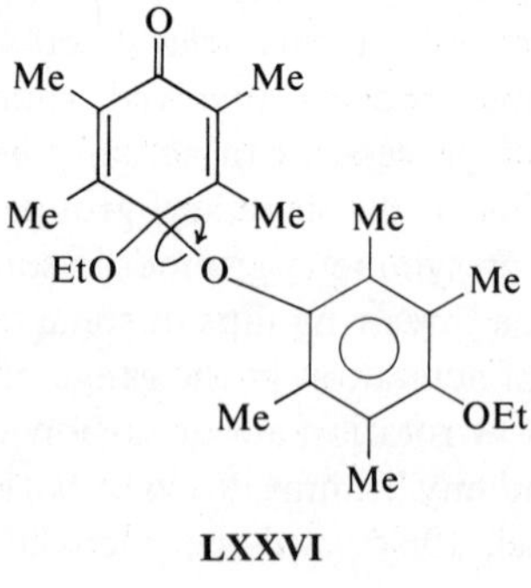

LXXVI

6. Rotation about Single Bonds

(Siddall *et al.*, 1967), and a similar steric effect can be seen in diphenyl ethers (**LXXIV**) (X = O), where the barrier to rotation rises from 10.6 kcal/mole for R = H to 17.8 kcal/mole for R = NO_2 (Bergman and Chandler, 1972). A steric effect has also been invoked to account for the reduction in the barrier to rotation in the corresponding diaryl sulfide (**LXXIV**, X = S, R = NO_2, $\Delta G^{\ddagger}_{12°C}$ = 15.0 kcal/mole) (Kessler *et al.*, 1968a). Another example (**LXXV**) of restricted rotation about an aryl–OAr bond was observed by Bolon (1966), while Huysmans *et al.* (1969) report somewhat lower rotational barriers ($\Delta G^{\ddagger}$ about 9 kcal/mole) in the mixed ketal **LXXVI**. Barriers to rotation about one, or both, of the indicated linkages have also been detected in diaryl disulfides (e.g., **LXXVII**), the highest recorded value being for R = CMe_3

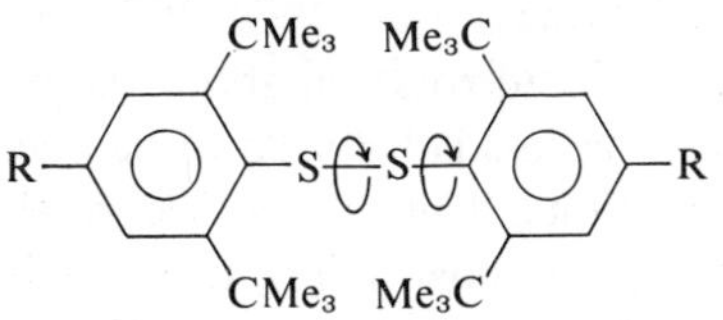

LXXVII

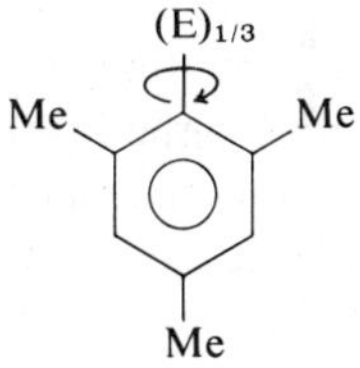

LXXVIII

($\Delta G^{\ddagger}_{54°C}$ = 16.3 kcal/mole). Significantly, the barrier to rotation in the corresponding diselenide is only 12.6 kcal/mole at $-19°C$ (Kessler *et al.*, 1968a; Kessler and Rundel, 1968). An analogous effect can be seen in a series of trimesityl derivatives of group VB elements (**LXXVIII**), where the free energy of activation to the rotation about the aryl–E bond decreases from 12.4 kcal/mole for E = phosphorus to 8.8 kcal/mole for E = arsenic and to undeterminable, but definitely lower, values for the corresponding antimony and bismuth derivatives. Clearly, a steric effect involving the increasing size of the central atom and hence a greater separation of the interacting aryl groups is involved (Rieker and Kessler, 1969).

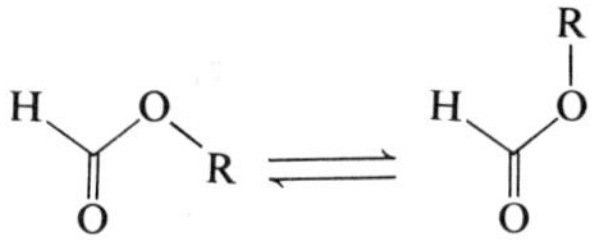

LXXIX

No barriers to rotation about the CO—O bond in esters have so far been determined by dnmr, but Ōki and Nakanishi (1970) reported that the S-trans and S-cis forms of some formates (**LXXIX**) give rise to separate sets of signals in their nmr spectra at about $-100°C$.

The rotation of the tertiary butyl groups in phosphines (**LXXX**) is associated with a free energy of activation of 6.4–6.5 kcal/mole for R = Cl and of 6.0 ± 0.3 kcal/mole for R = CMe₃ (Roberts and Roberts, 1972; Bushweller et al., 1972b). These results can be compared with the similarly substituted C—N and C—C bonds (cf. entry 3, Table VII, and entries 3, 14, 17, and 18, Table II, respectively). It can be seen that the longer C—P bond appears to be associated with a lower barrier to rotation (Robert and Roberts, 1972).

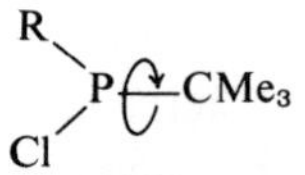

LXXX

A number of substituted hydrazines exhibit dynamic phenomena accessible to study by dnmr (Dewar and Jennings, 1969, 1970, 1973; Anderson et al., 1969), and while inversion and restricted rotation about the N—N bond have both been considered, the latter process is believed to be operative and to involve lone pair–lone pair interactions. The highest recorded barrier is 16.6 kcal/mole for **LXXXI** (Dewar and Jennings, 1973). Even higher barriers have been recorded for diacylhydrazines, e.g., $\Delta G^\ddagger_{147°C} = 23.5 \pm 0.5$ kcal/mole for **LXXXII** (Fletcher and Sutherland, 1969) and $\Delta G^\ddagger_{100°C} = 19.7$ kcal/mole for **LXXXIII** (Verma and Prasad, 1973). Substantial (up to 18.6 kcal/mole) barriers to rotation about the indicated bond have also been reported (Svanholm, 1971) for a series of carbazides, thiocarbazides, and selenocarbazides (**LXXXIV**, X = O, S, or Se).

The remaining examples of restricted rotation about N—N single bonds

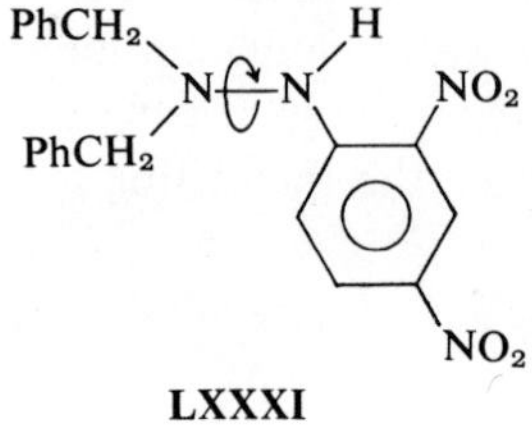

LXXXI 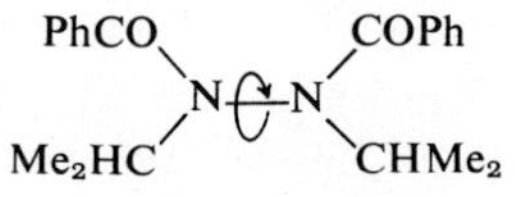   **LXXXII** 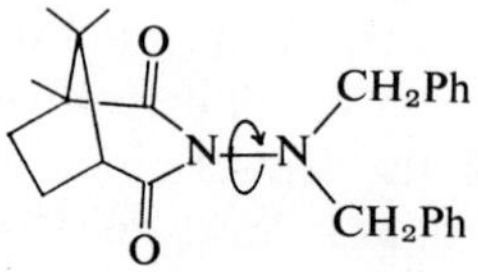   **LXXXIII**

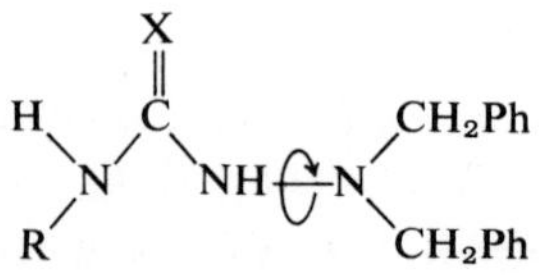

LXXXIV 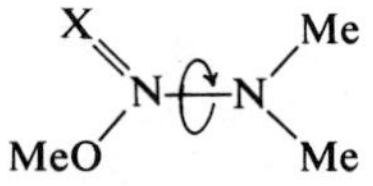   **LXXXV** 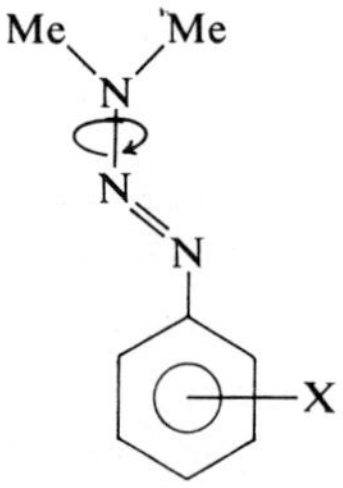   **LXXXVI**

TABLE VIII
Barriers to Rotation about N—X[a] Bonds

Entry	Compound	$\Delta G^{\ddagger}$ (kcal/mole)	(°C)[b]	Reference
1	LXXXVII, R_1 = CH_2Ph, R_2 = Me, R_3 = Me	12.3	(−25)	Raban and Kenney (1969)
2	LXXXVII, R_1 = CH_2Ph, R_2 = $CHMe_2$, R_3 = Me	12.8	(−15)	Raban and Kenney (1969)
3	LXXXVII, R_1 = CH_2Ph, R_2 = Me, R_3 = $CHMe_2$	12.8	(−15.5)	Raban and Kenney (1969)
4	LXXXVII, R_1 = H, R_2 = $CHMe_2$, R_3 = Me	8.2	(−115)	Walter and Schaumann (1971)
5	LXXXVII, R_1 = H, R_2 = CH_2Ph, R_3 = Me	<8.7	—	Walter and Schaumann (1971)
6	LXXXVIII	ca 10		Binsch (1968)
7	LXXXIX, R_1 = Ph, R_2 = Cl, R_3 = Me	10.9	(−50)	Cowley et al. (1968)[c]
8	LXXXIX, R_1 = Ph, R_2 = Cl, R_3 = $CHMe_2$	12.8	(−15)	Cowley et al. (1968)[c]
9	LXXXIX, R_1 = Ph, R_2 = Cl, R_3 = CHMeEt	14.6	(15)	Cowley et al. (1968)[c]
10	XC	8.5	(−100)	Cowley et al. (1968)[c]
11	XCI, R = Me	14.4	(17)	Raban et al. (1968, 1969)
12	XCI, R = $CHMe_2$	15.9	(45)	Raban et al. (1968, 1969)
13	XCI, R = adamantyl	17.0	(64)	Raban et al. (1968, 1969)
14	XCII	8.9	(−97)	Jackson and Kee (1972).

[a] Where X is any atom other than carbon or nitrogen.
[b] Free energy of activation to rotation about the bond indicated by an arrow on the appropriate structural formula at temperatures indicated in parentheses.
[c] See also Cowley et al. (1970).

involve electronic effects. These include the well-known (Binsch, 1968) example of nitrosamines, while restricted rotation in carbamates and thiocarbamates (**LXXXV**, X = O, S) are clearly analogous to amides (Inglefield and Kaplan, 1972). Two simultaneously appearing studies (Akhtar et al., 1968; Marullo et al., 1968) deal with the restricted rotation in aryl triazenes (**LXXXVI**), where the rate of rotation about the indicated bond varies with the electron withdrawing power of the group X, e.g., $\Delta G^{\ddagger}_{25°C}$ = 12.7 ± 0.4 for X = p-OMe and 15.7 ± 0.2 kcal/mole for X = p-NO$_2$).

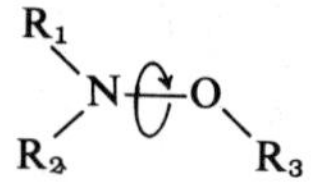

LXXXVII    LXXXVIII 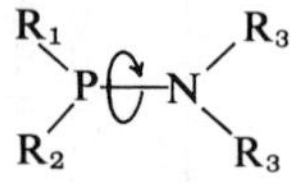   LXXXIX

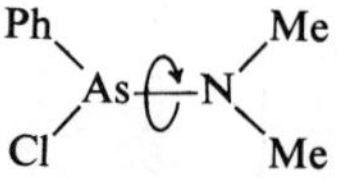

XC 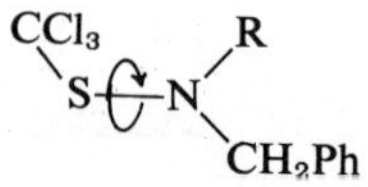   XCI 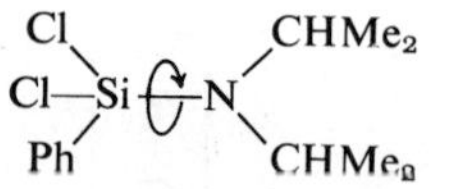   XCII

In hydrazones related to the imines (**LXVIII**) (X = N), the barriers to rotation remain substantial, although much below those for the corresponding analogs with X = CH (Mannschreck and Koelle, 1967, 1969).

Many examples of restricted rotation about N—X single bonds suitable for study by dnmr methods have been reported. In most cases the possibility of the dynamic process involving inversion about the nitrogen rather than rotation was considered, but rotation was favored as the origin of the energy barrier, because the processes tend to show steric deceleration (increase in barrier heights with increasing bulk of substituents) as expected for rotation but not for inversion. Typical examples and leading references are assembled in Table VIII. Good examples of steric deceleration can be seen by comparing the series of aminophosphines (entries 7, 8, and 9, Table VIII) and sulfenamides (entries 11, 12, and 13, Table VIII), while the conspicuous diminishing of the barrier between otherwise identical phosphorus and arsenic derivatives (entries 7 and 10, Table VIII) was ascribed to the diminishing lone pair-lone pair interection (Cowley et al., 1968).

REFERENCES

Abraham, R. J., and Bernstein, H. J. (1961). *Can. J. Chem.* **39**, 39.
Abraham, R. J., Cavalli, L., and Pachler, K. G. R. (1966). *Mol. Phys.* **11**, 471.
Akhtar, M. H., McDaniel, R. S., Feser, M., and Oehlschlager, A. C. (1968). *Tetrahedron* **24**, 3899.

6. Rotation about Single Bonds

Akkerman, O. S. (1970). *Rec. Trav. Chim. Pays-Bas* **89**, 673.
Alger, T. D., Gutowsky, H. S., and Vold, R. L. (1967). *J. Chem. Phys.* **47**, 3130
Alger, T. D., Grant, D. M., and Harris, R. K. (1972). *J. Phys. Chem.* **76**, 281.
Allen, P. S., and Cowking, A. (1968). *J. Chem. Phys.* **49**, 789.
Almog, J., Meyer, A. Y., and Shanan-Atidi, H. (1972). *J. Chem. Soc., Perkin Trans.* **2**, 451.
Anderson, J. E., and Pearson, H. (1971a). *J. Chem. Soc. B* p. 1209.
Anderson, J. E., and Pearson, H. (1971b). *Chem. Commun.* p. 871.
Anderson, J. E., and Pearson, H. (1972a). *Chem. Commun.* p. 908.
Anderson, J. E., and Pearson, H. (1972b). *Tetrahedron Lett.* p. 2779.
Anderson, J. E., Griffith, D. C., and Roberts, J. D. (1969). *J. Amer. Chem. Soc.* **91**, 6371.
Anderson, J. E., Pearson, H., and Rawson, D. I. (1973). *Chem. Commun.* p. 95.
Anet, F. A. L., and Ahmad, M. (1964). *J. Amer. Chem. Soc.* **86**, 119.
Anet, F. A. L., St. Jacques, M., and Chmurny, G. N. (1968). *J. Amer. Chem. Soc.* **90**, 5243.
Arlinger, L., Dahlquist, K. I., and Forsén, S. (1970a). *Acta Chem. Scand.* **24**, 662.
Arlinger, L., Dahlquist, K. I., and Forsén, S. (1970b). *Acta Chem. Scand.* **24**, 672.
Baas, J. M. A. (1972). *Rev. Trav. Chim. Pays-Bas* **91**, 1287.
Barber, B. H., and Schaefer, T. (1971). *Can. J. Chem.* **49**, 789.
Bartle, K. D., Bavin, P. M. G., Jones, D. W., and L'Amie, R. (1970). *Tetrahedron* **26**, 911.
Bartlett, P. D., and Tidwell, T. T. (1968). *J. Amer. Chem. Soc.* **90**, 4421.
Bates, R. B., Gosselink, D. W., and Kaczynski, J. A. (1967). *Tetrahedron Lett.* p. 205.
Bentley, K. W., Crocker, H. P., Walser, R., Fulmor, W., and Morton, G. O. (1969). *J. Chem. Soc., C* p. 2225.
Bentz, W. E., Colebrook, L. D., Fehlner, J. R., and Rosowsky, A. (1970). *Chem. Commun.* p. 974.
Bergman, J. J., and Chandler, W. D. (1972). *Can. J. Chem.* **50**, 353.
Binsch, G. (1968). *Top. Stereochem.* **3**, 138–139.
Bolon, D. A. (1966). *J. Amer. Chem. Soc.* **88**, 3148.
Bothner-By, A. A., Castellano, S. M., Ebersole, S. J., and Günther, H. (1966). *J. Amer. Chem. Soc.* **88**, 2466.
Breslow, R., Kaplan, L., and LaFollette, D. (1968). *J. Amer. Chem. Soc.* **90**, 4056.
Brewer, J. P. N., Heaney, H., and Marples, B. A. (1967). *Chem. Commun.* p. 27.
Brewer, J. P. N., Eckhard, I. F., Heaney, H., and Marples, B. A. (1968). *J. Chem. Soc., C* p. 664.
Brey, W. S., and Ramey, K. C. (1963). *J. Chem. Phys.* **39**, 844.
Brownstein, S., and Worsfold, D. J. (1972). *Can. J. Chem.* **50**, 1246.
Brownstein, S., Horswill, E. C., and Ingold, K. U. (1969). *Can. J. Chem.* **47**, 3243.
Brownstein, S., Horswill, E. C., and Ingold, K. U. (1970). *J. Amer. Chem. Soc.* **92**, 7217.
Burnell, E. E., and Diehl, P. (1972). *Mol. Phys.* **24**, 489.
Bushweller, C. H., and Anderson, W. G. (1972). *Tetrahedron Lett.* p. 1811.
Bushweller, C. H., and O'Neil, J. W. (1971). *Tetrahedron Lett.* p. 3471.
Bushweller, C. H., O'Neil, J. W., and Bilofsky, H. S. (1971a). *J. Amer. Chem. Soc.* **93**, 542.
Bushweller, C. H., O'Neil, J. W., and Bilofsky, H. S. (1971b). *Tetrahedron* **27**, 5761.
Bushweller, C. H., Dewlett, W. J., O'Neil, J. W., and Beall, H. (1971c). *J. Org. Chem.* **36**, 3782.
Bushweller, C. H., Rao, G. U., Anderson, W. G., and Stevenson, P. E. (1972a). *J. Amer. Chem. Soc.* **94**, 4743.

Bushweller, C. H., Brunelle, J. A., Anderson, W. G., and Bilofsky, H. S. (1972b). *Tetrahedron Lett.* p. 3261.
Bushweller, C. H., Anderson, W. G., O'Neil, J. W., and Bilofsky, H. S. (1973). *Tetrahedron Lett.* p. 717.
Cable, D. A., Ernst, J. A., and Tidwell, T. T. (1972). *J. Org. Chem.* **37**, 3420.
Calder, I. C., and Garratt, P. J. (1969). *Tetrahedron* **25**, 4023.
Calder, I. C.. Garratt P. J., and Sondheimer, F. (1967). *Chem. Commun.* p. 41.
Carter, R. E., and Berntsson, P. (1968). *Acta Chem. Scand.* **22**, 1047.
Carter, R. E., Marton, J., and Dahlquist, K. I. (1970). *Acta Chem. Scand.* **24**, 195.
Casy, A. F., and Ison, R. R. (1969). *Tetrahedron* **25**, 641.
Chambers, R. D., Corbally, R. P., Musgrave, W. K. R., Jackson, J. A., and Matthews, R. S. (1972). *J. Chem. Soc., Perkin Trans.* **1**, 1286.
Chandross, E. A., and Sheley, C. F. (1968). *J. Amer. Chem. Soc.* **90**, 4345.
Colebrook, L. D., and Jahnke, J. A. (1968). *J. Amer. Chem. Soc.* **90**, 4687.
Colebrook, L. D., Giles, H. G., and Rosowsky, A. (1972). *Tetrahedron Lett.* p. 5239.
Combrisson, S., Roques, B., and Basselier, J. J. (1971). *Can. J. Chem.* **49**, 904.
Cowley, A. H., Dewar, M. J. S., and Jackson, W. R. (1968). *J. Amer. Chem. Soc.* **90**, 4185.
Cowley, A. H., Dewar, M. J. S., Jackson, W. R., and Jennings, W. B. (1970). *J. Amer. Chem. Soc.* **92**, 1085.
Cupas, C. A., Bollinger, J. M., and Haslanger, M. (1968). *J. Amer. Chem. Soc.* **90**, 5502.
Dabrowski, J., and Kamienska-Trela, K. (1972). *Org. Magn. Resonance* **4**, 421.
Dahlquist, K. I. (1970). *Acta Chem. Scand.* **24**, 1941.
Dahlquist, K. I., and Forsén, S. (1965). *J. Phys. Chem.* **69**, 4062.
Dahlquist, K. I., and Forsén, S. (1970). *Acta Chem. Scand.* **24**, 2075.
Dewar, M. J. S., and Jennings, W. B. (1969). *J. Amer. Chem. Soc.* **91**, 3655.
Dewar, M. J. S., and Jennings, W. B. (1970). *Tetrahedron Lett.* p. 339.
Dewar, M. J. S., and Jennings, W. B. (1973). *J. Amer. Chem. Soc.* **95**, 1562.
Diehl, P., Kellerhals, H. P., and Niederberger, W. (1971a). *J. Magn. Resonance* **4**, 352.
Diehl, P., Henrichs, P. M., Niederberger, W., and Vogt, J. (1971b). *Mol. Phys.* **21**, 377.
Dix, D. T., Fraenkel, G., Karnes, H. A., and Newman, M. S. (1966). *Tetrahedron Lett.* p. 517.
Dunlop, R., Mackenzie, R. K., MacNicol, D. D., Mills, H. H., and Williams, D. A. R. (1971). *Chem. Commun.* p. 919.
du Preez, I. C., Rowan, A. C., Roux, D. G., and Feeney, J. (1971). *Chem. Commun.* p. 315.
Eliel, E. L. (1965). *Angew. Chem., Int. Ed. Engl.* **4**, 761.
Fletcher, J. R., and Sutherland, I. O. (1969). *Chem. Commun.* p. 706.
Fuhr, B. J., Goodwin, B. W., Hutton, H. M., and Schaefer, T. (1970). *Can. J. Chem.* **48**, 1558.
Gall, R. E., Landman, D., Newsoroff, G. P., and Sternhell, S. (1972). *Aust. J. Chem.* **25**, 109.
Gerteisen, T. J., Kleinfelter, D. C., Brophy, G. C., and Sternhell, S. (1971). *Tetrahedron* **27**, 3013.
Govil, G., and Bernstein, H. J. (1967). *J. Chem. Phys.* **47**, 2818.
Govil, G., and Bernstein, H. J. (1968a). *J. Chem. Phys.* **48**, 285.
Govil, G., and Bernstein, H. J. (1968b). *J. Chem. Phys.* **49**, 911.
Greenvald, A., and Rabinovitz, M. (1969). *Chem. Commun.* p. 642.
Gutowsky, H. S. (1963). *Pure Appl. Chem.* **7**, 93.

6. Rotation about Single Bonds

Gyulai, H. G., Fuhr, B. J., Hutton, H. M., and Schaefer, T. (1970). *Can. J. Chem.* **48**, 3877.
Hawkins, B. L., Bremser, W., Borcic, S., and Roberts, J. D. (1971). *J. Amer. Chem. Soc.* **93**, 4472.
Heatley, F., Allen, G., Hameed, S., and Jones, P. W. (1972). *J. Chem. Soc., Faraday Trans.* **2**, 1547.
Heidberg, J., Weil, J. A., Janusonis, G. A., and Anderson, J. K. (1964). *J. Chem. Phys.* **41**, 1033.
Heyd, W. E., and Cupas, C. A. (1971). *J. Amer. Chem. Soc.* **93**, 6086.
Hirsch, J. A. (1967). *Top. Stereochem.* **1**, 199.
House, H. O., Campbell, W. J., and Gall, M. (1970). *J. Org. Chem.* **35**, 1815.
Huysmans, W. G. B., Mijs, W. J., Westra, J. G., van den Hoek, W. J., Angad Gaur, H., and Smidt, J. (1969). *Tetrahedron* **25**, 2249.
Inglefield, P. T., and Kaplan, S. (1972). *Can. J. Chem.* **50**, 1594.
Jackson, W. R., and Kee, T. G. (1972). *Tetrahedron Lett.* p. 5051.
Jost, R., Rimmelin, P., and Sommer, J. M. (1971a). *Chem. Commun.* p. 879.
Jost, R., Rimmelin, P., and Sommer, J. M. (1971b). *Tetrahedron Lett.* p. 3005.
Jouanne, J., and Heidberg, J. (1971). *Ber. Bunsenges. Phys. Chem.* **75**, 261.
Jouanne, J., and Heidberg, J. (1973). *J. Amer. Chem. Soc.* **95**, 487.
Kamikawa, T., Nakatani, M., and Kubota, T. (1968). *Tetrahedron* **24**, 2091.
Kaplan, F., and Meloy, G. K. (1966). *J. Amer. Chem. Soc.* **88**, 950.
Kessler, H. (1968). *Tetrahedron* **24**, 1857.
Kessler, H. (1970a). *Chem. Ber.* **103**, 973.
Kessler, H. (1970b). *Angew. Chem., Int. Ed. Engl.* **9**, 219.
Kessler, H., and Leibfritz, D. (1970). *Tetrahedron Lett.* p. 1423.
Kessler, H., and Leibfritz, D. (1971). *Chem. Ber.* **104**, 2143.
Kessler, H., and Rundel, W. (1968). *Chem. Ber.* **101**, 3350.
Kessler, H., Rieker, A., and Rundel, W. (1968a). *Chem. Commun.* p. 475.
Kessler, H., Gusowski, V., and Hanack, M. (1968b). *Tetrahedron Lett.* p. 4665.
Kessler, H., Moosmayer, A., and Rieker, A. (1969). *Tetrahedron* **25**, 287.
Khetan, S. K., and George, M. V. (1969). *Tetrahedron* **25**, 527.
Kleinfelter, D. C., Aaron, R. W., Wilde, W. E., Bennett, T. B., Wei, H., and Wiechert, J. E. (1969). *Tetrahedron Lett.* p. 909.
Klinck, R. E., Marr, D. H., and Stothers, J. B. (1967). *Chem. Commun.* p. 409.
Kloosterziel, H. (1968). *Chem. Commun.* p. 1330.
Köbrich, G., Mannschreck, A., Misra, R. A., Rissmann, G., Rösner, M., and Zundorf, W. (1972). *Chem. Ber.* **105**, 3794.
Korver, P. K., van der Haak, P. J., and de Boer, T. J. (1966). *Tetrahedron* **22**, 3157.
Lauer, D., and Staab, H. A. (1969). *Chem. Ber.* **102**, 1631.
Mackenzie, R. K., and MacNicol, D. D. (1970). *Chem. Commun.* p. 1299.
McKinley, S. V., Grieco, P. A., Young, A. E., and Freedman, H. H. (1970). *J. Amer. Chem. Soc.* **92**, 5900.
MacNicol, D. D. (1969). *Chem. Commun.* p. 1516.
MacNicol, D. D., Wallace, R., and Brand, J. C. D. (1965). *Trans. Faraday Soc.* **61**, 1.
MacNicol, D. D., Porte, A. L., and Wallace, R. (1966). *Nature (London)* **212**, 1572.
Mannschreck, A., and Ernst, L. (1968). *Tetrahedron Lett.* p. 5939.
Mannschreck, A., and Ernst, L. (1971). *Chem. Ber.* **104**, 228.
Mannschreck, A., and Koelle, U. (1967). *Tetrahedron Lett.* p. 863.
Mannschreck, A., and Koelle, U. (1969). *Angew. Chem., Int. Ed. Engl.* **8**, 528.
Mannschreck, A., and Muensch, H. (1968). *Tetrahedron Lett.* p. 3227.

Mannschreck, A., Ernst, L., and Keck, E. (1970). *Angew. Chem., Int. Ed. Engl.* **9**, 806.
Mark, V., and Pattison, V. A. (1971). *Chem. Commun.* p. 553.
Martin, D. M. G., and Reese, C. B. (1967). *Chem. Commun.* p. 1275.
Martinson, P. (1969). *Acta Chem. Scand.* **23**, 751.
Marullo, N. P., Mayfield, C. B., and Wagener, E. H. (1968). *J. Amer. Chem. Soc.* **90**, 510.
Matsubayashi, G., Takaya, Y., and Tanaka, T. (1970). *Spectrochim. Acta, Part A* **26**, 1851.
Meyer, W. L., and Meyer, R. B. (1963). *J. Amer. Chem. Soc.* **85**, 2170.
Misiti, D., Settimj, G., Mantovani, P., and Chiaverelli, S. (1970). *Gazz. Chim. Ital.* **100**, 495.
Nakamura, N., and Ōki, M. (1972). *Bull. Chem. Soc. Jap.* **45**, 2565.
Nakanishi, H., Yamamoto, O., Nakamura, N., and Ōki, M. (1973). *Tetrahedron Lett.* p. 727.
Newmark, R. A., and Graves, R. E. (1968). *J. Phys. Chem.* **72**, 4299.
Newmark, R. A., and Sederholm, C. H. (1965). *J. Chem. Phys.* **43**, 602.
Newsoroff, G. P., and Sternhell, S. (1967). *Tetrahedron Lett.* p. 2539.
Nilsson, B., Carter, R. E., Dahlquist, K. I., and Marten, J. (1972). *Org. Magn. Resonance* **4**, 95.
Norris, R. K., and Sternhell, S. (1969). *Aust. J. Chem.* **22**, 935.
Ōki, M., and Nakamura, N. (1971). *Bull. Chem. Soc. Jap.* **44**, 1880.
Ōki, M., and Nakanishi, H. (1970). *Bull. Chem. Soc. Jap.* **43**, 2558.
Ōki, M., and Suda, M. (1971). *Bull. Chem. Soc. Jap.* **44**, 1876.
Ōki, M., and Yamamoto, G. (1971). *Bull. Chem. Soc. Jap.* **44**, 266.
Ōki, M., Akashi, K., Yamamoto, G., and Iwamura, H. (1971). *Bull. Chem. Soc. Jap.* **44**, 1683.
Peeling, J., Schaefer, T., and Wong, C. M. (1970). *Can. J. Chem.* **48**, 2839.
Peeling, J., Goodwin, B. W., Schaefer, T., and Wong, C. M. (1971). *Can. J. Chem.* **49**, 1489.
Pinkus, A. G., Riggs, J. I., and Broughton, S. M. (1968). *J. Amer. Chem. Soc.* **90**, 5043.
Price, B. J., Eggleston, J. A., and Sutherland, I. O. (1967). *J. Chem. Soc., B* p. 922.
Raban, M., and Kenney, G. W. (1969). *Tetrahedron Lett.* p. 1295.
Raban, M., Jones, F. B., and Kenney, G. W. (1968). *Tetrahedron Lett.* p. 5055.
Raban, M., Kenney, G. W., and Jones, F. B. (1969). *J. Amer. Chem. Soc.* **91**, 1969.
Rakshys, J. W., McKinley, S. V., and Freedman, H. H. (1971). *J. Amer. Chem. Soc.* **93**, 6522.
Reuvers, A. J. M., Sinnema, A., Nieuwstad, T. J., von Rantwijk, F., and von Bekkum, H. (1971). *Tetrahedron* **27**, 3713.
Reynolds, W. F., and Schaefer, T. (1964). *Can. J. Chem.* **42**, 2119.
Rieker, A., and Kessler, H. (1967). *Tetrahedron Lett.* p. 153.
Rieker, A., and Kessler, H. (1969). *Tetrahedron Lett.* p. 1227.
Rieker, A., Zeller, N., and Kessler, H. (1968). *J. Amer. Chem. Soc.* **90**, 6566.
Robert, J. B., and Roberts, J. D. (1972). *J. Amer. Chem. Soc.* **94**, 1972.
Roques, B., Jaureguiberry, C., Fournie-Zaluski, M. C., and Combrisson, S. (1971). *Tetrahedron Lett.* p. 2693.
Sandel, V. R., McKinely, S. V., and Freedman, H. H. (1968). *J. Amer. Chem. Soc.* **90**, 495.
Schuster, L. I., Colter, A. K., and Kurland, R. J. (1968). *J. Amer. Chem. Soc.* **90**, 4679.
Sergeyev, N. M., Abdulla, K. F., and Skvarchenko, V. R. (1972). *Chem. Commun.* p. 368.
Seymour, R. J., and Jones, R. C. (1967). *Tetrahedron Lett.* p. 2021.
Shoup, R. R., Miles, H. T., and Becker, E. D. (1972a). *J. Phys. Chem.* **76**, 64.

Shoup, R. R., Becker, E. D., and McNeel, M. L. (1972b). *J. Phys. Chem.* **76**, 71.
Shvo, Y., and Shanan-Atidi, H. (1969). *J. Amer. Chem. Soc.* **91**, 6683 and 6689.
Shvo, Y., Taylor, E. C., Mislow, K., and Raban, M. (1967). *J. Amer. Chem. Soc.* **86**, 4910.
Siddall, T. H., and Garner, R. H. (1966). *Tetrahedron Lett.* 3513.
Siddall, T. H., and Prohaska, C. A. (1966). *J. Amer. Chem. Soc.* **88**, 1172.
Siddall, T. H., and Stewart, W. E. (1968a). *Tetrahedron Lett.* p. 5011.
Siddall, T. H., and Stewart, W. E. (1968b). *Chem. Commun.* p. 1116.
Siddall, T. H., and Stewart, W. E. (1968c). *J. Chem. Phys.* **48**, 2928.
Siddall, T. H., and Stewart, W. E. (1969a), *J. Phys. Chem.* **73**, 40.
Siddall, T. H., and Stewart, W. E. (1969b). *J. Org. Chem.* **34**, 233.
Siddall, T. H., and Stewart, W. E. (1969c). *J. Org. Chem.* **34**, 2927.
Siddall, T. H., Stewart, W. E., and Good, M. L. (1967). *Can. J. Chem.* **45**, 1290.
Stewart, M. A. H., Schaefer, T., Hutton, H. M., and Wong, C. M. (1971). *Can. J. Chem.* **49**, 1085.
Stewart, W. E., and Siddall, T. H. (1970). *Chem. Rev.* **70**, 517.
Sutherland, I. O. (1971). *Annu. Rep. NMR (Nucl. Magn. Resonance) Spectrosc.* **4**, 71.
Svanholm, U. (1971). *Acta Chem. Scand.* **25**, 1166.
Theilacker, W., and Böhm, H. (1967). *Angew. Chem., Int. Ed. Engl.* **6**, 251.
Thompson, R. S., Jacques, D., Haslam, E., and Tanner, R. J. N. (1972). *J. Chem. Soc., Perkin Trans.* **2** 1387.
Verma S. M., and Prasad, R. (1973). *J. Org. Chem.* **38**, 1004.
Vigevani, A., Giola, B., Cavalleri, B., and Gallo, G. G. (1971). *Org. Magn. Resonance* **3**, 249.
Vold, R. L., and Gutowsky, H. S. (1967). *J. Chem. Phys.* **47**, 2495.
Walter, W., and Schaumann, E. (1971). *Justus Liebigs Ann. Chem.* **747**, 191.
Wasylishen, R., and Schaefer, T. (1971). *Can. J. Chem.* **49**, 3575.
Weigert, F. J., and Mahler, W. (1972). *J. Amer. Chem. Soc.* **94**, 5314.
Weigert, F. J., and Roberts, J. D. (1968). *J. Amer. Chem. Soc.* **90**, 3577.
Weigert, F. J., Winstead, M. B., Garrels, J. I., and Roberts, J. D. (1970). *J. Amer. Chem. Soc.* **92**, 7359.
Weinges, K., Marx, H. D., and Göritz, K. (1970). *Chem. Ber.* **103**, 2336.
Whitesides, G. M., Sevenair, J. P., and Goetz, R. W. (1967). *J. Amer. Chem. Soc.* **89**, 1135.
Woller, P. B., and Garbisch, E. W. (1972a). *J. Amer. Chem. Soc.* **94**, 5310.
Woller, P. B., and Garbisch, E. W. (1972b). *J. Org. Chem.* **37**, 4281.

7

Rotation about Partial Double Bonds in Organic Molecules

L. M. JACKMAN

I. Introduction 203
II. Amides and Related Systems (R^1R^2N—COX) 204
   A. Experimental Procedures........................... 204
   B. Relation between Structure and Barriers to Rotation about the Amide Bond 208
III. Amidines, Thionamides, Selenamides, and Related Compounds [R—C(=X)—N$\langle$; X = N—R′, S, Se]....................... 215
IV. Enamines.. 218
V. Enolates, Enol Ethers and Thio Enol Ethers, Acylides, Diazoketones, and Aminoboranes... 233
VI. Potentially Aromatic Systems............................... 237
VII. Cumulenes 243
VIII. Inversion versus Rotation—Imines and Related Systems.......... 244
   References 250

I. Introduction

The barriers to rotation about normal carbon–carbon and carbon–nitrogen double bonds are certainly in excess of 40 kcal mole^{-1} and correspond to rate processes well outside the nmr time scale. It is clear, therefore, that the examples considered in this chapter must embody special steric or electronic effects that either raise the ground state or lower the transition state energies (or both) by factors of substantial magnitude. Certain formal single bonds, notably the carbon–nitrogen bond in amides, possess sufficient double bond character to produce barriers in the range 10–25 kcal mole^{-1} and therefore fall within the purview of this chapter. As pointed out in the previous chapter, the division of systems between it and this chapter is arbitrary and some overlaps (but we hope few omissions) are inevitable.

The major class of compounds considered here is the amides. Indeed, the study of simple amides has played an important role in the development of dnmr techniques and this will be outlined in the next section. A later section dealing with a variety of enamines, including vinylogous amides, is particularly important in that it emphasizes the need for considering both the ground and transition states in rationalizing observed barriers.

In addition to the material referred to in the preceding paragraph, a fairly comprehensive compilation of barriers in a variety of other systems is presented and the data discussed. The attention of the reader, however, is directed to several excellent and very thorough reviews, notably those by Kessler (1970), Sutherland (1971), and Stewart and Siddall (1970) (on amides and thionamides).

II. Amides and Related Systems (R^1R^2N—COX)

A. Experimental Procedures

The phenomenon of the dependence of line shapes on rotation about bonds at rates commensurate with the nmr time scale was first observed in N,N-dimethylformamide by Phillips (1955) and was then treated quantitatively by Gutowsky and Holm (1956) by an approximation to the previously developed GMS equation (Gutowsky et al., 1953). Subsequently, numerous measurements of the barrier to rotation in this molecule using various approximations (Grunwald et al., 1957; Rogers and Woodbrey, 1962; Fryer et al., 1965) were reported (see Stewart and Siddall, 1970). Barriers were all determined by comparing the observed total line shapes (TLS) with those computed from the GMS equation. In general, reasonable agreement in $\Delta G^{\ddagger}$ (approx. 21 kcal mole^{-1}) but widely varying values of the activation parameters $\Delta H^{\ddagger}$ and $\Delta S^{\ddagger}$ were found.

The principal sources of errors in these measurements arose, not so much from the nature of the approximations to the GMS equation, but rather from the neglect of spin–spin coupling between the methyl protons and the formyl proton, the assumption of temperature invariance of the chemical shift and, in the case of $\Delta H^{\ddagger}$ and $\Delta S^{\ddagger}$, the rather narrow temperature range over which meaningful data could be acquired. The first two factors frequently limit the accuracy in the determination of rate processes by line-shape analysis (Gutowsky et al., 1967), although in several investigations some attempts to take partial account of them were made (Fraenkel and Franconi, 1960; Whittaker and Siegel, 1965; Fryer et al., 1965). Conti and von Phillipsborn (1967) used both homonuclear decoupling and deuterium replacement of the formyl proton to eliminate the effect of spin–spin coupling. However, the best approach by far is that used by Inglefield et al. (1968) and later by

7. Rotation about Partial Double Bonds

Rabinovitz and Pines (1969), which includes the spin–spin coupling explicitly in an iterative TLS analysis that determines both the chemical shift and the exchange lifetime. In this case, the spin system is essentially first order and the analysis is based on the method of McConnell (1958). This study undoubtedly provides the best kinetic parameters for pure N,N-dimethylformamide. The derivation of the Arrhenius parameters, however, have not been subjected to the weighted least squares and error analysis recommended in Chapter 3 (p. 76). It is possible that the protons of the two methyl groups are weakly coupled, although no resolvable fine structure arising from this source has been detected. Nevertheless, even a coupling of the order of 0.1 Hz could result in apparent differences in the effective $T_2{}^*$ for the absorptions under fast and slow exchange, which if not taken into account could introduce errors near the fast or slow exchange limits.

The problem of the small difference in chemical shift between the two methyl groups, and hence the limited temperature range over which rate data can be acquired, can be alleviated by using the higher magnetic fields provided by superconducting solenoids. The use of shift reagents to produce the same effect has been reported (Beauté *et al.*, 1971; Cheng and Gutowsky, 1972), but the method is fraught with danger, since not only does the addition of a shift reagent introduce a strong dependence of the chemical shift and line width on temperature, but there is also the possibility that complexation will effect the barrier under examination. It has been shown by Tanny *et al.* (1974) that, in the case of trimethylcarbamate, useful information is available from studies at constant temperature with varying concentrations of shift reagent. Investigation of ^{13}C resonances of the methyl groups might conceivably increase the range over which the exchange process can be studied. This approach has been investigated in the case of N,N-dimethyltrichloracetamide at natural abundance of ^{13}C by Gansow *et al.* (1971). These workers have also pointed out that while the chemical shift difference for the two methyl carbons of N,N-dimethylformamide of 5.3 ppm (i.e., 21 Hz at 21.1 kG) would permit an investigation of rates over a larger temperature range, the problems of sensitivity at or near coalescence is severe and ^{13}C-enriched samples would probably be required. It is also clear that, in principle, pulse methods (outlined in Chapters 4 and 5) in the case of N,N-dimethylformamide-d_1 could be used to extend the temperature range over which the rate process can be studied. In practice, however, the attempts by Inglefield *et al.* (1968) to extend the high temperature limit by spin echo experiments with N,N-dimethylformamide-d_1 were thwarted by interference associated with boiling of the substrate. In any case, the accuracy of this method over the accessible temperature range was found to be inferior to that of total line-shape analysis.

In all the experiments considered above, the rate of the exchange process has been studied by the exchange averaging of the chemical shifts associated with the two sites. Other nmr parameters can lead to the same information. In the particular case of N,N-dimethylformamide, the coupling constants between the formyl proton and the two methyl groups are too small to be of practical value in this context. On the other hand, the spin–lattice relaxation processes associated with the two methyl groups offer the possibility of investigating the exchange process in a different temperature range. The method of saturation transfer introduced by Forsén and Hoffman (1963) has not been applied to this system but might easily extend the lower temperature limit at which the rate process can be studied. The effect of temperature on the nuclear Overhauser enhancement of the formyl proton, which depends on the contribution of the methyl protons to its spin–lattice relaxation, has been determined (Saunders and Bell, 1970) and analyzed (Noggle and Schirmer, 1971) to yield rate data in the range 30°–90°C, substantially below the range for which line-shape analysis is applicable. Unfortunately, the activation energy thus obtained is clearly too low.

The studies of N,N-dimethylformamide just discussed provide an indicaation of the various levels of sophistication that may be brought to bear on such a problem. While the free energies of activation (at or near the coalescence temperature) thus determined (see Stewart and Siddall, 1970) agree reasonably well (± 1 kcal mole^{-1}) for all methods used, wide variations in the Arrhenius parameters are recorded, and only the values derived from the most detailed analysis are acceptable.

Detailed analyses have been reported for some other simple amides. Drakenberg and Forsén (1970) have determined the barrier to rotation in ^{15}N-formamide by comparing the observed spectra with those calculated for an ABCX spin system by the density matrix method (Chapter 3). The results for one solvent are presented in Fig. 1, from which it can be seen that, as stressed in Chapter 3. reliable rate data over a wide temperature range become available in cases involving complex spin systems. Similar studies have been carried out for N-acetylpyrrole and N,N-dimethylacetamide (Dahlqvist and Forsén, 1969). The barrier to rotation in the latter molecule was also carefully determined by Neuman and Jonas (1968) using N,N-dimethylacetamide-2-d_3, and by Reeves et al. (1971b), who used an iterative procedure based on their multi-site extension of the GMS equation, and which also included the temperature dependence of the chemical shift.

The systems thus far discussed involve exchange between equally populated pairs of sites. In the case of secondary amides and unsymmetrical tertiary amides, unequal population of sites will generally prevail. Since site populations will also be temperature dependent, the best approach involves iterative fitting of site populations, as well as chemical shifts and lifetimes

7. Rotation about Partial Double Bonds

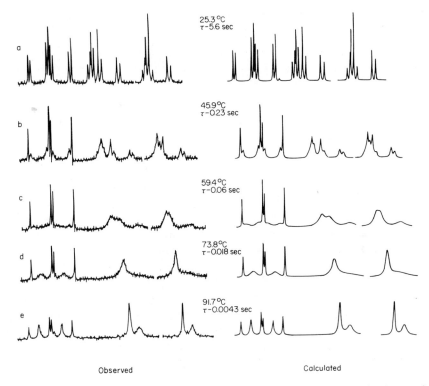

Fig. 1 Nuclear magnetic resonance spectra of ^{15}N-labeled formamide (Drakenberg and Forsén, 1970).

(cf. Jackman and Haddon, 1973). Note that for a two site problem this only requires the fitting of three parameters, viz. a chemical shift and two lifetimes.

Two further experimental factors require comment. First, the measurement of sample temperature in nmr experiments requires considerable care and is usually little better than ±1°C. Generally, temperatures are determined by peak separations for standard substances such as methanol and ethylene glycol, for which the most accurate calibrations have been provided by Van Geet (1968). Second, rotation about the amide bond is subject to acid catalysis (see below). It has been noted (Spaargaren *et al.*, 1971c) that minute quantities of hydrochloric acid in chloroform markedly effect line shapes and that this source of error can be eliminated by the addition of traces of pyridine.

In certain unsymmetrically *N,N*-disubstituted amides (see below) the barriers to rotation are sufficiently high to permit separation of geometric isomers. In some such cases dnmr methods cannot be used, although conventional nmr is useful for direct studes of the rates of equilibration of the pairs of isomers. In the case of N-benzyl-N-methylformamide, however, Gutowsky et al. (1967) were able to use both the direct kinetic method and dnmr lineshape analysis and thereby obtain rate data over a large temperature range.

B. Relation between Structure and Barriers to Rotation about the Amide Bond

An extremely comprehensive review of all work in this area up until 1969 has been published by Stewart and Siddall (1970). In this chapter, we will seek to develop the main concepts using selected examples. Most of the careful studies of amides in solution in reasonably inert solvents indicate that the entropy of activation for rotation is close to zero, and in view of the difficulty in obtaining accurate Arrhenius parameters, most of the ensuing discussion will be confined to considerations of the free energies ($\Delta G^{\ddagger}$) for rotation, on the assumption that this is the best available parameter for the comparison of substituent effects.

The parent system, formamide, has $\Delta G^{\ddagger}_{298}$ of 17.8 kcal mole^{-1} in both methyl propyl ketone and diethylene glycol dimethyl ether, to which $\Delta S^{\ddagger}$ makes only a small positive contribution (Drakenberg and Forsén, 1970). It is clear that this barrier is associated with the partial double bond character of the C—N bond. We will now examine the way in which substitution of formamide can influence this barrier.

Since few primary amides have been studied, the effect of substitution of the formyl proton is best seen in N,N-dimethylamides for which there are now abundant data, some of which are assembled in Table I. The structures **1** and **2** represent the ground and excited states for the rotational process.

It is a reasonable hypothesis that the electronic effects of the substituent R are likely to be greater in the transition than in the ground state, since the carbonyl carbon atom is clearly more electrophilic in the former. On the other hand, one can readily envisage the steric requirements of R destabilizing the ground state by causing out-of-plane deformation of the amide system. We will see later, however, that in certain systems, steric effects destabilize the transition state more than the ground state.

Alkylation of the carbonyl atom has the effect of decreasing the barriers in the other H, CH_3, CH_3CH_2, $(CH_3)_2CH$, $(CH_3)_3C$. Steric effects obviously

TABLE I

BARRIERS TO ROTATION ABOUT THE AMIDE BOND IN $RCON(CH_3)_2$

R	$\Delta G^{\ddagger}$ (kcal mole^{-1})	°K	Solvent	Reference
H—	20.9	392	None	Rabinovitz and Pines (1969)
CH_3—	17.3	298	CCl_4	Reeves et al. (1971b)
CH_3CH_2—	17.4	331	CCl_4	Isaksson and Sandström (1967)
$CH_3(CH_2)_2$—	18	330	None	Fryer et al. (1965)
$(CH_3)_2CH$—	16.2	299	o-$C_6H_4Cl_2$	Isaksson and Sandström (1967)
$(CH_3)_3C$—	11.9	298	$CHCl_3$	Walter et al. (1969)
Cyclopropyl	16.7	298	$CDCl_3$	Spaargaren et al. (1971c)
F_3C—	18.6	298	None	Abramson et al. (1966)
Cl_3C—	18.3	292	$(CBrF_2)_2$	Gansow et al. (1971)
$CH_2{=}CH$—	16.7	298	$CHCl_3$	Hobson and Reeves (1973b)
$CH_2{=}C(CH_3)$—	14.1	298	$(CHCl_2)_2$	Hobson and Reeves (1973b)
$(CH_3)_2C{=}CH$—	12.0	298	CCl_4	Hobson and Reeves (1973b)
$PhCH{=}CH$—	16.4	312	$CDCl_3$	Spaargaren et al. (1971b)
Ph—	15.5	298	CH_3CN	Jackman et al. (1969)
F—	18.1	298	CCl_4	Allan et al. (1972)
Cl—	16.5	298	CCl_4	Allan et al. (1972)
Br—	15.7	298	None	Allan et al. (1972)
N_3—	17.7	298	CCl_4	Allan et al. (1972)
SCN—	18.4	298	n-C_8H_{18}	Allan et al. (1972)
NC—	21.4	298	$(CHCl_2)_2$	Allan et al. (1972)
$PhCH_2O$—	15.9	270	Pyridine	Price et al. (1966)
$(CH_3)_2C{=}N{-}O$—	14.4	264	CS_2	Bushweller and Tobias (1968)
$\begin{array}{c}CH_2\\ \diagdown\\ N{-}\\ \diagup\\ CH_2\end{array}$	16.4	300	CCl_4	Anet and Osyany (1967)

play an important role by destabilizing the ground state, but almost certainly electronic stabilization of the excited state is also involved in the effect associated with substitution of hydrogen by an alkyl group. Table II shows the results of the analogous series of primary amides. Here it is seen that the greatest change occurs in going from formamide to acetamide, a situation in which a steric effect is not anticipated. In contrast, replacement of $(CH_3)_2CH$ by $(CH_3)_3C$ results in a diminution of only 0.5 kcal mole^{-1} compared with 3.9 in the N,N-dimethylamide series.

The two substituents CF_3 and CCl_3, which exert strong electron-withdrawing inductive effects, actually reduce the barrier, suggesting that steric destabilization of the ground state is more effective than inductive destabilization of the transition state. In the primary amide series (Table II), the CHF_2 group does increase the barrier relative to the CH_3 group. The nitrile group, which presumably provides little steric hindrance, but which is very

TABLE II
BARRIERS TO ROTATION IN SOME PRIMARY
ALIPHATIC CARBOXAMIDES (RCONH$_2$)

R	$\Delta G^{\ddagger}$ (kcal mole^{-1})	°K	Solvent
H	17.8[a]	298	2-Pentanone
CH$_3$	16.7[b]	333	Acetone
(CH$_3$)$_2$CH	16.2[b]	325	Acetone
(CH$_3$)$_3$C	15.7[b]	313	Acetone
Ph	15.7[b]	317	Acetone
CH$_2$F	16.4[c]	298	DMF
CHF$_2$	17.4[c]	298	DMF

[a] Drakenberg and Forsén (1970).
[b] Walter et al. (1973).
[c] Pendlebury and Philips (1972).

strongly electron withdrawing by both the mesomeric and inductive mechanisms, increases the barrier.

Substituents that can conjugate directly with the carbonyl group are generally highly effective in reducing the barrier to rotation. This is true of the vinyl group and the effect is dramatically enhanced by alkylation, particularly in the β-position. Similarly, the phenyl substituent produces a substantial lowering of the barrier that may be in part due to steric effects in the ground state but that also involves a substantial conjugative stabilization of the transition state, since a similar, although somewhat reduced trend, is observed in the primary amides (Table II). Furthermore, the effect of *m*- and *p*-substituents in the phenyl ring correlate well with Hammett's σ constants, or better, with the σ$^+$ constants with ρ values of +1.56 and +1.13, respectively (Jackman et al., 1969). Spassov et al. (1971) have established that similar correlations hold for *m*- and *p*-substituted *N,N*-dimethylcinnamides, the ρ values being smaller (+0.85 and +0.60, respectively) in this series. Spaargaren et al. (1971a, c) have also shown that the barriers for *m*- and *p*-substituted *N,N*-dimethylbenzamides and -cinnamides correlate well with the differences in π-delocalization energies calculated by an HMO method for the planar ground state versus the orthogonal transition state.

Substituents which can donate a lone pair of electrons to the carbonyl group can have profound effects. Ureas (R = $>\!\!\ddot{\text{N}}\!\!-$), for instance, evidently have very low barriers (Stewart and Siddall, 1970), a notable exception being the last entry in Table I, in which delocalization of the nitrogen lone pair from the aziridine ring is inhibited by the geometric restraint on sp^2 hybridization of the ring nitrogen atom. The barriers in urethans are also

7. Rotation about Partial Double Bonds

TABLE III
BARRIERS TO ROTATION ABOUT THE AMIDE BOND IN SOME SUBSTITUTED BENZAMIDES [RCON(CH$_3$)R']

R	R'	$\Delta G^{\ddagger}$ (kcal mole^{-1})	°K	Solvent
o-CH$_3$OC$_6$H$_4$	CH$_3$	18.4[a]	298	CH$_3$CN
p-CH$_3$OC$_6$H$_4$	CH$_3$	14.6[a]	298	CH$_3$CN
o-NO$_2$C$_6$H$_4$	CH$_3$	18.5[a]	298	CH$_3$CN
p-NO$_2$C$_6$H$_4$	CH$_3$	16.4[a]	298	CH$_3$CN
o-NH$_2$C$_6$H$_4$	CH$_3$	14.2[a]	298	CH$_3$CN
2,4,6-(CH$_3$)$_3$C$_6$H$_2$	PhCH$_2$	22.6[b] 22.9[b]	311	CCl$_4$
2,4,6-(t-Bu)$_3$C$_6$H$_2$	PhCH$_2$	30.3[c] 32.0[c]	393	—[d]

[a] Jackman et al. (1969).
[b] Mannschreck (1965).
[c] Staab and Lauer (1968).
[d] 1-Chloronaphthalene/benzotrichloride (1:1).

considerably reduced. The halogens and pseudohalogens are more complicated because of the usual conflict between inductive ($-I$) and mesomeric ($+M$) effects. For the halogens, the $+M$ effect becomes increasingly important with increasing atomic number, although steric destabilization of the ground state may also come into play.

Some special considerations prevail in certain 2,N,N-trisubstituted and 2,6,N,N-tetrasubstituted benzamides (Table III). The introduction of ortho substituents in the N,N-dimethylbenzamide series generally increases the rotational barriers presumably through steric inhibition of mesomerism in the transition state. An o-amino substituent, on the other hand, effects a dramatic reduction, which can be attributed to hydrogen bonding, since this is expected to enhance the conjugative stabilization of the transition state 3. The last two entries in Table III involve a steric destabilization of the transition

3

state rather than the ground state and result in barriers, which are so large that geometric isomers can be isolated and the kinetics of isomerization studied directly. In these systems, the planar amide bond is locked orthogonally to the benzene ring by the two ortho substituents (**4**). Rotation about

$$\underset{\mathbf{4}}{\text{R}\cdot\underset{\text{R}}{\overset{\text{R}\quad\text{CH}_3}{\bigcirc}}\underset{\text{O}}{\overset{\text{N}}{\text{C}}}\text{R}^1}$$

the amide bond can only take place by forcing the N-substituents past the ortho substituents and, in the case where these are tertiary butyl groups, the steric interaction is formidable.

Substituents on the nitrogen atom will also have an effect on the rotational barrier. Clearly, increased bulk will result in steric destabilization of the ground state, and any substituent that will delocalize the lone pair of electrons of the nitrogen atom will stabilize the transition state in preference to the ground state. Some data confirming these predictions are presented in Table IV. In the formamide series, steric effects are evidently minimal, since in N-t-butylformamide, the two geometric isomers differ by only 0.3 kcal mole^{-1} in free energy. The effect of alkylation must therefore be attributed to inductive stabilization of the mesomeric ground state. Conjugation, on the other hand, evidently lowers the barrier by stabilizing the transition state. A particularly interesting example is provided by N-trimethylsilylformamide in which $(p$–$d)\pi$ bonding has been postulated as stabilizing the transition state by as much as 3 kcal mole^{-1} (Komoriya and Yoder, 1972).

Similar trends are discernible in the acetamide series, although here steric destabilization of the ground state is evident, for example, in N,N-diisopropylacetamide. Again, conjugation with both N-vinyl and N-aryl groups results in lower barriers. Not unexpectedly, the barrier in N-acetylpyrrole is very low, since in the transition state, the pyrrole ring has full access to the nitrogen lone pair of electrons.

Solvents can have significant effects on barriers to rotation about the amide bond, but as pointed out by Drakenberg et al. (1972), there is a paucity of reliable data. Some data for N,N-dimethylamides are assembled in Table V. In general, the effects are not dramatic, except for water in which solvent barriers can be as much as 3.0 kcal mole^{-1} higher than found for nonpolar aprotic solvents. This can be attributed to stabilization of the transition state by hydrogen bonding, and the point has been made (Drakenberg et al., 1972) that, as the entropies of activation are approximately zero for both types of solvents, the hydrogen bond is also present in the ground state but is presumably weaker. It is difficult to tell whether the barriers are sensitive to changes in the bulk dielectric constant of the medium. More likely, the variations found are the results of specific solute–solvent interactions. Thus,

TABLE IV BARRIERS TO ROTATION ABOUT THE AMIDE BOND IN SOME MISCELLANEOUS AMIDES

Structure	$\Delta G^{\ddagger}$ (kcal mole^{-1})	°K	Solvent	Reference
H$_2$NCHO	17.8	298	2-Butanone	Drakenberg and Forsén (1970)
(CH$_3$)$_2$NCHO	20.9	392	None	Rabinovitz and Pines (1969)
(CH$_3$CH$_2$)NCHO	20.4	298	None	Hammaker and Gugler (1965)
(CH$_3$)$_3$CNHCHO	19.0, 19.3	298	PhCl	Komoriya and Yoder (1972)
CH$_2$=CH—N(CH$_3$)—CHO	18.3	372	None	Gehring et al. (1966)
[benzofused N(CH$_2$CH$_2$)—CHO]	15.5	342	Dioxane	Buchardt et al. (1969)
(CH$_3$)$_3$SiNHCHO	16.1, 16.5	298	PhCl	Komoriya and Yoder (1972)
H$_2$NCOCH$_3$	16.7	298	Acetone	Drakenberg (1972)
(CH$_3$)$_2$NCOCH$_3$	17.3	298	CCl$_4$	Reeves et al. (1971b)
(CH$_3$CH$_2$)$_2$NCOCH$_3$	16.9	298	None	Hammaker and Gugler (1965)
(CH$_3$)$_2$CH—NCOCH$_3$	15.7	298	None	Hammaker and Gugler (1965)
CH$_2$=CH—N(CH$_3$)—NCOCH$_3$	13.4	298	None	Hammaker and Gugler (1965)
CH$_3$O-aryl piperidine NCOCH$_3$	ca 14.0	280	CDCl$_3$	Monro and Sewell (1969)
pyrrolyl NCOCH$_3$	12.1	298	CD$_2$Cl$_2$	Dahlqvist and Forsén (1969)

TABLE V
EFFECT OF SOLVENT ON BARRIERS TO ROTATION IN SOME SIMPLE
N,N-DIMETHYLAMIDES

Solvent	D	$\Delta G^{\ddagger}$ (kcal mole^{-1})	$\Delta H^{\ddagger}_{298}$ (kcal mole^{-1})	$\Delta S^{\ddagger}$ (cal deg^{-1} mole^{-1})
		N,N-Dimethylformamide[a]		
None		20.9	20.5	−1.4 ± 1
CCl$_2$=CCl$_2$		20.7	20.8	0.4 ± 2
Decalin		20.3	19.6	−2.4 ± 2
		N,N-Dimethylacetamide[a]		
None		18.1	18.3	0.7 ± 1
Acetone-d_6		18.0	19.0	3.1 ± 2
DMSO-d_6		18.6	20.0	4.7
H$_2$O		19.3	19.1	−0.8 ± 1
		N,N-Dimethylbenzamide		
C$_6$H$_6$/CS$_2$(15:85)[b]	2.3	14.0	—	—
C$_6$H$_6$[b]	2.3	14.7	—	—
CDCl$_3$[c]	4.7	15.7	16.8	3.7 ± 0.2
Acetone[b]	20.7	14.8	—	—
CH$_3$CN[b]	37.5	15.5	16.7	3.8 ± 0.4
CH$_3$NO$_2$[c]	38.5	15.6	15.3	1 ± 0.4
H$_2$O[b]	78.5	18.1	18.4	4 ± 1.3

[a] Drakenberg et al. (1972).
[b] Jackman et al. (1969).
[c] Spaargaren et al. (1971b).

chloroform may enter into weak hydrogen bonding, whereas acetonitrile may stabilize the ground state by dipole–dipole interaction and benzene by the well-known charge transfer interaction (Hatton and Richards, 1960, 1962).

In primary amides, a somewhat different situation prevails in that the amide protons may be involved in hydrogen bonding. Evidently, the making or breaking of such hydrogen bonds accompanies the rotational process, as significant entropies of activation (9 and 5 cal/mole °K) have been observed for the barrier to rotation in acetamide in the solvents N,N-dimethylformamide and acetone, respectively (Drakenberg, 1972).

Acid catalysis of rotation about the amide bond has been unequivocally established (Berger et al., 1959; Fraenkel and Franconi, 1960; Bunton et al., 1962), even though it has been clearly demonstrated that oxygen is the preferred site of protonation (Berger et al., 1959; Gillespie and Birchall, 1963; Stewart and Siddall, 1970). That this catalysis is due to a small extent of N-protonation rather than to the incursion of 5 has been demonstrated by the H$_2$ ^{18}O exchange studies by Jackman et al. (1969). These workers also

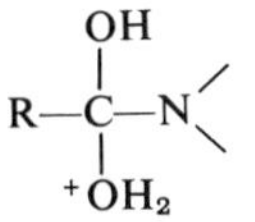

7. Rotation about Partial Double Bonds

$$R-\underset{\underset{+OH_2}{|}}{\overset{\overset{OH}{|}}{C}}-N\overset{\diagup}{\diagdown}$$

5

examined the effect of a wide range of acidities on the rotational barrier for N-(p-N,N-dimethylcarboxamidobenzyl)pyridinium bromide and showed that the dependence of the barrier on water activity is consistent with the involvement of a water molecule in the exchange between N- and O-protonated species. Thus, catalysis reaches a maximum value at a Hammett acidity function (H_0) of approximately -2.0. At very high acidities, the barrier increases sharply, presumably approaching the value for the O-protonated system, which evidently is the overwhelmingly predominant species. Lithium ions interact specifically with the oxygen atom of N,N-dimethylacetamide, since the barrier to rotation ($\Delta G^{\ddagger}$) increases by approximately 2 kcal mole^{-1} on the addition 1.0 M lithium chloride (Egan et al., 1972). In contrast, Ag$^+$ ions decreases the barrier in aqueous N,N-dimethylacetamide (Temussi et al., 1969) indicating a greater preference for interaction at nitrogen by this ion.

III. Amidines, Thionamides, Selenamides, and Related Compounds

$$\left[R-C(=X)-N\diagdown\!\!\!\!\diagup \; ; \; X = N-R', S, Se \right]$$

In this section, we examine the effect of replacing the amide oxygen atom by other heteroatoms. In the case of the amidines (**6**) two rotational barriers can be considered, namely that for rotation about the formal single bond, a process directly related to rotation about the amide bond, and rotation about the formal double bond. There is evidence that the latter process often involves inversion (lateral shift) of the trigonal nitrogen atom rather than rotation about the double bond. This problem will be considered more closely

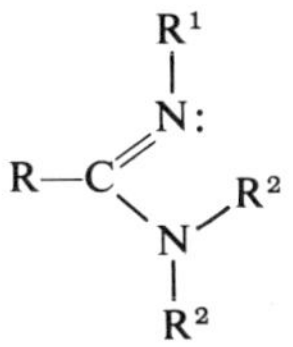

$$R-\overset{}{C}\underset{\underset{R^2}{\overset{|}{N}}}{\overset{\overset{\overset{R^1}{|}}{N:}}{\diagdown}}R^2$$

6

in Section VIII (p. 244). The barriers to rotation about the C—N bond in amidines (Table VI) are substantially lower than for analogously constituted amides, a fact that is doubtless due to decreased mesomeric stabilization of the ground state. An exception is provided by compound **11** in which the

TABLE VI
Barriers to Rotation in Amidines

Compound	$\Delta G^{\ddagger}$ (kcal mole^{-1})	°K	Solvent
7, R = t-Bu[a]	11.9	298	Toluene-d_8
7, R = p-NO$_2$C$_6$H$_4$[b]	14.1	298	CHCl$_3$
8, R = H[c]	12.6	250	CDCl$_3$
8, R = NO$_2$[c]	13.9	283	CDCl$_3$
9, R = H[c]	12.1	251	CDCl$_3$
9, R = CH$_3$O[c]	11.7	243	CDCl$_3$
10, R = NO$_2$[c]	13.1	267	CDCl$_3$
10, R = CH$_3$O[c]	12.3	256	CDCl$_3$
11[d]	17.6	326	CHBr$_3$

[a] Harris and Wellman (1968).
[b] Bertelli and Gerig (1967).
[c] Rappoport and Ta-Shma (1972).
[d] Jakobsen and Senning (1968).

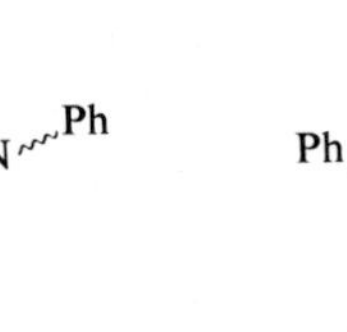

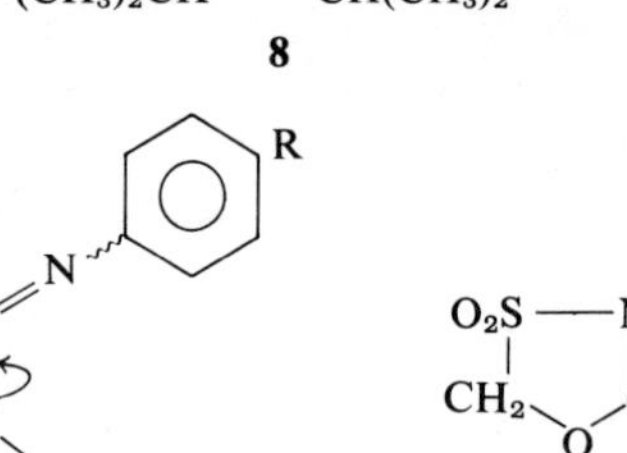

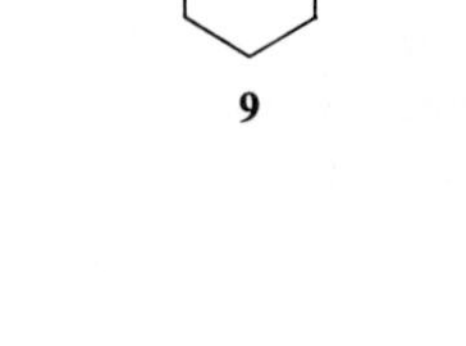

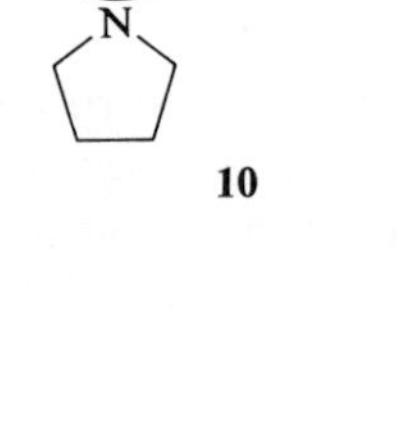

sulfone group attached to the trigonal nitrogen atom greatly enhances this stabilization. Similarly, protonation of the trigonal nitrogen atom also increases mesomeric stabilization. Thus, Neuman and Young (1965) were unable to detect nonequivalence of the N-methyl groups in N,N-dimethylacetamidine down to $-40°$C in chloroform but found a barrier ($\Delta G^{\ddagger}$) of 18.9 kcal mole^{-1} for N,N-dimethylacetamidinium chloride in formamide, a value which incidently must be regarded as a lower limit, since the mechanism could involve the amidine itself as an intermediate. Kessler and Leibriftz (1970) have found low barriers for the equivalencing of the four methyl groups in **12**, the rate-determining process of which they have shown to involve the inversion of the trigonal nitrogen, whereas barriers for rotation in the corresponding guanidinium salts (**13**) have $\Delta G^{\ddagger}_{298}$ of 15 and 21.2 for R = H and CH_3, respectively (Kessler and Leibfritz, 1969a). Note that rotation about all three carbon–nitrogen bonds is necessary to render equivalent the

12, R = H, CH_3

13, R = H, CH_3

four methyl groups comprising the N,N-dimethyl substituents, and in amidinium and guanidinium ions, it is not possible to recognize any one particular carbon–nitrogen bond as a "formal" double bond. In summary then, replacement of the oxygen of an amide by trigonal nitrogen results in a sharp decrease in the barrier to rotation. Protonation or quaternization of the imino nitrogen, however, raises the barriers to values that are probably substantially greater than analogously constituted amides.

In the thionamides and related systems, barriers ($\Delta G^{\ddagger}$) about the carbon–nitrogen bond are usually 2–5 kcal mole^{-1} higher than the oxo analogs. Some representative comparisons are provided in Table VII. The increased barrier is presumably due to destabilization of the transition state in which a full carbon–sulfur double bond must exist. The reluctance of second row elements to form $p\pi–p\pi$ double bonds has been attributed to inner shell penetration (Pitzer, 1948). The trend is apparently continued in the third row elements, since the barrier ($\Delta G^{\ddagger}_{298}$) to rotation in N,N-dimethylselenourea is 14.6 (Reeves et al., 1971a) compared with 13.9 kcal mole^{-1} (Table VII) for the corresponding thiourea. In making this comparison, it must be remembered that steric effects will destabilize the ground state of the seleno derivative more than for the thiourea. An interesting steric effect is seen (Table VII) in the comparison of the barriers for $(CH_3)_3CCON(CH_3)_2$ and $(CH_3)_3$-$CCSN(CH_3)_2$, in which a buttressing effect of the t-butyl group evidently results in increased steric destabilization of the ground state of the latter.

TABLE VII

COMPARISON OF THE BARRIERS TO ROTATION ABOUT THE
CARBON–NITROGEN BOND OF SOME AMIDES WITH
THIONAMIDES

$$\underset{R-C-Y}{\overset{X}{\|}}$$

		$\Delta G^{\ddagger}$ (kcal mole^{-1})		
Y	R	$X = O^a$	$X = S$	$\Delta\Delta G^{\ddagger}$
NH$_2$	H	17.8	>20.5^b	>2.7
NH$_2$	CH$_3$	16.4	20.2^b	3.8
NH$_2$	CH(CH$_3$)$_2$	16.1	19.7^b	3.6
NH$_2$	CH(CH$_3$)$_3$	15.7	18.6^b	2.9
NH$_2$	Ph	15.1	18.2^b	3.1
N(CH$_3$)$_2$	H	20.9	24.1^c	3.2
N(CH$_3$)$_2$	CH$_3$	17.3	21.8^c	4.5
N(CH$_3$)$_2$	CH(CH$_3$)$_2$	16.2	19.3^d	3.1
N(CH$_3$)$_2$	C(CH$_3$)$_3$	11.9	13.0^e	1.1
N(CH$_3$)$_2$	Ph	15.5	18.4^f	2.0
N(CH$_3$)$_2$	CN	21.4	23.5^g	2.1
N(CH$_3$)$_2$	F	18.1	20.7^g	2.6
N(CH$_3$)$_2$	Cl	16.5	18.7^h	2.2
N(CH$_3$)$_2$	NH$_2$	—	13.9^g	—

^{a} Data from Tables I and II.
^{b} In diethyleneglycol diethyl ether (Walter et al., 1973).
^{c} In o-dichlorobenzene (Siddall et al., 1970).
^{d} In o-dichlorobenzene (Isaksson and Sandström, 1967).
^{e} In chloroform (Walter et al., 1969).
^{f} In o-dichlorobenzene (Ramey et al., 1971).
^{g} In tetrachloroethane (Hobson et al., 1973).
^{h} In carbon tetrachloride (Hobson et al., 1973).

IV. Enamines

The degree of conjugation of an amino group with an unactivated carbon–carbon double bond is too small to give rise to sizable barriers about the carbon–nitrogen bond or to reduce the barrier about the carbon–carbon double bond to the point where rotation occurs with a rate comparable with the nmr time scale. If electron withdrawing substituents are present at the position β to the amino group, the mesomeric interaction becomes sufficient to move one or both of these barriers into the 10–25 kcal mole^{-1} range, and

7. Rotation about Partial Double Bonds

measurement of these barriers by nmr techniques becomes feasible. In addition, observable barriers may exist for the bond between the activating substituent and double bond, and rather complicated sets of equilibria and rate processes may result. This can be illustrated by considering the case of a simple vinylogous amide **14**, which can, in principle, exist on the nmr time scale as four isomers in each of which rotation about the carbon–nitrogen bond will also be commensurate with the nmr time scale. Two systems that

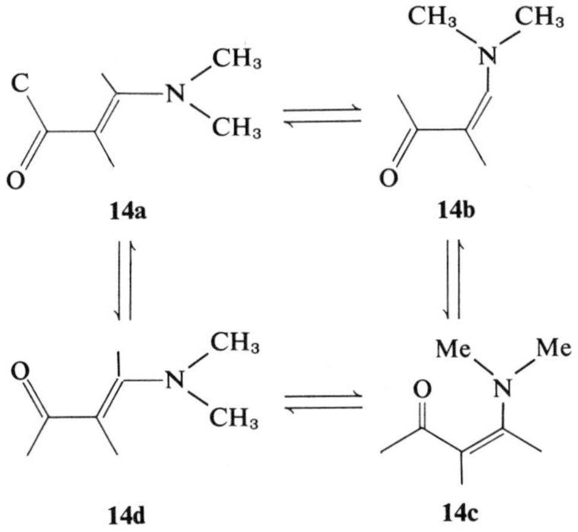

will exemplify the complexities associated with the existence of two or more rotational processes with rates comparable with the nmr time scale will now be considered.

The first example is provided by the study (Shvo and Shanan-Atidi, 1969b) of methyl-3-dimethylamino-2-cyanocrotonate (**15**), the temperature-dependent proton spectra of which are reproduced in Fig. 2. The low temperature ($-60°C$) spectra exhibit pairs of signals for each equivalent group of protons, corresponding to the isomers **15a** and **15b**. It should be noted that, at this temperature, rotation about the C—N bond is slow and the two N-methyl groups are nonequivalent. The relative populations of the two isomers are immediately obtained by integration of the two completely resolved C-methyl absorptions, the assignments of which are firmly based on the known deshielding effect by a *cis*-carbomethoxy group of the methyl

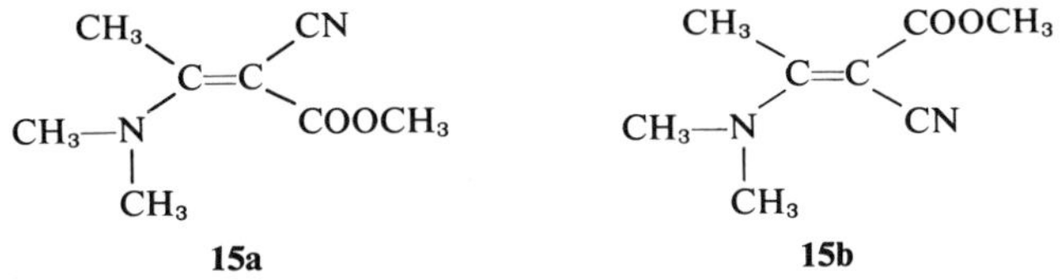

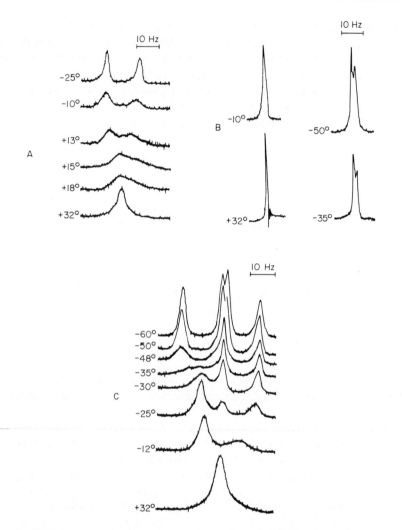

Fig. 2 Nuclear magnetic resonance spectra of methyl 3-dimethylamino 2-cyanocrotonate: (A) =CCH$_3$ signals. (B) COOCH$_3$ signals. (C) N(CH$_3$)$_2$ signals (Shvo and Shanan-Atidi, 1969b).

group in **15b** (Jackman and Wiley, 1960). As the temperature is raised, three distinct coalescences of the four N-methyl absorptions occur which must correspond to three different kinetic processes, namely rotation about the C—N bonds in **15a** and **15b**, and rotation about the carbon–carbon double bond. This last process is more conveniently studied by analysis of the C-methyl absorptions, and $\Delta G^{\ddagger}_{286}$ is found to be 14.8 and 14.9 kcal mole^{-1} for

7. Rotation about Partial Double Bonds

k_a and k_b, respectively. The barriers to rotation about the two C—N bonds are different, having values of 12.1 and 14.8 kcal mole^{-1} for **15a** and **15b**, respectively. As Shvo and Shanan-Atidi (1969b) have pointed out, these barriers may be in error because rotation about the carbon–carbon double bond interferes with the simple form of the analysis used. In principle, a more sophisticated multisite analysis would obviate this problem.

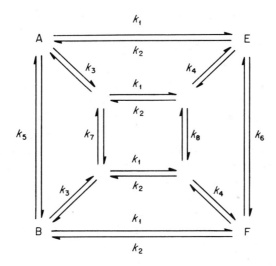

Fig. 3 Pathways for site exchange in **16a** and **16b** (Wennerbeck and Sandström, 1972)

The second example is provided by **16** and involves two isomers with four rotational processes in each and has been studied by Wennerbeck and Sandström, (1972). A total of eight rate constants are involved and the pathways by which the four N-methyl groups can undergo exchange are set out in Fig. 3.

Using the simple multisite GMS approach requires the following set of eight linear equations:

$$\alpha_A \cdot G_A = -i\gamma H_1 \cdot M_{OA} + k_2 \cdot G_E + k_3 \cdot G_C + k_5 \cdot G_B - G_A(k_1 + k_3 + k_5)$$
$$\alpha_B \cdot G_B = -i\gamma H_1 \cdot M_{OB} + k_2 \cdot G_F + k_3 \cdot G_D + k_5 \cdot G_A - G_B(k_1 + k_3 + k_5)$$
$$\alpha_C \cdot G_C = -i\gamma H_1 \cdot M_{OC} + k_2 \cdot G_G + k_3 \cdot G_A + k_7 \cdot G_D - G_C(k_1 + k_3 + k_7)$$
$$\alpha_D \cdot G_D = -i\gamma H_1 \cdot M_{OD} + k_2 \cdot G_H + k_3 \cdot G_B + k_7 \cdot G_C - G_D(k_1 + k_3 + k_7)$$
$$\alpha_E \cdot G_E = -i\gamma H_1 \cdot M_{OE} + k_1 \cdot G_A + k_4 \cdot G_G + k_6 \cdot G_F - G_E(k_2 + k_4 + k_6)$$
$$\alpha_F \cdot G_F = -i\gamma H_1 \cdot M_{OF} + k_1 \cdot G_B + k_4 \cdot G_H + k_6 \cdot G_E - G_F(k_2 + k_4 + k_6)$$
$$\alpha_G \cdot G_G = -i\gamma H_1 \cdot M_{OG} + k_1 \cdot G_C + k_4 \cdot G_E + k_8 \cdot G_A - G_G(k_2 + k_4 + k_8)$$
$$\alpha_H \cdot G_H = -i\gamma H_1 \cdot M_{OH} + k_1 \cdot G_D + k_4 \cdot G_F + k_8 \cdot G_C - G_H(k_2 + k_4 + k_8)$$

in which $\alpha_A = T_{2A}^{-1} - 2\pi i(\nu_A - \nu)$, $M_{OA} = M_{OB} = M_{OC} = M_{OD} = 0.25 p_1$ and $M_{OE} = M_{OF} = M_{OG} = M_{OH} = 0.25 p_2$, where p_1 and p_2 are the populations of the isomers **16a** and **16b**. The equations must be solved for the G's, which are then summed to give $G = u + iv$, the imaginary part of which yields the line shape as a function of the parameters. At high temperatures, a single N-methyl absorption was observed, and at $-100°C$ three strong and two weak signals appeared instead of a maximum of eight. Nevertheless, reasonable estimates of the various barriers were obtained.

The values for rotational barriers about the C—N bond, the C=C bond, and those involving the activating substituent for a variety of enamines are given in Tables VIII, IX, and X. Perhaps the most important feature of these data is that they bring into sharp focus the futility of using estimates of ground state delocalization (or π-bond orders) as guides to barriers in these systems. Such a procedure would suggest that the barrier about the C—N bond should increase at the expense of that for C=C, and this is certainly not always the case. Comparison of the barriers in $(CH_3)_2NCH{=}C(COOCH_3)_2$ and $(CH_3)_2NCCH_3{=}C(COOCH_3)_2$ (Table VIII) is instructive in this regard. The lower barrier in the C-methyl derivative is of course due to some steric destabilization of a coplanar conjugated system, which will result in a decrease in conjugation. However, in this system, the barrier about the carbon–carbon double bond is also greatly reduced, in spite of its predicted higher ground state bond order. The origin of these effects lies in the fact that a polar transition state **17** for rotation about the C=C bond is well stabilized by conjugation and little affected by steric interactions, whereas, in the

17

ground state, such stabilization is largely precluded by steric effects. An X-ray structure of $(CH_3)_2NCH=C(COOCH_3)_2$ reveals a fairly planar arrangement for the molecule except for the carbomethoxy group cis to the dimethylamino group, which is severely twisted out of plane. The carbon–carbon double bond length is 1.38 Å, suggestive of rather weak ground-state conjugation (Shmueli *et al.*, 1973), further substantiating the importance of transition-state stabilization in lowering the C=C barrier in this type of system.

In general, vinylogous amides have barriers about the C—N bond that are lower than the analogous amides. The ability of β-substituents to enhance these barriers shows the normal dependence on their electron-withdrawing power, provided the role of steric destabilization of the ground state is taken into account. An important observation (Shvo and Shanan-Atidi, 1969b) is that the conjugative ability, σ_R, of the carboxyl group relative to that of the nitrile group is largely responsible for the dramatic lowering (6–8 kcal mole^{-1}) in the barrier, effected by substituting the former for the latter. This is, of course, consistent with the nature of the transition state **17** and with the postulate that, in these systems, the stabilization of the transition state is dominant in determining barriers. We shall see presently that this latter situation does not always prevail (p. 233). Furthermore, it has been noted that, whereas the barrier for the C—N bond in $[(CH_3)_2N]_2C=CX_2$ is increased by 5.2 kcal mole^{-1} by exchanging X = CN for X = COOCH$_3$, the analogous change in substitution in $(CH_3)_2NCH=CX_2$ results in a decrease of 4.7 kcal mole^{-1} in the barrier (Wennerbeck and Sandström, 1972).

Some of the results presented in Table X deserve special comment. The origin of the difference between the "fixed" *S*-cis and *S*-trans models for vinylogous amides is not obvious (Dabrowski and Kozerski, 1972). The delocalization of the electron pair of the squaric acid moiety appears to be abnormally large, since the C—N bond barriers in its amide derivatives are among the largest found in vinylogous amides. Surprisingly, the benzene ring is almost as effective as a double bond (see Table VIII) in transmitting the conjugative interaction between a dimethylamino group and a carbonyl group. The origin of the large barrier recorded in the last entry of Table X is obscure.

The fulvene nucleus is an effective electron sink for the lone pair of electrons of a 6-amino substituent, and barriers to rotation about the C—N bond as well as the exocyclic carbon–carbon double bond have been determined by several groups of workers (Table XI). The effectiveness of the fulvene nucleus is, in fact, comparable with a geminal pair of dicarboxymethoxy groups and is considerably enhanced by electron withdrawing substituents in the ring. The corresponding indenylidene system is less effective.

Some quite large barriers have been observed in amino substituted, aromatic, heterocycles and heterocyclic cations, and these are recorded in Table

TABLE VIII
BARRIERS TO ROTATION IN N,N-DIMETHYL SUBSTITUTED ACYCLIC ENAMINES[a]

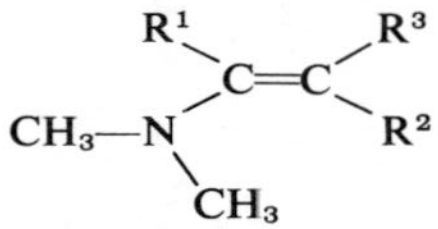

$$\begin{array}{c} R^1 R^3 \\ \diagdown C = C \diagup \\ \diagup \diagdown \\ CH_3 - N R^2 \\ | \\ CH_3 \end{array}$$

R^1	R^2	R^3	$\Delta G^{\ddagger}$ (C—N) (kcal mole^{-1})	°K	$\Delta G^{\ddagger}$ (C=C) (kcal mole^{-1})	°K	$\Delta G^{\ddagger}$ (C—R) (kcal mole^{-1})	°K	Solvent	Reference
H	H	CHO	14.6	292	—	—	—	—	CCl_2CCl_2	Dabrowski and Kozerski (1972)
H	H	$COCH_3$	13.4	267	—	—	—	—	CCl_2CCl_2	Dabrowski and Kozerski (1972)
H	H	$COCH_2CH_3$	13.2	265	—	—	—	—	CCl_2CCl_2	Dabrowski and Kozerski (1972)
H	H	$COCH(CH_3)_2$	13.2	264	—	—	—	—	CCl_2CCl_2	Dabrowski and Kozerski (1972)
H	H	$COC(CH_3)_3$	13.1	262	—	—	—	—	CCl_2CCl_2	Dabrowski and Kozerski (1972)
H	H	CH=CHCHO	13.6	268	—	—	—	—	CCl_4	Filleux-Blanchard et al. (1969)
H	H	$COOCH_2CH_3$	13.9	273	—	—	—	—	$CHCl_3$	Mannschreck and Koelle (1967)
H	H	NO_2	16.5	323	—	—	—	—	$CDBr_3$	Mannschreck and Koelle (1967)
H	H	CN	12.9	298	—	—	—	—	CCl_4	Hobson and Reeves (1973a)
CH_3	H	CHO	12.2	241	—	—	—	—	CCl_2CCl_2	Dabrowski and Kozerski (1972)
CH_3	H	$COCH_3$	11.3	221	—	—	—	—	CCl_2CCl_2	Dabrowski and Kozerski (1972)
H	$COOCH_3$	$COOCH_3$	13.3	264	15.6	292	—	—	CH_2Cl_2	Shvo and Shanan-Atidi (1969a)

7. Rotation about Partial Double Bonds

									Reference
CH₃	COOCH₃	~9	187	<9	178	—	—	CH₂Cl₂	Shvo and Shanan-Atidi (1969a)
H	CN	17.7	298	—	—	—	—	(CHCl₂)₂	Hobson and Reeves (1973a)
H	CN	17.6	331	—	—	—	—	C₆H₅Br	Shvo and Shanan-Atidi (1969b)
H	Ph	<12	218	—	—	—	—	Acetone-d	Mannschreck and Koelle (1967)
CH₃	COOCH₃	12.8[b]	238	14.9	287	—	—	CH₂Cl₂	Shvo and Shanan-Atidi (1969b)
CH₃	CN	12.1[b]	261	14.8	287	—	—	CH₂Cl₂	Shvo and Shanan-Atidi (1969b)
N(CH₃)₂	COOCH₃	14.0	273	—	—	10.2	208	C₆H₅F	Wennerbeck and Sandström (1972)
N(CH₃)₂	COOCH₃	15.5	283	<8	<150	—	157	CDCl₃:CS₂ (1:1)	Wennerbeck and Sandström (1972)
N(CH₃)₂	COCH₃	15.0	279	<8	<150	8.1		CS₂:CH₂Cl₂	Wennerbeck and Sandström (1972)
N(CH₃)₂	COCH₃	16.4	318	9.4	201	—	—	CH₂Cl₂:pyridine (1:1)	Wennerbeck and Sandström (1972)
N(CH₃)₂	COCH₃	15.2	295	9.8	193	12.9, 12.5	231	CDCl₃:CS₂ (1:1)	Wennerbeck and Sandström (1972)
N(CH₃)₂	COOCH₃	14.2, 13.7	266	15.2	266	—	—	C₆H₅F	Wennerbeck and Sandström (1972)
N(CH₃)₂	COPh	13.7, 13.5	253	14.7, 13.9	253	13.6, 13.2	266	CDCl₃:CS₂ (1:1)	Wennerbeck and Sandström (1972)
N(CH₃)₂	CN	10.3	203	—	—	—	—	CDCl₃	Wennerbeck and Sandström (1972)
N(CH₃)₂	p-NO₂C₆H₄	13.9, 14.1	274	14.3	281	—	—	C₆H₅F	Wennerbeck and Sandström (1972)
N(CH₃)₂	Ph	12.3, 13.0	298	20.1[c]	298	—	—		Wennerbeck and Sandström (1972)
SCH₃	COOCH₃	8.9	173	—	—	—	—	Acetone-d	Shvo and Belsky (1969)
SCH₃	CN	11.0	214	—	—	—	—	CH₂Cl₂	Shvo and Belsky (1969)

[a] Two entries occur when the rotation interconverts a pair of nondegenerate isomers.
[b] Subject to possible errors.
[c] o-C₆H₄Cl₂.

TABLE IX
BARRIERS TO ROTATION ABOUT CARBON–CARBON DOUBLE BONDS IN SOME MISCELLANEOUS ENAMINES

Compound		$\Delta G^\ddagger$ (kcal mole^{-1})	°K	Solvent	Reference
CH, RC$_6$H$_4$N(CH$_3$)–C=C(COOCH$_3$)(COOCH$_3$)	R=OCH$_3$ R=H R=NO$_2$	18.5 19.4 22.1	280 363 410	CHBr$_3$ CHBr$_3$ CHBr$_3$	Shvo et al. (1967) Shvo et al. (1967) Shvo et al. (1967)
H, RC$_6$H$_4$N(CH$_3$)–C=C(COOCH$_3$)(COOCH$_3$)	R=OCH$_3$ R=H R=NO$_2$	13.0 13.9 16.9	243 266 316	CH$_2$Cl$_2$ CH$_2$Cl$_2$ CH$_2$Cl$_2$	Shvo et al. (1967) Shvo et al. (1967) Shvo et al. (1967)
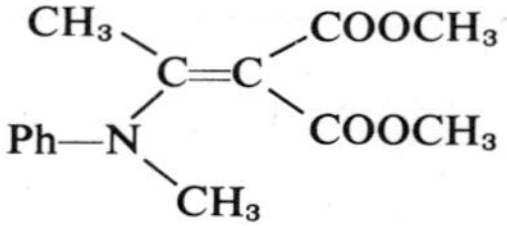 CH$_3$, Ph–N(CH$_3$)–C=C(COOCH$_3$)(COOCH$_3$)		10.0	197	Acetone-d_6	Shvo and Shanan-Atidi (1969a)
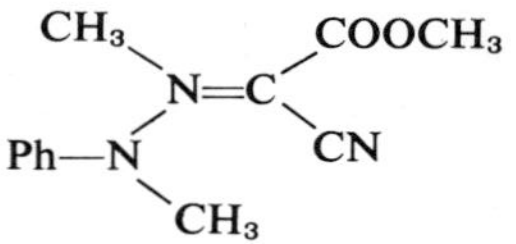 CH$_3$, Ph–N(CH$_3$)–C=C(COOCH$_3$)(CN)		18.3, 17.8	196	C$_6$H$_5$Br	Shvo and Shanan-Atidi (1969b)
pyrrolidine-C(=C(COOCH$_3$)(COOCH$_3$)), N–R	R=CH$_3$ R=CH$_2$Ph R=Ph	9.8 10.0 13.7	192 202 278	Acetone-d_6 Acetone-d_6 CH$_2$Cl$_2$	Shvo and Shanan-Atidi (1969a) Shvo and Shanan-Atidi (1969a) Shvo and Shanan-Atidi (1969a)
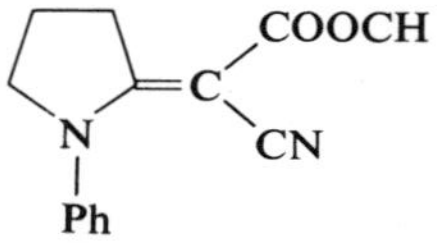 pyrrolidine-C(=C(COOCH)(CN)), N–Ph		18.6, 18.0	355	CDCl$_3$	Shvo and Shanan-Atidi (1969b)

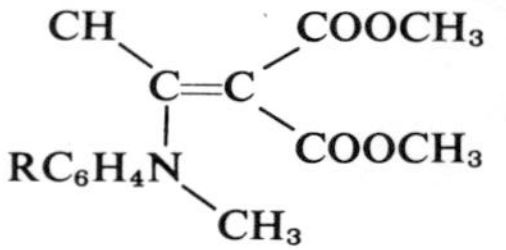

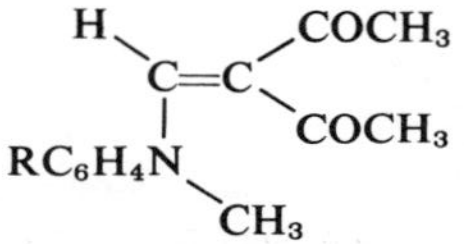

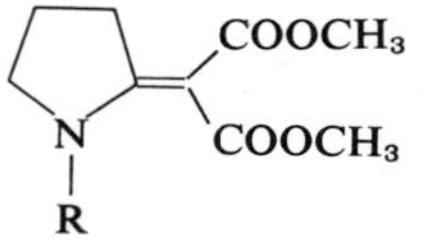

TABLE X
Barriers to Rotation about Carbon–Nitrogen Bonds in Some Miscellaneous Enamines

Compound	$\Delta G^{\ddagger}$ (kcal mole^{-1})	°K	Solvent	Reference
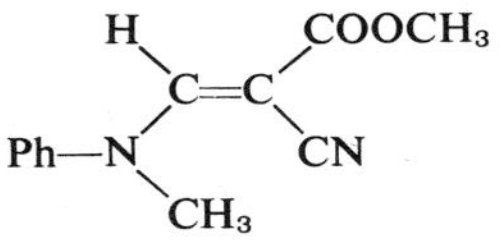 H, COOCH₃ / Ph–N, CN, CH₃	12.8, 12.2	263	Acetone-d_6	Shov and Shanan-Atidi (1969b)
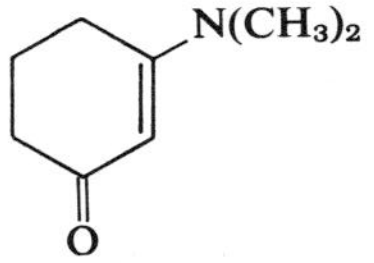 cyclohexenone with N(CH₃)₂	12.0	237	CCl₂CCl₂	Dabrowski and Kozerski (1972)
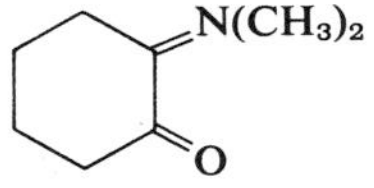 cyclohexanone with =N(CH₃)₂	9.2	178	CCl₂CCl₂	Dabrowski and Kozerski (1972)
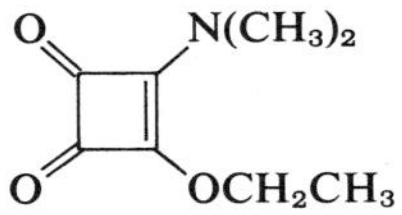 cyclobutenedione with N(CH₃)₂, OCH₂CH₃	16.9	233	CDCl₃	Thorpe (1968)
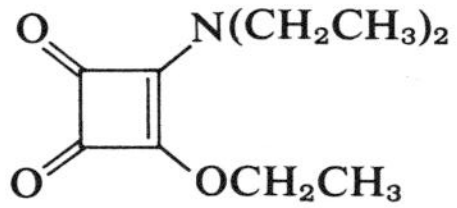 cyclobutenedione with N(CH₂CH₃)₂, OCH₂CH₃	17.3	234	CDCl₃	Thorpe (1968)

(continued)

TABLE X (continued)

Compound	$\Delta G^\ddagger$ (kcal mole^{-1})	°K	Solvent	Reference
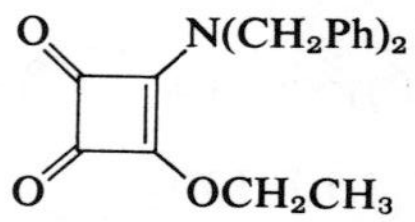 N(CH₂Ph)₂ / OCH₂CH₃ cyclobutenedione	16.7	234	CDCl₃	Thorpe (1968)
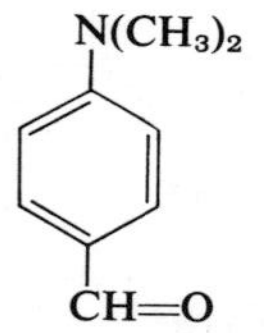 4-N(CH₃)₂-C₆H₄-CH=O	12.9	300	Toluene-d_8	Klinck et al. (1967)
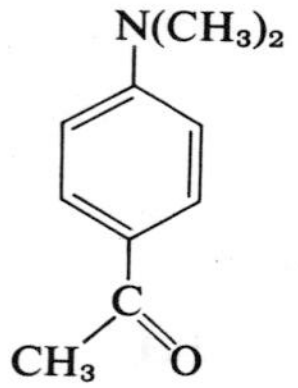 4-N(CH₃)₂-C₆H₄-C(=O)CH₃	12.4	300	Toluene-d_8	Klinck et al. (1967)
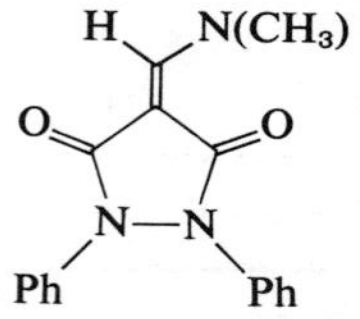 H,N(CH₃)-methylene pyrazolidinedione (N,N'-diphenyl)	21.5	433	Ph₂O	Mannschreck and Koelle (1967)

TABLE XI
BARRIERS TO ROTATION ABOUT C—N AND C=C IN 6-N,N-DIMETHYLAMINOFULVENES AND RELATED COMPOUNDS

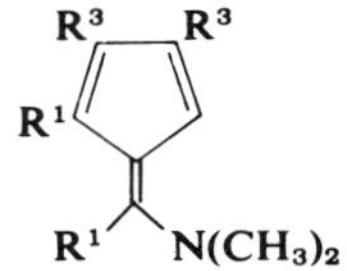

R^1	R^2	R^3	$\Delta G^{\ddagger}$ (C—N) (kcal mole^{-1})	°K	$\Delta G^{\ddagger}$ (C=C) (kcal mole^{-1})	°K	Solvent
H	H	H^{a,b}	13.5	273	22.1	421	DMSO
CH$_3$	H	H^a	—	—	15.5	293	DMSO–D$_2$O
			—	—	16.4	311	DMSO
			—	—	17.5	330	Acetone-d_6
Ph	H	H^a	10.7	208	17.5	330	CDCl$_3$
p-CH$_3$OC$_6$H$_4$	H	H^a	11.8	239	19.2	373	DMSO
H	CHO	H^{a,c}	12.1	245	18.8	364	DMSO
H	H	CHO^a	17.9	273	—	—	C$_2$HCl$_5$
H	H	CHO^a	21.5	273	—	—	C$_2$HCl$_5$
H	CH$_3$	CHO^a	20.3	273	—	—	C$_2$HCl$_5$

Compound	$\Delta G^{\ddagger}$ (C—N) (kcal mole^{-1})	°K	$\Delta G^{\ddagger}$ (C=C) (kcal mole^{-1})	°K	Solvent
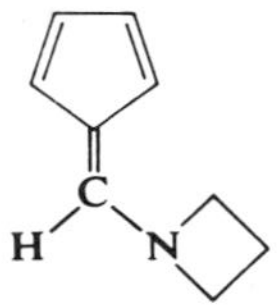	—	—	18.3^d	—	CH$_3$CN
	—	—	19.1^d	—	DMF
	—	—	19.6^d	—	Acetone

(continued)

TABLE XI (continued)

Compound		$\Delta G^{\ddagger}$ (C—N) (kcal mole^{-1})	°K	$\Delta G^{\ddagger}$ (C=C) (kcal mole^{-1})	°K	Solvent
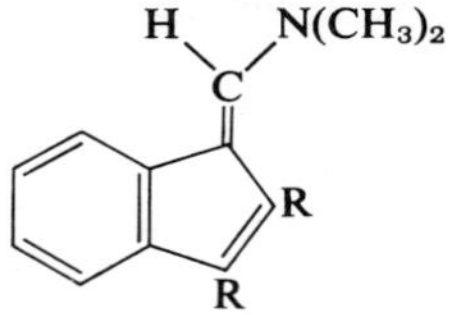 (indene-N(CH₃)₂ structure)	R=H[c]	10.4	213	—	—	Acetone-d_6
	R=CHO[a]	12.6	273	—	—	CDCl$_3$
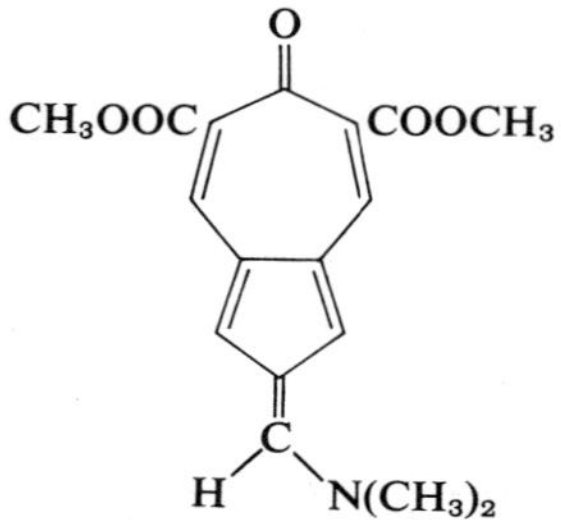 (azulene diester-N(CH₃)₂ structure)		20.6	281	—	—	Pyridine
		19.6	269	—	—	C$_2$HCl$_5$

[a] Downing et al. (1969).
[b] Crabtree and Bertelli (1967).
[c] Mannschreck and Koelle (1967).
[d] These values are activation energies.

TABLE XII
BARRIER TO ROTATION FOR C—N BONDS IN SOME N,N-DIMETHYLAMINO-SUBSTITUTED HETEROCYCLES

Compound	$\Delta G^{\ddagger}$ (kcal mole^{-1})	°K	Solvent	Reference
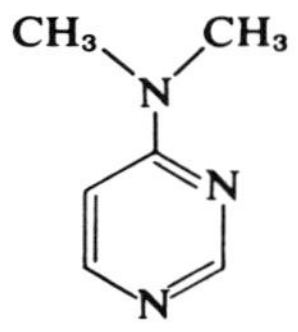 (CH$_3$)(CH$_3$)N–pyridyl	12.2	298	CHCl$_3$	Katritzky and Tiddy (1969)
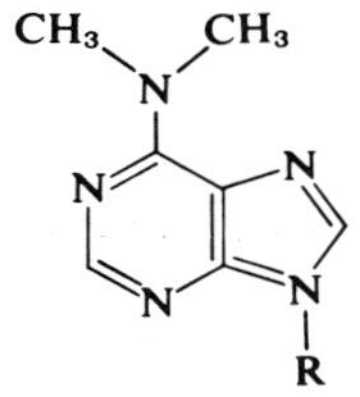 (CH$_3$)(CH$_3$)N–purinyl-R [a]	13.4	273	CDCl$_3$	Martin and Reese (1967)
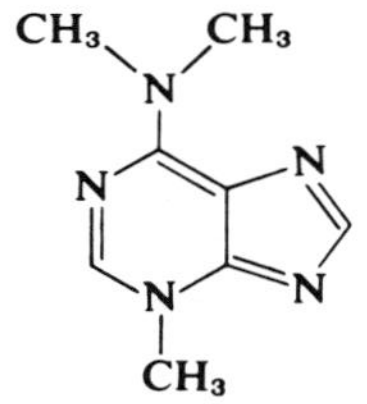 (CH$_3$)(CH$_3$)N–(N-CH$_3$)purinyl	15.3	303	CDCl$_3$	Neiman and Bergmann (1968)

(continued)

TABLE XII (continued)

Compound	$\Delta G^{\ddagger}$ (kcal mole^{-1})	°K	Solvent	Reference
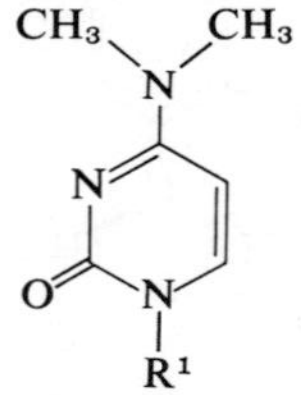 $R^1 = R^a$	15.5	303	CDCl$_3$	Martin and Reese (1967)
$R^1 = DH_3$	14.8	298	CDCl$_3$	Shoup et al. (1972)
$R^1 = X^b$	16.6	298	CDCl$_3$	Shoup et al. (1972)
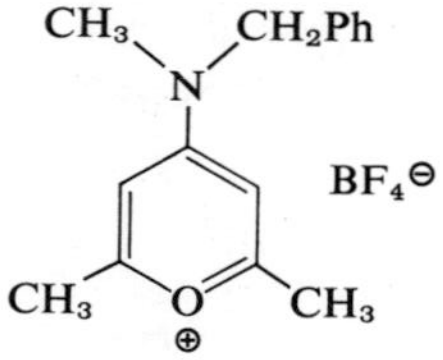 (2,6-dimethylpyrylium BF$_4^\ominus$, N(CH$_2$Ph)(CH$_3$))	22.5	298	(CH$_3$O)$_3$PO	J. P. Brown and L. M. Jackman (unpublished results, 1973)
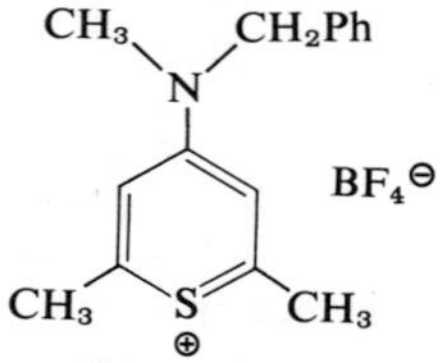 (2,6-dimethylthiopyrylium BF$_4^\ominus$, N(CH$_2$Ph)(CH$_3$))	19.5	298	(CH$_3$O)$_3$PO	J. P. Brown and L. M. Jackman (unpublished results, 1973)

[a] R = 2′,3′-O-isopropylideneribofuranosidyl.
[b] X = 2,3,4,6-O-tetraacetylglucopyranosyl.

7. Rotation about Partial Double Bonds

XII. In the first two entries, the existence of an amidine-like segment can be recognized. In the third, the increased barrier is presumably accounted for by the partial transfer of the lone pair of the dimethylamino group to the five-membered ring with consequent stabilization of the ground state. The cytosine derivatives may be regarded as vinylogous amides. The pyrilium and thiapyrilium salts evidently receive considerable stabilization of their ground states from canonical structures of the type **18**. This is not true for the

18

19, R = CH_3 or Ph

corresponding pyridinium salt (**18**, X = $NCH_2CH_2CH_2CH_3$), in which the barrier appears to be less than 10 kcal mole^{-1} (J. P. Brown and L. M. Jackman, unpublished results, 1973). Similarly the amine derivatives of the cyclopropenium cation (**19**) have barriers in the range 22–25 kcal mole^{-1} (Krebs and Breckwoldt, 1969).

V. Enolates, Enol Ethers and Thio Enol Ethers, Acylides, Diazoketones, and Aminoboranes

A number of highly activated enol and thio enol ethers have been investigated, and their barriers to rotation, together with those of the analogous dimethylamino compounds, are recorded in Table XIII. It is evident that the barriers to rotation for the oxygen and sulfur compounds are substantially higher as a consequence of the poorer ability of these elements to accommodate the positive charge in the polar transition state.

The last five entries in Table XIII illustrate the existence of steric destabilization of the ground state associated with the introduction of bulky groups α to the methoxy group. The effect of replacing hydrogen with methyl is presumably due to inductive and hyperconjugative stabilization of the polar transition state, but further decreases in the barriers in the series CH_3, CH_2CH_3, $CH(CH_3)_2$, $C(CH_3)_3$ must result in steric distortion of the planar double bond, which, of course, is relieved in the transition state.

The only reported (Jackman and Haddon, 1973) attempt to determine the barrier to rotation in an enolate ion is for lithioisobutyrophenone in triglyme. No evidence for exchange of the nonequivalent methyl groups was observed at 473°K, indicating a barrier in excess of 25 kcal mole^{-1}. Presumably, a

TABLE XIII
BARRIERS TO ROTATION ABOUT C=C IN SOME DIMETHYLAMINES,
ENOL METHYL ETHERS, AND THIO ENOL METHYL ETHERS IN
$R_1R_2C=CR_3R_4$

R_1	R_2	R_3	R_4	$\Delta G^\ddagger$ (kcal mole^{-1})	°K	Solvent
COOCH$_3$	COOCH$_3$	H	N(CH$_3$)$_2$	15.6	292	CH$_2$Cl$_2$[a]
COOCH$_3$	COOCH$_3$	H	OCH$_3$	> 27.7	479	C$_4$Cl$_6$[b]
COOCH$_3$	COOCH$_3$	CH$_3$	N(CH$_3$)$_2$	< 9.1	183	CH$_2$Cl$_2$[c]
COOCH$_3$	COOCH$_3$	CH$_3$	OCH$_3$	25.7	464	C$_4$Cl$_6$[b]
COOCH$_3$	COOCH$_3$	CH$_3$	SCH$_3$	> 27.5	473	C$_2$Cl$_6$[d]
COOCH$_3$	CN	N(CH$_3$)$_2$	N(CH$_3$)$_2$	14.0	260	CS$_2$-pyridine[e]
COOCH$_3$	CN	SCH$_3$	SCH$_3$	24.6	458	o-C$_6$H$_4$Cl$_2$[f]
COOCH$_3$	COOCH$_3$	H	OCH$_3$	> 27.7	479	C$_4$Cl$_6$[b]
COOCH$_3$	COOCH$_3$	CH$_3$	OCH$_3$	25.7	464	C$_4$Cl$_6$[b]
COOCH$_3$	COOCH$_3$	CH$_2$CH$_3$	OCH$_3$	24.7	435	C$_4$Cl$_6$[b]
COOCH$_3$	COOCH$_3$	CH(CH$_3$)$_2$	OCH$_3$	23.3[g]	425	C$_4$Cl$_6$[b]
COOCH$_3$	COOCH$_3$	C(CH$_3$)$_3$	OCH$_3$	18.3	334	C$_6$H$_5$Br[b]

[a] Shvo et al. (1967).
[b] Shvo (1968).
[c] Shvo and Shanan-Atidi (1969b).
[d] Shvo and Belsky (1969).
[e] Wennerbeck and Sandstrom (1972).
[f] Isaksson et al. (1967).
[g] 23.0 in C$_6$H$_5$Br.

number of more highly activated enolate ions would have barriers amenable to dnmr studies.

Nitrogen, sulfur, and phosphorus acylides (**20**), are, in a sense, related to enolate ions but might be expected to have lower barriers to rotation. Indeed, line shape phenomena explicable in terms of rate of rotation about the partial

20, X = $\equiv$N, $>$S, or $\equiv$P

carbon–carbon double bond commensurate with the nmr time scale have frequently been observed (Table XIV), but only one study in which actual barriers have been determined has been published. This is by Zeliger et al. (1969), who found barriers ($\Delta G^\ddagger$) of 18.4 (341°K) and 17 (335°K) kcal mole^{-1} for **21a** and **21b**, respectively, the two geometric isomers being almost equally

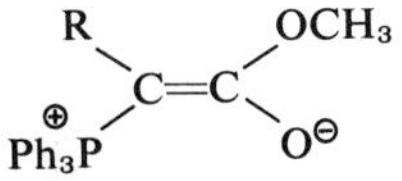

7. Rotation about Partial Double Bonds

$$\underset{\underset{Ph_3\overset{\oplus}{P}}{}}{R}C=C\underset{O^{\ominus}}{OCH_3}$$

21a, R = CH$_3$
21b, R = CH$_2$CH$_3$

populated. It would seem that, with the more sophisticated theoretical techniques currently available, examination of barriers in acylides could provide an interesting way of studying this group of synthetically important compounds.

Consideration of the various canonical structures (**22**), for diazoketones suggests that if **22c** makes a significant contribution, these compounds resemble acylides and could possibly exhibit cis–trans isomerism about the

TABLE XIV
ACYLIDES EXHIBITING LINE-SHAPE PHENOMENA DUE TO ROTATION ABOUT THE CARBON–CARBON DOUBLE BOND

Class	Reference
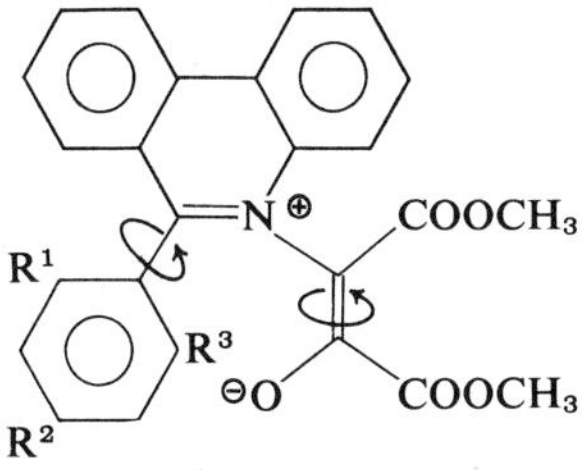 [structure with R^1, R^2, R^3, =N$^{\oplus}$, COOCH$_3$, COOCH$_3$, O$^{\ominus}$]	Acheson and Selby (1970) (These systems also exhibit slow phenyl rotation)
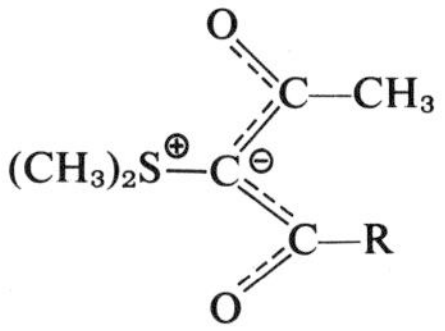 (CH$_3$)$_2$S$^{\oplus}$—C$^{\ominus}$ with C—CH$_3$ (=O) and C—R (=O)	Nozaki et al. (1967)
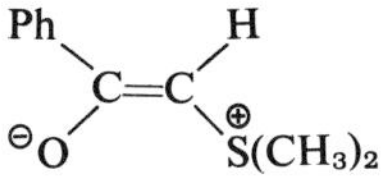 Ph, H, $^{\ominus}$O, S(CH$_3$)$_2^{\oplus}$ on C=C	Trost (1967)
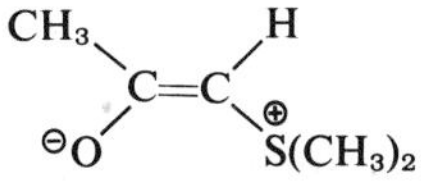 CH$_3$, H, $^{\ominus}$O, S(CH$_3$)$_2^{\oplus}$ on C=C	Casanova and Rutolo (1967); Smallcombe et al. (1968)
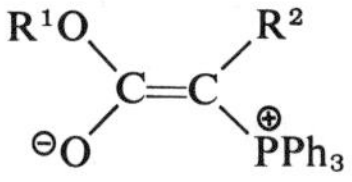 R^{1}O, R^2, $^{\ominus}$O, $\overset{\oplus}{P}$Ph$_3$ on C=C	Zeliger et al. (1969); Randall and Johnson (1968); Crouse et al. (1968)

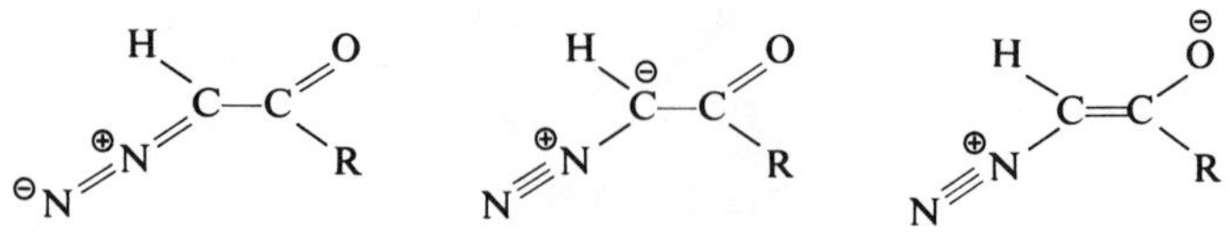

$$\underset{22a}{\overset{H}{\underset{\ominus N}{\overset{\oplus}{=}}}N\overset{}{-}\overset{H}{\underset{}{C}}\overset{O}{\underset{R}{=}}C} \quad \underset{22b}{\overset{H}{\underset{N}{\overset{\oplus}{=}}}N\overset{}{-}\overset{\ominus}{\underset{}{C}}\overset{O}{\underset{R}{=}}C} \quad \underset{22c}{\overset{H}{\underset{N}{\overset{\oplus}{=}}}N\overset{}{-}\overset{H}{\underset{}{C}}\overset{\ominus O}{\underset{R}{=}}C}$$

C—C bond. Kaplan and Meloy (1966) have demonstrated this to be the case and that the rates of isomerism at room temperature are comparable with the nmr time scale. More recently, Kessler and Rosenthal (1973) have provided additional examples. Of course, in some systems, one isomer is overwhelmingly favored, and the determination of barriers to rotation by dnmr is not

TABLE XV
BARRIERS TO ROTATION IN DIAZOKETONES, R^1CO—$C(N_2)R^2$, IN $CDCl_3$

R^1	R^2	$\Delta G^{\ddagger}_{298}$ $Z \to E$ (kcal mole^{-1})	$\Delta G^{\ddagger}_{298}$ $E \to Z$ (kcal mole^{-1})
H	CH_3	15.4	13.9[a]
H	CH_3CH_2	15.3	13.5[a]
H	$PhCH_2$	15.3	13.4[a]
H	CH_3O	12.8	12.7[a]
H	CH_3CH_2O	13.2	13.2[a]
H	2,4,6-$(CH_3)_3C_6H_2$	15.0[b]	—
H	2,4,6-(isopropyl)$_3C_6H_2$	15.5[b]	—
CH_3	2,4,6-(isopropyl)$_3C_6H_2$	16.7[b]	—

[a] Kaplan and Meloy (1966).
[b] Kessler and Rosenthal (1973).

feasible. Available data for barriers are assembled in Table XV. Extended Hückel MO calculations of some of these barriers give good agreement with the observed values (Csizmadia et al., 1969).

Aminoboranes evidently have an ylide-like structure (23) and exhibit

$$R^1R^2\ddot{N}\text{—}BR^3R^4 \longleftrightarrow R^1R^2\overset{\oplus}{N}\text{=}\overset{\ominus}{B}R^3R^4$$
23

barriers ranging from 10 to 25 kcal mole^{-1} (Imbery et al., 1970) depending on substitution (Table XVI). There has also been an extensive study (Totani et al., 1971) of solvent effects that indicates that the barriers are lowered by some lone-pair donating solvents. As expected, much lower barriers (~ 10 kcal mole^{-1}) prevail in diaminoboranes.

7. Rotation about Partial Double Bonds

TABLE XVI
BARRIERS TO ROTATION ABOUT THE BORON–NITROGEN BOND IN AMINOBORANES[a]
$R^1R^2N\text{—}BR^3R^4$

R^1	R^2	R^3	R^4	$\Delta G^{\ddagger}$ (kcal mole^{-1})	°K	Solvent
CH$_3$	CH$_3$	Cl	Ph	20.3	393	C$_4$Cl$_6$
CH$_3$CH$_2$	CH$_3$CH$_2$	Cl	Ph	19.2	373	C$_4$Cl$_6$
n-Pr	n-Pr	Cl	Ph	19.2	373	C$_4$Cl$_6$
n-Bu	n-Bu	Cl	Ph	19.7	383	C$_4$Cl$_6$
Isopropyl	Isopropyl	Cl	Ph	17.0	343	o-C$_6$H$_4$Br$_2$
sec-Bu	sec-Bu	Cl	Ph	17.9	365	C$_4$H$_6$
CH$_3$	CH$_3$	Ph	PhCH$_2$	22.3	423	m-C$_6$H$_4$Cl$_2$
sec-Bu	sec-Bu	Ph	CH$_3$	20.1	403	C$_4$Cl$_6$
CH$_3$	CH$_3$	Ph	N(CH$_3$)$_2$	10.5	213	CS$_2$
CH$_3$	PhCH$_2$	Ph	N(CH$_3$)CH$_2$Ph	10.2	203	CS$_2$
CH$_3$	Ph	CH$_3$	CH$_3$	18.9	368	C$_4$Cl$_6$
CH$_3$	PhCH$_2$	Cl	Ph	20.5	393	C$_4$Cl$_6$
CH$_3$	PhCH$_2$	Ph	PhCH$_2$	23.7, 22.6	453	m-C$_6$H$_4$Cl$_2$

[a] Imbery et al. (1970).

VI. Potentially Aromatic Systems

In systems in which one or both termini of the double bond are parts of potentially aromatic rings, stabilization of the transition state for rotation may be sufficient to result in rates comparable with the nmr time scale. Such a situation is well exemplified by the pentatriafulvene derivative (**23**), in which both the rings become aromatic in the transition state (Kende et al., 1966).

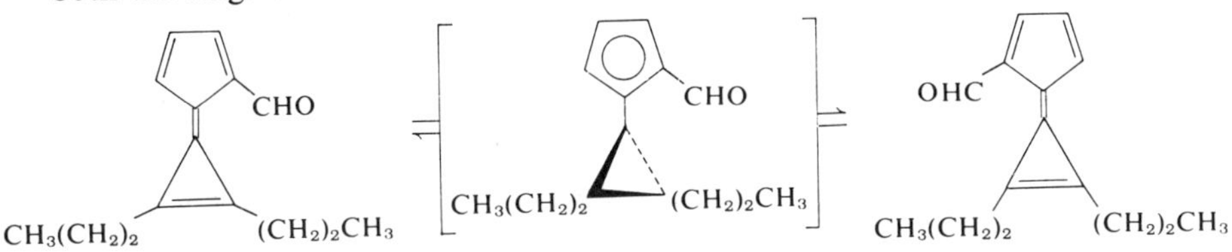

23

The iminium salts (**19**) discussed above provide another example if viewed as the canonical structure, **24**.

24

TABLE XVII
EFFECT OF SOLVENT ON THE BARRIER TO
ROTATION ABOUT THE PENTATRIAFULVENE (23)[a]

$\Delta G^{\ddagger}$ (kcal mole^{-1})	°K	Solvent
19.4	387	Ph$_2$O
19.2	378	o-C$_6$H$_4$Cl$_2$
18.4	356	PhCN
18.0	339	DMF

[a] Kende et al. (1966).

The barrier in **23** appears to show some dependence on solvent polarity (Table XVII) which is consistent with the polar transition state. Similar low barriers are observed for the quinone derivatives (**25** and **26**), in which the transition states involve the fully aromatic anthrolate anions. In compound **27**, the aromatic stabilization of the transition state is so effective that the barrier to rotation is apparently very low (Eichei and Pelz, 1970). Barriers to rotation have also been determined for a variety of diacylmethylenecyclopropenes, and these results are summarized in Table XVIII.

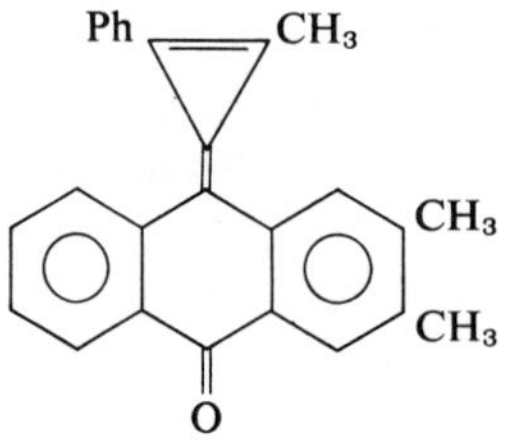

$\Delta G^{\ddagger}_{322} = 17.3$ (PhNO$_2$)
25

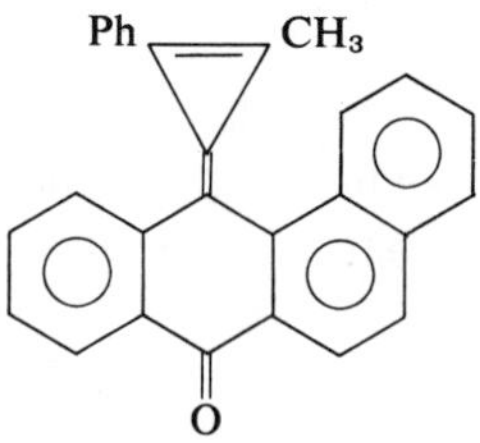

$\Delta G^{\ddagger}_{305} = 15.3$ (PhNO$_2$)
26

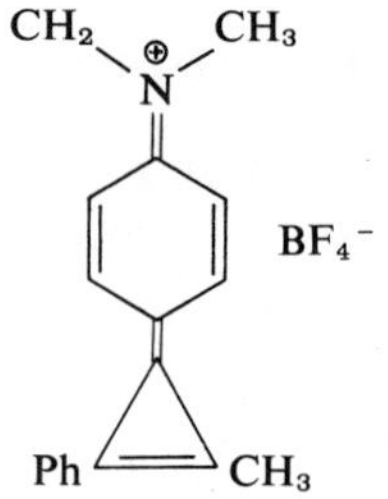

$\Delta G^{\ddagger}_{213} < 11$ (CH$_2$Cl$_2$)
27

A low barrier ($\Delta G^{\ddagger}_{291} = 14.5$ kcal mole^{-1}) has been observed in 2,6-di-*t*-butyl-4(*N,N*-dimethylaminomethylene)-2,5-cyclohexadienone (**28**). Here, the stabilizing influences in the transition state are the ability of the nitrogen atom to accommodate a positive charge and the formation of the aromatic phenolate ion. There is a significant barrier ($\Delta G^{\ddagger}_{227} = 12.4$ kcal mole^{-1}) to rotation about the C—N bond in this compound. The two corresponding barriers in the hydrazinoquinone (**29**) are, respectively, too high and too low to be determined (Mannschreck and Kolb, 1972).

Several potentially aromatic heterocyclic systems have been examined (Table XIX). Only in those cases in which the pyridinium cation is developed

TABLE XVIII
BARRIERS TO ROTATION ABOUT THE EXOCYCLIC DOUBLE BOND IN METHYLENE CYCLOPROPENES[a]

$$R^4CO\diagdown_{}\diagup COR^3$$
$$R^1\triangle R^2$$

R^1	R^2	R^3	R^4	$\Delta G^\ddagger$ (kcal mole^{-1})	°K	Solvent
H	Ph	CH$_3$	CH$_3$	19.0	360	PhNO$_2$
H	Ph	Ph	CH$_3$	22.6, 21.6	403	PhNO$_2$
H	Ph	Ph	PhCH$_2$	22.6, 21.8	404	PhNO$_2$
H	Ph	Ph	H	21.8, 21.2	421	PhNO$_2$
H	Ph	CH$_3$	PhNH	>23.7, >22.2	453	PhNO$_2$
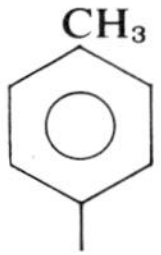 CH$_3$-C$_6$H$_4$	 CH$_3$-C$_6$H$_4$	Ph	CH$_3$	19.7	371	α-Chloronaphthalene
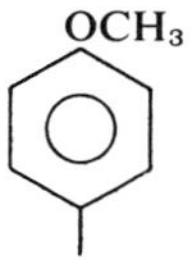 OCH$_3$-C$_6$H$_4$	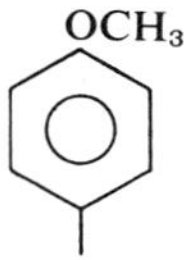 OCH$_3$-C$_6$H$_4$	Ph	CH$_3$	19.0	354	α-Chloronaphthalene

(continued)

TABLE XVIII (continued)

R¹	R²	R³	R⁴	$\Delta G^{\ddagger}$ (kcal mole^{-1})	°K	Solvent
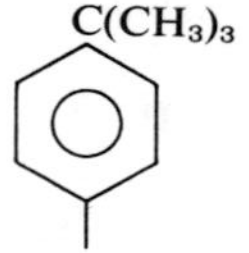 C(CH₃)₃-C₆H₄-	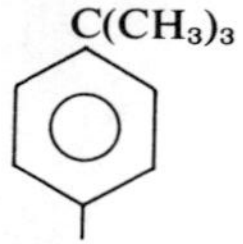 C(CH₃)₃-C₆H₄-	Ph	CH₃	18.5	343	PhNO₂
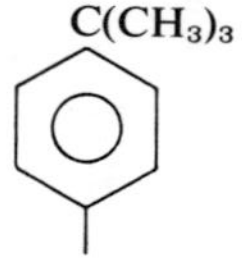 C(CH₃)₃-C₆H₄-	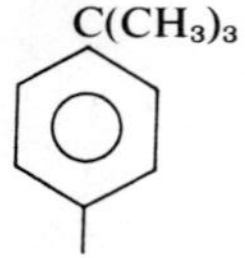 C(CH₃)₃-C₆H₄-	Ph	PhNH	20.9	382	PhNO₂
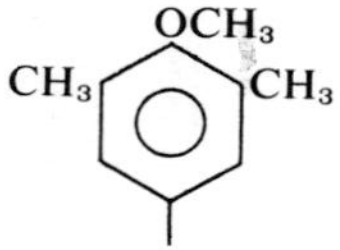 2,6-(CH₃)₂-3-OCH₃-C₆H₂-	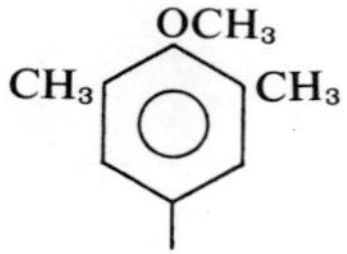 2,6-(CH₃)₂-3-OCH₃-C₆H₂-	Ph	CH₃	18.9	363	α-Chloronaphthalene
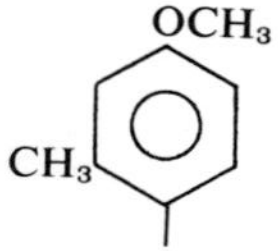 2-OCH₃-3-CH₃-C₆H₃-	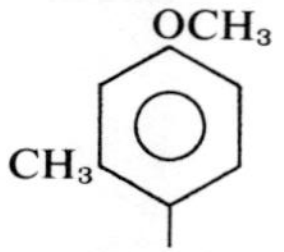 2-OCH₃-3-CH₃-C₆H₃-	Ph	CH₃	14.1	283	α-Chloronaphthalene

[a] Eicher and Pelz (1970).

TABLE XIX
Barriers to Rotation about Exocyclic Double Bonds in Potentially Aromatic Heterocycles[a]

Compound	ΔG_{298} (kcal mole^{-1})	Solvent
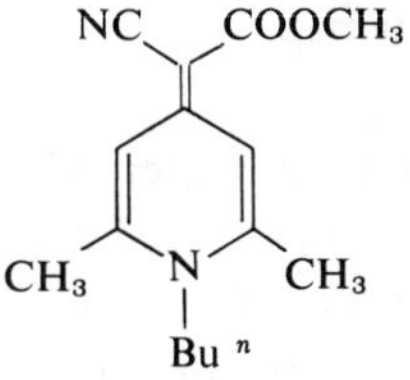 NC, COOCH₃ / CH₃, N(Bu^n), CH₃ pyridinylidene	20.1	Hexamethylphosphotriamide
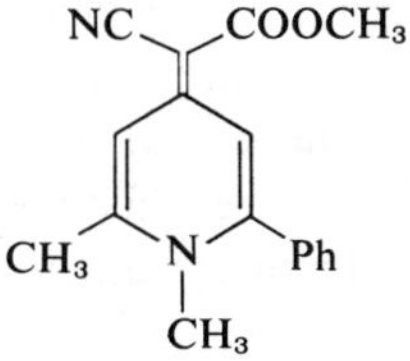 NC, COOCH₃ / CH₃, N(CH₃), Ph pyridinylidene	20.0	Hexamethylphosphotriamide
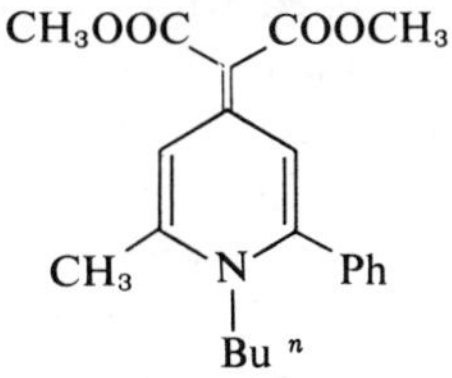 CH₃OOC, COOCH₃ / CH₃, N(Bu^n), Ph pyridinylidene	23.0	Hexamethylphosphotriamide
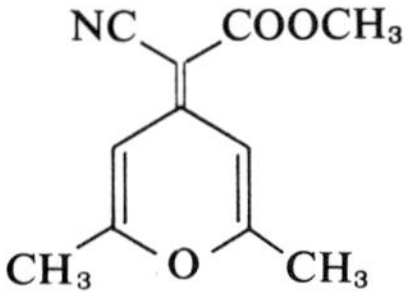 NC, COOCH₃ / CH₃, O, CH₃ pyranylidene	>27	Hexamethylphosphotriamide
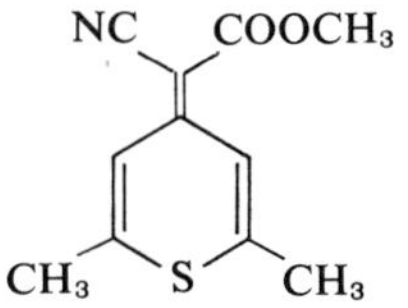 NC, COOCH₃ / CH₃, S, CH₃ thiopyranylidene	>27	Hexamethylphosphotriamide
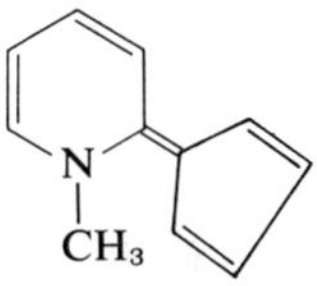 N-methylpyridinyl-cyclopentadienylidene[b]	11[c]	Acetone

[a] J. P. Brown and L. M. Jackman (unpublished results, 1973).
[b] Crabtree and Bertelli (1967).
[c] Assuming $\Delta S^{\ddagger} = 0$.

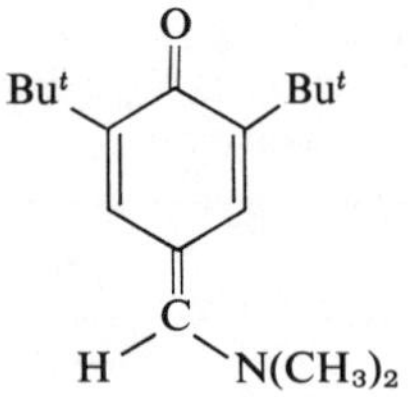

in the transition state have the barriers been accessible to measurement by the dnmr technique. These systems are interesting in two respects. First, comparison of the barriers (<9.1 and 23 kcal mole^{-1}) respectively, for $(CH_3)_2$-$NC(CH_3)$=$C(COOCH_3)_2$ (Table VIII) and, say, that for the third compound in Table XIX suggests that there must be considerable ground-state stabilization in the latter, since its transition state, involving as it does the pyridinium cation, is certainly effectively stabilized. The second feature of interest is the observation that =C(CN)COOCH$_3$ is more effective than =C(COOCH$_3$)$_2$ in lowering the barrier to rotation in the methylenedihydropyridine series, whereas the opposite trend has been observed for some acyclic enamines (p. 224). This is also possibly associated with the sizable degree of conjugation in the ground states of the former class of compounds.

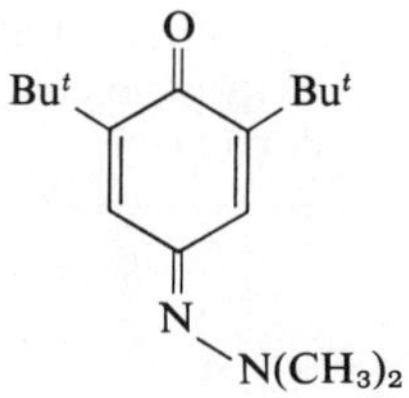

The rationalization of barriers in all compounds so far discussed in this chapter have been predicated on the assumption of the involvement of dipolar or polar transition states. We now consider several examples in which the transition states are presumably diradicals (singlets). The 2,4'-diphenoquinone (30) (in *trans*-decalin) has been shown to have a barrier ($\Delta G^\ddagger$) to rotation of 21 kcal mole^{-1} (Kessler and Rieker, 1966), and while steric destabilization of the ground state no doubt contributes to this low barrier, there must nevertheless be substantial stabilization, since the barriers in the

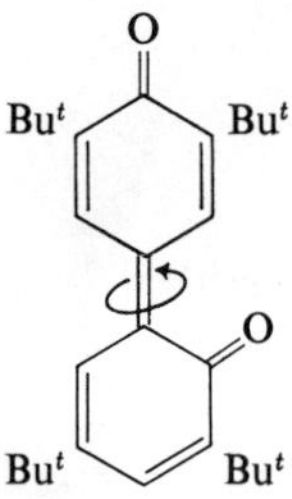

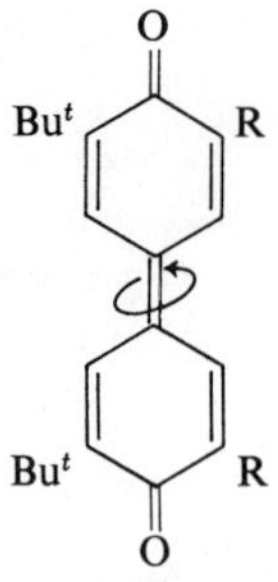

a, R = Ph; $\Delta G^\ddagger_{310}$ = 23.2 kcal mole^{-1}
b, R = OCH$_3$; $\Delta G^\ddagger_{386}$ = 22.0 kcal mole^{-1}

4,4'-diphenoquinones (30) (in C$_6$H$_3$Cl$_3$) (Rieker and Kessler, 1969) in 31 (in CHCl$_2$CHCl$_2$) (Boldt *et al.*, 1971) are only slightly higher. The transition state for this type of system is best envisaged as a pair of orthogonal phenoxyl radicals (Sutherland, 1971).

7. Rotation about Partial Double Bonds

Diradical transition states have also been postulated for rotation about the 9,9' double bond in bisfluorenylidenes (**32**) (Gault *et al.*, 1970). These systems are particularly interesting because it appears that the ground state is nonplanar and therefore chiral, so that both the cis and trans isomers exist as enantiomers (R ≠ R'). Interconversion of the diastereomers presumably involves the orthogonal diradical. Interconversion of enantiomers, on the other hand, is probably a more complex process, which at the simplest would involve a completely planar trans structure as the transition state for both diastereomers, but which in fact is believed to involve twisted (**33a**) and folded (**33b**) conformations and their interconversions. The problem has been approached in an interesting way by using R = CH_3 and R' = $CH(CH_3)_2$.

The inclusion of the prochiral isopropyl group permitted the determination of the barrier for interconversion of the enantiomers, the two "isopropyl" methyl groups being nonequivalent at low temperatures and equivalent at temperatures for which the rates of interconversions of the enantiomers is fast on the nmr time scale. The barriers ($\Delta G^\ddagger_{363}$) for interconversion of diastereomers and enantiomers are of the same order (21 kcal mole^{-1}).

VII. Cumulenes

One brief report (Kuhn *et al.*, 1966) of an nmr study of rotation about double bonds in cumulenes has been published. The barrier ($\Delta G^\ddagger_{373}$; in dimethylsulfoxide) for the interconversion of the cis and trans forms of 1,6-diphenyl-1,6-di-*t*-butylhexapentaene (**34a** and **34b**) was found to be 20 kcal mole^{-1}, in contrast to the corresponding butatriene for which the value is 30 kcal mole^{-1}.

VIII. Inversion versus Rotation—Imines and Related Systems

It has been pointed out by Curtin et al. (1966) that the rates of isomerization of imines (**35** and **37**) range over sixteen orders of magnitude. In general barriers ($\Delta G^{\ddagger}$) to isomerization of less than 22 kcal mole^{-1} are only observed if X is a conjugating substituent or if R^1 and R^2 have hetero atoms directly

35   **36**   **37**

attached to the imino carbon atom. Two extreme mechanisms have been proposed for this process. One is a rotation about the carbon–nitrogen double bond. The second is an inversion via the diagonal transition state (**36**) and is sometimes referred to as the lateral shift mechanism (Curtin et al., 1966). Of course, pathways having the character of both mechanisms are possible and have been considered (Raban and Carlson, 1971). There is, however, some controversy in this area as to the operative mechanisms in specific cases. Some representative data are assembled in Table XX.

The remarkably low barrier found in $(CH_3)_2C=NPh$ and related systems led Curtin et al. (1966) to evoke the lateral shift mechanism, since there was no known reason for such a low torsional barrier in a relatively unactivated double bond. On the other hand, the presence of the N-phenyl group could be expected to stabilize the diagonal transition state **36**. Furthermore, it has been pointed out that there is a striking parallelism between the dramatic effect of X on the rate of isomerization of **35** and **37** and its effect on the rate of inversion of the sp^3 hybridized nitrogen in aziridines (Kessler and Leibfritz, 1970). Kessler and his co-workers have also shown the existence of steric effects in the guanidines (**38**) (Table XXI) that argue in favor of the inversion mechanism. Thus, as the series R = H, CH$_3$, CH$_2$CH$_3$, CH(CH$_3$)$_2$ is ascended, the isomerization barrier decreases, in contrast to the guanidinium salts (**39**), for which substantial increases occur. The introduction of the o-substituents will cause the phenyl ring to be twisted out of the plane of the C=N bond. The out-of-plane conformation would allow the phenyl ring to stabilize the transition state (**36**) but not that for rotation. An earlier study by Wurmb-Gerlich et al. (1967) of steric effects on the rates of isomerization of acetone anils (Table XXII) revealed the same steric acceleration. Kessler and Leibfritz (1969b) have pointed out that good hydrogen bonding solvents (e.g., methanol) raise the barriers by approximately 1 kcal mole^{-1}, which is consistent with the lateral shift mechanism but not with the rotation mechanism.

TABLE XX
Barriers for the Degenerate Isomerization of Imines, $R_2C=NZ$

R	Z	$\Delta G^\ddagger$ (kcal mole^{-1})	°K	Solvent	Reference
CH_3	CH_2Ph	>23	443	Quinoline	Marullo and Wagener (1969)
CH_3	CN	18.9	358	Acetone-d_6	McCarty and Wieland (1969)
CH_3	Ph	21.0	413	Quinoline	Marullo and Wagener (1969)
CH_3	$CF(CF_3)_2$	ca. 13.0	—	None	Andreades (1962)
CF_3	Ph	15.45	298	Pyridine	Hall et al. (1971)
CF_3	Ph	19.4	378	Quinoline	Wurmb-Gerlich et al. (1967)
CH_3CO	Cl	>23	433	Ph_2O	Vögtle et al. (1967)
CH_3O	CH_2Ph	19.3	347	Acetone	Marullo and Wagener (1969)
CH_3O	CN	14.1	257	Acetone-d_6	McCarty and Wieland (1969)
CH_3O	Ph	14.4	274	Acetone	Marullo and Wagener (1969)
CH_3S	CH_3	18.6	346	Ph_2O	Vogtle et al. (1967)
CH_3S	CN	14.1	274	Acetone-d_6	McCarty and Wieland (1969)
CH_3S	Ph	13.8	251	Acetone	Marullo and Wagener (1969)
$(CH_3)_2N$	CN	<10	183	Acetone-d_6	McCarty and Wieland (1969)
$(CH_3)_2N$	Ph	12.1	256	CS_2—$CDCl_3$	Kessler and Leibfritz (1969b)

TABLE XXI
Barriers for Degenerate Isomerization of Some Guanidines[a] and Guanidinium Salts[b]

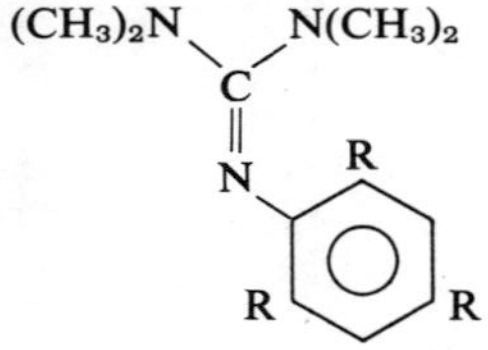

38: $(CH_3)_2N-C(=N-C_6H_3R_2)-N(CH_3)_2$

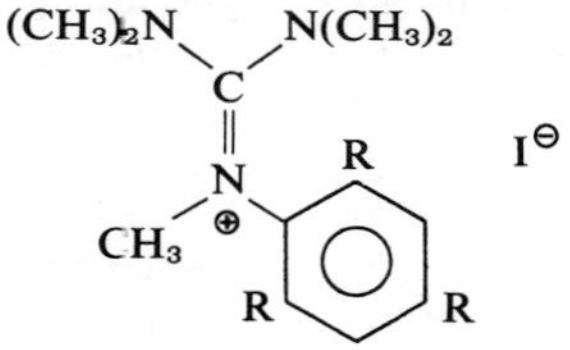

39: $(CH_3)_2N-C(N(CH_3)_2)=N^+(CH_3)(C_6H_3R_2) \; I^-$

R	$\Delta G^\ddagger$ (kcal mole^{-1})	°K	Solvent	$\Delta G^\ddagger$ (kcal mole^{-1})	°K	Solvent
H—	12.1	256	CS$_2$—CDCl$_3$	14.0	280	CD$_3$NO$_2$
CH$_3$—	12.6	272	CS$_2$—CDCl$_3$	21.1	420	1,2,4-C$_6$H$_3$Cl$_3$
CH$_3$CH$_2$—	11.9	266	CS$_2$—CDCl$_3$	22.5[c]	448	1,2,4-C$_6$H$_3$Cl$_3$
(CH$_3$)$_2$CH—	11.4	258	CS$_2$—CDCl$_3$	24.0	468	Formamide

[a] Kessler and Leibfritz (1969b).
[b] Kessler and Leibfritz (1971).
[c] 22.4 in formamide at 441°K.

TABLE XXII
Barriers for Degenerate Isomerization of a Series of Acetone Anils in Diphenyl Ether[a]

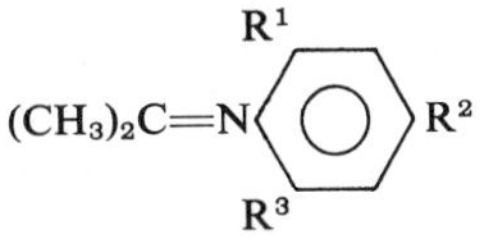

$$(CH_3)_2C=N\underset{R^3}{\overset{R^1}{\bigcirc}}R^2$$

R^1	R^2	R^3	$\Delta G^{\ddagger}$ (kcal mole^{-1})	°K
H	H	H	20.3	399
CH_3	H	CH_3	19.2	383
$CH(CH_3)_2$	$CH(CH_3)_2$	$CH(CH_3)_2$	18.8	373
CH_3	$C(CH_3)_3$	$C(CH_3)_3$	18.4	366

[a] Wurmb-Gerlich et al. (1967).

Marullo and Wagener (1966) have favored the rotation mechanism on the grounds that attachment of lone-pair-bearing substituents to the imino carbon atom produces a dramatic lowering of the barrier (Table XX). They further argue that the observation of positive ρ values (Table XXIII) for the effect of substituents in *N*-phenylimine derivatives is consistent with either mechanism (Marullo and Wagener, 1969). The Hammett ρ value for substitution in a phenyl group attached to the imino carbon atom in benzylidene anilines is only -0.4 (Wettermark et al., 1965) compared with the large

TABLE XXIII
Hammett ρ Values (25°C) for Degenerate Isomerization of Imine Derivatives

$$\underset{X^2}{\overset{X^1}{\diagdown}}C=NC_6H_4R$$

X^1	X^2	ρ	Reference
Ph	H	2.04	Wettermark et al. (1965)
p-$CH_3OC_6H_4$	p-$CH_3OC_6H_4$	1.88	Curtin et al. (1966)
CH_3O	CH_3O	1.44	Kessler et al. (1971)
CH_3S	CH_3S	1.30	Kessler et al. (1971)
$(CH_3)_2N$	$(CH_3)_2N$	2.21	Kessler et al. (1971)
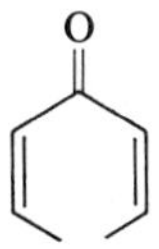 cyclohexadienone		1.90	Kessler and Leibfritz (1970)

values recorded in Table XXIII for N-phenyl substitution, a result which Kessler (1970) feels is better explained by the lateral shift mechanism.

Perhaps the most definitive evidence regarding the mechanism for the isomerization of imines has been provided by Hall *et al.* (1971) who have shown that, in the system $(CF_3)_2C\!\!=\!\!NC_6H_4R(p)$, the substituents R = CH_3O, CH_3, F, H, and Cl correlate well with σ^+ ($\rho^+ = -0.98$) but the nitro group deviates by about 1.7 kcal mole^{-1}, in the sense that it accelerates the degenerate isomerization. This certainly appears to indicate a change of mechanism from rotation to lateral shift on introduction of the nitro substituent. As the presence of the two trifluomethyl groups will stabilize the transition state for rotation, it is likely that the isomerizations of most simple imines involve the lateral shift mechanism. This line of argument cannot be extrapolated to systems with nitrogen, oxygen, and sulfur attached to the imino carbon atom, since in these cases the polarization of the double bond in the transition state for rotation will be in the opposite sense to the case just discussed.

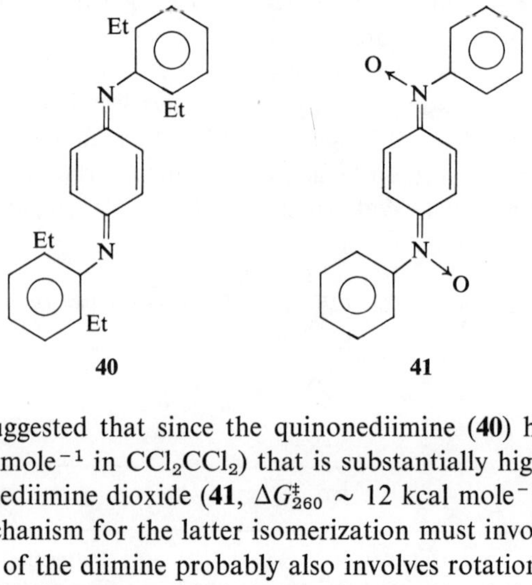

40   **41**

It has been suggested that since the quinonediimine (**40**) has a barrier ($\Delta G^{\ddagger}_{398} \sim 20$ kcal mole^{-1} in CCl_2CCl_2) that is substantially higher than the analogous quinonediimine dioxide (**41**, $\Delta G^{\ddagger}_{260} \sim 12$ kcal mole^{-1} in $CDCl_3$), and since the mechanism for the latter isomerization must involve rotation, the isomerization of the diimine probably also involves rotation (Layer and Carman, 1968). This is not a valid argument, however, as it is well known that nitroxyl radicals are relatively stable, so that the transition state for rotation involves a relatively stable diradical as well as a central aromatic nucleus.

Finally, the results of various quantum mechanical calculations (Table XXIV) suggest that the isomerization of simple imines involves the lateral shift mechanism but indicate that rotation may be a viable mechanism in certain guanidines.

TABLE XXIV
CALCULATED ISOMERIZATION BARRIERS FOR IMINE DERIVATIVES

Compound	$\Delta H^\ddagger$ (kcal mole^{-1}) lateral shift	$\Delta H^\ddagger$ (kcal mole^{-1}) rotation	Method	Reference
CH_2=NH	26–28	57	Ab initio	Lehn et al. (1970)
CH_2=NH	31	61	CNDO/2	Raban (1970)
HOCH=NH	32	51	CNDO/2	Raban (1970)
$(NH_2)_2C$=NH	36	28	CNDO/2	Raban (1970)
$(CH_3)_2C$=NO$^\ominus$	28	18	CNDO/2	Grubbs et al. (1973)
$(CH_3)_2C$=NO$^\ominus$	106	25	Ext. HMO	Grubbs et al. (1973)

REFERENCES

Abramson, K. H., Inglefield, P. T., Krakower, E., and Reeves, L. W. (1966). *Can. J. Chem.* **44**, 1685.
Acheson, R. M., and Selby, I. A. (1970). *Chem. Commun.* p. 62.
Allan, E. A., Hobson, R. F., Reeves, L. W., and Shaw, K. N. (1972). *J. Amer. Chem. Soc.* **94**, 6604.
Andreades, S. (1962). *J. Org. Chem.* **27**, 4163.
Anet, F. A. L., and Osyany, J. M. (1967). *J. Amer. Chem. Soc.* **89**, 352.
Beauté, C., Wolkowski, Z. W., and Thoai, N. (1971). *Chem. Commun.* p. 700.
Berger, A., Loewenstein, A., and Meiboom, S. (1959). *J. Amer. Chem. Soc.* **81**, 62.
Bertelli, D. J., and Gerig, J. T. (1967). *Tetrahedron Lett.* p. 2481.
Boldt, P., Michaelis, W., Lackner, H., and Krebs, B. (1971). *Chem. Ber.* **104**, 220.
Buchardt, O., Kumler, P. L., and Lohse, C. (1969). *Acta Chem. Scand.* **23**, 1155.
Bunton, C. A., Figgis, B. N., and Nayak, B. (1962). *Advan. Mol. Spectrosc.* **3**, 1209.
Bushweller, C. H., and Tobias, M. A. (1968). *Tetrahedron Lett.* p. 595.
Casanova, J., Jr., and Rutolo, D. A., Jr. (1967). *Chem. Commun.* p. 1224.
Cheng, H. N., and Gutowski, H. S. (1972). *J. Amer. Chem. Soc.*, **94**, 5505.
Conti, F., and von Philipsborn, W. (1967). *Helv. Chim. Acta* **50**, 603.
Crabtree, J. H., and Bertelli, D. J. (1967). *J. Amer. Chem. Soc.* **89**, 5384.
Crouse, D. M., Wehman, A. T., and Schweizer, E. E. (1968). *Chem. Commun.* p. 866.
Csizmadia, I. G., Houlden, S. A., Meresz, O., and Yates, P. (1969). *Tetrahedron* **25**, 2121.
Curtin, D. Y., Grubbs, E. J., and McCarty, C. G. (1966). *J. Amer. Chem. Soc.* **88**, 2775.
Dabrowski, J., and Kozerski, L. J. (1972). *Org. Magn. Resonance* **4**, 137.
Dahlqvist, K.-I., and Forsén, S. (1969). *J. Phys. Chem.* **73**, 4124.
Downing, A. P., Ollis, W. D., and Sutherland, I. O. (1969). *J. Chem. Soc.*, *B* p. 111.
Drakenberg, T. (1972). *Tetrahedron Lett.* p. 1743.
Drakenberg, T., and Forsén, S. (1970). *J. Phys. Chem.* **74**, 1.
Drakenberg, T., Dahlqvist, K.-I., and Forsén, S. (1972). *J. Phys. Chem.* **76**, 2178.
Egan, W., Bull, T. E., and Forsén, S. (1972). *Chem. Commun.* p. 1099.
Eicher, T., and Pelz, N. (1970). *Chem. Ber.* **103**, 2647.
Filleux-Blanchard, M. L., Clesse, F., Bignebat, J., and Martin, G. J. (1969). *Tetrahedron Lett.* p. 981.
Forsén, S., and Hoffman, R. A. (1963). *J. Chem. Phys.* **39**, 2892.
Fraenkel, G., and Franconi, C. (1960). *J. Amer. Chem. Soc.* **82**, 4478.
Fryer, C. W., Conti, F., and Franconi, C. (1965). *Ric. Sci.*, *Part 2 Sez. A* **35**, 788.
Gansow, O. A., Killough, J., and Burke, A. R. (1971). *J. Amer. Chem. Soc.* **93**, 4297.
Gault, I. R., Ollis, W. D., and Sutherland, I. O. (1970). *Chem. Commun.* p. 269.
Gehring, D. G., Mosher, W. A., and Reddy, G. S. (1966). *J. Org. Chem.* **31**, 3436.
Gillespie, R. J., and Birchall, T. (1963). *Can. J. Chem.* **41**, 148.
Grubbs, E. J., Parker, D. R., and Jones, W. D. (1973). *Tetrahedron Lett.* p. 3279.
Grunwald, E., Loewenstein, A., and Meiboom, S. (1957). *J. Chem. Phys.* **27**, 630.
Gutowsky, H. S., and Holm, C. H. (1956). *J. Chem. Phys.* **25**, 1228.
Gutowsky, H. S., McCall, D. W., and Slichter, C. P. (1953). *J. Chem. Phys.* **21**, 279.
Gutowsky, H. S., Jonas, J., and Sidall, T. H., III. (1967). *J. Amer. Chem. Soc.* **89**, 4300.
Hall, G. E., Middleton, W. J., and Roberts, J. D. (1971). *J. Amer. Chem. Soc.* **93**, 4778.
Hammaker, R. M., and Gugler, B. A. (1965). *J. Mol. Spectrosc.* **17**, 356.
Harris, D. L., and Wellman, K. M. (1968). *Tetrahedron Lett.* p. 5225.
Hatton, J. V., and Richards, R. E. (1960). *Mol. Phys.* **3**, 253.
Hatton, J. V., and Richards, R. E. (1962). *Mol. Phys.* **5**, 139.

7. Rotation about Partial Double Bonds

Hobson, R. F., and Reeves, L. W. (1973a). *J. Phys. Chem.* **77**, 419.
Hobson, R. F., and Reeves, L. W. (1973b). *J. Magn. Resonance* **10**, 243.
Hobson, R. F., Reeves, L. W., and Shaw, K. N. (1973). *J. Amer. Chem. Soc.* **77**, 1228.
Imbery, D., Jaeschke, A., and Friebolin, H. (1970). *Org. Magn. Resonance* **2**, 271.
Inglefield, P. T., Krakower, E., Reeves, L. W., and Stewart, R. (1968). *Mol. Phys.* **15**, 65.
Isaksson, G., and Sandström, J. (1967). *Acta Chem. Scand.* **21**, 1605.
Isaksson, G., Sandström, J., and Wennerbeck, I. (1967). *Tetrahedron Lett.* p. 2233.
Jackman, L. M., and Haddon, R. C. (1973). *J. Amer. Chem. Soc.* **95**, 3687.
Jackman, L. M., and Wiley, R. H. (1960). *J. Chem. Soc., London* pp. 2881 and 2886.
Jackman, L. M., Kavanagh, T. E., and Haddon, R. C. (1969). *Org. Magn. Resonance* **1**, 109.
Jakobsen, H. J., and Senning, A. (1968). *Chem. Commun.* p. 1245.
Kaplan, F., and Meloy, G. K. (1966). *J. Amer. Chem. Soc.* **88**, 950.
Katritzky, A. R., and Tiddy, G. J. T. (1969). *Org. Magn. Resonance* **1**, 57.
Kende, A. S., Izzo, P. T., and Fulmor, W. (1966). *Tetrahedron Lett.* p. 3697.
Kessler, H. (1970). *Angew. Chem., Int. Ed. Engl.* **9**, 219.
Kessler, H., and Leibfritz, D. (1969a). *Tetrahedron Lett.* p. 427.
Kessler, H., and Leibfritz, D. (1969b). *Tetrahedron* **25**, 5127.
Kessler, H., and Leibfritz, D. (1970). *Tetrahedron* **26**, 1805.
Kessler, H. and Leibfritz, D. (1971). *Chem. Ber.* **104**, 2158.
Kessler, H., and Rieker, A. (1966). *Tetrahedron Lett.* p. 5257.
Kessler, H., and Rosenthal, D. (1973). *Tetrahedron Lett.* p. 393.
Kessler, H., Bley, P. F., and Leibfritz, D. (1971). *Tetrahedron* **27**, 1687.
Klinck, R. E., Marr, D. H., and Stothers, J. B. (1967). *Chem. Commun.* p. 409.
Komoriya, A., and Yoder, C. H. (1972). *J. Amer. Chem. Soc.* **94**, 5285.
Krebs, A., and Breckwoldt, J. (1969). *Tetrahedron Lett.* p. 3797.
Kuhn, R., Schulz, B., and Jochims, J. C. (1966). *Angew. Chem., Int. Ed. Engl.* **5**, 420.
Layer, R. W., and Carman, C. J. (1968). *Tetrahedron Lett.* p. 1285.
Lehn, J. M., Munsch, B., and Millie, P. (1970). *Theor. Chim. Acta* **16**, 351.
McCarty, C. G., and Wieland, D. M. (1969). *Tetrahedron Lett.* p. 1787.
McConnell, H. M. (1958), *J. Chem. Phys.* **28**, 430.
Mannschreck, A. (1965). *Tetrahedron Lett.* p. 1341.
Mannschreck, A., and Koelle, U. (1967). *Tetrahedron Lett.* p. 863.
Mannschreck, A., and Kolb, B. (1972). *Chem. Ber.* **105**, 696.
Martin, D. M. G., and Reese, C. B. (1967). *Chem. Commun.* p. 1275.
Marullo, N. P., and Wagener, E. H. (1966). *J. Amer. Chem. Soc.* **88**, 5034.
Marullo, N. P. and Wagener, E. H. (1969). *Tetrahedron Lett.* p. 2555.
Monro, A. M., and Sewell, M. J. (1969). *Tetrahedron Lett.* p. 595.
Neiman, Z., and Bergmann, F. (1968). *Chem. Commun.* p. 1002.
Neuman, R. C., Jr., and Jonas, V. (1968). *J. Amer. Chem. Soc.* **90**, 1970.
Neuman, R. C., Jr., and Young, L. B. (1965). *J. Phys. Chem.* **69**, 2570.
Noggle, J. H., and Schirmer, R. E. (1971). "The Nuclear Overhauser Effect. Chemical Applications," p. 160. Academic Press, New York.
Nozaki, H., Tunemoto, D., Morita, Z., Nakamura, K., Watanake, K., Takaku, M., and Kondô, K. (1967). *Tetrahedron* **23**, 4279.
Pendlebury, M. H., and Philips, L. (1972). *Org. Magn. Resonance* **4**, 529.
Phillips, W. D. (1955). *J. Chem. Phys.* **23**, 1363.
Pitzer, K. S. (1948). *J. Amer. Chem. Soc.* **70**, 2140.
Price, B. J., Smallman, R. V., and Sutherland, I. O. (1966). *Chem. Commun.* p. 319.

Raban, M. (1970). *Chem. Commun.* p. 1415.
Raban, M., and Carlson, E. (1971). *J. Amer. Chem. Soc.* **93**, 685.
Rabinovitz, M., and Pines, A. (1969). *J. Amer. Chem. Soc.* **91**, 1585.
Ramey, K. C., Louick, D. J., Whitehurst, P. W., Wise, W. B., Mukherjee, R., Rosen, J. F., and Moriarty, R. M. (1971). *Org. Magn. Resonance* **3**, 767.
Randall, F. J., and Johnson, A. W. (1968). *Tetrahedron Lett.* p. 2841.
Rappoport, Z., and Ta-Shma, R. (1972). *Tetrahedron Lett.* p. 5281.
Reeves, L. W., Shaddick, R. C., and Shaw, K. N. (1971a). *J. Phys. Chem.* **75**, 3372.
Reeves, L. W., Shaddick, R. C., and Shaw, K. N. (1971b). *Can. J. Chem.* **49**, 3683.
Rieker, A., and Kessler, H. (1969). *Chem. Ber.* **102**, 2147.
Rogers, M. T., and Woodbrey, J. C. (1962). *J. Phys. Chem.* **66**, 540.
Saunders, J. K., and Bell, R. A. (1970). *Can. J. Chem.* **48**, 512.
Shmueli, U., Shanan-Atidi, H., Horwitz, H., and Shvo, Y. (1973). *J. Chem. Soc., Perkin, Trans.* **2**, 657.
Shoup, R. R., Miles, H. T., and Becker, E. D. (1972). *J. Phys. Chem.* **76**, 64.
Shvo, Y. (1968). *Tetrahedron Lett.* p. 5923.
Shvo, Y., and Belsky, I. (1969). *Tetrahedron* **25**, 4649.
Shvo, Y., and Shanan-Atidi, H. (1969a). *J. Amer. Chem. Soc.* **91**, 6683.
Shvo, Y., and Shanan-Atidi, H. (1969b). *J. Amer. Chem. Soc.* **91**, 6689.
Shvo, Y., Taylor, E. C., and Bartulin, J. (1967). *Tetrahedron Lett.* p. 3259.
Siddall, T. H., III, Stewart, W. E., and Knight, F. C. (1970), *J. Phys. Chem.* **74**, 3580.
Smallcombe, S. H., Holland, R. J., Fish, R. H., and Caserio, M. C. (1968). *Tetrahedron Lett.* p. 5987.
Spaargen, K., Korver, P. K., van der Haak, P. J., and de Boer, T. J. (1971a). *Org, Magn. Resonance* **3**, 604.
Spaargen, K., Korver, P. K., van der Haak, P. J., and de Boer, T. J. (1971b). *Org. Magn. Resonance* **3**, 615.
Spaargen, K., Korver, P. K., van der Haak, P. J., and de Boer, T. J. (1971c). *Org. Magn. Resonance* **3**, 639.
Spassov, S. L., Dimitrov, V. S., Agova, M., Kantschowska, I., and Todorova, R. (1971.) *Org. Magn. Resonance* **3**, 551.
Staab, H. A., and Lauer, D. (1968). *Chem. Ber.* **101**, 864.
Stewart, W. E., and Siddall, T. H., III. (1970). *Chem. Rev.* **70**, 517.
Sutherland, I. O. (1971). *Annu. Rep. NMR (Nucl. Magn. Resonance) Spectrosc.* **4**, 71.
Tanny S. R., Pickering, M., and Springer, C. S. (1973) *J. Amer. Chem. Soc.* **95**, 6227.
Temussi, P. A., Tancredi, T., and Quadrifoglio, F. (1969). *J. Phys. Chem.* **73**, 4227.
Thorpe, J. E. (1968). *J. Chem. Soc., B* p. 435.
Totani, T., Tori, K., Murakami, J., and Watanabe, H. (1971). *Org. Magn. Resonance* **3**, 627.
Trost, B. M. (1967). *J. Amer. Chem. Soc.* **89**, 138.
Van Geet, A. L. (1968). *Anal. Chem.* **40**, 2227.
Vögtle, F., Mannschreck, A., and Staab, H. A. (1967). *Justus Liebigs Ann. Chem.* **708**, 51.
Walter, W., Schaumann, E., and Paulsen, H. (1969). *Justus Liebigs Ann. Chem.* **727**, 61.
Walter, W., Schaumann, E., and Rose, H. (1973). *Org. Magn. Resonance* **5**, 191.
Wennerbeck, I., and Sandström, J. (1972). *Org. Magn. Resonance* **4**, 783.
Wettermark, G., Weinstein, J., Sousa, J., and Dogliotti, L. (1965). *J. Phys. Chem.* **69**, 1584.
Whittaker, A. G., and Siegel, S. (1965). *J. Chem. Phys.* **42**, 3320.
Wurmb-Gerlich, D., Vögtle, F., Mannschreck, A., and Staab, H. A. (1967). *Justus Liebigs Ann. Chem.* **708**, 36.
Zeliger, H. I., Snyder, J. P., and Bestmann, H. J. (1969). *Tetrahedron Lett.* p. 2199.

8

Dynamic Molecular Processes in Inorganic and Organometallic Compounds

J. P. JESSON and E. L. MUETTERTIES

I. Introduction.. 253
II. Recent Advances in Instrumentation and Technique 258
    A. Conventional Time Averaging 258
    B. Fourier Transform NMR 259
    C. Superconducting Systems.................................. 260
    D. Temperature Control and Measurement..................... 260
III. Applications... 261
    A. Intramolecular Exchange.................................. 261
    B. Intermolecular Exchange.................................. 299
IV. Conclusion 313
    References... 313

I. Introduction

The time scale of the frequency spectrum of most nuclear spin systems at normally accessible magnetic fields is in the range of 10^0 to 10^{-6} seconds, a range comparable to the lifetimes of many molecules in fast reactions (Pople et al., 1959). These fast reactions affect the resonance lines in a characteristic fashion which is amenable to kinetic analysis and, in some cases, to a relatively definitive mechanistic analysis (Jesson and Meakin, 1973a). In inorganic and organometallic systems, this facet of nuclear magnetic resonance is very important because a significant fraction of inorganic and organometallic molecules are not "well behaved" in solution (Muetterties and Phillips, 1962). They may undergo rapid intramolecular rearrangements, as well as intermolecular processes such as dissociation, ionization, and association. All these dynamic processes can significantly affect chemical behavior. Delineation of these dynamic processes not only can lead to an understanding of chemical behavior, but can also be used as a basis for further extension of the chemistry of the systems. It is the objective of this chapter to outline how the nmr technique can be exploited in probing dynamic molecular processes in inorganic and organometallic compounds. The power of the technique, and its limitations, for the study of intramolecular and intermolecular processes, especially in catalytic reactions, is illustrated primarily using examples from the literature. Coverage of the literature is not comprehensive; furthermore, some specific areas of dynamic molecular processes in inorganic molecules are treated in Chapters 2 ("Delineation of Nuclear Exchange Processes"), 9 ("Stereochemically Nonrigid Metal Chelate Complexes"), 10 ("Stereochemical Nonrigidity in Organometallic Compounds"), 11 ("Fluxional Allyl Complexes"), and 12 ("Stereochemical Nonrigidity in Metal Carbonyl Compounds").

The range of dynamic processes that can be envisaged for inorganic or organometallic molecules is very broad indeed. As a first step to classification we divide the dynamic processes into two main groups: (1) intramolecular rearrangements, and (2) intermolecular processes involving bond breaking and making. These are later subdivided into the more commonly encountered physical processes.

Intramolecular rearrangements (Muetterties, 1965, 1970b) are defined as processes in which nuclear positions in a molecule are permuted without bond breaking, and can sometimes be delineated by the fact that spin–spin coupling is maintained between the nuclei involved. Molecules undergoing these rearrangements are described as stereochemically nonrigid or fluxional. The following examples illustrate this type of process and some of the limitations in the nmr interpretation. In the slow exchange limit,* molecules of type **1** have the multiline proton spectrum for the hydride hydrogens expected for the A part of an $AA'XX'Y_2$ system, while in the fast exchange limit a binomial quintet is observed indicating magnetic equivalence within the set of phosphorus nuclei and within the set of hydrogen nuclei. Because hydrogen–phosphorus coupling is maintained in the fast exchange limit, the requisite conditions for definition of intramolecular ex-

---

* Throughout this chapter, the term "slow exchange limit" applies to the temperature range where the lifetime of a molecule with respect to a dynamic process is long relative to the time scale of the nmr experiment. The term "fast exchange limit" refers to the opposite situation. The transition region between these two extremes covers the temperature range in which the lifetime of the molecule is comparable to the time scale of the nmr experiment.

8. Dynamic Molecular Processes

change are met. Now, consider the case where one of the tricovalent phosphorus ligands is replaced by an atom, or a group of atoms, that has no detectable spin–spin coupling with the hydrogen or phosphorus nuclei. In this case, observation of a binomial quartet in the fast exchange limit does not in itself establish an intramolecular process, because magnetic equivalence could be achieved through a bond-breaking process involving the unique

1

atom or group. Of course, an intramolecular process should have a rate essentially independent of concentration and relatively insensitive to changes with solvent. Also, experiments can be carried out to determine whether the ligand with no spin can exchange with free ligand in solution. Thus, if the ligand in question were CO, the sample could be pressurized with ^{13}C-enriched CO and the exchange, if any, could be followed by ^{13}C nmr or direct chemical analysis of the gas after exchange. Hence, even in these cases, additional experiments can reasonably define an intramolecular rearrangement but without the rigor of the first example. It should also be noted that in the first example, if exchange occurs without the intermediacy of a second species of appreciable concentration, then the resonance lines representing transitions unaffected by the exchange process will usually remain sharp throughout the transition region. This is termed mutual exchange. If, however, there is a second species in substantial concentration, all of the lines broaden; the final value of the average chemical shift depends on the chemical shifts and the concentrations of the two species. This is nonmutual exchange (for an example, see p. 288) and can be discerned in some cases where the second or intermediate species is present in undetectable (nmr) amounts in the slow exchange limit. The extreme limit in this series of examples is in a molecule where all like nuclei are magnetically equivalent and the spin Hamiltonian is unaffected by permutation of nuclear positions in a mutual exchange process; e.g., in five-coordinate molecules of type **2** where permutation of axial and equatorial sites might be relatively fast, but the molecule spends nearly all its time in configuration **2**. Here, nmr obviously can be of no assistance as an experimental probe.

The rearrangements in stereochemically nonrigid molecules are often

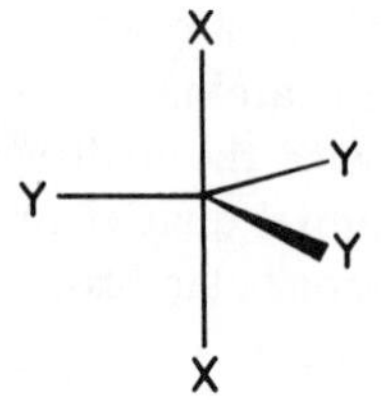

```
      X
      |
   Y--M--Y
      |    Y
      X

      2
```

referred to, with an element of imprecision, as nonbond-breaking processes simply because spin–spin coupling is retained in the fast exchange rearrangement limit. There is, in fact, at least one alternative process that is bond-breaking in character. Reconsider the first example, **1**, above. If an M—P bond is broken to give $H_2M(P\equiv)_3 + P\equiv$, and these fragments are protected from the rest of the species in solution by a solvent cage, then recombination may give a rearranged molecule, but one in which all spin–spin coupling is maintained, provided that the lifetime of separate species is short relative to the phosphorus nuclear relaxation times and to diffusion through the solvent cage. This alternative of a solvent cage is analogous to a problem encountered in the nmr study of chelates. If we alter structure **1** to the chelate analog **3**, then the nmr line shape changes, from the slow ex-

```
          \ /
           P
     H    /|    P<
      \  / |   /
       \/  |  /
       M   | /
      /|   |/
     / |   P<
    H  |  /
       P
      /\
```

       3

change multiplet to a fast exchange quintet proton nmr spectrum, are still characteristic of an intramolecular process, including one that comprises M—P bond-breaking through intermediates of the type shown in **4**. If an intermediate such as **4** has a lifetime that is short relative to the phosphorus

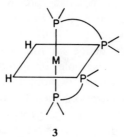

       4

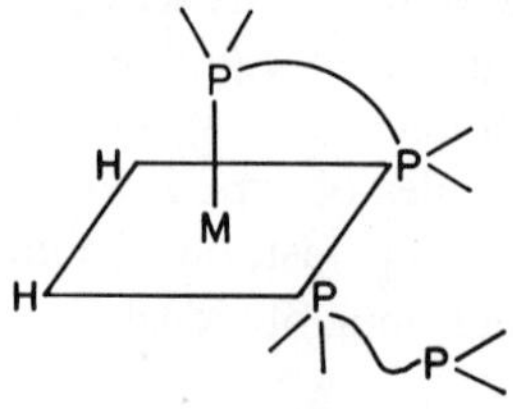

nuclear relaxation time, then M—P bond breaking and forming will not lead to loss of spin correlation. This problem in chelates of distinguishing between bond-breaking ("arm off") and nonbond-breaking processes is a classic one, and only recently have such distinctions been made with a reasonable degree of certainty for a few chelate structures in which diasterotopic groups (Eaton *et al.*, 1972, 1973; Edgar *et al.*, 1973) provided sufficient stereochemical information for mechanistic arguments. These studies are discussed in detail in Chapter 9.

A possible nonbond-breaking bimolecular process for rearrangement, which has been postulated although not as yet established, comprises formation of a dimer in which ligands serve as unsymmetrical bridges. For example, an MX_5 molecule could, in principle, form the dimer **5** in which an intra-

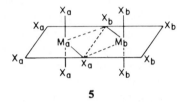

5

molecular process might occur followed by dissociation to the originally labelled MX_5 molecules without any M—X bond breaking and concomitant loss of spin correlation, but with the labeled nuclei having been permuted within a given molecule (Redington and Berney, 1965; Brownstein, 1967).

In intermolecular exchange processes, there are bond-breaking and bond-formation steps. Bond scission necessarily leads to loss of spin–spin coupling between nuclei in the two fragments generated in the scission process. This loss of spin correlation is the most definitive nmr test of an intermolecular exchange process and is extremely sensitive. The phenomenon can be observed for equilibria of the type

$$ML_x \underset{k_{-1}}{\overset{k_1}{\rightleftarrows}} ML_{x-1} + L \tag{1}$$

provided only that $1/k_1$ be comparable to the time scale of the resonance parameters in the experiment at some accessible temperature.* If there is in the nmr spectrum no discernible fine structure due to spin–spin coupling between nuclei in the two dissociated fragments, then the experiment will only sense a dissociation process if the values of k_1 and k_{-1} are comparable; under these circumstances, a marked temperature dependence of the chemical shift will be noted in the fast exchange limit, with the observed average chemical shift depending on the degree of dissociation and the true chemical

* Unless the dissociation involves short-lived ML_{x-1} and L protected by a solvent cage as discussed on page 256.

shifts of the individual species. Intermolecular exchange studies have the additional variables of dilution and relative concentrations of the species which may be used to bring the rate process into the appropriate range for study by nmr. For dilute solutions, the sensitivity of this application may be enhanced by pulse techniques and Fourier analysis (see below). In the slow exchange limit, separate resonances for ML_x, ML_{x-1}, and L may be detected if k_1/k_{-1} is in the appropriate range or if excess L is added.

In the past, one difficulty common to nmr experiments on exchange phenomena stemmed from the small concentration range that is practical in the experiment. Furthermore, low concentrations that approach the ideal solution range have been used only in the optimal cases where the nuclei had high sensitivities. This situation has made it relatively difficult to obtain kinetic data that can be assessed accurately, a critical point where the concentration effect is an important parameter for mechanistic interpretations. It has also left a "concentration gap" for comparison of data obtained by nmr with data obtained by such techniques as ultraviolet spectroscopy, molecular weight studies, conductivity, etc., which techniques normally deal with dilute solutions. One approach to nmr studies of dilute solutions has been computer averaging of transient signals, where on a time basis a practical gain in signal to noise is about ten. Recent advances in spectrometer design, especially in the use of pulse techniques with Fourier transform, now make it feasible, at least for some nuclei (e.g., ^{31}P, ^{13}C, ^{1}H, and ^{19}F), to study relatively dilute solutions. These advances are of such import to present day nmr studies that they serve as the basis of a brief discussion in the following section.

There is a final caveat in the use of dnmr for study of molecules in solution. Line shape studies identify the dominant dynamic processes. Other dynamic processes that may be only 10 to 100 times slower will go undetected, yet these can substantially modify the chemical nature of the molecule(s) in question. Dynamic nmr, although an extraordinarily effective probe of dynamic phenomena, is complementary to other techniques that probe other time scales (Muetterties, 1965) relevant to the structural and chemical properties of molecules.

II. Recent Advances in Instrumentation and Technique

A. *Conventional Time Averaging*

The use of digital storage devices (time-averaging computers) to enhance signal-to-noise has been an established technique in nmr for many years. However, it is only recently that spectrometer design has reached the point that full advantage may be taken of this approach. The signal-to-noise is enhanced over a single scan by the factor $n^{1/2}$ where n is the number of passes accumulated, provided the sweep through the spectrum is precisely reproduced for each of the scans. This latter requirement can be fulfilled in the following way

1. The magnetic field must be locked to a fixed rf signal so that any tendency of the field to drift is effectively eliminated. This feature has been standard for proton and fluorine nmr for some time, and was typically homonuclear; i.e., when observing a proton spectrum, the spectrometer was locked to a proton signal such as that of tetramethylsilane. Recently, systems with heteronuclear locks have been developed, so that the spectrometer can be stabilized with a strong fluorine or deuterium signal while observing nuclei of low abundance such as ^{13}C. The stabilization and observing frequencies must be tied together so that there is no drift of one with respect to the other (phase locking).

2. The spectrometer sweep must be precisely reproducible, and the sweep through the channels of the computer should be controlled by the same voltage as the spectral sweep.

3. The spectrometer should preferably have time-sharing capability such that the transmitter is on for only a small fraction of the time (typically one-tenth), and the receiver is on only when the transmitter is off. This eliminates the leakage problem (interaction between the transmitter and receiver) and gives rise to very stable baselines so that time averaging can be continued for days. The sloping or curved baselines associated with conventional spectrometers (non-time shared) at high rf power levels can make long-term time averaging very difficult.

B. Fourier Transform NMR

In the Fourier transform nmr experiment, the nuclear spin system is subjected to a sequence of identical pulse and delay steps. The pulses are sufficiently narrow such that the power distribution excites the whole spectral range of interest. The decay of the transverse magnetization—free induction decay (FID)—is recorded and digitized during all or part of the delay and coadded to the previously accumulated FIDs in a manner similar to the conventional time averaging described in the previous section (the increase in signal-to-noise over the FID acquired after a single pulse is again equal to the square root of the number of pulses assuming steady-state conditions for all pulses). At the end of the experiment, the time-averaged FIDs are Fourier transformed to obtain the familiar frequency spectrum. The gain in signal-to-noise over the conventional continuous wave experiment is often dramatic and arises from the fact that the whole spectrum is excited at one time. In principle, the saving in time is of the order of $(\Delta f/f)^{-1}$ where Δf is the average linewidth and f is the spectral width being covered. Heteronuclear lock capability as described in the previous section is essential, but one does not require a stable sweep system, since the pulse is applied at a fixed frequency. The fixed frequency must, however, be stable with respect to the lock frequency.

One must be careful in using the Fourier method for line-shape studies. Different T_1 relaxation terms at different sites can cause problems, since the spin system is, in general, not allowed to relax back to equilibrium between each pulse.

C. Superconducting Systems

Superconducting solenoids can produce large magnetic fields of high homogeneity (60–70 kG as opposed to the ~20 kG of conventional electromagnets). The two major advantages are increase in sensitivity and increase in chemical shift differences. This latter advantage, which is particularly important for proton spectra of complex molecules, can be a liability for exchange studies under some circumstances. Often crucial mechanistic information is contained in the behavior of the fine structure due to second-order effects as the lines start to exchange (see, for example, discussion of ML_5 complexes, p. 266). A spectrum that shows the requisite second-order effects at 20 kG may be first order at 60 kG and mechanistic information can no longer be extracted. In some cases, it can be advantageous to induce higher-order effects in a spectrum that is first order at 20 kG by examining the system at a much lower field.

D. Temperature Control and Measurement

These are particularly important factors in line-shape studies, especially if accurate thermodynamic data are required. The most reliable approach to temperature measurement is to use a thermocouple held coaxially in a spinning sample tube under identical conditions to those used for the spectral measurement. Temperature controllers in general control the temperature very accurately so that this does not represent a problem in most systems.

For mechanistic studies, the accessible temperature range is often more important than the accuracy of the temperature measurement. With transition metal complexes, it is desirable to be able to go as low as possible in temperature; temperatures in the $-160°C$ range should be attained routinely. Few solvents will dissolve the complexes of interest and at the same time remain sufficiently mobile at these temperatures to avoid line broadening due

8. Dynamic Molecular Processes

to viscosity effects. A notable exception is chlorodifluoromethane, which was used in many of the studies described in this chapter. In addition, the fluorine nuclei in the solvent can be used as a lock signal and to shim the magnet at very low temperatures. Shimming at low temperatures is quite difficult, for example, when using a deuterium lock.

III. Applications

A. Intramolecular Exchange

Inorganic and organometallic chemistry is blessed or cursed, depending on one's point of view, with a large number of stereochemically nonrigid molecules (Muetterties, 1965, 1970a,b; Cotton, 1968). This situation was not recognized until the 1960s, when nmr studies uncovered for many molecules spectral features explicable only in terms of rapid intramolecular rearrangements (Muetterties *et al.*, 1963, 1964). By the late 1960s, nmr studies of stereochemically nonrigid molecules became quite common. Now, a reasonably well-defined picture may be presented for these molecules. In the following sections, discussions of intramolecular rearrangements are organized by the coordination number of the molecule. Rearrangements in organometallic molecules and in chelate structures are discussed in Chapters 9, 10, 11, and 12, respectively.

1. The Quasi Four-Coordinate, Pyramidal Molecules

A classic fluxional molecule is ammonia, in which hydrogen nuclear positions move through an inversion process that involves a planar transition state with HNH angles of 120°. Quantum-mechanical tunneling is an important aspect of the rearrangement and gives rise to inversion doubling of the energy levels (Herzberg, 1951). These processes in ammonia are not only too fast for detection by the nmr technique but also effect no change in the spin Hamiltonian, and hence, are unobservable by nmr. This condition prevails for all ML_3 pyramidal molecules unless one of the L groups is a diastereotopic group, such as isopropyl or benzyl, and the other two L groups or atoms are inequivalent. In the pyramidal form of any molecule of form **6** the

$$L_1 \diagdown \underset{L_2}{\overset{M}{|}} \diagup C \diagup \overset{H}{\underset{CH_3}{\diagdown CH_3}}$$

6

methyl groups are environmentally nonequivalent, as can usually be shown by the observation of two chemically shifted methyl doublets in the proton nmr (doublets due to coupling with the CH proton). If inversion occurs at a rate fast with respect to the nmr time scale, then the methyl groups appear magnetically equivalent. The inversion process can be followed in the nmr experiment by the broadening of the two CH_3 doublets and final merging into a single doublet resonance. This technique has been employed by Mislow and co-workers to estimate the inversion barriers for a number of pyramidal molecules. A specific example is the phosphine **7** which has the surprisingly low barrier, $\Delta G^{\ddagger}_{25°C} = 16$ kcal/mole (Egan *et al.*, 1971). At 0°C, each isopropyl methyl group of **7** gives rise in the ^{1}H nmr spectrum to a doublet

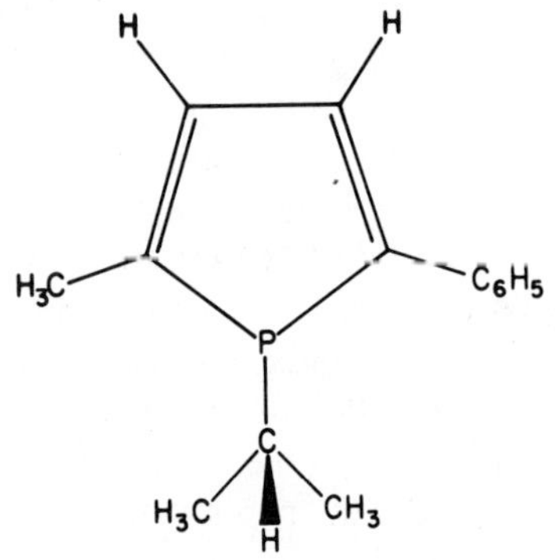

7

(J_{PH}) of doublets (J_{CH-CH_3}). Above 0°C, the inversion becomes rapid as evidenced by a broadening and merging of all the CH_3 resonances (Fig. 1), and finally, above 60°C, there is a single isopropyl CH_3 resonance which is a doublet (J_{PH}) of doublets (J_{CH-CH_3}).

Inversion can also be followed by nmr in pyramidal molecules where a saturated ring system comprises two of the ligand sites. For example, in **8** the α protons are inequivalent in the pyramidal molecule and appear magnetically

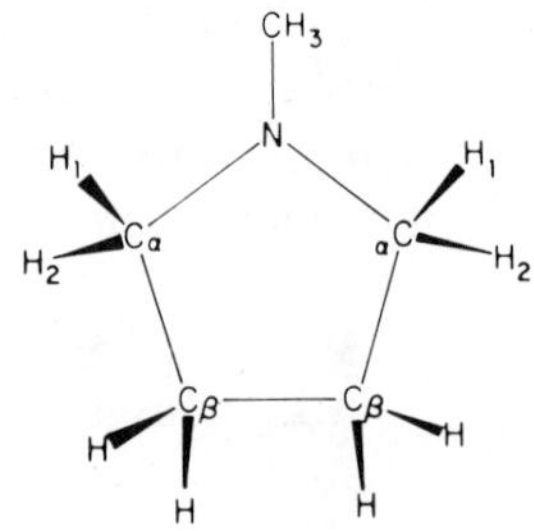

8

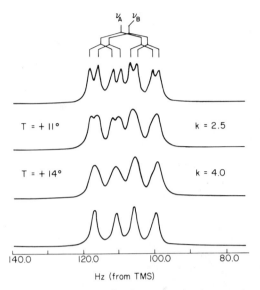

Fig. 1 Simulated dnmr proton spectra for **7** restricted to isopropyl methyl resonances. A and B denote diastereotopic methyl groups (Egan *et al.*, 1971.)

equivalent when inversion is rapid (ring pseudorotation is rapid with respect to the nmr time scale). Lambert and Oliver (1969) studied **8** and analogous cyclic amines by nmr. Spectral detail was simplified by irradiation at the resonance frequency of the β protons. Hence, in the slow exchange region there are well separated multiplets for the α hydrogen atoms, H_1 and H_2, and these collapse to a singlet (or complex triplet in the absence of irradiation) in the fast exchange limit.

Nuclear magnetic resonance studies of inversion in pyramidal molecules is limited, then, to those molecules having diastereotopic groups. Also, the nmr experiment for these molecules has no inherent mechanistic information in the line shape changes. There is only one subclass of permutations (Jesson and Meakin, 1973a) in this type of molecule that will affect the spin Hamiltonian; any kind of physical motion in the intramolecular exchange process would be consistent with the nmr line shape changes.

Factors that affect the barrier to inversion include the LML equilibrium angle (as the angle increases the barrier decreases), substituent electronic or steric effects (bulky groups generally destabilize the pyramidal state), and differences in ring strain between the pyramidal and planar states in cyclic species (see discussion by Rauk *et al.*, 1970). In Group V, the inversion barriers uniformly increase from N to P to As; i.e., for any fixed set of substituents the barrier is lowest for nitrogen and highest in arsenic. With fixed substituents which yield low inversion barriers, the order of inversion barriers

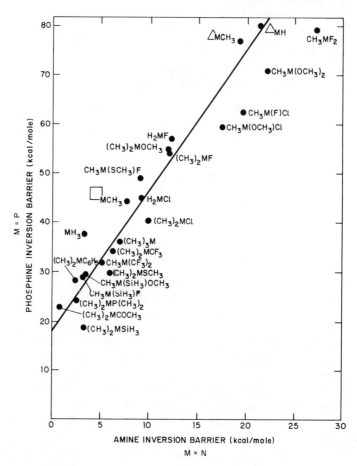

Fig. 2 Linear free energy relationship between inversion barriers for amines and phosphines. (Baehler et al., 1972.)

is $N < Si^- < P \sim As < S^+$. Mislow and co-workers have shown that the pyramidal inversion barriers of identically substituted amine phosphines, arsines, carbanions, silyl anions, sulfonium ions, and sulfoxides have a linear relationship as specifically shown in Fig. 2 for phosphines and amines (Baehler et al., 1972). They have also neatly summarized the relationship in a nomograph reproduced in Fig. 3. With this graph, it is possible to estimate fairly accurately (± 3 kcal) the inversion barrier for any $R_1R_2R_3M$ complex if the barrier is known for a specific $R_1R_2R_3M$ complex. Note that in the graph the slopes for some central atoms are such that they cross those for others; at these points the relative barriers invert. As reference points, the barriers for a few MR_3 molecules are noted on the graph.

8. Dynamic Molecular Processes

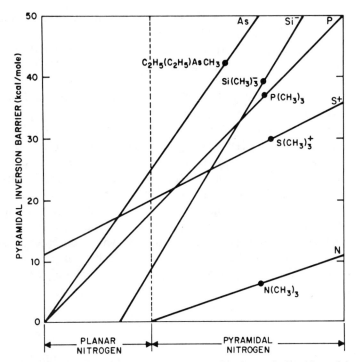

Fig. 3 Nomograph of Baehler et al. (1972) showing interrelationships of inversion barriers for five classes of ML$_3$ molecules and ions.

2. Four-Coordinate Molecules

In four-coordination, the favored polytopal form is the regular tetrahedron or a close approximation. The energy separation between tetrahedral and planar forms is so large for compounds in which the central atom has d^0 or d^{10} electronic configuration, that population of the planar state will probably be preceded by a dissociation of the molecule into neutral or ionic fragments (Muetterties, 1970a,b). On the other hand, if the central atom has partially-filled d orbitals, the energy levels of tetrahedral and planar forms often become of comparable magnitude for the lighter transition elements. In fact, tetrahedral and planar forms can coexist in solution for Ni^{2+} (d^8), and rates for interconversion of these polytopal forms commonly are very high, too high for detection of the transition or slow exchange regions by nmr as shown by Phillips and co-workers (Eaton et al., 1962). This is especially true for four-coordinate chelates (see Chapter 9). The only class of four-coordinate molecules in which the individual nmr resonances for planar and tetrahedral forms have been detected are bisphosphine complexes of nickel-(II) halides, (R$_3$P)$_2$NiX$_2$ (Pignolet et al., 1970; LaMar and Sherman, 1970).

The tetrahedral forms are triplet states, and the protons in these complexes are subject to contact shifts, which assist in distinguishing the nmr resonances associated with the tetrahedral form from those for the diamagnetic planar form. Analysis of the exchange regions for a series of bis(diarylalkylphosphine)nickel dihalides established an exchange order for the halides of I > Cl > Br (29, 2.6, and 0.45 × 10^5 sec^{-1} at 298°K for the $(C_6H_5)_2CH_3P$ derivatives). Representative exchange parameters are $\Delta H^{\ddagger} = 10$ and 13 kcal/mole and $\Delta S^{\ddagger} = 9$ and 3 kcal/mole for the bromide and chloride, respectively, of $[(p-CH_3OC_6H_4)_2CH_3P]_2NiX_2$. Geometries for these complexes are not precisely known, but comparison of electronic spectra for singlet and triplet complexes with those whose structures have been established indicate that interconversion does involve planar and tetrahedral forms. As in the case of pyramidal molecules, the nmr line shape changes provide, for four-coordinate molecules, no mechanistic information (Muetterties, 1972) about the physical details of the rearrangement process which conceptually is most simply viewed as a digonal twist, a process physically and permutationally indistinguishable from a tetrahedral compression to a square plane.

3. *Five-Coordination*

Stereochemical nonrigidity is a common feature of five-coordinate structures. If all ligating atoms or groups are equivalent or relatively similar in electronic and steric character, or if only one ligating atom or group is distinctively different, the five-coordinate structure is nonrigid. In these cases the barriers to intramolecular rearrangements are usually low, < 10 kcal/mole (Muetterties, 1970a,b, 1972). In ML_5E molecules* such as IF_5, the nonequivalence of the fluorine ligands can be observed using nmr; the barriers to rearrangement are high and the spectra are consistent with the idealized square pyramidal geometry. Molecules like IF_5 may be considered as quasi six-coordinate with the nonbonding electron pair acting as a phantom sixth ligand.

There is a class of cationic transition metal complexes in which five trivalent phosphorus ligands are bound to the central metal (Jesson and Meakin, 1973b,c). All these molecules are nonrigid and are unique among ML_5 species in that the rate of the exchange process can be reduced to a frequency that is low on the nmr time scale at low temperatures. In the slow exchange limit, the ^{31}P spectra are A_2B_3 or A_2B_3X patterns, which unambiguously define the solution state stereochemistry as trigonal bipyramidal. The pattern broadens in the transition region and becomes a single line (A_2B_3) or a doublet (A_2B_3X) in the fast exchange limit. The behavior is

---

* The E designation is for a valence shell lone electron pair.

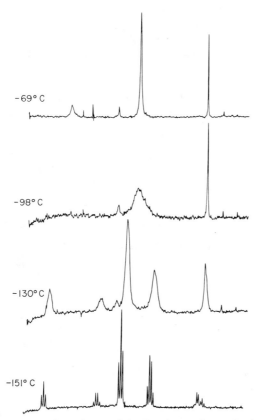

Fig. 4 Temperature dependence of the Fourier mode proton decoupled nmr spectrum assigned to $Pt[P(OCH_3)_3]^{2+}\ 2B(C_6H_5)_4^-$ in chlorodifluoromethane.

illustrated in Fig. 4 for $Pt[P(OCH_3)_3]_5^{2+}$ showing a series of ^{31}P spectra over a temperature range. Analogous behavior was observed for the corresponding complexes based on Co(I), Rh(I), Ir(I), Ni(II), and Pd(II). Exchange barriers proved relatively insensitive to increased positive charge on the metal or on going from one transition metal row to the next. There is, however, an increase in barrier with increasing steric bulk of the ligand, e.g., in going from trimethyl phosphite to triethyl phosphite to tri-n-butyl phosphite. Table I gives some free energies of activation for the exchange process (Jesson and Meakin, 1974).

The maintenance of metal-to-phosphorus coupling in the high-temperature limit for the Rh(I) and Pt(II) complexes indicates that the exchange process is intramolecular within our definition. Experiments with added ligand allow a similar conclusion for the other metals. Although in some of the complexes the chemical shift separation is temperature dependent, this does not appear to be due to ion-pairing effects; thus, the chemical shift separation

TABLE I
Activation Parameters for Intramolecular Rearrangement in d^8 ML_5 Cations

Complex	$\Delta G^{\ddagger}$ (kcal mole^{-1})	$\Delta H^{\ddagger}$ (kcal mole^{-1})	$\Delta S^{\ddagger}$ (calories mole^{-1} deg^{-1})	Temperature (°K)
Co$^+$				
$P(OCH_2)_3CCH_3$	10.0	12.1	10.5	200
Rh$^+$				
$P(OCH_3)_3$	7.5	6.35	−6.0	200
$P(OC_2H_5)_3$	9.9	—	—	208
$P(O\text{-}nC_4H_9)_3$	11.1	—	—	228
$P(OCH_2)_3CCH_3$	7.8	—	—	153
$P(OCH_2)_3CC_2H_5$	7.8	—	—	—
Ir$^+$				
$P(OCH_3)_3$	8.0	8.1	0.5	200
$P(OC_2H_5)_3$	10.4	—	—	215
$P(O\text{-}nC_4H_9)_3$	11.5	—	—	231
$P(OCH_2)_3CCH_3$	8.4	8.8	2.0	200
Ni^{2+}				
$P(OCH_2)_3CCH_3$	∼8.3	∼8.2	∼−0.5	200
$P(OCH_2)_3CC_2H_5$	8.3	8.2	−0.5	200
Pd^{2+}				
$P(OCH_3)_3$	6.2	—	—	138
$P(OC_2H_5)_3$	8.0	7.0	−5.0	200
Pt^{2+}				
$P(OCH_3)_3$	6.5	—	—	143
$P(OC_2H_5)_3$	9.2	—	—	199

is not appreciably dilution dependent and the height of the barrier is insensitive to the nature of the counterion.* Any interaction of this type must be averaged rapidly on the nmr time scale even at very low temperatures.

The character of the ^{31}P line-shape changes for these $M[P(OR)_3]_5^{x+}$ complexes contains mechanistic information (Meakin and Jesson, 1973). Assuming the "jump model," the nmr line-shape changes reflect the nuclear permutations which convert the initial labeled configuration 9 into the configuration after rearrangement. For these molecules, the number of permutations is 5!, or 120. Those permutations of the 120 that are symmetry operations of the D_{3h} point group of the molecule do not affect the spin Hamiltonian; the order of this subgroup is twelve. There are, in addition to the identity operation, two basic permutational sets or subclasses each of

* Earlier studies on HML_4 and HML_4^- complexes (see below) demonstrated that barriers for the isoelectronic pairs were similar, again indicating that ion pairing is not an important factor, the barriers being somewhat smaller for the charged relative to the neutral species.

8. Dynamic Molecular Processes

$$\begin{array}{c} 4 \\ 1 \overset{2}{\underset{3}{\rule{0pt}{0pt}}} \\ 5 \end{array}$$

which can lead to a line shape change distinguishable from the other (Jesson and Meakin, 1973a, 1974). These basic sets* are:

A	(15)(24)	B	(15)(23)
	(14)(35)		(12)(35)
	(25)(34)		(14)(23)
			(13)(24)
			(13)(25)
			(12)(34)

Using the spectral parameters derived from the slow exchange spectrum, the line shape behavior is calculated as a function of exchange rate for each of these mechanistic sets (Meakin and Jesson, 1973). These simulated spectra are illustrated in Figs. 5–7 for the $Rh[P(OCH_3)_3]_5^+$ case along with the observed spectra. Note for the rhodium case there is a spectral doubling because the rhodium nucleus has a spin of $\frac{1}{2}$. Only mechanistic set A yields spectra in close correspondence with the observed spectra. Thus, this permutational mechanism must be dominant. A physical motion that has this permutational character is the Berry rearrangement (Berry, 1960), schematically shown in Fig. 8, which traverses a square pyramidal intermediate or transition state wherein two axial positions are exchanged with two equatorial positions. Alternative, permutationally indistinguishable physical motions, such as the anti-Berry (Meakin et al., 1972b) and the turnstile mechanisms (Ugi et al., 1971), are discussed below. These experiments do incisively rule out such physical motions as axial–equatorial, axial–axial–equatorial, and axial–equatorial–equatorial exchanges. As noted earlier in the chapter, the possibility of extracting the mechanistic information depends critically on the observation of second-order splittings in the spectrum and would not have been possible if the chemical shift separation were larger or if the spectra were taken on a spectrometer operating at a much higher field than 21 kG.

All other ML_5 molecules studied to date exhibit magnetic equivalence of ligands down to the lowest temperatures accessible, as dictated by solubility characteristics and solvent freezing points. No transitional or slow exchange spectra have been observed for these other molecules. Presumably, rearrangement barriers are quite low, < 5 kcal/mole. The one ML_5 molecule that precipitated much of the interest in nonrigid five-coordinate molecules

* These operations and the identity operation plus the equivalent operations obtained by applying the symmetry operations of the point group will yield the 120 labeled configurations.

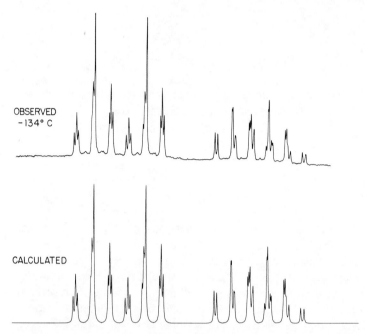

Fig. 5 Observed and simulated low temperature limit ($-134°C$) Fourier mode proton noise decoupled 36.43 MHz ^{31}P nmr spectrum assigned to $Rh[P(OCH_3)_3]_5{}^+ \ B(C_6H_5)_4{}^-$ in 90% chlorodifluoromethane–10% dichloromethane.

is PF_5, which has a doublet (PF spin coupling) ^{19}F spectrum down to temperatures below $-150°C$. Other ML_5 species examined by nmr include $P(C_6H_5)_5$(^1H) (Muetterties et al., 1963, 1964), $SiF_5{}^-$(^{19}F) (Klanberg and Muetterties, 1968), AsF_5(^{19}F), $Sb(CH_3)_5$(^1H) (Muetterties et al., 1963, 1964), $Co(CNR)_5{}^+$(^1H and ^{13}C) (Muetterties, 1973a), $Fe(CO)_5$(^{13}C) (Cotton et al., 1958; Bramley et al., 1962), $M(PF_3)_5$ (M = Fe, Ru, and Os) (^{31}P and ^{19}F) (Meakin et al., 1972b).

Five-coordinate structures with one unique ligand are also susceptible to rapid intramolecular rearrangements. If in an XML_4 structure the interbond angles are nearly 90° and 120° corresponding to a trigonal bipyramid, and if the X ligand is at an equatorial site, then the rearrangement barrier will be quite low, usually less than 10 kcal/mole. This is explicable in terms of a Berry rearrangement because the transition state or reaction intermediate (Fig. 8) may still have the C_{4v} square pyramidal geometry with the unique X ligand at the apical site. Actually, limiting slow exchange spectra have not been observed for most C_{2v}–XML_4 complexes, e.g., RPF_4, HPF_4, and $RSiF_4{}^-$. Higher rearrangement barriers are observed in phosphorus(V) compounds where the X ligand has π orbitals available for bonding, such as

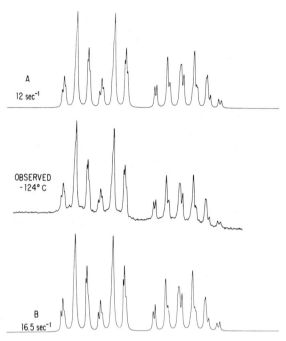

Fig. 6 The proton noise decoupled ^{31}P Fourier mode spectrum of $Rh[P(OCH_3)_3]_5{}^+$ $B(C_6H_5)_4{}^-$ at $-124°C$ in 10% dichloromethane–90% chlorodifluoromethane, together with spectra simulated for the two permutational mechanisms A and B defined in the text. The appropriate exchange rates are given for the calculated spectra. It is clear that permutational mechanism A, which corresponds to simultaneous exchange of the axial ligands with two equatorial ligands, gives the better fit.

R_2N, or RS, possibly due to the multiple bond character of the PX bond (Muetterties, 1970a,b, 1972; Muetterties et al., 1963, 1964). For example, in $(CH_3)_2NPF_4$, the rearrangement barrier is 9.5 kcal/mole. Whitesides and Mitchell (1969) showed that the character of line-shape changes in this molecule is consistent with a Berry-type of rearrangement. As in the ML_5 problem discussed above, there are, for a C_{2v}–XML_4 molecule, two distinguishable permutational mechanisms in a mutual exchange process (in addition to the identity operation) which have the character:

$$
\begin{array}{ll}
A \ (25)(34) & B \ (25) \\
\ \ \ (24)(35) & \ \ \ (35) \\
& \ \ \ (24) \\
& \ \ \ (34)
\end{array}
$$

for the labeled configuration **10**. It should be noted that irrespective of the nitrogen coordination geometry, i.e., planar or rapidly inverting pyramidal,

$$X \overset{5}{\underset{4}{-}} \hspace{-0.5em} \diagdown \hspace{-0.5em} \begin{matrix} 3 \\ 2 \end{matrix}$$

10

rotation about the PX bond must be fast to achieve fluorine equivalence via the Berry mechanism. These P—X bond rotational barriers can be large, as in $PF_3(NH_2)_2$ (Muetterties *et al.*, 1972), where E_a is 11.1 kcal/mole. PX multiple bond character is not the only factor contributing to the high barriers in NH_2 derivatives; the rearrangement barrier in H_2NPF_4 (Cowley and Schweiger, 1972) is very high, >15 kcal/mole. Barrier enhancement in this molecule may be due to proton–axial fluorine bonding interactions. In the R_2N and the RS derivatives of PF_5, the effective (averaged conformation if R_2N is pyramidal and inverting rapidly) plane of the substituent groups passes through the two axial sites and the phosphorus atom.

In C_{3v}–XML_4 structures with X at an axial position rearrangement. barriers are relatively high, provided that near trigonal bipyramidal angles are

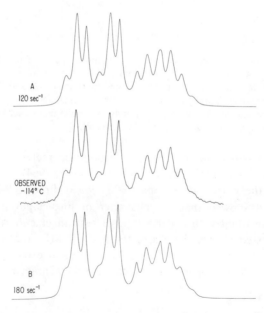

Fig. 7 Observed and calculated temperature-dependent proton noise decoupled ^{31}P Fourier mode nmr spectrum for a solution of $Rh[P(OCH_3)_3]_5^+ \; B(C_6H_5)_4^-$ in 10% dichloromethane–90% chlorodifluoromethane at $-114°$. The permutational mechanisms and exchange rates are given with the calculated spectra in the figure. Permutational mechanism A, which corresponds to simultaneous exchange of the axial ligands with two equatorial ligands, clearly gives the better fit.

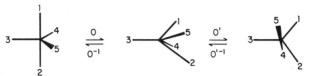

Fig. 8 Berry rearrangement that interrelates the two idealized forms for an ML_5 complex, namely the trigonal bipyramid and the square pyramid. This physical process can permute ligand atom positions. A one-step process for a labeled ML_5 complex is depicted to show the permutation of two axial (1,2) and two equatorial positions (4,5) in the trigonal bipyramid.

maintained and that the electronic or steric character of the X and L ligands are significantly different (Muetterties, 1970a,b). This behavior is again explained if the Berry rearrangement is the dominant rearrangement mechanism for trigonal bipyramidal molecules. Equilibration of L environments would require traverse, or population, of a high-energy C_{2v} state. On the other hand, if the C_{3v}–XML_4 molecule does not have the trigonal bipyramidal geometry, then the physical motions required in a Berry rearrangement are not feasible. A case in point is the highly fluxional class of transition metal hydrides of the composition HML_4 (Meakin et al., 1972b). Line-shape changes for one of these $HRh(PF_3)_4^-$, are shown in Fig. 9, and exchange

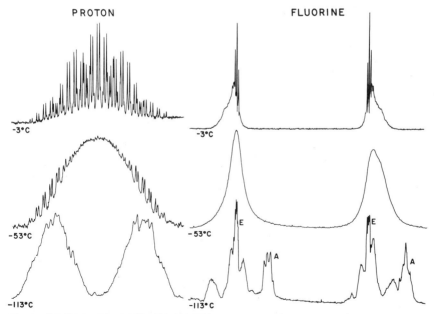

Fig. 9 ^1H (90 MHz) and ^{19}F (84.66 MHz) nmr spectra for $HRh(PF_3)_4$ as a function of temperature: first row, acetone solution; second and third rows, chlorodifluoromethane solution. A and E indicate axial and equatorial fluorine resonances.

TABLE II
EXCHANGE PARAMETERS OF HML_4 MOLECULES
DERIVED FROM NMR ANALYSIS[a]

Molecule	$\Delta G^{\ddagger}$ (kcal mole^{-1})	Low temperature range studied (°C)
$HFe(PF_3)_4^-$	< 5	−160
$HRu(PF_3)_4^-$	7.0	−160
$HOs(PF_3)_4^-$	8.0	−160
$HCo(PF_3)_4$	5.5	−160
$HRh(PF_3)_4$	9.0	−160
$HIr(PF_3)_4$	10.0	−100
$HCo[P(OC_2H_5)_3]_4$	< 6.0[b,c]	−160
$HCo[P(OCH_3)_3]_4$	< 6.0[b,c,d]	−160
$HRh[P(OC_2H_5)_3]_4$	7.25[b]	−150
$HRh[(C_6H_5)_2PCH_2CH_2P(C_6H_5)_2]_2$	< 7.5[b]	−110
$HRh[P(OC_2H_5)_2C_6H_5]_4$	< 8.5[b]	−50
$HIr[(C_6H_5)_2PCH_2CH_2P(C_6H_5)_2]_2$	< 7.5[b]	−110
$HNi[P(OC_2H_5)_3]_4^+$	< 5.0[b]	−160

[a] Meakin et al., 1972b.
[b] C_{3v} ground states assumed; limiting data from proton nmr studies.
[c] Limiting data from ^{31}P studies.
[d] Limiting data from ^{13}C studies (Muetterties and Hirsekorn, 1973).

parameters are listed in Table II. The limiting slow exchange spectra are consistent with a C_{3v} ground-state structure in which the unique hydride ligand is at an axial position. X-ray studies of four HML_4 hydrides establish the C_{3v} stereochemistry for the solid state and additionally, show that the structure is not trigonal bipyramidal. Rather, the four phosphorus atoms and the central metal atom describe a nearly regular tetrahedron—the hydride ligand then "lies" on a tetrahedral face. Unfortunately, there is no mechanistic information in the nmr line-shape changes for these molecules. In addition to the identity operation, there is only one basic mechanistic set, which has the permutational character (12), (13), and (14) for the labeled configuration **11**. Hence, any physical motion proposed for this rearrangement would be

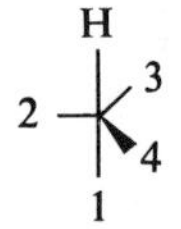

$$\begin{array}{c} H \\ 2 - \overset{3}{\underset{1}{\bigtriangleup}} 4 \\ \mathbf{11} \end{array}$$

consistent with the nmr data provided it did not involve H—M or P—M bond breaking. A physical picture of the rearrangement process presented by Meakin et al. (1972b) is the tetrahedral jump model illustrated in Fig. 10. As

8. Dynamic Molecular Processes

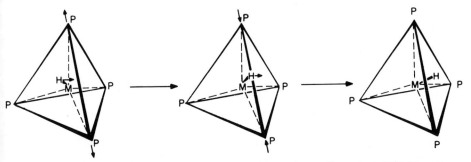

Fig. 10 The tetrahedral jump mechanism as applied to the quasi-tetrahedral HML_4 and H_2ML_4 complex in which hydrogen atoms are at tetrahedral facial positions in the ground state. In the concerted process, a PMP angle increases with concomitant transverse of a hydrogen atom to this affected tetrahedral edge, and then the PMP angle decreases as the hydrogen atom goes to a new tetrahedral face.

a PMP angle is increased, the M—H bond is bent to give a C_{2v} intermediate or transition state with the hydrogen on an edge of an MP_4 tetrahedron. Deuterium substitution only slightly affects the exchange rates, and apparently most of the reduced mass arises from the phosphorus atom motions. Similar processes, discussed below, have been proposed for H_2MP_4 and H_4MP_4 molecules.

It is unfortunate that in XML_4 molecules with C_{3v} symmetry, nmr lineshape methods using a jump model can provide no permutational mechanistic information. An important chemical modification of structure, use of diastereotopic groups, can extend the efficacy of the nmr technique in mechanistic distinctions. This has been little used for studies of five-coordinate molecules. One example is the work of Whitesides and Bunting (1967) on the phosphorane **12** shown in a possible trigonal bipyramidal ground state form with an equatorial isopropylphenyl substituent. At 25°, the axial and

12

equatorial methyl groups of the bitolyl ligands are nonequivalent, as are the methyl groups of the isopropyl substituent. The nonequivalence of the bitolyl methyl groups and that of the isopropyl methyls are independent probes of axial–equatorial positional exchange and interconversion of enantiomers. The observed rates of these two independent processes were identical within experimental error. This is consistent with a Berry rearrangement mechanism or a permutationally equivalent mechanism. Note that the nmr data are also consistent with a ground state square pyramidal form.

In C_{2v}–X_2ML_3 molecules, rearrangement barriers are relatively high, especially if the electronic characters of X and L are substantially different. The polarity rules (Muetterties et al., 1963, 1964) set the stereochemistry with the more electronegative ligands at the axial positions in a trigonal bipyramid (or the more bulky ligands at the equatorial sites). Thus, an X_2ML_3 molecule such as $(CH_3)_2PF_3$ has stereochemistry **13**. Equilibration of fluorine atom

13

environments via a Berry mechanism requires traverse of **14** and **15**, which states are clearly high-energy states. It has been found that $(CH_3)_2PF_3$

14 **15**

(Muetterties et al., 1963) and its analogs are rigid on the nmr time scale. On the other hand, Cl_2PF_3 and Br_2PF_3 are nonrigid (Mahler and Muetterties, 1965), reflecting the smaller electronegativity difference between fluorine and these halogens as compared to the situation in the alkylphosphoranes. In alkoxyphosphoranes, the barriers, due to intermediate (O versus C) electronegativity differences, are in the 10–17 kcal/mole range (Gorenstein, 1970).

More extensive discussions of five-coordinate intramolecular rearrangements may be found in reviews (Muetterties, 1970a,b, 1972; Mislow, 1970; Westheimer, 1968) and original literature. A concise summary of rearrangement mechanisms in five-coordinate molecules is as follows. The Berry mechanism is based on the close relationship of the two basic geometries found for five-coordinate molecules or ions. The trigonal bipyramidal form is clearly favored for most central atom electronic configurations, but the available experimental data and best possible theoretical calculations indicate that the energy difference between the trigonal bipyramidal and square pyramidal forms is very small. Clearly, this mechanism, which requires only small bending motions of the ligands, is a very attractive physical process. Most importantly, this mechanism, in conjunction with the polarity rule, can be

8. Dynamic Molecular Processes

used to rationalize a vast body of dynamic stereochemical data and has been utilized effectively in predicting stereochemical behavior in five-coordinate molecules (Muetterties, 1970a; Mislow, 1970; Westheimer, 1968). The Berry mechanism is not applicable to molecules whose structures do not reasonably approximate the trigonal bipyramid (or alternatively the square pyramid), as noted above for the quasi-tetrahedral HML_4 complexes; if fine distinctions are to be made concerning the physical motion, the Berry rearrangement can be strictly applied only to D_{3h}-, ML_5, and C_{2v}-XML_4 structures where a reaction path constrained to C_{4v} symmetry can be followed. Clearly, this is not feasible in C_{3v}-XML_4 species or any more complex molecules. One mechanism that has been proposed as an alternative to the Berry mechanism is the turnstile rotation (Ugi et al., 1971). This is permutationally and topologically identical to the Berry motion, but it does not possess the general predictive value and rationale of the Berry mechanism. The distinctions between the Berry mechanism and the turnstile mechanism are symmetry defined for the reaction path; these are lost if ground state geometry is not D_{3h}, C_{2v}, or C_{4v}. A more extensive discourse and critique of this alternative may be found in the articles by Ugi et al. (1971), by Meakin et al. (1972b), and by Muetterties and Guggenberger (1974).

4. Six-Coordination

Stereochemically nonrigid six-coordinate molecules are not very common. The octahedral form appears to represent such a deep well in the potential energy surface for six coordination, that alternative states, through which permutation of ligand sites can occur, are not readily attained. An extreme case is represented by $C_6H_5SF_5$ (Muetterties, 1970a). The ^{19}F nmr spectrum is an AB_4 pattern that does not change at temperatures as high as 220°C indicating a barrier to any kind of intramolecular (or intermolecular) rearrangement greater than 30 kcal/mole. On the other hand, many six-coordinate tris chelates are stereochemically labile. It is probable that the majority of these chelates rearrange through an internal dissociative process involving a five-coordinate species with one of the chelate ligands bound only as a unidentate ligand. As discussed earlier (p. 256), it is very difficult by nmr to distinguish in chelates between this internal dissociation and a true intramolecular process. However, by use of diastereotopic groups, a quite convincing argument based on nmr data has been presented for a fast intramolecular rearrangement in certain metal tristropolonates and tris(dithiocarbamates) (Eaton et al., 1972, 1973; Edgar et al., 1973). The rearrangement comprises a trigonal twisting about the threefold or pseudo threefold axis (Fig. 11). These studies are discussed in Chapter 9.

If a six-coordinate complex does not have regular octahedral geometry,

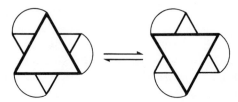

Fig. 11 Trigonal twist mechanism for octahedral structures illustrated for the specific case of a trischelate. This operation is performed about the threefold axis (pseudo threefold axis if the chelates are inequivalent or if the chelates are unsymmetrical and the complex is the trans isomer). This process effects a change in chirality at the metal center.

i.e., if it does not have near 90° interbond angles, the possibility of intramolecular rearrangements is increased. For example, in the chelates cited above, twist angles *significantly* less than the octahedral 60° twist angle should provide lower barriers to the twist rearrangement (Fig. 11). A classic example of ground state geometry other than the idealized regular octahedron providing a facile path to rearrangements is found in the six-coordinate metal dihydrides of H_2ML_4 form (Tebbe *et al.*, 1970; Meakin *et al.*, 1970, 1971, 1973a).

Metal dihydrides of the H_2ML_4 form with M = Fe, Ru, and L, a tricovalent phosphorus compound, exist in solution as a single stereochemical entity with but four exceptions discussed below (Meakin *et al.*, 1973a). In each case, slow exchange limit nmr data establish three magnetically nonequivalent sets and two environmentally nonequivalent sets of phosphorus atoms. These data are consistent with a cis-octahedral stereochemistry **16**. A representative limiting slow exchange spectrum for one complex is shown

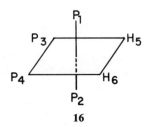

16

in Fig. 12 with a calculated spectrum and the associated nmr parameters referenced to the labeled configuration **16**. All the dihydrides are stereochemically nonrigid, and the iron complexes, which have lower barriers than the ruthenium analogs, have as limiting fast exchange ^1H and ^{31}P spectra binomial quintets and triplets, respectively. Thus, in the fast exchange limit, there is equivalence of all nuclei in the hydride hydrogen set and in the phosphorus set. Furthermore, retention of spin–spin coupling in this limit establishes the intramolecular character of the exchange. As a further

8. Dynamic Molecular Processes

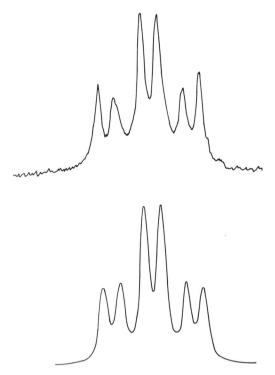

Fig. 12 Observed and calculated low-temperature limit ^{1}H 220 MHz hydride nmr spectrum for $H_2Fe[P(OC_2H_5)_3]_4$. The spectral parameters used in the calculation are

$$J_{15} = J_{16} = J_{25} = J_{26} = 66.5 \text{ Hz}; \quad J_{13} = J_{14} = J_{23} = J_{24} = 56.0 \text{ Hz}$$

$$\begin{Bmatrix} J_{36} = J_{45} \\ J_{35} = J_{46} \end{Bmatrix} = \begin{Bmatrix} 24 \text{ Hz} \\ -6 \text{ Hz} \end{Bmatrix}$$

$$J_{56} = -5 \text{ Hz}; \quad J_{34} \sim 0 \text{ Hz}; \quad \delta_{34} - \delta_{12} = 2.2 \text{ ppm} (-70°\text{C})$$

support of this last point, it was found that exchange rates were, within the experimental error, independent of hydride concentration and of the nature of the solvent.

The temperature dependent spectra of one hydride were studied in detail; this hydride, $H_2Fe[P(OC_2H_5)_3]_4$, serves as the main focal point in the discussion below although the other cis iron dihydrides behaved in an analogous fashion. In Figs. 13–15, the ^{1}H, ^{31}P, and ^{31}P{^{1}H} spectra of $H_2Fe[P(OC_2H_5)_3]_4$ are presented as a function of temperature together with calculated spectra (see below). Two points are to be made. First, note in the proton nmr spectra that the outer two lines stay sharp throughout the transition region. This is characteristic of a mutual exchange process. The invariant lines represent transitions unaffected by the permutations. In this particular case, they

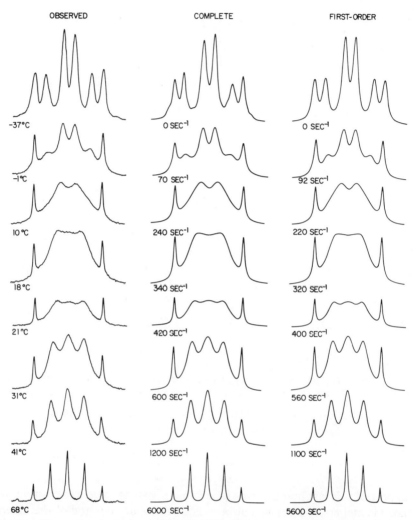

Fig. 13 Observed and calculated 220 MHz ^{1}H hydride nmr spectra of FeH$_2$[P(OC$_2$H$_5$)$_3$]$_4$ as a function of temperature. The results are shown for first-order and complete calculations using the basic permutational set IV.

correspond to transitions $|H(\alpha\alpha); P(\alpha\alpha\alpha\alpha)\rangle \rightarrow |H(\alpha\beta + \beta\alpha); P(\alpha\alpha\alpha\alpha)\rangle$ and $|H(\alpha\beta + \beta\alpha); P(\alpha\alpha\alpha\alpha)\rangle \rightarrow |H(\beta\beta); P(\alpha\alpha\alpha\alpha)\rangle$ for one and analogous functions with P($\beta\beta\beta\beta$) for the other. Second, and most important, there is potentially a wealth of mechanistic information in this six-spin system in contrast to little or no information in the four- and five-coordinate cases. The level of complexity is just about ideal for this iron dihydride, since it is approaching the limit of modern computational techniques.

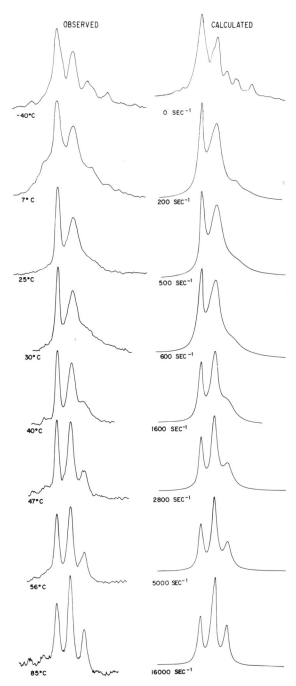

Fig. 14 Observed and calculated ^{31}P (36.4 MHz) nmr spectra of $FeH_2[P(OC_2H_5)_3]_4$ as a function of temperature. The calculated spectra are for the basic permutational set IV.

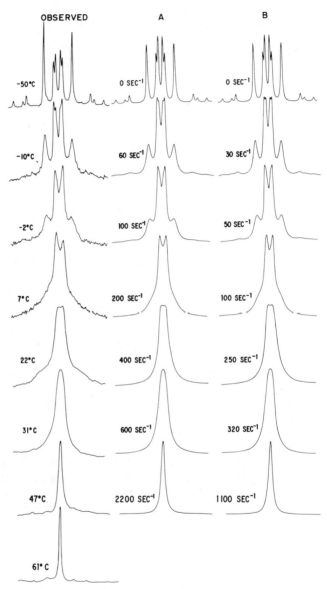

Fig. 15 Proton noise decoupled ^{31}P (36.4 MHz) spectrum for $FeH_2[P(OC_2H_5)_3]_4$ over a range of temperature together with simulated spectra for the basic permutational sets A and B.

8. Dynamic Molecular Processes

For the proton nmr spectra, the mutual exchange process may be represented by a group of labeled configurations (see **16**) for all possible cis molecules. The order of the group is $4! \times 2! = 48$. Decomposition of the group into basic mechanistic sets gives the following (Jesson and Meakin, 1973a; Meakin et al., 1971).

$$E \quad (1)(2)(3)(4)(5)(6)$$

$$\mathrm{I} \begin{pmatrix} 1 & 2 & 3 & 4 & 5 & 6 \\ 3 & 4 & 1 & 2 & 5 & 6 \\ 4 & 3 & 2 & 1 & 5 & 6 \end{pmatrix} = \begin{matrix} (1\ 3)(2\ 4) \\ (1\ 4)(2\ 3) \end{matrix}$$

$$\mathrm{II} \begin{pmatrix} 1 & 2 & 3 & 4 & 5 & 6 \\ 1 & 2 & 4 & 3 & 5 & 6 \end{pmatrix} = (3\ 4)$$

$$\mathrm{III} \begin{pmatrix} 1 & 2 & 3 & 4 & 5 & 6 \\ 1 & 3 & 4 & 2 & 5 & 6 \\ 1 & 4 & 2 & 3 & 5 & 6 \\ 3 & 2 & 4 & 1 & 5 & 6 \\ 4 & 2 & 1 & 3 & 5 & 6 \end{pmatrix} = \begin{matrix} (2\ 3\ 4) \\ (2\ 4\ 3) \\ (1\ 3\ 4) \\ (1\ 4\ 3) \end{matrix}$$

$$\mathrm{IV} \begin{pmatrix} 1 & 2 & 3 & 4 & 5 & 6 \\ 3 & 2 & 1 & 4 & 5 & 6 \\ 1 & 3 & 2 & 4 & 5 & 6 \\ 1 & 4 & 3 & 2 & 5 & 6 \\ 4 & 2 & 3 & 1 & 5 & 6 \end{pmatrix} = \begin{matrix} (1\ 3) \\ (2\ 3) \\ (2\ 4) \\ (1\ 4) \end{matrix}$$

together with the equivalent sets derived from these by the application of the operations of the point group (C_{2v}). It is immediately clear that permutation (II) cannot give rise to the fast exchange limit spectrum of the dihydride since this permutation cannot remove the environmental nonequivalence of the phosphorus atoms. The classical trigonal twist or Bailar mechanisms correspond to permutational mechanisms I and III for the two possible twists (Fig. 11); there are two types of pseudo threefold axes in cis-H_2ML_4.

Using the spectral parameters derived from the slow exchange spectrum, the line-shape behavior was calculated as a function of exchange rate for each of the mechanisms (first-order treatment). These first-order simulations are presented in Fig. 16. Spectra in column V are for random exchange—all permutations equally weighted. Of the basic mechanisms, only mechanism IV gave results closely in agreement with experiment. Similarly, linear combinations were examined and only those combinations that were predominantly IV gave reasonable fits. Side-by-side comparisons of observed and simulated

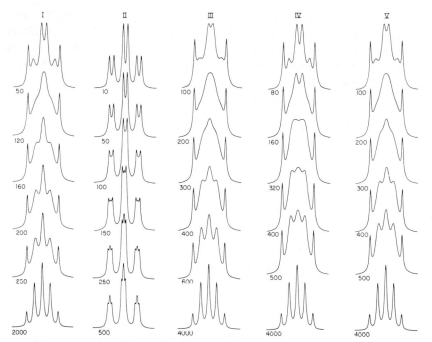

Fig. 16 Calculated first-order line shapes as a function of exchange rate for mechanisms I–IV together with a calculation for random exchange (V). The exchange rates are per second.

spectra based on mechanism IV are provided in Figs. 13 and 17 for 220 and 90 MHz spectra, respectively.

The simplest physical picture of the rearrangement that has been devised to date having the permutational character of IV is the *tetrahedral* jump mechanism as described earlier (Fig. 10) for the five-coordinate HML_4 complexes. Consider first the structure of a *cis*-H_2ML_4 complex. X-ray structural studies (Guggenberger *et al.*, 1972) of $H_2Fe[P(C_6H_5)(OC_2H_5)_2]_4$ show that the interbond angles depart substantially from the 90° regular octahedral angles. In fact, the PFeP angles more closely approach those for a regular MP_4 tetrahedron. Thus, the dihydride structure, shown in Fig. 18, may be viewed as a quasi-tetrahedral complex with the hydride ligands projecting from two tetrahedral faces. The physical picture of a concerted intramolecular rearrangement would be FeH bending so as to take a single hydride ligand from a tetrahedral face to a tetrahedral edge (transition state) with concomitant increase in the associated PFeP angle (Fig. 10). Deuteration does not affect the rate substantially; thus, it appears that the heavy atom motion makes the major contribution to the reduced mass. $\Delta G^\ddagger = 13.7$ kcal/mole, $\Delta H^\ddagger = 11.1$ kcal/mole, and $\Delta S^\ddagger = -8.8$ cal/mole/deg (298°K).

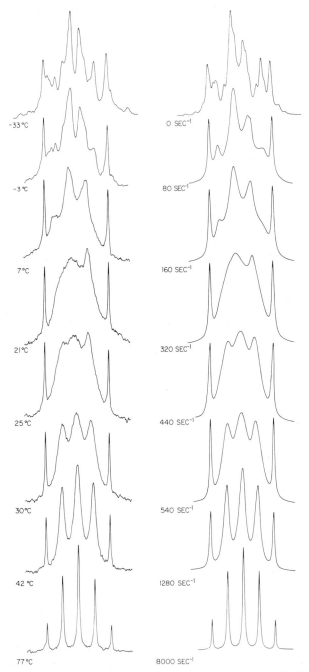

Fig. 17 Observed and calculated 90 MHz ^{1}H hydride nmr spectra for $FeH_2[P(OC_2H_5)_3]_4$. The calculated spectra are for the basic permutational set IV.

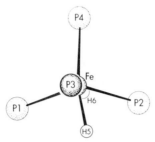

Fig. 18 Phosphorus–hydride–iron skeleton of $FeH_2[P(OC_2H_5)_2C_6H_5]_4$ taken from X-ray crystal structure data. The perspective looking down the P3–Fe–H6 axis was chosen to illustrate the postulated exchange mechanism. Phosphorus atoms P1, P2, and P4 are in a nearly trigonal array about this axis, and the mechanism involves the motion of H5, which is in the face between P1, P2, and P3, into the face between P1, P4, and P3 or between P2, P3, and P4. An essentially identical perspective could have been drawn looking down the P4–Fe–H5 axis to show the single hydrogen step for H6.

An alternative nonconcerted process would be movement of both hydrogen atoms to tetrahedral edges to give a trans structure as a short-lived reaction intermediate. From this intermediate, the hydrogen atoms may return to original faces (E), one may move on to a previously unoccupied face (IV) or both go to a previously unoccupied face (I). This process corresponds to the linear combination $2E + I + IV$ which is predominantly mechanism IV. This alternative process is attractive because three iron hydrides and two ruthenium hydrides exist in solution as equilibrium mixtures of cis and trans isomers.

We have found permutational analysis to be a very important adjunct to dynamic nmr studies. However, it should be used not just as an adjunct analysis, but to predesign the experiments and pinpoint the types of chemical structures necessary to establish mechanistic points. Unfortunately, there are limitations to this mathematically precise analysis of mechanism. There is an optimum range of spin complexities below which no distinctions (see HML_4 case above) can be made and above which the analysis of the nmr spectral changes becomes very difficult (see the eight-spin, H_4ML_4 case below). Spin complexity is not a sufficient basis alone for effective use of permutational analysis in nmr studies. The spectral parameters must be of certain relative magnitudes if distinctions between mechanisms are to be made. This is clearly demonstrated by the ruthenium dihydride analogs of the iron complexes. The signs and magnitudes for the coupling constants in the ruthenium hydrides are quite different than those for iron; this is illustrated in Fig. 19 for cis-$H_2Ru[P(OCH_3)_3]_4$, where observed and calculated slow exchange spectra are presented. Simulations were made of spectra as a function of

8. Dynamic Molecular Processes

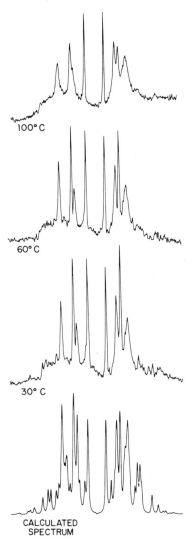

Fig. 19 Observed and calculated 90 MHz hydride region ^{1}H nmr spectra for $H_2Ru[P(OCH_3)_3]_4$. The low-field end of the spectrum is at the right side. The simulated slow exchange limit spectrum is shown on the bottom row.

exchange rate for this complex as dictated by the permutational character of the individual mechanisms (I–IV) (Fig. 20). The distinctions among the simulations for the individual mechanisms are too small for any realistic conclusions (except elimination of II) to be made.

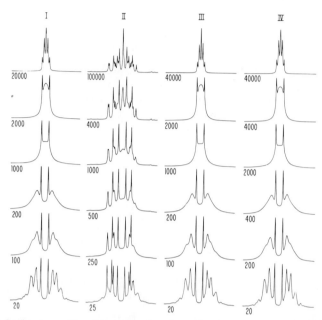

Fig. 20 Calculated nmr line shapes as a function of exchange rate for the basic permutational mechanisms I–IV. These simulations are intended to correspond to the hydride region ^{1}H nmr spectra of $H_2Ru[P(OCH_3)_3]_4$ taken at 90 MHz. The exchange rates are per second. The slow exchange limit spectra (identical in all cases) have been omitted from this figure and are all identical with that shown at the bottom of Fig. 19.

As mentioned above, several of the metal dihydrides yield solutions of cis and trans isomers. This is illustrated for $H_2Fe[P(OC_2H_5)_2C_6H_5]_4$ in Fig. 21. At low temperatures, individual resonances are observed for the trans (quintet) and cis (high field multiplet) forms. At elevated temperature exchange between isomers occurs, as can clearly be seen from the spectral changes in Fig. 21. The exchange is nonmutual and the primary change with temperature is a broadening and merging of the resonances due to the substantial chemical shift between cis and trans forms. This type of process gives rise to broad featureless spectra, and in general, there is no mechanistic information in the line-shape changes. Perhaps even more instructive from the standpoint of the importance of temperature-dependent nmr studies are the spectral changes in the proton spectrum for $H_2Fe\{o\text{-}C_6H_4[P(C_2H_5)_2]_2\}_2$ shown in Fig. 22. At low temperatures, the primary solution species is the trans isomer; cis is present but at too low a concentration to be detected. As the temperature is raised, the exchange between cis and trans forms becomes rapid with respect to the nmr time scale and there is a parallel increase

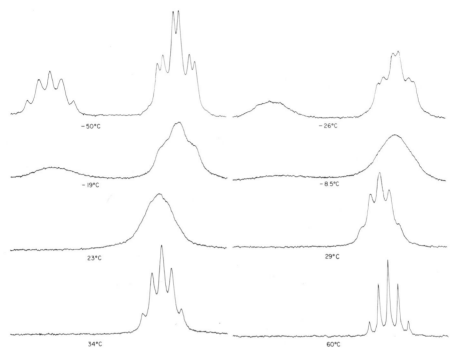

Fig. 21 Dynamic nmr proton spectrum (220 MHz) for $H_2Fe[C_6H_5P(OC_2H_5)_2]_4$. In the slow exchange limit ($-50°C$), the quintet at the left represents the trans isomer and the triplet of doublets at the right the cis isomer.

in the relative concentration of the cis isomer. Hence, the initial broadening of the quintet and, as the fast exchange limit is reached, the appearance of a new quintet whose position shifts to higher fields as the cis isomer concentration increases. Alternatively, the cis isomer might have been detected at low temperatures by Fourier transform spectroscopy; however, the temperature study clearly delineated the solution state stereochemistry and dynamics—a modestly complex solution phenomenon not suspected in earlier studies that relied on more classic analytical techniques.

5. *Seven-Coordination*

Seven-coordinate complexes with unidentate ligands should be stereochemically nonrigid (Muetterties, 1965, 1970a,b). In fact, intramolecular rearrangements in this coordination family should be less activated than in five-coordination because of the apparent close energies for the various

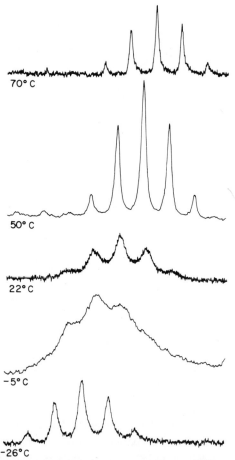

Fig. 22 The temperature-dependent 220 MHz hydride region ^{1}H nmr spectra of $H_2Fe[o-C_6H_4(P(C_2H_5)_2)_2]_2$ in toluene-d_8. The low field end of the spectrum is at the left side.

idealized geometries and the small nuclear displacements required for interconverting polytopal forms. The slow exchange spectrum for such seven-coordinate complexes has never been observed, e.g., IF_7 (Gillespie and Quail, 1964), IF_7, and ReF_7 (Muetterties and Packer, 1964) and $IrH_5[P(CH_3)_3]_2$ (E. L. Muetterties, unpublished data, 1973). Similarly, H_3ML_4 complexes should be susceptible to a rapid "tetrahedral jump" type of rearrangement.

Chelate structures may be the more fertile field of investigation in stereochemically nonrigid seven-coordinate structures, because chelate constraint may impose higher barriers to rearrangement than in unidentate complexes.

8. Dynamic Molecular Processes

In fact, all chelate structures examined to date show no nmr evidence of the necessary nonequivalence of ligand nuclei (facile bond-breaking processes also complicate the problem) with but two exceptions. The complex $HTa(CO)_2[(CH_3)_2PCH_2CH_2P(CH_3)_3]_2$ shows low-temperature-limit ^1H and ^{31}P nmr spectra consistent with the solid state geometry, which is a near capped octahedron with H as the capping ligand (Meakin et al., 1974). For $[(CH_3)_2NCS_2]_3MoNO$, there is in the proton nmr spectrum the CH_3 nonequivalence expected for the established solid state structure of near-pentagonal bipyramidal geometry **17** (Davis et al., 1971). In the latter case it is only at 60°–120° that the CH_3 groups become equivalent and then this most

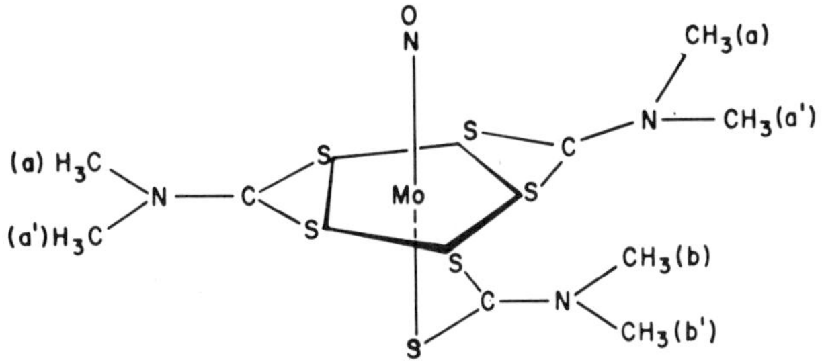

17

probably involves breaking of one Mo—S bond and rotation about the C—N axis. These are two interesting cases of relatively rigid seven-coordinate complexes.

6. Eight-Coordination

Eight-coordinate complexes are usually stereochemically nonrigid and most such complexes investigated to date, whether they have unidentate or polydentate ligands, do not show a limiting slow exchange spectrum. Thus, the rearrangement barriers for such species as dodecahedral $Mo(CN)_8^{4-}$ must be quite low, since the ^{13}C spectrum of $[(C_3H_7)_4N^+]_4[Mo(CN)_8]^{4-}$ is a single line down to $-130°C$, whereas the dodecahedral ground state demands two different cyanide environments (Muetterties, 1973c). Also, ^{13}C and ^1H nmr studies of $(R_2NCS_2)_4M$ and $(RR'NCS_2)_4M$ did not discern the limiting slow exchange spectra for these dodecahedral chelates at temperatures of $-130°$ to 160°C (Muetterties, 1973c). There is one recent exceptional class, the tetrahydrides of molybdenum and tungsten of the form H_4ML_4 with L a tricovalent phosphorus compound (Meakin et al., 1973b; Jesson et al., 1971; Bell et al., 1972).

An X-ray study of $H_4Mo[P(C_6H_5)_2CH_3]_4$ (Meakin et al., 1973b; Guggenberger, 1973) established a dodecahedral structure with the H atoms at A sites and P atoms at B sites (Fig. 23). Thus, the hydrogen set and the phosphorus set are environmentally equivalent but not magnetically equivalent. This is evident from the low temperature ^{1}H and ^{31}P spectra (Figs. 24 and 25); the former was shown to be invariant to magnetic field. The spectra are temperature dependent and in the fast exchange limit are binomial quintets which establish an intramolecular process that makes all hydride hydrogen atoms and all phosphorus atoms magnetically equivalent. Since the MP$_4$

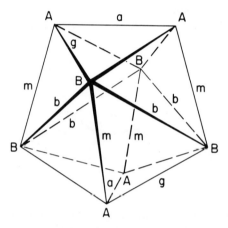

Fig. 23 The sites in the D_{2d} trigonal dodecahedron are labeled after Hoard and Silverton (1963), and in accord with their rules, the hydrogens occupy the A sites, and the phosphine ligands the B sites.

substructure is again tetrahedral (compressed from a regular tetrahedron), it has been proposed that the physical process comprises the tetrahedral jump mechanism with a concerted motion of three or four hydrogen atoms. For this molecule, the Hoard–Silverton (Hoard and Silverton, 1963) rearrangement mode, which permutes A and B sites, cannot alone effect magnetic equivalence. This eight-spin problem is very rich in mechanistic information. However, the practical problems to comprehensive permutational analysis and spectral simulations are nearly insurmountable. Even if an analysis of the limiting slow exchange spectra were achieved, it is doubtful that a complete density matrix calculation could be made for this eight-spin system [although approximate methods might be applied (Meakin and Jesson, 1973)]. The order of the permutational group for this mutual exchange problem is $4! \times 4! = 576$. Decomposition of the group into subclasses with respect to the subgroup representing the symmetry operations (Meakin et al., 1973b) for D_{2d} gives 20 permutational subclasses, including the identity set. Even

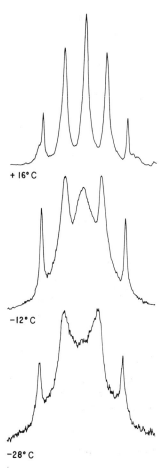

Fig. 24 Temperature dependence of the 90 MHz ^{1}H nmr spectrum for the hydride protons of $MoH_4[P(C_6H_5)_2CH_3]_4$ in CD_2Cl_2. The averaged P–H coupling constant is 32 Hz.

excluding those mechanistic sets that do not yield the proper fast exchange spectrum, the number of mechanistic possibilities is probably too large for a unique set to be selected, much less to consider the large number of linear combinations that would have to be explored. The problem is substantially simplified if the effective averaged symmetry is T_d in solution at the observed slow exchange limit, but even in this case the problem of complete density matrix calculations is difficult. Thus, we see the spin complexity is a necessary condition for mechanistic information to be obtained from nmr line shape changes, but practical considerations can set an upper limit to the size of the spin system which can be treated.

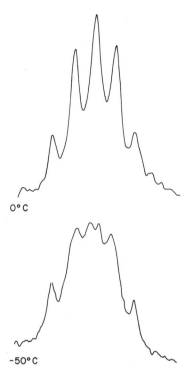

Fig. 25 Temperature dependence of the 36.4 MHz ^{31}P nmr spectrum of MoH$_4$-[P(C$_6$H$_5$)$_2$CH$_3$]$_4$ in dichloromethane. The spectrum is extensively time averaged and the ligand aromatic protons are selectively decoupled.

7. Nine- and Higher-Coordinate Structures

Rearrangement barriers drop rather rapidly for complexes of coordination number higher than eight,* because the extent of nuclear displacement required in polytopal rearrangement falls with increasing coordination number. In fact, all attempts to observe limiting slow exchange spectra in complexes of the type ReH$_9{}^2$, ReH$_8$PR$_3{}^-$, and ReH$_7$(PR$_3$)$_2$ have so far been unsuccessful (Muetterties, 1970a,b).

8. Boron Structures

Polytopal rearrangements occur in the polyhedral boranes (closoboranes) but the rates, at temperatures accessible with nmr spectrometers, are generally

---

* Excluding those somewhat artificial cases where ligands such as cyclopentadiene or benzene are considered to occupy five- or six-coordination sites, respectively.

8. Dynamic Molecular Processes

too low for mechanistic nmr studies of line shape changes (Lipscomb, 1966; Muetterties and Knoth, 1968; Muetterties, 1970a,b). An exception is the remarkable $B_8H_8^{2-}$ (Klanberg et al., 1967) which has been shown (Muetterties et al., 1973d) to exist in two polytopal forms, square antiprismatic and bicapped trigonal prismatic, in solution; both polytopal forms are fluxional. Other probable exceptions are $B_9C_2H_{11}$ and $B_{11}H_{11}^{-2}$, where the environmental nonequivalence of boron framework sites are *not* evident in the 25°C ^{11}B and ^1H spectra (Wiersema and Hawthorne, 1973; Tolpin and Lipscomb, 1973). On the other hand, there are a number of nidoboranes, boron frameworks that are fragments of the closoboranes, that show rapid intramolecular rearrangements. These are based on hydrogen exchange of two types: (1) exchange between a bridging and a terminal hydrogen analogous to carbonyl exchange in metal carbonyl clusters as discussed in Chapter 12, and (2) exchange of bridge hydrogen positions by a bridge hydrogen shift to an unoccupied B–B edge.

Covalent metal derivatives of the BH_4^- ion may have the bonding interactions shown in **18** through **20** of which the doubly and triply bridged structures have been unequivocally established by X-ray and neutron diffraction studies of metal borohydrides. The solution state structure has not

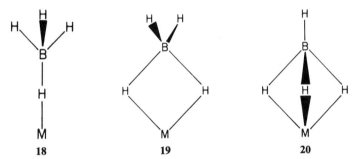

been established for any metalloborohydride, although strong claims have been advanced solely on the basis of infrared data. No limiting slow exchange nmr spectrum for a borohydride has been detected; in all instances spectroscopic equivalence of the BH protons is observed down to low temperatures. Facile exchange mechanisms are available in any proton bridge structure:

The top path is nonconcerted and the bottom is concerted (Muetterties, 1973d).

A classic case of hydrogen atom exchange in boron chemistry is the triborohydride ion which has structure **21** wherein there are two boron atoms

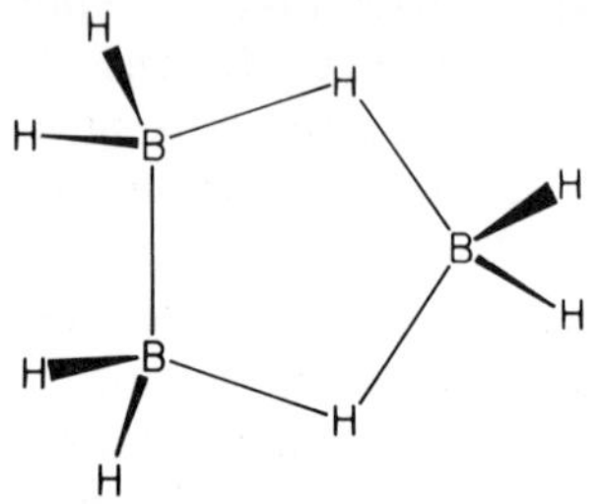

21

and three hydrogen environments (Phillips *et al.*, 1959; Lipscomb, 1963; Beall *et al.*, 1970; Bushweller *et al.*, 1971). The ^{11}B and ^{1}H nmr spectra (Fig. 26) of $B_3H_8^-$ are a binomial nonet and decet, respectively. These multiplets are unchanged over a wide temperature range, although there is broadening at low temperatures, ostensibly due to quadrupole relaxation effects; the two naturally occurring boron isotopes ^{10}B and ^{11}B have relatively large quadrupole moments. Bridge-terminal proton exchange requires only trivial B—B stretching and minor B—H bending:

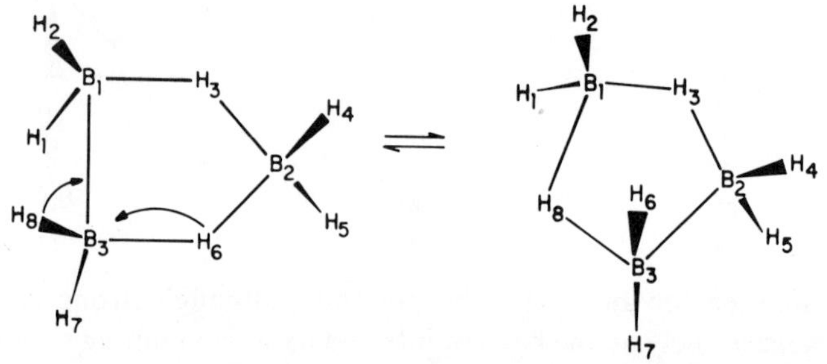

(3)

In $B_3H_8^-$ metalloborane complexes (Beall *et al.*, 1970; Bushweller *et al.*, 1971; Muetterties *et al.*, 1970; Muetterties and Alegranti, 1970), the hydrogen exchange is slowed and the expected nonequivalences of hydrogen and boron atoms are discernible. The four hydrogen atom nmr resonances expected for $(R_3P)_2CuB_3H_8$ complexes are observed at low temperatures ($-100°$C), but exchange broadening sets in slightly above this temperature (Beall *et al.*, 1970). In Fig. 27, the X-ray-established structure for $[(C_6H_5)_3P]_2CuB_3H_8$ is depicted. This stereochemistry requires nonequivalent phosphorus environments which have been detected by ^{31}P nmr. The ^{31}P AB resonances broaden at $-70°$C and merge into a single resonance (Muetter-

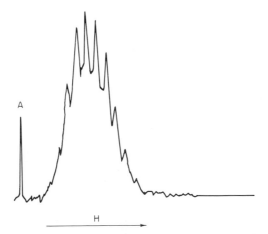

Fig. 26 Proton nmr spectrum of $Na^+B_3H_8^-$ in D_2O is a decet. Peak A is a water resonance.

ties *et al.*, 1970). This transition temperature is unaffected by added phosphine or $B_3H_8^-$ ion, indicating the exchange is an intramolecular process; line-shape changes due to added ligand are evident only at high concentrations of ligands and at temperatures of 20°C or above. Exchange activation parameters derived from the ^1H and ^{31}P spectra are comparable indicating a common process. A possible mechanism is an equilibrium between the ground state with a bidentate $B_3H_8^-$ ligand and an excited state with a unidentate $B_3H_8^-$ ligand in which fast hydrogen atom exchange similar to that in the free ion occurs. The line shape changes are in themselves not a source of mechanistic information. The solution state dynamics are substantially different for the pseudo octahedral $M(CO)_4B_3H_8^-$ (M = Cr, Mo, and W) complexes (structure shown in Fig. 28). Three ^{13}C carbonyl resonances are observed at 25°C as are two ^{11}B multiplets, which are the expected spectra

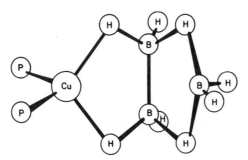

Fig. 27 Solid state structure of $[(C_6H_5)_3P]_2B_3H_8$. The substituent phenyl groups on the phosphorus atoms have not been included.

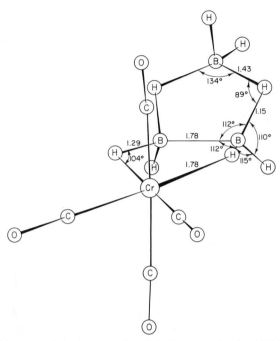

Fig. 28 Structure of the ion complex $Cr(CO)_4B_3H_8^-$ in the $(CH_3)_4N^+$ salt.

for the ground state structure (Fig. 28) (Klanberg *et al.*, 1968; P. Meakin and E. L. Muetterties, unpublished data, 1973). However, the proton spectra change in a complex and as yet uninterpretable fashion over a wide temperature range, $+60°$ to $-80°C$ (as do the fine structures for the two ^{11}B resonances) (P. Meakin and E. L. Muetterties, unpublished data, 1973). Full equivalence of all hydrogen atoms is not achieved at temperatures as high as $80°C$. Much better resolved ^1H and ^{11}B nmr data are required before the dynamic stereochemistry of this molecule is understood. Also complex and not understood are the nmr line-shape changes observed for $(CH_3)_2GaB_3H_7$ (Borlin and Gaines, 1972).

A number of boron hydride molecules and ions, $B_5H_8^-$, $CB_5H_7^-$, B_6H_{10}, and B_6H_9, show an nmr behavior that is consistent with a rapid shift of bridging hydrogen atom(s) to B–B edge positions not bridged in the ground-state structures without bridge-terminal hydrogen atom exchange as definitely established by the combination of nmr studies and chemical (deuterium) labeling of terminal sites (see review by Beall, 1973). Physically, the intramolecular process is visualized as a bridge hydrogen shift with minor, concomitant B–B bond stretching as shown below for $B_5H_8^-$ (Johnson *et al.*, 1970).

8. Dynamic Molecular Processes

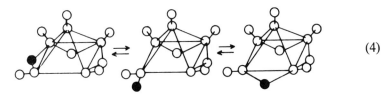

(4)

A common difficulty in dynamic nmr studies of boron hydrides is the small ^{11}B nmr chemical shift range and, critically, line broadening due to quadrupolar relaxation effects.

B. Intermolecular Exchange

The use of nmr to study intermolecular exchange has not yet reached the degree of sophistication characteristic of the latest intramolecular exchange investigations. This is partly due to the fact that the major effort has been concentrated on the intramolecular area and partly to the additional complexities inherent in intermolecular exchange systems. When a ligand, with a nucleus coupled to the rest of the molecule, dissociates, it will in general be replaced at the vacant coordination site by another ligand of the same type from the bulk of the solution. This means that spin correlation is usually lost for the nuclei involved. This, in turn, makes it more difficult to extract mechanistic information from the line shapes. Additionally the calculations are rendered more complex by the fact that the potential for symmetry factoring is greatly reduced.

Intermolecular exchange can be detected by the nmr technique in a number of ways. First, is the simplest and definitive experiment in which an ML_x complex with M—L spin–spin coupling dissociates (5).

$$ML_x \underset{k_{-1}}{\overset{k_1}{\rightleftharpoons}} ML_{x-1} + L \qquad (5)$$

The slow exchange limit is characterized by a multiplet, the transition region by a broadening of the multiplet, and the fast exchange limit by a single resonance. The ML_{x-1} and L dissociative species do not have to be present at detectable concentrations for a rigorous nmr characterization of this process. An alternative here to the ML_x case is an $ML_xL'_y$ in which there is spin–spin coupling between L and L' and loss of coupling is an unambiguous test of a bond-breaking process. If there is no such spin–spin coupling to probe the bond-breaking process, then a dissociative process (5) must yield substantial amounts of ML_{x-1} and L such that they can be detected in the slow exchange limit. Under these circumstances, an averaging process dependent on the parameters $K = k_1/k_{-1}$ will give, in the fast exchange limit, a single resonance that shifts with temperature in a manner that is determined by the variation of K. If the slow exchange region cannot be reached for these cases, evidence for the dissociation process must be derived from other kinds of measurements.

Exchange processes can also be detected by the nmr method utilizing special features of nuclear relaxation times. For example, the iodine nucleus has a large quadrupole moment and the nuclear resonance line for iodine is sharp enough for detection only in those instances where the field gradient is effectively zero at the nucleus (e.g., the iodide ion). This feature was employed to follow the very fast reaction between I^- and I_2.

$$I^- + I_2 \rightleftharpoons I_3^- \tag{6}$$

The degree of broadening is related to the average lifetime of an iodine nucleus as I_3^-. Analysis of line width variation, taking into account the relaxation processes, with I^- and I_2 concentrations yielded reaction rates as high as 10^8 sec^{-1}.

Electron exchange reactions can also be followed. For example, the rate of electron exchange between Co(I) and Co(II) was followed for $Co^{(I)}(CNR)_5^+$ complexes by the broadening of the Co(I) ligand proton resonance due to a relaxation mechanism associated with the Co(I)–Co(II) electron exchange. Rates were about 10^6 sec^{-1}. Similar studies have been carried out by ^{63}Cu and ^{65}Cu nmr in following Cu(I)–Cu(II) electron exchange.

Molecular exchange between diamagnetic and paramagnetic systems where the electron relaxation time is long give rise to broadened average resonance signals that result from a relaxation mechanism that depends on the hyperfine coupling interaction between the nuclear spin and the electron spin of the paramagnetic species. This hyperfine relaxation mechanism, which has the same origin as the isotropic nuclear resonance shift in paramagnetic systems, has been utilized in studying exchange of protons between the hydration sphere of a paramagnetic ion, e.g., Mn^{2+} and bulk water.

Similarly, the position of the averaged resonance in systems where rapid exchange is occurring between diamagnetic and paramagnetic species is becoming an increasingly important tool in studying dynamic molecular processes. A major area comprises the so-called "shift" reagents, and similar principles are involved in ion-pairing studies in which one member of the pair is diamagnetic and the other paramagnetic, as well as studies of second coordination sphere structure of paramagnetic species. For a general review of nmr exchange when paramagnetic species are involved (as in the previous three paragraphs) see LaMar et al. (1973).

In the following sections, specific applications of nmr to intermolecular exchange are presented to illustrate the variety of problems amenable to such studies. Our main interest will be in cases where quite detailed mechanistic information can be obtained for systems undergoing intermolecular exchange. The examples so far are few, but it is on these that we shall dwell at greatest length. The methods whereby kinetic and thermodynamic data are extracted in cases where no detailed mechanistic information is present in the line shapes are straightforward and will be treated briefly.

1. Calculational Aspects

Quite general programs using the density matrix approach have been written which will simulate mutual intramolecular exchange processes in non-first order systems. These have been used in a number of the detailed mechanistic studies described in the first part of this chapter. Programs for non-mutual intramolecular exchange in first order systems also exist. For intermolecular exchange it is much more difficult to write programs of any generality and particular systems or groups of clearly related systems are often best approached on an individual basis.

Computer programs that can simulate intermolecular exchange in first-order systems have been written for a few specific cases. The qualitative behavior can be intuitively predicted; sharp lines in the slow exchange limit broaden as the temperature is raised and finally sharpen in the high temperature limit into a line or group of lines reflecting the exchange between the various "sites."

More recently, programs have been written that can simulate intermolecular exchange in tightly coupled systems. The application of one of these programs to reactions of the type

$$AB \rightleftharpoons A + B \tag{7}$$

is shown in Fig. 29. As anticipated, in the low-temperature (slow exchange) limit an AB spectrum is predicted, whereas in the fast-exchange (high temperature) limit, two single resonances are calculated at the chemical shift positions of the A and B nuclei, respectively. However the dependence of the nmr line shapes at intermediate exchange rates on the ratio $J_{AB}/(\delta_A - \delta_B)$ shown in Fig. 29 is interesting because it is not intuitively obvious.

As more emphasis is placed on calculations in the intermolecular area, the programs will undoubtedly become more sophisticated, and the systems that can be examined will become more complex. In this section on intermolecular exchange, we now consider individual examples of particular interest, concluding with an analysis of an as yet hypothetical intermolecular exchange process in SF_4, which would be a problem of comparable complexity to the most complex case discussed under intramolecular exchange.

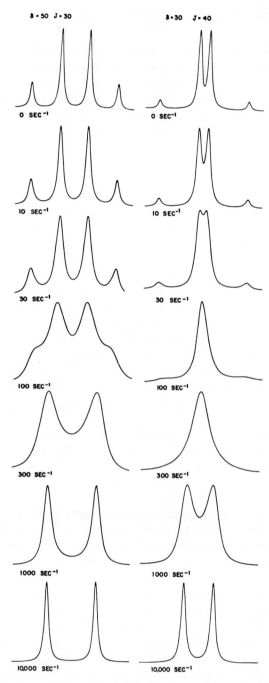

Fig. 29 Nuclear magnetic resonance line shapes simulated for the dissociation of an AB spin system AB ⇌ A + B. The degree of dissociation is assumed to be small.

2. Ligand Dissociation (Small Degree of Dissociation)

As indicated in the introduction, the loss of spin–spin coupling between a central metal atom and a ligand atom can be an extremely sensitive probe for ligand dissociations of the type

$$ML_x \underset{k_{-1}}{\overset{k_1}{\rightleftharpoons}} ML_{x-1} + L \qquad (8)$$

It should be emphasized at this point that we make the artificial distinction between ligand dissociation reactions and ligand association reactions simply on the basis of equilibrium constants. For the case of ligand dissociation, we include reactions where the equilibrium lies sharply to the left; in ligand association, we include reactions where the equilibrium lies far to the right. In the former case, in the absence of excess ligand, the species ML_x is usually the only one directly observable in solution by nmr; in the latter case, the species ML_{x-1} and L are usually the only ones directly observable, and the equilibrium is normally written in reverse form. Of course, if we start with only $(x - 1)$ ligands per metal in the associative case, there will be no free ligand observable. There is naturally a continuum between the two extremes and in cases where $k_{-1} \approx k_1$, it may be possible to detect all three species in the slow-exchange limit.

As a straightforward illustration of dissociations of the type shown in equilibrium (8) we may use the example (Meakin and Jesson, 1973, 1974)

$$Rh[P(OCH_3)_3]_5^+ \rightleftharpoons Rh[P(OCH_3)_3]_4^+ + P(OCH_3)_3 \qquad (9)$$

The $^{31}P\{^1H\}$ spectrum of this system in acetonitrile as a function of temperature is shown in Fig. 30. At 33°C, the spectrum is a simple doublet due to Rh–P coupling—the five phosphorus ligands in the trigonal bipyramidal ML_5 complex are rendered equivalent by rapid intramolecular exchange (see above). On raising the temperature the spectrum broadens and eventually sharpens to a single line at $\sim 82°C$, the Rh–P coupling having been lost. Since the chemical shift in the high-temperature limit is the average for all three species in (9) weighted by their concentrations, and since there is little variation of chemical shift with temperature, it is clear that the degree of dissociation is small.

In the above example we can extract quantitative data for the lifetime of a phosphorus ligand on the central metal as a function of temperature using one of the conventional line-shape programs, but the only mechanistic information we obtain is the fact that a ligand does dissociate.

We now present a more complex example of a ligand dissociation study (Meakin et al., 1972a,b; Tolman and Jesson, 1973; Tolman et al., 1974) that illustrates how, in a particular case, more detailed mechanistic information

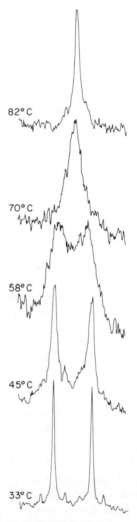

Fig. 30 Dissociative exchange in $Rh[P(OCH_3)_3]_5^+$ is established from the dnmr proton spectra, which shows loss of Rh–P coupling at the high temperature range. $^{31}P\{^1H\}$ spectrum (36.43 MHz) of $Rh[P(OCH_3)_3]_5^+\ B(C_6H_5)_4^-$ in methyl cyanide.

can be obtained and how resonance studies for one type of nucleus (^{31}P in this case) may yield this information, whereas studies involving other magnetic nuclei in the molecule (^1H in this case) may not. Hydrogen will react essentially quantitatively at 1 atm with $RhCl[P(C_6H_5)_3]_3$ to give a six coordinate dihydride according to the equation

$$RhCl[P(C_6H_5)_3]_3 + H_2 \rightleftharpoons H_2RhCl[P(C_6H_5)_3]_3 \qquad (10)$$

8. Dynamic Molecular Processes

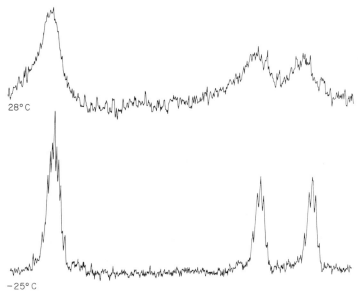

Fig. 31 Proton nmr spectra (90 MHz ^{1}H) of $H_2RhCl[P(C_6H_5)_3]_3$ at two different temperatures.

The ^{1}H hydride nmr of $H_2RhCl[P(C_6H_5)_3]_3$ at two temperatures is shown in Fig. 31. At $-25°C$, the spectrum is sharp, showing two nonequivalent hydrogen atoms (intensity ratio 1:1), only one of which has a large P–H coupling. The large coupling is characteristic of phosphorus trans to hydride hydrogen and the stereochemistry of the complex is unambiguously established as **22** with two magnetically equivalent phosphorus nuclei (P_2) and

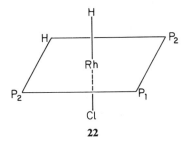

22

one unique phosphorus nucleus (P_1). On raising the temperature to 28°C, the hydride spectrum broadens reversibly, indicating a chemical exchange process but providing no further mechanistic information.

Figure 32 shows the $^{31}P\{^1H\}$ spectrum of the same complex at a similar set of temperatures. The low-temperature limit spectrum is a doublet of doublets for the phosphorus nuclei P_2 (split by Rh and by P_1) and a doublet

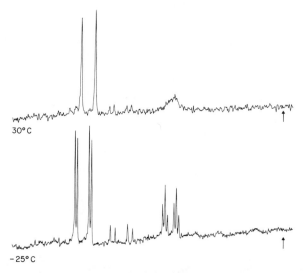

Fig. 32 $^{31}P\{^1H\}$ nmr spectra of $H_2RhCl[P(C_6H_5)_2]_3$ at two temperatures.

of triplets for the nucleus P_1 (split by Rh and $2P_1$) providing further confirmation of the stereochemistry.

At the higher temperature, the $^{31}P\{^1H\}$ spectrum yields quite detailed mechanistic information. The P_1–Rh and P_1–P_2 couplings are lost, while the P_2–Rh coupling is retained. This shows that only the P_1 ligand is dissociating at an appreciable rate at 30°C, providing a site for olefin coordination in hydrogenation catalysis. The nmr line shapes can be fitted in a straightforward manner for the situation described above and for the case where excess ligand is added. Simulations of this exchange, shown in Fig. 33, yield a forward rate constant for dissociation of ~ 400 sec^{-1} at 30°C. The simulations can also be used to confirm the deduction from the observed spectra that the degree of dissociation of the complex is small (Tolman *et al.*, 1974).

3. *Ligand Association (Small Degree of Association)*

Systems considered in Section A are, in general, stable five- and six-coordinate species with eighteen valence electrons (i.e., they are coordinatively saturated). The interpretation of the spectra is straightforward in that small amounts of free ligand impurity do not materially alter the results, and unambiguous simple mechanistic information is obtained. On the other hand, systems considered in this section are, in general, stable four- and five-coordinate species with sixteen valence electrons (i.e., they are coordinatively unsaturated), and small amounts of adventitious ligand together with

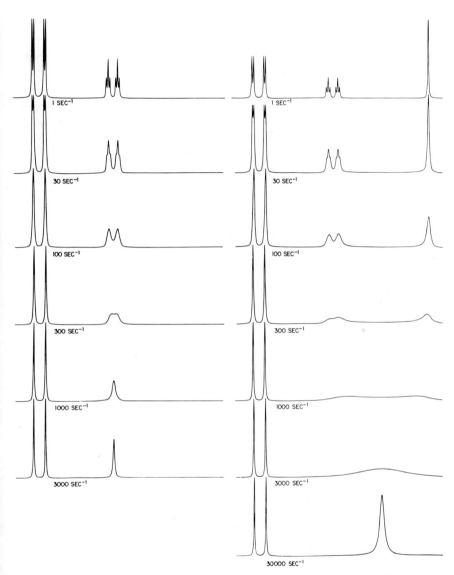

Fig. 33 Simulations of $^{31}P\{^1H\}$ dnmr spectra for $H_2RhCl[P(C_6H_5)_3]_3$.

subtleties of interpretation can qualitatively alter the conclusions as noted below.

Associative interactions are particularly important in the chemistry of square planar complexes of Pt(II), Pd(II), Rh(I), and Ir(I). Deeming and

Shaw (1969) have studied ligand exchange behavior in *trans*-RhCl(CO)-[P(CH$_3$)$_2$(C$_6$H$_5$)]$_2$ in benzene solution using proton nmr. The methyl resonance before addition of excess ligand is two triplets (triplet splitting arises from virtual coupling of the methyl protons to the two strongly coupled phosphorus nuclei ~ 6.4 Hz; doublet splitting arises from coupling to the rhodium ~ 1 Hz). Addition of 0.01 mole of free ligand per rhodium atom to a solution of the complex does not materially alter the pattern (slow exchange at this concentration). With 0.02–0.03 mole of added phosphine, the spectrum collapses to a singlet; on further addition, the spectrum broadens and starts to form a doublet such that with 0.08 mole of added phosphine, a well-defined 1:1 doublet is obtained. As more P(CH$_3$)$_2$(C$_6$H$_5$) is added, the doublet separation decreases, and at about 4–7 moles, a singlet is obtained. The resonance remains a singlet up to about 12 moles, when a narrow doublet is formed. Further addition causes the doublet separation to increase and approach the observed value for free P(CH$_3$)$_2$(C$_6$H$_5$) (~ 3.7 Hz). In this system we are looking at the equilibrium

$$\text{RhCl(CO)[P(CH}_3)_2(\text{C}_6\text{H}_5)]_2 + \text{P(CH}_3)_2(\text{C}_6\text{H}_5) \rightleftharpoons \text{RhCl(CO)[P(CH}_3)_2(\text{C}_6\text{H}_5)]_3 \tag{11}$$

with the position of the equilibrium lying far to the left.

Spectral behavior of this type was first interpreted by Fackler *et al.* (1969) and applied specifically to the above system by Fackler (1970). These workers noted that, in studies of methyldiphenylphosphine adducts of planar four coordinate group VIII transition metals, the doublet expected for methyl protons coupled to phosphorus was at times collapsed. In some cases the effect could be attributed to paramagnetism, while in others it was possible to show that the chemical shift and $J_{\text{P-CH}_3}$ were concentration-dependent and that the formation of a singlet was due to averaging of the coupling constants of the coordinated phosphine with the free phosphine. In the case of some cis palladium complexes, neither of these explanations appeared appropriate. Instead, it appeared that chemical exchange of the phosphine ligands with the diamagnetic complexes was producing the decoupling.

To explain the results, Fackler *et al.* (1969) used the density matrix approach to calculate nmr line shapes for the X part of the spectrum for systems of the type XAA' where X = H and A, A' = P (the approximation of ignoring X' is equivalent to ignoring X, X' coupling in the X$_n$AA'X$_n$' system and the approximation of setting $n = 1$ decreases the number of possible transitions). The effect of this latter approximation on the spectrum is particularly significant under conditions of no chemical exchange when $J_{\text{AA}'} \sim (J_{\text{XA}} - J_{\text{XA}'})$. To examine the generality of the conclusion for the XAA' fragment, the X$_2$AA system was also treated. No qualitative differences were noted; however, the full number of X nuclei must be included if the line shapes are to be used to extract rate information.

The theoretical results apply equally well to the equilibria

$$XAA' \rightleftharpoons XA + A'$$

$$XAA' + A^{*'} \rightleftharpoons XAA^{*'} + A$$

where $J_{AA'}$ may be large and virtual coupling effects can be observed. Spectra of this type may be simulated using a density matrix equation of motion very similar to that used in mutual exchange.*

$$\frac{d\rho}{dt} = 2\pi i[\rho, \mathcal{H}] + \left(\frac{d\rho}{dt}\right)_{\text{relax}} + \sum_i (P_i \rho P_i^\dagger - \rho)/\tau_i \tag{12}$$

where ρ is the spin density matrix and $\mathcal{H}$ is the high-resolution nmr Hamiltonian.

In this case, the P_i are spin-flip matrices rather than the spin exchange matrices used in mutual exchange calculations. Because of the similarity of equation to the equations used for simulating the nmr spectra for systems undergoing mutual exchange, existing programs for mutual exchange can readily be modified for this type of intermolecular exchange. This has been done and the calculations of Fackler et al. (1969) have been reproduced; furthermore, these same programs can be applied to a variety of other systems.

Figure 34 shows the calculated X spectrum for an XAA' system undergoing intermolecular exchange of the type $XAA' \rightleftharpoons A'$ or $XAA' + A^* \rightleftharpoons XAA^{*'} + A$. The nmr parameters are $J_{XA} = 66.6$ Hz, $J_{XA'} = 20$ Hz, and $J_{AA'} = 3330$ Hz, or in the notation of Fackler et al. (1969) $R = 50$, $S = 3.33$. The calculated spectra in Fig. 34 should be compared with those shown in Fig. 2 of this reference (note that in this reference the figure captions for Figs. 2 and 3 should be interchanged).

In the slow exchange limit, the spectrum is a triplet due to the large AA' coupling constant and in the fast exchange limit the spectrum is a doublet with separation J_{XA}. The A' spin has become "decoupled" by the exchange process. The narrow single-line spectrum at intermediate exchange rates first calculated by Fackler et al. (1969) is not easily understood in any intuitive fashion.

With the above theoretical basis, the results of Deeming and Shaw (1969) are interpretable in terms of the following stages:

I. Slow exchange doublet of triplets due to virtual coupling.

II. Intermediate exchange (small amount of added ligand) collapses the spectrum to a singlet as predicted by density matrix treatment.

* We are grateful to Dr. P. Meakin for allowing us to reproduce the following results prior to publication. This treatment is not rigorous and will be discussed in detail elsewhere.

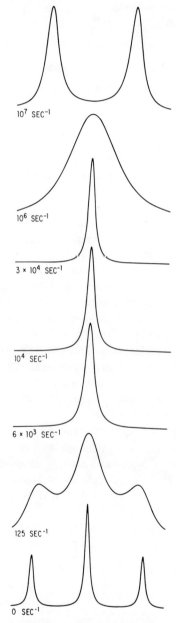

Fig. 34 Calculated X spectrum for an XAA′ spin system undergoing exchange of the type XAA′ ⇌ A′ or XAA′ + A* ⇌ XAA* + A′.

III. At faster exchange (high concentrations of added ligand), the spectrum broadens and then sharpens to a doublet as predicted by density matrix treatment.

IV. At still higher concentrations of added ligand, the doublet splitting gradually decreases to zero (averaging of H–P couplings of opposite sign in free and complexed ligand).

V. At still larger concentrations of excess ligand, the doublet reappears (reversal of sign of average P–H coupling).

VI. Finally, with very large excess of free ligand, the coupling constant and chemical shift approach those of free ligand as expected.

It is clear, as mentioned in the introduction, that great care must be exercised in interpreting ligand association spectra. One particularly easy trap to fall into when dealing with "pure" coordinatively unsaturated species is to assume that collapse of couplings is indicative of a dissociative process, since it may only take catalytic amounts of adventitious free ligand impurity to cause collapse by the associative route.

Finally, in this section we may consider the catalysis of cis ↔ trans isomerization in planar d^{10} systems by small amounts of added ligand. Haake and Pfeiffer (1970) found that when cis-L_2PtCl_2 (L = a phosphine) complexes are mixed in cyclohexane with a catalytic amount of a second phosphine L', that well-behaved first-order isomerization rates are observed both for L = L' and L ≠ L', and each system has a unique second-order rate constant over a range of L'. The data may be interpreted by assuming a five-coordinate intermediate with ligand association the rate determining step, there being rapid intramolecular rearrangement in the intermediate. The additional assumption has to be made in certain cases that when L ≠ L', L' is uniquely involved in the association and dissociation and L is not. Further study of this type of system is clearly warranted; detailed line-shape studies would be useful, and it is possible that the data are amenable to different interpretations.

4. General Intermolecular Exchange

By contrast to the cases considered in Sections A and B, intermolecular exchange systems can be extremely complex, involving many species and also several competing pathways. Few of these have been analyzed in any detail, and the relatively sophisticated types of calculation described above have not been applied. The approach, so far, has been to use simple site-exchange computer programs to investigate the kinetics and thermodynamics of the systems at equilibrium, to use diastereotopic or optically active groups as additional mechanistic probes, to vary concentrations of reactants and to study series in which there is a systematic variation of the structure of the reactants. These methods have been applied with success in some cases but do not fall within the intended scope of this chapter. The application of this type of approach and the possible complexities to be encountered in the general case of intermolecular exchange are elucidated in Chapter 11.

5. Sulfur Tetrafluoride

In Chapter 2, Klemperer has considered the permutational aspects of the dynamic processes which SF_4 might be expected to undergo. As a concluding section, we present a brief discussion of SF_4, since it represents one of the most complex cases of intermolecular exchange that could, potentially, be treated by a rigorous mathematical approach with the extraction of mechanistic information in the sense of the major bulk of the material in the present chapter.

The SF_4 molecule has C_{2v} symmetry and gives an A_2B_2 ^{19}F spectrum in the low-temperature limit. References of historical interest can be found in Chapter 2. In spite of the considerable number of spectroscopic studies of this molecule, it remains unclear as to whether the observed dynamic processes proceed by intramolecular exchange, by pure intermolecular exchange or by impurity catalyzed exchange. The potential mechanistic information available as determined by the methods of Klemperer (Chapter 2) may be summarized as follows

1. If the exchange involves catalysis by fluoride ion impurities:

  a. *Direct association to give D_{3h} intermediate.*—There are eleven differentiable permutation reactions. Three of these correspond to isomerization and eight to intermolecular exchange. Of the eight intermolecular processes two pairs must occur with equal probability since the reactant and product species have identical free energies. Thus only nine of the eleven processes imply distinct site exchange schemes.

  b. *Direct association to give a C_{4v} intermediate (attacking ion approaches "trans" to one of the ligands in the pair at approximately 180°).*—Here there are six different classes of associative mechanism that might be differentiated by the nmr approach.

  c. *Dissociation to give a cationic C_{3v} intermediate.*—There are four differentiable reactions, only two of which occur with significant probability since the remaining two involve the unlikely process of recoordination of the same fluoride ion which dissociated.

2. Bimolecular processes

A variety of bimolecular processes have been suggested for the exchange. The analysis in Chapter 2 presents six nmr differentiable reactions, three of which may be ruled out as being the sole processes responsible for fluorine exchange, since they do not permute axial and equatorial sites.

The ^{19}F spectrum of SF_4 does show small second order effects at conventional frequencies (60–100 MHz), but to provide a situation in which there is the highest probability of extracting mechanistic information, it is desirable to increase these higher-order effects until the individual transitions are as clearly resolved as possible. This can be done by running the spectra at much lower fields than normal. The best possibility of extracting mechanistic information is to examine the spectrum as a function of temperature at an rf frequency between about 6 and 20 MHz. Finally, it will be necessary to use SF_4 samples of higher purity in order to investigate both the pure material and the effect of adding catalytic amounts of fluoride ion.

IV. Conclusion

We have discussed a broad range of dynamic molecular processes in inorganic and organometallic compounds, excluding those systems considered in detail in other chapters. Our prime objective has been to emphasize the power of new theoretical and calculational approaches for extracting mechanistic information from nmr line shapes. These new approaches have so far been applied with maximum effect only to intramolecular processes, as illustrated by the discussion of H_2ML_4 and ML_5 species. The intermolecular area has not as yet been touched by these methods for reasons outlined in the text. We have closed the section on intermolecular exchange with the case of SF_4, which represents a challenge for the future. This challenge has already been taken up by a number of workers in the area and the results of their studies may well represent the next significant step in extending the scope of detailed mechanistic investigations.

We have deliberately excluded from the chapter any detailed discussion of cases where diamagnetic species are exchanging with paramagnetic species (ion-pairing, shift reagents, coordination number studies, outer sphere coordination studies, effects of electron relaxation processes, etc.). These, for the most part, are simple from the experimental nmr point of view (fast exchange averaged lines without fine structure), but the theory of the shift effects observed can be quite complex (see, for example, Jesson, 1973).

REFERENCES

Baehler, R. D., Andose, J. D., Stackhouse, J., and Mislow, K. (1972). *J. Amer. Chem. Soc.* **94**, 8060.
Beall, H. (1973). *Chem. Rev.* (in press).

Beall, H., Bushweller, C. H., Dewkett, H. J., and Grace, M. (1970). *J. Amer. Chem. Soc.* **92**, 3484.
Bell, B., Chatt, J., Leigh, G. J., and Ito, T. (1972). *Chem. Commun.* p. 34.
Berry, R. S. (1960). *J. Chem. Phys.* **32**, 933.
Borlin, J., and Gaines, D. F. (1972). *J. Amer. Chem. Soc.* **94**, 1367.
Bramley, R., Figgis, B. N., and Nyholm, R. S. (1962). *Trans. Faraday Soc.* **58**, 1893.
Brownstein, S. (1967). *Can. J. Chem.* **45**, 1711.
Bushweller, C. H., Beall, H., Grace, M., Dewkett, H. J., and Bilofsky, H. S. (1971). *J. Amer. Chem. Soc.* **93**, 2145.
Cotton, F. A. (1968). *Accounts Chem. Res.* **1**, 257.
Cotton, F. A., Danti, A., Waugh, J. S., and Fessenden, R. W. (1958). *J. Chem. Phys.* **29**, 1427.
Cowley, A. W., and Schweiger, J. R. (1972). *Chem. Commun.* p. 560.
Davis, R., Hill, M. N. S., Holloway, C. E., Johnson, B. F. G., and Al-Obaidi, K. H. (1971). *J. Chem. Soc., A* p. 994.
Deeming, A. J., and Shaw, B. L. (1969). *J. Chem. Soc., A* p. 597.
Eaton, D. R., Josey, A. D., Phillips, W. D., and Benson, R. E. (1962). *Discuss. Faraday Soc.* **34**, 77.
Eaton, S. S., Hutchison, J. R., Holm, R. H., and Muetterties, E. L. (1972). *J. Amer. Chem. Soc.* **94**, 6411.
Eaton, S. S., Eaton, D. R., Holm, R. H., and Muetterties, E. L. (1973). *J. Amer. Chem. Soc.* **95**, 1116.
Edgar, B. L., Duffy, D. J., Palazzotto, M. C., and Pignolet, L. H. (1973). *J. Amer. Chem. Soc.* **95**, 1125.
Egan, W., Tang, R., Zon, G., and Mislow, K. (1971). *J. Amer. Chem. Soc.* **93**, 6205.
Fackler, J. P., Jr. (1970). *Inorg. Chem.* **9**, 2625.
Fackler, J. P., Jr., Fetchin, J. A., Mayhew, J., Seidel, W. C., Swift, T. J., and Weeks, M. (1969). *J. Amer. Chem. Soc.* **91**, 1941.
Gillespie, R. J., and Quail, J. W. (1964). *Can. J. Chem.* **42**, 2671.
Gorenstein, D. (1970). *J. Amer. Chem. Soc.* **92**, 644.
Guggenberger, L. J. (1973).
Guggenberger, L. J., Titus, D. D., Flood, M. T., Marsh, R. E., Orio, A. A., and Gray, H. B. (1972). *J. Amer. Chem. Soc.* **94**, 1135.
Haake, P., and Pfeiffer, R. M. (1970). *J. Amer. Chem. Soc.* **92**, 4996.
Herzberg, G. (1951). "Molecular Spectra and Molecular Structure," Vol. II, p. 221. Van Nostrand-Reinhold, Princeton, New Jersey.
Hoard, J. L., and Silverton, J. V. (1963). *Inorg. Chem.* **2**, 235.
Jesson, J. P. (1973). In "Chemical Applications of Nuclear Magnetic Resonance in Paramagnetic Molecules" (G. N. LaMar, W. D. Horrocks, and R. H. Holm, eds.), Chapter 1, p.1. Academic Press, New York.
Jesson, J. P., and Meakin, P. (1973a). *Accounts Chem. Res.* **6**, 269.
Jesson, J. P., and Meakin, P. (1973b). *J. Amer. Chem. Soc.* **95**, 1344.
Jesson, J. P., and Meakin, P. (1973c). *Inorg. Nucl. Chem. Lett.* **9**, 1221.
Jesson, J. P., and Meakin, P. (1974) *J. Amer. Chem. Soc.* **96**, 5760.
Jesson, J. P., Muetterties, E. L., and Meakin, P. (1971). *J. Amer. Chem. Soc.* **93**, 5261.
Johnson, H. D., II, Geanangel, R. A., and Shore, S. G. (1970). *Inorg. Chem.* **9**, 908.
Klanberg, F., and Muetterties, E. L. (1968). *Inorg. Chem.* **7**, 155.
Klanberg, F., Eaton, D. R., Guggenberger, L. J., and Muetterties, E. L. (1967). *Inorg. Chem.* **6**, 1271.
Klanberg, F., Muetterties, E. L., and Guggenberger, L. J. (1968). *Inorg. Chem.* **7**, 2272.

LaMar, G. N., and Sherman, E. O. (1970), *J. Amer. Chem. Soc.* **92**, 2691.
LaMar, G. N., Horrocks, W. D., and Holm, R. H., eds. (1973). "Chemical Applications of Nuclear Magnetic Resonance in Paramagnetic Molecules." Academic Press, New York.
Lambert, J. B., and Oliver, W. L., Jr. (1969). *J. Amer. Chem. Soc.* **91**, 7774.
Lippard, S. J., and Melmed, K. M. (1969). *Inorg. Chem.* **8**, 2755.
Lipscomb, W. N. (1963). "Boron Hydrides." Benjamin, New York.
Lipscomb, W. N. (1966). *Science* **153**, 373.
Mahler, W., and Muetterties, E. L. (1965). *Inorg. Chem.* **4**, 1520.
Meakin, P., and Jesson, J. P. (1973). *J. Amer. Chem. Soc.* **95**, 7272.
Meakin, P., and Jesson, J. P. (1974). *J. Amer. Chem. Soc.* **96**, 5751.
Meakin, P., Guggenberger, L. J., Jesson, J. P., Gerlach, D. H., Tebbe, F. N., Peet. W. G., and Muetterties, E. L. (1970). *J. Amer. Chem. Soc.* **92**, 3482.
Meakin, P., Muetterties, E. L., Tebbe, F. N., and Jesson, J. P. (1971). *J. Amer. Chem. Soc.* **93**, 4701.
Meakin, P., Jesson, J. P., and Tolman, C.A. (1972a). *J. Amer. Chem. Soc.* **94**, 3240.
Meakin, P., Muetterties, E. L., and Jesson, J. P. (1972b). *J. Amer. Chem. Soc.* **94**, 5271
Meakin, P., Muetterties, E. L., and Jesson, J. P. (1973a). *J. Amer. Chem. Soc.* **95**, 75.
Meakin, P., Guggenberger, L. J., Peet, W. G., Muetterties, E. L., and Jesson, J. P. (1973b). *J. Amer. Chem. Soc.* **95**, 1467.
Meakin, P., Guggenberger, L. J., Tebbe, F. N., and Jesson, J. P. (1974). *Inorg. Chem.* **13**, 1025.
Mislow, K. (1970). *Accounts Chem. Res.* **3**, 321.
Muetterties, E. L. (1965). *Inorg. Chem.* **4**, 769.
Muetterties, E. L. (1970a). *Accounts Chem. Res.* **3**, 266.
Muetterties, E. L. (1970b). *Rec. Chem. Progr.* **31**, 51.
Muetterties E. L. (1972). *MTP Int. Rev. Sci.* Series 1, **9**, 39.
Muetterties, E. L. (1973a). *Chem. Commun.* p. 221.
Muetterties, E. L. (1973c). *Inorg. Chem.* **12**, 1963.
Muetterties, E. L., ed. (1975). "Boron Hydrides." Academic Press, New York.
Muetterties, E. L., and Alegranti, C. W. (1970). *J. Amer. Chem. Soc.* **92**, 4114.
Muetterties, E. L and Guggenberger, L. J. (1974) *J. Amer. Chem. Soc.* **96** 1748.
Muetterties, E. L., and Hirsekorn, F. J. (1973). *J. Amer Chem. Soc.* **95**, 5419.
Muetterties, E. L., and Knoth, W. H. (1968). "Polyhedral Boranes." Dekker, New York.
Muetterties, E. L., and Packer, K. J. (1964). *J. Amer. Chem. Soc.* **86**, 293.
Muetterties, E. L., and Phillips, W. D. (1962). *Advan. Inorg. Chem. Radiochem.* **4**, 231.
Muetterties, E. L., Mahler, W., and Schmutzler, R. (1963). *Inorg. Chem.* **2**, 613.
Muetterties, E. L., Mahler, W., Packer, K. J., and Schmutzler, R. (1964). *Inorg. Chem.* **3**, 1298.
Muetterties, E. L., Peet, W. G., Wegner, P. A., and Alegranti, C. W. (1970). *Inorg. Chem.* **9**, 2447.
Muetterties, E. L., Meakin, P., and Hoffmann, R. (1972). *J. Amer. Chem. Soc.* **94**, 5674.
Muetterties, E. L., Wiersema, R. V. and Hawthorne, M. F. (1974). *J. Amer. Chem. Soc.* **95**, 7520.
Phillips, W. D., Miller, H. C., and Muetterties, E. L. (1959). *J. Amer. Chem. Soc.* **81**, 4496.
Pignolet, L. H., Horrocks, W. D., Jr., and Holm, R. H. (1970). *J. Amer. Chem. Soc.* **92**, 1855.
Pople, J. A., Schneider, W. G., and Bernstein, H. J. (1959). "High-Resolution Nuclear Magnetic Resonance." McGraw-Hill, New York.

Rauk, A., Allen, L. C., and Mislow, K. (1970). *Angew. Chem., Int. Ed. Engl.* **9**, 400.
Redington, R. L., and Berney, C. V. (1965). *J. Chem. Phys.* **43**, 2020.
Tebbe, F. N., Meakin, P., Jesson, J. P., and Muetterties, E. L. (1970). *J. Amer. Chem. Soc.* **92**, 1068.
Tolman, C. A., and Jesson, J. P. (1973). *Science* **181**, 501.
Tolman, C. A., Meakin, P., Linder, D. L., and Jesson, J. P. (1974). *J. Amer. Chem. Soc.* **96**, 2762.
Tolpin, E. I., and Lipscomb, W. N. (1973). *J. Amer. Chem. Soc.* **95**, 2384.
Ugi, I., Marquarding, D., Klusacek, H., and Gillespie, P. (1971). *Accounts Chem. Res.* **4**, 288.
Westheimer, F. H. (1968). *Accounts Chem. Res.* **1**, 70.
Whitesides, G. M., and Bunting, W. M. (1967). *J. Amer. Chem. Soc.* **89**, 6801.
Whitesides, G. M., and Mitchell, H. L. (1969). *J. Amer. Chem. Soc.* **91**, 5384.
Wiersema, R. J., and Hawthorne, M. F. (1973). *Inorg. Chem.* **12**, 785.

9

Stereochemically Nonrigid Metal Chelate Complexes

R. H. HOLM

List of Abbreviations	317
I. Introduction	318
II. Four-Coordinate Systems	320
A. Bisbidentate Complexes	320
B. Monobidentate and Unidentate Phosphine Complexes	328
III. Six-Coordinate Systems	330
A. Monobidentate Complexes	331
B. Bisbidentate Complexes	333
C. Trisbidentate Complexes	338
D. Sexadentate Complexes	368
IV. Systems of Higher Coordination Number	369
V. Summary	372
References	374

List of Abbreviations

acac	Acetylacetonate	L–L'	Unsymmetrical bidentate ligand
bipy	2,2'-Bipyridyl		
bmp	6,6'-Dimethyl-2,2'-biphenylyl	LTP	Low-temperature process
		mhd	5-Methylhexane-2,4-dionate
bzac	Benzoylacetonate	mnt	Maleonitriledithiolene
C	Cis	phen	1,10-Phenanthroline
Cp	Pentahaptocyclopentadienyl	pmhd	1-Phenyl-5-methylhexane-2,4-dionate
dbm	Dibenzoylmethanide		
dibm	Diisobutyrylmethanide	pvac	Pivaloylacetonate
dpm	Dipivaloylmethanide	py$_3$tame	1,1,1-Tris(pyridine-2-carboxaldiminomethyl)ethane
hfac	Hexafluoroacetylacetonate		
HTP	High-temperature process		
L–L	Symmetrical bidentate ligand	R$_1$R$_2$-dtc	N,N-Disubstituted dithiocarbamate

R-SPhHR$_\alpha$	Substituted -aminothionate	tfac	Trifluoroacetylacetonate
α-RT	α-Substituted tropolonate	tfd	1,2-Bis(trifluoromethyl)-
SP	Square pyramid(al)		dithiolene
T	Trans; tropolonate	TP	Trigonal prism(atic)
TAP	Trigonal antiprism(atic)	X-R-sal	Substituted salicylaldiminate
TBP	Trigonal bipyramid(al)		

I. Introduction

Prior to the development of magnetic resonance techniques for the investigation of chemical kinetics, the rates of reaction of coordination compounds were determined by measurements on nonequilibrium systems. Consequently, the large majority of kinetic data were obtained on systems containing complexes of conventionally nonlabile metal ions such as Cr(III), Co(III), Rh(III), and Pt(II,IV). Of the reactions studied before the early 1960s, only a small fraction were intramolecular processes, which in most instances involved racemization of bis- or trischelate complexes (Basolo and Pearson, 1958, 1967). These processes were generally followed by spectrophotometric or spectropolarimetric techniques, sometimes accompanied by measurements of ligand exchange in order to establish the intramolecular nature of the reaction. Such techniques are inherently incapable of detecting intramolecular site interchanges, the knowledge of which, as will be demonstrated, is essential to deduction of reaction mechanism or to a significant limitation of the number of reaction pathways. At present only dynamic nuclear magnetic resonance (dnmr) spectroscopy provides a means of sensing molecular environmental interchanges indispensable to establishment of intramolecular rearrangement mechanisms that fall in the approximate rate domain of 10^{-2}–10^6 sec^{-1}.

This chapter* consists of a selective review and appraisal of the kinetics and proposed mechanisms of rearrangement reactions of metal complexes, some or all of whose ligating atoms are incorporated in chelate ring structures. Attention is restricted to those cases investigated by dnmr in which the stereochemical nonrigidity is known or may reasonably be assumed to occur in the course of strictly intramolecular processes. All such processes to be considered here may be described as metal-centered rearrangements in which the ligating framework is interchanged or permuted between or among magnetically inequivalent sites. On a microscopic physical basis, these rearrangements may be conceived to involve stretching or breaking of a metal–ligand (M—L) bond and/or deformation of L—M—L angles. Excluded from consideration are all intermolecular processes, such as ligand exchange (see, e.g,

* The material covered in this chapter is based on results published or available to the author through April, 1973.

Pinnavaia and Fay, 1966; Pinnavaia and Nweke, 1969) and dynamic conformational changes of chelate rings that occur without substantial motion of the ligating atoms (Hawkins, 1971; Holm and Hawkins, 1973).

For purposes of this chapter a stereochemically nonrigid complex is defined as one which manifests exchange line-broadening or achieves the fast exchange limit within the temperature interval of about $-100°$ to $200°C$ usually accessible in nmr measurements of adequately soluble and thermally stable compounds. Such complexes are also designated as "fast" to emphasize further their intramolecular lability, which under ordinary conditions prevents separation of isomeric forms. Their rearrangement kinetics are determined by measurement of rates of interconversion at equilibrium. Stereochemically rigid or "slow" complexes are those that do not or have not been shown to exhibit dnmr behavior in the above temperature interval. Consequently, their rearrangement kinetics are obtained by measurement of rates of approach to equilibrium. Geometrical isomers of slow complexes are often separable and optical isomers at least partially resolvable. Evidence is rapidly accumulating that classification of complexes into slow and fast categories on the basis of metal ion is not likely to be generally applicable. This situation is due in part to the availability of rate data only at ambient temperature for certain complexes that must currently be considered slow and, more frequently, to highly significant effects on rearrangement rates caused by variation in ligand structure. For examples, the xanthate complex $Co(S_2CO-l\text{-menthyl})_3$ undergoes slow mutarotation at $25°C$ ($k \sim 10^{-7}$ sec^{-1}) (Krebs and Schumacher, 1966), whereas the similar dithiocarbamate complex $Co[(PhCH_2)_2\text{-dtc}]_3$ experiences fast inversion of configuration above about $170°C$ ($k \sim 18$ sec^{-1}) (Pignolet et al., 1973). The half-life for inversion of $Co(acac)_3$ at $25°C$ is 14 years ($k \sim 10^{-9}$ sec^{-1}) (Fay et al., 1970), in stark contrast to the remarkably facile inversion of the tropolonate complex trans-$Co(\alpha\text{-}C_3H_5T)_3$ at this temperature ($k \sim 10^2$ sec^{-1}) (Eaton et al., 1972). Somewhat similar effects are encountered in the series $Ru(acac)_3$, $Ru(\alpha\text{-RT})_3$, and $Ru(Me, PhCH_2\text{-dtc})_3$. The first two complexes exhibit the usual slow behavior of Ru(III) species, even at elevated temperatures (Fay et al., 1970; Eaton et al., 1973), while the fast exchange limit for inversion of $Ru(Me, PhCH_2\text{-dtc})_3$ is reached at only $43°C$ (Pignolet et al., 1973). From these and other results considered subsequently, it is clear that the usual measures of metal ion lability such as, e.g., aquo ion–bulk water exchange rates (Basolo and Pearson, 1967, pp. 151-158; Cotton and Wilkinson, 1972), are not necessarily reliable guides to slow or fast intramolecular lability.

Any process that exchanges a given nucleus between two (or more) magnetically inequivalent sites is in principle detectable by dnmr, provided the exchange frequency is compatible with the time scale of the method and the slow exchange chemical shift difference between the sites ($\Delta \nu$) is resolvable. Successful analysis of dnmr line-broadening depends on the condition that $\tau \sim \Delta\nu^{-1}$, where τ is the mean lifetime of the site and $\Delta\nu$ is in hertz. This is frequently the case for some of the most fundamental intramolecular processes of many diamagnetic and paramagnetic metal chelates. The following processes have been detected and investigated by dnmr methods: (a) inversion of chiral molecular configurations; (b) geometrical isomerization; (c) interconversion of structures with different coordination polyhedra. In addition, there exist cases in which site interchange may occur without alteration of geometrical or absolute configurations. In the sections which follow, stereochemically nonrigid complexes are discussed in terms of their coordination number, and for each coordination number, in terms of the number and functionality of their chelate ligands. Five-coordinate complexes containing chelate rings are treated in Chapter 8 and will not be dealt with here. Throughout, L–L symbolizes a symmetric bidentate ligand whose chelate ring possesses a perpendicular plane of symmetry; an unsymmetrical bidentate ligand is represented by L–L'. Emphasis is placed on those systems whose dnmr properties have yielded mechanistic information. Consequently, trischelate complexes are examined in greatest detail. In a perceptive paper, Fay and Piper (1964) reported the first attempt to deduce intramolecular rearrangement pathways of trischelates from variable-temperature nmr data. This work is now well recognized as the starting point for subsequent increasingly detailed and sophisticated experimental approaches directed toward elucidation of rearrangement mechanisms for this group of complexes, and, indeed, for metal chelate complexes in general.

The dynamic stereochemistry of chelate complexes has been reviewed in some detail elsewhere (Basolo and Pearson, 1967, pp. 300–334; Holm, 1969; Muetterties, 1970, 1972; Fortman and Sievers, 1971; Holm and O'Connor, 1971; Holm and Hawkins, 1973; Pignolet and LaMar, 1973; Serpone and Bickley, 1973). The treatments by Holm and O'Connor (1971) and Serpone and Bickley (1973) for coordination numbers four and six, respectively, are quite comprehensive and should be consulted for additional information that cannot be included here.

II. Four-Coordinate Systems

A. Bisbidentate Complexes

Most divalent ions of the first transition series, Cd(II), and Hg(II) form discrete complexes of the types $M(L-L)_2$ and $M(L-L')_2$. These complexes approach or possess the limiting planar and tetrahedral stereochemistries,

9. Nonrigid Chelate Complexes

although in a few cases, primarily Cu(II) complexes, structures intermediate between these extremes are stable (Holm and O'Connor, 1971). Bisbidentate complexes exhibit two types of nonrigidity, which are considered next.

1. Planar–Tetrahedral Interconversion

A substantial number of complexes are known for which the free energy differences between planar and tetrahedral forms are sufficiently small that both are detectably populated at or near room temperature. Nearly all such cases involve Ni(II) complexes, for which the structural equilibrium (1) is well established in noncoordinating solvents. This equilibrium has been most

$$\text{Planar } (S = 0) \rightleftharpoons \text{tetrahedral } (S = 1) \qquad (1)$$

extensively investigated in the aminotroponeimine (**1**, R = Me, aryl), salicylaldimine (**2**, R = *sec*-alkyl), and β-ketoamine (**3**, R = Me, *n*-alkyl, aryl) groups of complexes but exists for a number of other bischelate Ni(II) complexes as well (Holm and O'Connor, 1971). The relative energies of the two forms can be adjusted by variation of the nitrogen substituent. Thus, the complexes **1** (R = H), **2** (R = H, Me, *n*-alkyl, Ph), and **3** (R = H) exist only in the planar form in solution at room temperature, whereas **1** (R = *n*-alkyl), **2** (R = *t*-Bu), and **3** (R = *sec*-alkyl) are essentially completely tetrahedral under these conditions. Without exception the pmr spectra of these complexes (obtained at 40–100 MHz) display one set of signals whose chemical shifts are the weighted averages of those of the diamagnetic planar and paramagnetic tetrahedral forms. The signals are usually substantially displaced from normal diamagnetic resonance positions due to the paramagnetism of the tetrahedral isomer, which causes isotropic contributions to the chemical shifts [Eq. (2)] (Holm, 1969; Jesson, 1973). On the assumption that the isotropic shifts are predominantly or completely contact in origin, their

$$(\Delta\nu_i^{iso}/\nu_0) = (\Delta\nu_i^{obs}/\nu_0) - (\Delta\nu_i^{dia}/\nu_0) \qquad (2)$$

temperature dependencies in many cases have been fit to Eq. (3) (Horrocks, 1965; Jesson, 1973) in which the bracketed term is the mole fraction of

$$\frac{\Delta\nu_i^{iso}}{\nu_0} = -A_i\left(\frac{\gamma_e}{\gamma_H}\right)\frac{g\beta S(S+1)}{6SkT}[\exp(\Delta G^0/RT) + 1]^{-1} \qquad (3)$$

tetrahedral form and the remaining symbols have their usual meanings. [For a discussion of the validity of Eq. (3) when applied to Ni(II) systems with equilibrium (1), see Holm and Hawkins (1973).] Averaged signals persist at temperatures as low as about −70° to −90°C, and in no case have separate signals for the two forms been frozen out nor has convincing evidence been presented that a condition other than the fast exchange limit been attained.

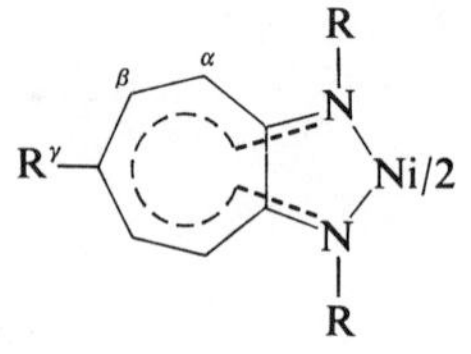

Recent ^{13}C nmr investigations of several aminotroponeimine complexes involved in the structural equilibrium also reveal averaged signals at ambient temperature (Doddrell and Roberts, 1970). While the isotropic ^{13}C shifts at up to ten to fifteen times larger than ^{1}H shifts in parts per million, they are less than four times larger in hertz due to the lower resonance frequency (15 MHz) employed in the ^{13}C measurements. Consequently, ^{13}C nmr does not afford an important extension of the range of detectable lifetimes of the two forms. Because the pmr chemical shift differences between these forms may in some cases be about 10,000 Hz or greater and both forms are detectable in the electronic spectra of equilibrium mixtures, it has been estimated that the upper and lower lifetimes of each isomer are about 10^{-4} to 10^{-6} and 10^{-13} seconds, respectively (LaMar, 1965; Eaton, 1968).

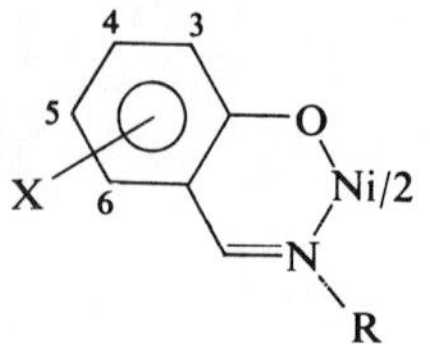

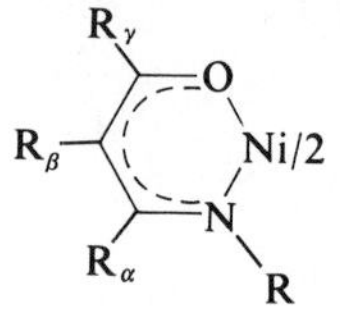

The origin of the extremely low barrier for the planar-tetrahedral interconversion of Ni(II) complexes is not understood, nor is the mechanism of this interconversion known. The rapidity of the structural change implies a nonbond rupture pathway in which case the relatively small nuclear displacements required may be visualized in terms of the digonal twist process (Eq. 4). Here, motion of the chelate rings affords a dihedral angle between them requisite to stabilization of the spin-triplet ground state of the tetrahedral isomer. From solid state structural data this angle would appear to be

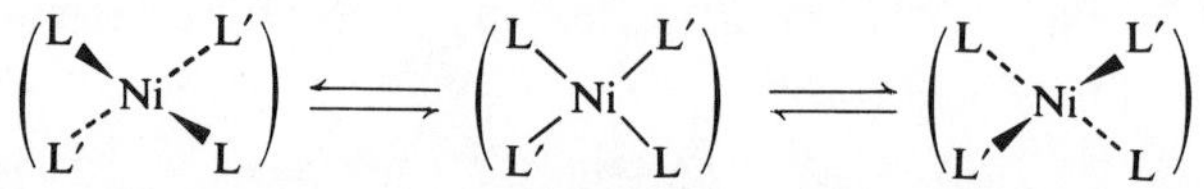

about 80° or more for salicylaldimine complexes (Holm and O'Connor, 1971). In the event of a strictly planar configuration **4**, this process should lead to facile racemization ($\Delta \rightleftharpoons \Lambda$) of any tetrahedral Ni(L–L')$_2$ species involved in equilibrium (1). This appears to be the case for a number of diastereoisomeric complexes of types **2** and **3** prepared from optically active amines. Such complexes contain chiral (+ or −) R substituents, leading to the following diastereomers. The isotropically shifted pmr spectra of species

$$\Delta(+,+) \equiv \Lambda(-,-) \quad \biggr\} \text{ active}$$
$$\Lambda(+,+) \equiv \Delta(-,-)$$
$$\Delta(+,-) \equiv \Lambda(+,-) \quad \text{meso}$$

obtained from racemic amines exhibit only *two* sets of signals, consistent with equilibria (5) and (6) (Holm *et al.*, 1964; Ernst *et al.*, 1967; see also Ernst

Planar[(+, +) or (−, −)] ⇌ tetrahedral[(+, +) or (−, −)]   (5)

Planar(+, −) ⇌ tetrahedral(+, −)   (6)

et al., 1968; Gerlach and Holm, 1969a). Only one set of signals is found for complexes prepared from optically pure amines. Barring failure to observe chemical shift differences between species variant only in overall molecular chirality, these results indicate that structural interconversion is accompanied by racemization of the tetrahedral forms. Supporting evidence for this interpretation is afforded by a pmr study of the quadridentate Ni(II) salicylaldimine complexes 7 (O'Connor *et al.*, 1968). These complexes are also nonrigid and exist in about 6–15% tetrahedral form in chloroform solution at 30°C. The spectrum of Ni(3-*s*-Bu-5-Me-sal)$_2$bmp, prepared from racemic ligand components, is given in Fig. 1. Three 5-Me and azomethine signals are clearly resolved and have been proven to arise from the meso and two active diastereomers. In this case, the complex is resistant to racemization of the absolute configuration because of the high activation energy (45 kcal/mole) of the parent diamine 2,2′-diamino-6,6′-dimethylbiphenyl. Consequently, structural interconversion (8 ⇌ 9) occurs without racemization of the tetrahedral form, at least up to 90°C. Inversion reactions of tetrahedral complexes not involved in equilibrium (1) are described in the following section.

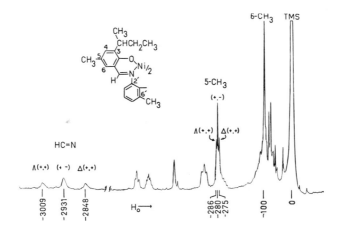

Fig. 1 Pmr spectrum (60 MHz) of a mixture of diastereomers of Ni(3-*s*Bu-5-Me-sal)$_2$-bmp (7) in CDCl$_3$ solution at ∼27°C (O'Connor *et al.*, 1968). Chemical shifts are given in hertz relative to tetramethylsilane.

7   **8 Δ**   **9 planar**

The picture that has emerged for the Ni(II) chelate complexes described above is one of extraordinarily rapid interconversion between planar and tetrahedral isomers ($k \geqslant 10^5$–10^6 sec^{-1}) via a low barrier pathway that is as yet unestablished. Rate measurements on these systems presumably can only be accomplished by recourse to faster kinetic techniques such as ultrasonic absorption or dielectric relaxation. In addition to Ni(II) systems, this type of structural interconversion has been established only for several bis(β-ketoaminato) (**3**, R = H) and bis(β-mercaptoketonato)Co(II) complexes (Everett and Holm, 1966, 1968; Gerlach and Holm, 1969b). Investigations of their interconversion rates have not been carried out. In assessing the generality of the occurrence of planar and tetrahedral isomers (and in the present context the possibility of seeking examples of nonrigid chelates amenable to kinetic study), the following conclusions have been reached by Gerlach and Holm (1970):

> There is no evidence at the present time that there exist for a given complex well defined energy minima corresponding to planar and tetrahedral isomers unless the isomers have different spin multiplicities. In those cases where the intrinsic ligand field strengths are insufficient to cause a spin change in passing from one limiting structure to another, the system may respond by populating one of these structures only or by progressively, rather than discontinuously, distorting toward the other structure to an extent expected to be mainly dependent upon steric factors and temperature.

2. Inversion of Tetrahedral Complexes

Stereochemical nonrigidity of bischelate complexes that exist only in the tetrahedral form necessarily results from inversion of configuration ($\Delta \rightleftharpoons \Lambda$). This process is restricted to M(L–L')$_2$ complexes (C$_2$ symmetry); M(L–L)$_2$ species are achiral. [Other chiral four-coordinate chelates are of course possible but only those of the M(L–L')$_2$ type have been investigated.] Nearly all inversion reactions that have been studied are of slow complexes and rate data, which are not extensive and seldom quantitative, are summarized elsewhere (Basolo and Pearson, 1958, p. 284; Eaton and Holm, 1971a).

9. Nonrigid Chelate Complexes

The only investigation of stereochemically nonrigid, exclusively tetrahedral species is that by Eaton and Holm (1971a), who analyzed the proton dnmr spectra of bis(β-aminothionato) Zn(II) and Cd(II) complexes **10**, M(R-SPhHR$_\alpha$)$_2$, by a total line shape procedure. Inasmuch as the spectra of the two enantiomers are identical in achiral media, complexes with the substituent R = isopropyl were employed. The methyl groups (1 and 2) of each substituent are diastereotopic in these chiral molecules, and interconversion between different magnetic environments, arbitrarily designated as r and s, is illustrated by the process **11** $\rightleftharpoons$ **12**. Rates of environmental exchange and inversion of configuration are identical. Ligand exchange between Zn(II)

complexes is slow compared to inversion below about 130°C, and it was found that r–s chemical shift differences are larger in mixed than in pure ligand complexes at slow exchange. Methyl pmr spectra of the equilibrium system (7) are shown in Fig. 2. The diastereotopic chemical shift differences M_1–M_2 and P_1–P_2 are averaged to zero in the fast exchange limit at which two spin doublets are observed. At higher temperatures the spectra

$$\text{Zn(isopropyl-SPhHH)}_2 + \text{Zn(PhCH}_2\text{-SPhHH)}_2 \rightleftharpoons \\ 2\text{Zn(isopropyl-SPhHH)(PhCH}_2\text{-SPhHH)} \quad (7)$$

show evidence of intermolecular ligand exchange. For related bis(β-aminothionato) Cd(II) complexes diastereotopic averaging and ligand exchange were observed at substantially lower temperatures. Rates and activation parameters are collected in Table I.* Possible rearrangement pathways are depicted in Fig. 3. Slow ligand exchange between complex and free ligand or between two different complexes establishes an intramolecular pathway for Zn(II) species, which may involve a twist (a) or bond-rupture (b) mechanism. Both are consistent with the very small activation entropies. The twist mechanism has been favored primarily because diastereotopic averaging in the β-ketoaminato complex Zn(isopropyl-PhHMe)$_2$ occurs at much higher

---

* Here and elsewhere, the full body of available kinetic data is not necessarily included. Values of the preexponential factor A and $\Delta G^\ddagger$ are not quoted. An extensive tabulation of kinetic parameters for six-coordinate complexes is given by Serpone and Bickley (1973).

TABLE I

KINETIC PARAMETERS FOR DIASTEREOTOPIC AVERAGING OF ISOPROPYL METHYL GROUPS AND LIGAND EXCHANGE OF BIS(β-AMINOTHIONATO)-Zn(II) AND Cd(II) COMPLEXES[a]

Complex	Solvent	$\Delta H^{\ddagger}$ (kcal/mole)	$\Delta S^{\ddagger}$ (eu)	E_a (kcal/mole)	$k^{298°C}$ (sec^{-1})
Zn(isopropyl-SPhHMe)(PhCH$_2$-SPhHH)	PhCl	20.6 (1.3)	0.2 (4.0)	21.4 (1.4)	0.007
Zn(isopropyl-SPhHH)(PhCH$_2$-SPhHH)[b]	PhCl	20.2 (1.5)	0.1 (4.0)	21.0 (1.5)	0.0065
Cd(isopropyl-SPhHMe)$_2$	CDCl$_3$–PhCl	12.3 (1.5)	−7.5 (4.0)	12.8 (1.5)	130
Cd(isopropyl-SPhHH)$_2$	CDCl$_3$	11.0 (1.5)	−10.1 (4.0)	11.5 (1.0)	290
Cd(isopropyl-SPhHMe)$_2$–Cd(isopropyl-SPhHMe)(PhCH$_2$-SPhHH)[c]	PhCl	13.7 (1.5)	−9.5 (4.5)	14.3 (1.6)	4.0

[a] Eaton and Holm (1971a); estimated errors given in parentheses.
[b] Kinetic data are also applicable to Zn(isopropyl-SPhHH)$_2$ in PhCl.
[c] Data refer to ligand exchange.

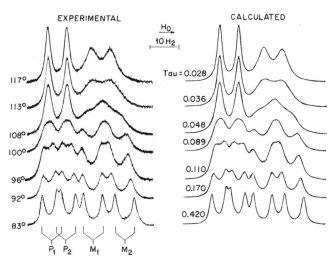

Fig. 2 Experimental and calculated isopropyl methyl spectra (100 MHz) of Zn(II) complexes 10 in the intermediate exchange region for a chlorobenzene solution of Zn(isopropyl-SPhHH)$_2$ and Zn(PhCH$_2$-SPhHH)$_2$: P$_1$, P$_2$, Zn(isopropyl-SPhHH)$_2$; M$_1$, M$_2$, Zn(isopropyl-SPhHH)(PhCH$_2$-SPhHH) (Eaton and Holm, 1971a); methyl spin doublets are indicated; τ values are in seconds.

temperatures (fast exchange limit about 170°C). The slower rate of diastereotopic averaging in the Zn–N$_2$O$_2$ compared to the Zn–N$_2$S$_2$ case is consistent with a planar transition state that is stabilized to a greater extent by sulfur than by oxygen donor atoms. Preferential stabilization of this sort has been demonstrated for bis(β-aminothionato) and bis(thiosalicylaldiminato)Ni(II) complexes compared to their oxygen-donor analogs 3 and 2, respectively (Gerlach and Holm, 1969a; Bertini et al., 1972). For Cd(II) complexes, kinetic parameters for diastereotopic averaging and Cd(isopropyl-SPhHMe)$_2$–Cd(PhCH$_2$-SPhHH)$_2$ ligand exchange are closely comparable (Table I), and the ligand exchange process has been found to have an overall second-order dependence on complex concentration. These results suggest that inversion occurs via bond rupture [pathway (b), Fig. 3] accompanied by ligand exchange [pathway (c)] at a somewhat slower rate. The data provide no information on whether a Cd—S (as shown) or a Cd—N bond is broken. Attempts to relate more directly the mechanism(s) of the inversion and ligand-exchange processes were thwarted by the low solubilities of the two Cd(isopropyl-SPhHR$_\alpha$)$_2$ complexes investigated.

The increased rates of rearrangement with increasing ionic radius [Shannon-Prewitt radii are quoted throughout (see Jesson and Muetterties, 1969)] (Zn^{2+}, 0.64 Å; Cd^{2+}, 0.84 Å) is frequently encountered in d^0 and d^{10} chelate complexes and appears to be a general property. Additional examples

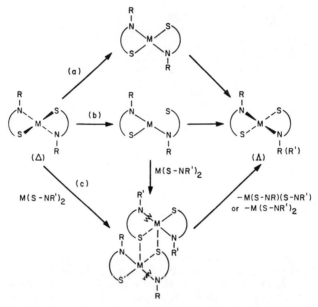

Fig. 3 Possible inversion pathways for bis(β-aminothionato)M(II) complexes (**10**): (a) digonal twist; (b) bond rupture (arbitrarily shown as M–S rupture); (c) ligand exchange via a bimolecular transition state (Eaton and Holm, 1971a).

of this behavior will be discussed in Section III. The failure to observe separate signals for active diastereomers of the Ni(II) complexes **10** (Gerlach and Holm, 1969a) implies rapid inversion of the tetrahedral form via a planar intermediate (see above). If correct, the apparent order of inversion is Ni(II) > Cd(II) > Zn(II), at least for β-aminothionato complexes where a full comparison can be made. Insufficient data for inversion of slow complexes are available to allow formulation of a metal ion rate order for inversion. The most optically stable tetrahedral bischelate thus far described appears to be bis(d-benzoylcamphorato)Be(II), which has been separated into its diastereomers and requires a catalyst for mutarotation in pure and dry solvents at ambient temperature (Lowry and Traill, 1931).

B. Monobidentate and Unidentate Phosphine Complexes

Other than the complexes described in the preceding section and in greater detail elsewhere (Holm and O'Connor, 1971), the only chelate species known to exhibit equilibrium (1) are the monobidentate Ni(II) complexes **13** (X = Br$^-$, I$^-$) (Van Hecke and Horrocks, 1966). Averaged pmr signals were observed down to $-30°$C, the lowest temperature of observation. The re-

9. Nonrigid Chelate Complexes

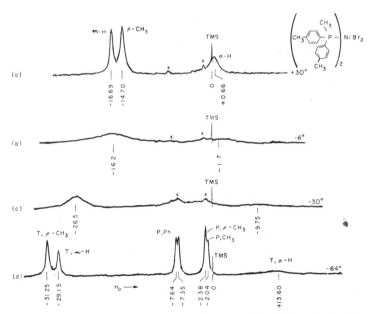

lated complexes derived from 1,2-bis(diphenylphosphino)ethane displayed no isotropic shifts indicative of the presence of a paramagnetic form. Evidently, the longer carbon chain in **13** is required to decrease the energy difference between the planar and tetrahedral isomers.

Although not chelate species the dihalobis(phosphino)Ni(II) complexes **14** are relevant because they are the only examples in which the planar–tetrahedral interconversion has been frozen out. All of these complexes display averaged signals at or near room temperature, but at lower temperatures undergo spectral changes consistent with an intramolecular two-site exchange process (Pignolet and Horrocks, 1968; Pignolet et al., 1970; LaMar and Sherman, 1969, 1970; Que and Pignolet, 1973). Variable temperature pmr spectra are typified by those of $Ni[(p\text{-tol})_2MeP]_2Br_2$ shown in Fig. 4. In the $-64°C$ spectrum the p-tolylmethyl signals of the two isomers

Fig. 4 Pmr spectra (100 MHz) of $Ni[(p\text{-tol})_2MeP]_2Br_2$ in $CDCl_3$ solution illustrating completely (a) averaged, (b,c) intermediate, and (d) frozen out cases (Pignolet et al., 1970). X indicates an impurity, P = planar; T = tetrahedral.

TABLE II
Kinetic Parameters for the Tetrahedral → Planar Conversion of $Ni(R_1R_2R_3P)_2X_2$ Complexes

Phosphine	X	$\Delta H^{\ddagger}$ (kcal/mole)	$\Delta S^{\ddagger}$ (eu)	$k^{298°C}$ ($\times 10^{-5}$ sec)	Reference[c]
Ph_2MeP	Cl^a	~9	~−4	2.6	1
	Br^a	9.3	−6.1	0.45	1
	Br^b	11 ± 4	6 ± 10	8.5	2
	I^a	8.5	−0.6	29	1
Ph_2EtP	Cl^a	9.0	−1.0	9.2	1
	Br^a	9.0	−4.7	1.6	1
$Ph_2(n\text{-}Pr)P$	Cl^a	9.4	−0.6	6.6	1
	Br^a	9.8	−2.4	1.3	1
$Ph_2(n\text{-}Bu)P$	Cl^a	9.1	−1.3	7.5	1
	Br^a	9.2	−4.3	1.4	1
$(p\text{-}C_6H_4Cl)_2MeP$	Br^b	11 ± 4	6 ± 10	8.5	2
$(p\text{-}C_6H_4OMe)_2MeP$	Cl^b	10 ± 4	5 ± 10	1.6	2
	Br^b	13 ± 4	9 ± 10	1.4	2
$(C_3H_5)_3P$	Cl^a	9.2	0	12	3
$Ph(C_3H_5)_2P$	Br^a	10.7	1	1.5	3
$Ph_2(C_3H_5)P$	Br^a	9.5	−3	1.5	3

[a] CD_2Cl_2 solution.
[b] $CDCl_3$ solution.
[c] References: (1) LaMar and Sherman (1970); (2) Pignolet et al. (1970); (3) Que and Pignolet (1973).

are particularly clearly evident. The fast exchange spectrum at 30°C contains signals whose shifts are intermediate between those of pairs of resonances at lower temperatures. Analysis of temperature dependencies of linewidths has yielded rate constants and activation parameters for the structural change. The rate constants are approximately 10^2–10^3 sec^{-1} at −50°C and 10^5–10^6 sec^{-1} at 25°C in chloroform or dichloromethane solution. The available kinetic data, most of which are collected in Table II, indicate the rate order Br < Cl < I for constant alkyl substituent R and the order R = Me < n-Pr < n-Bu < Et at parity of halide. There is at present no satisfactory interpretation of these rate orders nor of the apparently higher barrier to structural change for **14** compared to **13** and bischelate Ni(II) complexes.

III. Six-Coordinate Systems

Nonrigid chelate complexes with coordination number six are at present mainly limited to species containing bidentate ligands, the number of which

9. Nonrigid Chelate Complexes

provides a convenient classification of complexes of this type. As will be seen firm mechanistic information is available only for certain trisbidentate chelates $M(L-L')_3$.

A. Monobidentate Complexes

The only complexes of this type whose nonrigid behavior has been demonstrated are the Ti(IV) species $RTiCl_3(L-L)$ and $RTiCl_3(L-L')$ (R = Me, OMe) (Clark and McAlees, 1970, 1972a,b). The bidentate ligands are of the type XCH_2CH_2Y with X, Y = OMe, SMe, NMe_2. Complexes with unsymmetrical ligands can exist as three geometrical isomers: two meridional forms (mer, **15, 16**) and one facial form (fac, **17**). The dnmr behavior of these complexes, which has been examined in the range $-90°C$ to $27°C$, appears to

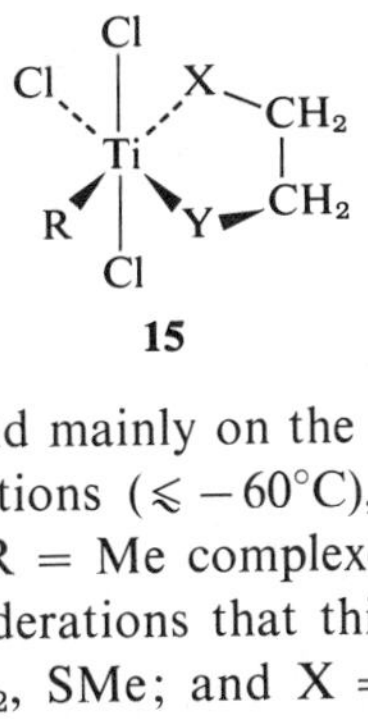

<pre>
         Cl   Cl   R
    Cl. | .X   Cl. | .Y   Cl. | .X
       `.|.´ `CH₂   `.|.´ `CH₂   `.|.´ `CH₂
         Ti | Ti | Ti |
       .´|`. .CH₂               .´|`. .CH₂   .´|`. .CH₂
    R´ | `Y                R´   |  `X   Cl´ | `Y
         Cl   Cl   Cl
         15   16   17
</pre>

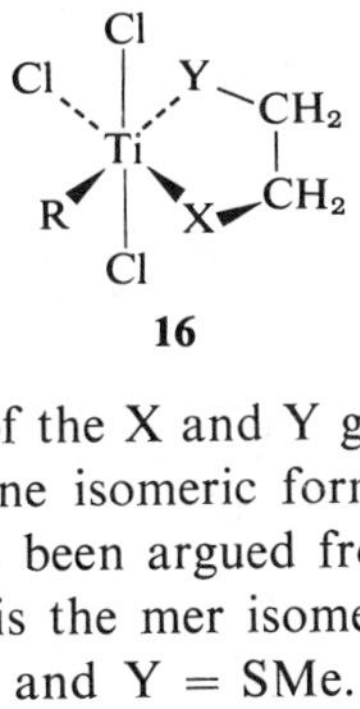

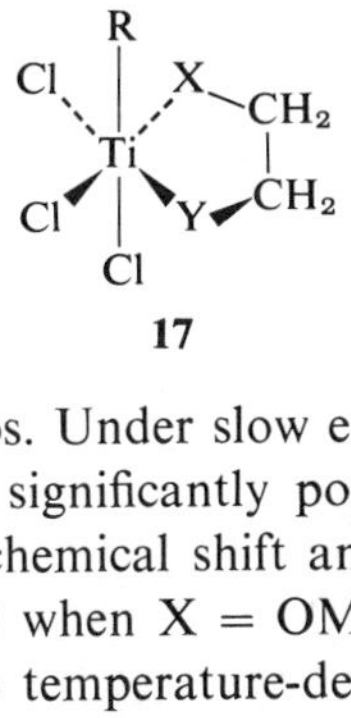

depend mainly on the nature of the X and Y groups. Under slow exchange conditions ($\leq -60°C$), only one isomeric form is significantly populated. For R = Me complexes it has been argued from chemical shift and other considerations that this form is the mer isomer **15** when X = OMe, Y = NMe_2, SMe; and X = NMe_2 and Y = SMe. The temperature-dependent pmr spectra for $MeTiCl_3(MeOCH_2CH_2NMe_2)$ are given in Fig. 5 as an example of spectral changes encountered in this group of complexes. At slow and fast exchange, one methyl signal of each type is observed. However, their chemical shifts are somewhat different, and at intermediate temperatures all signals undergo broadening. The single NMe_2 resonance at $-90°C$ is suggestive of the presence of one of the two mer isomers, inasmuch as the two methyl groups are inequivalent in the fac form **17**. The nonrigid behavior of this and related methyltitanium trichloride complexes has been interpreted in terms of the dynamic equilibrium mer-(**15**) $\rightleftharpoons$ mer-(**16**), which is shifted toward **16** as the temperature is increased. The process observed is one of methyl site exchange and has been proposed to occur by the twisting process depicted below, which takes place about an axis bisecting two opposite faces of the pseudooctahedron (Clark and McAlees, 1972a). For similar methoxytitanium trichloride complexes a mer-(**15**) $\rightleftharpoons$ fac-(**17**) interconversion by a twist about a different axis has been proposed as the source of nonrigidity (Clark and McAlees, 1972b). While apparently consistent with the dnmr results, these mechanisms are in no sense required by them and

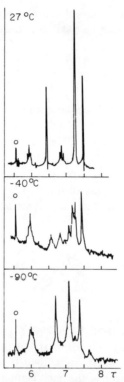

Fig. 5 Pmr spectra (100 MHz) of MeTiCl$_3$(MeOCH$_2$CH$_2$NMe$_2$) illustrating slow (−90°C) and fast exchange (27°C) conditions (Clark and McAlees, 1972a). The circle denotes a ^{13}C—H side band arising from the dichloromethane solvent. The signals at slow exchange are Ti-Me (7.38τ), N-Me$_2$ (7.09τ), O-Me (6.71τ), and CH$_2$CH$_2$ (6τ).

must be regarded as tentative proposals. A prerequisite to further mechanistic interpretation would appear to be the attainment of equilibrium systems, which under slow exchange conditions contain significant amounts of identifiable stereoisomers whose interconversion can be demonstrated at higher temperatures. At present, there is no direct evidence for the implication of forms such as **16** or **17** in the exchange processes.

B. Bisbidentate Complexes

Four classes of complexes of this type are readily recognized: $M(L-L)_2X_2$, $M(L-L')_2X_2$, $M(L-L)_2XY$, and $M(L-L')_2XY$; X and Y represent monodentate ligands. Examples of each class are known, and the large majority of these are neutral complexes containing two β-diketonates and X, Y = halide. The static stereochemistry and/or nonrigid behavior of complexes with M = Ti(IV), Zr(IV), Hf(IV), Mo(VI), Si(IV), Ge(IV), and Sn(IV) have been investigated by nmr methods. Assignment of stereochemistry can in principle be made from the number and relative intensities of ring signals under slow exchange conditions. For the largest class of complexes investigated, $M(L-L)_2X_2$, differentiation of the cis (**18a**) and trans (**18b**) isomers is trivial provided slow exchange can be reached and chemical shift differences between inequivalent sites are resolvable. This is frequently the case, and nmr

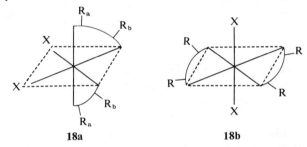

18a   **18b**

results together with a smaller body of dipole moment data indicate that the cis configuration is usually the more stable. $M(L-L')_2X_2$ complexes can exist as five geometric isomers [the designation of isomers is the same as that employed elsewhere (Serpone and Fay, 1967)]: cis-cis-cis (**19**), cis-trans-cis (**20**), cis-cis-trans (**21**), trans-cis-cis (**22**), and trans-trans-trans (**23**). Consequently, assessment of equilibrium isomer distribution is somewhat more involved. In favorable cases, all isomers present can be detected, as illustrated by the pmr spectrum of $Ge(pvac)_2Cl_2$ (Fig. 6). Resolution of six $R_1 = t$-Bu and $R_2 = $ Me signals demonstrates that all five isomers are present in benzene solution at 40°C (Pinnavaia et al., 1970b), although it has not been possible to assign signals to particular isomers. The stereochemical results for neutral bisbidentate complexes (Fortman and Sievers, 1971; Serpone and Bickley, 1973) reveals that both cis and trans forms are detectably populated in a significant proportion of Group IVb complexes with the cis form generally favored. Of the complexes derived from Group IVa metal ions, the trans isomer has been detected only for $Ti(acac)_2I_2$ (Fay and Lowry, 1970).

The nonrigid behavior of cis-$M(L-L)_2X_2$ complexes is typified by $Ti(acac)_2X_2$ (Fay and Lowry, 1967) and $MoO_2(dpm)_2$ (Pinnavaia and Clements, 1971). A simple exchange of the chelate ring substituents R in **18a**

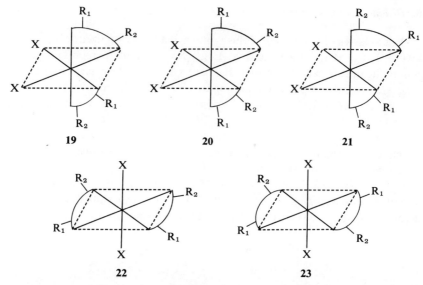

between sites a and b is observed (Fig. 7). Selected kinetic parameters for site interchange of methyl and t-Bu groups are collected in Table III. Data for the Ti(IV) alkoxide complexes are presumably the least reliable due to the approximate nature of the line-shape analysis employed, which can lead to systematically low activation energies and excessively negative activation entropies (Allerhand et al., 1966; Serpone and Bickley, 1973). The intramolecular nature of the averaging processes is supported by observation of slow intermolecular ligand exchange and, in the case of cis-$Sn(acac)_2Cl_2$, by retention of Sn-ring CH coupling under intermediate and fast exchange conditions (Faller and Davison, 1967). The dnmr behavior of $M(L-L')_2X_2$

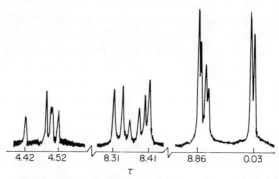

Fig. 6 Pmr spectrum (60 MHz) of $Ge(pvac)_2Cl_2$ in benzene at 40°C (Pinnavaia et al., 1970b). The three spectral regions from low to high field contain —CH=, CH_3, and $C(CH_3)_3$ signals, respectively.

species has been investigated in detail only for Ti(bzac)$_2$X$_2$ complexes (X = F Cl, Br) (Serpone and Fay, 1967). Fluorine and methyl spectra over the slow to fast exchange temperature intervals are set out in Fig. 8. The six ^{19}F resonances at slow exchange have been assigned to the four nonequivalent fluorine atoms of the three cis isomers **19, 20,** and **21**, with signals 1,2,5, and 6 deriving from the AB spin system in the cis-cis-cis isomer **19**. The four methyl signals observed at slow exchange are consistent with the presence of the

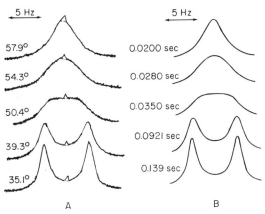

Fig. 7 (A) Temperature-dependent pmr spectra (60 MHz) of t-Bu signals of MoO$_2$-(dpm)$_2$ in dichloromethane. The weak central peak is free ligand impurity. (B) Calculated spectra for the mean preexchange lifetimes indicated. (Pinnavaia and Clements, 1971.)

three cis isomers. As the temperature is increased, all fluorine and methyl signals appear to undergo exchange broadening at approximately the same rates, and a single averaged resonance is observed in the fast exchange limit.

There is insufficient information contained in the two-site or multisite dnmr spectra of the types shown in Figs. 7 and 8 to allow deduction of the mechanism(s) responsible for the site interchanges or significant restriction of the number of reasonable mechanistic possibilities. The first essential question, i.e., whether or not site interchange occurs in the cis isomers with retention or inversion of configuration, is largely unanswered. The very limited evidence suggests that inversion and interchange may occur simultaneously. This is the indicated situation for Ti(dibm)$_2$Cl$_2$ (Serpone and Bickley, 1973), whose isopropyl groups can sense both processes. These groups in Ti(acac)$_2$(O-isopropyl)$_2$ are also diastereotopic and the collapse of four methyl spin doublets below 17°C to one doublet at higher temperatures concomitant with methyl site averaging might result from inversion instead of restricted rotation as originally proposed (Bradley and Holloway, 1969). The

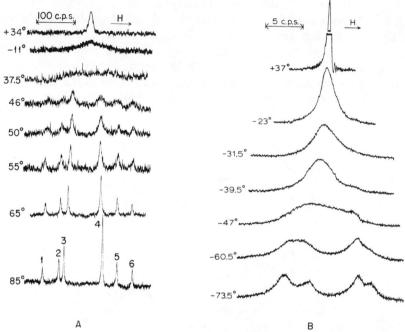

Fig. 8 Temperature-dependent spectra of Ti(bzac)$_2$F$_2$ in dichloromethane solution: (A) ^{19}F spectra; (B) Me pmr spectra. (Serpone and Fay, 1967.)

dnmr spectra of Ti(bzac)$_2$F$_2$ (Fig. 8) as well as those of the chloride and bromide analogs appear to result from the exchange of fluorine atoms and methyl groups among the four nonequivalent sites of each in the three cis isomers. Possible mechanistic pathways involving twist and bond rupture mechanisms have been considered in some detail elsewhere (Serpone and Fay, 1967; Fortman and Sievers, 1971; Serpone and Bickley, 1973), and a topological representation of twist processes has been presented (Muetterties, 1968). Because mechanistic arguments are indecisive at this stage, they are not assessed here. The representations of mechanistic pathways in terms of twist and bond rupture processes are analogous to those which are employed in the following section dealing with trisbidentate complexes, whose probable rearrangement mechanisms can be more fully examined.

Although the kinetic parameters in Table III cannot be associated with definite rearrangement mechanisms, they do serve to define a metal ion-dependent rate order for the same exchange process in a series of closely related complexes. For bis(β-diketonates) the rate order (Eq. 8) is apparent,

Zr(IV)(0.72), Hf(IV)(0.71) > Ti(IV)(0.61) > Mo(VI)(0.60) ∼ Sn(IV)(0.69) ≫
$\qquad$ Ge(IV)(0.54) > Si(IV)(0.40) (8)

TABLE III
SELECTED KINETIC PARAMETERS FOR SITE INTERCHANGE OF RING SUBSTITUENTS IN $M(L-L)_2X_2$ [a]

Complex	Solvent	E_a (kcal/mole)	$\Delta S^{\ddagger}$ (eu)	$k^{298°C}$ (seconds)	Reference[b]
$Ti(acac)_2F_2$	CD_2Cl_2	11.6 (0.5)	−2.4 (2.3)	1.6×10^4	1
$Ti(acac)_2Cl_2$	CD_2Cl_2	11.2 (0.6)	−10.0 (2.3)	6.7×10^2	1
$Ti(acac)_2Br_2$	CD_2Cl_2	11.6 (0.4)	−6.3 (1.6)	2.3×10^3	1
$Ti(acac)_2(OMe)_2$	CCl_4	10.1 (0.4)	−19 (2)	—	2
$Ti(acac)_2(O\text{-isopropyl})_2$[c]	CCl_4	10.8 (0.5)	−19 (2)	—	2
$Ti(acac)_2(O\text{-}t\text{-Bu})_2$	CCl_4	13.2 (0.5)	−14 (2)	—	2
$MoO_2(dpm)_2$	CD_2Cl_2	17.0 (1.0)	—	1.4	3
$Ge(dpm)_2Cl_2$	$Ph_2CH_2\text{-}m\text{-dimethoxybenzene}$	25.2 (1.4)	−1.1 (3.0)	3×10^{-6}	4
$Ge(dpm)_2Br_2$	$Ph_2CH_2\text{-}m\text{-dimethoxybenzene}$	25.1 (2.2)	−0.7 (4.8)	5×10^{-6}	4
$Ge(dpm)_2I_2$	$Ph_2CH_2\text{-}m\text{-dimethoxybenzene}$	13.8 (0.4)	−8.7 (1.2)	15	4
$Sn(dpm)_2F_2$	$Ph_2CH_2\text{-}m\text{-dimethoxybenzene}$	12.8 (0.3)	−13.5 (1.1)	8.2	4
$Sn(dpm)_2Cl_2$	$Ph_2CH_2\text{-}m\text{-dimethoxybenzene}$	14.6 (0.4)	−12.6 (1.2)	0.58	4
$Sn(dpm)_2Br_2$	$Ph_2CH_2\text{-}m\text{-dimethoxybenzene}$	15.0 (0.5)	−10.9 (1.3)	0.64	4
$Sn(dpm)_2I_2$	$Ph_2CH_2\text{-}m\text{-dimethoxybenzene}$	14.4 (0.4)	−10.6 (1.3)	2.5	4

[a] Estimated errors in parentheses. For a more extensive body of data, see Serpone and Bickley (1973).
[b] References: (1) Fay and Lowry (1967); (2) Bradley and Holloway (1969); (3) Pinnavaia and Clements (1971); (4) Jones (1971).
[c] For additional data, see Harrod and Taylor (1971).

in which the ionic radius (angstroms) for each metal ion is shown in parentheses. The leading position of Zr(IV) and Hf(IV) in the series follows from the observation of single methyl resonances in M(acac)$_2$X$_2$ (X = Cl, Br) at ambient temperature and in Zr(acac)$_2$Cl$_2$ at temperatures as low as $-130°$C (Pinnavaia and Fay, 1968). The cis structure of these complexes in solution is supported by infrared and Raman evidence (Fay and Pinnavaia, 1968). The terminal position of Si(IV) is based on the finding that Si(acac)$_2$(OAc)$_2$ is a rigid or slow complex* at ambient temperature, with trans $\rightleftharpoons$ cis equilibration occurring at a measurable rate (Holloway et al., 1966). For these closed-shell ions, increasing rate is paralleled by increasing ionic size, with the exception of Sn(IV). In addition to Sn(dpm)$_2$X$_2$, Sn(acac)$_2$X$_2$ and Sn(acac)$_2$RX (R = Me, Ph) rearrange at rates intermediate between those of Ti(IV) and Ge(IV) complexes (Jones, 1971; Serpone and Bickley, 1973) and thus do not follow a simple ionic size relationship. Nonetheless, the influence of metal ion radius is unmistakable and is also encountered in the rearrangements of trischelate complexes (Section III,C).

C. Trisbidentate Complexes

The three most important classes of nonrigid complexes are M(L L)$_3$, M(L–L')$_3$, and M(L–L')$_2$(L–L). When the twist angle ϕ,† defined generally by **24**, is different from zero the three types of complexes are chiral. Of these only M(L–L)$_3$ does not possess geometrical isomers. Consequently, its only source of nonrigidity is inversion of configuration, $\Delta \rightleftharpoons \Lambda$. As will become evident, the dnmr features of M(L–L)$_3$ complexes constitute an underdetermined

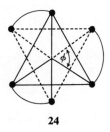

24

---

* Note that fast stereochemical lability has been observed for R$_3$Si(acac) complexes. However, the stable forms of these species are not chelated, and the dynamic process observed is the interchange of nonequivalent acac methyl groups for which a chelated six-coordinate transition state has been proposed (Pinnavaia et al., 1970a).

† This angle is defined by projection of metal–ligand vectors of the same chelate ring on a plane normal to a real or pseudo-threefold axis. When $\phi = 0°$ and 60° the coordination sphere is an idealized trigonal prism (TP) and trigonal antiprism (TAP), respectively, with the latter forming an octahedron when all ligand–metal–ligand angles are 90°.

9. Nonrigid Chelate Complexes

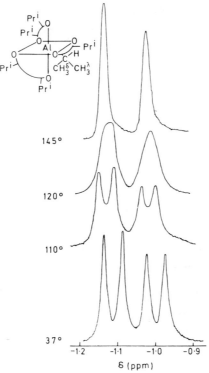

Fig. 9 Temperature-dependent methyl pmr spectra (60 MHz) of Al(dibm)$_3$ in chlorobenzene (Jurado and Springer, 1971).

situation for deduction of mechanism, and only the occurrence of the inversion process itself can be detected. Because knowledge of the presence or absence of inversion is critical to mechanistic arguments for other classes of trisbidentates, it is instructive to examine briefly evidence for the occurrence of this process in a case uncomplicated by geometrical isomerization. Such a case is furnished by Al(dibm)$_3$, whose methyl dnmr spectra are shown in Fig. 9 (Jurado and Springer, 1971). The methyl groups of the isopropyl ring substituents are diastereotopic with environments designated here as δ and λ. Upon inversion of the molecular configuration by any means, a given methyl group changes environments. Thus, the passage from two methyl spin doublets at 37°C to one doublet at fast exchange signifies the site averaging $\delta \rightleftharpoons \lambda$ and, therewith, inversion. The use of diastereotopic probes for inversion in chelate complexes has become commonplace and, coupled with information relating to site interchanges effected by isomerization reactions, can in favorable circumstances lead to a specification of mechanism.

1. The $M(L-L')_3$ Class

Trischelate complexes containing three identical unsymmetrical bidentate ligands can exist as cis (C) and trans (T) isomers, each of which is enantiomeric. The cis-Λ and trans-Λ forms are illustrated by **25** and **26**, respectively. The three chelate rings in **25** are equivalent, whereas those in **26** are inequivalent, thus allowing detection of one isomer in the presence of the other by nmr, provided chemical shift differences are large enough. This method of stereochemical assay was first recognized by Fay and Piper (1962, 1963) and

25   **26**

has since been applied to a variety of diamagnetic and paramagnetic complexes (see, e.g., Holm, 1969; Fortman and Sievers, 1971; Gordon *et al.*, 1971; Springer *et al.*, 1971; Holm and Hawkins, 1973).

Complexes of the $M(L-L')_3$ type can exhibit two types of intramolecular rearrangement processes which result in a change in molecular stereochemistry: inversion ($\Delta \rightleftharpoons \Lambda$), and isomerization (C $\rightleftharpoons$ T). When viewed on a physical basis, these processes can proceed by two types of reaction pathways, which are distinguished by the effective coordination number of the transition state. If no metal–ligand bonds are broken, the transition state is an idealized trigonal prism (TP) which may be achieved by means of various twist motions of the chelate rings with respect to real (r-C_3) or artificial [pseudo (p-C_3) and imaginary (i-C_3)] threefold symmetry axes of the cis and trans isomers. Twists about the r- or p-C_3 axes have been referred to as trigonal or Bailar twists (Bailar, 1958) and those about i-C_3 axes are rhombic or Rây-Dutt twists (Rây and Dutt, 1943). Bond rupture leads to a five-coordinate transition state with an idealized trigonal-bipyramidal (TBP) or square-pyramidal (SP) geometry. A detailed analysis of the various twist and bond-rupture [Springer (1973) has recently presented a detailed analysis of the role of five-coordinate transition states in dissociative reactions of octahedral complexes] pathways has been given by Gordon and Holm (1970, see also Muetterties, 1968). Twist mechanisms have also been analyzed by Springer and Sievers (1967), who first provided the useful demonstration that relative displacements of chelate rings not involving bond rupture may be envisaged as proceeding by means of twist motions of the three rings with respect to real or artificial threefold axes of the two geometric isomers. Illustrations of twist and TBP pathways are depicted in Fig. 10, and correla-

9. Nonrigid Chelate Complexes

Twist Mechanism

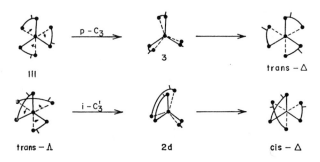

Bond Rupture – TBP Transition State

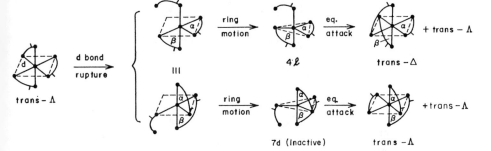

Fig. 10 Illustration of the formation of products in reactions of trans-Λ M(L–L′)$_3$ proceeding by twist and bond rupture mechanisms. TBP transition states 4*l* and 7*d* possess dangling axial and equatorial ligands, respectively. α and β designate chelate ring planes that have different relative orientations in the two types of TBP transition states (Gordon and Holm, 1970).

tion diagrams and representations of transition states are set out in Figs. 11 and 12. The SP pathway is somewhat more complicated and is dealt with elsewhere (Gordon and Holm, 1970). The advantage of M(L–L′)$_3$ species as vehicles for studying rearrangement mechanisms is evident from the correlation diagrams, in which stable isomers are placed at the vertices of the figures and transition states on the edges. These diagrams show that constraints are placed on one-step interconversions by the different mechanisms. For example, a trigonal twist can effect inversion but not isomerization (CΔ ⇌ CΛ, TΔ ⇌ TΛ), and bond rupture through a TBP-axial transition state will lead

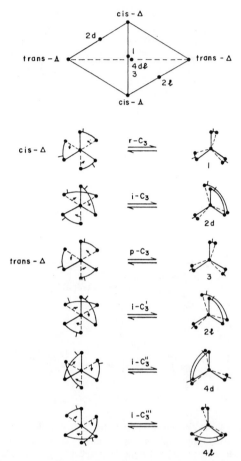

Fig. 11 Topological correlation diagram for the interconversion of M(L–L′)₃ isomers by twist mechanisms. The six possible TP transition states produced by ring motion relative to the real (*r*), pseudo (*p*), and imaginary (*i*) threefold axes are illustrated. (Gordon and Holm, 1970.)

to simultaneous inversion and isomerization but only isomerization if a TBP-equatorial transition state is involved. Using this type of approach, a complete treatment of the rearrangement kinetics of slow M(L–L′)₃ complexes has been developed (Gordon and Holm, 1970; Girgis and Fay, 1970).

The preceding treatment of rearrangements is based on the conception of transition states in terms of idealized stereochemistries (TP, TBP, SP), which have predictable mechanistic consequences and may be considered physically reasonable inasmuch as a number of stable metal complexes approach or

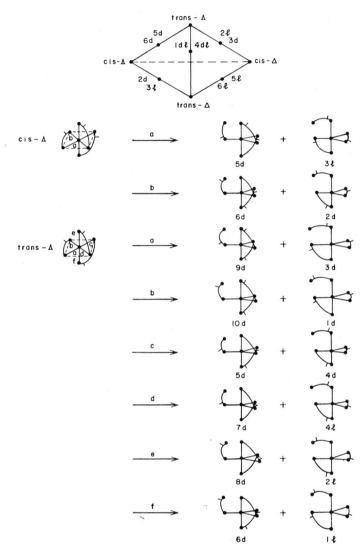

Fig. 12 Topological correlation diagram for the interconversion of M(L–L')$_3$ isomers through TBP transition states. The transition states produced by rupture of the eight distinguishable metal–ligand bonds of cis- and trans-Λ are shown (Gordon and Holm, 1970).

possess these structures. These transition states have also been frequently involved in mechanisms of other types of reactions of six-coordinate complexes (Basolo and Pearson, 1967). However, the dnmr behavior of nonrigid complexes reflects the consequences of site interchange only, and any mechanistic approach founded on such physically reasonable transition states does not necessarily insure that all interchanges and, hence, all feasible rearrangements, have been included. This problem is best attacked by permutational analysis of site interchanges and several alternate approaches have recently been described for the M(L–L')$_3$ class. Musher (1972) has grouped various sets of site permutations into modes or split modes of rearrangement. A general mathematical treatment of the problem has been developed by Klemperer (1973). Eaton et al. (1972) and Eaton and Eaton (1973) have derived permutational sets using the symmetry groups of nonrigid molecules originally introduced by Longuet-Higgins (1963). This rather straightforward procedure is briefly summarized.

For an M(L–L')$_3$ complex, the magnetically inequivalent sites of a ring substituent in the trans isomer are designated as x, y, z. If this substituent contains diastereotopic groups, their environments are denoted by r, s. Thus a cis–trans mixture constitutes an eight-site problem. The notation is illustrated in Fig. 13 for α-substituted tropolonate complexes whose rearrangements are described below. The set of all permutations and permutation–inversions of the six ligating nuclei is a group of order 384, which factors into a group of order 16 (consisting of the rearrangements) and a group of order 24 (which can be regarded as consisting of the rigid-body rotations of the molecule). The group of order 16, which is of primary interest here, is an Abelian group so that each operation is in a separate class. These sixteen operations constitute the complete group of rearrangements of stereochemically nonrigid trischelate complexes. It is important to note that the sixteen rearrangements are expressed as permutations; i.e., they describe only the net change and not the pathway by which the atoms move from

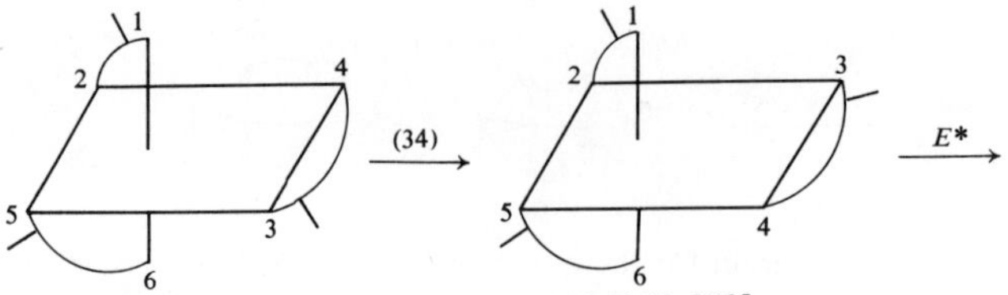

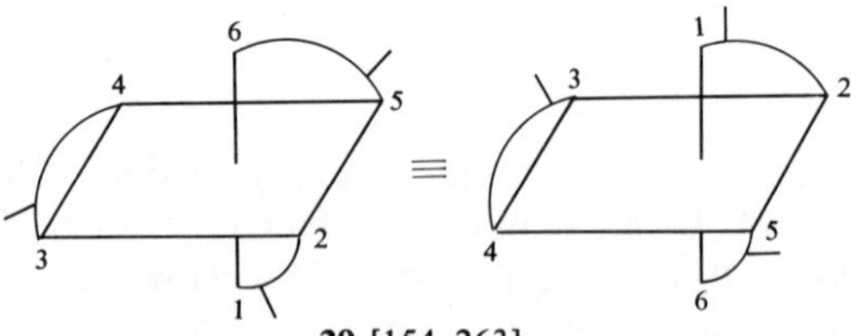

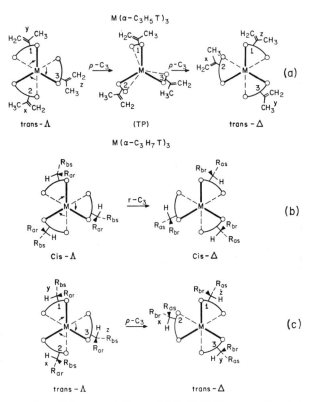

Fig. 13 (a) Inversion of the trans-Λ form of M(α-C₃H₅T)₃ proceeding by a twist about the p-C_3 axis. (b,c) Inversion of the cis-Λ and trans-Λ forms of M(α-C₃H₇T)₃ (R = Me) proceeding by twists about the r-C_3 and p-C_3 axes, respectively. Substituent groups are labeled numerically and group magnetic environments are designated by x, y, z. Methyl groups of the isopropyl substituents are labeled a and b, and their individual magnetic environments are arbitrarily designated by r and s (Eaton et al., 1972).

initial to final positions. The nmr site averaging for each operation is presented in Table IV.

The notation is illustrated in **27** for the cis isomer [135–462]. As the complex is viewed down the r-C_3 (or p-C_3) axis with the triangular face containing vertex 1 upward, the labels are given clockwise for this face starting with 1 and similarly for the lower face starting with the vertex to the right of 1. Square brackets denote isomers and parentheses denote permutations. The permutation (34) is illustrated by **27** → **28** and takes cis to trans. The conversion **28** → **29** is effected by E^*, the inversion of all ligands through the metal center. The operation that takes **27** to **29** is the permutation–inversion (34)*. The effect of the sixteen operations on [135–462] is given in Table IV.

TABLE IV

PERMUTATIONAL ANALYSIS OF $M(L–L')_3$ NMR SPECTRA[a]

Operation	Resulting isomer[b]	Averaging set	Net configurational change	Net site interchanges (trans)
E	[135–462]	A_1	None	None
(12)(34)(56)	[153–642]	A_2	None	(yz)
(12)	[164–532]	A_3	cis → trans	$(xy), (xz)$
(34)	[145–362]		[trans → $\frac{1}{3}$cis + $\frac{2}{3}$trans]	
(56)	[136–452]			
(12)(34)	[163–542]	A_4	cis → trans	$(xzy), (xyz)$
(34)(56)	[146–352]		[trans → $\frac{1}{3}$cis + $\frac{2}{3}$trans]	
(12)(56)	[154–632]			
E*	[153–264]	A_5	$\Lambda \rightleftarrows \Delta$	(rs)
(12)(34)(56)*	[135–246]	A_6	$\Lambda \rightleftarrows \Delta$	$(yz), (rs)$
(12)*	[146–235]	A_7	$\Lambda \rightleftarrows \Delta$	$(xy), (xz), (rs)$
(34)*	[154–263]		cis → trans	
(56)*	[163–254]		[trans → $\frac{1}{3}$cis + $\frac{2}{3}$trans]	
(12)(34)*	[136–245]	A_8	$\Lambda \rightleftarrows \Delta$	$(xyz), (xzy), (rs)$
(34)(56)*	[164–253]		cis → trans	
(12)(56)*	[145–236]		[trans → $\frac{1}{3}$cis + $\frac{2}{3}$trans]	

[a] Eaton et al. (1972); Eaton and Eaton (1973).
[b] Isomers resulting from performing the indicated operations on [135–462] are given as examples. Columns 3, 4, and 5 summarize net effects of operations on all sixteen isomers.

9. Nonrigid Chelate Complexes

Because all sixteen isomers generated in this fashion contribute to the pmr spectra, the net effect of each rearrangement on the pmr spectra is derived by summing the changes in ligand environments produced by a particular operation. These results are compiled in Table IV. Permutations which give the same net averaging pattern for the sites x, y, z, r, and s are placed together in averaging sets A_i, although the permutations within each set may have different effects on different isomers. Certain of these permutations cause the same site interchanges as do the physically based mechanisms. The following correspondences can be made: p-C_3 or r-C_3 twists, A_6; i-C_3 twists, A_8; TBP-axial, A_7; TBP-equatorial, A_3; SP-axial, $A_3 + A_6 + A_8$. These correspondences can be verified by reference to the site exchange tabulations for the physical mechanisms derived by Hutchison et al. (1971).

Virtually all of the detailed mechanistic studies on the dynamic stereochemistry of $M(L-L')_3$ complexes involve species that are ligated exclusively by oxygen or sulfur. Results for these two groups of complexes are considered in the following sections.

a. *Complexes with the* M—O_6 *Core.* The important complexes of this group are the β-diketonates (**30**), and the α-isopropenyl- and α-isopropyltropolonates (**31** and **32**, respectively). Tris(β-diketonate) complexes and related species derived from acetylcamphor have been extensively investigated, and sufficient information is at hand to allow division into rigid and

nonrigid categories in terms of metal ion as shown in Table V. The classification is based on observation of slow exchange spectra at or above room temperature and in some cases on the separation of geometrical isomers and (partial) resolution of optical isomers. In certain of the latter cases, measurements of isomerization and racemization rates have been carried out. The most detailed work of this sort on rigid complexes has involved two Co(III) β-diketonates of type **30**: Co(mhd)$_3$(R_1 = isopropyl, R_2 = Me; Gordon and Holm, 1970) and Co(bzac)$_3$ (R_1 = Ph, R_2 = Me; Girgis and Fay, 1970). In chlorobenzene solution, these complexes isomerize and racemize simultaneously but at slightly different rates, which fall within the interval of about 10^{-4}–10^{-5} sec^{-1} at 90°–96°C. Activation energies (30–33 kcal/mole)

TABLE V
NMR KINETIC CLASSIFICATION OF TRIS(β-DIKETONATE) AND
TRIS(ACETYLCAMPHORATE) COMPLEXES

Metal ion	Classification	References[a]
Al(III)	Nonrigid	1–4, 6, 17
Ga(III)	Nonrigid	1, 3, 5, 6
In(III)	Nonrigid	1
Si(IV)	Rigid	16
Sc(III)	Nonrigid	3
V(III)	Rigid	7–9
Cr(III)	Rigid	1, 10
Mn(III) (high-spin)	Nonrigid	8, 11
Fe(III) (high-spin)	Nonrigid	8
Ru(III)	Rigid	8, 10, 15
Co(III)	Rigid	3, 8, 10, 12, 13, 14, 17
Rh(III)	Rigid	10

[a] References: (1) Fay and Piper (1964), (2) Jurado and Springer (1971); (3) Hutchison et al. (1971); (4) Fortman and Sievers (1967); (5) Pinnavaia et al. (1969); (6) Case and Pinnavaia (1971); (7) Röhrscheid et al. (1967); (8) Gordon et al. (1971); (9) Chen and Everett (1968); Everett and Chen (1970); (10) Fay et al. (1970); (11) Everett and Johnson (1972); (12) Gordon and Holm (1970); (13) Girgis and Fay (1970); (14) Springer et al. (1971); (15) Everett and King (1972); (16) Larsen et al. (1966); (17) Palmer et al. (1964).

and other activation parameters are closely similar for both processes, indicating that they proceed through the same transition state(s). The two independent kinetic analyses concur in concluding that the reaction pathway involves bond rupture generating a high percentage of TBP-axial transition states. Rates and activation parameters for isomerization of Co(tfac)$_3$ (Fay and Piper, 1964) and racemization of Co(acac)$_3$ (Fay et al., 1970) are comparable to those for the rearrangements of Co(mhd)$_3$ and Co(bzac)$_3$, suggesting that the same mechanism obtains for all four complexes.

Considerable effort has been expended in attempts to establish rearrangement mechanisms of nonrigid β-diketonates, especially those of Al(III) and Ga(III) (Fortman and Sievers, 1971; Serpone and Bickley, 1973). Unfortunately, no unique mechanisms have been established from dnmr results, but in some cases it has been possible to limit the number of possible reaction pathways. The initial investigation was that of Fay and Piper (1964), who found that coalescence of the four ^{19}F signals of a cis–trans mixture of Al(tfac)$_3$ (**30**, $R_1 = CF_3$, $R_2 = Me$) were not consistent with the simple site interchange pattern generated by a trigonal twist (see below). This mechanism

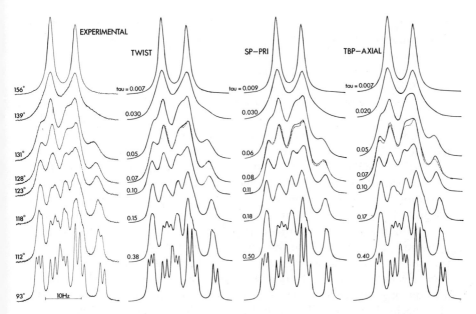

Fig. 14 Experimental methyl pmr spectra (100 MHz) of Al(pmhd)$_3$ in chlorobenzene and spectra calculated for various rearrangement mechanisms: twist, 100% rotation about the i-C$_3$ axes of cis and trans isomers; SP-PRI = SP-primary process; dotted line indicates experimental spectrum at 128°C; τ values are in seconds. (Hutchison et al., 1971.)

also cannot account for the ^{19}F dnmr spectra of a system containing the three geometrical isomers of Al(tfac)$_2$(acac) (Palmer et al., 1964).

The most detailed attempt to determine rearrangement mechanisms of tris(β-diketonate) species is that of Hutchison et al. (1971), who examined the dnmr behavior of M(pmhd)$_3$ (**30**, M = Al(III), Ga(III); R$_1$ = isopropyl, R$_2$ = CH$_2$Ph). In these complexes the isopropyl groups act as probes for isomerization and inversion. Under slow exchange conditions in chlorobenzene, thirteen and nine of the sixteen possible methyl signals (eight inequivalent methyl groups each split into a doublet by coupling with the methine proton) were resolved for cis–trans mixtures of Al(pmhd)$_3$ and Ga(pmhd)$_3$, respectively. Dynamic nmr methyl spectra of the Al(III) complex, shown in Fig. 14, reveal a series of line-shape changes that are too complicated to be analyzed visually. Spectra of the Ga(III) analog are qualitatively similar and in both cases the fast exchange spectrum consists of a single methyl spin-doublet. The spectra were analyzed by a total line-shape method using site exchange matrices for the various twist and bond-rupture mechanisms as well as for completely random exchange. The line-shape analysis demonstrated that the rearrangements of Al(pmhd)$_3$ and Ga(pmhd)$_3$ involve both isomerization and inversion and led to exclusion of the following pathways as the sole reaction mechanism for either complex: twists about the r-C_3 and p-C_3 axes only; bond rupture producing TBP-equatorial states or a 1:1 mixture of TBP-axial and TBP-equatorial transition states; random scrambling of ligand environments; for Ga(pmhd)$_3$, bond rupture producing TBP-axial transition states. Computed line shapes were in satisfactory agreement with experimental spectra for the following mechanisms: twists involving less than about 50% and from about 25 to 75% rotation about the r-C_3 and p-C_3 axes of Al(pmhd)$_3$ and Ga(pmhd)$_3$, respectively; bond rupture with formation of SP-axial transition states; for Al(pmhd)$_3$, bond rupture producing TBP-axial transition states. Comparisons of experimental and certain calculated spectra are given in Fig. 14. The kinetic analysis on which the line-shape calculations were based included transformations proceeding through TP, TBP, and SP transition states. Within the framework of the physically acceptable set of transition states of these types, the line-shape approach allowed rejection of some mechanisms as decidedly less probable than others, but did not afford a unique mechanism for either complex. Activation parameters for the probable mechanisms are collected in Table VI. The analysis of M(pmhd)$_3$ rearrangements involved, in effect, the computation of line shapes as a function of mechanism. This is a necessarily laborious procedure and involves rather detailed problems of assigning slow exchange resonances to specific sites, as well as the extrapolation of temperature-dependent linewidths and chemical shifts from the slow to the intermediate exchange region. These difficulties may in part be responsible for the indecisive mechanistic conclusions summarized above, although the possibility of multiple rearrangement pathways for M(pmhd)$_3$ and other complexes cannot be discounted.

The α-substituted tropolonates **31** and **32** have proven more amenable to detailed mechanistic analysis than have the β-diketonates. The two types of complexes have the M—O$_6$ core in common and can be examined under similar experimental conditions. The initial dnmr studies of these complexes were reported by Muetterties and Alegranti (1969), who recognized the use of isopropyl methyl resonances in M(α-C$_3$H$_7$T)$_3$ as probes from isomerization and inversion and offered a qualitative interpretation of exchange-broadened spectra which subsequent work has shown to be essentially correct. Methyl pmr spectra of the series M(α-RT)$_3$ [R = C$_3$H$_5$, isopropyl; M = Al(III), Ga(III), Ge(IV), V(III), Mn(III), Ru(III), Rh(III)] have been studied in considerable detail and have provided rather definite mechanistic information in some cases (Eaton and Holm, 1971b; Eaton *et al.*, 1972, 1973). The dnmr behavior of the Al(III), Ga(III), and Co(III) complexes reveals two partially overlapping kinetic regions. Rearrangements that achieve these fast exchange limits in the lower and higher temperature intervals are referred to as the

TABLE VI
Kinetic Parameters for Rearrangements of M—O$_6$ Complexes[a]

Complex	Solvent	Process	E_a (kcal/mole)	$\Delta S^{\ddagger}$ (eu)	$k^{298°C}$ (sec)	Reference[c]
Al(pmhd)$_3$	PhCl	Twist[b]	27.6 (1.4)	12.9 (3.5)	10^{-4}	1
	PhCl	TBP-axial[b]	30.2 (1.4)	19.5 (3.5)	$\sim 10^{-4}$	1
	PhCl	SP-primary[b]	29.6 (1.4)	17.6 (3.5)	10^{-4}	1
Al(acac)$_2$(hfac)	CH$_2$Cl$_2$	Site exchange	18.4 (0.7)	10.0 (2.4)	80	2
Al(acac)(hfac)$_2$	CH$_2$Cl$_2$	Site exchange	21.3 (0.7)	10.7 (2.2)	0.86	2
	C$_6$H$_6$	Site exchange	19.0 (1.3)	2.9 (4.2)	0.84	2
Al(acac)$_2$(dbm)	o-C$_6$H$_4$Cl$_2$	Site exchange	22.0 (0.6)	0.9 (1.5)	0.0022	2
Ga(pmhd)$_3$	PhCl	Twist[b]	20.1 (1.2)	2.9 (3.4)	~ 8	1
	PhCl	SP-primary[b]	20.5 (1.2)	3.6 (3.4)	~ 8	1
Ga(acac)$_2$(hfac)	CH$_2$Cl$_2$	Site exchange	15.3 (1.2)	3.9 (4.6)	770	2
Ga(acac)$_2$(dbm)	C$_6$H$_6$	Site exchange	20.2 (3.2)	2.2 (9.3)	0.077	2
Al(α-C$_3$H$_5$T)$_3$	C$_2$H$_2$Cl$_4$	C $\rightarrow$ T	16.9 (1.1)	-4.3 (3.2)	1.0	3
	CH$_2$Cl$_2$	T$\Delta \rightarrow$ TΛ	13.2 (1.4)	-7.4 (4.8)	80	3
Al(α-C$_3$H$_7$T)$_3$	C$_2$H$_2$Cl$_4$	C$\Delta \rightarrow$ CΛ	10.5 (3.6)	-16 (12)	120	3
	C$_2$H$_2$Cl$_4$	T$\Delta \rightarrow$ TΛ	12.3 (2.0)	-11 (7)	75	3
	C$_2$H$_2$Cl$_4$	C $\rightarrow$ T	25.8 (4.0)	21 (12)	0.04	3
Co(α-C$_3$H$_5$T)$_3$	CDCl$_3$	T$\Delta \rightarrow$ TΛ	15.1 (1.1)	-5.3 (3.7)	4.0	3
	CDCl$_3$	C$\Delta \rightarrow$ CΛ	16.7 (0.9)	5.4 (3.9)	100	3
Co(α-C$_3$H$_7$T)$_3$	CHCl$_3$	T$\Delta \rightarrow$ TΛ	14.3 (1.0)	-2.1 (3.5)	210	3
	CHCl$_3$	T$\Delta \rightarrow$ TΛ	16.1 (1.3)	4.8 (3.9)	350	3
	CHCl$_3$	C $\rightarrow$ T	15.7 (1.4)	-5.1 (4.2)	5.0	3

[a] Estimated errors in parentheses.
[b] Isomerization and inversion; mechanisms consistent with calculated line shapes (see text).
[c] Reference: (1) Hutchison et al. (1971); (2) Case and Pinnavaia (1971); (3) Eaton et al. (1972).

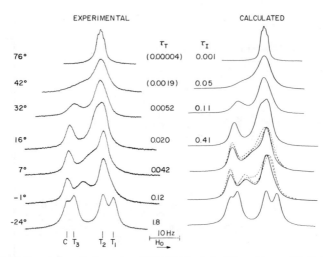

Fig. 15 Experimental methyl spectra (100 MHz) of cis (C) and trans (T) Co(α-C$_3$H$_5$T)$_3$ in chloroform solution, and spectra calculated for inversion (LTP, p-C$_3$ twist) below 20°C with a random isomerization process (HTP) superimposed above 20°C. Spectra calculated using different τ values for an exchange process involving all three trans sites are indicated by the dotted lines. τ values are in seconds: τ_T, inversion of T; τ_I, C → T isomerization. Extrapolated values are in parentheses. (Eaton et al., 1972.)

low temperature process (LTP) and high temperature process (HTP), respectively. Complexes of V(III) and Mn(III) do not exhibit two distinct kinetic regions, and as for the HTP of other complexes, the spectral changes indicate cis–trans isomerization. As will be shown, the LTP involves inversion of configuration. On the basis of the methyl dnmr spectra, M(α-RT)$_3$ complexes can be grouped as follows: nonrigid complexes that attain the fast exchange limit of inversion and/or isomerization (i) below about 0°C [V(III), Mn(III), Ga(III)], (ii) below about 100°C [Al(III), Co(III)], (iii) above about 100°C [Ge(IV)]. Complexes of Si(IV)*, Rh(III), and Ru(III) are rigid or slow. Line-shape changes due to isomerization have provided no mechanistic information and in several cases could be simulated by assuming random site interchange (Eaton et al., 1972). Therefore, attention is confined to the LTP.

The dnmr properties of tropolonate complexes are best approached by considering first the methyl spectra of the α-isopropenyl species **31**. LTP spectra of the Al(III), Ga(III), and Co(III) complexes are similar and those of Co(α-C$_3$H$_5$T)$_3$ are given in Fig. 15. The slow exchange (-24°C) spectrum clearly shows the presence of cis and trans isomers. The key spectral feature

---

* The classification of Si(IV) species as rigid is based on the resolution of [SiT$_3$]$^+$ (Ito et al., 1969) and the partial separation of [Si(α-C$_3$H$_7$T)$_3$]$^+$ into cis and trans forms (Muetterties and Alegranti, 1969).

9. Nonrigid Chelate Complexes

in the $-24°$ to $16°C$ region is the averaging of two of the three methyl signals of the trans isomer without simultaneously affecting the remaining trans and the cis resonance. This process is consistent with the site interchange Eq. (9), i.e., inversion without isomerization. The $C\Delta \rightleftharpoons C\Lambda$ inversion

$$T\Lambda \begin{Bmatrix} 1-y \\ 2-x \\ 3-z \end{Bmatrix} \rightleftharpoons T\Delta \begin{Bmatrix} 1-z \\ 2-x \\ 3-y \end{Bmatrix} \qquad (9)$$

is doubtless also occurring but is undetectable by pmr. Spectra of $M(\alpha\text{-}C_3H_5T)_3$ serve to define clearly the substituent group site interchange in the LTP and accord with the averaging sets A_2 and A_6 obtained from permutational analysis (Table IV). (Set A_5 results in inversion but does not average the trans x, y, z sites). Distinction between the sets A_2 and A_6 is afforded by the methyl spectra of the closely related α-isopropyl complexes **32**, in which the methyl groups are diastereotopic (Fig. 13). The spectra of $Co(\alpha\text{-}C_3H_7T)_3$ (Fig. 16) show a series of line shape changes for the LTP (about $-17°$ to $30°C$), which could be accurately simulated on the basis of site interchanges Eqs. (10) and (11). Spectra of Al(III) and Ga(III) analogs can be similarly

$$C\Lambda \begin{Bmatrix} a-r \\ b-s \end{Bmatrix} \rightleftharpoons C\Delta \begin{Bmatrix} a-s \\ b-r \end{Bmatrix} \qquad (10)$$

$$T\Lambda \begin{Bmatrix} 1-yr \\ 1-ys \\ 2-xr \\ 2-xs \\ 3-zr \\ 3-zs \end{Bmatrix} \rightleftharpoons T\Delta \begin{Bmatrix} 1-zs \\ 1-zr \\ 2-xs \\ 2-xr \\ 3-ys \\ 3-yr \end{Bmatrix} \qquad (11)$$

interpreted. Interchanges (10) and (11) involve the pairwise averaging of eight methyl doublets to four, and the latter features are clearly evident in the fast exchange LTP spectrum of $Ga(\alpha\text{-}C_3H_7T)_3$ (Eaton *et al.*, 1973). Extensive computer analyses of the LTP spectra of $Al(\alpha\text{-}C_3H_7T)_3$ and $Co(\alpha\text{-}C_3H_7T)_3$ (Eaton *et al.*, 1972) demonstrated that the experimental spectra could only be adequately reproduced assuming (10) and (11). The latter conforms to averaging set A_6 inversion without isomerization and simultaneous exchange of trans sites y and z. In particular, the spectra could not be fit to set A_2, which corresponds to trans y, z interchange without inversion. The most satisfactory physical conception of the site interchanges is considered to be the trigonal twist. This mechanism is set out in Fig. 13, which provides a simple illustration of the concordance of interchanges (9)–(11) with this reaction pathway

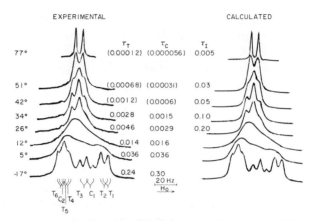

Fig. 16 Experimental methyl spectra (100 MHz) of cis (C) and trans (T) Co(α-C_3H_7T)$_3$ in deuteriochloroform solution, and spectra calculated for inversion [LTP = r-C_3 twist (C), p-C_3 twist (T)] below 20°C with a random isomerization process (HTP) superimposed above 20°C. τ values are in seconds; τ_T, inversion of T; τ_C, inversion of C; τ_I, C $\rightarrow$ T isomerization. Extrapolated values are in parentheses. (Eaton et al., 1972.)

for the M(α-RT)$_3$ series. Kinetic parameters for the rearrangements of these complexes obtained by full line-shape analysis are collected in Table VI. The investigations of the tropolonate complexes provided the first incisive demonstration of the twist mechanism, which, however, is not confined to this group of complexes (see Section III,C,1b).

In addition to the operation of the trigonal twist mechanism in their inversion reactions, the most significant kinetic features of the M(α-RT)$_3$ complexes are the unexpectedly low barriers and consequently high rates of rearrangement. Rearrangement rates and activation energies for the tropolonates and various groups of β-diketonate complexes are compared in Table VII (Eaton et al., 1973). As may be seen, tropolonates invariably rearrange faster than β-diketonate complexes containing the same metal ion. This situation includes fluorinated β-diketonates for which the data shown reveal that CF_3 groups have an accelerating influence on rate. The relative rate enhancement is most dramatically illustrated by Co(α-RT)$_3$. These were the first nonrigid Co(III) complexes discovered (Eaton and Holm, 1971b), and at 25°C they invert $\sim 10^{11}$ times faster than does Co(acac)$_3$ and $\sim 10^9$ times faster than Co(tfac)$_3$ isomerizes. Similar situations are encountered upon comparison of the kinetic classification in Table V with the data of Table VII. As additional examples the kinetic behavior of V(tfac)$_3$ (Gordon et al., 1971) and V(α-C_3H_7T)$_3$ (Eaton et al., 1973) may be contrasted. There is no exchange broadening in the pmr spectrum of the former up to 100°C, whereas there is coalescence of the α-H and γ-H signals of the tropolonate at about

−40°C. Similar relations hold for Mn(III) complexes with coalescence temperature differentials in the 80°–100°C range. Data for complexes of Al(III), Ga(III), V(III), Mn(III), and Co(III) are fully consistent with ligand kinetic series (12). In the Al(III) and Ga(III) cases, rates become comparable to isomerization rates of the tropolonates only when at least one hfac ligand (30, $R_1 = R_2 = CF_3$) is present.

$$\alpha\text{-RT} > \text{tfac} > \text{bzbc} \sim \text{acac} \sim \text{pmhd} \sim \text{mhd} \qquad (12)$$

Inversion of Al(α-RT)$_3$ also proceeds at rates comparable to site exchange in Al(acac)$_2$(hfac), but the inversion of Ga(α-RT)$_3$ is substantially faster than site exchange in Ga(acac)$_2$(hfac). It is not known whether site exchange in M(acac)$_2$(hfac) proceeds with inversion of configuration.* Like Co(tfac)$_3$, Co(tfac)$_2$(acac) is stereochemically rigid up to at least 115°C (Palmer et al., 1964). The fast exchange limit for isomerization (HTP) of Co(α-RT)$_3$ is reached at about 70°–80°C. From the results at hand it is concluded that rate enhancement of tris(tropolonate) compared to M(tfac)$_3$ and nonfluorinated tris(β-diketonate) complexes is a general phenomenon.

The kinetic results for the tropolonates conform to certain regularities in metal ion dependence of rearrangement rates. The collective rate data for the $d^{0,10}$ complexes of the M—O$_6$ type reveals the metal kinetic series (Eq. 13) in which ionic radii are given in parentheses. Pinnavaia and Clements (1971)

$$\text{In(III) (0.79), Sc(III) (0.73)} \gg \text{Ga(III) (0.62)} > \text{Al(III) (0.53)} \qquad (13)$$

have proposed a more extensive rate order by combining series (8) and (13). Decreasing intramolecular lability with decreasing ionic radius is clearly evident in these kinetic series and, despite several apparent exceptions† to this trend, ionic size must be considered the dominant parameter determining relative rearrangement rates in series of $d^{0,10}$ complexes with invariant ligands. In contrast, rates for d^2–d^6 complexes show no correlation with ionic radius. Further the rate orders Fe(III) (0.65) > Ru(III) (0.70) and Co(III) (0.53) > Rh(III) (0.67) are the reverse of those found in descending the Ti, Al, and Si vertical groups.

The origin of the rearrangement barriers in the metal kinetic series (8) and (13) remains unclear. The problem of assessing these barriers is compounded

---

* Kinetic parameters for diastereotopic averaging in Al(hfac)$_2$(dibm) ($E_a = 21.5$ kcal/mole, log $A = 15.6$) and CF$_3$ site exchange in Al(hfac)$_2$(acac) ($E_a = 21.3$ kcal/mole, log $A = 15.6$) strongly imply that the latter process occurs with inversion of configuration (unpublished results of C. S. Springer et al., quoted by Serpone and Bickley, 1973).

† As mentioned earlier, bis(β-diketonato)SnX$_2$ appear to be slower than anticipated on the basis of the size of Sn(IV). In addition, the partial resolution of Y(acac)$_3$ [Y(III), 0.89 Å] has been claimed (Moeller et al., 1959). The sample used in this case was apparently hydrated or impure and, as already suggested (Fortman and Sievers, 1971), a reinvestigation of these results would be desirable.

TABLE VII
RELATIVE REARRANGEMENT RATES OF M—O$_6$ COMPLEXES

Ligand	Process	Rate order[a]	References
α-RT[b]	Inversion	Ga > Co $\gtrsim$ Al > Ge > Si, Rh, Ru [$k \sim 10^2 (-65°)$] ($k = 10^2, E_a = 17$) ($k = 30, E_a = 13$)	c, h, i, j
	Isomerization	V ~ Mn ~ Ga > Co[d] $\gtrsim$ Al[d] > Ge > Si, Rh, Ru ($k = 4.0, E_a = 16$) ($k = 1.0, E_a = 17$)	c, h, i, j
tfac	Isomerization	Fe, In > Mn > Ga [$k > 36 (-57°C), E_a < 14$] [$k \sim 10^3 (70°C)$] [$k = 38 (62°C), E_a = 21$] Al > V > Co[d] [$k = 34 (103°C), E_a = 24$] ($k = 5 \times 10^{-8}, E_a = 31$) Ru > Rh ($k = 2 \times 10^{-12}, E_a = 33$) [$k < 10^{-8} (163°C), E_a > 42$]	k, l
bzac	Isomerization	Mn > V > Co[d] [$k = 8 \times 10^2 (80°C)$] ($k = 5 \times 10^{-9}, E_a = 33$)	k, m
pmhd	Isomerization + inversion	Sc > Ga[e] > Al[e] ($k \sim 8, E_a = 20$) ($k \sim 10^{-4}, E_a \sim 29$)	n

		Co	Rh	Cr	Ru	
acac	Inversion	$(k = 8 \times 10^{-10}, E_a = 35)$ ≳	[$k < 2 \times 10^{-6}$ (165°C), $E_a > 42$] ≳	$(k = 2 \times 10^{-11}, E_a = 35)$ >	[$k < 3 \times 10^{-6}$ (135°C), $E_a > 39$]	o
		Ga	>	Al		p
acac-hfac[f]	Site exchange	$(k = 8 \times 10^{2}, E_a = 15)$		$(k = 80, E_a = 18)$		
		Ga	>	Al		p
acac-dbm[g]	Site exchange	$(k = 8 \times 10^{-2}, E_a = 20)$		$(k = 2 \times 10^{-3}, E_a = 22)$		

[a] Measured or extrapolated value at 25°C unless otherwise noted; $k(\text{sec}^{-1})$, $E_a(\text{kcal/mole})$.
[b] Data for R = C_3H_5 complexes.
[c] Eaton et al. (1973). $k(\text{sec}^{-1})$, $E_a(\text{kal/mole})$.
[d] C → T; T → C kinetic parameters similar.
[e] Calculated using average values of the kinetic data obtained by lineshape analysis.
[f] M(acac)₂(hfac).
[g] M(acac)₂(dbm).
[h] Eaton et al. (1972).
[i] Eaton and Holm (1971b).
[j] Ito et al. (1969).
[k] Gordon et al. (1971).
[l] Fay and Piper (1964).
[m] Girgis and Fay (1970).
[n] Hutchison et al. (1971).
[o] Fay et al. (1970).
[p] Case and Pinnavaia (1971).

by the lack of any definitive mechanistic information for fast bis- and tris-(β-diketonates). The findings that Co(α-RT)$_3$ and Co(III) β-diketonates rearrange by twist and bond rupture pathways, respectively, suggest that rate differences for tropolonate and β-diketonate complexes of other metal ions may be due to similar mechanistic differences. With the present state of knowledge it is possible only to summarize those features of tropolonates that appear to favor the twist mechanism for which the transition state is an idealized trigonal prism (Fig. 13) (Eaton et al., 1973).

1. The rigid, planar nature of the tropolonate ligand should tend to suppress a bond rupture mechanism, since additional energy in the form of M—O or M—O—C bond deformations would be required to remove one end of the ligand from bonding distance to the metal compared to the case for the internally flexible β-diketonates.

2. The relatively short bite distance of the tropolonate ligand, about 2.5 Å (Muetterties and Guggenberger, 1972), leads to polyhedron radius to polyhedron edge ratios that are close to the ideal value of 0.76 given by Day and Hoard (1970) for TP geometry. The following ratios have been calculated from X-ray data assuming bite and M—O distances are the same in the TP as in the ground state trigonal antiprismatic (TAP) configuration: CoT$_3$ 0.74; Co(acac)$_3$, 0.66; AlT$_3$, 0.76; Al(acac)$_3$, 0.69; GaT$_3$, 0.78 (estimated); MnT$_3$, ~0.78.

3. Ligand electrostatic repulsion energy differences in TP and TAP geometries decrease with increasing ratio of M—O to bite distance (Kepert, 1972); these energy differences are smaller in the tropolonates.

An additional factor that could promote the probability of a trigonal twist pathway is a ground state distortion toward TP geometry. The twist angle ϕ is 49° in AlT$_3$ (Muetterties and Guggenberger, 1972), not the octahedral value of 60°, which is closely approached by tris(β-diketonates). This change from the octahedral toward the trigonal prismatic value of 0° may have a favorable effect on the energetics for twisting about an r-C$_3$ or p-C$_3$ axis. The corresponding angle in CoT$_3$ is ~55° (R. Eisenberg and A. P. Gaughan unpublished results, 1972) and in MnT$_3$ ~ 49° (Fackler et al., 1972). This structural factor alone is insufficient to explain the fast rates and the operation of the twist mechanism in the M(α-RT)$_3$ series.

b. *Complexes with the* M—S$_6$ *Core.* Nonrigid complexes of this type are at present confined to the tris(N,N-disubstituted dithiocarbamates) 33, with R$_1$, R$_2$ = alkyl, aryl, and M = Fe(III, IV), Ru(III), and Co(III). From a variety of diffraction and spectroscopic studies, it has been well established that the C—N bonds in dithiocarbamate complexes possess a degree of double bond character resulting from a contribution of the ligand VB struc-

9. Nonrigid Chelate Complexes

$$\underset{33}{\underset{R_1}{R_2}}N=\!\!\!\!\underset{S}{\overset{S}{\diamondsuit}}M/3 \qquad \underset{34}{\underset{-S}{\overset{-S}{\diamondsuit}}C=\!\overset{+}{N}\underset{R_1}{\overset{R_2}{\diamondsuit}}}$$

ture **34**. Consequently, there are two potential sources of nonrigidity in tris(dithiocarbamates): restricted rotation about C—N bonds and metal-centered rearrangements resulting in inversion and/or isomerization. Geometrical isomers are possible in complexes with $R_1 \neq R_2$ and will be detectable by nmr only when C—N bond rotation is slow. In a number of M—S_6 complexes, including several of the mixed ligand type (Section III,C,2) two distinct kinetic regions are observable in the dnmr spectra. In several cases the averaging patterns demonstrate that the LTP involves inversion. Accordingly, the HTP has been assigned to bond rotation, although other types of rearrangements such as cis–trans isomerization cannot be eliminated (Musher, 1972).

A particularly clear example of the type and mechanism of a low-temperature rearrangement reaction is afforded by the Fe(IV) complex [Fe-(Me, PhCH$_2$-dtc)$_3$]$^+$ (Duffy and Pignolet, 1972). This species has a spin–triplet ground state and, as is the case with many paramagnetic complexes (Holm, 1969; Holm and Hawkins, 1973), isotropic (paramagnetic) contributions to the total chemical shifts [Eq. (2)] produce a nonlinearly expanded shift scale such that signals of most or all of the nonequivalent sites are resolvable. The pmr spectra of this complex below $-100°C$ (Fig. 17) reveal four methyl and seven of the eight resonances of the diastereotopic methylene protons, observations which require that bond rotation and metal-centered processes are slow under these conditions. At the fast exchange limits of the LTP, three methyl ($-77°C$) and four methylene signals ($-47°C$) are found. Spectra near the slow exchange limit signify an approximately statistical isomeric mixture (3 trans:1 cis), and the coalescence of four methyl signals to three indicates y, z site averaging in trans [site interchange (9)]. Collapse of the methylene spectra to four resonances conforms to site interchanges (10) and (11) and thus to the averaging set A_6 in Table IV. These spectral changes are entirely equivalent to those of the LTP of Al(III), Ga(III), and Co(III) α-substituted tropolonates. As in those cases, site permutation corresponding to set A_2 may be eliminated, since this requires coalescence of only two pairs of methylene signals of the trans isomer with four trans resonances unaffected. Hence, the LTP of [Fe(Me, PhCH$_2$-dtc)$_3$]$^+$ results in inversion without isomerization, which may reasonably be conceived in terms of a trigonal twist (Fig. 18). Kinetic parameters for this process were not evaluated, because the slow exchange spectrum of the LTP was not completely frozen out. The HTP (Fig. 17) is consistent with any rearrangement that randomly

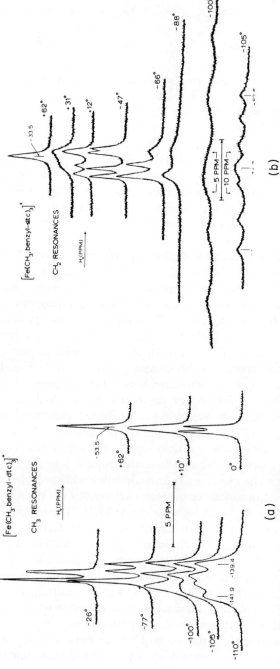

Fig. 17 Pmr spectra (100 MHz) of [Fe(Me, PhCH$_2$-dtc)$_3$](BF$_4$) in CD$_2$Cl$_2$ solution: (a) methyl spectra. (b) methylene spectra. Chemical shifts are in parts per million relative to CHDCl$_2$ internal standard. (Duffy and Pignolet, 1972.)

9. Nonrigid Chelate Complexes

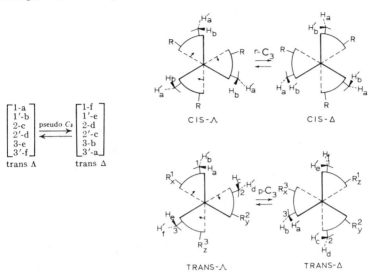

Fig. 18 Inversion of the cis-Λ and trans-Λ isomers of $[Fe(Me, PhCH_2\text{-}dtc)_3]^+$ proceeding by a trigonal twist about the real and pseudo C_3 axes, respectively. R refers to the methyl group and H to the methylene protons. Numbers and primes refer to substituent groups, while letters refer to magnetic environments. The site interchange pattern consistent with the methylene spectra in Fig. 17 is shown. (Duffy and Pignolet, 1972.)

scrambles all cis and trans environments, and it has been proposed that the process involved is C—N bond rotation (Duffy and Pignolet, 1972).

A discovery of considerable importance is the nonrigidity of Ru(III) dithiocarbamates (Pignolet et al., 1973). This is the first instance of fast kinetic behavior of a chelate complex of any second or third transition series d^n ion ($n \neq 0, 10$). Methyl and methylene dnmr spectra of $Ru(Me, PhCH_2\text{-}dtc)_3$ are shown in Fig. 19. The slow exchange spectra resolve four methyls and seven methylene signals and demonstrate the presence of cis and trans isomers. Coalescence behavior is similar to that of $[Fe(Me, PhCH_2\text{-}dtc)_3]^+$, except that the collapse of the methylene resonances from eight to four is not well defined. At the fast exchange limit of the LTP (43°C) the two methyl trans resonances (T_2, $T_{1,3}$) are accidentally degenerate; the anticipated four methylene signals are possibly broadened due to the onset of C—N bond rotation. The dnmr spectra are consistent with inversion of the cis and trans isomers by the trigonal twist pathway. $Fe(R_1R_2\text{-}dtc)_3$ complexes are also nonrigid (Palazzotto and Pignolet, 1972), and detailed studies of the type just outlined provide evidence for low temperature rearrangements by means of trigonal twists (Palazzotto et al., 1973). Nonrigidity of $Co[(PhCH_2)_2\text{-}dtc]_3$ has been detected at elevated temperature. This complex is of the $M(L-L)_3$ type and C—N bond rotation has no effect on the nmr spectrum. The AB spectrum

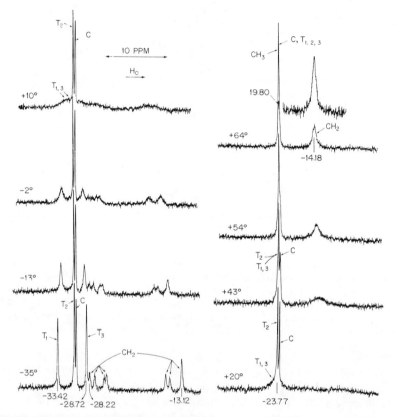

Fig. 19 Pmr spectra (100 MHz) of the methyl and methylene regions of Ru(Me, PhCH$_2$-dtc)$_3$ in CD$_2$Cl$_2$ solution. Chemical shifts are in parts per million relative to CHDCl$_2$ internal standard. (Pignolet et al., 1973.)

of the diastereotopic methylene protons coalesces to a singlet in nitrobenzene above 180°C (Pignolet et al., 1973; Palazzotto et al., 1973), indicating inversion. No mechanistic information is available for this process.

A detailed body of kinetic data for dithiocarbamate rearrangements has not yet been obtained. However, the following rate constants (sec^{-1}) and activation enthalpies (kcal/mole) have been reported (Pignolet et al., 1973; Palazzotto et al., 1973): Fe(R$_1$R$_2$-dtc)$_3$, —, 7.6–10.3; [Fe(R$_1$R$_2$-dtc)$_3$]$^+$, —, 8.4; Ru(Me, PhCH$_2$-dtc)$_3$, 78(10° C), ~13; Co[(PhCH$_2$)$_2$-dtc]$_3$, 18(168°C), 26; Rh[(PhCH$_2$)$_2$-dtc]$_3$, —, ≥27. The Rh(III) complex is rigid, as indicated by retention of the AB methylene spectrum up to 200°C in nitrobenzene. These data lead to the metal kinetic series in (14) for the complexes **33**.

$$\text{Fe(IV)} \gtrsim \text{Fe(III)} > \text{Ru(III)} \gg \text{Co(III)} > \text{Rh(III)} \quad (14)$$

This rate order does not follow an ionic size relationship and the only evident regularities are the slower rates of the diamagnetic complexes and of second row complexes compared to their first row analogs. Iron(III) and cobalt(III) dithiocarbamates and xanthates are decidedly distorted from the TAP limit with twist angles of 33°–41° and 42°–43° (Healy and White, 1972). This structural effect may reduce the barrier for the twist mechanism of $Fe(R_1R_2\text{-dtc})_3$, but cannot account for large differences in rates and activation enthalpies between Fe(III) and Co(III) complexes. Cobalt(III) xanthates have inversion rate constants of about 10^{-5}–10^{-7} sec^{-1} at or below 60°C (Krebs and Rasche, 1954; Krebs and Schumacher, 1966; Krebs et al., 1956). The complexity of metal kinetic series is emphasized by inversion of Ru(III) and Co(III) in (14) compared to the rate orders for $M(\alpha\text{-RT})_3$, $M(\text{tfac})_3$, and $M(\text{acac})_3$ (Table VII). These series are a consequence of a complicated interplay of metal ion and ligand structural and electronic effects in the ground and transition states that cannot yet be adequately explained.

2. Mixed Ligand Complexes

Dynamic nmr behavior of mixed ligand M—O_6 complexes have been referred to in the preceding sections, and kinetic data are summarized in Table VI. Fuller descriptions of the nmr features of such species are available elsewhere (Fortman and Sievers, 1971; Serpone and Bickley, 1973). In addition, the dnmr spectra of $Co(\text{acac})_2(4,7\text{-Me}_2\text{phen})$ has been investigated (LaMar, 1970). Site exchange is observed in these cases but in the absence of diastereotopic groups it cannot be established whether the rearrangement proceeds with inversion of configuration (Musher, 1972). Inversion has been demonstrated for $Al(\text{acac})_2(\text{dibm})$ (Jurado and Springer, 1971), but the mechanism of the reaction is unknown. High-spin Fe(II) complexes of the type $Fe(R_1R_2\text{-dtc})_2(\text{phen})$ and $Fe(R_1R_2\text{-dtc})_2(\text{bipy})$ have also been examined (Palazzotto et al., 1973). Pmr shifts exhibit Curie behavior over the range −100° to 30°C. Low-temperature signal multiplicities have been interpreted in terms of slow C—N bond rotations and fast metal-centered rearrangement involving inversion via a trigonal twist.

The most detailed study of the dynamic stereochemistry of a mixed ligand complex is that of the bis(dithiocarbamato)dithioleneiron species **35** (R = CN, CF$_3$; R$_1$, R$_2$ = alkyl, Ph) (Pignolet et al., 1971, 1972). Like the tris-(dithiocarbamates) **33**, these complexes have the potential of exhibiting nonrigidity as a consequence of metal-centered rearrangements and restricted C—N bond rotation. Two kinetic regions are observed in the dnmr spectra of these complexes, as illustrated for $Fe(\text{Et}_2\text{-dtc})_2(\text{mnt})$ in Fig. 20. The extremely large chemical shifts and signal separations are due to isotropic interactions.

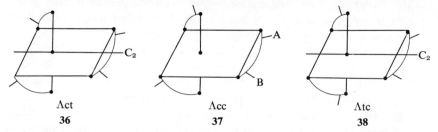

35

Complexes of type **35** are thermally distributed over singlet and triplet spin states, the interconversion between which is sufficiently fast that averaged shifts are observed. The presence of four signals from the diastereotopic methylene protons at slow exchange ($-80°C$) demonstrates slow metal-centered and bond rotation processes. The averaging of these signals to yield two resonances at the fast exchange limit of the LTP (about $-16°C$) can be accommodated by any process that effects pairwise equilibration of the four inequivalent protons. The two physical processes considered initially were inversion, which averages H_a with H_b and H_a' with H_b', and bond rotation, which averages H_a with H_a' and H_b with H_b'. Coalescence of the two methyl resonances is also consistent with either process.

A distinction between bond rotation and other processes is afforded by the dnmr spectra of the related complex $Fe(Me, Ph\text{-}dtc)_2(tfd)$ (**35**, $R = CF_3$) (Pignolet et al., 1971). Under conditions of slow bond rotation, this complex is of the $M(L-L')_2(L-L)$ class and can exist as three geometrical isomers. These are the cis–trans (ct, **36**), cis–cis (cc, **37**), and trans–cis (tc, **38**) forms (shown in the Λ configuration). Of these, cc lacks a twofold axis and thus contains two inequivalent methyl and CF_3 groups. The metal-centered rearrangements of this general type of complex have been described in terms

Λct	Λcc	Λtc
36	**37**	**38**

of topological correlation diagrams for the various physical mechanisms (Pignolet et al., 1971) analogous to those for the $M(L-L')_3$ class. In addition,

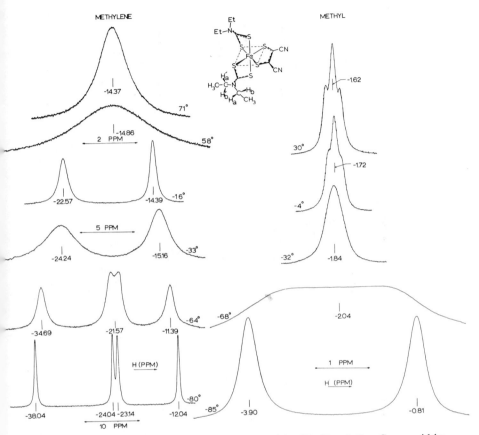

Fig. 20 Pmr spectra (100 MHz) of Fe(Et$_2$-dtc)$_2$(mnt) in CD$_2$Cl$_2$ solution. Sweep widths vary and are indicated below each series of spectra. Chemical shifts are downfield of tetramethylsilane reference. (Pignolet et al., 1972.)

permutational analyses of site interchanges have been performed (Musher, 1972; Eaton and Eaton, 1973). The slow exchange spectra of Fe(Me, Ph-dtc)$_2$(tfd) (Fig. 21) reveal the presence of all three isomers. Both the methyl and ^{19}F spectra demonstrate the occurrence of ct ⇌ tc interconversion by a process that does not simultaneously equilibrate these isomers with cc. The pmr spectra show that the two methyl sites of the cc isomer are averaged in the LTP, whose fast exchange limit is reached below 0°C. Retention of ^{19}F spin–spin multiplet components provides the further constraint that the inequivalent CF$_3$ groups of cc are not averaged in this process. These results are inconsistent with C—N bond rotation, which would equilibrate all three isomers, but can be rationalized by a trigonal twist pathway. The process **39 ⇌ 40 ⇌ 41** for the cc isomer averages methyl environments a and b

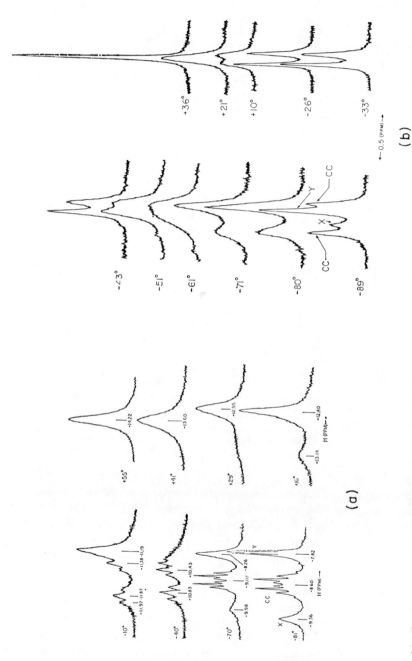

Fig. 21 Dnmr spectra of Fe(Me, Ph-dtc)₂(tfd) in CD₂Cl₂ solution. Signals x, y refer to ct, tc isomers. (a) ^{19}F spectra; sweep widths are not the same at each temperature. (b) Methyl spectra. Chemical shifts move progessively to lower field as the temperature is raised due to increasing population of the triplet spin state. (Pignolet et al., 1971.)

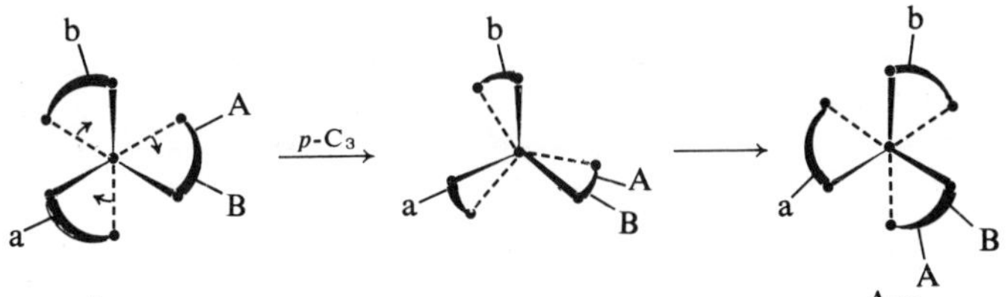

9. Nonrigid Chelate Complexes

Λcc
39

40

Δcc
41

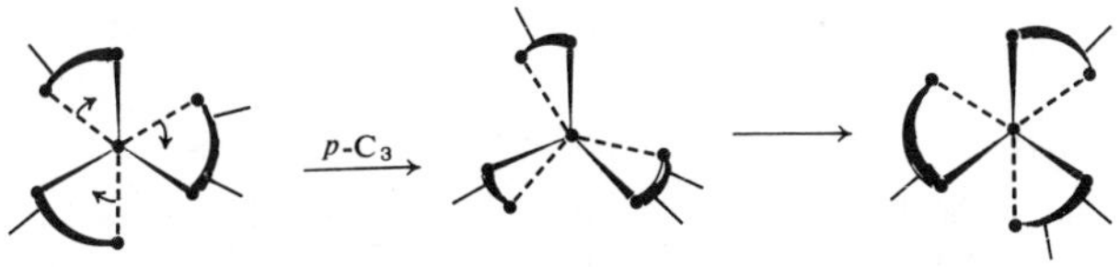

Λct
42

43

Δtc
44

without affecting CF$_3$ environments A and B, as required by the dnmr spectra. The ct ⇌ tc interconversion can be interpreted by a similar mechanism (**42** ⇌ **43** ⇌ **44**), which does not involve the cc isomer. Hence, the LTP of this and similar complexes is consistent with inversion of configuration by means of a trigonal twist. No other physical mechanism of the types considered earlier can sensibly accommodate the averaging features of the cc form. An ambiguity remains, however. Because only nondiastereotopic groups are present in Fe(Me, Ph-dtc)$_2$(tfd), there is no direct proof that its LTP is one of inversion. As recognized explicitly by permutational analysis (Musher, 1972; Eaton and Eaton, 1973), inversion of the cc isomer of this complex cannot be distinguished from any other process that results in methyl a ⇌ b site interchange without inversion. The problem is similar to distinguishing between the M(L–L')$_3$ averaging sets A_2 and A_6, and requires the use of diastereotopic probes. In the complexes **35**, the diastereotopic groups must be incorporated in the dithiolene ligand, a nontrivial synthetic task that has not yet been accomplished.

While the trigonal twist cannot be considered proven for bis(dithiocarbamato)dithioleneiron complexes, there are several additional factors that might be considered to support it. Examination of processes that result in site interchange without inversion or bond rupture suggest that a greater extent of ligand motion is required to achieve the several types of transition states that might be involved than is the case for the trigonal twist. The ground state structure of Fe(Et$_2$-dtc)$_2$(tfd) (Johnston et al., 1971) is severely distorted from the TAP configuration with twist angles of 33° (dtc) and 41° (tfd), a situation that should reduce the activation energy required to reach the TP transition states **40** or **43**.

TABLE VIII
KINETIC PARAMETERS FOR METAL-CENTERED REARRANGEMENTS OF $Fe(R_1R_2\text{-dtc})_2(S_2C_2R_2)$ IN CD_2Cl_2 SOLUTION[a]

R	R_1, R_2	$k^{233°C}$ (sec^{-1})	$\Delta H^{\ddagger}$ (kcal/mole)	$\Delta S^{\ddagger}$ (eu)
CF_3	Me, Me	1.89×10^2	9.2 (0.9)	−6.1 (3.9)
CF_3	Et, Et	7.85×10^2	8.3 (0.6)	−7.5 (3.7)
CN	Et, Et	3.11×10^3	8.6 (1.5)	−3.4 (5.0)

[a] Pignolet et al. (1972). Estimated errors in parentheses.

Rate constants and activation parameters for the LTP of the complexes **35**, assigned as a trigonal twist (Pignolet et al., 1971, 1972) prior to full permutational analysis of the $M(L-L')_2(L-L)$ cases, are summarized in Table VIII. The HTP of these complexes has been postulated to be C—N bond rotation. Activation enthalpies are 3–8 kcal/mole larger than for the LTP and are tabulated elsewhere (Pignolet et al., 1972).

D. Sexadentate Complexes

Nonrigidity has been detected in the high-temperature dnmr spectra of the complexes **45** derived from the ligand py$_3$tame (Wandiga et al., 1972). These species contain diastereotopic methylene protons and chiral structures are indicated by the AB patterns observed for diamagnetic [Fe(py$_3$tame)]$^{2+}$ and [Co(py$_3$tame)]$^{3+}$ at ambient temperature. In D_2O solution, the spectrum of the Fe(II) complex collapses to a broad singlet at $\sim 95°C$, while the spectrum of the Co(III) complex in dimethylsulfoxide coalesces to a broad feature at $\sim 145°C$. These spectral changes are consistent with inversion of configuration and the following kinetic data [$k^{353°C}$ (sec^{-1}), E_a(kcal/mole)] have been obtained: Fe(II), 123, 18.4; Co(III), 678, 10.1. The reaction pathway is unknown. It has been postulated that these complexes rearrange by a trigonal twist, inasmuch as the rather rigid nature of the sexadentate ligand should

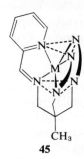

45

tend to suppress outright bond rupture. In the solid state, [Fe(py$_3$tame)]$^{2+}$ has an average twist angle of 43° (Fleischer et al., 1972). Other complexes of py$_3$tame and similar sexadentates have considerably smaller twist angles (Fleischer et al., 1972; Larsen et al., 1972), indicating that a TP transition state is not unreasonable for such systems. The structure of [Co(py$_3$tame)]$^{3+}$ is not known. The smaller size of the Co(III) ion will presumably result in a larger twist angle (Larsen et al., 1972; Zakrewski et al., 1971). If so, the rate order Co(III) > Fe(II) is not simply a consequence of ground state structures, a matter emphasized by the large difference in preexponential factors (Wandiga et al., 1972).

IV. Systems of Higher Coordination Number

A rather large number of coordination polyhedra with different idealized symmetries are possible for molecules with coordination number > 6. For a given CN, certain of these geometries may be interconverted by relatively small nuclear displacements (Lippard, 1966; Muetterties and Wright, 1967; Muetterties, 1970, 1972). This situation could lead to facile thermally induced rearrangements involving coordination polyhedra of comparable energies that may represent stable molecular configurations or readily accessible transition states for rearrangement. With the exception of several organometallic chelate complexes described below, this situation appears to prevail for species with coordination numbers 7 and 8. Thus far only the fast exchange limiting nmr spectra for such molecules have been observed. For the seven-coordinate species M(acac)$_3$X, at least two inequivalent methyl sites should exist in each of the idealized coordination geometries, yet at temperatures down to −130°C, only a single methyl resonance is found in the Zr(IV) and Hf(IV) complexes (Pinnavaia and Fay, 1968). The structures of most chelates of the types M(L–L)$_4$ and M(L–L')$_4$ approach square antiprismatic, dodecahedral, or bicapped trigonal prismatic forms. There are, for example, three and six geometrical isomers possible for antiprismatic and dodecahedral M(L–L)$_4$, respectively, and nearly all of these forms have ring substituent site inequivalencies. The ^{19}F spectra of M(tfac)$_4$ [M = Zr(IV), Hf(IV), Ce(IV), Th(IV)] indicate rapid rearrangement, since at −105°C in acetone, only a single resonance is found (Pinnavaia and Fay, 1966). Tetrakis-β-diketonate complexes undergo intermolecular ligand exchange that is slow on the nmr time scale at ambient temperature. Signal multiplicities in the systems M(acac)$_{4-n}$(tfac)$_n$ (Adams and Larsen, 1966; Pinnavaia and Fay, 1966) and Y(hfac)$_{4-n}$(tfac)$_n$ (Cotton et al., 1966) are consistent with the presence of mixed ligand complexes, but site inequivalencies within each were not detectable. Spectra of tetrakisdithiocarbamate, β-diketonate, and tropolonate chelates of lanthanides, Th(IV), and U(IV) similarly imply rapid intramolecular averaging if structures of D_2, C_2, and lower symmetry found for symmetrical chelates in the solid state persist in solution (Holm and Hawkins, 1973).

The full range of dnmr spectra of chelate complexes with coordination number >6 has been observed for CpM(acac)$_2$X [M = Zr(IV), Hf(IV); X = Cl, Br] (Pinnavaia and Lott, 1971) and CpZr(β-diketonate)$_3$ containing hfac, tfac, and dbm (Howe and Pinnavaia, 1970). X-ray structural investigations of CpZr(acac)$_2$Cl (Stezowski and Eick, 1969) and CpZr(hfac)$_3$ (Elder et al., 1969) have established the structures represented schematically by **46** and **47**. Each contains a pentahaptocyclopentadienyl group. Structure **46** indicates the presence of two and four inequivalent —CH= and methyl groups, respectively, and the slow exchange spectrum (42°C) is consistent with C_1 symmetry. The spectrum of CpZr(bzac)$_2$Cl indicates the presence of four geometrical isomers based on **46** with the chlorine cis to the Cp group. The dnmr spectra of CpM(acac)$_2$X reveal coalescence of the two methine and four methyl signals to a single resonance above about 80°C. Methyl site averaging appears to be random and a mechanism cannot be established from these results. Spectra of CpZr(hfac)$_3$ are somewhat more complicated and indicate the operation of two kinetic processes. The presence of four ^{19}F and two methine signals at slow exchange (−43°C) accords with structure **47** (R = CF$_3$). The LTP interconverts the pairs of nonequivalent CF$_3$ groups A and B with fast exchange being reached at 50°C. At higher temperatures, the unique hfac ligand exchanges with the two equivalent equatorial ligands, resulting in simultaneous averaging of the two methine and all nonequivalent CF$_3$ environments. The dnmr spectra of CpZr(dbm)$_3$ are similar but indicate larger activation energies for the two processes, consistent with ligand kinetic series (12). Intermolecular ligand exchange cannot be ruled out in the HTP.

TABLE IX

KINETIC PARAMETERS FOR REARRANGEMENTS OF CYCLOPENTADIENYL-β-DIKETONATE Zr(IV) AND Hf(IV) COMPLEXES[a]

Complex	Solvent	E_a (kcal/mole)	$\Delta S^\ddagger$ (eu)	$k^{298°C}$ (sec^{-1})	Reference
CpZr(acac)$_2$Cl[b]	C$_6$H$_6$	23.1 (2.0)	7.1 (5.4)	0.0068	Pinnavaia and Lott (1971)
CpZr(acac)$_2$Br[b]	C$_6$H$_6$	22.8 (1.8)	5.7 (4.8)	0.0061	Pinnavaia and Lott (1971)
CpHf(acac)$_2$Cl[b]	C$_6$H$_6$	24.2 (2.2)	7.5 (6.0)	0.0013	Pinnavaia and Lott (1971)
CpZr(hfac)$_3$[c]	(Isopropyl)$_2$O	15.7 (0.5)	5.1 (1.7)	646	Howe and Pinnavaia (1970)
CpZr(dpm)$_3$[c]	o-xylene[e]	21.3 (1.9)	3.8 (5.5)	0.0284	Howe and Pinnavaia (1970)
CpZr(hfac)$_3$[d]	(Isopropyl)$_2$O	26.4 (1.9)	17.2 (5.2)	0.0045	Howe and Pinnavaia (1970)

[a] Estimated errors in parentheses.
[b] Methyl site exchange.
[c] Averaging of CF$_3$ or t-Bu groups A and B in 47.
[d] Axial–equatorial ligand interchange (possibly an intermolecular process).
[e] Parameters similar in diisopropyl ether and toluene.

Additional studies with CpZr(tfac)$_3$ suggest that the LTP proceeds by either a bond rupture or digonal twist mechanism involving the equatorial ligands (Howe and Pinnavaia, 1970). Kinetic parameters for the rearrangements of complexes **46** and **47** are set out in Table IX.

V. Summary

The investigations described in this chapter and elsewhere (Fortman and Sievers, 1971; Serpone and Bickley, 1973; Pignolet and LaMar, 1973) provide ample testimony that the ability to detect and kinetically characterize the intramolecular rearrangement processes of nonrigid metal chelate complexes is now rather highly developed. These advances in inorganic stereochemistry and reaction kinetics have evolved in the last decade and result from perspicacious utilization of the theory and practice of dnmr spectroscopy by a variety of workers. Particularly contributory to the current state of the art in this area of inorganic reaction chemistry are computer line-shape programs capable of handling efficiently multisite exchange problems, evolvement of methods for permutational analysis of site exchange, frequent facile inclusion of diastereotopic probes within molecular structure allowing detection of chirality changes, and exploitation of isotropic interactions in paramagnetic molecules, which in favorable cases can allow visual determination of site exchange without recourse to computer simulation of spectra. It is abundantly clear that the capacity to determine accurate kinetic parameters has in general outstripped the ability to provide convincing mechanistic arguments for the rearrangement reactions in question, a situation indigenous to virtually all forms of chemical kinetics.

For stereochemically nonrigid molecules of the types considered here, it is manifest that secure mechanistic deductions can follow only if site interchange patterns, once established, are consistent with a small number or, ideally, one set of the total array of site exchanges generated by permutational analysis. At this point, the experimental observables and theoretical constraints afforded by the dnmr method are exhausted and the chemist must turn to subjective criteria consistent with microscopic reversibility, such as the principle of least motion and the physical reasonableness of intermediates or transition states, in framing mechanistic propositions. In this light, the trigonal twist rearrangement pathway for species such as Al(α-RT)$_3$, Co(α-RT)$_3$, [Fe(Me, PhCH$_2$-dtc)$_3$]$^+$, and Ru(Me, PhCH$_2$-dtc)$_3$ may be considered as "established" with, however, recognition of the proviso that the nmr data provide no direct evidence that metal–ligand bonds remain intact during the inversion process. Indeed, a twist-with-rupture mechanism (Eaton et al., 1972) is equally consistent with the observed site interchanges. This mechanism and the conventional visualization of the trigonal twist (Figs. 10, 13, and 18) are limiting representations of the same overall transformation—inversion without isomerization—and are not subject to direct experimental differentiation in those cases where the trigonal twist has been proposed.

Equally vexing at present is the problem of establishing those structural and electronic features responsible for a particular rearrangement pathway. While metal and ligand kinetic series have been defined, it is clear that these may be of limited generality, a point illustrated by the rate order Co(III) > Ru(III) for tropolonates and β-diketonates and the reverse for dithiocarbamates. Delineation of these factors is best attempted for those complexes which ostensibly rearrange by the same mechanism. Attempts to correlate activation enthalpies of $M(R_1R_2\text{-dtc})_3$ complexes with calculated values of TP–TAP ligand field stabilization energies has been only partly successful (Palazzotto et al., 1973). Further, the presence of a significant ligand field barrier to formation of a TP structure on the one hand and the absence of such an effect on the other obviously does not dominate the inversion kinetics of the Co(III) and Al(III) tropolonates (Table VI). Attempts to relate rearrangement rates and operation of the twist mechanism to significant ground state distortions, first suggested by studies of $Fe(R_1R_2\text{-dtc})_2(\text{tfd})$ (Pignolet et al., 1971), are not uniformly convincing.

While all complexes with twist angles substantially less than 60° for which mechanistic information is available appear to adopt the twist pathway, such angular distortions are not a sufficient criterion of mechanism. Cobalt(III) tropolonates invert by a twist yet for CoT_3 $\phi \sim 55°$. Further, Co(III) and predominantly low-spin Fe(III) dithiocarbamates have nearly identical twist angles, but the former rearrange only at elevated temperatures (by an as yet unestablished mechanism). Stereochemical stabilization of the assumed TP transition state for the twist is also presumably significant. As already discussed, tris(tropolonates) appear to conform rather closely to the ideal radius–edge ratio of 0.76, which is one purely geometrical criterion of stability. Values for Fe(III) dithiocarbamates (0.82–0.85) do not as closely fit this criterion.

These attempted relationships, which are part of a larger body of similarly unclear mechanistic and structure-electronic variables, are cited here in order to emphasize that the problem of deciphering the mechanisms of chelate rearrangements does not end with deduction of the most probable mechanism. It can be anticipated that the factors that contribute to a particular mechanistic pathway, like previous and future mechanistic studies themselves, will come under increasingly sophisticated and rigorous scrutiny. By these means, it is hoped that a satisfactory explanation of the factors that promote the various rearrangement pathways of both rigid and nonrigid chelate complexes will ultimately be evolved. Certainly, one of the most interesting problems to be addressed in the near future concerns the generality of the trigonal twist pathway and, specifically, whether or not this concerted mechanism is confined to reactions with the relatively low barriers characteristic of nonrigid complexes.

REFERENCES

Adams, A. C., and Larsen, E. M. (1966). *Inorg. Chem.* **5**, 228.
Allerhand, A., Gutowsky, H. S., Jonas, J., and Meinzer, R. A. (1966). *J. Amer. Chem. Soc.* **88**, 3185.
Bailar, J. C. (1958). *J. Inorg. Nucl. Chem.* **8**, 165.
Basolo, F., and Pearson, R. G. (1958). "Mechanisms of Inorganic Reactions," 1st ed. Wiley, New York.
Basolo, F., and Pearson, R. G. (1967). "Mechanisms of Inorganic Reactions," 2nd ed. Wiley, New York.
Bertini, I., Sacconi, L., and Speroni, G. P. (1972). *Inorg. Chem.* **11**, 1323.
Bradley, D. C., and Holloway, C. E. (1969). *J. Chem. Soc., A* p. 282.
Case, D. A., and Pinnavaia, T. J. (1971). *Inorg. Chem.* **10**, 482.
Chen, Y. T., and Everett, G. W. (1968). *J. Amer. Chem. Soc.* **90**, 6660.
Clark, R. J. H., and McAlees, A. J. (1970). *J. Chem. Soc., A* p. 2026.
Clark, R. J. H., and McAlees, A. J. (1972a). *Inorg. Chem.* **11**, 342.
Clark, R. J. H., and McAlees, A. J. (1972b). *J. Chem. Soc., Dalton Trans.* p. 640.
Cotton, F. A., and Wilkinson, G. (1972). "Advanced Inorganic Chemistry," 3rd ed., pp. 656–658. Wiley (Interscience), New York.
Cotton, F. A., Legzdins, P., and Lippard, S. J. (1966). *J. Chem. Phys.* **45**, 3461.
Day, V. W., and Hoard, J. L. (1970). *J. Amer. Chem. Soc.* **92**, 3626.
Doddrell, D., and Roberts, J. D. (1970). *J. Amer. Chem. Soc.* **92**, 4484 and 5255.
Duffy, L. J., and Pignolet, L. H. (1972). *Inorg. Chem.* **11**, 2843.
Eaton, D. R. (1968). *J. Amer. Chem. Soc.* **90**, 4272.
Eaton, S. S., and Eaton, G. R. (1973). *J. Amer. Chem. Soc.* **95**, 1459.
Eaton, S. S., and Holm, R. H. (1971a). *Inorg. Chem.* **10**, 1446.
Eaton, S. S., and Holm, R. H. (1971b). *J. Amer. Chem. Soc.* **93**, 4913.
Eaton, S. S., Hutchison, J. R., Holm, R. H., and Muetterties, E. L. (1972). *J. Amer. Chem. Soc.* **94**, 6411.
Eaton, S. S., Eaton, G. R., Holm, R. H., and Muetterties, E. L. (1973). *J. Amer. Chem. Soc.* **95**, 1116.
Elder, M., Evans, J. G., and Graham, W. A. G. (1969). *J. Amer. Chem. Soc.* **91**, 1245.
Ernst, R. E., O'Connor, M. J., and Holm, R. H. (1967). *J. Amer. Chem. Soc.* **89**, 6104.
Ernst, R. E., O'Connor, M. J., and Holm, R. H. (1968). *J. Amer. Chem. Soc.* **90**, 5735.
Everett, G. W., and Chen, Y. T. (1970). *J. Amer. Chem. Soc.* **92**, 508.
Everett, G. W., and Holm, R. H. (1966). *J. Amer. Chem. Soc.* **88**, 2442.
Everett, G. W., and Holm, R. H. (1968). *Inorg. Chem.* **7**, 776.
Everett, G. W., and Johnson, A. (1972). *J. Amer. Chem. Soc.* **94**, 6397.
Everett, G. W., and King, R. M. (1972). *Inorg. Chem.* **11**, 2041.
Fackler, J. P., Avdeef, A. V., and Costamagna J. (1972). *Proc. 14th Int. Conf. Coord.* p. 589.
Faller J. W. and Davison A. (1967). *Inorg. Chem.* **6**, 182.
Fay, R. C., and Lowry, R. N. (1967). *Inorg. Chem.* **6**, 1512.

Fay, R. C., and Lowry, R. N. (1970). *Inorg. Chem.* **9**, 2048.
Fay, R. C., and Pinnavaia, T. J. (1968). *Inorg. Chem.* **7**, 508.
Fay, R. C., and Piper, T. S. (1962). *J. Amer. Chem. Soc.* **84**, 2303.
Fay, R. C., and Piper, T. S. (1963). *J. Amer. Chem. Soc.* **85**, 500.
Fay, R. C., and Piper, T. S. (1964). *Inorg. Chem.* **3**, 348.
Fay, R. C., Girgis, A. Y., and Klabunde, U. (1970). *J. Amer. Chem. Soc.* **92**, 7056.
Fleischer, E. B., Gebala, A. E., Swift, D. R., and Tasker, P. A. (1972). *Inorg. Chem.* **11**, 2775.
Fortman, J. J., and Sievers, R. E. (1967). *Inorg. Chem.* **6**, 2022.
Fortman, J. J., and Sievers, R. E. (1971). *Coord. Chem. Rev.* **6**, 331.
Gerlach, D. H., and Holm, R. H. (1969a). *J. Amer. Chem. Soc.* **91**, 3457.
Gerlach, D. H., and Holm, R. H. (1969b). *Inorg. Chem.* **8**, 2292.
Gerlach, D. H., and Holm, R. H. (1970). *Inorg. Chem.* **9**, 588.
Girgis, A. Y., and Fay, R. C. (1970). *J. Amer. Chem. Soc.* **92**, 7061.
Gordon, J. G., and Holm, R. H. (1970). *J. Amer. Chem. Soc.* **92**, 5319.
Gordon, J. G., O'Connor, M. J., and Holm, R. H. (1971). *Inorg. Chim. Acta* **5**, 381.
Harrod, J. F., and Taylor, K. (1971). *Chem. Commun.* p. 696.
Hawkins, C. J. (1971). "Absolute Configuration of Metal Complexes." Wiley (Interscience), New York.
Healy, P. C., and White, A. H. (1972). *J. Chem. Soc., Dalton Trans.* p. 1163.
Holloway, C. E., Luongo, R. R., and Pike, R. M. (1966). *J. Amer. Chem. Soc.* **88**, 2060.
Holm, R. H. (1969). *Accounts Chem. Res.* **2**, 307.
Holm, R. H., and Hawkins, C. J. (1973). *In* "NMR of Paramagnetic Molecules: Theory and Applications" (G. N. LaMar, W. D. Horrocks, and R. H. Holm, eds.), Chapter 7. Academic Press, New York.
Holm, R. H., and O'Connor, M. J. (1971). *Progr. Inorg. Chem.* **14**, 241.
Holm, R. H., Chakravorty, A., and Dudek, G. O. (1964). *J. Amer. Chem. Soc.* **86**, 379.
Horrocks, W. D. (1965). *J. Amer. Chem. Soc.* **87**, 3779.
Howe, J. J., and Pinnavaia, T. J. (1970). *J. Amer. Chem. Soc.* **92**, 7342.
Hutchison, J. R., Gordon, J. G., and Holm, R. H. (1971). *Inorg. Chem.* **10**, 1004.
Ito, T., Tanka, N., Hanazaki, I., and Nakagura, S. (1969). *Inorg. Nucl. Chem. Lett.* **5**, 781.
Jesson, J. P. (1973). *In* "NMR of Paramagnetic Molecules: Theory and Applications" (G. N. LaMar, W. D. Horrocks, and R. H. Holm, eds.), Chapter 1. Academic Press, New York.
Jesson, J. P., and Muetterties, E. L. (1969). "Chemist's Guide," pp. 4–6. Dekker, New York.
Johnston, D. L., Rohrbaugh, W. L., and Horrocks, W. D. (1971). *Inorg. Chem.* **10**, 1474.
Jones, R. W., Jr. (1971). Ph.D. Thesis, Cornell University, New York (quoted by Serpone and Bickley, 1973).
Jurado, B., and Springer, C. S. (1971). *Chem. Commun.* p. 85.
Kepert, D. L. (1972). *Inorg. Chem.* **11**, 1561.
Klemperer, W. G. (1973). *J. Amer. Chem. Soc.* **95**, 2105.
Krebs, H., and Rasche, R. (1954). *Naturwissenschaften* **41**, 63.
Krebs, H., and Schumacher, W. (1966). *Z. Anorg. Allg. Chem.* **344**, 187.
Krebs, H., Diewold, J., Arlitt, H., and Wagner, J. A. (1956). *Z. Anorg. Allg. Chem.* **287**, 98.
LaMar, G. N. (1965). *J. Amer. Chem. Soc.* **87**, 3567.
LaMar, G. N. (1970). *J. Amer. Chem. Soc.* **92**, 1806.

LaMar, G. N., and Sherman, E. O. (1969). *Chem. Commun.* p. 809.
LaMar, G. N., and Sherman, E. O. (1970). *J. Amer. Chem. Soc.* **92**, 2691.
Larsen, E., Mason, S. F., and Searle, G. H. (1966). *Acta Chem. Scand.* **20**, 191.
Larsen, E., LaMar, G. N., Wagner, B. E., Parks, J. E., and Holm, R. H. (1972). *Inorg. Chem.* **11**, 2652.
Lippard, S. J. (1966). *Progr. Inorg. Chem.* **8**, 109.
Longuet-Higgins, H. C. (1963). *Mol. Phys.* **6**, 445.
Lowry, T. M., and Traill, R. C. (1931). *Proc. Roy. Soc., Ser. A* **132**, 398.
Moeller, T., Gulyas, E., and Marshall, R. H. (1959). *J. Inorg. Nucl. Chem.* **9**, 82.
Muetterties, E. L. (1968). *J. Amer. Chem. Soc.* **90**, 5097.
Muetterties, E. L. (1970). *Accounts Chem. Res.* **3**, 266.
Muetterties, E. L. (1972). *MTP Int. Rev. Sci.* **9**, Chapter 2.
Muetterties, E. L., and Alegranti, C. W. (1969). *J. Amer. Chem. Soc.* **91**, 4420.
Muetterties, E. L., and Guggenberger, L. J. (1972). *J. Amer. Chem. Soc.* **94**, 8046.
Muetterties, E. L., and Wright, C. M. (1967). *Quart. Rev., Chem. Soc.* **21**, 109.
Musher, J. I. (1972). *Inorg. Chem.* **11**, 2335.
O'Connor, M. J., Ernst, R. E., and Holm, R. H. (1968). *J. Amer. Chem. Soc.* **90**, 4184.
Palazzotto, M. C., and Pignolet, L. H. (1972). *J. Chem. Soc., Chem. Commun.* p. 6.
Palazzotto, M. C., Duffy, D. J., Edgar, B. L., Que, L., and Pignolet, L. H. (1973). *J. Amer. Chem. Soc.* **95**, 4537.
Palmer, R. A., Fay, R. C., and Piper, T. S. (1964). *Inorg. Chem.* **3**, 875.
Pignolet, L. H., and Horrocks, W. D. (1968). *J. Amer. Chem. Soc.* **91**, 3976.
Pignolet, L. H., and LaMar, G. N. (1973). *In* "NMR of Paramagnetic Molecules: Theory and Applications" (G. N. LaMar, W. D. Horrocks, and R. H. Holm, eds.), Chapter 8. Academic Press, New York.
Pignolet, L. H., Horrocks, W. D., and Holm, R. H. (1970). *J. Amer. Chem. Soc.* **92**, 1855.
Pignolet, L. H., Lewis, R. A., and Holm, R. H. (1971). *J. Amer. Chem. Soc.* **93**, 360.
Pignolet, L. H., Lewis, R. A., and Holm, R. H. (1972). *Inorg. Chem.* **11**, 99.
Pignolet, L. H., Duffy, D. J., and Que, L. (1973). *J. Amer. Chem. Soc.* **95**, 295.
Pinnavaia, T. J., and Clements, W. R. (1971). *Inorg. Nucl. Chem. Lett.* **7**, 127.
Pinnavaia, T. J., and Fay, R. C. (1966). *Inorg. Chem.* **5**, 233.
Pinnavaia, T. J., and Fay, R. C. (1968). *Inorg. Chem.* **7**, 502.
Pinnavaia, T. J., and Lott, A. L. (1971). *Inorg. Chem.* **10**, 1388.
Pinnavaia, T. J., and Nweke, S. O. (1969). *Inorg. Chem.* **8**, 639.
Pinnavaia, T. J., Sebeson, J. M., and Case, D. A. (1969). *Inorg. Chem.* **8**, 644.
Pinnavaia, T. J., Collins, W. T., and Howe, J. J. (1970a). *J. Amer. Chem. Soc.* **92**, 4544.
Pinnavaia, T. J., Matienzo, L. J., and Peters, Y. A. (1970b). *Inorg. Chem.* **9**, 993.
Que, L., and Pignolet, L. H. (1973). *Inorg. Chem.* **12**, 156.
Rây, P., and Dutt, N. K. (1943). *J. Indian Chem. Soc.* **20**, 81.
Röhrscheid, F., Ernst, R. E., and Holm, R. H. (1967). *Inorg. Chem.* **5**, 1315.
Serpone, N., and Bickley, D. G. (1973). *Progr. Inorg. Chem.* **17**, 391.
Serpone, N., and Fay, R. C. (1967). *Inorg. Chem.* **6**, 1835.
Springer, C. S. (1973). *J. Amer. Chem. Soc.* **95**, 1459.
Springer, C. S., and Sievers, R. E. (1967). *Inorg. Chem.* **6**, 852.
Springer, C. S., Sievers, R. E., and Feibush, B. (1971). *Inorg. Chem.* **10**, 1242.
Stezowski, J. J., and Eick, H. A. (1969). *J. Amer. Chem. Soc.* **91**, 2890.
Van Hecke, G. R., and Horrocks, W. D. (1966). *Inorg. Chem.* **5**, 1968.
Wandiga, S. O., Sarneski, J. E., and Urbach, F. L. (1972). *Inorg. Chem.* **11**, 1349.
Zakrewski, G. A., Ghilardi, C. A., and Lingafelter, E. C. (1971). *J. Amer. Chem. Soc.* **93**, 4411.

10

Stereochemical Nonrigidity in Organometallic Compounds

F. A. COTTON

Introduction	377
I. Cyclopolyenylmetal Systems	378
A. Monohaptocyclopentadienyl and 1-Indenyl Compounds	378
B. Cycloheptatrienyl Compounds	400
II. Cyclopolyenemetal Systems	403
A. Cyclooctatetraenemetal Compounds	403
B. (Cycloocta-1,3,5-triene)diiron Hexacarbonyl and Related Molecules	413
C. Cyclohexatrienemetal Compounds	416
III. Systems with Two Independent Metal Atoms on One Ring	417
IV. Scrambling of Differently Bonded Rings	419
A. Interchange of Mono- and Pentahaptocyclopentadienyl Rings	419
B. Interchange of Tetra- and Hexahaptocyclooctatetraene Rings	424
C. Interchange of Hexa- and Tetrahaptohexatriene Rings	425
V. Fluxionality in the Solid State	425
VI. Miscellaneous Systems	427
A. Allene Complexes	427
B. Rotation of Coordinated Monoolefins	428
C. Biscyclohexadienyliron Compounds	431
D. Nonconjugated Diene Ligands	432
E. Some Metallocyclic Compounds	433
F. Polynuclear Benzyne Complexes	434
VII. Some Nonfluxional Molecules, or the Dog in the Night-Time	435
References	437

Introduction

The term "stereochemically nonrigid" will be used for all molecules that undergo intramolecular rearrangements rapidly enough to influence nmr line shapes at temperatures within the practical range of experimentation. The term "fluxional" will be used for that large and important subclass of stereochemically nonrigid molecules in which all of the interconverting species that are observable are chemically and structurally equivalent.

The first comprehensive review (Cotton, 1968) of fluxional and other stereochemically nonrigid organometallic molecules is now quite out of date. A more recent, but fairly condensed review is also available (Vrieze and van Leeuwen, 1971). This chapter will present an up-to-date review, with emphasis on work published since the earlier reviews. However, for the sake of coherence and readability, and to make clear the basis for our beliefs regarding certain fundamental points, some material in the earlier reviews will be covered again here. This chapter is intended to be readable independently of the earlier reviews.

Metal allyl compounds are excluded, except for occasional passing references, from this chapter. The volume of significant results on them is so large as to merit a separate review by Professor Vrieze which forms Chapter 11 of this book.

I. Cyclopolyenylmetal Systems

A. *Monohaptocyclopentadienyl and 1-Indenyl Compounds*

These compounds occupy an especially important place in the field for several reasons. For one thing, they were the first fluxional organometallic molecules to be discovered (Piper and Wilkinson, 1956a,b). One of them, $(\eta^5\text{-}C_5H_5)Fe(CO)_2(\eta^1\text{-}C_5H_5)$, was then the first one to be subjected to a complete variable-temperature nmr study, followed by a line-shape analysis leading to the characterization of the rearrangement pathway (Bennett *et al.*, 1966). Finally, $(\eta^1\text{-}C_5H_5)M$ compounds have been more extensively studied since 1966 than any other type of fluxional organometallic molecule.

Within the entire class of $(\eta^1\text{-}C_5H_5)M$-containing species, the molecule $(\eta^5\text{-}C_5H_5)Fe(CO)_2(\eta^1\text{-}C_5H_5)$ and some isoelectronic or congeneric ones, e.g., $(\eta^5\text{-}C_5H_5)Cr(NO)_2(\eta^1\text{-}C_5H_5)$ and $(\eta^5\text{-}C_5H_5)Ru(CO)_2(\eta^1\text{-}C_5H_5)$, or derivatives, e.g., $(\eta^5\text{-}C_5H_5)Fe(CO)_2(1\text{-indenyl})$, have been the objects of so much study that it is convenient to give first a detailed account of their behavior. Following this, the data and conclusions for other $(\eta^1\text{-}C_5H_5)M$ species can be more concisely reviewed.

1. $(\eta^5\text{-}C_5H_5)Fe(CO)_2(\eta^1\text{-}C_5H_5)$ *and Its Closer Relatives*

One of the earliest pmr spectra ever recorded for an organometallic compound was that of $(\eta^5\text{-}C_5H_5)Fe(CO)_2(\eta^1\text{-}C_5H_5)$. Based on the inert gas

10. Nonrigidity in Organometallic Compounds

rule and the chemistry then known (about 1955) for compounds of its type, the presence of one pentahapto and one monohapto ring (one π ring and one σ ring, to use the terminology of the time), i.e., structure **1** was anticipated.

There was chemical and infrared evidence to support this. For this structure, the pmr spectrum should consist of a singlet of relative intensity 5 in the range of $\tau 5$ to $\tau 7$ for η^5-C_5H_5, an AA'BB' multiplet of relative intensity 4 due to the olefinic protons of η^1-C_5H_5 and a resonance of relative intensity 1 at a higher τ value for the proton on C-1 of the η^1-C_5H_5 ring. The observation that the pmr spectrum consisted of only two singlets of equal intensity was therefore highly surprising. It is remarkable that Wilkinson had the perspicacity and daring to propose an idea which was then unprecedented: that there was indeed an η^1-C_5H_5 ring, but that it is engaged in continual shifting of the Fe—C σ bond among all five of the ring carbon atoms, with concomitant rearrangements of the double bond system, the rapidity of these shifts being sufficient to cause spectrum averaging (Piper and Wilkinson, 1956a,b). Similar observations and a similar proposal were made with respect to $(C_5H_5)_2Hg$ (see Section I.A.3). A handy informal expression for a motion of this sort, which we shall find it convenient to use at times, is "ring whizzing."

It was ten years before any further report appeared in the literature bearing on this problem. It was then shown (Bennett et al., 1966) that in the crystal the pentahapto/monohapto structure, **1**, was indeed correct and that a pmr spectrum consistent with this structure could be observed using a solution cooled below $-80°C$. As the temperature was raised, the separate resonances for the two types of olefinic protons and the aliphatic proton all collapsed, and eventually the single line, first observed by Piper and Wilkinson, appeared for the monohapto ring. This was the first step of the work by Cotton, Davison, et al.,* and it confirmed Wilkinson's proposal.

The next step taken by Cotton, Davison, et al. was a major turning point in the study of all fluxional molecules. This step was the first mechanistic analysis of line shape changes that occur in the range of intermediate exchange rates. Except in trivial cases, e.g. exchange between just two equally populated sites, information about the mechanism—or "pathway" as Cotton, Davison, et al. proposed to call it—can be obtained by observing which, if any, of the three or more lines broaden most rapidly. Because the treatment by Cotton, Davison, et al. of the η^1-C_5H_5 ring in $(\eta^5$-$C_5H_5)Fe(CO)_2$-$(\eta^1$-$C_5H_5)$ was the first published example of such an analysis, and because it

* See Bennett et al. (1966) for all Cotton, Davison, et al. references.

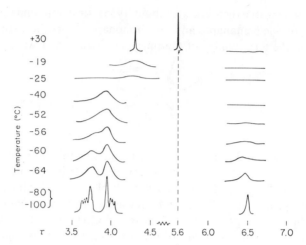

Fig. 1 The proton nmr spectra (60 MHz) of $(\eta^5\text{-}C_5H_5)Fe(CO)_2(\eta^1\text{-}C_5H_5)$ in CS_2 at various temperatures. The line for the $\eta^5\text{-}C_5H_5$ protons is shown only on the 30°C spectrum, the amplitude of which is 0.1 times that of the others. (Reproduced by permission from Bennett et al., 1966.)

constitutes an entirely representative case, it will be reviewed here in detail. The spectra at various temperatures are shown in Fig. 1. The line due to the $\eta^5\text{-}C_5H_5$ protons is invariant with temperature and is omitted from the traces at lower temperatures.

It is immediately obvious that one part of the AA'BB' multiplet centered at about $\tau 3.8$ broadens faster than the other as the temperature is raised. The first thing that can be inferred from this is that the rearrangement pathway does not permute the two kinds of olefinic nuclei at the same rate. Therefore, all pathways that would cause random exchange, such as dissociation followed by recombination, or movement of the iron atom to a position equidistant from all five carbon atoms, whence it might move with equal probability to any one of them, are rigorously eliminated. The problem of determining the pathway is, then, reduced to that of distinguishing between just two possibilities: 1,2 shifts in which the Fe—C bond moves to one of the two nearest carbon atoms, or 1,3 shifts, in which the bond is reformed at one of the next-nearest carbon atoms of the ring.

To pursue the argument further, the five ring proton sites are labeled thus:

10. Nonrigidity in Organometallic Compounds

A column matrix can be written listing these sites in the order in which they are initially occupied by protons 1–5:

$$\begin{bmatrix} H \\ A \\ B \\ B' \\ A' \end{bmatrix}$$

After a 1,2 or a 1,3 shift occurs, the sites are differently occupied, as follows, where the five rows of the matrix continue to correspond to the list of protons from 1 to 5:

$$\begin{bmatrix} A \\ H \\ A \\ B \\ B \end{bmatrix} \xleftarrow{1,2 \text{ shifts}} \begin{bmatrix} H \\ A \\ B \\ B \\ A \end{bmatrix} \xrightarrow{1,3 \text{ shifts}} \begin{bmatrix} B \\ A \\ H \\ A \\ B \end{bmatrix}$$

We then observe that for each 1,2 shift, both protons initially in A sites change to other (H or B) sites, whereas only half of the initial B protons change. In other words, for 1,2 shifts, protons initially in A sites have shorter residence times (by one-half) than those initially in B sites. For 1,3 shifts, the situation is reversed; protons in B sites have a mean residence time of one-half that for protons in A sites. Hence, in either case, more rapid broadening of one-half of the AA′BB′ multiplet is expected—and this is what is observed. To decide which pathway is correct, it is necessary to know how to assign the AA′BB′ multiplet. Cotton, Davison, et al., presented arguments for assigning the lower-field portion, which broadens faster, to the A sites. They concluded, therefore, that the molecule rearranges by 1,2 shifts.

Clearly, the argument used is an entirely rigorous one, provided that the assignment is correct. If the assignment is incorrect, i.e., if it should be reversed, then the conclusion would have to be reversed. The correctness of the assignment thus becomes a crucial question. A completely rigorous criterion for the assignment itself is not available, and thus as many different lines of independent supporting evidence as possible for the assignment were sought.

The initial criterion for assigning the low-field portion of the multiplet to the A sites (predominantly) was that the structure here was less sharp than in the high-field portion in the limiting low-temperature spectrum. This could be attributed to greater coupling of A protons to the proton at the H site. It was assumed that $|J_{HA}| > |J_{HB}|$; this is likely but not certain. In view of the small chemical shift difference between the A and B sites, no argument based on chemical shifts was considered reliable, and none was used. It must be borne in mind that with such magnetically anisotropic groups as η^5-C_5H_5 and CO also bound to the iron atom, as well as possible high-order magnetic effects from the iron atom itself, no small chemical shift difference such as that observed could be trusted to indicate which site is which.

Good supporting evidence for the assignment first proposed (and hence for the final conclusion that 1,2 shifts are occurring) was obtained soon after by study of the indenyl analog (η^5-C_5H_5)Fe(CO)$_2$(1-indenyl) (Cotton et al., 1967a). Two lines of argument were used in this work. First, it was noted that if direct 1,3 shifts, perhaps via a (η^3-allyl)-metal intermediate or transition

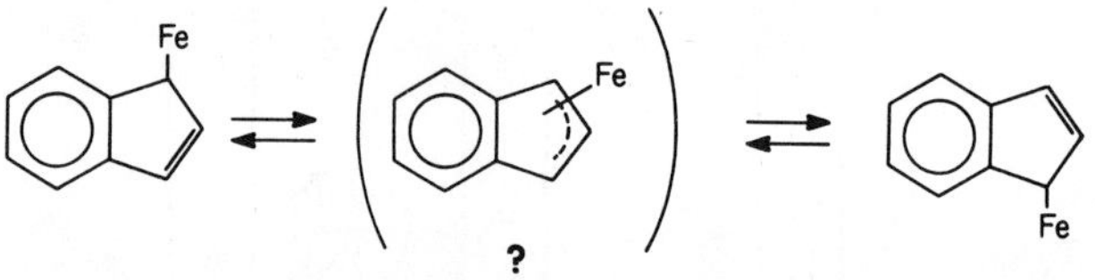

Fig. 2 Schematic representation of a hypothetical 1,3 shift in (η^5-C_5H_5)Fe(CO)$_2$(1-indenyl) via a possible η^3-allyl transition state. This process has not been observed.

state (see Fig. 2) occurred for η^1-C_5H_5, then the indenyl system could probably rearrange quite rapidly in the same way. However, 1,2 shifts would be suppressed in the indenyl system, since passage through a high-energy intermediate with an o-xylylene structure would be required. Thus, the occurrence of fluxionality in a 1-indenyl compound would imply that, contrary to the earlier conclusion, 1,3 shifts are the preferred pathway, whereas observation of a rigid 1-indenyl molecule would support the proposal of 1,2 shifts. The 1-indenyl molecule was observed to be nonfluxional.

In addition, by study of appropriately deuteriated derivatives of the 1-indenyl compound, it was shown that for this rather close analog of the original η^1-C_5H_5 compound, the chemical shifts for the A and B sites were in the same order as that required for the 1,2 shift mechanism.

There have since been several other investigations that have confirmed the assignment and hence the 1,2 shift mechanism. It has been shown (Cotton and Marks, 1969c) that coupling constants chosen to provide an accurate fitting of a computed spectrum to that observed in the slow exchange limit for (η^5-C_5H_5)Ru(CO)$_2$(η^1-C_5H_5) were qualitatively unreasonable for that assignment which would correspond to 1,3 shifts, but reasonable for that corresponding to 1,2 shifts. The terms reasonable and unreasonable in this context have to do with conformity to semiquantitative, semiempirical rules relating coupling constants to torsional and dihedral angles.

In the compound (C_5H_5)$_3$MoNO, to be discussed in some detail later with respect to ring interchange (Section IV,A), there is one observation of considerable importance with respect to the question of 1,2 versus 1,3 shifts (Cotton and Marks, 1969c). One of the rings in this compound is a monohapto type, and at temperatures between $-20°C$ and $-50°C$, the slow-exchange spectrum consisting of a well-defined AA'BB' multiplet and a singlet for the H proton forms and sharpens (Cotton and Legzdins, 1968). As with the iron and ruthenium compounds of the type $(\eta^5\text{-}C_5H_5)M(CO)_2$-$(\eta^1\text{-}C_5H_5)$ just discussed, the low-field half of the AA'BB' multiplet broadens first as the temperature is raised from about $-50°C$.

What is particularly interesting—and significant—is that this same portion of the AA'BB' multiplet also broadens below $-50°C$ so that at about $-95°C$ it has totally disappeared into the base line. At about $-110°C$, this signal, originally at about $\tau 3.2$ with relative intensity 2, has reappeared as two signals, each of relative intensity 1 at about $\tau 2.7$ and $\tau 3.7$. At the same time the other part of the AA'BB' multiplet merely undergoes moderate broadening and a small shift (about 0.2τ units) to lower frequency. These changes can be seen in Fig. 3.

The proposed explanation for them (Cotton and Marks, 1969c) is that below about $-50°C$, a particular rotational configuration with respect to the Mo—C σ bond, namely that shown as **2**, is being frozen in. Since the A protons

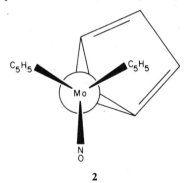

2

are considerably closer than the B protons to the $(C_5H_5)_2(NO)Mo$ group, they would sense the difference in their environments far more than the B protons. Hence, the signal that not only broadens but splits below $-50°C$ must be due to the A protons. Since this was the same signal that broadened most rapidly above $-50°C$, as the fluxional behavior of the $\eta^1\text{-}C_5H_5$ ring accelerated, the latter behavior is thus shown to occur via 1,2 shifts.

Other persuasive lines of evidence for 1,2 shifts in $(\eta^5\text{-}C_5H_5)Fe(CO)_2$-$(\eta^1\text{-}C_5H_5)$ were provided by observation of the ^{13}C satellites of the H_A and H_B resonances, and by examination of the ruthenium analog (Campbell and Green, 1970). Owing to the large values of the $^{13}C-^1H$ coupling constants, the

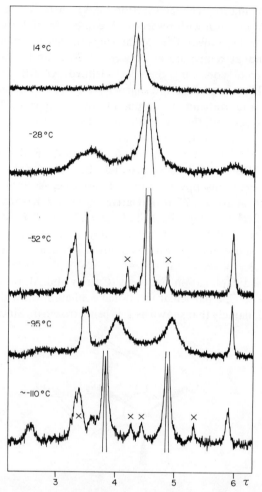

Fig. 3 Proton nmr spectra (60 MHz) of $(C_5H_5)_3MoNO$ (~0.2 M in CS_2) at various temperatures. Gain varies from trace to trace. Spinning side bands are marked ×. (Reproduced, by permission, from Cotton and Legzdins, 1968.)

fine structure in the ^{13}C shifted components of the AA'BB' multiplet caused by coupling to other ^1H nuclei is essentially first order. It was thus possible for Campbell and Green to make assignments with considerable certainty, and their assignments were such as to require 1,2 shifts. In addition, these workers noted that change of the metal atom from Fe to Ru produces a larger shift in the position of the low field portion of the AA'BB' multiplet than in the high field portion, as well as a still larger shift for the unique

10. Nonrigidity in Organometallic Compounds

proton on the carbon atom bonded to the metal atom. Since it seems reasonable that the simple change of metal atom should affect the shifts of the protons in the order $\Delta\delta_H > \Delta\delta_A > \Delta\delta_B$, this observation favors that assignment of the AA'BB' multiplet which leads to 1,2 shifts.

Quite recently ^{13}C spectra of $(\eta^5\text{-}C_5H_5)Fe(CO)_2(\eta^1\text{-}C_5H_5)$ have been recorded at various temperatures from $-88°C$ to $55°C$ (Ciappenelli et al., 1972). These data provide further support for the 1,2 shift mechanism. The resonances were assigned by comparison with the chemical shifts observed for the allyl compound, $(\eta^5\text{-}C_5H_5)Fe(CO)_2CH_2CH=CH_2$, as shown in Fig. 4. With the assignment so established, the different rates of broadening

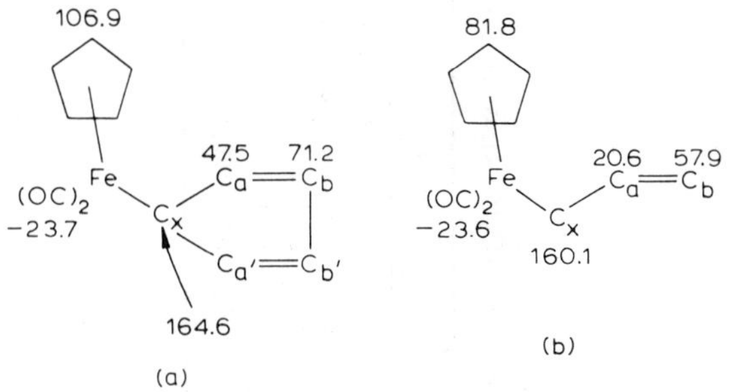

Fig. 4 Carbon-13 chemical shifts (ppm versus CS_2) in the allyl reference compound (b) and, by inference therefrom, in $(\eta^5\text{-}C_5H_5)Fe(CO)_2(\eta^1\text{-}C_5H_5)$(a). (Reproduced by permission from Ciappenelli et al., 1972.)

for C_a and C_b signals, i.e., $C_a > C_b$, lead unequivocally to the 1,2 shift mechanism.

The ^{13}C study also afforded the most accurate Arrhenius activation parameters for $(\eta^5\text{-}C_5H_5)Fe(CO)_2(\eta^1\text{-}C_5H_5)$. These and some figures obtained earlier from pmr spectra are shown in Table I.

The compound $(\eta^5\text{-}C_5H_5)Cr(NO)_2(\eta^1\text{-}C_5H_5)$ is isoelectronic and isostructural with $(\eta^5\text{-}C_5H_5)Fe(CO)_2(\eta^1\text{-}C_5H_5)$ and has very similar dynamic behavior (Cotton et al., 1967a).

2. Cyclopentadienyl Compounds of Copper and Gold

a. *Copper.* The cyclopentadienyl copper compounds, of general formula $(C_5H_5)CuL$, where L is some neutral ligand such as R_3P, have had a surprisingly troubled history, considering how simple they appear to be. Their existence was first satisfactorily demonstrated in 1956 when Piper and Wilkinson described the compound with $L = (C_2H_5)_3P$ (1956a,b). These workers commented on the remarkable stability of the compound in comparison to known alkyl and aryl copper(I) compounds; nevertheless they

TABLE I

ARRHENIUS ACTIVATION PARAMETERS FOR "WHIZZING" OF η^1-C_5H_5 RINGS IN SEVERAL COMPOUNDS

Compound	Type of data	E_a (kcal)	log A	Reference
$(\eta^5$-$C_5H_5)Fe(CO)_2(\eta^1$-$C_5H_5)$	pmr	9.8 ± 0.1	11.7 ± 0.1	Cotton and Marks (1969c)
	pmr	8.5 ± 0.8	9.8 ± 0.7	Campbell and Green (1970)
	^{13}C-nmr	10.7 ± 0.5	12.6 ± 0.5	Ciappenelli et al. (1972)
$(\eta^5$-$C_5H_5)Ru(CO)_2(\eta^1$-$C_5H_5)$	pmr	10.3 ± 0.3	11.2 ± 0.2	Cotton and Marks (1969c)
	pmr	9.6 ± 1.0	10.0 ± 0.7	Campbell and Green (1970)

proposed that it be considered a η^1-C_5H_5 species and, in order to account for the presence of only a single, sharp pmr line for the ring protons, they further proposed that it is what we should now call a fluxional molecule. The presence of a η^1-C_5H_5 ring was later alleged to be supported by its infrared spectrum and the compound was cited as a reference example of such a metal-to-ring bond (Fritz, 1964).

After Cotton, Davison et al. had, in October 1966, *published* their report showing how the detailed dynamic behavior of $(\eta^5$-$C_5H_5)Fe(CO)_2(\eta^1$-$C_5H_5)$ could be elucidated by proper analysis of low-temperature pmr spectra, a report of a similar investigation of $(C_5H_5)CuPEt_3$ was *submitted* in December 1966 (Whitesides and Fleming, 1967). Liquid sulfur dioxide was used as the solvent. In this work it was found the single C_5H_5 pmr line observed at room temperature was again transformed into an AA'BB'X type spectrum at lower temperature, and again, the width of the halves of the AA'BB' multiplet varied differently with temperature. In this case, it was the up-field half that collapsed more rapidly. Whereas Cotton, Davison, et al. had refrained from any attempt to use the very small chemical shift difference between halves of the AA'BB' multiplet to make an assignment of it, believing that this would be dangerously unreliable, and had instead employed an argument based on the expectation that J_{HA} should be larger than J_{HB}, Whitesides and Fleming attempted a chemical shift argument. This led them to an assignment that required the conclusion that 1,3 shifts are occurring. The basis for the assignment of Whitesides and Fleming was that for a series of $(\eta^1$-$C_5H_5)M$ compounds, they expected the chemical shift of the B protons to be substantially constant while that of the A protons would be more variable, because the former are separated by four bonds and the latter by only three bonds from the metal atom. Quite apart from what one might think *a priori* about the reasonableness of this argument, it has been demonstrated to be unsafe through the agency of a blatant counterexample (Cotton and Marks, 1969a). In monohapto-1-indenyl compounds, the resonances of A and B protons can be unequivocally assigned. For two such compounds, as shown in Table II, the behavior is exactly the reverse of what Whitesides and Fleming assumed;

TABLE II
CHEMICAL SHIFTS OF PROTONS α AND β TO THE 1-POSITION IN TWO INDENYL–METAL COMPOUNDS[a]

Compound	A protons	B protons
$(\eta^1$-$C_5H_5)(CO)_2Fe(C_9H_7)$	$\tau 3.28$	$\tau 3.47$
$Hg(C_9H_7)_2$	$\tau 3.32$	$\tau 3.09$

[a] 60 MHz in CCl_3D solution.

the shift of the B protons varies greatly while that of the A protons is fairly constant.

The reason Whitesides and Fleming's argument is so unreliable is that it assumes (a) that the chemical shifts depend simply on the inductive effects of the metal atoms, and (b) that these effects will then be attenuated inversely as the number of bonds between the metal atom and the proton in question. However, in the molecules of interest, the diamagnetic anisotropies of other ligands, such as η^5-C_5H_5, CO, and R_3P, can make important contributions to the chemical shifts; thus a different set of ligands [e.g., $(C_2H_5)_3P$ instead of η^5-C_5H_5 and 2CO] can, and apparently does, alter the chemical shifts of η^1-C_5H_5 protons in ways quite beyond the scope of Whitesides and Fleming's assumptions.

It has later been shown, however, that when the coupling constant argument of Cotton, Davison, et al. is applied, it requires the opposite assignment from that of Whitesides and Fleming and thus leads to the conclusion that here, too, 1,2 shifts are occurring.

While $C_5H_5CuPEt_3$ does undoubtedly have a fluxional η^1-C_5H_5 type structure in liquid sulfur dioxide and, according to Whitesides and Fleming, also in propionitrile and petroleum ether (although no details concerning solutions in these solvents were actually reported), infrared studies (Cotton and Marks, 1969b) on both $C_5H_5CuP(C_2H_5)_3$ and $C_5H_5CuP(C_4H_9)_3$ indicated that in the solid state, at least, these molecules contain η^5-C_5H_5 rings. Subsequently, X-ray crystallographic studies of $C_5H_5CuP(C_6H_5)_3$ (Cotton and Takats, 1970) and $C_5H_5CuP(C_2H_5)_3$ (Delbaere et al., 1970) showed that both contain η^5-C_5H_5 rings. The remarkable stability of these compounds, at least as solids, is thus understandable. What is not clear at present, and certainly deserving of the appropriate investigation, is dependence of structure on the solvent in which the C_5H_5CuL compound may be dissolved. Apparently, the $(\eta^1$-$C_5H_5)CuL$ and $(\eta^5$-$C_5H_5)CuL$ structures do not differ greatly in inherent stability. That being so, however, it is interesting that in liquid sulfur dioxide the fluxional rearrangement of the η^1-C_5H_5 ring proceeds primarily, and perhaps exclusively through a mechanism that causes unsymmetrical line shape changes rather than through a $(\eta^1$-$C_5H_5) \rightarrow (\eta^5$-$C_5H_5)$ $\rightarrow (\eta^1$-$C_5H_5)$ sequence, which would cause symmetrical ones.

b. *Gold.* There is no conclusive information as to the structural or dynamic character of cyclopentadienyl gold compounds. $Au(PPh_3)(C_5H_5)$ gives a single sharp nmr line for C_5H_5 down to $-50°C$, below which it splits into a doublet. The doublet shows no broadening even at $-100°C$. The doublet structure is due to coupling with ^{31}P, and it was shown that the doublet collapses above $-50°C$ due to exchange of the PPh_3 ligand. Whether the C_5H_5 ring is pentahapto or whether it is monohapto but still rapidly whizzing even at $-100°C$ cannot be inferred from the data (Campbell and Green, 1971).

The compound $Au(PPh_3)(CH_3C_5H_4)$ has been examined (Huttel et al., 1967; Campbell and Green, 1971). The earlier study was inconclusive, but the latter one afforded evidence that in solution in 2-methyltetrahydrofuran this molecule has a σ-bonded ring, and that the isomer **3** is the predominant one:

Ph₃PAu—[cyclopentadiene ring with CH₃]

3

3. Mercury Compounds

A major breakthrough with respect to cyclopentadienylmercury compounds came in 1969 (West et al., 1969). Prior to this work there had been infrared evidence (e.g., Cotton and Marks, 1969b; Maslowsky and Nakamoto, 1969), some suggestive but not fully conclusive nmr evidence (Maslowsky and Nakamoto, 1968), and some chemical evidence (e.g., Piper and Wilkinson, 1956a,b) pointing to $(\eta^1\text{-}C_5H_5)Hg$ bonding in C_5H_5HgX, X = Cl, Br, I, C_5H_5 compounds, but nmr spectra of solutions had not yielded any positive structural information; only a single sharp line for all protons had been observed in each case. Rausch and coworkers were able to obtain low-temperature spectra for C_5H_5HgX molecules with X = Cl, Br, and I, which are consistent only with the instantaneous structure $(\eta^1\text{-}C_5H_5)HgX$. The chloro compound had the highest coalescence temperature and was studied in greatest detail. Unfortunately, the details of the collapse of the spectra to the single line spectrum at higher temperatures could not be discerned well enough to yield any conclusion as to the rearrangement pathway (i.e., to discriminate between 1,2, 1,3, or random shifts). The observation of satellites due to $^{199}Hg\text{-}^1H$ coupling shows that the process is intramolecular below $0°C$. Interestingly, the satellites begin to broaden between $0°C$ and room temperature, indicating that intermolecular exchange is becoming important in that temperature range.

For $(C_5H_5)_2Hg$, the slow exchange limit was not reached, although at or below $-100°C$ the line is observably broadened. This suggests that the molecule is fluxional and thus not $(\eta^5\text{-}C_5H_5)_2Hg$, but tells nothing positive about its structure or the rearrangement pathway.

Another line of evidence bearing on the nature of cyclopentadienylmercury compounds derives from studies of the methylcyclopentadienyl mercury species, $Hg(CH_3C_5H_4)_2$ (Campbell and Green, 1971). The results obtained could only be interpreted reasonably by postulating that in solution there exists the rapidly interconverting set of isomers, **4, 5,** and **6**. It was assumed, in the absence of any evidence to the contrary, that the two rings behave independently of each other in their bonding relationship to the mercury atom. The alternate postulate, involving pentahapto rings, was

[Structures 4, 5, 6 showing Hg-cyclopentadienyl isomers with different positions of CH$_3$, H$_\alpha$, H$_\beta$ labels]

shown to be inconsistent with the data. The data and interpretation thereof may be summarized as follows:

Relative intensity	−10°C		−95°C	
	τ	J_{Hg-H}	τ	J_{Hg-H}
2	4.20	48	4.09	31
2	4.74	100	4.97	120
3	7.91	38	7.90	46

Even at −95°C, the lowest temperature at which spectra were recorded, spectra assignable to the individual isomers **4**, **5**, and **6** were not observed. At all temperatures from 30° to −95°C, the spectrum consisted of three peaks; data for −10°C and −95°C are tabulated above. While the appearance of two resonances for ring protons would be expected for a pentahapto ring, a vast disparity in their ^{199}Hg–H couplings would not be expected. Thus, the η^5-C$_5$H$_5$ structure is not in agreement with the data. The rapidly interconverting set, **4**, **5**, and **6**, appears to fit all the facts, provided that **4** is a minor constituent, and the ratio **5/6** varies from 0.3 at 30°C to 0.1 at −95°C. The ring resonance with the large value of ^{199}Hg–H coupling, which increases as the temperature drops, must be due to the β protons; and the other ring resonance, with a smaller ^{199}Hg–H coupling, which decreases as the temperature drops, must be due to the α protons. An alternative postulation of **6/5** < 1 with reversal of the H$_\alpha$ and H$_\beta$ assignments can be ruled out, since it leads to unacceptable values of the coupling constants; the original article must be consulted for the full argument.

The fully methylated compound (C$_5$Me$_5$)$_2$Hg has a proton nmr spectrum at room temperature characteristic of a bismonohapto structure in the slow exchange limit (Flores et al., 1969). The great effect of complete methylation on the rate of rearrangement is notable.

Finally, it is convenient to note here that bisindenylmercury has been characterized structurally and dynamically (Cotton and Marks, 1969a), and the results are consistent with the best information on the cyclopentadienyl compounds. At $-41°C$, the proton nmr spectrum shows that the structure is bis(1-*monohapto*-indenyl)mercury. As the temperature is raised, the spectrum changes in a manner indicative of increasingly rapid occurrence of some unimolecular process that exchanges the environments of the 1 and 3 protons of the indenyl groups. The rapid occurrence of what appear to be net 1,3 shifts in this case, whereas they are not observed for $(\eta^5\text{-}C_5H_5)(CO)_2Fe\text{-}$ (1-indenyl), may be attributed to the fact that the mercury atom can employ additional $6p$ orbitals to allow it to form some kind of a delocalized bond to three, or perhaps even five, of the indenyl carbon atoms in the transition state. Clearly, the same thing might equally well happen with a cyclopentadienylmercury compound. Thus, in the absence of any explicit experimental evidence as to the pathway for rearrangement of $(\eta^1\text{-}C_5H_5)HgX$ compounds (see above), 1,2, 1,3, or random shifts must all be considered as real possibilities at this time.

4. Other $(\eta^1\text{-}C_5H_5)M$ Compounds of Transition Metals

These are several more transition metal complexes containing $\eta^1\text{-}C_5H_5$ rings that should be cataloged here for the sake of completeness. For the most part their "ring-whizzing" behavior has not been studied in great detail, and in no case does it appear to have any special or unusual features.

There are a few $(\eta^1\text{-}C_5H_5)Pt$ complexes. The first ones reported were *cis*- and *trans*-$Pt(\eta^1\text{-}C_5H_5)_2(Me_2S)_2$, which decompose within a few minutes at 25°C where they have a single sharp signal for the ring protons; the structures assigned, while probably correct, lack firm support (Fritz and Schwarzhans, 1966). Five rather more stable complexes of the type $(\eta^1\text{-}C_5H_5)PtLL'X$ (where L and L' may be R_3P and/or CO, or a biarsine, and X is Cl or C_6H_5) were later reported (Cross and Wardle, 1971). Again, the presence of the $\eta^1\text{-}C_5H_5$ group, while most likely correct, lacks firm support. All of these compounds have a single sharp signal (complicated by Pt–H and P–H splittings, however), for the cyclopentadienyl protons at room temperature; two of the compounds were examined at $-90°C$ and no broadening was detected.

The molecule $(C_5H_5)_2MoNO(S_2CNMe_2)$ has one pentahapto and one monohapto ring in the crystal (N. Bailey, personal communication, 1972).* This structure is retained in solution, and at $-80°C$, the monohapto C_5H_5 ring gives rise to a pattern of signals of the ABCDX type. These broaden and coalesce, giving a single line at 20°C. The compound $(C_5H_5)_2MoNO\text{-}[S_2C_2(CN)_2]$ behaves similarly (Kita *et al.*, 1971). The first of these molecules will be discussed again later (Section IV,A,3) since at higher temperatures $(\eta^1\text{-}C_5H_5)/(\eta^5\text{-}C_5H_5)$ ring interchange occurs.

* Work done in the Department of Chemistry, Sheffield University.

The molecule $(C_5H_5)_4Mo$ would be expected to have a $(\eta^5\text{-}C_5H_5)_2$-$(\eta^1\text{-}C_5H_5)_2Mo$ structure, since there are many species of the type $(\eta^5\text{-}C_5H_5)_2$-MoX_2 known, where X is a univalent group such as Cl, and the 18-electron configuration is thereby achieved. The compound has two sharp proton resonance signals of equal intensity at room temperature (Calderon and Cotton, 1971) and thus presumably has rapidly rearranging $\eta^1\text{-}C_5H_5$ rings.

The $(C_5H_5)_4Ti$ molecule has the $(\eta^5\text{-}C_5H_5)_2(\eta^1\text{-}C_5H_5)_2Ti$ structure in the crystal (Calderon et al., 1971b). Its most interesting dynamical behavior (Calderon et al., 1971a) is the interchange of monohapto and pentahapto rings (see Section IV,A,1). The $\eta^1\text{-}C_5H_5$ signal is a sharp line at $\tau 4.1$ in toluene at $-27°C$, at $\tau 4.1$ at $-36°C$ in toluene–ether, 1:1, and at $\tau 5.25$ at $-36°C$ in ethyl ether. Viscosity broadening was a severe problem in the solvents containing toluene at lower temperatures. However, in both the mixed solvent and in pure ether, this resonance collapsed and a coalescence temperature of about $-93°C$ was observed. In pure ether, between $-116°C$ and $-138°C$, the AA'BB' multiplet due to the $\eta^1\text{-}C_5H_5$ rings developed.* Thus, the $\eta^1\text{-}C_5H_5$ rings in this compound definitely undergo rapid reorientation. However, the activation energy is quite low as compared to the iron, ruthenium and copper species already discussed. It must also be noted that the poor signal to noise ratio did not allow any information to be obtained about the pathway. The possibility certainly exists here for 1,3 shifts via an intermediate or transition state in which the ring interacts with two metal valance shell orbitals using three π electrons. This possibility is much more plausible here, because the titanium atom in $(\eta^5\text{-}C_5H_5)_2(\eta^1\text{-}C_5H_5)_2Ti$ has a formal electron configuration numbering only 16, whereas in all of the other cases the count is 18.

5. Cyclopentadienyl and Indenyl Compounds of Group IV Elements

A review of this subject can appropriately begin with the work of Davison and Rakita, since earlier work in the field is marred by misconceptions and misinterpretations. Davison and Rakita first studied the following series of molecules: $(CH_3)_3MR$ where $M = Si$, Ge, or Sn and $R = C_5H_5$, $CH_3C_5H_4$, or $C_5(CH_3)_5$ (1968, 1970b). They found that all such molecules are fluxional, and in the same general way (i.e., by ring whizzing) as are the transition metal monohaptocyclopentadienyl compounds. Quite generally, for the series of three molecules with a given R group, the rates with the three group IV elements are in the order $Sn \gg Ge > Si$. Usually, the tin compounds are too fast and the silicon compounds too slow to allow convenient study of the entire rate range, from the slow- to the fast-exchange limits. The germanium

* Unfortunately, in the paper by Calderon et al. (1971a), Fig. 2 and Fig. 3a were interchanged.

species are usually very convenient subjects for study. Excessive heating of the silicon compounds, in an effort to reach the high-temperature limit, leads to irreversible isomerizations due to hydrogen shifts, giving one or both of the two vinylsilyl isomers. It may be explicitly noted here that hydrogen migration is some 10^6 times slower than trimethylsilyl migration, so that the observable fluxional behavior is not (contrary to what some have suggested) due to movements of hydrogen atoms.

The spectra of $C_5H_5Ge(CH_3)_3$, Fig. 5, show features typical for this class of compounds. The most notable is the unsymmetrical collapse of the AA'BB' multiplet, with the high-field side collapsing fastest. In exactly the same manner as before with the iron compounds (Cotton et al., 1967a), a comparison with the 1-indenyl analog (Davison and Rakita, 1970a) was used to determine the assignment. It was found that for these Group IV molecules the up-field resonance is due to the A protons and thus again, 1,2 shifts are operative.

Support for an assignment leading to 1,2 shifts in another Group IV compound, C_5H_5-$SiCH_3Cl_2$, was later provided by a spin tickling experiment, though not without some rather unfortunate confusion along the way. A spin tickling experiment consists in irradiating certain transitions within a spin multiplet at a low power level and observing which transitions within the same (or another) multiplet are perturbed. Only transitions that have at least one energy level in common with the irradiated transition will be affected. The results can help in sorting out the relative signs of coupling constants in complex spectra. A knowledge of the relative signs, in conjunction with semi-empirical rules governing the signs and magnitudes of couplings through various numbers of bonds, can lead to assignments that have a high probability, though not absolute certainty, of being correct.

The paper reporting the experimental results (Sergeyev et al., 1970a) contained an interpretation and thence an assignment from which the prevalence of 1,3 shifts had to be inferred. This interpretation was incorrect, however, and soon afterward the correct interpretation was given (Sergeyev et al., 1970b; Cotton and Marks, 1970). From the correct interpretation, together with some well-founded and secure assumptions as to the signs which would be expected for $J_{AX}(>0)$ and $J_{BX}(\leq 0)$, the assignment of the AA'BB' multiplet follows rigorously and from that the conclusion that 1,2 shifts predominate follows rigorously in turn.

Because cyclopentadienyltin compounds so often give single line cyclopentadienyl spectra even at quite low temperatures (due to the extreme ease of ring whizzing), it has on occasion been proposed that the tin compounds, in contrast to their Si and Ge analogs, contain η^5-C_5H_5 rings. Firm evidence against this has been obtained by pmr study of $Sn(CH_3C_5H_4)_4$ (Campbell and Green, 1971) and from infrared spectra of various cyclopentadienyltin

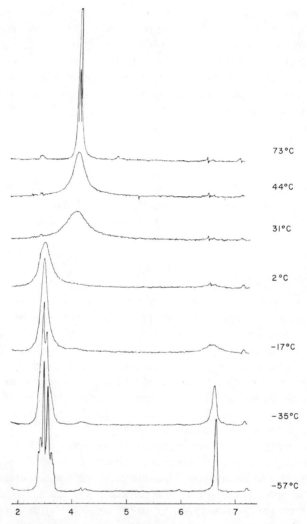

Fig. 5 The nmr spectra of $(\eta^1\text{-}C_5H_5)Ge(CH_3)_3$ at various temperatures. (Reproduced by permission from Davison and Rakita, 1970b.)

compounds (Davison and Rakita, 1970b). The tin compounds are important in showing that the rearrangement process must be an intramolecular one, since Sn–H and Sn–C couplings can be observed in the fast exchange limit.

Group IV indenyl compounds provide very valuable evidence in favor of 1,2 shifts (Rakita and Davison, 1969; Davison and Rakita, 1970a). As already noted, the signals due to the protons in the five-membered ring of indene can be unambiguously assigned. When this assignment is carried over to the cyclopentadienyl compounds, it provides the basis for concluding that 1,2 shifts occur. However, in the indenyl compounds per se, the decision cannot be directly made between 1,2 and 1,3 shifts. All that is observed is line-averaging consonant with the overall process **7 ⇌ 7′**:

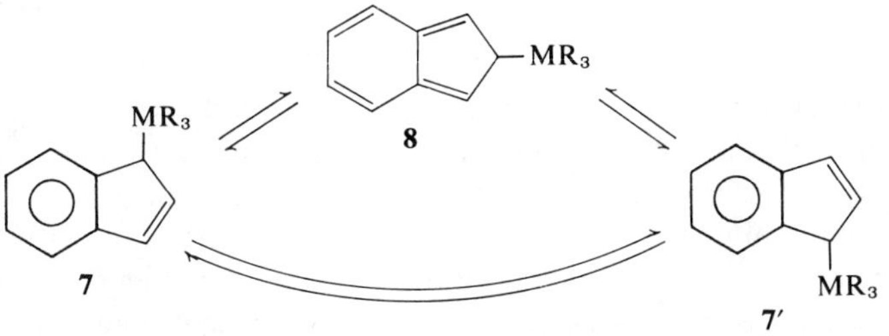

There is no way to tell whether this exchange occurs in one step (a 1,3 shift) or as a result of passage through the unstable *o*-xylylidine-like 2-indenyl intermediate, **8**, in other words, in two steps, each of which is a 1,2 shift.

Further insight was provided by a study (Davison and Rakita, 1970a) of $C_9H_7SnMe_2Ph$, in which the methyl groups are diastereotopic. It was found that the overall shift of the $SnMe_2Ph$ between the 1 and 3 carbon atoms was accompanied by site exchange of the methyl groups. Assuming first that only suprafacial shifts need be considered, the authors distinguished four possible pathways, as shown in Eqs. (1)–(4). Of these, Eq. (2) is inconsistent

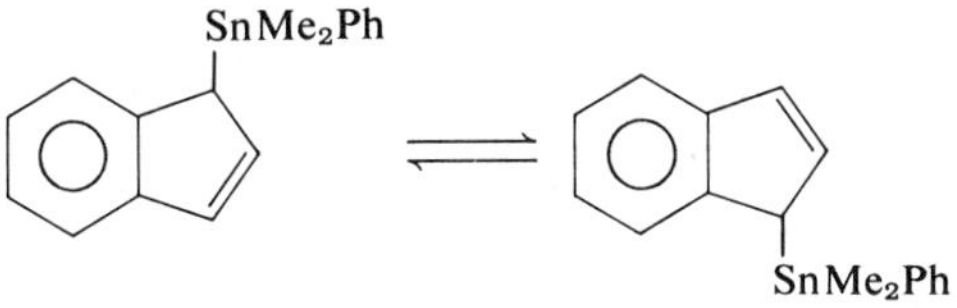

Direct 1,3 shift without inversion (1)

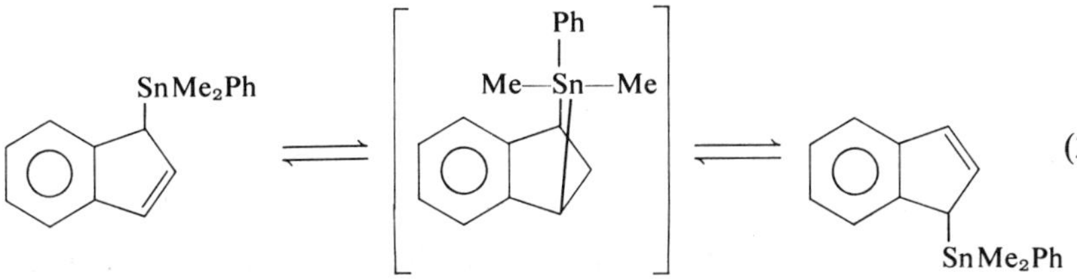

Direct 1,3 shift, through symmetrical intermediate, with inversion at Sn (2)

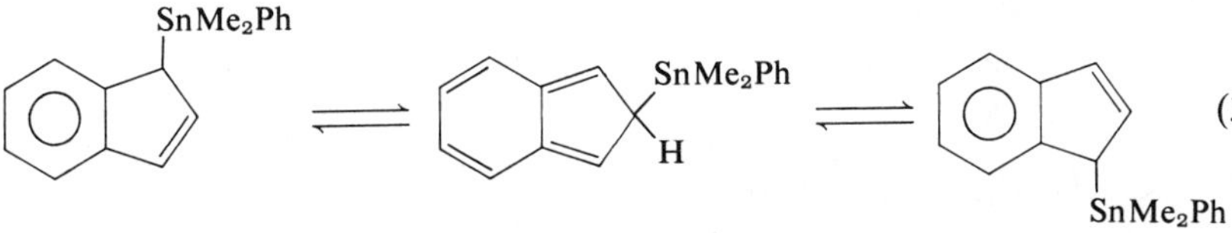

Sequence of 1,2 shifts (3)

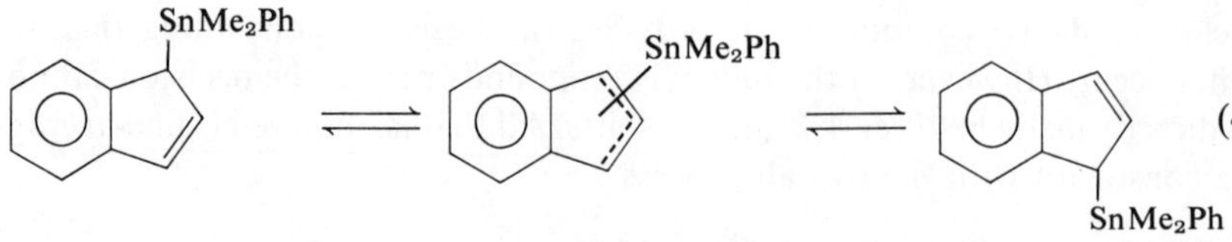

$$\text{(4)}$$

η^3-Allyl type intermediate

with the data, since simultaneous inversion of the overall molecular configuration and inversion of the local configuration at the tin atom would not lead to interchange of the environments of the methyl groups.

Perhaps the most mechanistically informative experiments on Group IV compounds were those done on 1,2-bis(trimethylsilyl)indene. In the slow exchange limit, the two trimethylsilyl groups are nonequivalent, as expected. At higher temperatures the trimethylsilyl groups become equivalent at the same time as do the 1 and 3 protons of the indene moiety. This clearly and unequivocally rules out direct 1,3 shifts, Eq. (5), and is consistent with the

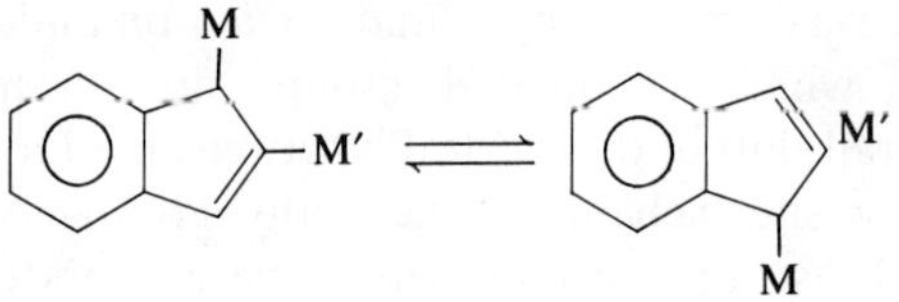

$$\text{(5)}$$

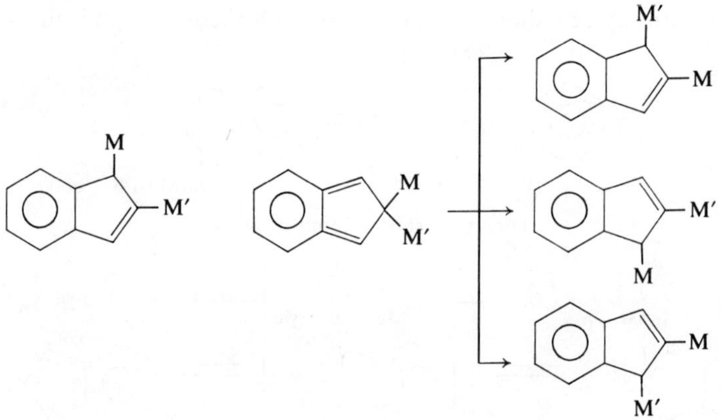

$$\text{(6)}$$

consequences of only 1,2 shifts, Eq. (6). The observations are also consistent with the process depicted in Eq. (7), a pair of simultaneous, concerted 1,2 shifts with no o-xylylidine-like intermediate.

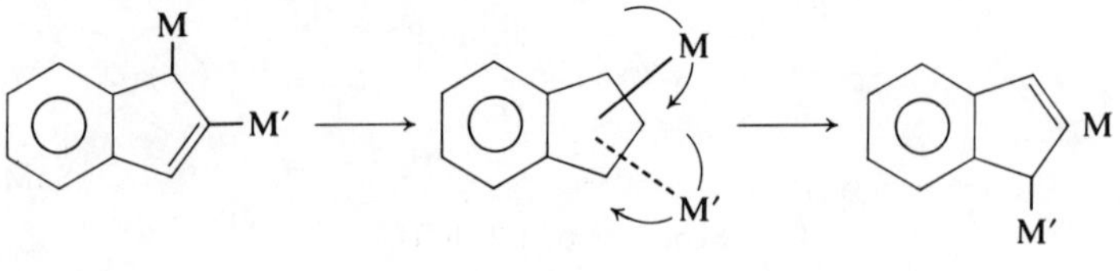

$$\text{(7)}$$

TABLE III
ACTIVATION PARAMETERS FOR SEVERAL INDENYL COMPOUNDS OF SILICON AND TIN[a]

	Compound	E_a	log A
1.	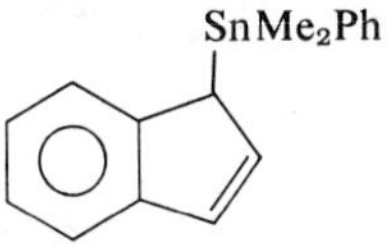 SnMe$_2$Ph (indenyl)	14.1	11.9
2.	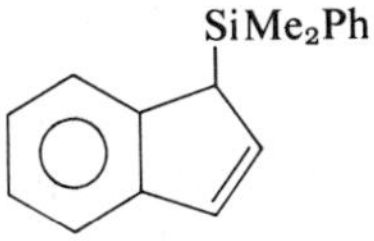 SiMe$_2$Ph (indenyl)	23.0	11.5
3.	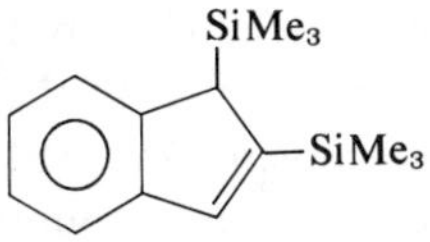 SiMe$_3$ (indenyl)—SiMe$_3$	26.1	14.4
4.	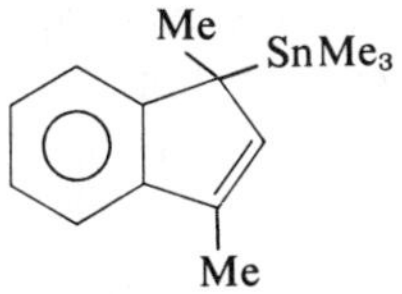 Me, SnMe$_3$ (indenyl), Me	14.2	12.5

[a] Davison and Rakita (1970a).

Detailed kinetic analysis of some of the processes occurring in indenyl compounds of Group IV elements lead to the activation parameters in Table III. Several important conclusions can be drawn from these quantities. First, it is seen that activation energies are primarily a function of whether Si or Sn is involved and depend little on details of how the ring or the Group IV element are substituted. Comparison of compounds 2 and 3 reveals sufficient similarity in their activation parameters to suggest that their rearrangement pathways are essentially the same. This, in turn, leads to the conclusion that these pathways are best represented by Eqs. (3) and (6) for compounds 2 and 3, respectively. The slightly higher activation energy for 3 is not unreasonable in view of probable steric destabilization of the 2,2-disubstituted indene intermediate in Eq. (6). In summary then, it seems that the study of the 1,2-disubstituted compound does afford evidence that 1,2 shifts occur in the singly substituted compounds as well.

Yet another line of evidence favoring 1,2 shifts in the Group IV compounds was provided by study of trimethylgermyl-N-pyrazolides and -N-imidazolides (Cotton and Ciappenelli, 1972). The idea is simply that if 1,2 shifts are allowed and 1,3 shifts are not, then the pyrazolides will be fluxional (Eq. 8) and the imidazolides will not, whereas, if the reverse is true, the imidazolides will be fluxional (Eq. 9) and the pyrazolides will not. The ex-

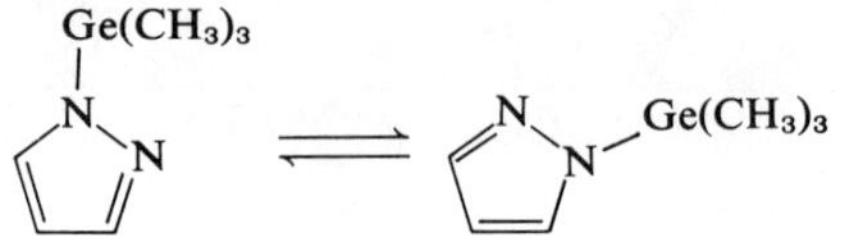

$$\underset{\substack{|\\ \text{Ge(CH}_3)_3}}{\underset{N-N}{\triangle}} \rightleftharpoons \underset{N-N}{\triangle}\text{—Ge(CH}_3)_3 \qquad (8)$$

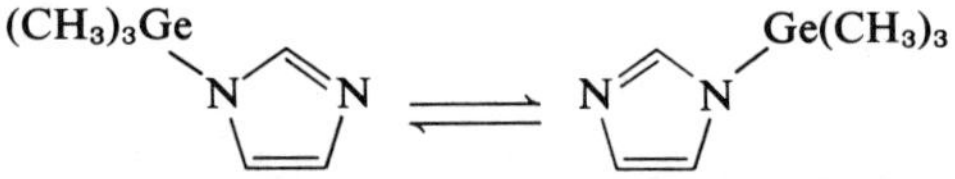

$$(\text{CH}_3)_3\text{Ge}-\underset{N\frown N}{\triangle} \rightleftharpoons \underset{N\frown N}{\triangle}-\text{Ge(CH}_3)_3 \qquad (9)$$

perimental data, while not as cleancut as had been hoped, definitely favored 1,2 shifts.

The results of ^{13}C nmr study of several Group IV cyclopentadienyl compounds (Grishin et al., 1972) are consistent with the earlier pmr investigations in respect to both rates and mechanism. Observation of ^{13}C offers the advantage that the more rapidly fluxional tin compounds, such as C_5H_5Sn-$(CH_3)_3$, can be more easily studied in the slow exchange limit, since the ^{13}C chemical shift differences are much larger than those for ^{1}H.

One more significant experiment bearing on this point must be mentioned. It has been shown by chemically trapping the o-xylylidine intermediate **9** as **10** (Ashe, 1970) that the activation energy for the 1,2 shift in $Me_3Si(1$-indenyl) is about 22.5 kcal/mole. Since this agrees very well with the value for the overall 1,3 shift in $PhMe_2Si(1$-indenyl), Table III, it affords good evidence that the overall 1,3 rearrangement proceeds stepwise via two 1,2 shifts.

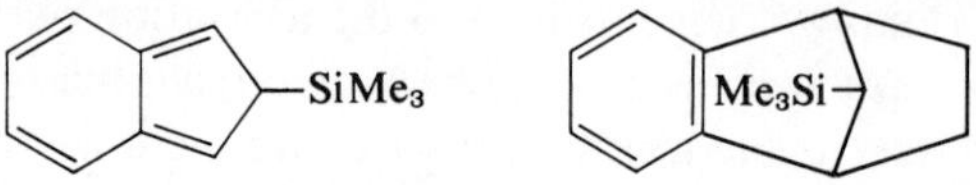

9   **10**

6. *Monohaptocyclopentadienyl Compounds of Other Nontransition Elements*

In view of the fluxional character of η^1-C_5H_5 and 1-indenyl derivatives of the Group IV elements Si, Ge, and Sn, it would not be surprising to find that other nontransition elements also form roughly similar fluxional compounds. A few have been reported and the data will be summarized here.

a. *Group III.* Several $R_2AlC_5H_5$ compounds have been studied by proton nmr. In each case [R = CH_3, C_2H_5, $(CH_3)_2CHCH_2$] the cyclopentadienyl resonance was a single line at ambient temperature and down to $-91°$C for the isobutyl and $-60°$C for the ethyl compounds (Kroll and

Naegele, 1969). These authors proposed that these compounds contain η^1-C_5H_5 rings that undergo 1,2 shifts with a low activation energy. Later, Kroll (1969) examined adducts of the $R_2AlC_5H_5$ molecules with Lewis bases and found that these too show only a single sharp line for the C_5H_5 group. Interestingly, the $R_2AlC_5H_5$ compounds in inert solvents do not disproportionate, whereas the adducts do at room temperature. It was proposed that the adducts also contain fluxional η^1-C_5H_5 rings.

The observation of a single line for the C_5H_5 groups at all temperatures could be explained in either of at least two other ways, and it must be stressed that no evidence was provided against either of these nor was there any positive evidence in favor of fluxionality. Thus, (a) intramolecular exchange of η^1-C_5H_5 groups might be occurring; or (b) the ring might be of the η^5-C_5H_5 type at least in the $R_2AlC_5H_5$ molecules. The fact that these do not disproportionate readily, in contrast to most $R_2R'Al$ type molecules could be explained in terms of (b). The infrared spectra of some $R_2AlC_5H_5$ compounds have since been interpreted in terms of η^5-C_5H_5 structures (Haaland and Weidlein, 1972), although this is by no means conclusive evidence.

The compound $(CH_3)_2TlC_5H_5$ has been examined (Lee and Sheldrick, 1969). The best interpretation of the results (single line for C_5H_5 protons with no splitting due to the thallium isotopes, both of which have $S = \frac{1}{2}$, though the methyl resonance is split) is that there is a monomer–dimer equilibrium and that, presumably via the dimers, the C_5H_5 groups pass from one metal atom to another.

b. *Group V.* The compound $C_5H_5PF_2$ has been clearly shown to be fluxional (Bentham *et al.*, 1972). Its ir spectrum is typical for a η^1-C_5H_5 compound. The ^{19}F resonance is a doublet ($J_{P-F} = 1170$ Hz) of doublets ($J_{H-F} = 11.5$ Hz) from $-85°$ to $-30°C$. The J_{H-F} splitting then collapses, and beginning at 25°C, a sextet pattern develops and becomes quite sharp at 50°C ($J_{H-F} = 2.5$ Hz). The temperature dependence of the proton spectrum is consistent. No facts bearing on the rearrangement pathway (i.e., 1,2 shifts, 1,3 shifts, random shifts, etc.) were reported.

The proton nmr spectra of the tricyclopentadienyl compounds of As, Sb, and Bi have all been reported from 35° to about $-70°C$ (Deubzer *et al.*, 1970). For $(C_5H_5)_3Bi$, only a single line, which does not change width, is observed throughout the temperature range. For $(C_5H_5)_3Sb$, there is also only a single line, but between $-45°$ and $-60°C$ it broadens. For $(C_5H_5)_3As$, there is a single line ($\tau 3.80$) at 35°C, which broadens and separates into two lines ($\tau 3.62$, $4H$ and $\tau 6.45$, $1H$) between 0° and $-35°C$ and gives a limiting spectrum consistent with an $As(\eta^1$-$C_5H_5)_3$ structure at $-70°C$. It seems clear that the As molecule is a true fluxional η^1-C_5H_5 type, but unfortunately no observations bearing on the detailed rearrangement pathway were reported.

The Sb and Bi compounds may well be homologous, since in Group IV the activation energies were seen to drop off sharply in descending the group.

B. Cycloheptatrienyl Compounds

The C_7H_7 ring can function as a mono-, tri-, or pentahapto ligand, and fluxional compounds of all three types are known. Trihapto binding is overwhelmingly the most common type.

1. *Monohapto(cycloheptatrienyl)metal Compounds*

Actually, only one of these is known, but it is of considerable importance for the light it throws on mechanistic principles. The compound in question is $Ph_3Sn(\eta^1\text{-}C_7H_7)$, **11**, and it has been shown crystallographically that the

11

structure is indeed that depicted (Weidenborner *et al.*, 1972). This molecule is fluxional, and it has been shown (Larrabee, 1971) that the main pathway is by 1,5 shifts (or, by 1,4 shifts, the two being equivalent and indistinguishable if omnidirectionality is assumed).

In the case of fluxional $(\eta^1\text{-}C_5H_5)M$ compounds, the "least motion" pathway, namely 1,2 shifts, and the pathway that would be expected according to the ordinary rules (Woodward and Hoffmann, 1969) for sigmatropic shifts, namely 1,5 shifts, are identical (omnidirectionality assumed). Hence, we are left perfectly ignorant as to whether "least motion" or "orbital symmetry" is the controlling factor in sigmatropic reactions involving metal atoms if we have only the results for $(\eta^1\text{-}C_5H_5)M$ compounds to go on. Larrabee's results for $Ph_3Sn(\eta^1\text{-}C_7H_7)$ clearly show that orbital symmetry considerations are dominant in this case. The possibility is thus raised that this factor is dominant in many if not all other cases.

However, there is a need for caution before generalizing from this tin compound to the transition metal compounds. The "ordinary" rules for sigmatropic shifts were developed assuming that the atom that moves presents only a single, essentially invariant, sigma orbital for interaction with the polyenyl residue. It appears that, despite the conceivable availability of

10. Nonrigidity in Organometallic Compounds

outer d (or other) orbitals of the tin atom, it substantially satisfies the assumption. Quite probably, other nontransition elements, such as silicon and germanium, will do so as well. However, it seems entirely uncertain whether this is a safe assumption for transition metals which have five d orbitals as well as the s and p orbitals in their valence shells. It seems quite possible that with the flexibility thus afforded, the transition metal atom may be able to present orbitals of varied symmetry properties to the polyenyl residue over which it is to move. Thus, orbital symmetry restrictions might not limit the possible motions to only one, and some other factor, or factors, might become controlling.

At this point, we simply do not have the necessary data to tell what controls the behavior of the transition metal systems. There has not been a successful synthesis of a transition metal $M(\eta^1\text{-}C_7H_7)$ compound, the one purported case having later been shown to be an $M(\eta^3\text{-}C_7H_7)$ compound (Ciappenelli and Rosenblum, 1969).

2. Trihapto(cycloheptatrienyl)metal Compounds

Compounds containing the $(\eta^3\text{-}C_7H_7)M$ moiety are the commonest type formed by the cycloheptatrienyl group, and all of those known are fluxional. The molybdenum compound, **12**, is representative and has been studied in greatest detail. Its structure has been established crystallographically (Cotton *et al.*, 1971a), and its proton nmr spectrum has been recorded and interpreted from 25° to −120°C (Faller, 1969; Bennett *et al.*, 1969). The coalescence temperature is around −50°C and the low-temperature limiting spectrum is reached at about −110°C.

It has been shown that rearrangement proceeds by 1,2 shifts. Using the site labeling shown below, it can easily be seen that for 1,2, 1,3, and 1,4 shifts

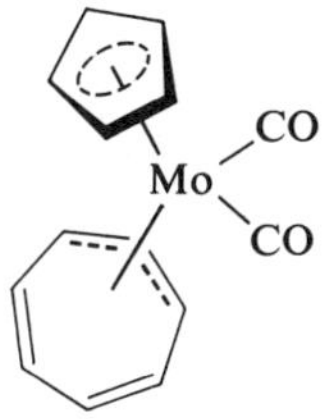

12 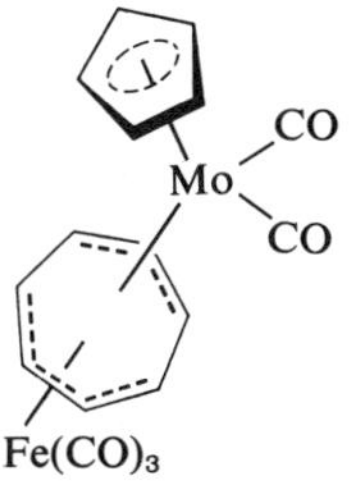   13

it will be the resonances for sites d, b, or c, respectively, that will broaden more slowly than the others, since the mean residence times in those sites will be twice as great as for the others. It was possible to assign the limiting spectrum unambiguously, and it was found that it is the d resonance that broadens most slowly as the temperature is raised. This is shown in Fig. 6, which shows the behavior of the c ($\delta = 6.5$) and d ($\delta = 5.5$) signals from $-102°$ to $-76°C$. Thus 1,2 shifts are conclusively established. The Arrhenius activation energy was found to be about 12.5 kcal/mole.

The molecule **13**, which can be obtained from **12**, is also fluxional; it is discussed later (Section III).

The compound $C_7H_7Co(CO)_3$ has been studied by nmr (Bennett *et al.*, 1969). The results indicate that, as expected, it contains a trihapto C_7H_7 group, and that it too rearranges by 1,2 shifts. The activation energy here is appreciably lower than for $(\eta^5-C_5H_5)Mo(CO)_2(\eta^3-C_7H_7)$ discussed above, since even at $-140°C$ a limiting spectrum showing resolved spin–spin splittings is not reached.

Again, in $(\eta^5-C_5H_5)Fe(CO)(\eta^3-C_7H_7)$ the proton nmr spectrum as a function of temperature (Ciappenelli and Rosenblum, 1969) clearly shows that (a) the C_7H_7 ring is trihapto, and (b) rearrangement occurs by 1,2 shifts. Possibly because of steric hindrance, the activation energy here must be

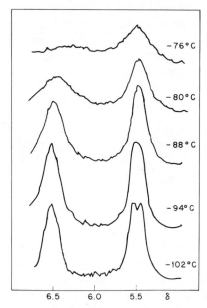

Fig. 6 The behavior of the *c* and *d* resonances of $(\eta^3-C_7H_7)Mo(CO)_2(\eta^5-C_5H_5)$. (Reproduced by permission from Faller, 1969.)

considerably higher than in the other two cases, since the major collapse of lines occurs between $-20°$ and $-10°C$.

Finally, the pyrazolylborate complex $H_2B(3,5\text{-dimethyl-l-pyrazolyl})_2Mo(CO)_2(C_7H_7)$ has been shown to contain a trihapto C_7H_7 group (Cotton et al., 1972), which is fluxional (Calderon et al., 1972). Although this was not studied in detail, the pattern of line-shape collapse is again indicative of 1,2 shifts.

3. Pentahapto(cycloheptatrienyl)metal Compounds

Only two species of this type appear to have been characterized, namely, $[(\eta^5\text{-}C_7H_7)Fe(CO)_3]^+$ (Pettit, 1964; Whitesides and Budnik, 1971) and $(\eta^5\text{-}C_7H_7)Mn(CO)_3$ (Whitesides and Budnik, 1971). The cation has a limiting slow-exchange spectrum at $-80°C$, and a coalescence temperature of about $-50°C$, whereas the neutral molecule has a limiting spectrum at or below $-47°C$ and a coalescence temperature of $27° \pm 10°C$. The limiting spectra support the pentahapto formulations. There is not as yet any report of a detailed line-shape analysis leading to a specification of the rearrangement pathway (e.g., 1,2-shifts).

The comparison afforded here between two isostructural and isoelectronic species differing only in charge is, thus far, unique. It shows that charge alone does have a substantial effect on the lability of the fluxional behavior. Whitesides and Budnik suggest that this is because the degree of backbonding is an important determinant of the rate of rearrangement. They suggest that the cationic species is faster because there is less backbonding.

II. Cyclopolyenemetal Systems

A cyclic conjugated polyolefin, such as cyclooctatetraene, with n double bonds can often be bonded to a metal atom, or to several metal atoms, in such a way that only a sequence of $1, 2, \ldots, n-1$ of the π electron pairs are employed, thus leaving a sequence of $n-1, n-2, \ldots, 1$ double bonds free. In such cases, there will be n equivalent positions for the metal atom or group of metal atoms. Rapid circulation among these n sites constitutes a form of fluxionality.

A. Cyclooctatetraenemetal Compounds

1. *Cyclooctatetraene Bound to One Metal Atom*

a. *Iron, Ruthenium, and Osmium Tricarbonyl Compounds.* The first such molecule to be reported was $(C_8H_8)Fe(CO)_3$ (Manuel and Stone, 1959, 1960;

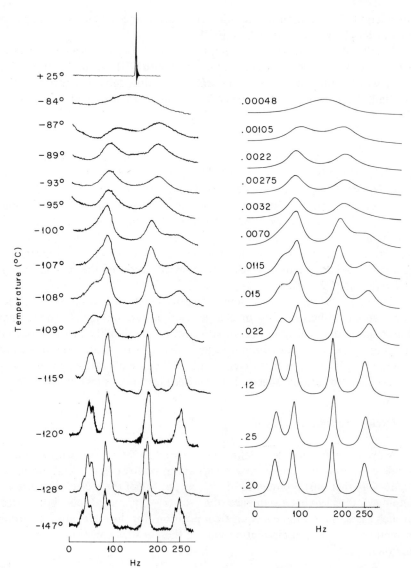

Fig. 7 Observed proton nmr spectra (left) for $(C_8H_8)Ru(CO)_3$ at various temperatures. Computed spectra (right) for various mean residence times based on 1,2 shifts. (Reproduced by permission from Cotton *et al.*, 1969.)

Rausch and Schrauzer, 1959; Nakamura and Hagihara, 1959). Its proton nmr spectrum at 25°C was found to be a single sharp line. The possibility that this is a fluxional molecule first became evident (Dickens and Lipscomb, 1962) when an X-ray crystallographic study showed that, at least in the crystal, the ring is attached to the iron atom in a 1,2,3,4-tetrahapto fashion. The early efforts to determine whether this structure persists in solution and thence to develop in detail a picture of the fluxional behavior of the molecule have previously been reviewed (Cotton, 1968). Difficulties arise from the low activation energy that makes the limiting low-temperature spectrum inaccessible for $C_8H_8Fe(CO)_3$. However, the activation energies increase for the Ru and Os analogs, with the result that for $C_8H_8Ru(CO)_3$ a limiting spectrum can be reached, or very nearly reached, at $-145°C$ and for $C_8H_8Os(CO)_3$, a very sharp, limiting spectrum is observed at $-100°C$ (Cooke et al., 1970). It is clear from the limiting spectra of the Ru and Os compounds that these two definitely have the 1,2,3,4-tetrahapto structure, and thus it seems a fair assumption that the iron analog does also. It is also very probable that the mechanism of rearrangement will be qualitatively the same in all three cases, and thus it is only necessary to establish it for one. Naturally, each case would have to be studied separately to obtain quantitative activation data, but from the approximate coalescence temperatures, it can be estimated that the activation energies increase by 2–3 kcal at each step in passing from $C_8H_8Fe(CO)_3$ to $C_8H_8Ru(CO)_3$ to $C_8H_8Os(CO)_3$. In any event, only an approximate activation energy has been obtained in each case from the proton spectrum because of the complexity of the spectrum and the small but not entirely negligible deviations from first order character.

The $C_8H_8Ru(CO)_3$ case has been examined in detail (Bratton et al., 1967; Cotton et al., 1969). Five pathways of migration were considered: 1,2 shifts, 1,3 shifts, 1,4 shifts, 1,5 shifts, and random shifts. Spectra recorded from $-147°$ to $+25°C$ are shown in Fig. 7. A diagram of the molecular structure defining labels for the proton sites is shown in Fig. 8. It is clear that random shifts can be at once ruled out, because the line-shape changes are not uniform, even though all four sites are equally populated. Moreover, 1,5 shifts can also be ruled out, since these would average only α with δ and β with γ,

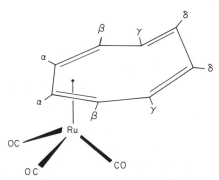

Fig. 8 A schematic representation of the structure of the $(C_8H_8)Ru(CO)_3$ molecule, showing how the proton sites are labeled.

thus leading to a two-line high-temperature limiting spectrum. The limiting spectrum could be assigned, and the pattern of spectral change expected for each of the remaining nonrandom pathways (1,2, 1,3, and 1,4) was computed. For all but the 1,2 pathway, there were gross qualitative discrepancies between the observed and calculated spectra. Computed spectra for the 1,2 shift pathway are also shown in Fig. 7. From the matched set of residence times and temperatures, an Arrhenius plot yielded $E_a \approx 9.5$ kcal/mole and $\log A \approx 14$. The former is probably accurate to ± 2 kcal/mole, the uncertainty arising chiefly from the unknown effect of spin–spin coupling in violating the assumption of independent spins, which is inherent in the computation of line shapes. However, there is no doubt that the mechanism is qualitatively correct.

b. *Chromium, Molybdenum and Tungsten Tricarbonyl Compounds.* Another important set of cyclooctatetraene complexes are those formed by the Group VI metals, Cr, Mo, and W, which also have the composition $C_8H_8M(CO)_3$ (Kreiter *et al.*, 1966; Winstein *et al.*, 1965; King, 1967). In these cases, six of the carbon atoms (and six π electrons) are engaged in bonding to the metal atom, the structure (McKechnie and Paul, 1966) being

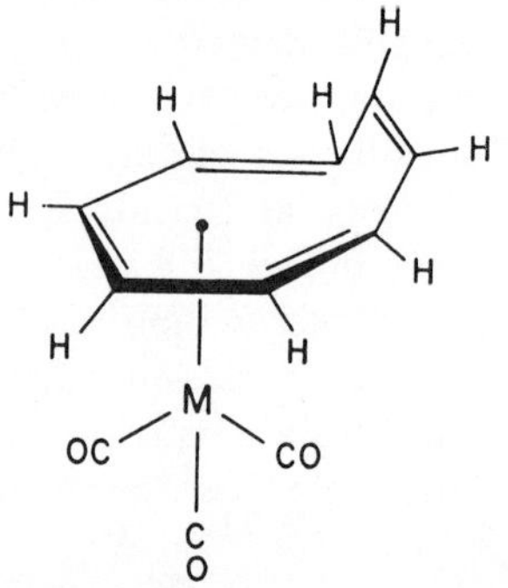

Fig. 9 A schematic representation of the structure of the $(C_8H_8)M(CO)_3$ molecules, M = Cr, Mo, W.

as shown schematically in Fig. 9. All three compounds have sharp complex spectra at $-40°C$ consistent with this structure. These spectra broaden and collapse as the temperature is raised, and at temperatures of 50° to 100°C, single line spectra are observed. Thus, the occurrence of a rearrangement process that leads to time-averaging of all proton environments is demonstrated, but because of extensive spin–spin coupling, the spectra are too complex for a dynamical analysis. A neat solution to this problem would be to study the ^{13}C spectra at the natural abundance level, but this was not possible until very recently (cf. p. 409).

Using 1,3,5,7-tetramethylcyclooctatetraene (TMCOT), three compounds analogous to the $C_8H_8M(CO)_3$ compounds were prepared and studied (Cotton et al., 1968). These derivatives were chosen because it was expected that proton–proton coupling would be negligible, thus allowing a detailed line-shape analysis. It was hoped that from a knowledge of the rearrangement pathway for the $(\eta^6\text{-TMCOT})M(CO)_3$ compounds the pathway for the $(\eta^6\text{-COT})M(CO)_3$ compounds could be inferred. As indicated later, this does not prove to be the case. The structure of the chromium compound (Bennett et al., 1968a) was shown to be essentially the same as that of $C_8H_8Mo(CO)_3$. The spectrum in the low temperature limit consisted of four singlets of relative intensity 1 due to the ring protons and four singlets of relative intensity 3 due to the methyl group protons.

The line-shape changes as the temperature is raised are very selective, as shown in Fig. 10. Between $-23°$ and about $+50°C$, two of the ring–proton resonances collapse and coalesce, while the other two do not change; simultaneously, all of the methyl resonances broaden and then coalesce pairwise.

It was shown that there is a simple and unique explanation for these changes, namely, that the relative movement of the metal atom is that shown in the upper left of Fig. 11. Clearly, this leads to averaging of two ring-proton sites (B and D), while the other two (A and C) remain entirely unaffected. At the same time, the two methyl groups flanking A and the two flanking C become time-averaged. The 1,4 shift shown at the upper right would also produce the same qualitative set of spectral changes, whereas none of the other five possibilities shown is qualitatively acceptable.

The choice between the two qualitatively acceptable types of shift was made in favor of the 1,2 shift on the basis of a partial assignment of the ring-proton resonances. According to this assignment, one of the two lines that collapse has to be due to proton B and one of the two lines that remain unchanged has to be due to proton A. This agrees with the predicted effect of 1,2 shifts but is opposite to the result expected for 1,4 shifts.

Above about 50°C, further changes begin to occur, and it appears that these are leading, finally, to single resonances for the ring and methyl protons, although the fast exchange limit cannot be fully attained. It was not possible to demonstrate rigorously from the experimental data what pathway causes this new set of changes, and there are a number of plausible, or at least credible, possibilities. However, the simplest, and for that reason, perhaps, the most reasonable assumption is that the alternative type of 1,2 shift, that shown as $S \leftrightarrow Z$ in the lower left diagram of Fig. 11 becomes activated. When this is added to the $S \leftrightarrow T$ type, the metal atom is able to circulate completely around the ring in either direction, which would time-average all the ring–proton resonances and all the methyl resonances.

By the usual method of fitting computed spectra at various temperatures,

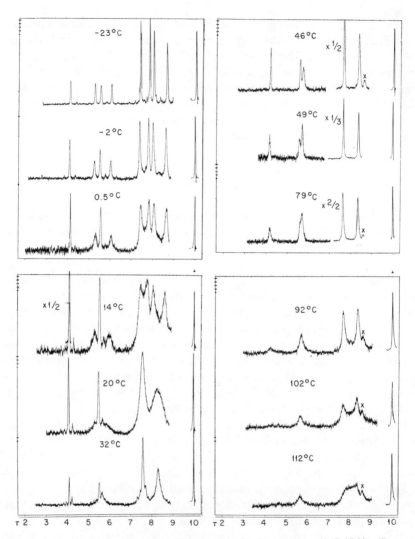

Fig. 10 Proton nmr spectra of (1,3,5,7-tetramethylcyclooctatetraene) Cr(CO)$_3$. (Reproduced by permission from Cotton *et al.*, 1968.)

a kinetic analysis was carried out for the first stage process for the three compounds. The results are given in the tabulation. The results for the

Compound	E_a(kcal/mole)	log A
(TMCOT)Cr(CO)$_3$	15.8 ± 1.0	13.8 ± 0.7
(TMCOT)Mo(CO)$_3$	16.0 ± 1.0	13.5 ± 0.5
(TMCOT)W(CO)$_3$	19 ± ?	16 ± ?

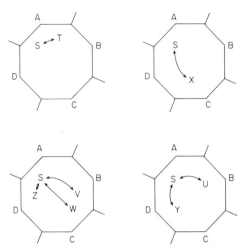

Fig. 11 Diagrams showing the possible shift pathways for a $[1,3,5,7\text{-}(CH_3)_4C_8H_4]M\text{-}(CO)_3$ molecule.

tungsten compound are least reliable, being derived from the line-shape variation of only one resonance over a limited temperature range. The high value of log A arouses suspicion. However, even if log A were reduced to 13.5, the value of E_a would still be high, about 18.4. Thus, it appears that for the three compounds, the activation energies are in the order Cr $\approx$ Mo < W.

From the clear results obtained for the $(\eta^6\text{-TMCOT})M(CO)_3$ molecules, where a restricted 1,2 shift was found to be the lowest energy process, there was a temptation to extrapolate to an assumed pathway for the $(\eta^6\text{-COT})\text{-}M(CO)_3$ molecules. The extrapolation would be based on the plausible assumption that, in the absence of methyl substituents on the ring, 1,2 shifts would occur in both directions cyclically. However, direct study of $(\eta^6\text{-COT})\text{-}Mo(CO)_3$ by ^{13}C nmr has shown that this is not so (Cotton, F. A., Hunter, D. L., and Lahuerta, P., unpublished results, 1973). It was found that the four ring-carbon resonances collapse at the same rate. This unequivocally rules out 1,2 shifts as the main pathway of rearrangement, and is consistent only with (a) 1,3 shifts, or (b) a random process. The 1,3 shift pathway is improbable since the relative ring-to-metal motion required could scarcely occur without passage through an intermediate configuration so close to the immediately adjacent minimum-energy configuration as to cause the system to drop into that configuration, i.e., to execute a 1,2 shift.

To explain the uniform line broadening, the metal may be assumed to pass to a position over the center of a nearly flat ring from which it then has equal access to all 8 bonding sites which allow it to recover a hexahapto structure. The reason why this could occur with the COT compound but not with the TMCOT analog is to be found in the steric effect of the four methyl groups in destabilizing a planar ring. The reality of that effect, and its very substantial magnitude, is evident from a comparison of the known activation energies for bond shift, via a planar activated state, for COT and TMCOT. These energies are *ca.* 13 kcal mole^{-1} and *ca.* 21 kcal mole^{-1}, respectively. Such a difference could easily explain the observations.

c. $(C_8H_8)_2Fe$. This molecule contains both a tetrahapto and a hexahapto ring (Carbonaro *et al.*, 1968; Allegra *et al.*, 1968). The former executes a ring-whizzing motion with significantly lower activation energy than the latter, which is in accord with the observations discussed above on the separate sets of $(\eta^4\text{-}C_8H_8)M(CO)_3$ and $(\eta^6\text{-}C_8H_8)M(CO)_3$ compounds. This compound is of greater interest because of the interchange of the $\eta^4\text{-}C_8H_8$ and $\eta^6\text{-}C_8H_8$ rings and is discussed further in that connection in Section IV,B.

d. *Other Compounds.* It seems likely that $(\eta^4\text{-}C_8H_8)M$ systems are characteristically fluxional. An example in support of this is the cobalt compound **14**, which shows a single sharp resonance for the C_8H_8 protons at 0°C in C_6D_6 (Greco *et al.*, 1971).

Finally, some interesting observations have been made on benzo- and naphthocyclooctatetraene complexes of $Fe(CO)_3$ (Whitlock and Stucki, 1972). While, as stated earlier, $C_8H_8Fe(CO)_3$ has a barrier to rearrangement of only about 7 kcal/mole, the benzo compound executes the **15a** to **15b** rearrangement with an activation energy of about 18.6 kcal/mole, and the analogous **16a** to **16b** rearrangement has an E_a of about 31 kcal/mole. Whitlock and Stucki make the very reasonable suggestion that these increases are due to the increasing instability of the ortho quinoidal intermediates. The phenomenon is essentially the same as that involved in going from $M\text{-}C_5H_5$ to $M(1\text{-indenyl})$ species, as already noted in Section 1. According to HMO calculations, as well as the observed behavior of the C_5H_5 versus 1-indenyl species, increases in E_a of the order of 14–15 kcal/mole are to be expected.

2. *Cyclooctatetraene Bound to Two Connected Metal Atoms*

The first examples in this class were $C_8H_8Fe_2(CO)_5$ and $(TMCOT)Fe_2(CO)_5$. The $C_8H_8Fe_2(CO)_5$ was shown to have only a sharp single line in its proton nmr spectrum (Keller *et al.*, 1965), and later the structure was shown to be **17a** (Fleischer *et al.*, 1966). Later, the TMCOT analog was found to have an nmr spectrum at both 25° and −60°C consisting of one signal for the ring protons and one for the methyl groups (Cotton and Musco, 1968), and the structure was shown to be **17b** (Cotton and LaPrade, 1968). The structure in each case has effective C_{2v} symmetry, and the nmr results must be attrib-

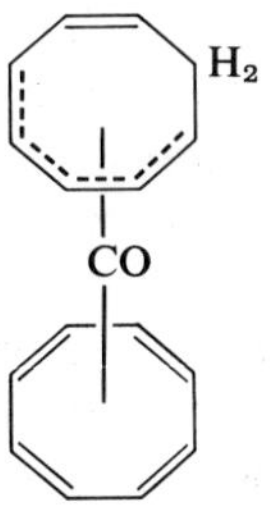

10. Nonrigidity in Organometallic Compounds

14

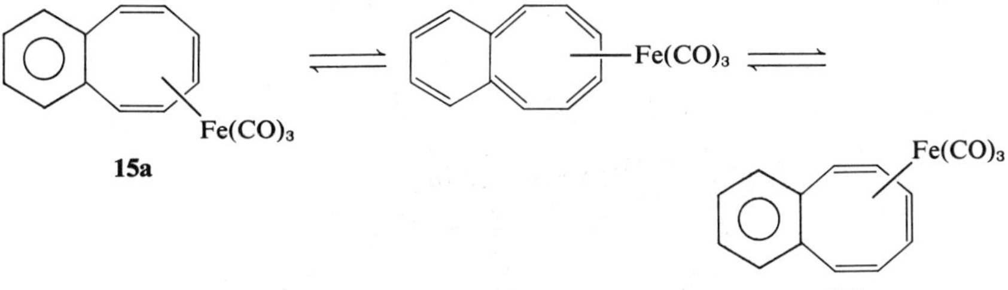

15a **15b**

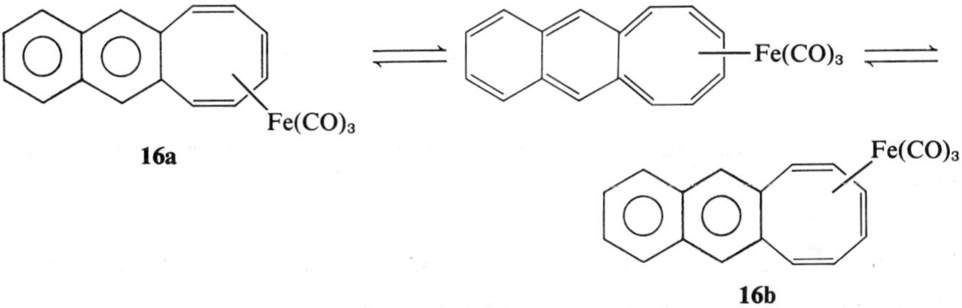

16a **16b**

uted to rapid rotational reorientations of the $Fe_2(CO)_5$ group with respect to the rings. In **17a**, these must be either 1,2 or 1,4 shifts, since 1,3 or 1,5 would fail to average all ring positions; it seems safe to assume 1,2 rather than 1,4 shifts.

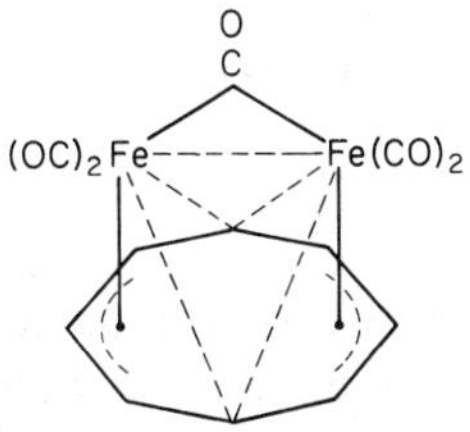

17a 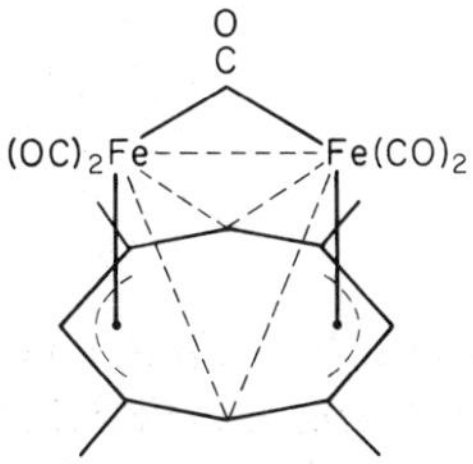 **17b**

A similar and yet interestingly different situation is found in $(C_8H_8)_2$-$Ru_3(CO)_4$ (Cotton et al., 1967b; Bennett et al., 1968b). The molecular structure is shown in Fig. 12. Only a single pmr signal is seen for this molecule at 25°C; it is only sparingly soluble and therefore low-temperature study was not practicable.

Close inspection of the orientation of each Ru–Ru axis with respect to its C_8H_8 ligand shows that it does not have the exact orientation found for the Fe–Fe axes in the two $Fe_2(CO)_5$, namely that shown schematically in **18a**.

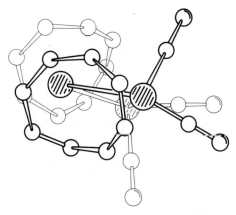

Fig. 12 The molecular structure of $(C_8H_8)_2Ru_3(CO)_4$. (Reproduced from Bennett et al., 1968b.)

Now, an alternative bonding scheme for all the systems being considered in this section, and the one actually proposed before the X-ray data on the iron systems disproved it is that shown in **18b**. In $(C_8H_8)_2Ru_3(CO)_4$, the orientation is approximately halfway between the two. This serves to point up the fact that a fluxional molecule may be expected to be very readily deformable in a manner corresponding to progress along the reaction coordinate of its fluxional process.

      18a   **18b**

The compound $(C_8H_8)Fe_2(CO)_6$ will be discussed in Section II,B, since, as will be seen there, the $Fe_2(CO)_6$ group interacts with only three of the double bonds and the system has qualitatively the same behavior as the corresponding cyclooctatriene complex $(C_8H_{10})Fe_2(CO)_6$.

3. Cyclooctatetraene Bound to Three Connected Metal Atoms

The well-known metal atom cluster systems of the type $XCCo_3(CO)_9$ (Penfold and Robinson, 1973) react with polyolefins to displace two or more CO groups. An interesting special case is the reaction of $C_6H_5CCo_3(CO)_9$ with cyclooctatetraene to give $C_6H_5C[Co(CO)_2]_3C_8H_8$, in which each cobalt atom loses its axial CO group, its place being taken by one double bond of the cyclooctatetraene ring. The resulting molecule is fluxional (Robinson and Spencer, 1971). The triangular set of three Co atoms lies over the C_8H_8 ring in the manner shown schematically in Fig. 13. There are four different types

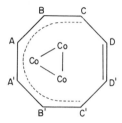

Fig. 13 A schematic representation of how the triangle of cobalt atoms in $C_6H_5CCo_3$-$(CO)_6C_8H_8$ lies over three of the double bonds of C_8H_8. The four nonequivalent types of hydrogen atom are denoted A, B, C, and D.

of proton (distinguished by A, B, C, and D in Fig. 13). At temperatures of $-20°C$ and below, four separate proton signals are observed. At about 30°C, there is only one broad signal, which sharpens as the temperature is raised further. Analysis of the intermediate spectra to yield mechanistic information was carried out straightforwardly using quantitative arguments directly adapted from earlier work, especially the analysis of the $[C_8H_4$-$(CH_3)_4]Mo(CO)_3$ (Cotton et al., 1968). The analysis led unambiguously, as in the earlier instance, to the conclusion that the main rearrangement pathway is via 1,2 shifts.

B. (Cycloocta-1,3,5-triene)diiron Hexacarbonyl and Related Molecules

According to generally accepted ideas concerning valence in olefinmetal carbonyl molecules, each of the metal atoms in the moieties $(OC)_3Fe$–$Fe(CO)_3$ and $(OC)_3Ru$–$Ru(CO)_3$ lacks three electrons of a closed-shell configuration. Therefore, such $M_2(CO)_6$ moieties would normally be expected to interact with olefinic systems capable of supplying a total of six electrons, or with portions of more extensive olefinic systems, so as to leave any π electrons beyond the requirement of six not engaged in metal–carbon bonding. Naturally, the olefin must also meet certain steric requirements.

The first such compound to be described was $C_8H_{10}Fe_2(CO)_6$ (**19**) where the C_8H_{10} originates as 1,3,5-cyclooctatriene (though it need not necessarily remain as such in the complex). The structure **19a** was suggested for it (King, 1963). This structure has a plane of symmetry and nonequivalent iron atoms.

A little later, Emerson *et al.* (1964) reported the isolation of $C_7H_8Fe_2$-$(CO)_6$, **20**. The Mössbauer spectra (Fe) of both **19** and **20** were studied and the conclusion was drawn that the two iron atoms are chemically equivalent in

19a

19b

19c

20a

20b

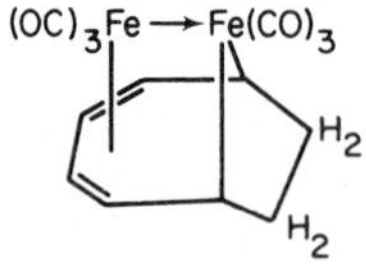

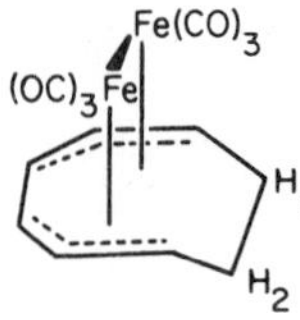

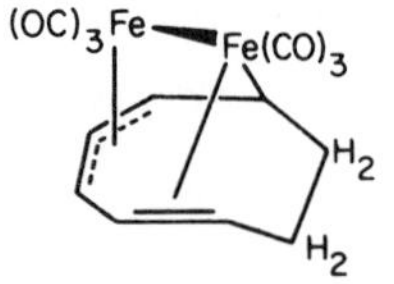

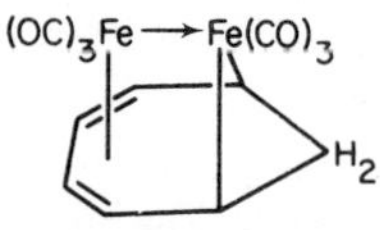

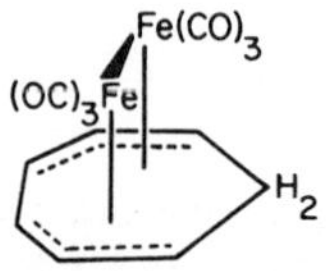

each of these complexes. Since King-type structures **19a** and **20a** place the metal atoms in different environments, Pettit and his co-workers proposed structures **19b** and **20b**. The two compounds **19** and **20** were reported to have similar infrared and nmr absorption patterns, though these have never been described in detail. Later, Keller *et al.* (1965) reported the preparation of $C_8H_8Fe_2(CO)_6$, **21**, to which they assigned structure **21b** on the basis of its Mössbauer, infrared, and nmr spectra. The nmr spectrum was complex but consistent with the ring protons being divided into two sets of four by a plane of symmetry.

21a, M=Fe
22a M=Ru

21b, M=Fe
22b M=Ru

21c M=Fe
22c M=Ru

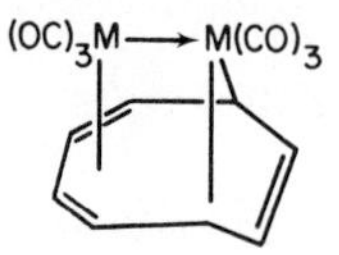

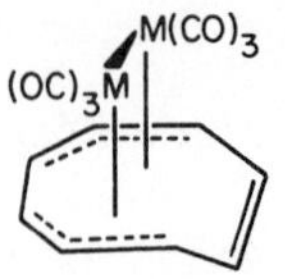

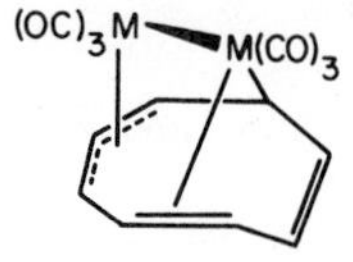

Then, an X-ray study of the ruthenium analog of **21**, namely $C_8H_8Ru_2$-$(CO)_6$, **22** (Cotton and Edwards, 1968), showed that neither a King-type **22a** nor a Pettit-type **22b** structure, but rather structure **22c** (or its enantiomorph) was correct—at least in the crystal. Compound **22** has an nmr spectrum very similar to that of **21**, but it is temperature dependent (Cotton *et al.*, 1969), and evidently the two enantiomorphous forms of **22c** interconvert rapidly at room temperature. It was suggested that this interconversion might proceed via **22a**.

It was then shown that compound **19** also has a skew structure, **19c**, in the crystal (Cotton and Edwards, 1969) and that this molecule, too, is fluxional in solution, possibly involving, again, interconversion of the enantiomorphs of **19c**, via the intermediate **19a** (Cotton and Marks, 1969d). At this point it seemed reasonable to suppose that skew structures such as **19c** and **22c** were probably correct for all such compounds, i.e., specifically, **20**, from which it would follow that the nmr spectrum of **20** should be temperature dependent. However, it was found that the spectrum of **20** remains unchanged down to $-100°C$. In order to find out whether this meant that **20** is fluxional but with a very low activation energy, or whether **20** has a structure with a static plane of symmetry, an X-ray investigation was carried out (Cotton *et al.*, 1971b).

It was found that compound **20** has structure **20b**. This result prompted some reconsideration of the earlier view that a structure of type **a** was the "obvious" intermediate in the interconversion of enantiomorphous structures of type **c**. The actual observation of a **b**-type structure as the stable one in this case requires one to be more cautious than previously in speculating about the rearrangement pathway. So long as the **b**-type structure remained purely hypothetical, it seemed logical that the **a** structure was the most probable intermediate structure in the interconversion of the enantiomorphs. This was chiefly because the $c \rightleftharpoons a \rightleftharpoons c'$ process appears to involve only slight nuclear and electronic shifts. However, the fact that a **b** structure had now been observed in a molecule closely similar in type to **19**, **21**, and **22**, whereas no **a** structure had yet been observed, obviously makes the $c \rightleftharpoons b \rightleftharpoons c'$ process a viable alternative.

In principle, there is some possibility of determining experimentally which process is operative, since one of them leads to site exchange and hence to time-average equivalence of the two $Fe(CO)_3$ groups, whereas the other does not. Study of ^{13}C or ^{17}O nmr spectra at various temperatures might therefore resolve the question, but so far the necessary data have not been published.

$(OC)_3Fe\text{—}Fe(CO)_3$

23

Quite recently, still another fluxional molecule, **23**, of this sort has been examined (Cotton *et al.*, 1973). In this case, a cyclobutane ring is fused to the cyclooctatriene ring at the 7,8 positions. The structure is essentially identical to that of **19**, but the ring fusion has the effect of lowering the activation energy enough so that line broadening becomes noticeable only at about $-65°C$, whereas for **1**, it is observable already at about $-15°C$.

The preparation and nmr study of compounds **24** and **25** has been briefly described (Whitlock and Stucki, 1972). It was assumed that they have the same type of instantaneous structure as do **19, 21, 22,** and **23**.

From the temperature dependence of the proton spectra, which can be rigorously analyzed as a system of two uncoupled sites, it was concluded that a $c \to a \to c'$ mechanism, with an E_a of 8.1 kcal/mole, is compatible with the data for **24**, although there is not any positive evidence against the rotatory, i.e., $c \to b \to c'$ mechanism. For **25**, however, the persistence of P–H coupling

to H-5 and H-6, while there is none to H-3, and H-8 eliminates the rotatory mechanism in that case. Whether it is then safe to consider the rotatory mechanism eliminated in all comparable cases on the basis of this result is debatable.

C. Cyclohexatrienemetal Compounds

There are a few metal–olefin complexes in which one of the ligands may be considered formally as a substituted benzene ring but is more realistically treated as a cyclohexatriene, since the ring is nonplanar and the π system is dismembered into coordinated and uncoordinated portions. The first such compound (Dickson and Wilkinson, 1964; Churchill and Mason, 1966) was **26**, $(\eta^5\text{-}C_5H_5)[\eta^4\text{-}C_6(CF_3)_6]Rh$. Later a similar one, **27**, in which pentamethylcyclopentadienyl replaced C_5H_5 and $C(O)OCH_3$ groups replaced the CF_3 groups was prepared and its dynamical behavior investigated (Kang *et al.*, 1970).

At room temperature, the proton nmr spectrum of this compound shows three peaks of equal intensities for the —$C(O)OCH_3$ protons, in accord with the $[\eta^4\text{-}\mathbf{C_6}(COOCH_3)_6]$ Rh structure. As temperature is raised, these lines broaden; they coalesce at about 120°C and the single line becomes sharp at 155°C. One of the peaks broadens more rapidly than the other two, which shows that the shifts are not random. 1,3 shifts would also lead to equal

10. Nonrigidity in Organometallic Compounds

[Structures 26 and 27 shown: compound 26 is a Rh complex with cyclopentadienyl ring and hexakis(trifluoromethyl)benzene; compound 27 is a Rh complex with pentamethylcyclopentadienyl and hexakis(methoxycarbonyl)benzene.]

26 **27**

rates of broadening and are ruled out as well. 1,4 shifts need not be considered, since they would not result in a single line in the fast exchange limit, but rather in a two-line (intensities 1:2) spectrum. Thus, by a process of elimination, rearrangement must take place by 1,2 shifts.

Compound **26** has a three-line ^{19}F nmr spectrum at room temperature (Dickson and Wilkinson, 1964) and so also does the isoelectronic compound [(CF$_3$)C]$_6$Ru[(MeO)$_3$P](CO)$_2$ (Burt et al., 1970). There has been no report as to whether either of these compounds shows evidence of nonrigidity at higher temperatures.

There are several nickel compounds, namely, [(CF$_3$)C]$_6$NiL$_2$, where L$_2$ = C$_8$H$_{12}$, (R$_3$As)$_2$, or (R$_3$P)$_2$, which have only a single fluorine resonance at ambient temperature (Brown et al., 1971). It was cautiously suggested that they might be fluxional molecules with η^4-cyclohexatriene ligands.

III. Systems with Two Independent Metal Atoms on One Ring

The first such system to be examined, *trans*-C$_8$H$_8$[Fe(CO)$_3$]$_2$, showed no evidence of fluxionality at room temperature or even at 160°C, where it begins to decompose rapidly (Cotton et al., 1971a). It was suggested (Dickens and Lipscomb, 1962) that this nonfluxional character might be due to the nearly zero overlap between the π systems of the two (diene)Fe(CO)$_3$ systems that make up the molecule (Fig. 14).

It was not, in fact, until 1969 that a fluxional molecule consisting of two separate metal-containing moieties bonded to different portions of a single cyclic polyene or polyenyl ring was prepared (Cotton and Reich, 1969). This was the molecule *trans*-[(η^5-C$_5$H$_5$)Mo(CO)$_2$][Fe(CO)$_3$]C$_7$H$_7$, Fig. 15, in which the molybdenum atom is attached to three carbon atoms of the C$_7$H$_7$ ring and the iron atom to the other four, the two metal atoms being on opposite sides of the mean plane of the ring (Cotton et al., 1971a). At 25°C, the seven protons of the C$_7$H$_7$ ring give rise to a single sharp resonance. The coalescence

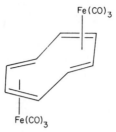

Fig. 14 A schematic representation of the structure of *trans*-$C_8H_8[Fe(CO)_3]_2$, stressing the relative independence of the two butadiene-$Fe(CO)_3$ parts of the molecule.

temperature is roughly 0°C and a spectrum showing well-resolved spin–spin splittings is observed around −70°C. Detailed analysis led to the conclusion that the rearrangement mechanism consists mainly (and, very probably, solely) of 1,2 shifts of the ring relative to the essentially immobile metal atoms. The activation parameters were evaluated as: $E_a = 13 \pm 1$ kcal mole^{-1} and log $A = 12.7 + 0.8$. This molecule rearranges slightly less facilely than $(\eta^5\text{-}C_5H_5)Mo(CO)_2C_7H_7$, for which a coalescence temperature of about −45°C and a 1,2 shift mechanism were reported (Faller, 1969).

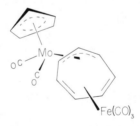

Fig. 15 A schematic representation of the structure of *trans*-$[(\eta^5\text{-}C_5H_5)Mo(CO_2)][Fe(CO)_3]C_7H_7$.

It has been shown (Cotton *et al.*, 1971a) that the fluxionality of *trans*-$[(\eta^5\text{-}C_5H_5)Mo(CO)_2][Fe(CO)_3]C_7H_7$ in contrast to the nonfluxionality of *trans*-$[Fe(CO)_3]_2C_8H_8$, can be correlated with the fact that in the former, the two π systems (i.e., the C_3 system bound to Mo and the C_4 system bound to iron) have an overlap angle of 61° (cos 61° = 0.48), while in the diiron molecule, the angle is 98° (cos 98° = −0.14).

More recently, several molecules of similar type, in which the η^5-C_5H_5 ring is replaced by a trispyrazolylborate ligand, have been prepared and found to have fluxional character very similar to that of their $[(\eta^5\text{-}C_5H_5)Mo(CO)_2][Fe(CO)_3]C_7H_7$ prototype (Calderon *et al.*, 1973).

IV. Scrambling of Differently Bonded Rings

The type of process under discussion in this section is geometrically possible whenever there are two or more ligands of the same intrinsic identity not all bound in the same way to a metal atom or cluster of metal atoms. This situation will typically arise when a metal atom, due to its inherent electron population as well as the influence of some ligands to which it may already be bound, finds itself able to accommodate more than the total number of electrons that can be donated by one organic moiety but less than the total number which can be donated by two. Thus, two such organoligands may be bound, but in different ways, one having a higher "hapticity" than the other. An example is $(C_5H_5)_2Fe(CO)_2$. The iron atom in $Fe(CO)_2$ already has twelve valence shell electrons; it needs more than the five electrons that a single η^5-C_5H_5 group can supply, but less than the ten electrons that two such rings could provide. In fact, six electrons are required and these are provided by one η^5-C_5H_5 and one η^1-C_5H_5 ring. However, in this case, and several others where one η^5-C_5H_5 and one η^1-C_5H_5 ring are present, these rings never interchange their structural roles rapidly (on the nmr time scale) even at the decomposition temperature. In other cases, e.g., $(\eta^1$-$C_5H_5)_2(\eta^5$-$C_5H_5)_2Ti$, they do so at room temperature.

A. Interchange of Mono- and Pentahaptocyclopentadienyl Rings

1. $(\eta^1$-$C_5H_5)_2(\eta^5$-$C_5H_5)_2Ti$

Interchange of η^5-C_5H_5 and η^1-C_5H_5 rings was first observed and studied in detail in $(\eta^1$-$C_5H_5)_2(\eta^5$-$C_5H_5)_2Ti$ (Calderon et al., 1970, 1971a,b). The structure of the molecule in the crystal was firmly established crystallographically as $(\eta^1$-$C_5H_5)_2(\eta^5$-$C_5H_5)_2Ti$.

The proton nmr spectra at various temperatures are shown in Fig. 16. At room temperature, two very broad, partially coalesced lines are seen; complete coalescence occurs on raising the temperature to $\sim 36°C$ and the single peak narrows as the temperature is raised further. At $-27°C$, the two peaks are quite sharp and of equal intensity. The spectra in this temperature range can be understood in terms of $(\eta^5$-$C_5H_5)/(\eta^1$-$C_5H_5)$ interchanges, which are too slow to cause line broadening at $\sim -27°C$ and become rapid enough to cause collapse to a single line at temperatures above $\sim 36°C$. The η^1-C_5H_5 rings give rise to a sharp singlet at $-27°C$ because ring whizzing is occurring rapidly at this temperature. When the temperature is lowered sufficiently, this resonance broadens, collapses, and reforms in the manner expected for slowing of ring whizzing.

The question naturally arises, why is $(\eta^1$-$C_5H_5)/(\eta^5$-$C_5H_5)$ ring interchange

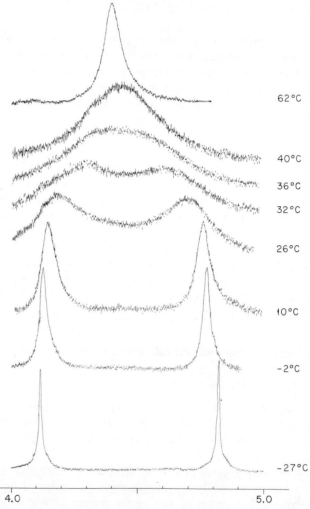

Fig. 16 Proton nmr spectra of $(\eta^1\text{-}C_5H_5)_2(\eta^5\text{-}C_5H_5)_2Ti$ in toluene at various temperatures.

facile in this case whereas in $(\eta^1\text{-}C_5H_5)(\eta^5\text{-}C_5H_5)(CO)_2Fe$ it is not observed up to the temperature (about 140°C) of decomposition? The answer to that question can best be given after the properties of $(C_5H_5)_3MoNO$ have been examined.

2. $(C_5H_5)_3MoNO$

This unusual molecule has proven very informative as to the probable transition state for $(\eta^1\text{-}C_5H_5)/(\eta^5\text{-}C_5H_5)$ ring interchange. To begin with, its structure, shown in Fig. 17, is unique. Ring 1 is a normal η^1-C_5H_5 group and the NO ligand is of the standard linear three-electron-donor type. Thus, from these two ligands the Mo atom acquires four electrons; together with its own six electrons, it then has a total of ten electrons. The other two rings must therefore be contributing a total of eight on the assumption that a final total of eighteen electrons is to be achieved. Structurally, the two remaining C_5H_5 rings (2 and 3) in Fig. 17 are equivalent; the formal possibility

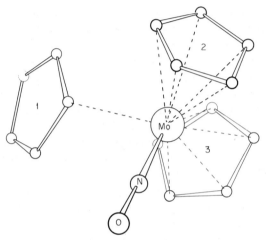

Fig. 17 The molecular structure of $(C_5H_5)_3MoNO$.

of one η^5-C_5H_5 and one η^3-C_5H_5 is ruled out (Calderon et al., 1969). Each of these rings is related geometrically to the metal atom in such a way that there are two short, one intermediate and two long C–Mo distances, of lengths approximately 2.64 ± 0.05 Å, 2.44 ± 0.02 Å, and 2.34 ± 0.03 Å. The same pattern was found in $Mo(C_5H_5)_2(CH_3)NO$ (Cotton and Rusholme, 1972), thus showing that it is truly characteristic. It is not possible to describe the bonding of such a ring to the metal atom in any simple and unique way. However, the important thing to be noted is that the strength and extent of attachment of the ring to the metal atom is intermediate between the well-defined limiting cases of $(\eta^1$-$C_5H_5)M$ and $(\eta^5$-$C_5H_5)M$. Since the two rings jointly are interacting equivalently with a metal atom so as to donate eight electrons to its valence shell, it could be said that each ring is, formally, a four-electron donor.

The proton nmr spectrum of $(C_5H_5)_3MoNO$ varies enormously with temperature, as illustrated in Fig. 3. Three different dynamic processes are responsible for the complete pattern of variation. At very low temperatures ($< -50°C$), the rate of rotation of the $(\eta^1$-$C_5H_5)$ ring about the C—Mo σ bond goes from fast to slow, thus causing the other two rings to have nonaveraged, different environments and the η^1-C_5H_5 ring itself to loose its plane of symmetry. At intermediate temperatures, the η^1-C_5H_5 ring to Mo bond executes increasingly rapid 1,2 shifts in the manner typical of (η^1-C_5H_5)M systems in general, as discussed in Section I. However, before this process reaches a rate at which a single discrete line can be observed for the five η^1-C_5H_5 protons, separate from the line for the other ten C_5H_5 protons, the third process supervenes. This third process is exchange of the structural roles of the three C_5H_5 rings.

The relative ease of the ring interchange process in this molecule may reasonably be attributed to the fact that two of the rings have, as noted above, an intermediate degree of bonding to the metal. Hence, the attainment of a transition state in which the η^1-C_5H_5 ring and one of the other two become equivalent, possibly a structure approximating (η^1-C_5H_5)$_2$(η^5-C_5H_5)Mo(NO) is energetically facilitated.

With the concept of this intermediately bonded (four-electron donor) ring, which may be denoted formally, by η^4-C_5H_5, it is possible to understand why (η^1-C_5H_5)/(η^5-C_5H_5) ring interchange occurs readily in some cases but is not observed in others. A pair of compounds that illustrate the problem clearly are (η^1-C_5H_5)$_2$(η^5-C_5H_5)$_2$Ti and (η^1-C_5H_5)$_2$(η^5-C_5H_5)$_2$Mo. The former, as already noted, begins to undergo rapid ring interchange around room temperature, whereas the latter (Calderon and Cotton, 1971) does not. The titanium compound has only a sixteen-electron configuration in its ground state structure. Thus, there is one valence shell orbital of the metal available for use in forming a suitable transition state, and the following process can occur.

$$(\eta^1\text{---}C_5H_5^*)_2(\eta^5\text{---}C_5H_5)_2\text{Ti} \rightleftharpoons (\eta^1\text{---}C_5H_5^*)(\eta^4\text{---}C_5H_5^*)(\eta^4\text{---}C_5H_5)(\eta^5\text{---}C_5H_5)\text{Ti}$$
16 electrons $\qquad\qquad\qquad\qquad\qquad$ 18 electrons

$$\rightleftharpoons (\eta^1\text{---}C_5H_5^*)(\eta^1\text{---}C_5H_5)(\eta^5\text{---}C_5H_5^*)(\eta^5\text{---}C_5H_5)\text{Ti}$$
16 electrons

In the case of the molybdenum compound, which already has an eighteen-electron configuration in its ground state structure, this process is impossible. Similarly, most other molecules with eighteen-electron structures are unable to interchange η^1-C_5H_5 and η^5-C_5H_5 rings. The only exceptions might—and apparently do—occur if there is another ligand present that can switch temporarily from being an n-electron donor to being an $(n-2)$-electron donor. In the next section this possibility will be illustrated with an example.

Finally, we note that for the $(C_5H_5)_4$Zr and $(C_5H_5)_4$Hf molecules all ring protons are equivalent even at $-150°C$ (Calderon et al., 1971b). It has been suggested by the authors just cited that these molecules may well have ground state structures of the type (η^1-C_5H_5)(η^4-C_5H_5)$_2$(η^5-C_5H_5)M. Ring interchange would thus become very facile indeed, since the ground state is virtually a transition state for the interchange process. At the very least, the energy of such a structure is probably much lower for the Zr and Hf compounds, even if it is not actually the most stable structure. An X-ray study of $(C_5H_5)_4Zr$ is sufficiently crude as to be consistent with a $(\eta^1\text{-}C_5H_5)(\eta^4\text{-}C_5H_5)_2(\eta^5\text{-}C_5H_5)Zr$ structure, whereas the proposed $(\eta^1\text{-}C_5H_5)(\eta^5\text{-}C_5H_5)_3Zr$ structure is not acceptable, since it implies a twenty-electron configuration.

3. $Mo(\eta^1\text{-}C_5H_5)(\eta^5\text{-}C_5H_5)(NO)(S_2CNMe_2)$

The molecule $Mo(\eta^1\text{-}C_5H_5)(\eta^5\text{-}C_5H_5)(NO)(S_2CNMe_2)$ (Kita et al., 1971) presents an interesting—and, in its details, unique—example of $(\eta^1\text{-}C_5H_5)/(\eta^5\text{-}C_5H_5)$ ring interchange. The structure of the molecule in the crystal is known (N. Bailey, private communication, 1972).* The Mo—N—O group is essentially linear with a short (1.79 Å) Mo—N distance, and NO must therefore be regarded as a three-electron donor. The Me_2NCS_2 group is bonded symmetrically to the molybdenum atom, and it too, as a neutral radical, is to be regarded as a three-electron donor. Thus, with the $\eta^1\text{-}C_5H_5$ and $\eta^5\text{-}C_5H_5$ rings being one- and five-electron donors, respectively, the valence shell of the molybdenum atom is populated by a total of $3 + 3 + 1 + 5 + 6 = 18$ electrons. It might, therefore, have been expected that while the $\eta^1\text{-}C_5H_5$ ring would undergo the usual ring whizzing, there would be no scrambling of the two rings.

Scrambling, however, does occur. At 30°C, each ring gives a single proton resonance signal. Between 30° and 60°C, these signals broaden and collapse. At 70°C, they coalesce, and above this temperature, this single peak sharpens. In the view of this writer, it is necessary to postulate that one of the two three-electron donors, NO or Me_2NCS_2, opens a pathway for ring scrambling by detaching itself from the metal to the degree required to become, temporarily, a one-electron donor. It is then possible for the $\eta^1\text{-}C_5H_5$ ring to move into a more intimate bonding posture toward the metal atom. A very attractive possibility is the same as that postulated in the case of $(\eta^1\text{-}C_5H_5)_2(\eta^5\text{-}C_5H_5)_2Ti$, in which a pair of rings, one initially $\eta^1\text{-}C_5H_5$ and the other initially $\eta^5\text{-}C_5H_5$ become equivalent in the same sense as are the two rings in $(C_5H_5)_2Mo(NO)R$ molecules, i.e., become $\eta^4\text{-}C_5H_5$ rings. Then, restoration of the ground-state geometry and bonding can occur with ring interchange.

Which of the two three-electron ligands is the "gate keeper" for this process? There are certain data that suggest that it is the NO group. As the crystal structure shows, the two methyl groups are structurally nonequivalent. McCleverty and his co-workers observed that in the pmr spectrum at lower temperatures they are also magnetically nonequivalent. Furthermore, they remain magnetically nonequivalent, giving a sharp doublet separated by

* Work done in the Department of Chemistry, Sheffield University.

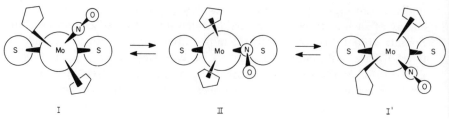

Fig. 18 The pathway for $(\eta^1\text{-}C_5H_5)/(\eta^5\text{-}C_5H_5)$ ring scrambling that is being proposed here. II represents a structure with two $\eta^4\text{-}C_5H_5$ groups and a bent (NO$^-$) type nitrosyl ligand.

3.5 Hz even after the two cyclopentadienyl resonances have coalesced. Only at ~90°C do the methyl signals begin to broaden, by which temperature the single cyclopentadienyl resonance has already begun to sharpen. It therefore seems unlikely (though one cannot say rigorously that it is impossible) that the dithiocarbamate ligand is gate keeper for the ring interchange. By a process of elimination, then, the NO ligand must play that role.

The NO group is known to function either as a three-electron donor (sometimes described as NO$^+$ coordination) to give a linear M—N—O group, or as a one-electron donor (describable as NO$^-$ coordination) to give a bent M—N—O chain. Even the suggestion that it can pass from one of these stereoelectronic functions to the other is not new, having been made (Basolo and Pearson, 1967) to account for the relatively fast S_N2 reactions of Co(NO)(CO)$_3$ and Fe(NO)$_2$(CO)$_2$ with other ligands, such as R$_3$P, RNC, and py. Thus, the ring-scrambling pathway shown in Fig. 18 appears to be the most reasonable one in view of all the evidence.

B. Interchange of Tetra- and Hexahaptocyclooctatetraene Rings

There appears to be only one reported example of this, namely, in the molecule (C$_8$H$_8$)$_2$Fe (Carbonaro *et al.*, 1968; Allegra *et al.*, 1968). It was shown crystallographically that the structure of the molecule is the expected one, $(\eta^4\text{-}C_8H_8)(\eta^6\text{-}C_8H_8)$Fe, **28**. As noted previously, Section II,A, both of the

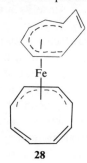

28

rings are individually fluxional. In addition, at a temperature of about $-50°C$, ring interchange begins to occur about as rapidly as eccentric rotation of the η^6-C_8H_8 ring. Therefore, this ring never gives an observable single signal of its own. Instead, beginning at about $-35°C$ a single broad signal is given by all of the sixteen protons, and this narrows with increasing temperature until, at room temperature, a sharp line is observed.

C. Interchange of Hexa- and Tetrahaptohexatriene Rings

There seems to be only one reported example of this phenomenon (Fischer and Elschenbroich, 1970; Huttner and Lange, 1972), namely in bis-(hexamethylbenzene)ruthenium. X-ray work shows that the structure consists of an η^6-C_6Me_6 ring and a 1,2,3,4-η^4-C_6Me_6 ring. At a temperature of 5°C, the proton nmr spectrum consists of four signals: $\tau 7.95$ (6H), $\tau 8.37$ (6H), $\tau 8.60$ (6H), and $\tau 8.14$ (36H), which is consistent with the crystal structure. At 30°C, this spectrum has collapsed, and there is only a broad singlet overlapping a sharp one. The former is presumably due to averaging of the three signals from the η^4-C_6Me_6 ring. At 70°C, there is only one sharp signal at $\tau 8.20$; presumably (η^6-C_6Me_6/(η^4-C_6Me_6) interchange is occurring rapidly at this temperature.

Unfortunately, no other spectra were reported, and thus the fluxional processes cannot be more explicitly characterized. There is no experimental indication reported as to whether the rotation of the η^4 ring, which collapses part of the spectrum between 5° and 30°C proceeds by 1,2 or 1,3 shifts. A closer study of this novel system would be desirable.

V. Fluxionality in the Solid State

The observation of fluxionality in the solid state by wide-line proton nmr measurements was first reported by Fyfe and co-workers (Campbell et al., 1971, 1972a) for $C_8H_8Fe(CO)_3$ and $C_8H_8Fe_2(CO)_5$. For the former, a line with a second moment of 7.9 G^2 is observed at temperatures below 204°K, which agrees well with a value of 7.7 G^2 calculated from the crystal structure (Dickens and Lipscomb, 1962). On raising the temperature, this line narrows markedly, and above 300°K has a second moment of only 1.0 G^2, which may be compared with a value of 0.7 G^2 calculated for rapid cycling of the protons around the set of crystallographic proton positions. The activation energy was estimated as 8.3 ± 0.4 kcal/mole. For $C_8H_8Fe_2(CO)_5$ a narrow line, $\langle M^2 \rangle = 0.5$ G^2, is observed at all temperatures above 77°K, whereas for the rigid structure (Fleischer et al., 1966), a second moment of 7 ± 1 G^2 would be expected. It was concluded that the site exchange process in this case remains very rapid down to 77°K.

It has also been shown (Cottrell et al., 1972) that solid $(C_8H_8)_2Ru_3(CO)_4$ has a markedly temperature-dependent line width. A comparison of theoretical second moments—calculated for various site-exchange sequences over the temperature range studied—with the experimental results led to best agreement when the set of eight protons in each ring was postulated to move sequentially over the eight crystallographic proton positions. An activation energy of 5.2 kcal/mole was estimated.

Studies of several compounds known or presumed to contain $\eta^1\text{-}C_5H_5$ rings have also been reported (Campbell et al., 1972b). For $(\eta^5\text{-}C_5H_5)$Fe-$(CO)_2(\eta^1\text{-}C_5H_5)$, it appears that reorientation of the $\eta^1\text{-}C_5H_5$ ring never becomes rapid enough to have its full effect on the line width even at the melting point, although there is a marked line-width change at about 125°K, which the authors attribute to accelerated rotation of the $\eta^5\text{-}C_5H_5$ ring. For the mercury compounds, C_5H_5HgX, $X = Cl$, Br, I, C_5H_5, results indicative of ring reorientation were obtained, with activation energies decreasing in the order $X = Cl$, Br, I, C_5H_5, which is in accord with the results of solution studies (West et al., 1969). The crystal structures of these mercury compounds are not known. Thus, thorough interpretation of the data could not be carried out, and even the assignment of $\eta^1\text{-}C_5H_5$ binding is speculative although probably correct.

As noted earlier (p. 424), $(C_8H_8)_2$Fe has quite elaborate fluxional behavior in solution since both the η^4- and η^6-C_8H_8 rings engage in eccentric rotation and at higher temperatures ring interchange also occurs. In the solid state, there are also rearrangements, but the observations are quite complex and have not been definitely interpreted (Chierico and Mognaschi, 1973). The crystals contain two crystallographically independent sets of molecules, one set occupying general positions and crystallographically ordered, the other set on special positions and crystallographically disordered. Below $-185°C$ the observed second moment of the wide proton resonance line agrees with the value calculated for the static structure. Above $-185°C$, this line begins to narrow, and also eventually develops a very narrow (1–2 G) central spike. The authors contend that the broad band behaves as though a process with Arrhenius parameters of $E_a = 2.6 \pm 0.1$ kcal mole^{-1} and $A = 6 \times 10^7$ sec^{-1} were occurring, and conclude from this that simple ring reorientation such as occurs in $C_8H_8Fe(CO)_3$ must be ruled out because the latter (see above) was found to have $E_a = 8.3 \pm 0.4$ kcal mole^{-1}. They finally conclude that the broad and narrow contributions must be due to the ordered and disordered molecules, respectively, but that the motions which cause the line narrowing cannot be specified.

This writer takes the liberty of doubting the interpretation in view of the extremely low value of the reported frequency factor. In order to obtain the direct input for the Arrhenius plot, an approximate equation was used to derive rates from the measured line widths at various temperatures. It seems that this may be a source of enough systematic error to negate the resulting Arrhenius plot. If a more reasonable value of A, say about 10^{13}, were used, the value of E_a would be much higher and perhaps not greatly different from 8.3 kcal mole^{-1}. It would then be possible to invoke simple eccentric ring rotations to explain the line narrowing. Such motions seem very plausible and very probable.

Studies of fluxional behavior in the solid state are of interest for at least two reasons. First, the fact that solid-state fluxionality can occur is informative. It provides strong support for the 1,2-shift pathway, since this appears to be the only one compatible with the occurrence of site exchange under the restrictions imposed by the crystal packing. The crystallographic positions of the metal atoms presumably are invariant. The rings rotate (and, to a degree, simultaneously bend) about a point close to their own centers of gravity. In this way, the point of attachment of the ring to the metal necessarily changes from an initial one to the next nearest equivalent one. Movement to the second or third nearest equivalent one could not occur without passing through the configuration of the nearest one.

Second, the temperature range available for study is greatly extended to lower temperatures. Solution studies are limited by solvent properties to temperatures above about 115°K, whereas the wide-line work can be done even at liquid helium temperatures.

The disadvantage of the solid state studies is, of course, that motions requiring significant movement of the center of gravity of the organic moiety relative to the position of the metal atom will presumably not occur in the crystal, even though they might have quite high rates in solution. It would be interesting to know, for example, whether $(C_6H_5)_3SnC_7H_7$ would show any line narrowing at higher temperatures in the solid state, since the ring reorientation proceeds by 1,5 shifts in solution (see Section I,B).

VI. Miscellaneous Systems

A. Allene Complexes

Fluxional behavior of an allene–metal complex was first reported for $Fe(CO)_4(Me_2C=C=CMe_2)$ (Ben-Shoshan and Pettit, 1967). This compound, for which the schematic structure **29** would be expected, exhibits only a single proton nmr peak at $+30°C$, whereas three resonances with relative intensities 1:1:2, for methyl groups (1), (2), and (3) + (4), respectively, would

$$\underset{29}{\underset{H_3C^{(4)}}{\overset{H_3C^{(3)}}{>}}C=\underset{|}{\underset{Fe(CO)_4}{C}}=\overset{{}^{(1)}CH_3}{\underset{{}^{(2)}CH_3}{<}}C}$$

be expected. On lowering the temperature, however, this single resonance collapsed, and at $-60°C$, a spectrum consisting of three sharp peaks in the proper intensity ratio was observed. The process responsible for the site exchange was shown to be nondissociative, and it was proposed that the point of attachment of the iron atom can change rapidly among the four equivalent ones, the path of the iron atom relative to the C=C=C group being a kind of double helix.

A number of other allene complexes have been prepared and X-ray studies (Kashiwagi et al., 1969; Hewitt et al., 1969; Racanelli et al., 1969) have shown that the bonding is essentially as assumed by Ben-Shoshan and Pettit except that the C=C=C chain is distinctly bent, with angles ranging from 147° to 158°C. Vrieze et al. (1970) have shown that many allene complexes of platinum and rhodium are also fluxional, with frequency factors of $\sim 10^{10}$ and activation energies from 7.8 to 11.4 kcal/mole for the platinum complexes. A transition state in which the central allene carbon atom alone is bound to the metal atom has been depicted by these workers.

B. *Rotation of Coordinated Monoolefins*

It is generally accepted that the binding of monoolefins, or individual olefinic groups in nonconjugated polyolefins to metal atoms can be described by the Dewar–Chatt scheme, shown in Fig. 19. The bond consists of two components, one of which (called the σ, or sometimes μ, component) would not be particularly restrictive of rotation about the bond axis. Bond axis in this context refers to a line from the metal atom to the midpoint of the double bond. The strength of the other (π component) maximizes sharply for that orientation of the olefin in which the C–C axis lies in, or close to, the plane of the metal d orbital, which overlaps with the π orbitals of the olefin. This feature tends to restrict rotation.

It was, therefore, at first surprising to learn (Cramer, 1964) that nmr spectra over a temperature for $(C_5H_5)Rh(C_2H_4)_2$ (**a**) range showed line-shape changes that could most reasonably be explained by rotation of each ethylene ligand about the axis from Rh to the midpoint of the olefin. The assumed instantaneous structure of the molecule, Fig. 20, has four endo protons and four exo protons. At low temperatures ($< -20°C$) each set gives its own sharp multiplet signal, but as the temperature is raised, these broaden and

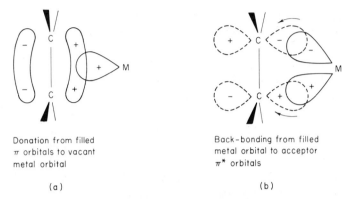

(a) Donation from filled π orbitals to vacant metal orbital

(b) Back-bonding from filled metal orbital to acceptor π* orbitals

Fig. 19 The Dewar–Chatt scheme of bonding in a simple metal–olefin complex. [Reproduced by permission from "Advanced Inorganic Chemistry," 3rd ed., by F. A. Cotton and G. Wilkinson, Wiley (Interscience), New York, 1972.]

coalesce, finally giving a single broad peak at 57°C. In $C_5H_5Rh(SO_2)(C_2H_4)$ (**b**) (Cramer, 1967) a similar temperature dependence of the spectrum was observed, but rotation was less hindered, since the limiting low-temperature spectrum was reached only at about $-50°C$ and collapse to a single line had already occurred at $-2°C$.

A detailed analysis of the spectra for **a**, **b**, and $C_5H_5Rh(C_2H_4)(C_2F_4)$ (**c**) (Cramer et al., 1969) showed that the activation energies are 15.0 ± 0.2, 12.2 ± 0.8 and 13.6 ± 0.6 kcal/mole for **a**, **b**, and **c**, respectively. The C_2F_4 in **c** does not rotate rapidly enough to influence ^{19}F line shapes up to 100°C, where decomposition beings.

It was later observed (Moseley et al., 1970) that $(Me_5C_5)Rh(C_2H_4)_2$ behaves similarly to $(C_5H_5)Rh(C_2H_4)_2$ but has a somewhat higher coalescence temperature. The iridium compound $(Me_5C_5)Ir(C_2H_4)_2$, however, gave the

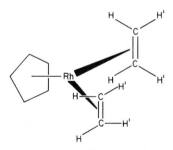

Fig. 20 The assumed structure of $(C_5H_5)Rh(C_2H_4)_2$. The exact relationship of the C_5H_5 ring to the metal atom (i.e., whether it is truly pentahapto or more tilted) is unknown, but in any case it reorients sufficiently rapidly at all temperatures above $-20°C$ to produce only one sharp ^1H resonance. The endo olefin protons are denoted H'.

slow-exchange spectrum up to 110°C; the higher barrier here is presumably of electronic origin, since steric factors should be less restrictive about the larger metal atom.

More recently a variety of platinum(II) olefin complexes have been studied by Lewis and Johnson and co-workers (Ashley-Smith et al., 1972; Holloway et al., 1969, 1970) with the objective of sorting out the relative importance of electronic and steric factors on the rates of rotation. In systems of the type Pt(olefin)X(diketo), where X = Cl, Br, and diketo = acetylacetonate or a substituted acetylacetonate, no marked variations clearly attributable to either factor were seen. In complexes of the type $Pt(C_2H_4)XYL$, X, Y = Cl, Br, or CF_3CO_2 and L = R_3P, $(RO)_3P$, R_3As, or RNH_2, with L cis to C_2H_4, steric contributions to the rotation barrier were found.

Olefin rotation has been reported in some cationic Pt(II) complexes, where the steric factor seems to be very important in controlling the rate (Chisholm and Clark, 1973).

Some evidence for rotation of the coordinated styrene ligands (including substituted styrenes) in dinuclear complexes of the type $[(styrene)MCl_2]_2$, M = Pt, Pd has been presented (Iwao et al., 1972). Rotation appears to be more rapid in the palladium compounds than the platinum ones, which is parallel with the Rh versus Ir case discussed above.

A singularly important investigation in this area (Johnson and Segal, 1972) has established directly that the nature of the rotatory process is indeed that postulated originally by Cramer, Fig. 21(a), and not the alternative one, Fig. 21(b), which is equally in accord with the experimental data

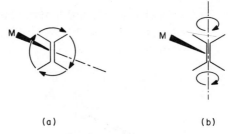

Fig. 21 The two *a priori* possible types of rotation a μ-bonded olefin might undergo; (a) is that originally postulated by Cramer and recently proved by Johnson and Segal.

previously available. This was done by designing and studying a system wherein each end of the coordinated olefin is in a different environment, namely, the five-coordinate complex $[Os(CO)(NO)(PPh_3)_2C_2H_4]^+PtF_6^-$. The ^{13}C nmr spectrum at 20°C has but a single resonance for the olefinic carbon atoms that collapses at lower temperatures and reforms as two signals of equal intensity at −80°C. The proton spectrum was also studied; it

10. Nonrigidity in Organometallic Compounds

showed the expected temperature dependence and also confirmed the nondissociative nature of the process because $^{31}P-^{1}H$ coupling was preserved in the high-temperature limit.

It would certainly not be surprising to find that similar rotational processes might occur for acetylenes appropriately bound (through one pair of π electrons) to a metal atom. For some complexes of the type $[Os(CO)(NO)(R_3P)_2RC\equiv CR']^+PF_6^-$, it has been found (Ashley-Smith et al., 1973) the R and R' sites are nonequivalent at lower temperatures, but that they become indistinguishable at higher temperatures. While no evidence was presented to rule out the occurrence of a dissociative process, it is probable that, as the authors suggest, a rotatory process is the cause of the spectral averaging.

It is important to emphasize that the rotation of the type under discussion is possible because, for any given d orbital that interacts with the π^* orbital of the olefin in the ground state conformation, there is another d orbital at 90° to it, which can give a similar interaction when the olefin is rotated 90°. Because of the properties of d orbitals the olefin π^* orbital can interact with both d orbitals in continuously varying proportion as the angle of rotation increases. If there were no steric factor, and if the two d orbitals were equivalent in the molecular environment, there would be no barrier to rotation. The electronic contribution to the barrier stems from the nonequivalence of the two d orbitals in the molecular environment.

C. Biscyclohexadienyliron Compounds

In 1970 several biscyclohexadienyliron compounds, e.g., **30**, were reported to have temperature-dependent proton nmr spectra (Helling and

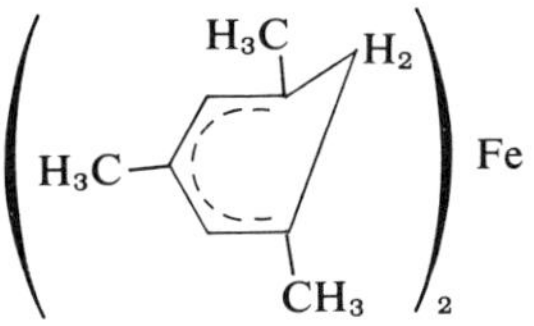

$$\left(\begin{array}{c} H_3C \\ H_3C \end{array} \right)_2 Fe$$

30

Braitsch, 1970). At lower temperatures, the protons at positions 2 and 4 and the methyl groups at positions 1 and 5 gave separate resonances that coalesced at higher temperatures. It was suggested that the process responsible was that in Eq. (10). Since it is improbable in the extreme that the bonding would be as depicted rather than maximally delocalized so that each ring would have a plane of symmetry (the whole molecule having C_{2h} symmetry when in the antirotational conformation), this explanation of the temperature dependence

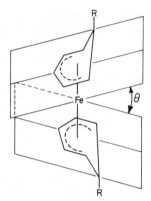

Fig. 22 An *a priori* plausible structure for a biscyclohexadienyl iron compound in solution. This type of structure is actually found in the crystal.

was not very attractive. An obvious alternative explanation is that the structure is as shown in Fig. 22 and that interconversion by internal rotation of the enantiomorphous forms of this structure causes spectrum averaging. The recent X-ray crystallographic demonstration (Mathew and Palenik, 1972) that the molecule does possess a structure of the type shown in Fig. 22 ($\theta \equiv 60°$) appears to establish the latter explanation firmly.

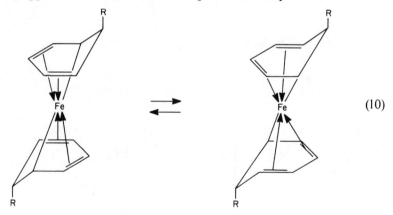

(10)

D. Nonconjugated Diene Ligands

Nonconjugated diolefins, especially 1,5-cyclooctadiene and norbornadiene, form a host of important organometallic compounds in which they function as chelating diolefins, filling two adjacent coordination positions on a metal atom. Quite often the plane of symmetry bisecting the two double bonds that exists in the free diolefin is nullified by the lower symmetry of the entire complex; thus, the four olefinic protons are no longer structurally equivalent. In many cases, however, they are nmr equivalent at room temperature (and below) and this is thought to be due to internal rotation about an axis coincident with the twofold symmetry axis of the free olefin. In no case, however, had this motion been slowed sufficiently, by lowering of the temperature, to obtain direct evidence for it from line-shape changes, until a study of the compound $CH_3CCo_3(CO)_7C_7H_8$ was reported (Elder and Robinson, 1972).

The norbornadiene, C_7H_8, occupies two coordination positions on one cobalt atom; the limiting slow exchange spectrum, which is observed below about $-60°C$, is consistent with an orientation of C_7H_8, relative to the rest of the molecule, which allows it to retain that plane of symmetry that passes through the bridgehead carbon atoms but not the one that bisects each of the double bonds. Thus, the olefinic protons form two equivalent pairs and the two bridgehead protons are nonequivalent. Above about $-50°C$, two single lines in an intensity ratio of 2:1 are observed for the four olefinic and two bridgehead protons, respectively. The increasing rapidity of rotation about the quasi twofold axis presumably accounts for these line-shape changes. Steric hindrance from the CO groups on the cobalt atoms (both that on the Co to which the C_7H_8 is bound and at least two on the adjacent cobalt atoms) may be responsible for making this particular case of rotation detectable.

E. Some Metallocyclic Compounds

Several molecules of the type **31** have long been known. Their proton nmr spectra show no temperature dependence over the temperature ranges that have been studied. However, it has been established chemically that they

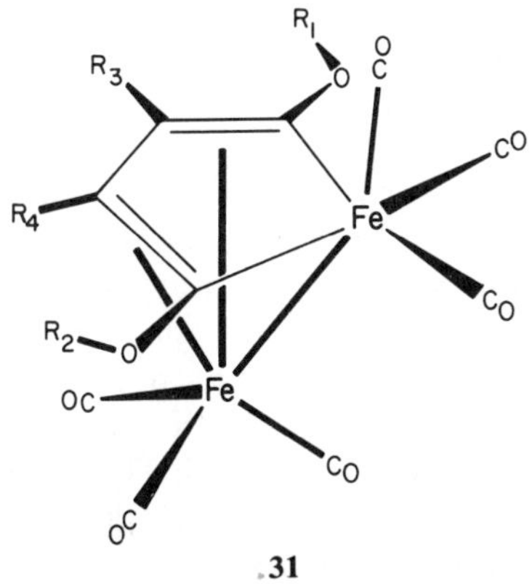

31

do undergo a rearrangement in which the iron atoms and their attached carbonyl groups exchange their roles (Case et al., 1962). This was demonstrated by the observation that the optically active compound in which $R_1 = R_2 = CH_3$, $R_3 = C(CH_3)_3$, and $R_4 = H$ undergoes rapid racemization above room temperature.

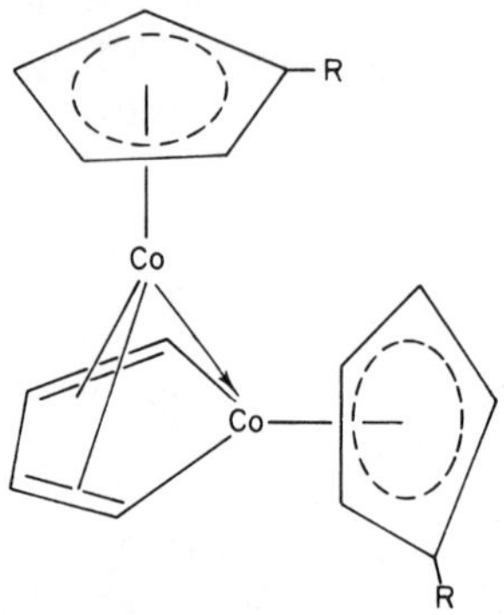

32

Related (i.e., isoelectronic) systems that rearrange rapidly enough to influence nmr line shapes are **32a** ($R = H$) and **32b** ($R = COOCH_3$) (Rosenblum et al., 1971). The two cyclopentadienyl rings are nonequivalent, but in the approximate temperature range 90°–160°C, the signals for **32a** broaden, coalesce, and finally appear as a narrow line. Essentially similar but more complex spectral changes occur for **32b**. Since the signals for the α and β protons of the cobaltocyclic ring are not averaged, the rearrangement process appears to involve a simple flip of the central C_4H_4 chain so as to transfer the σ and π bonding functions from one cobalt atom to the other. The activation energies are approximately 25 kcal/mole.

F. Polynuclear Benzyne Complexes

It has been shown crystallographically that a compound of composition $Os_3(C_6H_4)(PPh_2)(CO)_7$ has the structure **33** (Bradford et al., 1972). A proton nmr study of the PMe_2 analog has shown it to be nonrigid (Deeming et al., 1972). The structure is such that the four protons of C_6H_4 should constitute an ABXY set; at $-60°C$, they do. As the temperature is raised, the appropriate signals broaden, and at $+60°C$, a sharp spectrum in which these protons appear as an A_2X_2 set is observed. The authors proposed the reasonable explanation that **33** is in rapid exchange with **33'**, perhaps through the intermediacy of **34**. A related molecule, $Os_3(C_6H_4)(PMe_2)(CO)_9$, of unknown structure, was reported to behave similarly.

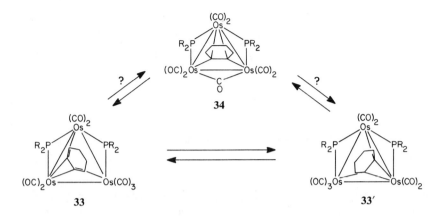

VII. Some Nonfluxional Molecules, or the Dog in the Night-Time

Those familiar with literature in the English language will have read the following quintessentially Holmesian passage in A. Conan Doyle's *Silver Blaze*. It begins with a question to Sherlock Holmes.

"Is there any point to which you wish to draw my attention?"
"To the curious incident of the dog in the night-time."
"The dog did nothing in the night-time."
"That was the curious incident," remarked Sherlock Holmes.

While the greatest knowledge of the "ground rules" governing behavior of fluxional organometallic molecules is, obviously, to be derived from study of those molecules that are fluxional, insights and confirmation of insights, which are not to be despised, are obtained by examining certain molecules that are not fluxional. A few such cases will be cited here. Several have already been mentioned, namely, various M(1-indenyl) compounds and *trans*-$[Fe(CO)_3]C_8H_8$, which confirm, respectively, that 1,2 shifts are allowed while 1,3 shifts are forbidden, and that a significant degree of overlap between all π orbitals may be important in promoting fluxionality.

Two compounds that show also that jumps that are not 1,2 shifts and that cannot be executed without traversing an intermediate that must be of high energy because it has no satisfactory electronic structure are $C_7H_8Fe(CO)_3$ and (azulene)$Fe_2(CO)_5$. $C_7H_8Fe(CO)_3$, **35**, was reported many years ago (Burton *et al.*, 1961) and shown to be rigid. The only rearrangement that might have been considered plausible is that shown as **35a** to **35b**. The overall process is a 1,3 shift. If we assume that this cannot occur as a single step, but only as a succession of two 1,2 shifts, we see that traversal of a very unsatisfactory intermediate, **35c**, which has an insufficiently coordinated iron atom, would be required. A formally acceptable electronic structure for such an intermediate might be **35d**, but the added energy of forming the norcaradiene structure in the ring system is evidently prohibitive.

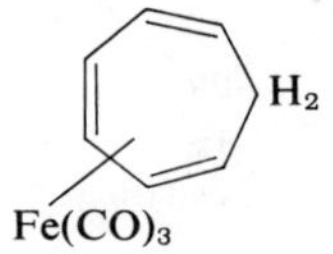

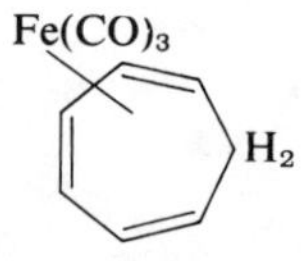

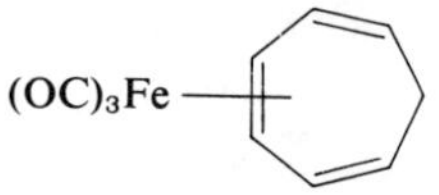

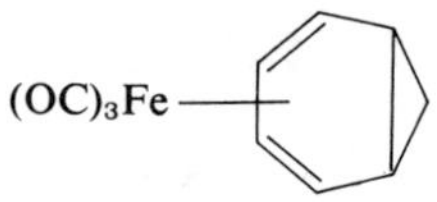

In the case of (azulene)Fe$_2$(CO)$_5$ (Churchill, 1967), the rearrangement **36a** to **36b** is also unobserved. Again, this can be explained by assuming that this direct 1,3 shift process is not allowed and observing that no satisfactory intermediate structure can be written. Excluding diradicals, there is only **36c**, which is quite obviously not likely to be of low energy.

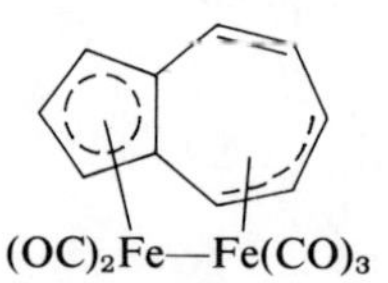

An informative comparison of two compounds in which concerted movements of two separate metal carbonyl moieties attached to a bicyclic system might be considered possible is presented by the compounds **37** and **38** (Aumann, 1971a,b). The molecule **37** is fluxional. It executes the enantiomer interconversion depicted rapidly on the nmr time scale above 0°C ($\Delta G^{\ddagger}$ at ~25°C is ~16 kcal/mole). The enantiomers of molecule **38**, however, do not interconvert rapidly enough to affect the proton nmr spectrum at 25°C.

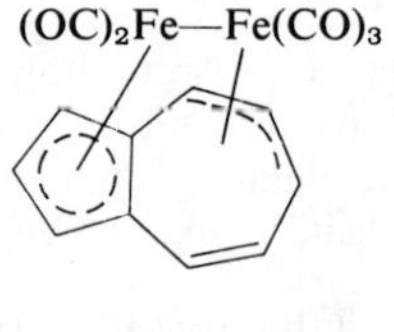

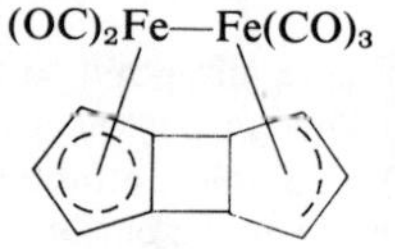

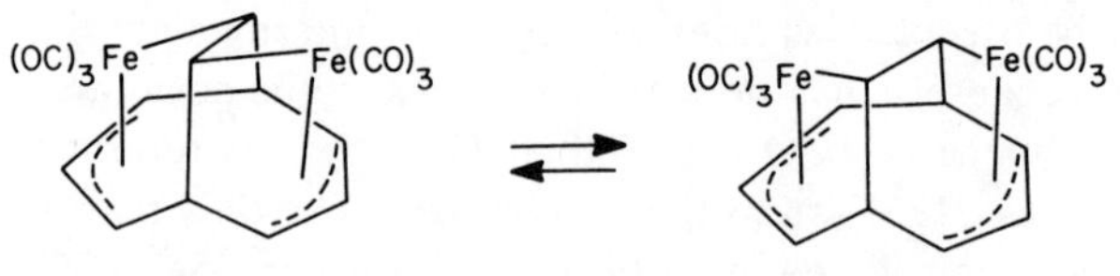

While the two processes appear similar, there is a crucial difference that accounts for the difference in behavior. In compound **37**, the process involves concerted 1,2 shifts of the Fe—C σ bonds, a process with a relatively low

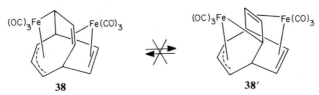

38 **38'**

barrier, as in numerous other cases. In compound **38**, however, one of the iron atoms would be required to execute a 1,3 metallotropic shift; this process is subject to a high barrier, as might be expected from observations on other systems cited elsewhere in this review.

REFERENCES

Allegra, G., Colombo, A., Immirzi, A., and Bassi, I. W. (1968). *J. Amer. Chem. Soc.* **90**, 4455.
Ashe, A. J. (1970). *Tetrahedron Lett.* p. 2105.
Ashley-Smith, J., Douek, I., Johnson, B. F. G., and Lewis, J. (1972). *J. C. S. Dalton* p. 1776.
Ashley-Smith, J., Johnson, B. F. G., and Segal, J. A. (1973). *J. Organometal. Chem.* **49**, C38.
Aumann, R. (1971a). *Angew. Chem., Int. Ed. Engl.* **10**, 188.
Aumann, R. (1971b). *Angew. Chem., Int. Ed. Engl.* **10**, 189.
Basolo, F., and Pearson, R. G. (1967). "Mechanisms of Inorganic Reactions," 2nd ed., p. 571 *et seq.* Wiley, New York.
Bennett, M. A., Bramley, R., and Watt, R. (1969). *J. Amer. Chem. Soc.* **91**, 3089.
Bennett, M. J., Cotton, F. A., Davison, A., Faller, J. W., Lippard, S. J., and Morehouse, S. M. (1966). *J. Amer. Chem. Soc.* **88**, 4371.
Bennett, M. J., Cotton, F. A., and Takats, J. (1968a). *J. Amer. Chem. Soc.* **90**, 903.
Bennett, M. J., Cotton, F. A., and Legzdins, P. (1968b). *J. Amer. Chem. Soc.* **90**, 6335.
Ben-Shoshan, R., and Pettit, R. (1967). *J. Amer. Chem. Soc.* **89**, 2231.
Bentham, J. E., Ebsworth, E. A. V., Moretto, H., and Rankin, D. W. H. (1972). *Angew. Chem.* **84**, 683.
Bradford, C. W., Nyholm, R. S., Gainsford, G. J., Guss, J. M., Ireland, P. R., and Mason, R. (1972). *J. Chem. Soc., Chem. Commun.* p. 87.
Bratton, W. K., Cotton, F. A., Davison, A., Musco, A., and Faller, J. W. (1967). *Proc. Nat. Acad. Sci. U.S.* **58**, 1324.
Brown, J., Cundy, C. S., Green, M., and Stone, F. G. A. (1971). *J. Chem. Soc., A* p. 448.
Burt, R., Cooke, M., and Green, M. (1970). *J. Chem. Soc., A* p. 2981.
Burton, R., Pratt, L., and Wilkinson, G. (1961). *J. Chem. Soc., London* p. 594.
Calderon, J. L., and Cotton, F. A. (1971). *J. Organometal. Chem.* **30**, 377.
Calderon, J. L., Cotton, F. A., and Legzdins, P. (1969). *J. Amer. Chem. Soc.* **91**, 2528.
Calderon, J. L., Cotton, F. A., DeBoer, B. G., and Takats, J. (1970). *J. Amer. Chem. Soc.* **92**, 3801.
Calderon, J. L., Cotton, F. A., and Takats, J. (1971a). *J. Amer. Chem. Soc.* **93**, 3587.
Calderon, J. L., Cotton, F. A., DeBoer, B. G., and Takats, J. (1971b). *J. Amer. Chem. Soc.* **93**, 3592.
Calderon, J. L., Cotton, F. A., and Shaver, A. (1972). *J. Organometal. Chem.* **42**, 419.
Calderon, J. L., Cotton, F. A., and Shaver, A. (1973). *J. Organometal. Chem.*, **57**, 121.

Campbell, A. J., Fyfe, C. A., and Maslowsky, E., Jr. (1971). *Chem. Commun.* p. 1032.
Campbell, A. J., Fyfe, C. A., and Maslowsky, E., Jr. (1972a). *J. Amer. Chem. Soc.* **94**, 2690.
Campbell, A. J., Fyfe, C. A., Goel, R. G., Maslowsky, E., Jr., and Senoff, C. V. (1972b). *J. Amer. Chem. Soc.* **94**, 8387.
Campbell, C. H., and Green, M. L. H. (1970). *J. Chem. Soc., A* p. 1318.
Campbell, C. H., and Green, M. L. H. (1971). *J. Chem. Soc., A* p. 3282.
Carbonaro, A., Segré, A. L., Greco, A., Tosi, C., and Dall'Asta, G. (1968). *J. Amer. Chem. Soc.* **90**, 4453.
Case, R., Jones, E. R. H., Schwartz, N. V., and Whiting, M. C. (1962). *Proc. Chem. Soc., London* p. 256.
Chierico, A., and Mognaschi, E. R. (1973). *J. Chem. Soc., Faraday Trans.* **2**, 433.
Chisholm, M. H., and Clark, H. C. (1973). *Inorg. Chem.* **12**, 991.
Churchill, M. R. (1967). *Inorg. Chem.* **6**, 190.
Churchill, M. R., and Mason, R. (1966). *Proc. Roy. Soc., Ser. A* **292**, 61.
Ciappenelli, D., and Rosenblum, M. (1969). *J. Amer. Chem. Soc.* **91**, 3673 and 6873.
Ciappenelli, D. J., Cotton, F. A., and Kruczynski, L. (1972). *J. Organometal. Chem.* **42**, 159.
Cooke, M., Goodfellow, R. J., Green, M., Maher, J. P., and Yondle, J. R. (1970). *Chem. Commun.* p. 565.
Cotton, F. A. (1968). *Accounts Chem. Res.* **1**, 257.
Cotton, F. A., and Ciappenelli, D. J. (1972). *Syn. Inorg. Metal.-Org. Chem.* **2**, 197.
Cotton, F. A., and Edwards, W. T. (1968). *J. Amer. Chem. Soc.* **90**, 5412.
Cotton, F. A., and Edwards, W. T. (1969). *J. Amer. Chem. Soc.* **91**, 843.
Cotton, F. A., and LaPrade, M. D. (1968). *J. Amer. Chem. Soc.* **90**, 2026.
Cotton, F. A., and Legzdins, P. (1968). *J. Amer. Chem. Soc.* **90**, 6232.
Cotton, F. A., and Marks, T. J. (1969a). *J. Amer. Chem. Soc.* **91**, 3178.
Cotton, F. A., and Marks, T. J. (1969b). *J. Amer. Chem. Soc.* **91**, 7281.
Cotton, F. A., and Marks, T. J. (1969c). *J. Amer. Chem. Soc.* **91**, 7523.
Cotton, F. A., and Marks, T. J. (1969d). *J. Organometal. Chem.* **19**, 237.
Cotton, F. A., and Marks, T. J. (1970). *Inorg. Chem.* **9**, 2802.
Cotton, F. A., and Musco, A. (1968). *J. Amer. Chem. Soc.* **90**, 1444.
Cotton, F. A., and Reich, C. R. (1969). *J. Amer. Chem. Soc.* **91**, 847.
Cotton, F. A., and Rusholme, G. A. (1972). *J. Amer. Chem. Soc.* **94**, 402.
Cotton, F. A., and Takats, J. (1970). *J. Amer. Chem. Soc.* **92**, 2353.
Cotton, F. A., Musco, A., and Yagupsky, G. (1967a). *J. Amer. Chem. Soc.* **89**, 6136.
Cotton, F. A., Davison, A., and Musco, A. (1967b). *J. Amer. Chem. Soc.* **89**, 6796.
Cotton, F. A., Faller, J. W., and Musco, A. (1968). *J. Amer. Chem. Soc.* **90**, 1438.
Cotton, F. A., Davison, A., Marks, T. J., and Musco, A. (1969). *J. Amer. Chem. Soc.* **91**, 6598.
Cotton, F. A., DeBoer, B. G., and LaPrade, M. D. (1971a). *Int. Congr. Pure Appl. Chem. 23rd. 1971 Spec. Lect.* Vol. 6, p. 1.
Cotton, F. A., DeBoer, B. G., and Marks, T. J. (1971b). *J. Amer. Chem. Soc.* **93**, 5069.
Cotton, F. A., Jeremic, M., and Shaver, A. (1972). *Inorg. Chim. Acta* **6**, 543.
Cotton, F. A., Frenz, B. A., Deganello, G., and Shaver, A. (1973). *J. Organometal. Chem.* **50**, 227.
Cottrell, C. E., Fyfe, C. A., and Senoff, C. V. (1972). *J. Organometal. Chem.* **43**, 203.
Cramer, R. (1964). *J. Amer. Chem. Soc.* **86**, 217.
Cramer, R. (1967). *J. Amer. Chem. Soc.* **89**, 5377.
Cramer, R., Kline, J. B., and Roberts, J. D. (1969). *J. Amer. Chem. Soc.* **91**, 2519.

Cross, R. J., and Wardle, R. (1971). *J. Chem. Soc., A* p. 2000.
Davison, A., and Rakita, P. E. (1968). *J. Amer. Chem. Soc* **90**, 4479.
Davison, A., and Rakita, P. E. (1970a). *J. Organometal. Chem.* **23**, 407.
Davison, A., and Rakita, P. E. (1970b). *Inorg. Chem.* **9**, 289.
Deeming, A. J., Nyholm, R. S., and Underhill, M. (1972). *J. Chem. Soc., Chem. Commun.* p. 224.
Delbaere, L. T. J., McBride, D. W., and Ferguson, R. B. (1970). *Acta Crystallogr., Sect. B* **26**, 518.
Deubzer, B., Elian, M., Fischer, E. O., and Fritz, H. P. (1970). *Chem. Ber.* **103**, 799.
Dickens, B., and Lipscomb, W. N. (1962). *J. Chem. Phys.* **37**, 2084.
Dickson, R. S., and Wilkinson, G. (1964). *J. Chem. Soc.* p. 2699.
Elder, P. A., and Robinson, B. H. (1972). *J. Organometal. Chem.* **36**, C45.
Emerson, G. F., Mahler, J. E., Pettit, R., and Collins, R. (1964). *J. Amer. Chem. Soc.* **86**, 3590.
Faller, J. W. (1969). *Inorg. Chem.* **8**, 767.
Fischer, E. O., and Elschenbroich, C. (1970). *Chem. Ber.* **103**, 162.
Fleischer, E. B., Stone, A. L., Dewar, R. B. K., Wright, J. D., Keller, C. E., and Pettit, R. (1966). *J. Amer. Chem. Soc.* **88**, 3158.
Flores, B., Illuminati, G., and Ortaggi, G. (1969). *Chem. Commun.* p. 492.
Fritz, H. P. (1964). *Advan. Organometal. Chem.* **1**, 239.
Fritz, H. P., and Schwarzhans, K. E. (1966). *J. Organometal. Chem.* **5**, 181.
Greco, A., Green, M., and Stone, F. G. A. (1971). *J. Chem. Soc., A* p. 285.
Grishin, Y. K., Sergeyev, N. M., and Ustynyuk, Y. A. (1972). *Org. Magn. Resonance* **4**, 377.
Haaland, A., and Weidlein, J. (1972). *J. Organometal. Chem.* **40**, 29.
Helling, J. F., and Braitsch, D. M. (1970). *J. Amer. Chem. Soc.* **92**, 7207.
Hewitt, T. G., Anzenhofer, K., and de Boer, J. J. (1969). *Chem. Commun.* p. 312.
Holloway, C. E., Hulley, G., Johnson, B. F. G., and Lewis, J. (1969). *J. Chem. Soc., A* p. 53.
Holloway, C. E., Hulley, G., Johnson, B. F. G., and Lewis, J. (1970). *J. Chem. Soc., A* p. 1658.
Hüttel, R., Raffay, V., and Reinheimer, H. (1967). *Angew. Chem., Int. Ed. Engl.* **6**, 862.
Huttner, G., and Lange, S. (1972). *Acta Crystallogr., Sect. B* **28**, 2049.
Iwao, T., Saika, A., and Kinugasa, T. (1972). *Inorg. Chem.* **11**, 3106.
Johnson, B. F. G., and Segal, J. A. (1972). *J. Chem. Soc., Chem. Commun.* p. 1312.
Kang, J. W., Childs, R. F., and Maitlis, P. M. (1970). *J. Amer. Chem. Soc.* **92**, 721.
Kashiwagi, T., Yasuoka, N., Kasai, N., and Kukudo, M. (1969). *Chem. Commun.* p. 317.
Keller, C. E., Emerson, G. F., and Pettit, R. (1965). *J. Amer. Chem. Soc.* **87**, 1388.
King, R. B. (1963). *Inorg. Chem.* **2**, 807.
King, R. B. (1967). *J. Organometal. Chem.* **8**, 1967.
Kita, W. G., Lloyd, M. K., and McCleverty, J. A. (1971). *Chem. Commun.* p. 420.
Kreiter, C. G., Maasbol, A., Anet, F. A. L., Kaesz, H. D., and Winstein, S. (1966). *J. Amer. Chem. Soc.* **88**, 3444.
Kroll, W. R. (1969). *Chem. Commun.* p. 844.
Kroll, W. R., and Naegele, W. (1969). *Chem. Commun.* p. 246.
Larrabee, R. B. (1971). *J. Amer. Chem. Soc.* **93**, 1510.
Lee, A. G., and Sheldrick, G. M. (1969). *Chem. Commun.* p. 441.
McKechnie, J. S., and Paul, I. C. (1966). *J. Amer. Chem. Soc.* **88**, 5927.
Manuel, T. A., and Stone, F. G. A. (1959). *Proc. Chem. Soc., London* p. 90.
Manuel, T. A., and Stone, F. G. A. (1960). *J. Amer. Chem. Soc.* **82**, 366.

Maslowsky, E., Jr., and Nakamoto, K. (1968). *Chem. Commun.* p. 257.
Maslowsky, E., Jr., and Nakamoto, K. (1969). *Inorg. Chem.* **8**, 1108.
Mathew, M., and Palenik, G. J. (1972). *Inorg. Chem.* **11**, 2809.
Moseley, K., Kang, J. W., and Maitlis, P. M. (1970). *J. Chem. Soc., A* p. 2875.
Nakamura, A., and Hagihara, N. (1959). *Bull. Chem. Soc. Jap.* **32**, 880.
Penfold, B. R., and Robinson, B. H. (1973). *Accounts Chem. Res.* **6**, 73.
Pettit, R. (1964). *J. Amer. Chem. Soc.* **86**, 2589.
Piper, T. S., and Wilkinson, G. (1956a). *J. Inorg. Nucl. Chem.* **2**, 32.
Piper, T. S., and Wilkinson, G. (1956b). *J. Inorg. Nucl. Chem.* **3**, 104.
Rakita, P. E., and Davison, A. (1969). *Inorg. Chem.* **8**, 1165.
Racanelli, P., Pantini, G., Immirzi, A., Allegra, G., and Porri, L. (1969). *Chem. Commun.* p. 361.
Rausch, M. D., and Schrauzer, G. N. (1959). *Chem. Ind. (London)* p. 957.
Robinson, B. H., and Spencer, J. (1971). *J. Organometal. Chem.* **33**, 97.
Rosenblum, M., Giering, W. P., North, B., and Wells, D. (1971). *J. Organometal. Chem.* **28**, C17.
Sergeyev, N. M., Avramenko, G. I., and Ustynyuk, Y. A. (1970a). *J. Organometal. Chem.* **22**, 79.
Sergeyev, N. M., Avramenko, G. I., and Ustynyuk, Y. A. (1970b). *J. Organometal. Chem.* **24**, C39.
Vrieze, K., and van Leeuwen, P. W. N. M. (1971). *Progr. Inorg. Chem.* **14**, 1.
Vrieze, K., Volger, H. C., and Praat, A. P. (1970). *J. Organometal. Chem.* **21**, 467.
Weidenborner, J. E., Larrabee, R. B., and Bednowitz, A. L. (1972). *J. Amer. Chem. Soc.* **94**, 4140.
West, P., Woodville, M. C., and Rausch, M. D. (1969). *J. Amer. Chem. Soc.* **91**, 5649.
Whitesides, G. M., and Fleming, J. S. (1967). *J. Amer. Chem. Soc.* **89**, 2855.
Whitesides, T. H., and Budnik, R. A. (1971). *Chem. Commun.* p. 1514.
Whitlock, H. W., Jr., and Stucki, H. (1972). *J. Amer. Chem. Soc.* **94**, 8594.
Winstein, S., Kaesz, H. D., Kreiter, C. G., and Friedrich, E. C. (1965). *J. Amer. Chem. Soc.* **87**, 3267.
Woodward, R. B., and Hoffmann, R. (1969). "The Conservation of Orbital Symmetry," p. 114. Academic Press, New York.

11

Fluxional Allyl Complexes

K. VRIEZE

I. Introduction. 441
II. Bonding and Movements of the Allyl Group in Metal–Allyl Compounds. 442
   A. Metal–π-Allyl Bonding. 442
   B. Metal–σ-Allyl Bonding. 447
   C. "Ionic" Metal–Allyl Bonding. 448
   D. Fluxional Movements of Metal–π-Allyl Compounds 449
   E. Fluxional Movements of Metal–σ-Allyl Compounds 456
   F. Fluxional Behavior of Ionic Metal–Allyl Compounds 457
III. Survey of Reactions Occurring for Metal–π-Allyl Compounds. 457
   A. Intramolecular Reactions Involving Syn–Syn, Anti–Anti Exchange of the Allylic Protons 457
   B. Intramolecular Reactions Involving Syn–Anti Proton Exchange . 462
   C. Intermolecular Reactions Involving Syn–Syn, Anti–Anti Proton Exchange 466
   D. Intermolecular Reactions Involving Syn–Anti Proton Exchange 476
IV. Fluxional Metal–σ-Allyl and Ionic Metal–Allyl Compounds. 480
   A. Metal–σ-Allyl Compounds 481
   B. Ionic Metal–Allyl Compounds 482
V. Concluding Remarks 483
References. 483

I. Introduction

Until about 1960, investigations into the nature of metal–allyl compounds were restricted almost exclusively to allylic compounds of the main group and subgroup elements, while particular attention was paid to allylic Grignard reagents (Kharasch and Reinmuth, 1954; De Wolfe and Young, 1956). Relatively little information had been obtained about the nature of the metal–allyl bond. It was, however, the application of nuclear magnetic resonance techniques, that on the one hand led to the discovery of the metal-π-allyl bond (Heck and Breslow, 1960; Moiseev et al., 1958; Smidt and Hafner, 1959) and on the other hand yielded much information about the fluxional nature of the allyl group in metal–allyl compounds. The fluxional character of the allyl group in $(C_3H_5)MgBr$ was reported in 1959 (Nordlander and Roberts, 1959). Shortly afterwards, Dehm and Chien (1960) and Powell et al. (1965) showed that the π-allyl group in allyl–palladium compounds may become fluxional under certain circumstances.

Since about 1965, precise kinetic nmr studies and the use of double irradiation techniques have afforded a much deeper insight into the intra- and intermolecular reactions of metal–allyl compounds and into the types of fluxional motion of the allyl group and/or of the whole metal–allyl complex.

At the time of writing, however, there is still controversy about some aspects of the fluxional motion of allyl groups and about the metal–allyl bond in compounds of the main group and subgroup elements, while very little is known about the factors which promote the various fluxional motions of the allyl group. This chapter attempts to present a unified view of the present state of the subject and to indicate the gaps in our knowledge.

II. Bonding and Movements of the Allyl Group in Metal–Allyl Compounds

From the available experimental data, one may at present distinguish between roughly three types of metal–allyl bonding, namely; metal–π-allyl bonding (π-allyl = η^3-C_3H_5), which is found in the case of transition metal derivatives, metal–σ-allyl bonding (σ-allyl = η^1-C_3H_5), which has been observed across almost the whole periodic system, and finally "ionic" compounds, e.g., $M^+C_3H_5^-$ (M = Li, Na, K). [N.B. The nomenclature used here was proposed by Cotton (1968); e.g., η^1-C_3H_5 is monohapto allyl, η^3 = C_3H_5 is trihapto allyl.]

In Sections A, B, and C, the various types of metal–allyl bonding are discussed, while the possible movements are considered in the remainder of Section II.

A. Metal–π-Allyl Bonding

Metal–π-allyl bonding has been extensively studied by various methods (Lobach et al., 1967), such as proton magnetic resonance (Fedorov, 1970; Vrieze et al., 1969; Vrieze and van Leeuwen, 1971), ^{13}C magnetic resonance (Mann et al., 1971a; Sokolov et al., 1972), infrared and Raman spectroscopy (Davidson, 1972), and ultraviolet (Hartley, 1970; see also comments by

Vrieze and van Leeuwen, 1971) and photoelectron spectroscopy (D. R. Lloyd, private communication to Brown and Owens, 1971). The enthalpy of formation of $[(\eta^3\text{-}C_3H_5)PdCl]_2$ and the energy of the palladium–allyl bond was measured by differential scanning calorimetry (Ashcroft and Mortimer, 1971).

The main characteristic for the metal–π-allyl bond, as was deduced from infrared, Raman, nmr, and X-ray measurements, is that the three carbon atoms of the allyl group are all bonded to the metal atom. For example, the crystal structure of $[(\eta^3\text{-}C_3H_5)PdCl]_2$ (Smith, 1965; Oberhansli and Dahl, 1965) shows that the allyl group is symmetrically bonded to the metal atom, which according to nmr measurements (Shaw and Sheppard, 1961) is also the case for the compound in solution. The bonding may be considered formally as an interaction between an allyl anion with bivalent palladium, but may equally well be described as an interaction of either an allyl radical with monovalent palladium or of an allyl cation with zero valent palladium. The actual situation depends very much on the metal atom and in particular on the other ligands, as will be seen later.

Typical for palladium–π-allyl compounds, in which the palladium atom is surrounded by the bidentate π-allyl group and two other ligands, is that the dihedral angle between the allyl plane and the plane of the palladium atom and the other ligands lies in the range of 110°–125° (Fig. 1). This tilting of the allyl group has been proved to occur in a wide range of four-coordinate palladium–π-allyl compounds and was rationalized by the following arguments.

Kettle and Mason (1966) used as interacting orbitals the $4d$ orbitals of palladium and the allyl Hückel orbitals $\psi_1 = \frac{1}{2}(\phi_1 + \phi_3 + \sqrt{2}\phi_2)$, $\psi_2 = 1/\sqrt{2}(\phi_1 - \phi_3)$, and $\psi_3 = \frac{1}{2}(\phi_1 + \phi_3 - \sqrt{2}\phi_2)$, in order of increasing energy. They calculated the overlap integrals of ψ_1, ψ_2, and ψ_3 with the relevant $4d$

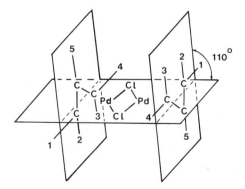

Fig. 1 Structure of $[(\eta^3\text{-}C_3H_5)PdCl]_2$.

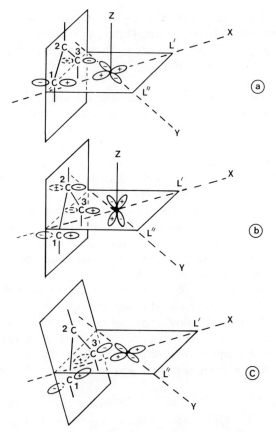

Fig. 2 Interactions of metal orbitals and allyl orbitals. (See text for details.)

orbitals for two extreme situations, i.e., with the plane of the allyl group respectively perpendicular to, and coinciding with, the plane of the palladium atom and the two ligands. In both extreme cases, the strongest bonding involved the lowest lying ψ_1 orbital, closely followed by the next strongest interaction involving the ψ_2 orbital. Maximization of the total energy with respect to a varying dihedral angle for the interactions of ψ_1 and of ψ_2 separately yielded predicted values for the dihedral angle of 114° and 102° respectively, which seems reasonably close to the experimental values.

Van Leeuwen and Praat (1970b) used a more qualitative and somewhat different approach, while they took into account also the 5s and 5p orbitals of palladium. In Fig. 2, the choice of coordinate axes is shown. Overlap may occur between ψ_1 and the palladium 5s orbital and a combination of $5p_x$ and $5p_y$, between ψ_2 and $4d_{x^2-y^2}$ and a combination of $4d_{xz}$ and $4d_{zy}$.

In Fig. 2a, it is shown that the ψ_2 orbital overlaps most efficiently if the terminal carbon atoms of the allyl group are in the PdL_2 plane and if the allyl group is perpendicular to the PdL_2 plane. From Fig. 2b, however, it may be concluded that the overlap situation for ψ_3 is more favorable if the nodal planes of the allyl group and of the PdL_2 unit are coincident; i.e., the terminal carbon atoms will lie below the PdL_2 plane. A compromise may be reached by tilting the allyl group, as shown in Fig. 2c, which gives better overlap for ψ_2 and still good overlap for ψ_1 and ψ_3 with their respective metal orbitals.

The application of this bonding scheme is very useful for the discussion of asymmetric palladium–π-allyl compounds, such as $(\eta^3\text{-}2\text{-}CH_3 \cdot C_3H_4)PdCl(PPh_3)$ (Mason and Russell, 1966) and $(\eta^3\text{-}C_3H_5)Pd(SnCl_3)(PPh_3)$ (Mason and Whimp, 1969). In both cases it appeared that the C—C bond trans to the strongest trans ligand; i.e., PPh_3 is shorter than the C—C bonds trans to the weaker trans ligands, e.g., Cl^- and $SnCl_3^-$. However, the difference between the C—C bond lengths trans to PPh_3 and trans to $SnCl_3$ was much less than the difference between the C—C bond lengths trans to PPh_3 and Cl^-. This is understandable, as the difference in trans influences of $SnCl_3$ and PPh_3 is much less than between Cl and PPh_3. Furthermore it appeared that the Pd—C bond lengths trans to the ligands with the strongest trans effect is larger than the Pd—C bonds trans to the weaker trans ligands.

If $L' = PPh_3$ and $L'' = Cl^-$, it may be reasoned that the p_x contribution of the Pd—C-1 bond is less than the p_y contribution to the Pd—C-3 bond. Consequently, the electron density in the allyl orbital ψ_2 will be larger on the carbon atom trans to the chloride ligand. It was pointed out that one may envisage in an extreme situation—i.e., with a large difference between the *trans* influences of L' and L''—that the highest occupied allyl orbital (ψ_2) acts as a σ-bond to the metal atom trans to Cl, while ψ_1 becomes a π orbital trans to PPh_3.

Fogleman *et al.* (1969) carried out semiempirical molecular orbital calculations for several π-allyl nickel systems. Their calculations indicate, in agreement with the above arguments, that for $(\eta^3\text{-}C_3H_5)NiCl(PPh_3)$, the overlap populations for the C—C bond trans to PPh_3, is larger than the C—C bond trans to Cl^-, while the value for the C—C bonds of the $C_3H_5^-$ anion was intermediate between both values of the Ni complex.

On basis of the above considerations, one may further infer that in an allyl–palladium complex, the allyl group or part of it may act as a good electron donor if a Cl atom is trans coordinated. This seems to be borne out by theoretical work of Brown and Owens (1971), who calculated overlap populations for $(\eta^3\text{-}C_3H_5)_2M$ and $[(\eta^3\text{-}C_3H_5)MCl]_2$ (M = Ni, Pd, and Pt) and discussed their results in the light of photoelectron spectroscopy measurements and calculations by Veillard (1969a,b) and by Hillier and Canadine

(1969). The metal–carbon overlap populations increase for $(\eta^3\text{-}C_3H_5)_2M$ in the order Ni ~ Pd < Pt and for $[(\eta^3\text{-}C_3H_5)MCl]_2$ in the order Ni < Pd < Pt, while the overlap populations of the dimers are always higher than for the corresponding $(\eta^3\text{-}C_3H_5)_2M$ compounds. From the calculations, it was apparent that the donating properties of the allyl group are increased and the electron density on the Pd atom is decreased if in $(\eta^3\text{-}C_3H_5)_2M$ the other allyl group is replaced by a Cl atom. It was pointed out that the ψ_1 and ψ_2 orbitals of the allyl group act as electron donor and electron acceptor orbitals, respectively. However, the ψ_2 orbital seems to act as an acceptor in $(\eta^3\text{-}C_3H_5)_2M$ and as donor in $[(\eta^3\text{-}C_3H_5)MCl]_2$. In other words, the allyl group behaves more as an anion in $(\eta^3\text{-}C_3H_5)_2M$ and more as a cation in $[(\eta^3\text{-}C_3H_5)MCl]_2$.

Finally, it seems useful to consider the bonding in $[(\eta^3\text{-}C_3H_5)PdCl]_2$ in a somewhat different way by putting the coordinate axes, as shown in Fig. 3, in

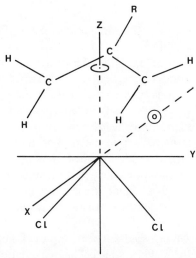

Fig. 3 A possible choice of coordinate axes for $[(\eta^3\text{-}C_4H_7)PdCl]_2$.

which an idealized $(\eta^3\text{-}C_3H_5)Pd(Cl/2)_2$ has been drawn with the two Cl/2 ions and a vacant position on the lower triangle of an octahedron (Vrieze et al., 1966). The allyl plane has been drawn parallel to this triangle and therefore makes an angle of about 125° with the $PdCl_2$ plane, which value is also near to the experimental values. The situation is thus similar to half sandwich compounds such as $(\eta^6\text{-}C_6H_6)Mo(CO)_3$ and $(\eta^3\text{-}C_3H_5)Co(CO)_3$ (Andrews and Davidson, 1972). The ψ_1 orbital again acts as a donor and overlaps with a combination of $4d_{z^2}$, $5s$ and $5p_z$.

The ψ_3 acceptor orbital interacts with $4d_{xz}$, while ψ_2 overlaps with mainly

$4d_{yz}$ and may act as a donor orbital in accordance with the suggestion of Brown and Owens (1971). Of interest is to see what will happen if another ligand is attached to the Pd atom in the place of the drawn empty coordination site. The structure of the resulting five-coordinate complex seems reasonable in view of the structure of the half-sandwich compound $(\eta^3\text{-}C_3H_5)\text{-}Co(CO)_3$ (Andrews and Davidson, 1972). In this configuration, the d_{yz} and d_{xz} orbitals will become more equivalent than in the four-coordinate situation. In an extreme situation one may envisage a diabolo-shaped electron density as in ferrocene and dibenzene chromium (Hillier and Canadine, 1969). It is therefore not unreasonable to suggest that the barrier to rotation of the π-allyl group in its own plane will be lower in five-coordinate compounds than in four-coordinate compounds. Later on, this model will be used to indicate that such a rotation, which is identical to an angular twist mechanism, may be one of the ways in which fluxional motion may occur in five coordinate species.

B. Metal–σ-Allyl Bonding

Metal–σ-allyl bonding occurs in various parts of the periodic system, but is more widespread for compounds of the nontransition metals. Clear-cut examples are $(\eta^1\text{-}C_3H_5)Co(CN)_5^{3-}$ (Kwiatek and Seyler, 1965), $(\eta^1\text{-}C_3H_5)\text{-}Mn(CO)_5$ (Kaesz et al., 1960; McClellan et al., 1961; Clarke and Fitzpatrick, 1972), $(\eta^1\text{-}C_3H_5)_4M$ with M = Si, Ge, Sn (Fishwick and Wallbridge, 1970), and $(\eta^1\text{-}C_3H_5)_2Hg$ (Zieger and Roberts, 1969a). The metal–σ-allyl bond M—CH$_2$—C(H)=CH$_2$ has been deduced from a combination of infrared, Raman, and nmr data. The M—C bond is thus similar to that in metal–alkyl compounds.

The occurrence of metal–σ-allyl bonding in the case of transition metal compounds is no doubt mainly due to the absence of empty coordination sites. However, upon irradiation in solution, the σ-allylcobalt and -manganese complexes were converted into the π-allyl compounds $[(\eta^3\text{-}C_3H_5)Co(CN)_4]^{2-}$ and $(\eta^3\text{-}C_3H_5)Mn(CO)_4$, respectively (Kaesz et al., 1960). For nontransition metal compounds, the absence of energetically available d orbitals is apparently an important cause for the nonexistence of π-allyl compounds.

For the allyl compounds $(C_3H_5)_2M$ (M = Mg, Zn, Cd), it was impossible, even at low temperatures, to obtain nmr evidence for the type of bonding owing to very fast allyl rearrangement reactions. In these cases, conclusions have been made on the basis of fairly slender evidence; i.e., of infrared and Raman data. In Table I, some infrared and nmr data have been given for a range of allyl compounds. It is clear that for all these compounds with a metal–σ-allyl bond the C=C-stretching frequency lies above about 1600 cm^{-1}.

TABLE I
INFRARED AND ^{1}H NMR^a DATA FOR METAL–ALLYL COMPOUNDS

Compound	Infrared data			NMR data		
	νC = C (cm^{-1})		Ref.	δH$_X$ (ppm) from TMS)	δH$_A$ (ppm from TMS)	Ref.^{b}
C$_3$H$_5$Cl	1640–1650		1	—	ABCX$_2$ spectrum	1
(OC)$_5$Mn(η^1-C$_3$H$_5$)	1617		2	—	ABCX$_2$ spectrum	2
Si(η^1-C$_3$H$_5$)$_4$	1629		3	—	ABCX$_2$ spectrum	3
Ge(η^1-C$_3$H$_5$)$_4$	1631		3	—	ABCX$_2$ spectrum	3
Sn(η^1-C$_3$H$_5$)$_4$	1625		3	—	ABCX$_2$ spectrum	3
Hg(η^1-C$_3$H$_5$)$_2$	1620		4	—	ABCX$_2$ spectrum	4
ClHg(η^1-C$_3$H$_5$)	1620–1635		5	3.64	6.02	5
Cd(η^1-C$_3$H$_5$)$_2$	—			2.86	6.14	6
Zn(η^1-C$_3$H$_5$)$_2$	1633–1622		7	2.96	6.30	7
[(C$_2$H$_5$)$_2$N]$_3$Ti(η^1-C$_3$H$_5$)	1602–1607		8	3.31	6.45	8
[(η^3-C$_3$H$_5$)PdCl]$_2$	—		3	4.77	6.82	9
Mg(η^1-C$_3$H$_5$)$_2$	1575		4	2.45	6.30	4
Li(C$_3$H$_5$)	1525–1540		4	2.17	6.38	10
Na(C$_3$H$_5$)	1535		11	—	—	

^{a} Only AX$_4$ spectra are reported.

^{b} References: (1) Thompson and Torkington (1966); (2) Clarke and Fitzpatrick (1972); (3) Fishwick and Wallbridge (1970); (4) Zieger and Roberts (1969a); (5) Kitching et al. (1972); (6) Thiele and Köhler (1967); (7) Thiele and Zdunneck (1965); (8) Neese and Bürger (1971); (9) Chien and Dehm (1961); (10) West et al. (1968); (11) Lanpher (1957).

Therefore, it seems reasonable that metal–σ-allyl bonding is predominant for Zn(C$_3$H$_5$)$_2$ and Cd(C$_3$H$_5$)$_2$. Mg(C$_3$H$_5$)$_2$ is a borderline case, as both infrared and Raman data indicate that this compound is intermediate between metal–σ-allyl and "ionic" metal–allyl compounds such as LiC$_3$H$_5$.

C. *"Ionic" Metal–Allyl Bonding*

Examples of a very interesting type of compound are Li(C$_3$H$_5$) (West et al., 1968), Na(C$_3$H$_5$) (Lanpher, 1957), Li(1,3-Ph$_2$C$_3$H$_3$) (Freedman et al., 1967), and M(3-Ph-C$_3$H$_4$) with M = Li, Na, K (Sandel et al., 1968), for which nmr measurements at suitable temperatures have shown that the thermodynamically most stable structure involves very likely a polar metal–allyl bond, while the allyl group may be regarded as an allylic carbanion with delocalized bonding (Fig. 4). The configurational stability of the allyl anion

11. Fluxional Allyl Complexes

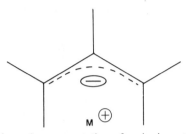

Fig. 4 Schematic representation of an ionic metal–allyl bond.

increases with increasing electropositivity of the metal atom, i.e., in the order Li < Na < K. From nmr and conductance measurements, it was concluded that there was no appreciable dissociation into ions, but a large bond polarity seems likely in view of the strongly shielded terminal and central protons of the allyl group (Table I). Additional evidence was provided by the low C═C stretching frequencies for $Li(C_3H_5)$ and $Na(C_3H_5)$ (Lanpher, 1957) (see Table I) and by ultraviolet spectra of $Li(C_3H_5)$ (West et al., 1968; Waack and Doran, 1963; Kuwata, 1960), which showed absorptions in accordance with theoretical estimates for an allyl carbanion. Many problems have to be solved, however, as little is known about the influence of aggregation and solvent and the influence of substituents on the allyl group (Fedorov, 1970).

Interestingly, in the series Na–Li–Mg–Zn–Cd–Hg "ionic" metal–allyl compounds are formed for the electropositive metals, while for the electronegative metals (e.g., Hg), stable metal–σ-allyl compounds are the rule. It is therefore clearly of great interest to study in the case of the borderline metal compounds the influence of various factors on the type of bonding.

D. Fluxional Movements of Metal–π-Allyl Compounds

The rearrangements taking place in fluxional metal–π-allyl compounds may be due either to movements of the allyl group relative to the other part of the complex or to the fluxional behavior of the whole complex (e.g., pseudorotations). At present, it is for most cases impossible to determine the precise mechanism. It seems useful, therefore, to discuss the possible mechanisms in the light of the observed nmr coalescence patterns.

In principle one may distinguish three cases. The first case involves interchange of syn and anti protons, which may take place either at one side of the allyl group (Fig. 5) or at both sides. We will call these interchanges "syn–anti exchange." The second case is an interchange between the syn protons 1 and 4 and simultaneously an interchange between the trans protons 2 and 3 (Fig. 6), the socalled "syn–syn, anti–anti exchange." This interchange can be observed by nmr if the complex is asymmetric, as is shown in Fig. 6; i.e., if protons 1

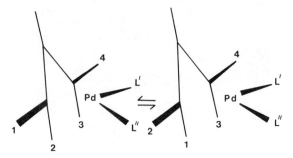

Fig. 5 Syn–anti interchange of $(\eta^3\text{-}C_4H_7)PdL'L''$.

and 4 and also 2 and 3 have different chemical shifts. Another possibility is the interconversion between conformers in solution if the allyl group in each conformer has a different surrounding. One may then study exchange rates between conformers for both symmetrically and asymmetrically bonded allyl groups, as will be discussed later. This third case will be discussed in Section II,D,1.

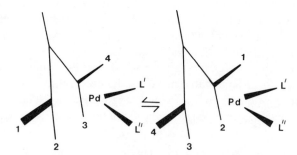

Fig. 6 Syn–syn, anti–anti interchange of $(\eta^3\text{-}C_4H_7)PdL'L''$.

1. *Syn–Syn, Anti–Anti Exchange*

The reaction depicted in Fig. 5 may take place in several ways. One of these, which is encountered most frequently, involves a ligand exchange via a five-coordinate intermediate (Vrieze *et al.*, 1969) and is essentially a cis–trans isomerization with chemically identical starting and end products.

In this context, it is significant that for a large number of isoelectronic planar four-coordinate compounds of Pt(II), slow ligand exchange reactions occur with retention of configuration (Basolo and Pearson, 1967). In Fig. 7, the proposed mechanism, which involves a trigonal bipyramidal intermediate, is shown.

Recently, however, it was found for the reaction of *cis*-$[(n\text{-}C_3H_7)_3P]_2PtCl_2$

11. Fluxional Allyl Complexes

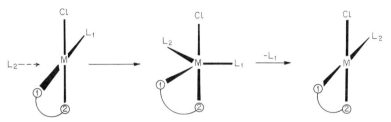

Fig. 7 The Basolo mechanism for retention of configuration.

with $(n\text{-}C_4H_9)_3P$, that in addition to a ligand exchange reaction, there occurred a very rapid cis–trans isomerization that was dependent on the concentration of the added ligand. However, the ligands did not mix during isomerization, which indicated a five-coordinate intermediate with $(n\text{-}C_4H_7)_3P$ in a unique position. Pseudorotational movements may then interchange Cl and $(n\text{-}C_3H_7)_3P$. The ligand displacement reaction seemed to occur via a much higher-energy five-coordinate species (Haake and Pfeiffer, 1970). Similar cis–trans isomerizations, which were catalyzed by other species, but also without chemical exchange occurring, have been found for (COD)IrCl(AsPh$_3$) (Volger et al., 1968), with COD = 1,5-cyclooctadiene, for 3-thiolo-1,3-diphenylprop-2-en-1-onebicyclo[2.2.1]hepta-2,5-dienerhodium(I) [in short, (NOR)Rh(SDBM)] (Heitner and Lippard, 1972) and possibly for (SDBM)Pd(η^3-C_4H_7) (Lippard and Morehouse, 1972a,b).

Recently, very interesting observations have been made for five-coordinate compounds (diene)M(PPhMe$_2$)$_2$(CH$_3$), in which the CH$_3$ and one of the olefinic bonds occupy the axial positions, while the other olefinic bond and the two phosphine ligands are situated in the equatorial positions of a trigonal bipyramid (Churchill and Bezman, 1971, 1972). An intramolecular exchange of the ends of the bidentate diene ligand without interchange of the other ligands was noted for (COD)Ir(PPhMe$_2$)$_2$(CH$_3$) (Shapley and Osborn, 1970). However, intermolecular phosphine exchange without interchange of the ends of the diene ligand was observed for (COD)Rh(PPhMe$_2$)$_2$(CH$_3$) (Rice and Osborn, 1971), i.e., a retention of configuration. As possible mechanisms, Shapley and Osborn (1970) suggested in the first place a Berry pseudorotational mechanism and secondly an angular twist mechanism.

In view of the above, it is worthwhile to realize that in the case of planar four-coordinate compounds, intramolecular isomerizations and intermolecular ligand exchange processes may also occur independent or dependent of each other. Furthermore, if the intramolecular process is dependent of the intermolecular exchange, the incoming ligand may or may not be the outgoing one.

The cis–trans isomerizations of (η^3-C_4H_7)PdCl(PPh$_3$) (Vrieze et al., 1968a) and of (diene)MCl(ligand) (Vrieze et al., 1968c,d,e), which are

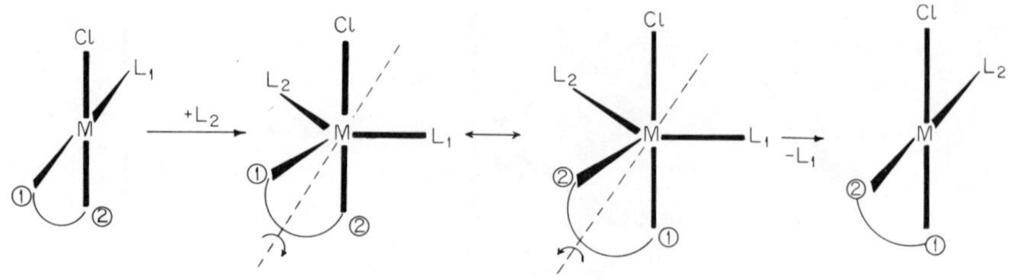

Fig. 8 The angular twist mechanism for cis–trans isomerization.

dependent on the ligand exchange process, have been explained using an angular twist mechanism, as is shown in Fig. 8. In principle, cis–trans isomerization may also occur without ligand exchange; i.e., the incoming ligand L_2 is also the outgoing ligand. The ligand exchange (L_2 in and L_1 out) must predominate, however, as loss of ^{31}P coupling was observed for L = phosphine.

A second possible mechanism, which again could take place with and, to a lesser extent, without ligand exchange, involves a number of Berry pseudorotational movements. For example, if one uses the five-coordinate structure in Fig. 9, which is similar to the configuration of the five-coordinate (COD)-Ir(PPhMe$_2$)(CH$_3$) (Shapley and Osborn, 1970), the mechanism shown in Fig. 10 involves cis–trans isomerization with ligand exchange. On the other hand, use of the structure in Fig. 11 may lead to cis–trans isomerization without ligand exchange (Fig. 12).

Later on it will become clear that very little information is as yet available about the precise mechanisms of each case. The above discussion indicates clearly, however, that one may expect a rich variety of mechanisms.

The existence of conformers each with a symmetrically bonded allyl group, has been found for a range of molecules, of which $(\eta^3\text{-}C_3H_5)(OC)_2Mo\text{-}(\eta^5\text{-}C_5H_5)$ is an example (Davison and Rode, 1967; Faller and Incorvia,

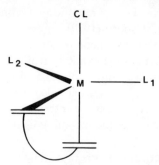

Fig. 9 Schematic configuration of a possible intermediate (diene)ML$_2$Cl.

11. Fluxional Allyl Complexes

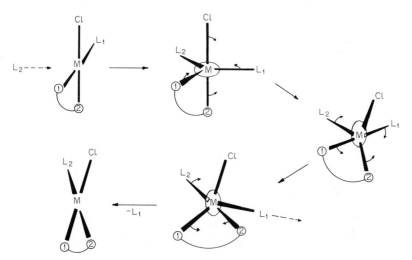

Fig. 10 Cis–trans isomerization by means of two Berry pseudorotations (L_2 in and L_1 out).

1968). At temperatures between $-10°$ to $-50°C$, two allyl groups are distinguished in the nmr spectrum in an unequal ratio, which become magnetically equivalent in both $CDCl_3$ and C_6H_6 at high temperatures. No evidence was found for interchange of syn and anti protons. The proposed mechanism, which interconverts the conformers, involved the rotation of the π-allyl group in its own plane, while the possibility of pseudorotational movements was not investigated but may be possible.

Very recently, Cooke *et al.* (1971) carried out double irradiation experiments on $(\eta^3\text{-}C_3H_5)_2Ru(CO)_2$ and showed that both allyl groups are equivalent, but that they are asymmetrically bonded to the metal atom. The experiments further showed a syn–syn and anti–anti coalescence process, which was explained by an intramolecular twist mechanism (Ray and Dutt, 1943),

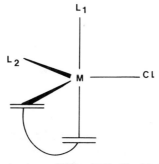

Fig. 11 Schematic structure of (diene)ML_2Cl with Cl in an axial position.

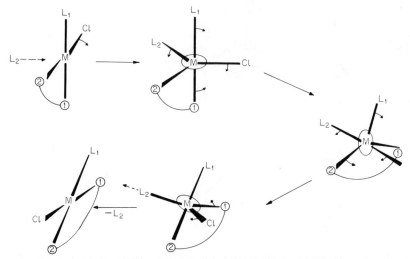

Fig. 12 Cis–trans isomerization by means of two Berry pseudorotations (L_2 in and L_2 out).

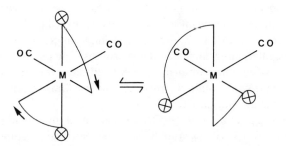

Fig. 13 The Ray–Dutt mechanism for interchange in $(\eta^3\text{-allyl})_2\text{Ru(CO)}_2$.

which involves a movement of the π-allyl group in its own plane as is shown in Fig. 13.

2. Syn–Anti Exchange

In Fig. 5, an interchange of syn and anti protons on one side of the allyl group is shown, which causes the coalescence of the appropriate proton signals. The most widely accepted mechanism that explains the observed coalescence is the π–σ–π movement (Vrieze et al., 1966; Cotton et al., 1967). In Fig. 14, the π–σ–π mechanism is depicted for $(\eta^3\text{-}2\text{-CH}_3 \cdot \text{C}_3\text{H}_4)\text{PdCl-}(\text{PPhMe}_2)$. Van Leeuwen and Praat (1970d) demonstrated that the coalescence of the protons 3 and 4 cis to the phosphine occurred at the same rate as the

11. Fluxional Allyl Complexes

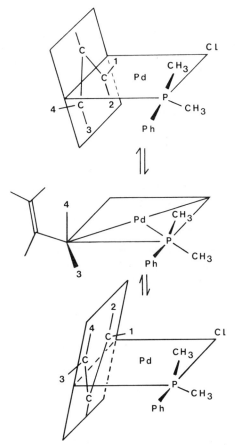

Fig. 14 The mechanism of syn–anti interchange of $(\eta^3\text{-}C_4H_7)PdCl(PPhMe_2)$.

coalescence of methyl signals of the phosphine ligand, which are inequivalent due to the absence of a plane of symmetry in the coordination plane. The π–σ–π movement explains this simultaneous collapse, as the starting and end products, although chemically identical, are mirror images. If the π–σ–π movement is rapid enough, there is effectively, i.e., in the time scale of the experiment, a plane of symmetry.

A corroborating evidence for this mechanism was achieved by substitution of the 2-position of the allyl group by an isopropyl group (van Leeuwen et al., 1970, 1971); the nonequivalent methyl groups of the isopropyl substituent again become magnetically equivalent simultaneously with the protons 3 and 4 in $(\eta^3\text{-}2\text{-isopropyl-}C_3H_4)PdCl(PPh_3)$. This simultaneous collapse can only occur if during the process the palladium atom becomes

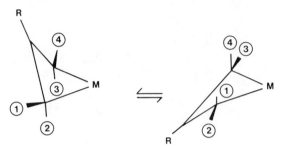

Fig. 15 The allyl flip mechanism for syn–anti interchange.

attached to different faces of the allyl plane, as is shown in Fig. 14. Recently π–σ–π movements were proved to take place in dimeric compounds with an isopropyl group substituted on the 1-position (Alexander et al., 1970), e.g., $[\eta^3$-syn-1-isopropylallyl-PdCl$]_2$.

For symmetric π-allyl compounds one could imagine a flip of the allyl group via a σ–σ bonded allyl group (Fig. 15). In view of the available evidence in particular for asymmetric π-allyl compounds, this mechanism seems rather improbable, as was pointed out by Vrieze et al. (1969).

A third mechanism (Becconsall and O'Brien, 1966a, 1967a,b) involves a rotation about the C—C bonds, while the allyl group remains π-bonded with the same face to the allyl group. Because of the above arguments, it is clear that this mechanism is impossible for palladium compounds and in general, in our view, for compounds with an interaction between metal d orbitals and the appropriate allyl orbitals.

However, the mechanism could occur for π-allyl compounds of Ti(IV) and Zr(IV), which have no d electrons for backbonding, and thus the π-allyl group may be intermediate between normal metal–π-allyl bonding and ionic metal allyl bonding. For compounds of the last type [e.g., Li(C$_3$H$_5$)] rotation about the carbon–carbon bond is a likely possibility, as will be discussed later.

E. Fluxional Movements of Metal–σ-Allyl Compounds

In the case of B(η^1-C$_3$H$_5$)$_3$, the allyl groups are clearly σ-bonded to the boron atom, as was shown by the ABCX$_2$ pattern of the nmr spectrum at temperatures below $-40°$C (Bogdanov et al., 1967, 1968). At higher temperatures, the ABCX$_2$ pattern coalesced to an AX$_4$ spectrum. It was proved that the 1,3 shift of the boron atom is an intramolecular process, which probably proceeds via a π-allyl type of intermediate.

Such a σ–π–σ mechanism seems certainly the most likely one to explain the coalescence to an AX$_4$ spectrum in the cases of (η^1-C$_3$H$_5$)Pd(O$_2$CC$_5$H$_4$N)(PPhMe$_2$) and (η^1-C$_3$H$_5$)Pd(S$_2$CNMe$_2$)(PPhMe$_2$), which are both four-coordinate σ-allyl metal compounds (Powell and Chan, 1972) and for the tetrahedrally coordinated [(CH$_3$)$_2$N]$_3$Ti(η^1-C$_3$H$_5$) (Neese and Bürger, 1971).

Slow intermolecular allyl exchange has been found to take place for Hg(η^1-C$_3$H$_5$)$_2$ (Zieger and Roberts, 1969a,b) in addition to intramolecular exchange. This intermolecular exchange was found, however, to be very rapid for mixtures of (η^1-C$_3$H$_5$)HgCl and mercuric chloride. The occurrence of intermolecular allyl exchange, although interesting in itself, makes the study of fluxional movements of the metal–σ-allyl compounds a difficult one. In any case it seems not unlikely that in addition to an intermediate of the metal–π-allyl type, intermediates also may be formed, in particular for Zn(η^1-C$_3$H$_5$)$_2$ and Mg(η^1-C$_3$H$_5$)$_2$, which have an ionic metal–allyl type of bond.

F. Fluxional Behavior of Ionic Metal–Allyl Compounds

The low-temperature nmr spectra of Li(C$_3$H$_5$) (West et al., 1968), (1,3-Ph$_2$·C$_3$H$_3$)Li (Freedman et al., 1967), and (3-Ph·C$_3$H$_4$)Li (Sandel et al., 1968) clearly indicate a structure of the type shown in Fig. 4. The nmr spectrum of Li(C$_3$H$_5$) is of the A$_2$B$_2$X type; i.e., the syn protons are equivalent as are also the anti protons. At higher temperatures an interchange of syn and anti protons causes the coalescence to an AX$_4$ pattern. It was found that the barrier to this interchange was increased with increasing electropositivity of the metal and therefore with increasing negative charge on the allyl group.

The interchange may be explained by the formation of a short-lived metal–σ-allyl intermediate. A more likely possibility seems at present the rotation around the carbon–carbon bonds in the allyl carbanion. This seems reasonable, as even rotations about a carbon–carbon double bond may occur in the nmr time scale, i.e., with $\Delta G^\ddagger$ values lower than about 20 kcal (Shvo and Shanan-Atidi, 1969).

Complicating factors are, as was pointed out before, that very little is known about the influence of aggregation and solvent, while again, intra- and intermolecular processes may proceed simultaneously.

III. Survey of Reactions Occurring for Metal–π-Allyl Compounds

A. Intramolecular Reactions Involving Syn–Syn, Anti–Anti Exchange of the Allylic Protons

In this section, mainly intramolecular reactions in which the π-allyl group remains π-bonded to the metal during the reaction are discussed.

1. Compounds of Cr, Mo, and W

The well-studied stereochemically nonrigid compounds $(\eta^5\text{-}C_5H_5)M\text{-}(\eta^3\text{-}C_3H_4R)(CO)_2$ with M = Mo or W and R = H, CH_3 were prepared by Cousins and Green (1963), Green and Stear (1964), Murdoch (1965), Murdoch and Henzi (1966), and Hayter (1968). King (1966) suggested, on the basis of the observation of four infrared carbonyl stretching frequencies, the possibility of cis and trans isomers. However, Davison and Rode (1967) and Hayter (1968) proposed the existence of two conformers (Fig. 16), which at low

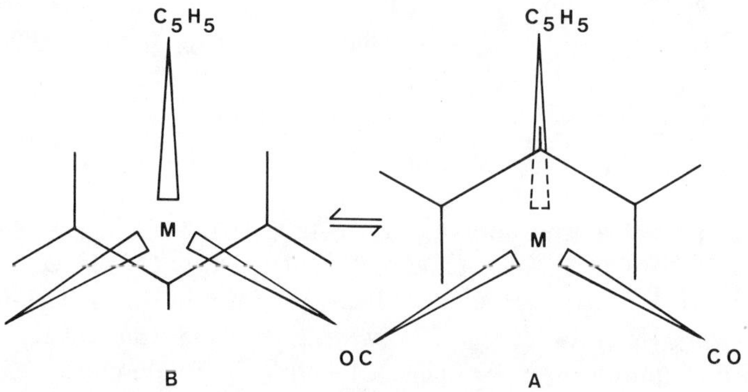

Fig. 16 Rotation of the allyl group in its own plane for $(C_5H_5)(CO)_2M(\eta^3\text{-allyl})$.

temperature occur in uneven quantities and interconvert at higher temperatures in the nmr time scale. For R = H, the concentration of isomer A is higher than that of B, while the opposite was found for R = CH_3. The reversal is no doubt due to the nonbonded interactions of the cyclopentadienyl and methyl groups (Faller and Incorvia, 1968; Faller and Jakubowski, 1971) (Figure 16). The reader is also referred to the work on stereochemical nonrigid poly(1-pyrazolyl)boratodicarbonylmolybdenum π-allyls (Meakin et al., 1972) and on cyclopentadienyl–molybdenum compounds of the type $(\eta^5\text{-}C_5H_5)Mo(CO)_2LR$ (Faller and Anderson, 1970), from which one may also deduce the possibility of pseudorotational movements for $(\eta^5\text{-}C_5H_5)M\text{-}(\eta^3\text{-}C_3H_4R)(CO)_2$.

A similar equilibrium between two conformers was found for $(OC)_4\text{-}W(\eta^3\text{-}C_3H_5)X$ with X = Br, I (Holloway et al., 1969). Double resonance experiments indicated exchange rates at room temperature greater than the inverse of spin–lattice relaxation time, but smaller than about 1 sec^{-1}, which is thus much lower than the rates for the above cyclopentadienyl compounds.

2. Compounds of Fe and Ru

Iron–π-allylcarbonyl chlorides $(\eta^3\text{-}2\text{-}R\text{-}C_3H_4)Fe(CO)_3X$ have been reported by several groups (Impastato and Ihrman, 1961; Emerson and Pettit,

11. Fluxional Allyl Complexes

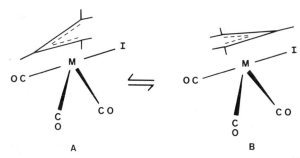

Fig. 17 A possible mechanism for interchange in $(\eta^3\text{-allyl})\text{Fe(CO)}_3\text{I}$.

1962; Plowman and Stone, 1962; Murdoch and Weiss, 1962; Heck and Boss, 1964; Nesmeyanov et al., 1966). The structure of $(\eta^3\text{-}C_3H_5)\text{Fe(CO)}_3\text{I}$ was determined by Minasyan et al. (1966) and shown as isomer A in Fig. 17. Nesmeyanov and Kritskaya (1968) and Nesmeyanov et al. (1968) showed for each compound the existence of two rotational isomers in ratios varying between 1:1 and 1:8 depending upon the nature of the group R and the halogen X. The temperature dependence of the interconversion of the isomers could not be determined due to decomposition above $+25°C$.

More extensively studied are the fluxional bis-π-allyl complexes $(\eta^3\text{-}C_3H_5)_2\text{Fe(CO)}_2$ (Nesmeyanov et al., 1968) and $(\eta^3\text{-}C_3H_5)\text{Ru(CO)}_2$ (Cooke et al., 1971). In the case of the iron compound, the spectrum at $-67°C$ was reported to consist of two A_2M_2X patterns which coalesced to one A_2M_2X pattern at $+20°C$ with an energy barrier of 4.6 ± 0.8 kcal. The proposed equilibrium is shown in Fig. 18.

Some doubt has been cast on these spectral assignments by Cooke et al. (1971), who showed conclusively that for $(\eta^3\text{-}C_3H_5)_2\text{Ru(CO)}_2$, which at $+33°C$ has an nmr spectrum closely resembling that of the analogous iron compound at $-67°C$, the allyl groups are equivalent but asymmetrically bonded, and not inequivalent and symmetrically bonded as proposed for the iron compound. The spectral assignment also agreed with the structures suggested for $(\eta^3\text{-allyl})_2\text{Ru(diene)}$ (Powell and Shaw, 1968a; Cooke et al., 1971). Double irradiation experiments on solutions of $(\eta^3\text{-}C_3H_5)_2\text{Ru(CO)}_2$ indicated a slow syn–syn and anti–anti proton exchange. At $+120°C$, the spectrum consists of one A_2M_2X pattern analogous to the situation for

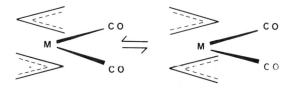

Fig. 18 A possible mechanism for interchange in $(\eta^3\text{-allyl})_2\text{Fe(CO)}_2$.

(η^3-C$_3$H$_5$)$_2$Fe(CO)$_2$ at +33°C. Cooke *et al.* (1971) suggested as a possible mechanism the Ray–Dutt process (Fig. 13), which involves a simultaneous movement of the π-allyl groups in their own planes (Ray and Dutt, 1943). Other twist mechanisms, however, were not excluded.

When we consider the influence of the metal and the other ligands on the rates of the fluxional process of (η^3-C$_3$H$_5$)$_2$ML$_2$, it is clear that the energy barrier for the iron carbonyl compound is much lower than for the ruthenium carbonyl compound. Substitution of the two carbonyl groups in the ruthenium complex by two trimethyl phosphite ligands again raises the energy barrier (Cooke *et al.*, 1971), while no fluxional behavior has been reported for (η^3-C$_3$H$_5$)$_2$Ru(diene). Obviously, a systematic study of the influence of the type of ligands on the fluxional behavior seems appropriate. In this respect, it is also of interest to point out that the interconversion rates for the isomers of (η^3-C$_3$H$_5$)M(CO)$_2$(η^5-C$_5$H$_5$) were higher for the tungsten compound than for the molybdenum compound (Davison and Rode, 1967), which in view of the known trend of metal–carbon bond strengths, seems not the expected order.

3. *Compounds of* Rh *and* Ir

A rather unusual fluxional behavior was observed for (η^3-C$_3$H$_5$)$_3$Rh (Becconsall and O'Brien, 1966b; Powell and Shaw, 1966, 1968b). At −74°C, all three allyl groups are inequivalent but symmetrically bonded to the rhodium atom. At +34°C, two of the allyl groups become magnetically equivalent, probably due to a rotation of the third allyl group in its own plane, as is schematically shown in Fig. 19. The activation energy for this

Fig. 19 Rotation of the allyl group in its own plane in (η^3-allyl)$_3$Rh.

process, in which the allyl groups remain linked to the metal atom, is about 9.5 kcal/mole. This situation was also found for (η^3-C$_3$H$_5$)$_3$Ir at room temperature (Chini and Martinengo, 1967) but was unfortunately not further studied. At higher temperatures all three allyl groups of the rhodium compound start to move in their own planes. The process, however, occurs in addition to a π–σ–π reaction, and could therefore not be studied properly.

The bis-π-allyl compounds [(η^3-C$_3$H$_5$)RhCl]$_2$ are isoelectronic with (η^3-C$_3$H$_5$)Ru(CO)$_2$ and are structurally very similar (Powell and Shaw, 1966,

1968b). The X-ray analysis (McPartlin and Mason, 1967) and nmr measurements at $-20°C$ clearly showed that all allyl groups are equivalent but unequally bonded (Powell and Shaw, 1968b; Ramey et al., 1968). At higher temperatures, line broadening results indicated both syn–syn, anti–anti and syn–anti exchange, but decomposition hindered a precise investigation. Preliminary double irradiation results (Cooke et al., 1971) on the dimeric rhodium compounds showed conclusively only syn–syn, anti–anti exchange at $+50°C$, and at higher temperatures, also syn–anti exchange. An obvious mechanism to account for the syn–syn, anti–anti exchange is the Ray–Dutt process or a similar twist mechanism (Cooke et al., 1971).

4. *Compounds of* Ni, Pd, *and* Pt

Two isomeric forms in various ratios and each with an A_2M_2X spectrum were observed for solutions of $Ni(\eta^3\text{-}C_3H_5)_2$, $Ni(\eta^3\text{-}C_4H_7)_2$ (Keim, 1963; Bönneman et al., 1967), and $Pd(C_3H_5)_2$ (Keim, 1963; Becconsall and O'Brien, 1967a,b; Faller and Incorvia, 1968). An X-ray analysis of $Ni(\eta^3\text{-}C_4H_7)_2$ (Uttech and Dietrich, 1965) showed that in the solid state only one isomer is found which has parallel allyl planes, while the methyl groups are trans to each other. The other isomer, which is formed in solution, probably has a configuration in which the methyl groups are cis to each other, while the allyl planes may be parallel or slightly staggered.

The platinum compounds $Pt(\eta^3\text{-}C_3H_5)_2$ and $Pt(\eta^3\text{-}C_4H_7)_2$ were reported by Keim (1963), but nmr provided no conclusive evidence about the structure. Reexamination with a 220 MHz nmr spectrometer (O'Brien, 1970; Mann et al., (1971b) gave strong evidence that the structure is similar to that of the nickel and palladium compounds. In solution, two isomers, each with symmetrically bonded allyl groups, were observed.

For all of the above compounds, syn–syn, anti–anti exchange was noted and also syn–anti exchange at higher temperatures. In the case of $Pt(\eta^3\text{-}C_4H_7)_2$, Mann et al. (1971b) pointed out that also intermolecular exchange of the allyl groups was observed. For none of the compounds in the Ni column were precise kinetic measurements carried out, which precludes conclusions about the trends in the case of fluxional movements. The exchange rates seem to decrease in going from Ni to Pt.

Intramolecular movements of the complex as a whole have been observed for the dimeric allylic palladium carboxylates (Powell, 1969, 1971; van Leeuwen and Praat, 1970 a,b; van Leeuwen et al·, 1972). At $-80°C$ the nmr spectrum of $[(\eta^3\text{-}2\text{-}CH_3\text{-}C_3H_4)Pd(OAc)]_2$ shows that there are three species. The species with the highest concentration (80–90%) possesses a symmetric structure (isomer A of Fig. 20). The species with the lower concentrations are isomers B and C (Fig. 21), which occur in equal quantities. Of interest is that in

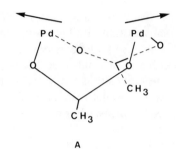

Fig. 20 Structure of isomer A of $[(\eta^3\text{-}C_4H_7)Pd(OAc)]_2$.

the solid state the asymmetric structure is the most favored one (Churchill and Mason, 1964). Between $-80°$ and $-40°C$, the signals of the low-concentration species coalesce, which was found to be an intramolecular process. Models indeed show that the interconversion of isomers B and C is via twisting of the acetate group without breaking of any bond. Above $-40°C$ there is also an exchange involving isomer A. This process requires bond breaking of the Pd—O bond(s).

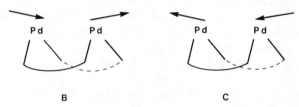

Fig. 21 Structures of the asymmetric isomers B and C of $[(\eta^3\text{-}C_4H_7)Pd(OAc)]_2$.

In accordance with the above interpretation, it was found that in the case of the bis(π-methallyl)palladium carboxylates, in which there is a bridge between the carboxylate groups, there is no interconversion between the two low-concentration isomers, as this is sterically impossible. At higher temperatures, an exchange takes place between symmetric and asymmetric species due to bond breaking of the Pd—O bond(s) (van Leeuwen et al., 1972).

B. Intramolecular Reactions Involving Syn–Anti Proton Exchange

1. *Compounds of* Zr, Hf, *and* Th

The tetra-π-allyl compounds $Zr(\eta^3\text{-}C_3H_5)_4$ and $Th(\eta^3\text{-}C_3H_5)_4$ show at $-74°C$ an A_2M_2X pattern, which indicates four equivalent symmetrically bonded π-allyl groups. The hafnium derivative $Hf(\eta^3\text{-}C_3H_5)_4$ shows even at

−74°C an AX_4 spectrum, which is observed for the Zr and Th compounds at −10°C and +80°C, respectively. The activation energies are about 10 and 15 kcal/mole, respectively, while the activation energy for this syn–anti exchange must be very low indeed for the hafnium compound (Becconsall et al., 1967b; Wilke et al., 1966). The mechanism probably involves a π–σ–π movement. However, as the above metal atoms are rather electropositive and have no filled d orbitals, and as the chemical shifts of the X signal of AX_4 spectrum of the Zr, Hf, and Th compounds—2.63, 2.91, and 2.82 respectively—are between $Mg(C_3H_5)_2$ and $Zn(C_3H_5)_2$ (see Table I), a rotation about the C—C bonds of a carbanion type of allyl group might be a possibility. As long as definitive evidence has not been given about the true mechanism, the π–σ–π movement is preferred by the author.

2. Compounds of Mo and W

Detailed data have been obtained for the trihapto benzyl compounds $(\eta^5\text{-}C_5H_5)(CO)_2M(1,2,7\text{-}\eta^3\text{-}4CH_3\text{-}C_6H_4CH_2)$ (M = Mo, W) and $(\eta^5\text{-}C_5H_5)(CO)_2Mo(1,2,7\text{-}\eta^3\text{-}3,5\text{-}[(CH_3)_2CH]_2C_6H_2CH_2)$ (King and Fronzaglia, 1966; Cotton and Marks, 1969). The X-ray analysis of $(\eta^5\text{-}C_5H_5)(CO)_2Mo(1,2,7\text{-}\eta^3\text{-}4\text{-}CH_3\text{-}C_6H_4CH_2)$ by Cotton and LaPrade (1968) showed that the benzyl group is bonded to the metal atom in a trihapto fashion (Fig. 22).

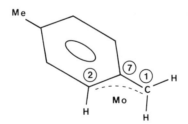

Fig. 22 Schematic representation of a part of $(\text{benzyl})Mo(CO)_2(\eta^5\text{-}C_5H_5)$.

A line-shape analysis of the spectral changes as a function of temperature showed that the metal atom can move to all four equivalent positions of the benzyl group at the place of the C-1, C-7, C-2 unit. Cotton and Marks (1969) proposed as the most probable pathway π–σ–π movement via a monohapto (σ) bonded intermediate. This intermediate, which has a sixteen-electron configuration, may be stabilized by the resonance stabilization of the benzene ring, which will be larger for the monohapto-bonded species than for the trihapto-bonded benzyl compound, as is clear in the last case from the strongly differing carbon–carbon distames in the benzene ring (Cotton and

Marks, 1969). In view of this, it is of interest to point out, that π–σ–π movements have not been observed for the analogous simple π-allyl compounds $(\eta^5\text{-}C_5H_5)(CO)_2M(\eta^3\text{-}C_3H_5)$ (Chapter III, A,1).

The activation enthalpies for the fluxional motion for all molybdenum compounds are about 18 kcal/mole, while the value is about 15 kcal/mole in the case of the tungsten compound. Although this fluxional behavior and the interconversion between conformers of $(C_5H_5)(CO)_2M(\eta^3\text{-}C_3H_5)$ are not strictly comparable, it is of interest to note that in both cases the rates are higher for the tungsten than for the molybdenum compounds.

3. *Compounds of* Fe, Ru, Rh, *and* Ir

Syn–anti exchange has been observed in the cases of $Rh(\eta^3\text{-}C_3H_5)_3$ (Becconsall and O'Brien, 1966b; Powell and Shaw, 1966), $[(\eta^3\text{-}C_3H_5)_2RhCl]_2$ (Powell and Shaw, 1968b; Ramey *et al.*, 1968), and of $(\eta^3\text{-}C_3H_5)M(CO)_2$ with M = Fe, Ru (Cooke *et al.*, 1971; Nesmeyanov *et al.*, 1968). These π–σ–π movements could not be studied properly owing to the occurrence of other reactions and/or decomposition.

More suitable compounds of the type $L_2Cl_2Rh(\eta^3\text{-allyl})$ were prepared by Lawson *et al.* (1966) and Volger and Vrieze (1967). Volger and Vrieze (1967) reported a whole series of compounds $L_2Cl_2Rh(\eta^3\text{-}2C\text{-}H_3C_3H_4)$ with L = Ph_3P, $(4\text{-}Br\text{-}C_6H_4)Ph_2P$, $[4\text{-}(CH_3)_2NC_6H_4]_3As$, Ph_2MeAs, Ph_3As, and Ph_3Sb and showed by means of nmr, infrared, and dipole-moment studies that the ligands L are cis to each other, while they are trans to the π-methallyl group (Fig. 23). This structure was confirmed by an X-ray analysis of the triphenylarsine derivative (Hewitt and de Boer, 1968) and is analogous to the structure suggested for $L_2Cl_2Ir(\eta^3\text{-}C_4H_7)$ with L = PEt_3, PMe_2Ph (Powell and Shaw, 1968c). The nmr spectra of the arsine and stibine derivatives consist of an $A_2M_2X_3$ spectrum due to protons 1 and 4, 2 and 3, and the methyl protons. In the case of the phosphine compounds, three signals were observed

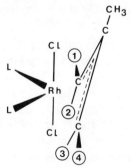

Fig. 23 Structure of $L_2Cl_2Rh(\eta^3\text{-}C_4H_7)$.

TABLE II
Transition State Parameters for the Syn–Anti Exchange in $L_2Cl_2Rh(\eta^3\text{-}2\text{-}CH_3\text{—}C_3H_4)$

L	$\Delta H^{\ddagger}$ (kcal/mole)	$\Delta S^{\ddagger}$ (eu)
Ph_3Sb	> 20	—
Ph_3As	17.3 ± 1.0	+3 ± 3
Ph_2MeAs	13.2 ± 1.0	−14 ± 3
{4-$(CH_3)_2NC_6H_4$}$_3$As	11.3 ± 1.0	−14 ± 3
(4-Br-C_6H_4)Ph_2As	9.8 ± 1.0	−12 ± 3
Ph_3P	8.5 ± 1.0	−15 ± 3

in the ratio 2:1:1 due to protons 1,4 and 2 and 3, respectively. This splitting pattern was also observed for the analogous phosphine derivatives of iridium (Powell and Shaw, 1968c). The two signals of intensity 1 were separated by 0.1 ppm, which may be caused by phosphor coupling (Powell and Shaw, 1968c) or a slight asymmetry in the compound (Volger and Vrieze, 1967).

At elevated temperatures, the signals of protons 1, 2, 3, and 4 coalesced for all compounds excepting the triphenylstibine derivative. The transition state parameters for this π–σ–π process are shown in Table II.

From Table II it is clear that the π–σ–π process, which is intramolecular (Vrieze and Volger, 1967), is favored by strongly electron-donating ligands. A similar trend was observed in the case of the asymmetric compounds (η^3-allyl)PdCl(L), but the picture is obscured by intermolecular ligand exchange reactions.

Brown and Owens (1971) concluded that the metal–π-allyl bond is more stable if the trans groups are electron attracting, as has been discussed before. This view agrees with our results, if it is assumed, that the increase in the rate of the π–σ–π process with increasing donor properties of the ligand L is mainly due to the destabilization of the initial π-allyl state. The influence of the ligands on the stability of the σ-allyl intermediate that probably resembles the transition state is not known but is not expected to vary very much along the above mentioned series.

Comparison of the above phosphine rhodium compound with analogous iridium compounds shows that the π–σ–π process is much faster for the rhodium than for the iridium compounds. This trend will not be discussed further, as not enough information is known.

4. Compounds of Ni, Pd, and Pt

Processes involving intramolecular syn–anti proton exchange, owing probably to π–σ–π movements, have been suggested for the bis(π-allyl)metal compounds of Ni, Pd, and Pt (Becconsall and O'Brien 1967a,b; O'Brien 1970). Even the more stable [(η^3-C$_4$H$_7$)PdCl]$_2$ was shown by means of nuclear magnetic double resonance experiments to undergo a slow π–σ–π movement at +90°C at a rate of about 0.2 sec^{-1}, which could be observed, as the longitudinal relaxation time was about 10 seconds (van Leeuwen and Praat, 1970c; Hoffmann and Forsén, 1966). It is not certain, however, if this process is intramolecular or is caused by an intermolecular reaction with other species.

Van Leeuwen and Praat (1970b) and Powell and Chan (1972) reported intramolecular syn–anti proton interchanges, but they are more conveniently discussed in a later section (III,D,1), as will be also systems investigated by Guthrie et al. (1971), Faller and Mattina (1972), and Faller et al. (1972).

C. Intermolecular Reactions Involving Syn–Syn, Anti–Anti Proton Exchange

Interchange of the syn protons and simultaneously also of the anti protons in the case of the asymmetric planar four-coordinate π-allyl compounds of Ni(II), Pd(II), and Pt(II) is, as discussed in Section II,D,1, essentially a cis–trans isomerization. This isomerization is known to occur for (η^3-allyl)NiCl(PPh$_3$) (Heck et al., 1961), (η^3-allyl)PtCl(L) with L = pyridine or isoquinoline (Mann et al., 1971b), (η^3-C$_4$H$_7$)PdCl(L) with L = AsPh$_3$, PPh$_3$, CO (Vrieze et al., 1969; Volger et al., 1970), N-donor ligands (Faller and Mattina, 1972), (η^3-C$_4$H$_7$)Pd(CN)(L) (Shaw and Shaw, 1971), and (η^3-C$_4$H$_7$)Pd(OAc)(L) (van Leeuwen and Praat, 1970b,c). Furthermore, cis–trans isomerization has also beeen found for (η^3-C$_4$H$_7$)Pd(asymmetric bidentate) (Ban et al., 1972; Lippard and Morehouse, 1972a,b). For almost all cases, bimolecular reactions of the above species with other molecules were the cause of the isomerization.

Dimeric compounds such as the asymmetric [(η^3-C$_4$H$_7$)Pd(SCN)]$_2$ (Tibbetts and Brown, 1970) and the symmetric [(η^3-allyl)PdCl]$_2$ (van Leeuwen et al., 1972) may be involved in direct dimer–dimer reactions. In the last case, the dimer occurs as two conformers, which recently was also reported for [(η^3-C$_4$H$_7$)PtCl]$_2$ (Mann et al., 1971b). The interchange between the conformers may proceed via dimer–dimer reactions or via dissociation of the dimeric species (van Leeuwen et al., 1972).

To keep the discussion of this complicated subject as simple as possible, we will consider fairly extensively systems containing monomer (η^3-C$_4$H$_7$)PdCl(L) (= ML) and dimer [(η^3-C$_4$H$_7$)PdCl]$_2$ (= M$_2$) and discuss also as a special case systems containing only monomer ML with and without excess ligand L. Attention will be paid in particular to phosphine, arsine, and amine systems, which have been the most thoroughly studied.

11. Fluxional Allyl Complexes

In Fig. 24, the monomer ML and dimer M_2 have been drawn symbolically. They may represent not only the above π-allyl systems, but also the diene systems (diene)MCl(L) and [(diene)MCl]$_2$ [M = Rh(I), Ir(I)], which are very similar to the palladium compounds (Vrieze et al., 1968c,d,e; Volger et al., 1968; van Leeuwen et al., 1969a,b,c). The sites A and B of ML in Fig. 24 may therefore represent either protons 1 and 4 (or 2 and 3) of $(\eta^3\text{-}C_4H_7)\text{PdCl(L)}$ or the olefinic protons of (diene)MCl(L) trans and cis to the ligands L and Cl,

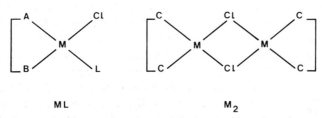

Fig. 24 Schematic structures of for example $(\eta^3\text{-}C_4H_7)\text{PdCl(L)}$ and $[(\eta^3\text{-}C_4H_7)\text{PdCl}]_2$.

respectively. The sites C of M_2 similarly represent the protons (1, 4) (or 2 and 3) of $[(\eta^3\text{-}C_4H_7)\text{PdCl}]_2$ or the diene protons of the dimer [(diene)MCl]$_2$. In some instances, one may also consider the methyl protons of the methallyl compounds or the bridgehead protons of the norbornadiene systems, however this makes no essential difference for the treatment given below.

In this section we will restrict ourselves to left–right interchange (i.e., a proton jumps from A to B and vice versa), so that π–σ–π movements are excluded. A similar reaction scheme may be written down for syn–anti exchange.

The monomer ML may dissociate:

$$\text{ML} \rightleftharpoons \text{M} + \text{L} \tag{1}$$

Reaction (1) will be first order if cis–trans isomerization occurs. Retention of configuration will not be observed, unless one studies for example ligand signals. The dimer may also dissociate into "half-dimers":

$$M_2 \rightleftharpoons 2M \tag{2}$$

This equilibrium lies very much to the left for solvents such as $CDCl_3$. The forward and backward reactions are very rapid. The signals of M_2 and M will therefore always be in the fast exchange limit.

The monomer ML may react in a bimolecular reaction with ligand L formed in reaction (1):

$$\text{ML} + \text{L}^* \rightleftharpoons \text{MLL}^* \rightleftharpoons \text{ML}^* + \text{L} \tag{3}$$

This reaction will be observed if there is cis–trans isomerization. There may or may not be a ligand exchange. Generally, ligand exchange is found. Fast ligand exchange reactions with retention of configuration will not be observed unless ligand signals themselves are studied.

The dimer may react in two steps with ligand to form monomer:

$$M_2 + L \rightleftharpoons M_2L \tag{4}$$

$$M_2L + L \rightleftharpoons 2ML \tag{5}$$

This reaction is measured using the signals of M_2 and will be first order in L if M_2L (intermediate which is not observed) and ML are in the fast-exchange limit, while a second-order behavior in L will be observed if M_2 and M_2L are in the fast-exchange limit. This last case was always found, but it must be borne in mind that one may find with chemical reaction kinetics that the reaction is first order in L, i.e., if reaction (4) is rate determining (Basolo and Pearson, 1967), while with nmr kinetics, the same reaction may be second order in L. The reason is, as was pointed out by Vrieze et al. (1969), that with nmr kinetics the chemical shift difference of the species involved determines to a great extent the rate expressions.

The monomer ML may interact with a half dimer species. If this involves chemical exchange the monomer and dimer signals will coalesce:

$$ML + M^* \rightleftharpoons M^*L + M \tag{6}$$

Without chemical exchange, signals A and B will coalesce if cis–trans isomerization takes place:

$$ML + M^* \rightleftharpoons M^*L + M \tag{7}$$

Direct reactions of monomer and dimer may or may not involve exchange of ligand L, as is shown in reactions (8) and (9), respectively.

$$M^*L + M_2 \rightleftharpoons ML + M^*M \tag{8}$$

$$M^*L + M_2 \rightleftharpoons M^*L + M_2 \tag{9}$$

Reaction (8), consequently, will involve coalescence of monomer and dimer signals, while reaction (9) will be observed on the signals of ML is cis–trans isomerization takes place. In the last case, chloride exchange between ML and M_2 may well occur, but it cannot be observed here with nmr.

Finally, the monomer ML may react with a similar species, involving exchange of A and B:

$$ML + M^*L^* \rightleftharpoons ML + M^*L^* \tag{10}$$

Dimer–dimer reactions also takes place, but, as the interconversion reaction of the two conformers of $[(\eta^3\text{-}C_4H_7)PdCl]_2$ is very rapid and the chemical shift differences between the proton signals of the two conformers is very

11. Fluxional Allyl Complexes

small (~ 0.02 ppm), only the averaged dimer signals occur in the temperature range of interest. Consequently, dimer–dimer reactions will not show up in the systems discussed below.

The rate equations for the line broadening of monomer ML and dimer M_2 are given in Eqs. (11) and (12), respectively:

$$1/\tau(\text{ML}) = k_1 + k_3[\text{L}] + 2k_{-5}[\text{ML}] + (k_6 + k_7)[\text{M}]$$
$$+ (k_8 + k_9)[\text{M}_2] + k_{10}[\text{ML}] \quad (11)$$

$$1/\tau(\text{M}_2) = k_{-1}\frac{[\text{M}][\text{L}]}{2[\text{M}_2]} + k_5\frac{[\text{L}][\text{M}_2\text{L}]}{[\text{M}_2]} + k_6\frac{[\text{ML}][\text{M}]}{2[\text{M}_2]} + \tfrac{1}{2}k_8[\text{ML}] \quad (12)$$

In Eqs. (13) and (14), the concentrations of L and M are given.

$$[\text{L}] = K_1[\text{ML}]/K_2^{1/2}[\text{M}_2]^{1/2} \quad (13)$$

$$[\text{M}] = K_2^{1/2}[\text{M}_2]^{1/2} \quad (14)$$

Eqs. (13) and (14) substituted in (11) and (12) yield (15) and (16, respectively:

$$1/\tau(\text{ML}) = k_1 + k_3\frac{K_1}{K_2^{1/2}}\frac{[\text{ML}]}{[\text{M}_2]^{1/2}} + 2k_{-5}[\text{ML}]$$
$$+ (k_6 + k_7)K_2^{1/2}[\text{M}_2]^{1/2} + (k_8 + k_9)[\text{M}_2] + k_{10}[\text{ML}] \quad (15)$$

$$1/\tau(\text{M}_2) = k_1\frac{[\text{ML}]}{2[\text{M}_2]} + k_{-5}\frac{[\text{ML}]^2}{[\text{M}_2]} + k_6 K_2^{1/2}\frac{[\text{ML}]}{2[\text{M}_2]^{1/2}} + \tfrac{1}{2}k_8[\text{ML}] \quad (16)$$

One may now distinguish between several possibilities, assuming that in general the amount of dissociated L is so small that the weighed-in concentration of ML and M_2 are also the actual concentrations present in solution.

For the triphenylarsine systems sufficient free arsine is formed in reaction (1) to give rapid reactions of ligand with monomer and dimer, respectively. The rate equations become:

$$\frac{1}{\tau}(\text{ML}) = k_1 + \frac{k_3(K_1/K_2^{1/2})[\text{ML}]}{[\text{M}_2]^{1/2}K_2^{1/2}} + 2k_{-5}[\text{ML}] \quad (17)$$

and

$$\frac{1}{\tau}(\text{M}_2) = k_{-5}\frac{[\text{ML}]^2}{[\text{M}_2]} \quad (18)$$

When the concentration of the free ligand L is very small, it turns out that one may observe reactions of monomer with the half dimer species M. As M_2 and M are in the fast-exchange limit, the rate expressions are:

$$1/\tau(\text{ML}) = (k_6 + k_7)K_2^{1/2}[\text{M}_2]^{1/2} + (k_8 + k_9)[\text{M}_2] \quad (19)$$

and

$$1/\tau(\text{M}_2) = k_6 K_2^{1/2}[\text{ML}]/2[\text{M}_2]^{1/2} \quad (20)$$

This behavior has been found for all phosphine systems and in addition also for the system with L = {4-$(CH_3)_2NC_6H_4$}$_3$As in diene compounds (van Leeuwen et al., 1969a). The term $k_i[M_2]$ was found for triphenylphosphine systems involving π-allyl compounds of palladium and the norbornadiene system of rhodium.

Rather particular rate expressions were found for mixtures of (COD)-RhCl(AsPh$_3$) and [(COD)RhCl]$_2$. They are:

$$1/\tau(ML) = k_1 \qquad (21)$$

and

$$1/\tau(M_2) = \tfrac{1}{2}k_1[ML]/[M_2] \qquad (22)$$

These concentration dependencies indicate a very fast dissociation reaction (1) and a rapid reaction (2). The equilibrium concentration of ligand L is, however, very small (Vrieze et al., 1969). This situation has not been observed in the case of metal–π-allyl systems and will not be further discussed in this chapter.

A relatively simple case exists when only monomer is present in solution with or without added ligand L. Monomer–monomer reactions sometimes appear to occur [Reaction (6)], but in general reactions of monomer with ligand L are dominant, i.e.,

$$1/\tau(ML) = k_3[L] \qquad (23)$$

1. *Reactions Occurring in Solutions of* (η^3-Allyl)PdX(L) *with or without added ligand* L

Addition of small quantities of triphenylarsine or triphenylphosphine to CDCl$_3$ solutions of (η^3-C$_4$H$_7$)PdCl(AsPh$_3$) or (η^3-C$_4$H$_7$)PdCl(PPh$_3$), respectively, causes a very rapid cis–trans isomerization, i.e., a coalescence of the signals of the syn protons and simultaneously also of the signals of the anti-protons (Vrieze et al., 1966, 1967, 1968a, 1969). A similar cis–trans isomerization was observed for the planar isoelectronic compounds (diene)MCl(L) of monovalent rhodium and iridium. In all cases, the interchange was found to be first order in concentration of ligand L. A five-coordinate intermediate for these d^8 complexes seems a reasonable suggestion. Five-coordinate compounds (diene)MCl(L)$_2$ were indeed isolated from suitable solvents.

Similar reactions were also reported for π-allyl amine palladium compounds (Ganis et al., 1969; Maglio et al., 1970; Faller et al., 1969; Faller and Incorvia, 1969; Faller and Mattina, 1972). An additional insight into the exchange mechanism was given by the use of optically active amines, which are readily available.

For example the nmr spectrum at $-60°$C of a solution of (η^3-C$_3$H$_5$)-

PdCl(α-phenylethylamine), which was prepared from the optically active R-α-phenylethylamine, consisted of two ABCDX patterns in one-to-one ratio, which correspond to the two expected diastereoisomers (Faller *et al.*, 1969). At +30°C, the spectrum has changed into two AA′BB′X patterns. On the other hand, if racemic α-phenylethylamine is used, coalescence to only one AA′BB′X spectrum was observed. These observations can only be explained by an intermolecular exchange of the amine ligands. In the temperature range of −60° to +35°C, no syn–anti proton interchange took place, which means that the palladium atom remains bonded to the same face of the allyl group (see also Vrieze and van Leeuwen, 1971).

A more detailed picture was obtained for $(\eta^3\text{-crotyl})\text{PdX}(2\text{-picoline})$ (X = Cl, Br) by Faller and Incorvia (1969) and Faller and Mattina (1972). At −100°C, four isomers could be discerned, namely the endo and exo isomers

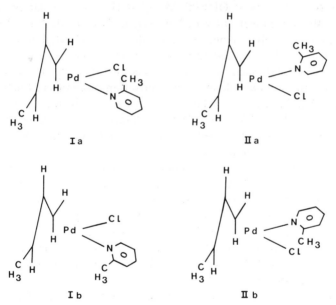

Fig. 25 Structures of the exo and endo isomers of both the cis and trans isomers of $(\eta^3\text{-crotyl})\text{PdCl}(2\text{-picoline})$. (See text for details.)

of both cis and trans isomers. The names "endo" and "exo" are used to indicate whether the methyl group is pointing in the same or in the opposite direction of the central allylic proton (Fig. 25). The cis and trans isomers are the species I and II, respectively. Between −100° and −30°C, the endo isomers are converted into the exo isomers and vice versa; i.e. Ia ↔ Ib and IIa ↔ IIb (Fig. 25). The activation energies are 11.0 and 8.0 kcal/mole, respectively for the chloride compound, while values of 11.3 and 8.3 kcal/mole were derived for the bromide derivative. The interconversion involves clearly a rotation of the 2-picoline ligand about the Pd—N bond. At temperatures between $-35°$ and $+30°C$, interconversion of the cis and trans isomers takes place via a bimolecular ligand exchange reaction, although a ligand dissociation reaction, such as described in the introduction of Section III,C, i.e., reaction (1), is also important. At temperatures above $+30°C$, syn–anti proton exchange was observed, which will be discussed below.

Several mechanisms may explain the cis–trans isomerization. The necessary rearrangement may take place by means of an angular twist mechanism (π-rotation) (Vrieze et al., 1966) or by means of a number of Berry pseudorotations (Figs. 8 and 10). The last mechanism was favored by Shapley and Osborn (1970) and by Rice and Osborn (1971) to rationalize the intramolecular interchange of the ends of the diene ligand in five-coordinate compounds of the type $(COD)(PPhMe_2)_2Ir(CH_3)$. An angular twist mechanism for this rearrangement was, however, not excluded. A third possible mechanism may involve an ionic intermediate, e.g. $[(\eta^3-C_4H_7)Pd(PR_3)_2]^+Cl^-$, which for $L = PR_3$ is indeed formed in solvents such as dichloromethane (Vrieze et al., 1968a). Such species are formed relatively slowly and therefore do not play an essential role in the isomerization mechanism. Furthermore, they can never be used to explain the isomerization in the five-coordinate compounds $(COD)(PPhMe_2)_2Ir(CH_3)$, which seems a good model compound for the intermediates proposed in the isomerization reactions of four-coordinate planar d^8 complexes.

In this respect, it is of interest to point out that cis–trans isomerization in the case of $(NOR)RhCl(AsPh_3)$ can also be achieved by the addition of chloride ions, while it could be proved that no free $AsPh_3$ was formed in the temperature range studied (Vrieze et al., 1968d). In general, chloride ions were found to be ineffective to give cis–trans isomerization.

For example cis–trans isomerization was not observed in the nmr time scale for $(\eta^3-C_4H_7)PdCl(L)$ with $L = AsPh_3$ and PPh_3, $(COD)RhCl(PPh_3)$ and $(COD)IrCl(PPh_3)$, when chloride ions were added.

Cis–trans isomerization occurs more easily in the presence of species such as the dimer itself. For example, isomerization was observed for mixtures of $(\eta^3-C_4H_7)PdCl(PPh_3)$ and $[(\eta^3-C_4H_7)PdCl]_2$ and for mixtures of $(NOR)RhCl(PPh_3)$ and $[(NOR)RhCl]_2$. In both cases (Vrieze et al., 1968a,d), the rate of the cis–trans isomerization is strictly proportional to the dimer concentration (between $-20°$ and $+20°C$). No exchange at all of triphenylphosphine between monomer and dimer was measured, although probably chloride exchange did occur.

Recently it was found that cis–trans isomerization could also be catalyzed by ligands such as pyridine and triphenylarsine in the case of π-allyl palladium chelate compounds. Examples are $(\eta^3$-allyl$)Pd($oxinate$)$, $(\eta^3$-allyl$)Pd$-

(picolinate) (Ban et al., 1972) and, for comparison, also (COD)Rh(DBSM), which was reported by Heitner and Lippard, 1972). In the latter case, pseudo-rotational movements were mentioned as a possible mechanism. Partial loosening of the anionic bidentate ligand, followed by temporary pyridine substitution and subsequent rearrangements, were proposed by Ban et al. (1972) to rationalize the behavior of the palladium compounds.

A further complicating factor is that in addition to cis–trans isomerization also ligand exchange reactions may occur with retention of configuration. For example, addition of $AsPh_3$ to $(\eta^3\text{-}C_4H_7)Pd(OAc)(AsPh_3)$ leads to cis–trans isomerization. However, for mixtures of this compound and $(\eta^3\text{-}C_4H_7)PdCl(PPh_3)$ (van Leeuwen and Praat, 1970c), it was demonstrated that the two $ABCDX_3$ patterns coalesced to one $ABCDX_3$ pattern instead of to one $A_2B_2X_3$ spectrum. Similarly, acetate ligand rapidly exchanges between $(\eta^3\text{-}C_4H_7)Pd(OAc)(PPh_3)$ and $PhHg(OAc)$ without change in the $ABCDX_3$ spectrum of the allylic group (van Leeuwen and Praat, 1970c).

Reactions with retention of configuration may be faster or slower than cis–trans isomerization. For example, isomerization is faster than retention of configuration for L_2PtCl_2 (Haake and Pfeiffer, 1970), while the opposite seems to be the case for the above mentioned acetate compounds. If both cis–trans isomerization and retention of configuration reactions are caused by the same ligand, they may be independent of each other; i.e., they may proceed by different intermediates. The independence of both processes was also observed for five-coordinate complexes (Rice and Osborn, 1971). On the other hand, there is reason to assume that there are cases in which both processes take pathways via a common intermediate.

From the available evidence, it may be concluded that we are still far away from a clear systematic picture of the behavior of the above type of compounds. The picture is, however, a very complicated one, as it seems now clear that the mechanism of the cis–trans isomerization may differ from one system to the other, depending on the type of ligands attached to the metal atom and the type of ligand used as attacking agent.

2. *Reactions Occurring in Mixtures of* $(\eta^3\text{-}C_4H_7)PdCl(AsPh_3)$ *and* $[(\eta^3\text{-}C_4H_7)PdCl]_2$

A relatively simple situation exists for systems containing mixtures of $(\eta^3\text{-}C_4H_7)PdCl(AsPh_3)$ and $[(\eta^3\text{-}C_4H_7)PdCl]_2$, and for comparison's sake, also for mixtures of $(NOR)RhCl(AsPh_3)$ and $[(NOR)RhCl]_2$ and of $(COD)IrCl(AsPh_3)$ and $[(COD)IrCl]_2$ (Vrieze et al., 1967; Volger et al., 1968).

Reactions (1–5) dominate the reaction behavior. The reaction of dimer is second order in ligand L, which shows that the M_2L intermediate is in the fast exchange with M_2, and in the slow exchange ML; i.e., step (4) is faster than step (3) with respect to the chemical shift differences between M_2L and M_2 and ML, respectively. In the actual situation, it may well be that $k_5[L] > k_{-4}$, so that one would measure first-order kinetics if one used "normal" kinetic measurements.

Monomer and dimer react with ligand L formed by the dissociation reaction (1). Reaction (1) is, however, unimportant relative to the ligand exchange reaction (3), which is extremely fast. Similarly, the term $k_{-5}[ML]$ is relatively unimportant.

3. *Reactions Occurring in Mixtures of* $(\eta^3\text{-}C_4H_7)PdCl(PPh_3)$ *and* $[(\eta^3\text{-}C_4H_7)PdCl]_2$.

As mentioned previously, below $-20°C$, reactions occur between monomer and dimer species without exchange of ligand L. Illustrations of this are the systems containing mixtures of $(\eta^3\text{-}C_4H_7)PdCl(PPh_3)$ and $[(\eta^3\text{-}C_4H_7)PdCl]_2$ and of $(NOR)RhCl(PPh_3)$ and $[(NOR)RhCl]_2$ (Vrieze et al., 1968a,d). Reaction (9) is first order in monomer and dimer. This reaction was not found for the analogous cyclooctadiene system (Volger et al., 1968). The intermediate probably has a trinuclear composition M_3L. This association compound seems to be stabilized by a more positive charge on the metal atom of the monomer. Indeed, the tendency to go from four to five coordination increases in the order COD < NOR for diene compounds, as was also found by Heitner and Lippard (1972).

Between $-20°C$ and $+20°C$, all phosphine systems—i.e., $(\eta^3\text{-}C_4H_7)PdCl(PPh_3)$, $(NOR)RhCl(PPh_3)$, and $(COD)RhCl(PPh_3)$—in the presence of the parent dimer compounds, showed the same kinetic behavior, namely a fast dissociation of dimer [Reaction (2)] and an exchange reaction of triphenylphosphine occurring in a dinuclear intermediate M_2L of reaction (6).

The expected and measured concentration dependencies are given by

$$1/\tau(ML) = k_6 K_2^{1/2}[M_2]^{1/2} \tag{24}$$

and Eq. (20). These reactions were not expected in the first instance, but it is very understandable that they do occur, because the Pd—P bond is very strong, and rarely is any free ligand formed to give rise to the reactions observed for the amine and arsine systems. The intermediate, M_2L, of reaction (6) probably has a dimer configuration with the phosphine bonded to one of the metal atoms. The nearness of the two metal atoms in M_2L probably provides a reatively low-energy pathway for the phosphine ligand to jump from one metal atom to the other.

Van Leeuwen et al. (1969b) were able to unravel to a satisfactory degree the influence of the ligand L on the cis–trans isomerization by studying reactions of $(\eta^3\text{-}C_4H_7)PdCl(L)$ with $[(COD)RhCl]_2$. It was found that the reaction increased in the order: $PEt_2Ph < PEtPh_2 < PPh_3 < P(C_6H_4Cl)_3$, which is the order of decreasing donor capacity. From the concentration dependence, it seems very likely that the reaction occurs between $(\eta^3\text{-}C_4H_7)\text{-}PdCl(L)$ and the half-dimer compound (COD)RhCl. In the intermediate species, M_2L, the palladium atom was assumed to be five-coordinated by the bidentate allyl group, the ligand L and two Cl atoms. The mechanism of the cis–trans isomerization may involve a π-rotation or a number of Berry pseudorotations. In any case, it is clear that the reaction is favored if the ligand L is more electron attracting, which is analogous to the reaction of monomer with the complete dimer compound. Stabilization of a five-coordinate intermediate probably plays a major role.

4. *Reactions Occurring in Mixtures of* $(\eta^3\text{-allyl})PdCl(amine)$ *and* $[(\eta^3\text{-allyl})PdCl]_2$

When $(\eta^3\text{-}C_4H_7)PdCl(\alpha\text{-phenylethylamine})$ is dissolved in chloroform, some weak resonances due to the dimer are observed also. Obviously, the dissociation reaction (1) lies so much to the right, that dimer is formed from two half-dimer species in reaction (2). Cis–trans isomerization is accompanied by reaction of dimer with ligand, as the dimer signals coalesce with the monomer signals.

Faller and Incorvia (1969) and Faller and Mattina (1972) investigated by means of nuclear magnetic double resonance the pathway of amine exchange. They found that for mixtures of $(\eta^3\text{-crotyl})PdCl(2\text{-picoline})$ and $[(\eta^3\text{-crotyl})PdCl]_2$, the cis–trans isomerization proceeds mainly via dimer and less so via the direct ligand exchange reaction (3).

$$\begin{array}{c} A \\ | \\ B \end{array}\!\!\!\!>\!\!\text{ML} \underset{c}{\rightleftarrows} \begin{array}{c} B \\ | \\ A \end{array}\!\!\!\!>\!\!\text{ML}$$

$$\searrow \qquad \swarrow$$

$$M_2 + 2L$$

This behavior is in sharp contrast with that of the arsine systems, in which reaction (3) is much more important. Concentration-dependence studies were difficult to carry out, but at high concentrations of dimer or complex bimolecular processes appear to be dominant.

The above indicates that reactions (1), (2), (4), (5), and probably also (6) are important. In reaction (6) the species M_2L is produced, which is also the proposed intermediate in the reaction (4) and (5).

From the available information it seems reasonable to conclude that for both the π-allylpalladium and the diene metal systems the tendency for the ligand L in (η^3-allyl)PdCl(L) to dissociate decreases in the order amine > arsine > phosphine. The strongly varying kinetic behavior is due mainly to the amount of free ligand present.

D. Intermolecular Reactions Involving Syn–Anti Proton Exchange

Interchange of syn and anti protons, which is caused by intermolecular reactions, has been studied extensively for a large variety of π-allylpalladium complexes. Some relevant examples will be given in order to indicate the present state of the art.

1. π-Allyl and π-Methallyl Compounds

Syn–anti exchange has been observed to be promoted by ligands such as group V donor ligands (Vrieze et al., 1969), carbon monoxide (Volger et al., 1970), dienes (Hughes and Powell, 1972; Sokolov et al., 1971), and even dimer species.

In Fig. 14, a drawing has been given of the π–σ–π movement, which explains the syn–anti proton exchange. In the example given in Fig. 14 the end of the allyl group, which is trans to the ligand PPhMe$_2$ with the largest trans effect, leaves the palladium atom so that protons 3 and 4 cis to the phosphine ligand may interchange in the short-lived σ-allyl intermediate. Protons 1 and 2, of course, do not exchange positions, as is indeed observed with nmr. This type of interchange is most common for the asymmetric compounds (η^3-C$_4$H$_7$)PdCl(PPh$_3$) (Vrieze et al., 1966, 1968a,b; Powell and Shaw, 1967; Cotton et al., 1967; Statton and Ramey, 1966; Ramey and Statton, 1966). The rate of the syn–anti interchange of protons 3 and 4 is proportional to the concentration of the dimer [(η^3-C$_4$H$_7$)PdCl]$_2$ in the temperature range of about $-40°$ to $-10°$C (Vrieze et al., 1968b). The phosphine ligand remained coordinated to the same metal atom, as could be deduced from the phosphor coupling on both protons 1 and 2 and from the continuing sharpness of these signals in the mentioned temperature range. The proposed intermediate involves a trinuclear intermediate, M$_3$L. The important part of this intermediate may well consist of a palladium atom coordinated by a σ-allyl group, a phosphine ligand, and two halide ligands which are all held together as a bidentate by the dimer molecule. Van Leeuwen and Praat (1970b) indicated that the first step may involve an attachment of the electron-accepting dimer to the chloride atom of the monomer (η^3-C$_4$H$_7$)PdCl(PPh$_3$), whereby the difference in trans effect between the ligands opposite to the allyl group is increased. The second step may then be the coordination of a chloride atom of the dimer to the palladium atom of the monomer, while

11. Fluxional Allyl Complexes

simultaneously the allyl group is converted from a π-allyl to a σ-allyl type of bonding.

Of great interest is that a formally similar process appears to occur in a monomolecular reaction, which was observed for compounds such as $(\eta^3\text{-}C_4H_7)Pd(OAc)PPh_3$ (van Leeuwen and Praat, 1970b) and $(\eta^3\text{-}C_4H_7)Pd\text{-}(xanthate)PPhMe_2$ (Powell and Chan, 1972). At about $-30°$ and $-80°C$, the syn and anti protons cis to the phosphine ligand started to interchange. A four-coordinate short-lived σ-allyl intermediate in which the xanthate acts as a bidentate ligand was proposed. An analogous monomolecular reaction was reported by Guthrie et al. (1971) for the conversion of the syn and anti isomers of the product from $[(\eta^3\text{-2-chloroallyl})PdCl]_2$ and isoprene (Fig. 26). At 73°C, the methyl signals of the isomers A and B coalesced with a $\Delta G^{\ddagger}$ of 17.3 ± 0.2 kcal/mole. The proposed intermediate C, which involves a σ-allyl species at the most substituted carbon atom, has again a four-coordinate structure with the olefin function of one of the substituents temporarily acting as the fourth ligand. These monomolecular reactions are therefore very similar to the bimolecular reactions of monomer with dimer, which were discussed previously.

Reactions with ligands such as dimethylsulfoxide, triphenylarsine, and triphenylphosphine also promote syn–anti proton exchange, but at much higher temperatures ($+30°$ to $+40°C$) than in the case of the above mentioned examples (Chien and Dehm, 1961; Statton and Ramey, 1966; Vrieze

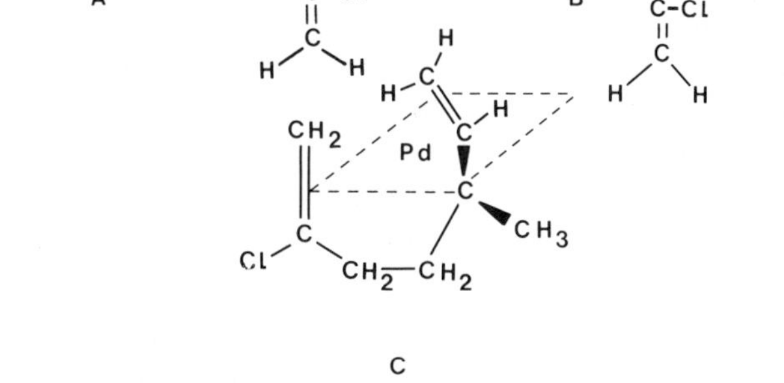

Fig. 26 The structures of the isomers A and B of the product of $[(2\text{-chloroallyl})PdCl]_2$ and isoprene and the structure C of the proposed intermediate for the syn–anti interconversion.

et al., 1966, 1967, 1968a; Lippard and Morehouse, 1972a,b; Tibbetts and Brown 1970). The reaction is first order in added ligand and very likely involves four-coordinate short-lived σ-allyl intermediates of the type $(\eta^1\text{-}C_4H_7) \times PdCl(L)_2, [(\eta^1\text{-}C_4H_7)Pd(diphos)(pyridine)]PF_6$ (Tibbetts and Brown, 1970), or $[(\eta^1\text{-}C_4H_7)Pt(PPh_3)_3]Cl$ (Volger and Vrieze, 1967; Vrieze and Volger, 1967). The added ligand occupies the fourth coordination position, but it does not increase the difference in trans effect of the ligand opposite the allyl group, as does the dimer. As a result the π–σ–π reaction occurs at higher temperatures than in the case of dimer promoted π–σ–π reactions.

A further point of interest is that in the case of the monomer–dimer reactions, the syn–anti exchange occurs at much lower temperatures than the cis–trans isomerization, while the opposite was observed for the monomer-ligand exchange reactions, for which cis–trans isomerization was already found to take place from about $-80°$ to $-40°C$ onward. This reversal in trends will not be discussed any further here, as little is as yet known about the actual mechanism of the cis–trans isomerization. However, stabilization of the four-coordinate σ-allyl species in the case of the π–σ–π reaction and the stabilization of the five-coordinate π-allyl intermediate for the cis–trans isomerization should lie at the center of this dicussion.

It is rather enigmatic that the amine systems behave in several respects rather differently from the arsine and phosphine systems. First, the syn–anti proton exchange of, for example, $(\eta^3\text{-}C_3H_5)PdCl(R\text{-}\alpha\text{-phenylethylamine})$ (Faller et al., 1969, 1971; Ganis et al., 1969), starts at around the same temperatures as the arsine and phosphine systems, while higher temperatures would be expected, as the amine ligands have lower trans effects than arsines and phosphines. Second, it was found in this compound and also for example in $(\eta^3\text{-crotyl})PdCl(2\text{-picoline})$ (Faller and Mattina, 1972) and $(\eta^3\text{-1-acetyl-2-methallyl})PdCl(S\text{-}\alpha\text{-phenylethylamine})$ (Faller et al., 1971) that the π–σ–π reactions appeared to be monomolecular, as neither excess dimer nor excess amine had any influence on the rates of the processes, which is in contrast to the cis–trans isomerizations.

There seems little point in discussing these observations in any depth, as investigations are being carried out to unravel the reactions occurring in these complicated systems (Faller and Mattina, 1972). One comment the author would like to make is that one may propose again an intermediate of the type $(\eta^1\text{-allyl})PdCl(\text{amine})_2$. In order to rationalize the monomolecular nature of the reaction, one may then assume as a possibility that there is a rapid reaction:

$$(\eta^3\text{-Allyl})PdCl(\text{amine}) + \text{amine} \rightleftharpoons (\eta^3\text{-allyl})PdCl(\text{amine})_2 \qquad (25)$$

which is followed by a slow rate determining π–σ–π reaction:

$$(\eta^3\text{-Allyl})PdCl(\text{amine})_2 \rightleftharpoons (\eta^1\text{-allyl})PdCl(\text{amine})_2 \qquad (26)$$

The amine may be formed by dissociation of (η^3-allyl)PdCl(amine) or may be added. In the case of the reactions of monomer (η^3-allyl)PdCl(L) with L = AsPh$_3$ or PPh$_3$, the first reaction may be slow, while the second reaction is fast.

2. Substituted Metal Allyl Compounds

Palladium–π-allyl compounds with substituents on the allyl group have received much attention, as one may study the influence of the substituents on the relative stabilities of the π- and σ-allylic species. On the other hand, substituted allyl compounds may also be used to investigate the detailed mechanism of the syn–anti group exchanges.

Powell and Chan (1972) and Hughes and Powell (1972) reported that in general the relative stability of the σ-allyl intermediate appeared to decrease in the order: 2-chloroallyl > allyl > 2-methylallyl > 2-ter-butylallyl > 2-neopentylallyl. It was suggested that electron-withdrawing groups enhance the stability, while large bulky groups, because of steric interactions with the cis ligands, decrease the stability of the σ-allyl intermediate relative to the initial π-allyl state.

Faller et al. (1971) carried out precise line-width measurements on the complex (η^3-C$_3$H$_5$)PdCl(S-α-phenylethylamine) above $+35°$C and showed conclusively that the syn–anti exchange proceeded via a π–σ–π movement and not via a flip mechanism.

More detailed mechanistic information about the π–σ–π movement was obtained on compounds such as (η^3-1-acetyl-2-methallyl)PdCl(amine) in which the amine may be pyridine or an optically active amine, e.g., S-α-phenylethylamine (Faller et al., 1971; Faller and Thomsen, 1969; De Candia et al., 1968). In the case of the pyridine compound, two geometric isomers were formed, i.e., with the acetyl group R, syn, or trans with respect to the methyl substituent on the 2-position. As the temperature was raised, three independent processes were observed. The two processes that were observed first involved syn–anti proton exchange at the unsubstituted end of the allyl group in both the syn and anti acetyl isomer. The third process is the isomerization of the syn acetyl isomer to the anti acetyl isomer and vice versa (Faller et al., 1971). When the optically active amine is used, each resonance of the pyridine system is split into two signals owing to the formation of epimers. By line-broadening and nuclear magnetic double resonance experiments, it can be shown conclusively that the epimerization (i.e., A ↔ B and C ↔ D in Fig. 27 if, for R' = H, syn H interchanges with anti H) occurs via a σ-bonded intermediate at the unsubstituted end of the allyl group, while isomerization (i.e., A ↔ D and B ↔ C in Fig. 27 if R' = H) takes place at the acetyl-substituted end of the allyl group via a σ-bonded intermediate.

Epimerization can be prevented by using (η^3-1-acetyl-2,3-dimethylallyl)-PdCl(S-α-phenylethylamine). When the crystalline compound is dissolved at $-20°C$ only one epimer of the anti acetyl isomer is formed with the 3-methyl group in the syn position, which is sterically much more favorable (Fig. 27). At higher temperature, this epimer A is converted to only one of the epimers D of the syn acetyl isomer. If, on the other hand, the optically active amine is

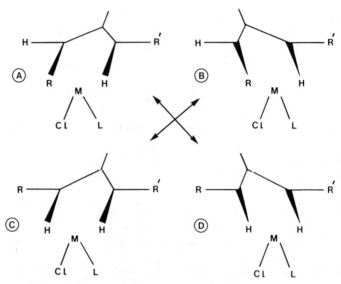

Fig. 27 The isomerization of (η^3-allyl)PdCl(S-α-phenylethylamine). (See text for details.)

added to the parent dimer the species A, B, C, and D are present, but epimerization can of course again not occur (Faller and Tully, 1972).

Steric interactions may further be used to prepare specific π-allyl compounds, as was discussed extensively by Faller *et al.* (1971) and Lukas *et al.* (1969, 1970). This subject will not be further treated here.

IV. Fluxional Metal–σ-Allyl and Ionic Metal–Allyl Compounds

In Sections II,B, C, E, and F, the bonding and possible movements of the allyl group in metal–σ-allyl and ionic metal–allyl compounds were discussed in short. The emphasis of the chapter has been mainly on metal–π-allyl compounds, which is understandable, as much information has become available. Nonetheless, the subject would not be complete without a discussion of metal–σ-allyl and ionic metal–allyl compounds, as they may be good models of intermediates occurring in reactions of metal–π-allyl compounds. In Sections IV,A and IV,B some relevant examples of these compounds will be discussed.

A. Metal–σ-Allyl Compounds

Only a few transition metal–σ-allyl compounds are known that contain at ambient temperatures fluxional allyl groups. Powell and Chan (1972) reported the four coordinate compounds (η^1-allyl)Pd($O_2CC_5H_4N$)($PPhMe_2$) and (η^1-2-CH_3-C_3H_4)Pd(S_2CNMe_2)($PPhMe_2$), in which the picolinate and carbamate ligands behave as bidentates. At room temperature, a rapid σ–π–σ movement takes place with probably a four-coordinate π-allyl species, in which the picolinate and carbamate groups temporarily act as monodentate ligands. This situation, for example, is therefore the reverse of the analogous acetate and xanthate compounds.

Recently, Neese and Bürger (1971) isolated [(CH_3)$_2$N]$_3$Ti(η^1-allyl) (allyl = C_3H_5 and 2-CH_3-C_3H_4), [(C_2H_5)$_2$N]$_3$Ti(η^1-2-CH_3-C_3H_4) and [(C_2H_5)N]$_2$Ti(η^1-2-CH_3-C_3H_4)$_2$, which all showed AX_4 type of nmr spectra. The nmr spectra did not show changes in the temperature range of $-65°$ to $+80°C$. From the carbon–carbon double bond-stretching frequencies, which lie around 1605 cm^{-1}, it was concluded that the compounds contain σ-allyl groups. Rapid σ–π–σ movements via, e.g., five-coordinate π-allyl intermediates, seem likely.

In the case of the four-coordinate compounds Si(η^1-C_3H_5)$_4$ and Ge(η^1-C_3H_5)$_4$ the allyl groups are nonfluxional in the temperature range studied (Fishwick and Wallbridge, 1970). Some reasons may be the strong bond between the allyl groups and the electronegative central atoms and the nonavailability of energetically favorable d orbitals, which could stabilize, e.g., five-coordinate π-allyl intermediates. In the case of mixtures of Sn(η^1-C_3H_5)$_4$ and $SnCl_4$ intermolecular exchange of allyl groups and the formation of mixed allyltin halide complexes was observed (Fishwick and Wallbridge, 1970).

Intramolecular rearrangements were observed for B(η^1-C_3H_5)$_3$ by Bogdanov et al. (1967, 1968). The spectrum changed from an $ABCX_2$ pattern at $-65°C$ to an AX_4 spectrum at $+135°C$. The acceptor properties of trivalent boron, and the formation of a four-coordinate intermediate probably facilitate these 1,3-metallotropic shifts.

Intermolecular exchange of allyl groups appear to take place, although rather slowly, in Hg(η^1-C_3H_5)$_2$ (Zieger and Roberts, 1969b). Allyl exchange was also observed for allylmercuric halides (Kitching et al., 1972). A very large increase in the exchange rate occurred on addition of mercuric halides.

Both intra- and intermolecular movements of the allyl groups seem to take place, but no definite mechanism could be derived from the available information. The allyl compounds $M(\eta^1-C_3H_5)_2$ of the more electropositive metals Cd (Thiele and Köhler, 1967), Zn (Thiele and Zdunneck, 1965), and Mg (Zieger and Roberts, 1969a), and also $Zn(\eta^1-2-CH_3-C_3H_4)_2$ (Thiele et al., 1967) showed even at very low temperatures a very rapid intermolecular and possibly also an intramolecular 1,3-metallotropic exchange of the allyl groups. From the infrared and Raman spectra, it may be tentatively deduced that the allyl groups are σ-bonded to the metal atom in the initial state, although the carbon–carbon double bond stretching frequency of $Mg(\eta^1-C_3H_5)_2$ is between those of ionic metal–allyl compounds and metal–σ-allyl compounds (see Table I). A certain degree of bond polarity is implied. In the case of (dimethylallyl)MgBr (Whitesides et al., 1962a,b), the methyl signals could be distinguished at $-40°C$. The interpretation of the experimental data involved a rapid equilibrium at room temperature between α,α-dimethylallyl-MgBr and γ,γ-dimethylallyl-MgBr, which is the major species in solution. An activation energy of 7 ± 3 kcal/mole was derived for the interchange of the cis and trans methyl groups of γ,γ-dimethylallyl-MgBr.

B. Ionic Metal–Allyl Compounds

Johnson et al. (1961) and West et al. (1968) reported the nmr spectra of $Li(C_3H_5)$. West et al. (1968) found at $-87°C$ an A_2B_2C spectrum which was converted to an AX_4 spectrum at $+37°C$. The activation frequency and frequency factor are 10.5 ± 2.0 kcal/mole and $10^{12.5 \pm 1.5}$ sec^{-1}, respectively. The equivalence at low temperature on the one hand of the syn protons and on the other hand of the anti protons strongly indicates a symmetric allylic species with a delocalized carbanion type of bond, which agrees with the ultraviolet spectrum and the carbon–carbon stretching frequencies (1520–1540 cm^{-1}), which are analogous to that of $Na(C_3H_5)$ (Lanpher, 1957). The exact nature of the bonding is still uncertain, as $Li(C_3H_5)$ is aggregated in tetrahydrofuran and diethyl ether (West et al., 1968), which is not unexpected in view of the reported aggregation for, e.g., $LiCH_3$ (McKeever et al., 1969; McKeever and Waack, 1969).

On the basis of the nmr pattern and the high upfield shifts of the phenyl resonances Freedman et al. (1967) suggested for (1,3-diphenylallyl)lithium a delocalized structure with a large negative charge delocalized into the phenyl rings.

Rotational barriers about the carbon–carbon bonds were measured by Sandel et al. (1968) for the phenylallyllithium, -sodium, and -potassium compounds. The $\Delta G^\ddagger$ values for the rotation at the unsubstituted end are

17.0, >17.8, and >20.1 kcal/mole, respectively, for the Li, Na, and K compounds (in THF). Rotation at the phenyl-substituted end occurred at much lower temperatures ($-30°$ to $-15°C$) with lower energy barriers, i.e., 11.9, 12.9, and 12.9 kcal/mole respectively. Interchange via a metal–σ-allyl intermediate was not excluded, but seemed unlikely. The lower energy barriers for rotation at the substituted end of the allyl group may be due to delocalization of negative charge into the phenyl ring. The rotational barriers clearly increase with the electropositivity of the metal, because more negative charge is situated on the allyl group.

V. Concluding Remarks

It has been attempted to give in this chapter an impression of the development in the area of metal–allyl compounds, with particular emphasis on the bonding aspects and fluxional nature of these complexes. In the case of metal–π-allyl complexes, there is conclusive evidence, that the syn–anti exchange, which has been extensively studied on palladium compounds, can be satisfactorily explained by π–σ–π movements. Less well understood is the origin of the syn–syn, anti–anti exchange, which may be the result of both monomolecular and bimolecular processes. Again, the palladium systems were the most thoroughly investigated, but there are as yet no definitive conclusions about the nature of the mechanisms occurring. Indeed, it seems that the type of mechanism alternates from one system to the other. It may be further concluded that there is still no clear picture of the factors that influence the movements occurring in metal–π-allyl complexes. Obviously, more work is needed to elucidate the various processes in well chosen systems.

Comparatively little is known about the fluxional behavior of metal–σ-allyl and ionic metal–allyl compounds. More attention is needed, particularly in the area of the ionic metal–allyl compounds, with respect to the structure and bonding aspects.

The author wants to point out that sometimes great caution is needed in the interpretation of the kinetic results in particular of the translation of concentration dependencies measured with nmr into reaction sequences. The reader is referred to a few publications (Vrieze *et al.*, 1968a, 1969; Vrieze and van Leeuwen, 1971) in which several pitfalls are mentioned.

REFERENCES

Alexander, C. W., Jackson, W. R., and Spratt, R. (1970). *J. Amer. Chem. Soc.* **92**, 4990.
Andrews, D. C., and Davidson, G. (1972). *J. Chem. Soc., Dalton Trans.* p. 1381.

Ashcroft, S. J., and Mortimer, C. T. (1971). *J. Chem. Soc., A* p. 781.
Ban, E., Chan, A., and Powell, J. (1972). *J. Organometal. Chem.* **34**, 405.
Basolo, F., and Pearson, R. G. (1967). "Mechanisms of Inorganic Reactions," 2nd ed. Wiley, New York.
Becconsall, J. K., and O'Brien, S. (1966a). *Chem. Commun.* p. 302.
Becconsall, J. K., and O'Brien, S. (1966b). *Chem. Commun.* p. 720.
Becconsall, J. K., and O'Brien, S. (1967a). *J. Organometal. Chem.* **9**, P27.
Becconsall, J. K., and O'Brien, S. (1967b) *J. Chem. Soc., A* p. 423.
Bogdanov, V. S., Pozdnev, V. F., Lagodzinskaya, G. V., and Mikhailov, B. M. (1967). *Teor. Eksp. Khim.* **3**, 488.
Bogdanov, V. S., Budnov, Yu. N., Frolov, S. I., and Mikhailov, B. M. (1968). *Izv. Akad Nauk SSSR* p. 307.
Bönnemann, H., Bogdanovic, B., and Wilke, G. (1967). *Angew. Chem.* **79**, 817.
Brown, D. A., and Owens, A. (1971). *Inorg. Chim. Acta* **5**, 675.
Chien, J. C. W., and Dehm, H. C. (1961). *Chem. Ind. (London)* p. 745.
Chini, P., and Martinengo, S. (1967). *Inorg. Chem.* **6**, 837.
Churchill, M. R., and Bezman, S. A. (1971). *J. Organometal. Chem.* **31**, C43.
Churchill, M. R., and Bezman, S. A. (1972). *Inorg. Chem.* **11**, 2243.
Churchill, M. R., and Mason, R. (1964). *Nature (London)* **204**, 777.
Clarke, H. L., and Fitzpatrick, N. J. (1972). *J. Organometal. Chem.* **40**, 379.
Cooke, M., Goodfellow, R. J., and Green, M. (1971). *J. Chem. Soc., A* p. 16.
Cotton, F. A. (1968). *J. Amer. Chem. Soc.* **90**, 6230.
Cotton, F. A., and LaPrade, M. D. (1968). *J. Amer. Chem. Soc.* **90**, 5418.
Cotton, F. A., and Marks, T. J. (1969). *J. Amer. Chem. Soc.* **91**, 1339.
Cotton, F. A., Faller, J. W., and Musco, A. (1967). *Inorg. Chem.* **6**, 179.
Cousins, M., and Green, M. L. H. (1963). *J. Chem. Soc., London* p. 889.
Davidson, G. (1972). *Organometal. Chem. Rev. Sect. A* **8**, 303.
Davison, A., and Rode, W. C. (1967). *Inorg. Chem.* **6**, 2124.
De Candia, F., Maglio, G., Musco, A., and Paiaro, G. (1968). *Inorg. Chim. Acta* **2**, 233.
Dehm, H. C., and Chien, J. C. W. (1960). *J. Amer. Chem. Soc.* **82**, 4429.
De Wolfe, R. H., and Young, W. G. (1956). *Chem. Rev.* **56**, 753.
Emerson, G. F., and Pettit, R. (1962). *J. Amer. Chem. Soc.* **84**, 4591.
Faller, J. W., and Anderson, A. S. (1970). *J. Amer. Chem. Soc.* **92**, 5852.
Faller, J. W., and Incorvia, M. J. (1968). *Inorg. Chem.* **7**, 840.
Faller, J. W., and Incorvia, M. J. (1969). *J. Organometal. Chem.* **19**, P13.
Faller, J. W., and Jakubowski, A. (1971). *J. Organometal. Chem.* **31**, C75.
Faller, J. W., and Thomsen, M. E. (1969). *J. Amer. Chem. Soc.* **91**, 6871.
Faller, J. W., and Mattina, M. J. (1972). *Inorg. Chem.* **11**, 1296.
Faller, J. W., and Tully, M. T. (1972). *J. Amer. Chem. Soc.* **94**, 2676.
Faller, J. W., Incorvia, M. J., and Thomsen, M. E. (1969). *J. Amer. Chem. Soc.* **91**, 518.
Faller, J. W., Thomsen, M. E., and Mattina, M. J. (1971). *J. Amer. Chem. Soc.* **93**, 2642.
Faller, J. W., Tully, M. T., and Laffey, K. J. (1972). *J. Organometal. Chem.* **37**, 193.
Fedorov, L. A. (1970). *Russ. Chem. Rev.* **39**, 655.
Fishwick, M., and Wallbridge, M. G. H. (1970). *J. Organometal. Chem.* **25**, 69.
Fogleman, W. W., Cusachs, L. C., and Jonassen, H. B. (1969). *Chem. Phys. Lett.* **3**, 52.
Freedman, H. H., Sandel, V. R., and Thill, B. P. (1967). *J. Amer. Chem. Soc.* **89**, 1762.
Ganis, P., Maglio, G., Musco, A., and Segré, A. L. (1969). *Inorg. Chim. Acta* **3**, 266.
Green, M. L. H., and Stear, A. N. (1964). *J. Organometal. Chem.* **1**, 230.
Guthrie, D. J. S., Spratt, R., and Nelson, S. M. (1971). *Chem. Commun.* p. 935.
Haake, P., and Pfeiffer, R. M. (1970). *J. Amer. Chem. Soc.* **92**, 4996.
Hartley, F. R. (1970). *J. Organometal. Chem.* **21**, 227.

11. Fluxional Allyl Complexes 485

Hayter, R. G. (1968). *J. Organometal. Chem.* **13**, P1.
Heck, R. F., and Boss, C. R. (1964). *J. Amer. Chem. Soc.* **86**, 2580.
Heck, R. F., and Breslow, D. S. (1960). *J. Amer. Chem. Soc.* **82**, 750.
Heck, R. F., Chien, J. C. W., and Breslow, D. S. (1961). *Chem. Ind. (London)* p. 986.
Heitner, H. I., and Lippard, S. J. (1972). *Inorg. Chem.* **11**, 1447.
Hewitt, T. G., and de Boer, J. J. (1968). *Chem. Commun.* p. 1413.
Hillier, I. H., and Canadine, R. M. (1969). *Discuss. Faraday Soc.* **47**, 27.
Hoffmann, R. A., and Forsén, S. (1966). In "Progress in NMR Spectroscopy" (J. W. Emsley, J. Feeney, and L. H. Sutcliffe, eds.), Vol. 1, p. 173. Pergamon, Oxford.
Holloway, C. E., Kelley, J. D., and Stiddard, M. H. B. (1969). *J. Chem. Soc., A* p. 931.
Hughes, R. P., and Powell, J. (1972). *J. Amer. Chem. Soc.* **94**, 7723.
Impastato, F. J., and Ihrman, K. G. (1961). *J. Amer. Chem. Soc.* **83**, 3726.
Johnson, C. S., Jr., Weiner, M. A., Waugh, J. S., and Seyferth, D. (1961). *J. Amer. Chem. Soc.* **83**, 1306.
Kaesz, H. D., King, R. B., and Stone, F. G. A. (1960). *Z. Naturforsch. B* **15**, 682.
Keim, W. (1963). Ph.D. thesis, Max Planck Institut für Kohlenforschung.
Kettle, S. F. A., and Mason, R. (1966). *J. Organometal. Chem.* **5**, 573.
Kharasch, M. S., and Reinmuth, O. (1954). In "Grignard Reactions of Nonmetallic Substances" (Wendell M. Latimer, ed.), Chapter XVII, p. 1133. Prentice-Hall, Englewood Cliffs, New Jersey.
King, R. B. (1966). *Inorg. Chem.* **5**, 2242.
King, R. B., and Fronzaglia, A. (1966). *J. Amer. Chem. Soc.* **88**, 709.
Kitching, W., Bullpitt, M. L., Sleezer, P. D., Winstein, S., and Young, W. G. (1972). *J. Organometal. Chem.* **34**, 233.
Kuwata, K. (1960). *Bull. Chem. Soc. Jap.* **33**, 1091.
Kwiatek, J., and Seyler, J. K. (1965). *J. Organometal. Chem.* **3**, 421.
Lanpher, E. J. (1957). *J. Amer. Chem. Soc.* **79**, 5578.
Lawson, D. N., Osborn, J. A., and Wilkinson, J. (1966). *J. Chem. Soc., A* p. 1733.
Lippard, S. J., and Morehouse, S. M. (1972a). *J. Amer. Chem. Soc.* **94**, 6949.
Lippard, S. J., and Morehouse, S. M. (1972b). *J. Amer. Chem. Soc.* **94**, 6956.
Lobach, M. I., Babitskii, B. D., and Kormer, S. (1967). *Russ. Chem. Rev.* **36**, 476.
Lukas, J., Coren, S., and Blom, J. E. (1969). *Chem. Commun.* p. 1303.
Lukas, J., van Leeuwen, P. W. N. M., Volger, H. C., and Kramer, P. (1970). *Chem. Commun.* p. 799.
McClellan, W. R., Hoehn, H. H., Cripps, H. N., Muetterties, E. L., and Howk, B. W. (1961). *J. Amer. Chem. Soc.* **83**, 1603.
McKeever, L. D., and Waack, R. (1969). *Chem. Commun.* p. 750.
McKeever, L. D., Waack, R., Doran, M. A., and Baker, E. B. (1969). *J. Amer. Chem. Soc.* **91**, 1057.
McPartlin, M., and Mason, R. (1967). *Chem. Commun.* p. 16.
Maglio, G., Musco, A., and Palumbo, R. (1970). *Inorg. Chim. Acta* **4**, 153.
Mann, B. E., Pietropaolo, R., and Shaw, B. L. (1971a). *Chem. Commun.* p. 790.
Mann, B. E., Shaw, B. L., and Shaw, G. (1971b). *J. Chem. Soc., A* p. 3536.
Mason, R., and Russell, D. R. (1966). *Chem. Commun.* p. 26.
Mason, R., and Whimp, P. O. (1969). *J. Chem. Soc., A* p. 2709.
Meakin, P., Trofimenko, S., and Jesson, J. P. (1972). *J. Amer. Chem. Soc.* **94**, 5677.
Minasyan, M. Kh., Struchkov, Y. T., Kritskaya, I. I., and Avoyan, R. L. (1966). *Zh. Strukt. Khim.* **7**, 903.
Moiseev, I. I., Federoskaya, E. A., and Syrkin, Ya. K. (1958). *Russ. J. Inorg. Chem.* **3**, 1218.
Murdoch, H. D. (1965). *J. Organometal. Chem.* **4**, 119.

Murdoch, H. D., and Henzi, R. (1966). *J. Organometal. Chem.* **5**, 552.
Murdoch, H. D., and Weiss, E. (1962). *Helv. Chim. Acta* **45**, 1927.
Neese, H. J., and Bürger, H. (1971). *J. Organometal. Chem.* **32**, 213.
Nesmeyanov, A. N., and Kritskaya, I. I. (1968). *J. Organometal. Chem.* **14**, 387.
Nesmeyanov, A. N., Rubezhov, A. Z., and Gubin, S. P. (1966). *Izv. Akad. Nauk SSSR, Ser. Khim.* p. 194.
Nesmeyanov, A. N., Ustynyuk, Y. A., Kritskaya, I. I., and Shcembelov, G. A. (1968). *J. Organometal. Chem.* **14**, 395.
Nordlander, J. E., and Roberts, J. D. (1959). *J. Amer. Chem. Soc.* **81**, 1769.
Oberhansli, W. E., and Dahl, L. F. (1965). *J. Organometal. Chem.* **3**, 43.
O'Brien, S. (1970). *J. Chem. Soc., A* p. 9.
Plowman, R. A., and Stone, F. G. A. (1962). *Z. Naturforsch.* **B 17**, 575.
Powell, J. (1969). *J. Amer. Chem. Soc.* **91**, 4311.
Powell, J. (1971). *J. Chem. Soc., A* p. 2233.
Powell, J., and Chan, A. W. L. (1972). *J. Organometal. Chem.* **35**, 203.
Powell, J., and Shaw, B. L. (1966). *Chem. Commun.* p. 323.
Powell, J., and Shaw, B. L. (1967). *J. Chem. Soc., A* p. 1839.
Powell, J., and Shaw, B. L. (1968a). *J. Chem. Soc., A* p. 159.
Powell, J., and Shaw, B. L. (1968b). *J. Chem. Soc., A* p. 583.
Powell, J., and Shaw, B. L. (1968c). *J. Chem. Soc., A* p. 780.
Powell, J., Robinson, S. D., and Shaw, B. L. (1965). *Chem. Commun.* p. 78.
Ramey, K. C., and Statton, G. L. (1966). *J. Amer. Chem. Soc.* **88**, 4387.
Ramey, K. C., Lini, D. C., and Wise, W. B. (1968). *J. Amer. Chem. Soc.* **90**, 4275.
Ray, P., and Dutt, N. K. (1943). *J. Indian Chem. Soc.* **90**, 81.
Rice, D. P., and Osborn, J. A. (1971). *J. Organometal. Chem.* **30**, C84.
Sandel, V. R., McKinley, S. V., and Freedman, H. H. (1968). *J. Amer. Chem. Soc.* **90**, 495.
Shapley, J. R., and Osborn, J. A. (1970). *J. Amer. Chem. Soc.* **92**, 6976.
Shaw, B. L., and Shaw, G. (1971). *J. Chem. Soc., London* p. 3533.
Shaw, B. L., and Sheppard, N. (1961). *Chem. Ind. (London)* p. 517.
Shvo, Y., and Shanan-Atidi, H. (1969). *J. Amer. Chem. Soc.* **91**, 6683.
Smidt, J., and Hafner, W. (1959). *Angew. Chem.* **71**, 284.
Smith, A. E. (1965). *Acta Crystallogr.* **18**, 331.
Sokolov, V. N., Khvostic, G. M., Poddubnyi, I. Ya., and Kondratenkov, G. P. (1971). *J. Organometal. Chem.* **29**, 313.
Sokolov, V. N., Khvostic, G. M., Poddubnyi, I. Ya., and Kondratenkov, G. P. (1972). *Dokl. Chem.* **204**, 408.
Statton, G. L., and Ramey, K. C. (1966). *J. Amer. Chem. Soc.* **88**, 1327.
Thiele, K. H., and Köhler, J. (1967). *J. Organometal. Chem.* **7**, 365.
Thiele, K. H., and Zdunneck, P. (1965). *J. Organometal. Chem.* **4**, 10.
Thiele, K. H., Engelhardt, G., Köhler, J., and Arnstedt, M. (1967). *J. Organometal. Chem.* **9**, 385.
Thompson, H. W., and Torkington, P. (1966). *Trans. Faraday Soc.* **42**, 432.
Tibbetts, D. L., and Brown, T. L. (1970). *J. Amer. Chem. Soc.* **92**, 2031.
Uttech, R., and Dietrich, H. (1965). *Z. Kristallogr., Kristallgeometrie, Kristallphys., Kristallchem.* **122**, 112.
van Leeuwen, P. W. N. M., and Praat, A. P. (1970a). *Rec. Trav. Chim. Pays-Bas* **89**, 321.
van Leeuwen, P. W. N. M., and Praat, A. P. (1970b). *J. Organometal. Chem.* **21**, 501.
van Leeuwen, P. W. N. M., and Praat, A. P. (1970c). *J. Organometal. Chem.* **22**, 483.
van Leeuwen, P. W. N. M., and Praat, A. P. (1970d). *Chem. Commun.* p. 365.

van Leeuwen, P. W. N. M., Praat, A. P., and Vrieze, K. (1969a). *J. Organometal. Chem.* **19**, 181.
van Leeuwen, P. W. N. M., Vrieze, K., and Praat, A. P. (1969b). *J. Organometal. Chem.* **20**, 219.
van Leeuwen, P. W. N. M., Vrieze, K., and Praat, A. P. (1969c). *J. Organometal. Chem.* **20**, 277.
van Leeuwen, P. W. N. M., Praat, A. P., and van Diepen, M. (1970). *J. Organometal. Chem.* **24**, C31.
van Leeuwen, P. W. N. M., Praat, A. P., and van Diepen, M. (1971). *J. Organometal. Chem.* **29**, 433.
van Leeuwen, P. W. N. M., Lukas, J., Praat, A. P., and Appelman, M. (1972). *J. Organometal. Chem.* **29**, 433.
van Leeuwen, P. W. N. M., Lukas, J., Praat, A. P., and Appelman, M. (1972). *J. Organometal. Chem.* **38**, 199.
Veillard, A. (1969a). *Chem. Commun.* p. 1022.
Veillard, A. (1969b). *Chem. Commun.* p. 1427.
Volger, H. C., and Vrieze, K. (1967). *J. Organometal. Chem.* **9**, 527.
Volger, H. C., Vrieze, K., and Praat, A. P. (1968). *J. Organometal. Chem.* **14**, 429.
Volger, H. C., Vrieze, K., Lemmers, J. W. F. M., Praat, A. P., and van Leeuwen, P. W. N. M. (1970). *Inorg. Chim. Acta* **4**, 435.
Vrieze, K., and van Leeuwen, P. W. N. M. (1971). *Prog. Inorg. Chem.* **14**, 1.
Vrieze, K., and Volger, H. C. (1967). *J. Organometal. Chem.* **9**, 537.
Vrieze, K., MacLean, C., Cossee, P., and Hilbers, C. W. (1966). *Rec. Trav. Chim. Pays-Bas* **85**, 1077.
Vrieze, K., Cossee, P., MacLean, C., Hilbers, C. W., and Praat, A. P. (1967). *Rec. Trav. Chim. Pays-Bas* **86**, 769.
Vrieze, K., Cossee, P., Praat, A. P., and Hilbers, C. W. (1968a). *J. Organometal. Chem.* **11**, 353.
Vrieze, K., Praat, A. P., and Cossee, P. (1968b). *J. Organometal. Chem.* **12**, 533.
Vrieze, K., Volger, H. C., and Praat, A. P. (1968c). *J. Organometal. Chem.* **14**, 185.
Vrieze, K., Volger, H. C., and Praat, A. P. (1968d). *J. Organometal. Chem.* **15**, 195.
Vrieze, K., Volger, H. C., and Praat, A. P. (1968e). *J. Organometal. Chem.* **15**, 447.
Vrieze, K., Volger, H. C., and van Leeuwen, P. W. N. M. (1969). *Inorg. Chim. Acta Rev.* **3**, 109.
Waack, R., and Doran, M. A. (1963). *J. Amer. Chem. Soc.* **85**, 1091.
West, P., Purmort, J. I., and McKinley, S. V. (1968). *J. Amer. Chem. Soc.* **90**, 797.
Whitesides, G. M., Nordlander, J. E., and Roberts, J. D. (1962a). *Discuss. Faraday Soc.* **34**, 185.
Whitesides, G. M., Nordlander, J. E., and Roberts, J. D. (1962b). *J. Amer. Chem. Soc.* **84**, 2010.
Wilke, G., Bogdanovic, B., Hardt, P., Heimbach, P., Keim, W., Krömer, M., Oberkirch, W., Tanaka, K., Steinrücke, E., Walter, D., and Zimmermann, H. (1966). *Angew. Chem.* **78**, 157.
Zieger, H. E., and Roberts, J. D. (1969a). *J. Org. Chem.* **34**, 1976.
Zieger, H. E., and Roberts, J. D. (1969b). *J. Org. Chem.* **34**, 2826.

12

Stereochemical Nonrigidity in Metal Carbonyl Compounds

R. D. ADAMS and F. A. COTTON

I. Introduction. 489
II. Permutational and Geometrical Isomerizations 491
III. Bridge-Terminal Exchange and Scrambling of CO Groups 500
    A. Bridged and Nonbridged Structures. 500
    B. Binuclear Systems 503
    C. Polynuclear Systems 517
    Note Added in Proof 519
    References. 520

I. Introduction

Of all ligands known to the organometallic chemist, carbon monoxide is undoubtedly the most ubiquitous. This is due to the inherent stability of metal to carbonyl bonds under a great range of conditions and the compatibility of the CO group with a great variety of organo groups and other ligands. There are four main structural functions of the carbonyl ligand:

1. *The terminal, essentially linear* M—C—O *arrangement.* This was the first one to be recognized and is certainly the most common one. Its great stability is best explained in terms of a dual mode of interaction between the metal atom and the CO. There is a σ interaction which consists essentially of overlap of a carbon σ orbital with a metal σ orbital, with the pair of electrons originally in the carbon σ orbital occupying the region of overlap. Second, there is overlap between one or (usually) two mutually perpendicular π type orbitals of the metal, which usually have mainly d character, and the $p\pi^*$ MOs of the CO group. Electron density initially in the metal π orbitals flows

into the overlap region. These two modes of bonding engender opposite polarities and tend to be mutually reinforcing, i.e., synergic. A considerable variation in both the absolute and relative amounts of σ and π interaction is possible, thus accounting for the environmental versatility of the terminal CO group.

Strict linearity of the M—C—O group can be expected only in the ideal case, where two mutually perpendicular π interactions are exactly equal in strength and there are no steric effects due to nonbonded contacts. In practice, most terminal M—C—O groups are slightly bent, with angles in the range of 170°–179°.

2. *Doubly bridging arrangement.* This is the second most common type. The entire set of four atoms, M—C(O)—M, is essentially planar and symmetrical. The bonding has usually been described in terms of two M—C single bonds and a C=O double bond; hence the appellation "ketonic." However, it has been argued (Braterman, 1972), quite persuasively, that this is an inadequate approximation. Instead, a more flexible scheme involving multicenter bonds should be used; such a scheme is also capable of natural extension to cover the triply bridging case (see below).

3. *The triply bridging arrangement.* This is a relatively rare situation in which a CO group lies along the threefold symmetry axis of an equilateral triangle of three metal atoms, which are themselves joined by M—M bonds. It appears that the bonding can only be satisfactorily described from a molecular orbital or multicenter bond point of view. If the carbon atom were to form a full two-center two-electron single bond to each metal atom, the valence of the oxygen atom could not be satisfied.

4. *Grossly unsymmetrical* CO *bridges.* This arrangement might equally well be considered as a severely perturbed terminal arrangement. It consists of an M—C—O chain that "leans" toward a neighboring metal atom, M'. The C···M' distance is short enough to imply some bonding interaction (e.g., ~2.3–2.4 Å), while the M—C distance is about as short as that for an ordinary terminal group. The M—C—O angle is about 165°–170°, and the M—C(O)—M' set is approximately planar. There are only about ten reported examples of this arrangement [see Cotton *et al.* (1973a) for leading references], but it is of interest here in connection with the pathway of bridge-terminal exchanges with which much of this article will be concerned.

A fuller discussion of structure, bonding and spectra of metal carbonyls can be found elsewhere (Cotton and Wilkinson, 1972; Braterman, 1972).

12. Nonrigidity in Metal Carbonyl Compounds

Stereochemical nonrigidity of metal carbonyl compounds takes two basic forms. First, there are the interconversions of permutational isomers and geometric isomers of mononuclear molecules, presumably involving what Muetterties has called polytopal rearrangements. Second, there is the interconversion of bridge and terminal arrangements and the migration and scrambling of CO groups in di- and polynuclear molecules. Both of these forms will be reviewed here in the order mentioned, but greater emphasis will be given to the second one.

Activity in both areas has been relatively restricted, since ^{13}C nmr spectroscopy is the most potent tool, and suitable instrumentation, especially pulse and Fourier transform circuitry (Farrer and Becker, 1971), has only recently become fairly generally available. However, by the same token, the field is now ripe for rapid advances in the next few years. Until now, most studies have been conducted on molecules that could be examined via the ^1H resonance(s) of other ligands, such as C_5H_5 and CNR, and thus the activities of the CO groups themselves were inferred at second hand.

II. Permutational and Geometrical Isomerizations

In the first ^{13}C investigation of a metal carbonyl molecule (Cotton et al., 1958), the spectrum of pure liquid $Fe(CO)_5$, with ^{13}C at the natural abundance level, was recorded in the dispersion mode at room temperature. Only a single resonance was observed, although there is no reasonable structure for a five-coordinate complex that makes all ligands equivalent. It was generally supposed at that time that the molecule is trigonal bipyramidal, and this has since been proved. The nmr observation was later verified using an enriched sample (Bramley et al., 1962). Subsequently, the sharp signal has been found at $-110°C$ (Gansow et al., 1972) and even $-170°C$ (Meakin and Jesson, 1973).

In 1960, the first and most celebrated of all polytopal rearrangements (Muetterties, 1970), the Berry pseudorotation was proposed to explain the occurrence of only one doublet in the ^{19}F spectrum of the trigonal bipyramidal molecule PF_5 (Berry, 1960). The Berry rearrangement of a trigonal bipyramid proceeds (Fig. 1) by closing the 180° angle between the two axial ligands and opening that between two equatorial ones until a square pyramidal configuration is attained. A continuation of these motions eventually carries the two ligands initially in axial positions into equatorial ones and the two initially equatorial groups finally become axial. The process can be repeated so that the equatorial ligand that remained as such in the first cycle is now transferred to an axial site.

It is not known (and cannot be determined) if the Berry mechanism is

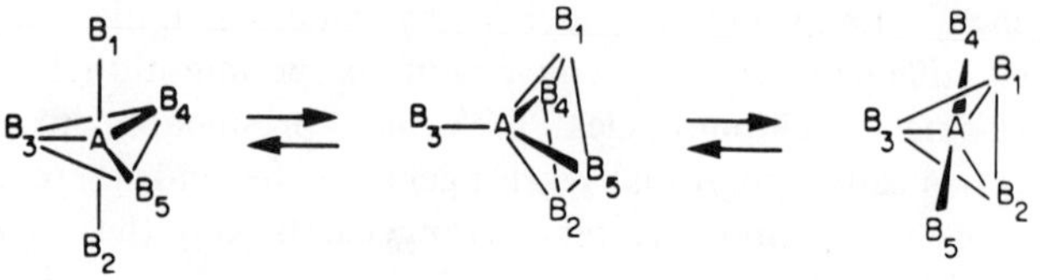

Fig. 1 The Berry pseudorotation mechanism for interchange of axial and equatorial ligands in trigonal bipyramidal molecules (1–3).

actually the correct one for Fe(CO)$_5$, since all reported observations are confined to the fast exchange limit, although it has been shown conclusively that a mechanism that either is the Berry mechanism or is permutationally indistinguishable from it operates in other cases (Whitesides and Mitchell, 1969; Meakin and Jesson, 1973). It could even be argued that there is no direct evidence that Fe(CO)$_5$ is stereochemically nonrigid and that there is some other explanation, such as those considered by Cotton et al., for the appearance of only one line. It seems to be generally accepted, however, that Fe(CO)$_5$ is a nonrigid molecule.

There is evidence to show that other five-coordinate, d^{10}, metal carbonyl, isonitrile, and phosphine species are also stereochemically nonrigid. Thus, ir studies of PF$_3$-substituted derivatives of Fe(CO)$_5$ and RCo(CO)$_4$ show that they exist as mixtures of isomers (Udovich and Clark, 1969; Udovich et al., 1969), which cannot, however, be distinguished in the nmr spectra. Partial methanolysis of the PF$_3$ group yielded a mixture of product isomers with a different distribution, but left the equilibrium distribution in the unreacted starting material unchanged. All these systems appear to rearrange so rapidly that no direct nmr proof of stereochemical nonrigidity, nor any mechanistic information can be obtained.

The Co(CNCH$_3$)$_5^+$ ion has been shown to have a trigonal bipyramidal structure in Co(CNCH$_3$)$_5$ClO$_4$ (Cotton et al., 1965), and it is reasonable to assume that this structure will occur in Co(CNR)$_5^+$ species generally. The (t-ButNC)$_5$Co$^+$ ion and a number of substitution products with mono- and polydentate phosphines and mixed arsine–phosphine ligands have been investigated (King and Saran, 1972). It was reported that the proton nmr spectra of (t-BuNC)$_5$Co$^+$ at $-60°$C and at $25°$C correspond to the slow and fast exchange limits, respectively, for axial–equatorial site exchange. This was, of course, very surprising and puzzling, in view of the characteristically low activation energies for rearrangement of most five-coordinate species (Jesson and Meakin, 1973). The claim has since been shown to be erroneous (Muetterties, 1973). Indeed the only transition metal systems in which it has thus far been possible to observe slow exchange nmr spectra and to carry out mechanistic analysis are {Rh[P(OR)$_3$]$_5$}$^+$ complexes (Meakin and Jesson,

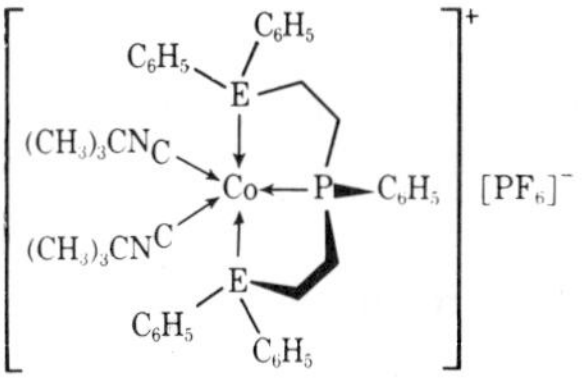

12. Nonrigidity in Metal Carbonyl Compounds

$$\left[\begin{array}{c} \text{C}_6\text{H}_5 \\ \text{C}_6\text{H}_5 \diagdown \underset{\text{E}}{\text{P}} \diagdown \\ (\text{CH}_3)_3\text{CNC} \downarrow \\ \text{Co} \leftarrow \text{P} \blacktriangleright \text{C}_6\text{H}_5 \\ (\text{CH}_3)_3\text{CNC} \uparrow \\ \diagup \underset{\text{E}}{\diagdown} \\ \text{C}_6\text{H}_5 \quad \text{C}_6\text{H}_5 \end{array}\right]^+ [\text{PF}_6]^-$$ $$\left[\begin{array}{c} \text{C}_6\text{H}_5 \\ \text{C}_6\text{H}_5 \diagdown \underset{\text{E}}{\text{P}} \diagdown \\ \text{C}_6\text{H}_5 \diagup \text{Co} \leftarrow \text{E}-\text{C}_6\text{H}_5 \\ (\text{CH}_3)_3\text{CNC} \uparrow \quad | \\ \text{C} \quad \text{C}_6\text{H}_5 \\ \text{N} \\ (\text{CH}_3)_3\text{C} \end{array}\right]^+ [\text{PF}_6]^-$$

4 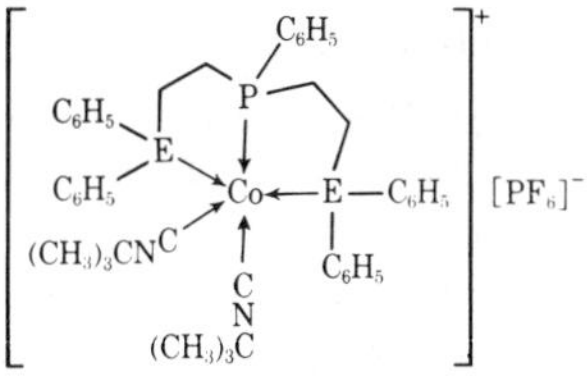   5

Fig. 2 Two possible structures (4, 5) for the cation in [(t-ButNC)$_2$Co(Pf-Pf-Pf)]PF$_6$. (Reproduced by permission from King and Saran, 1972.)

1973). The free energies of activation, which increase with ligand size, range from 7.5 to 12.0 kcal/mole. A detailed analysis of the complex with R = CH$_3$. for which $\Delta G^\ddagger$ = 7.5 kcal/mole at 200°K, showed that the rearrangement process is one in which the two axial and two of the equatorial ligands of a trigonal bipyramid exchange simultaneously. This is consistent with the Berry mechanism, but also with others, such as turnstile rotation. This case is discussed further in Sec. III.A.3 of Chapter 8.

Evidence for stereochemical nonrigidity was presented for the molecule [(t-ButNC)$_2$Co(Pf-Pf-Pf)][PF$_6$]. (See Fig. 2. (Pf-Pf-Pf) = tridentate [(C$_6$H$_5$)$_2$PCH$_2$CH$_2$]$_2$PC$_6$H$_5$.) Two separate t-butyl methyl resonances were observed at 30°C, which coalesced at about 70°C. Because the separation of the two signals was very small (0.08 ppm), the authors eschewed the obvious possibility that there are axial and equatorial t-ButNC groups in a roughly trigonal bipyramidal molecule and suggested instead that both t-ButNC groups occupy equatorial positions. The nonequivalence of the t-ButNC groups would then be due to the unsymmetrical conformation of the tridentate ligand. No suggestion was offered as to how this conformation could flip so as to interchange the sites of the isonitrile groups.

There have been several studies of molecules of the type Fe(CO)$_3$(diphos), where diphos represents a ligand of the type R$_2$P(CH$_2$)$_n$PR$_2$, in which R may be Me or Ph and n may be 1, 2, or 3. In the series with R = Ph and n = 1, 2, 3, ^{13}C spectra at room temperature contain only a single line for the carbonyl carbon atoms (F. A. Cotton and L. Kruczynski, unpublished, 1971). Stereochemical nonrigidity was assumed to account for this. Before undertaking low-temperature ^{13}C nmr studies, structural information was sought crystallographically. Thus far, only the structure of the compound with diphos = Ph$_2$PCH$_2$PPh$_2$ has been determined, but it is quite revealing (Cotton et al., 1973b). The ligand arrangement is intermediate between the two limiting ones, namely, trigonal bipyramidal and square pyramidal. All three CO groups are nonequivalent. However, only slight bends or twists (angle changes of about 15° or shifts of about 0.2 Å in atomic positions) are necessary to carry the structure into configurations in which one or another pair of CO groups become equivalent. The small shifts required might be expected to occur very rapidly at ambient temperature and to have such low activation energies as to remain rapid even at the lowest practical temperatures for high-resolution nmr measurements. An interesting thing about this molecule is that because of the pronounced deviations from both trigonal bipyramidal and square pyramidal geometries, the "canonical" mechanisms, such as Berry pseudorotation, turnstile rotation and others, need not be considered; relatively minor distortion can accomplish all that needs to be done.

The compound with diphos = $Me_2PCH_2CH_2PMe_2$ has also been studied (Akhtar et al., 1972). Its infrared spectrum was said to be consistent with a trigonal bipyramidal structure with the phosphorus atoms occupying one axial and one equatorial position. Actually, very marked distortions from such a regular structure would not be detectable in the infrared spectrum, and this statement must be interpreted accordingly. The ^{13}C spectrum showed only a single triplet (J_{P-C} = 6.9 Hz) from room temperature to $-80°C$. Thus, again the molecule is apparently stereochemically nonrigid, but there is no basis for a discussion of the rearrangement pathway.

The dynamical stereochemistries of the compounds $(\eta^5\text{-}C_5H_5)Mo(CO)_2LR$ have been thoroughly investigated by variable temperature nmr methods (Faller and Anderson, 1970). In these compounds, L is generally a tertiary phosphine or phosphite and R is an alkyl group, halide, or hydride. The molecular geometry can be usefully described as square pyramidal with the center of the η^5-cyclopentadienyl ring occupying the apical position. These molecules exist as cis and trans isomers, where the notation describes the relative positioning of the two carbonyl groups in the basal plane. Variable temperature pmr investigations have shown that the $\eta^5\text{-}C_5H_5$ resonances of a mixture of these isomers broaden and coalesce when the samples are sufficiently heated. Ample evidence was provided to demonstrate that the rearrangement processes are truly intramolecular. Equation (1) shows one mechanism that has been proposed. The square pyramidal cis form **6** converts through angle bending to the trigonal bipyramidal intermediate **7**.

$$\begin{array}{ccc} \mathbf{6} & \mathbf{7} & \mathbf{8} \end{array} \tag{1}$$

Conversion to the square pyramidal trans form **8** completes the rearrangement. The process is different from the Berry rearrangement in that the apical position in the square pyramidal form is converted into an axial position in the trigonal bipyramidal form.

12. Nonrigidity in Metal Carbonyl Compounds

For seventeen systems of the type (η^5-C_5H_5)Mo(CO)$_2$LR, in which L was one of five PR$_3$ or P(OR)$_3$ ligands and R was H, I, Cl, Br, CH$_3$, or PhCH$_2$, free energies of activation ranged from 12.8 to 25.2 kcal/mole at 25°C; entropies of activation ranged from -7 to $+1$ eu, corresponding to log A values in the Arrhenius equation from 11.7 to 13.4.

In addition to the cis–trans isomerization just described, special cases provide evidence for a cis–cis rearrangement in which enantiomeric cis isomers are interconverted. Cis configurations of these molecules are chiral, and when they contain a ligand such as dimethylphenylphosphine, the methyl groups are diastereotopic (nonequivalent in the nmr). They can only be interchanged by inversion of the molecular configuration. Variable temperature pmr investigations of these compounds showed that the methyl resonances in the cis isomers broadened and coalesced before cis–trans isomerization occurred at an appreciable rate. It was then necessary to propose that there must be a cis–cis rearrangement, attributable to the conversion of isomer **6** to **7**, whence it may return to the enantiomer of the original isomer. Such a process would exchange the environments of the phosphine methyl groups and give no exchange between the cis and trans isomers. It was generally observed that cis–cis rearrangement barriers were about 3 kcal/mole less than the limiting cis–trans rearrangement barrier.

Further evidence concerning the cis–cis rearrangement was obtained from similar complexes in which the cyclopentadienyl ligand was replaced with an indenyl group (Faller et al., 1971). In the cis isomers of these compounds, the protons which are equivalent in the free indenyl ligand are diastereotopic in the complex, due to the chirality around the molybdenum atom. As with the methyl groups on the phosphine ligand in the previously discussed complex, the nonequivalent protons on the indenyl ligand in these complexes exchange environments with cis–cis isomerization rearrangements. It was shown for the complex (η^5-C_9H_7)Mo(CO)$_2$[P(CH$_3$)$_2$C$_6$H$_5$] that the nonequivalent protons on the indenyl group averaged at the same rate as the methyl groups on the phosphorus ligand. This observation thus provides strong evidence that the spectral nonequivalences are genuinely caused by the chirality of the molecule as a whole and supports the cis–cis exchange mechanism.

Except for the hydrides, the rearrangement barriers in all these dicarbonyl molybdenum complexes are relatively high (i.e., 20–25 kcal/mole) and coalescence temperatures are usually $> 100°C$. As a result, the report of a very facile equilibrium in the diphosphine complexes (η^5-C_5H_5)Mo(CO)L$_2$Cl [L = P(CH$_3$)$_2$C$_6$H$_5$ and P(CH$_3$)(C$_6$H$_5$)$_2$] is unusually interesting. (Wright and Mawby, 1971). Only at $-62°C$ in the latter complex can the rearrangement rate be slowed sufficiently to observe separate signals for both the cis and trans isomers. Although this result could readily be attributed to steric effects, it is also clear that the electronic environment in these complexes will be significantly changed, and it seems likely that both factors will be of importance.

This section will conclude with a discussion of some nonrigid carbonyl molecules in which the rearrangement pathway is still subject to some controversy, as will be seen. They are discussed here because the authors of this chapter favor a pathway based on polytopal rearrangement rather than bridge-terminal shifts. The compounds in questions are of the type $[\mu\text{-}(R_nE)M(CO)_3]_2$ (9), where, for example, M = Fe and E = P, As; M = Fe and E = S; or M = Co and E = Ge or Sn, to choose representative examples. Crystallographic studies of $[\mu\text{-}(PhMeP)Fe(CO)_3]_2$ (L. F. Dahl,

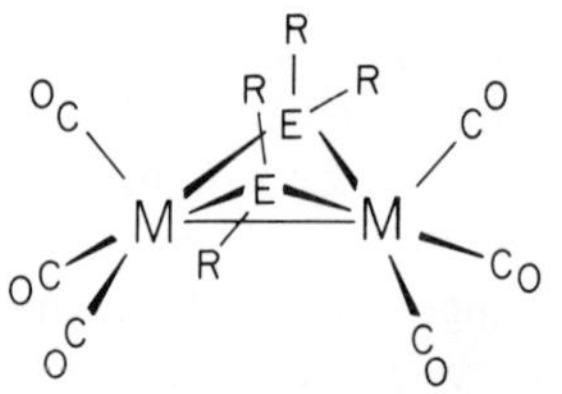

Fig. 3 A structural diagram of molecules of the type $[(\mu\text{-}R_2E)M(CO)_3]_2$ (9), which possess an M—M bond.

unpublished results, 1972, cf. Treichel et al., 1972) and $[\mu\text{-}(C_2H_5S)Fe(CO)_3]_2$ (Dahl and Wei, 1963) show that these particular molecules have a folded four-membered ring with a metal–metal bond, as shown schematically in Fig. 3. The presence of an Fe—Fe bond is indicated by the short metal–metal distance and is necessary to give a diamagnetic system with closed valence shell electron configurations on each metal atom. Compounds with two Co atoms and two R_2Si, R_2Ge or R_2Sn bridges are isoelectronic with the RS and R_2P bridged diiron compounds, and are diamagnetic; the same folded M—M bonded structure can safely be assumed for them. It is worth noting explicitly that this structure can be viewed as a pair of square pyramids sharing a basal edge. Its point symmetry is C_{2v}.

The observations and discussion reported for $[\mu\text{-}(Me_2Ge)Co(CO)_3]_2$ (Adams and Cotton, 1970) provide a good illustration of this type of problem. Proton nmr spectra (the methyl groups) of this compound are shown for various temperatures in Fig. 4. The sharp singlet that appears at room temperature collapses at lower temperatures and eventually (about $-45°C$) two peaks of equal intensity appear and become sharp at temperatures below 80°C. The following explanation has been proposed (Adams and Cotton, 1970). As shown by **10** and **11** (Fig. 5), the presumed structure has methyl groups in two different environments. The spectra at low temperatures are in

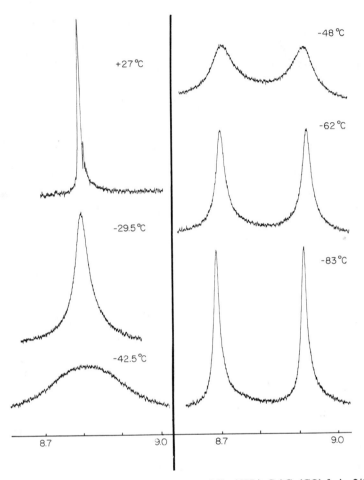

Fig. 4 The variable temperature pmr spectra of $[(\mu\text{-}(CH_3)_2Ge)Co(CO)_3]_2$ in 2/1, v/v, CF_2Cl_2/CH_2Cl_2 solvent. Temperatures have been corrected from the original report. (Reproduced by permission, from Adams and Cotton, 1970.)

accord with this. Obviously, at higher temperatures, the methyl groups undergo site exchange; in effect, **11** is reversibly transformed into **12** and **13**.

The proposed pathway is essentially the Berry pseudorotation in reverse. The square-pyramidal coordination at either cobalt atom (or perhaps both simultaneously) rearranges to trigonal bipyramidal, as indicated in **14**. From configuration **14**, there is an equal probability of recovering **11** or going on to **13**. Any two of the configurations, **11**, **12**, and **13** can be interconverted in one such step, and any one such configurational interconversion interchanges the methyl group sites.

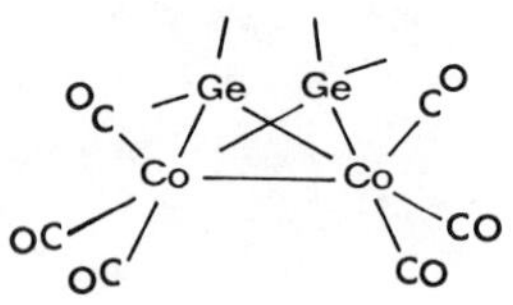

Fig. 5 Schematic representations of the various conformers and proposed rearrangement intermediates for $[(\mu\text{-}(CH_3)_2Ge)Co(CO)_3]_2$ (**10–15**).

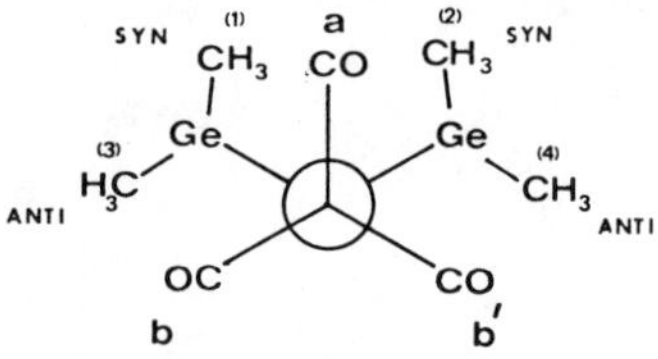

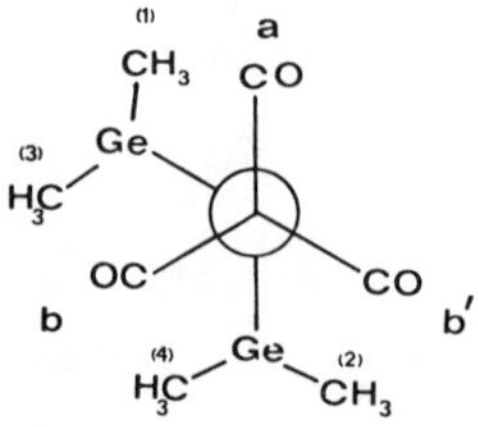

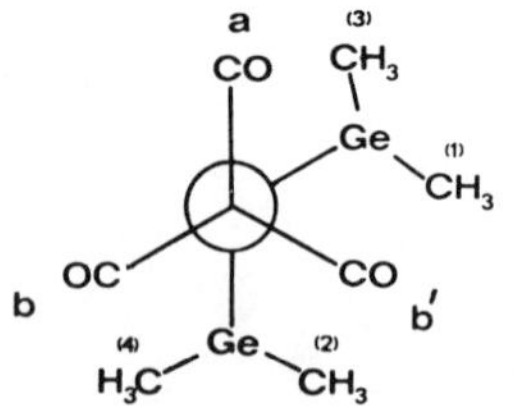

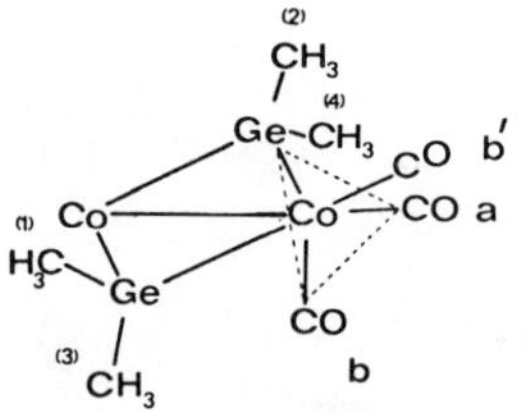

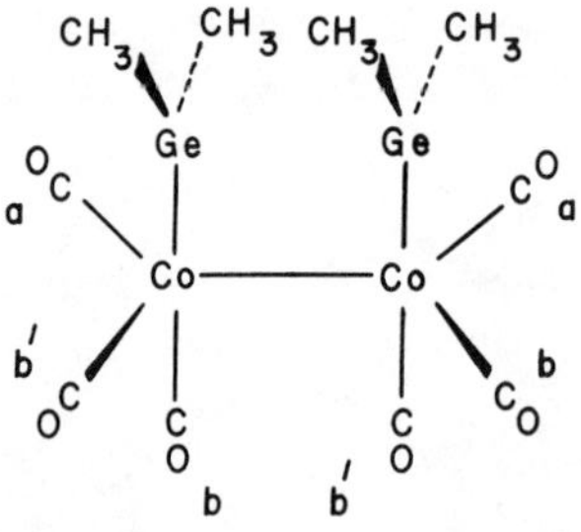

A relatively obvious alternative pathway, which traverses structure **15**, attained from **11**, **12**, or **13** by opening the $(CH_3)_2Ge$ bridges, is analogous to the pathways to be discussed in Section III for CO-bridged molecules. However, this pathway was considered less likely (Adams and Cotton, 1970) because there is no precedent for the carbenoid-like R_2Ge—M groups that are involved. Even true carbenoid complexes (Cotton and Lukehart, 1972) of the type XX'C—M are stable only when either X or X' is a group such as RO or R_2N which can form a partial dative π bond to the carbon atom; no R_2C—M systems are known. More recently (Marks, 1971) has shown that the attempted preparation of R_2Sn— and R_2Ge— analogs of XX'C—Cr(CO)$_5$ species gives stable products only when a molecule of donor solvent such as THF or pyridine is strongly held in the product. An X-ray study of a repre-

12. Nonrigidity in Metal Carbonyl Compounds

sentative compound (Brice and Cotton, 1973), $t\text{-But}_2\text{SnCr(CO)}_5 \cdot \text{py}$, shows that the pyridine is coordinated to the tin atom; thus, no genuine $t\text{-But}_2\text{Sn}$—Cr nor any other R_2Ge—M or R_2Sn—M grouping has yet been shown to exist.

The compound $[\mu\text{-}(CF_3)_2\text{PFe(CO)}_3]_2$ (Grobe, 1968), which is isoelectronic and presumably isostructural with $[\mu\text{-Me}_2\text{GeCo(CO)}_3]_2$, appears to behave similarly but with a higher activation energy. It shows separate ^{19}F doublet resonances for the two types of CF_3 groups even at 40°C but only one doublet at 120°C. These observations were reported without discussion of the pathway for site exchange.

More recently (Dessy et al., 1972), it has been shown that $[\mu\text{-Me}_2\text{PFe-}(CO)_3]_2^{2-}$ also undergoes methyl-group site exchange with a coalescence temperature of about $-65°\text{C}$. Since the effect of the added electrons upon the structure is unknown, speculation on the reason for this increased lability would be premature.

Another case in which site exchange may occur via polytopal rearrangement with retention of the bridge has recently been reported (Bennett et al., 1972), although again, it is not absolutely certain that this is so. The reaction of acetylenes with dicobalt octacarbonyl produces molecules in which the two bridging CO groups are replaced by a π-bonded acetylene, which bridges the two metal atoms (Sly, 1959). In a similar manner, $\text{Rh}_2(\text{PF}_3)_8$ also reacts with acetylenes to give complexes of the form $(\text{acetylene})\text{Rh}_2(\text{PF}_3)_6$ in which the acetylene is also π bonded to the two metal atoms, **16**. As emphasized in

16   **17**

structure **17** for the molecule $(\mu\text{-}C_6H_5C_2CH_3)\text{Rh}_2(\text{PF}_3)_6$, there are three chemically nonequivalent PF_3 groups. The room temperature ^{19}F nmr spectrum shows only a single resonance with appropriate ^{31}P splitting. When cooled, however, the resonance collapses and reforms as three peaks of equal intensity. The spectral changes are unaffected by the presence of uncoordinated PF_3 in the samples. The evidence led Bennett et al. to suggest that at ambient temperatures the PF_3 groups are rapidly and intramolecularly interconverted between the three different coordination sites. One possible mechanism would interchange the PF_3 groups by propellorlike rotations around the pseudo threefold axis shown in **17**. A rotation of the $\text{Rh}_2(\text{PF}_3)_6$ unit as a whole relative to the acetylene would produce the same effect.

III. Bridge-Terminal Exchanges and Scrambling of CO Groups

A. Bridged and Nonbridged Structures

In a binuclear metal carbonyl it is in general possible to write both bridged and nonbridged structures (**18, 19**) or structures with one (**21**) or three (**20**)

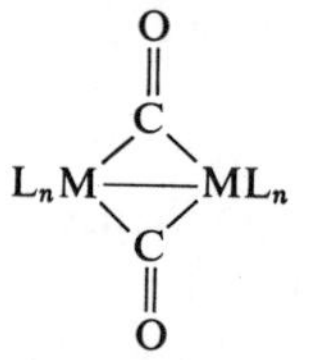

$$
\begin{array}{cc}
\text{18} & \text{19} \\
\text{20} & \text{21}
\end{array}
$$

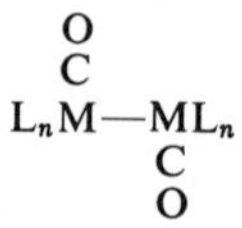

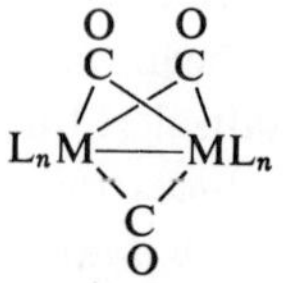

bridges. In each one, the effective atomic number of the metal atoms is the same, and it seems likely that the aggregate bond energies would also be similar. Thus the question of why some molecules such as $Mn_2(CO)_{10}$ and $[(\eta^5\text{-}C_5H_5)Mo(CO)_3]_2$ have nonbridged structures, while others, such as $Fe_2(CO)_9$, $[(\eta^5\text{-}C_5H_5)Fe(CO)_2]_2$, and $Co_2(CO)_8$, have bridged ones framed itself quite naturally. For the case of one versus three bridges, specific examples are provided by $(OC)_3Fe(\mu\text{-}CO)_3Fe(CO)_3$ and $(OC)_4Os(\mu\text{-}CO)Os(CO)_4$ (Moss and Graham, 1970) and by the stability of $[(\eta^5\text{-}C_5H_5)Rh(CO)]_2(\mu\text{-}CO)$, as opposed to the instability of $(\eta^5\text{-}C_5H_5)Rh(\mu\text{-}CO)_3(\eta^5\text{-}C_5H_5)Rh$. A corollary question is whether there might not be cases in which the two structures have sufficiently similar stabilities to exist together in equilibrium.

In 1963, it was shown for the first time that this does actually happen (Noack, 1963; Bor, 1963) for $Co_2(CO)_8$ in solution, though the one crystalline form known contains only bridged molecules. In order to interpret the variation in the infrared spectrum of $Co_2(CO)_8$ with temperature, it was necessary to postulate that a nonbridged isomer (**23**), whose structure has not been unequivocally established, is in equilibrium with a bridged isomer (**22**), presumably having the same structure as that found in the crystal (Sumner et al., 1964). This is shown in Fig. 6. In hydrocarbon solvents at about 25°C, the two tautomers are present in comparable quantities. Another case in which there are substantial quantities of both forms is $[(\eta^5\text{-}C_5H_5)Ru(CO)_2]_2$ (Cotton and Yagupsky, 1967).

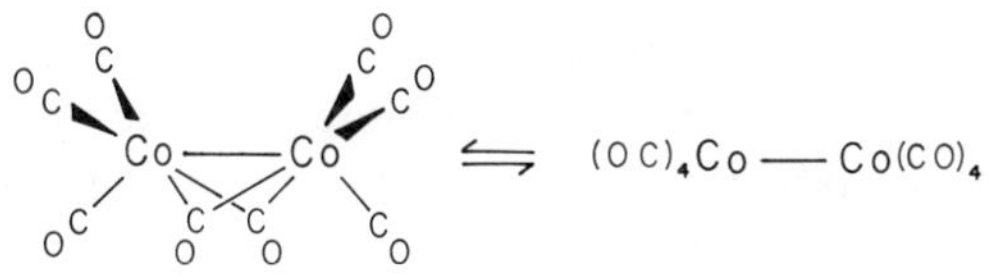

12. Nonrigidity in Metal Carbonyl Compounds

Fig. 6 The bridged/nonbridged (**22, 23**) tautomeric equilibrium for $Co_2(CO)_8$.

In other instances, efforts to detect a second tautomer have failed. For example, work in this laboratory has revealed no bridged species for $Mn_2(CO)_{10}$, $[(\eta^5\text{-}C_5H_5)Cr(CO)_3]_2$ and $[(\eta^5\text{-}C_5H_5)Mo(CO)_3]_2$, whereas for $[(\eta^5\text{-}C_5H_5)Ni(CO)]_2$ and $[(\eta^5\text{-}C_5H_5)Ni(CNCH_3)]_2$ there are no detectable quantities of the nonbridged isomers.

There appears to be some correlation between the relative stabilities of the bridged and nonbridged structures for a given compound and the size of the metal atoms. With the largest metal atoms, only nonbridged structures are observed (or singly bridged instead of triply bridged), while with the smallest metal atoms only the bridged structures occur. The reason for this correlation may be that when the metal atoms are too large, the long M—M distance strains the M—C(O)—M angles too much, and the nonbridged structure becomes preferable. It is noteworthy that the M—C(O)—M angles tend to be quite small, 80°–90°. Thus, as any group is descended, nonbridged structures supersede bridged ones, and in a given row of the periodic table, nonbridged structures are found to the left and bridged ones to the right. See Cotton and Wilkinson (1972) for further examples and references.

At its most primitive level, the principle underlying the fact that structures for binuclear species that differ by two bridges (i.e., 0 versus 2, or 1 versus 3) are both viable alternates is simply that in each pair, each of the two alternatives allows both metal atoms to attain the same number of electrons in its valence shell orbitals. Deletion of two bridges deprives each metal atom of two electrons; attachment of one terminal CO group to each metal atom restores two electrons to each metal atom.

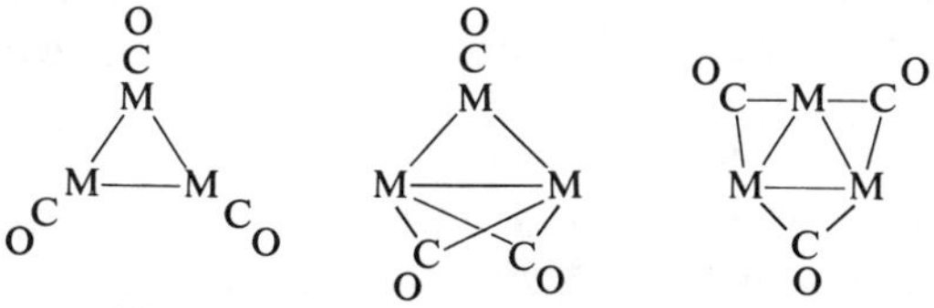

This principle may be extended to discuss alternative structures for tri- and tetranuclear metal carbonyls. For trinuclear species, the three obvious possibilities are **24, 25,** and **26**. Structure **24** is found in $Ru_3(CO)_{12}$ and $Os_3(CO)_{12}$, while **25** occurs in crystalline $Fe_3(CO)_{12}$ (another illustration of the occurrence of bridges being favored by smaller metal atoms). The **24** versus **25** comparison is in one sense no more than another example of the 0 versus 2 bridge situation already discussed for binuclear systems. However, it acquires extended significance here, since repeated $24 \rightarrow 25 \rightarrow 24$ type interconversions, with different edges of the M_3 triangle being used to form different **25** structures would afford a pathway for scrambling CO groups all around the triangle. It is obvious that ^{13}C studies of the $M_3(CO)_{12}$ systems (M = Fe, Ru, Os) would be of interest, to see if such scrambling really does occur and if so, how rapidly. Structure **26** has not been observed as a stable one among the $M_3(CO)_{12}$ compounds, although its participation as an intermediate in scrambling processes is a possibility.

A structure of type **26** has been observed in one of the crystalline isomers of $(\eta^5-C_5H_5)_3Rh_3(CO)_3$, while a structure of type **25** has been found in another crystalline form (Paulus et al., 1967; Mills and Paulus, 1967). There is no information at all as to the possible dynamic behavior of these isomers in solution.

For tetranuclear species of the type $M_4(CO)_{12}$ many structures, such as **27–30** can be written. There are clearly several more structures obtainable by

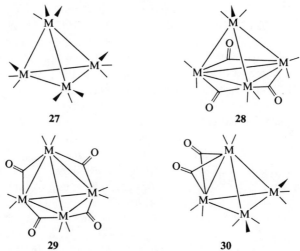

building into **30** still more doubly-bridged edges. Structure **27** is found for crystalline $Ir_4(CO)_{12}$ and **28** for crystalline $Co_4(CO)_{12}$ and $Rh_4(CO)_{12}$ (Wei and Dahl, 1966; Wei et al., 1967). Structure **29** was suggested as the solution structure for $Co_4(CO)_{12}$ (Smith, 1965), but this was later shown to be incorrect. Thus, neither **29**, **30**, nor any other more highly bridged structure has yet been observed. However, as early as 1966, it was realized (Cotton, 1966) that structures such as **27**, **28** and **29** are readily interconvertible by concerted opening and closing of bridges.

12. Nonrigidity in Metal Carbonyl Compounds

In summary, it is clear that interconversion of structures with different numbers of bridging CO groups can afford possible pathways for permuting CO groups among all metal atoms and all stereochemically nonequivalent sites in any one structure in polynuclear metal carbonyls. We shall now review the results available that show that such scrambling processes actually do occur, and at rates that influence nmr line shapes at temperatures of interest.

B. Binuclear Systems

1. $(\eta^5\text{-}C_5H_5)_2Fe_2(CO)_4$

This molecule—or, better, the set of interconverting molecules that all have this formula—provides a good starting point. It and some of its derivatives have been studied in more detail than any comparable system or set of related systems. It is now reasonably well understood due to the efforts of many workers over a considerable period of time (Stammreich et al., 1959; Noack, 1967; Cotton and Yagupsky, 1967; Manning, 1968; McArdle and Manning, 1970; Bryan and Greene, 1970; Bryan et al., 1970; Bullitt et al., 1970, 1972; Gansow et al., 1972; Adams and Cotton, 1973b).

Three structures (isomers or tautomers) are important (Fig. 7), namely, the cis bridged, **31**, trans bridged, **32**, and the nonbridged, **33**; the last has

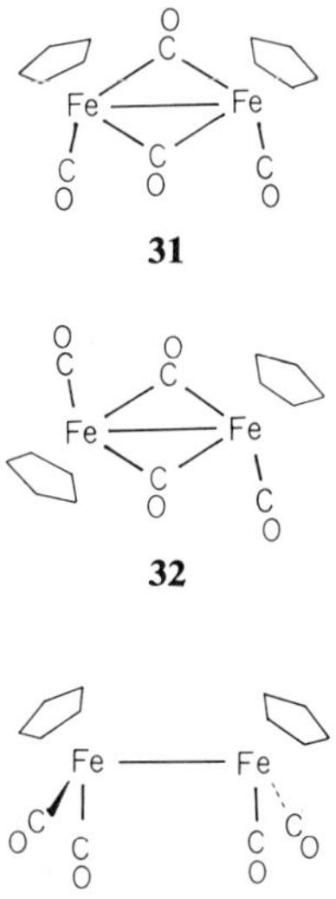

Fig. 7 The three tautomers (**31–33**) of $(\eta^5\text{-}C_5H_5)_2Fe_2(CO)_4$.

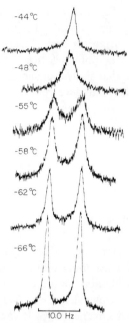

Fig. 8 The proton nmr spectra of $(\eta^5\text{-}C_5H_5)_2Fe_2(CO)_4$ at various temperatures. The two signals are for the cis and trans isomers with the cis–trans ratio slightly >1. (Reproduced by permission from Bullitt et al., 1972.)

several possible rotational conformations, a point that will be of major importance later. Tautomers **31** and **32** have both been obtained in crystalline form and are well defined as to structural details (Bryan et al., 1970). In nonpolar solvents, **31** and **32** are present in comparable quantities and are accompanied by a small proportion (about 2%) of **33**. An increase in the solvent polarity increases the ratio of **31** to **32**; because **31** is somewhat more stable than **32** in polar solvents, the **31/32** ratio increases as the temperature of such a solution is lowered. The foregoing facts can be established from infrared spectra alone (Manning, 1968; McArdle and Manning 1970) and provide a necessary background for interpretation of nuclear resonance data which will now be considered.

It was first shown, by observing the cyclopentadienyl proton resonances, that the cis and trans bridged structures interconvert rapidly at room temperature but slowly enough to give separate proton nmr signals at temperatures below about $-50°C$ (Bullitt et al., 1970, 1972). Figure 8 shows the data for a solution in toluene–CS_2, which is sufficiently nonpolar that the **31/32** ratio is close to unity at all temperatures. The addition of even 10% of CD_2Cl_2 causes this ratio to change to 1.5 at $-70°C$ and in pure CD_2Cl_2

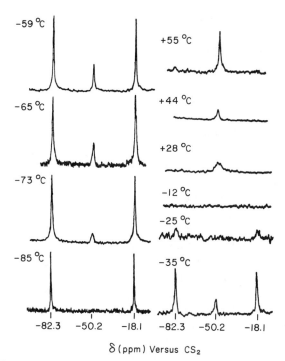

Fig. 9 The ^{13}C spectra at various temperatures of the CO groups of $(\eta^5\text{-}C_5H_5)Fe_2(CO)_4$. (Reproduced by permission, from Gansow *et al.*, 1972.)

(F. A. Cotton and T. J. Marks, unpublished work) at $-70°C$ the ratio is about 12.

It was proposed (Bullitt *et al.*, 1970) that **31** and **32** interconvert by opening of the CO bridges, giving **33**. Rotation could then occur about the Fe—Fe single bond in **33** to produce rotational conformations from which either **31** or **32** could be recovered by reclosing of the bridges. This was the first explicit proposal of rapid opening and reclosing of carbonyl bridges as a basis for stereochemically nonrigid behavior and, simultaneously, the first proven example (unless, of course, one is willing to believe that, somehow, **31** and **32** can interconvert without dismantling the bridging system).

However, from the proton spectra of the cyclopentadienyl groups, the finer details of the process cannot be ascertained. The results of a ^{13}C study (Gansow *et al.*, 1972) provide the basis for a detailed mechanistic analysis (Adams and Cotton, 1973b). The ^{13}C spectra are shown in Fig. 9.

There appears to be but one acceptable assignment of these spectra, which is that proposed by Gansow *et al.* (1972). At $-85°C$, the observed signals are those of the bridging and terminal CO groups of **31**. No signal for

32 is seen for two reasons. First, at this low a temperature in a polar solvent the proportion of **32** present is quite small. Second, the coalescence temperature for **32** is about $-85°C$.

As the temperature is raised, a signal for all the CO groups of **32** appears at a position appropriate for the average of bridge and terminal chemical shifts, and sharpens. It also increases in relative intensity since the **32**/**31** equilibrium ratio is increased at higher temperatures. Thus, at about $-60°C$. the solution contains about 80% **31** and 20% **32** tautomers. Interchange of bridging and terminal CO groups is proceeding so slowly in the **31** tautomers that sharp signals are observed for each type of ligand, while it is proceeding so rapidly in the **32** tautomers that only a single sharp signal is seen for all the CO ligands. Beginning at about $-40°C$, all signals begin to broaden at what appears to be the same rate, coalescing around $-12°C$ and eventually reforming as one signal (about $20°C$). This becomes sharp as the temperature is raised further. Evidently, at about $-40°C$, both bridge-terminal exchange in the cis isomer, **31**, and interconversion of the isomers become rapid enough at the same time to begin to broaden all the signals. The salient points are: (a) that bridge-terminal exchange requires much less activation energy in **32** than in **31**; and (2) bridge-terminal exchange in **31** and **31** $\rightleftharpoons$ **32** isomerization proceed at essentially the same rate.

The following analysis of these observations has been proposed (Adams and Cotton, 1973b). First, it is postulated that the only important pathway for the interchange of bridging and terminal CO groups, for either **31** or **32**, is by way of the nonbridged tautomer, **33**; this is the original idea of Bullitt *et al.* (1970). Other possibilities, such as the formation of a quadruply-bridged intermediate, or the occurrence of concerted interchange of one bridging CO with one terminal CO are excluded.

The second postulate, which is vital, is that as bridge opening of **31** or **32** occurs there are small concerted shifts and twists in other parts of the molecule so that **33** is generated only in that one (or two) of its staggered rotameric conformations which is (or are) most directly accessible from the initial bridged structure. This means, for example, that **31** can open only to **34** and **35**, while **32** can open only to **36**, as shown in Fig. 10. Then, by the principle of microscopic reversibility, only **34** and **35** (but not **36**) can directly generate **31** and only **36** (but not **34** or **35**) can directly generate **32**.

The third and final postulate is that internal rotations, such as those connecting **34**, **35**, and **36** may have barriers of the same order as, or even greater than, the activation energies for opening and closing of pairs of bridges. Indirect experimental evidence for the third postulate is provided by a study of $[(\eta^5\text{-}C_5H_5)_2Mo(CO)_3]_2$ (Adams and Cotton, 1973a), in which a similar kind of barrier is shown to have a magnitude of 15.3 ± 1.0 kcal/mole.

The foregoing postulates are then employed in the following way. Let us use labels to mark the CO groups initially in bridging positions. The asterisks

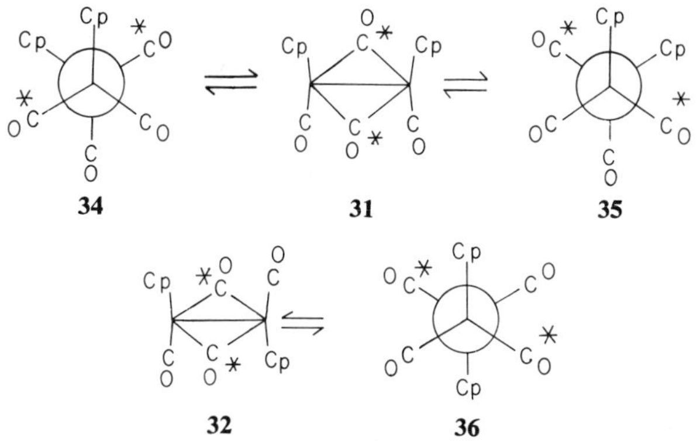

Fig. 10 Diagrams showing which rotational conformations (**34–36**) of **33** are accessible directly from **31** and **32**. The conformations of **33** are represented by Newman-like projections, in which the view is such as to superpose the two iron atoms.

in Fig. 10 serve this purpose. In **36**, all four CO groups have become equivalent and **36** may generate directly not only the particular **32** tautomer from which it was formed, but another in which bridging and terminal CO groups have changed places. Interchange of bridging and terminal CO groups can occur for **32** merely by opening and reclosing bridges, without any internal rotation of the **33** type intermediate.

For **31**, the situation is different. Both **34** and **35** regenerate **31** with the same pair of bridging ligands. Only after **34** or **35** has undergone a rotation by $2\pi/3$, of the type which carries one Cp group past the other, or a rotation by $4\pi/3$, in the course of which they pass through a transoid rotamer, can a **31** type structure be recovered in which bridge and terminal ligands have changed places.

Therefore, interchange of bridging and terminal CO ligands in the cis isomer and interconversion of the cis and trans isomers are both processes that require the same hindered rotation in the nonbridged intermediate. On the other hand, interchange of bridging and terminal CO ligands in the trans isomer does not require this step. This explains why the latter process can occur rapidly at low temperatures where neither of the former two processes are very rapid. It also explains why the first two processes reach the critical rate for line broadening at the same temperature. A convenient way to summarize the essential features of this explanation is in a potential energy diagram such as that shown in Fig. 11 (Adams and Cotton, 1973a).

2. $(\eta^5\text{-}C_5H_5)_2Fe_2(CO)_2(CNCH_3)_2$

The system now to be described provides very strong support for the mechanistic analysis advanced to account for the behavior of $(\eta^5\text{-}C_5H_5)_2Fe_2(CO)_4$

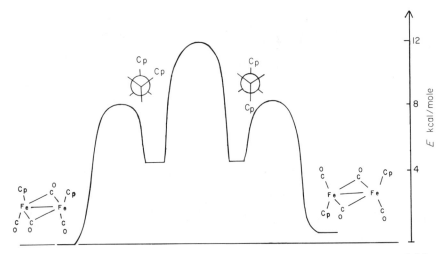

Fig. 11 A potential energy diagram for the intramolecular rearrangements of $(\eta^5\text{-}C_5H_5)_2Fe_2(CO)_4$. (Adapted, by permission, from Adams and Cotton, 1973a.)

(Adams and Cotton, 1973b). The substance with the above composition, prepared by reaction of CH_3NC with the tetracarbonyl, is found by both infrared and nmr spectroscopy at 25°C to consist entirely of the cis isomer with two bridging CH_3NC ligands (isomer **37**, Fig. 12) and one other isomer, which has one bridging and one terminal CH_3NC ligand, i.e., either isomer **39** or **40**. The presence of isomer **37** has been confirmed by isolation of it from the mixture in crystalline form and X-ray crystallographic structure determination (Cotton and Frenz, 1974).

There are a total of six isomeric bridged structures (not counting optical isomers) for a molecule of this composition. They are shown in Fig. 12; isomers **39** and **40** have enantiomers. It can be shown that according to the postulates of Adams and Cotton, supplemented by the further one that there be no intermediate nonbridged structure with both CH_3NC ligands on the same metal atom, isomers **37**, **40**, and **41** are interconvertible with each other, and so are isomers **38**, **39**, and **42**. However, it is not possible to convert any member of one set to any member of the other. Thus, since one of the isomers present is definitely **37**, the other must be **40**. Isomer **41** is mechanistically accessible; its absence is presumably attributable to intrinsically low thermodynamic stability.

The effect of temperature on the proton nmr spectrum is shown in Fig. 13. The signals observed in the room temperature spectrum can be assigned as follows: $\tau 5.33$ and $\tau 6.37$ with a 5:3 intensity ratio, to the cyclopentadienyl and methyl protons of **37**; $\tau 5.38$, $\tau 5.47$, $\tau 6.23$, $\tau 7.48$, with relative intensities of 5:5:3:3, to cyclopentadienyl, cyclopentadienyl, bridging CH_3NC and

12. Nonrigidity in Metal Carbonyl Compounds

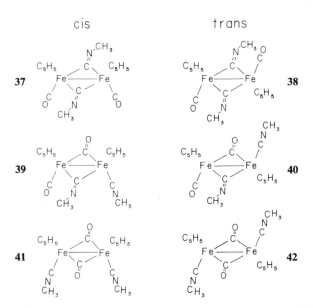

Fig. 12 The six isomeric structures (**37–42**) for $(\eta^5\text{-}C_5H_5)_2Fe_2(CO)_2(CNCH_3)_2$ omitting the enantiomers of **39** and **40**. (Reproduced, by permission, from Adams and Cotton, 1973b.)

terminal CH$_3$NC protons of isomer **40**. As the temperature is raised the signals from isomer **40** broaden first. This means that bridge-terminal exchange of CH$_3$NC ligands (accompanied by site exchange of the cyclopentadienyl groups) occurs with lower activation energy than the interconversion of the isomers.

These observations are exactly what the Adams and Cotton mechanism predicts. Collapse of both C$_5$H$_5$ and CH$_3$ resonances in isomer **40** can occur solely by opening and reclosing of bridges without internal rotation in the nonbridged intermediate. Interconversion of isomers **37** and **40**, however, requires internal rotation.

It is also important to note that after the solution containing isomers **37** and **40** has been held at a temperature of 95°–100°C for about 15 minutes—in the normal course of recording the nmr spectrum at that temperature—and then returned to room temperature, no new signals that might be due to other isomers are observed. Thus, no mechanism of rearrangement that could allow crossover between the **37**, **40**, **41** and the **38**, **39**, **42** isomer sets is detectable even under conditions where transformations which are permitted within the Adams and Cotton scheme are proceeding very rapidly.

This comparison between the rates of Adams-and-Cotton-allowed (ACA) and Adams-and-Cotton-forbidden (ACF) processes may be expressed

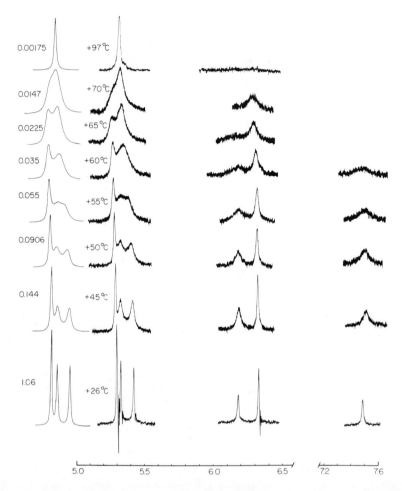

Fig. 13 The proton nmr spectra at various temperatures of $(\eta^5\text{-}C_5H_5)_2Fe_2(CO)_2\text{-}(CNCH_3)_2$ in o-dichlorobenzene. At the left are computer-simulated spectra in the cyclopentadienyl region for the proposed rearrangement mechanism as discussed in the text. (Reproduced, by permission, from Adams and Cotton, 1973b.)

quantitatively. Assuming that there can be no more than 5% conversion of the original isomer mixture (**37, 40, 41**) into one or more members of the other set (**38, 39, 42**), the rate constant k_{ACF} at about 100°C of any process capable of effecting such a conversion (an ACF process) must be $\leq 5 \times 10^{-5}$ mole sec^{-1}. At 97°C, the rate constant of isomer interconversion, the slower of the two ACA processes, is equal to $(0.006)^{-1} \simeq 1.5 \times 10^2$. Thus, even the fastest ACF process must be at least 10^7 times slower than the slowest ACA process.

12. Nonrigidity in Metal Carbonyl Compounds

Assuming approximate equality of frequency factors for ACF and ACA processes, this means that the activation energy of the most facile ACF process must be at least 22 kcal/mole greater than that for the least facile ACA process observed.

3. Other Isonitrile Derivatives of $(\eta^5\text{-}C_5H_5)_2Fe_2(CO)_4$

Because the case of $(\eta^5\text{-}C_5H_5)_2Fe_2(CO)_2(CNCH_3)_2$ just discussed possesses certain inherent features and is understood in sufficient detail, it provides cogent support for the detailed mechanism of rearrangement of doubly-bridged binuclear carbonyls that we proposed first for $(\eta^5\text{-}C_5H_5)_2Fe_2(CO)_4$. Quite apart from the finer mechanistic details, however, the behavior of the compound shows that CH_3NC ligands can pass rapidly between terminal and bridging positions and, in this way, from one metal atom to the other.

There are several other bridged compounds for which similar observations have been made, thus indicating the generality of the exchange processes under discussion, although similarly detailed analyses of the observations are not at present available.

The following three molecules (or, as already noted, sets of interconverting molecules with the composition given) have been prepared and studied by proton nmr spectroscopy at various temperatures (Adams, 1973): $(\eta^5\text{-}C_5H_5)_2\text{-}Fe(CO)_{4-n}L_n$, where $n = 1$ and $L = (CH_3)_3CNC$; $n = 1$ and $L = CH_3NC$; $n = 3$ and $L = CH_3NC$. In the first, $(\eta^5\text{-}C_5H_5)_2Fe_2(CO)_3(Me_3CNC)$, there is no detectable concentration of any isonitrile-bridged isomer; nevertheless, at room temperature, the $(CH_3)_3CNC$ ligand is passing rapidly from one iron atom to the other, since only a single sharp C_5H_5 resonance is seen. Separate sharp resonances are observed for the two rings at $-23°C$ and below.

For the second compound, $(\eta^5\text{-}C_5H_5)_2Fe_2(CO)_3(CH_3NC)$, at $-44°C$, the pmr spectrum shows that both CH_3NC-bridged and CH_3NC-terminal isomers are present. Line broadening begins at about $-30°C$, and by $50°C$, a single sharp signal is observed for the cyclopentadienyl protons.

The third compound, $(\eta^5\text{-}C_5H_5)_2Fe_2(CO)(CNCH_3)_3$, exists entirely in a form that has two CH_3NC bridges; the C_5H_5 rings are therefore nonequivalent and give rise to separate signals. Rearrangement is relatively slow, however, and at $95°C$, only line broadening, but not actual collapse, is observed. Unfortunately, measurements could not be made at higher temperatures due to the formation of paramagnetic decomposition products.

It should also be noted that for compounds containing bridging isonitrile groups, a further source of isomerism and stereochemically nonrigid behavior is the orientation of the alkyl or aryl group. That such groups are bent is well known (Joshi et al., 1965; Adams et al., 1971; Cotton and Frenz, 1974); thus when two metal atoms, M and M', are not chemically equivalent, isomers

43 and **44** may arise. In doubly bridged systems, syn and anti isomers, **45** and **46** are possible.

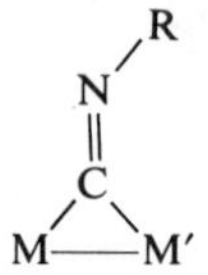

Information as to the occurrence of such isomers and the barriers to their interconversion is vital to a complete understanding of how isonitrile ligands pass from a terminal position on one metal to a terminal position on another via a bridged intermediate. The principal of microscopic reversibility demands that an inversion at nitrogen occur in this bridged intermediate according to one of the following two alternative pathways, Eq. (2). Thus, in order to

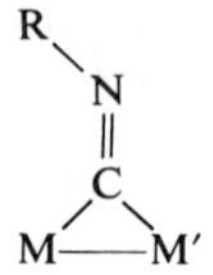

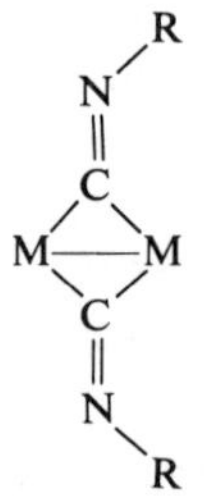

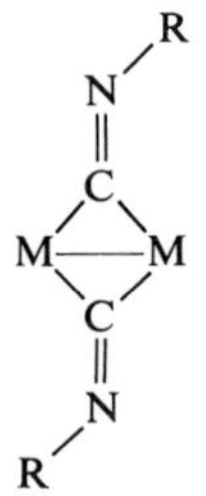

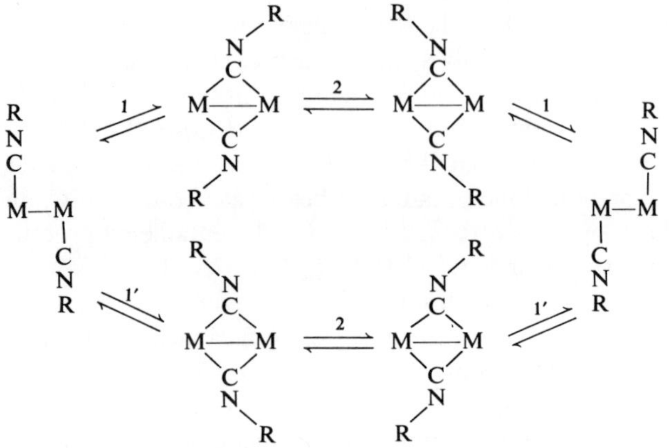

understand fully the processes of bridge-terminal exchanges and scrambling of isonitrile ligands, the inversion step must be understood.

A detailed investigation (Adams and Cotton, 1974) of such isomers and their rates of interconversion has been carried out for the two molecules $(\eta^5\text{-}C_5H_5)_2Fe_2(CO)(CNCH_3)_3$ and $(\eta^5\text{-}C_5H_5)_2Fe_2(CO)_2(CNCH_3)_2$. In both cases it was shown that the expected isomers do occur and can be observed at low temperatures. The barriers to inversion at nitrogen are relatively low (~10 kcal/mole), and it does not appear that this step is in itself rate determining in any of the systems so far investigated.

4. The $[(\eta^5\text{-}C_5H_5)Mn(CO)(NO)]_2$ System

It has been reported (Marks and Kristoff, 1972) that this compound exists in solution as a mixture of cis and trans isomers, in each of which there is one bridging and one terminal NO group. The proton nmr spectra at temperatures between $-62°$ and 40°C indicate that within each of these two isomers rearrangements occur that render the two sets of ring protons nmr equivalent at room temperature. Above room temperature, interconversion of the cis and trans isomers begins to be rapid. End-to-end scrambling occurs more rapidly in the trans isomer than in the cis isomer.

Marks and Kristoff pointed out that in all likelihood interconversion of isomers can occur only via a nonbridged intermediate in which both NO ligands are attached to the same metal atom. On the other hand, opening of bridges to give nonbridged isomers with one NO group on each metal atom can be followed by reclosing of one CO and one NO bridge in such a way as to result in a terminal NO ligand now residing on the other metal atom. In the case of the trans isomer, this can occur without internal rotation, as shown in Fig. 14; in the case of the cis isomer, internal rotation is necessary.

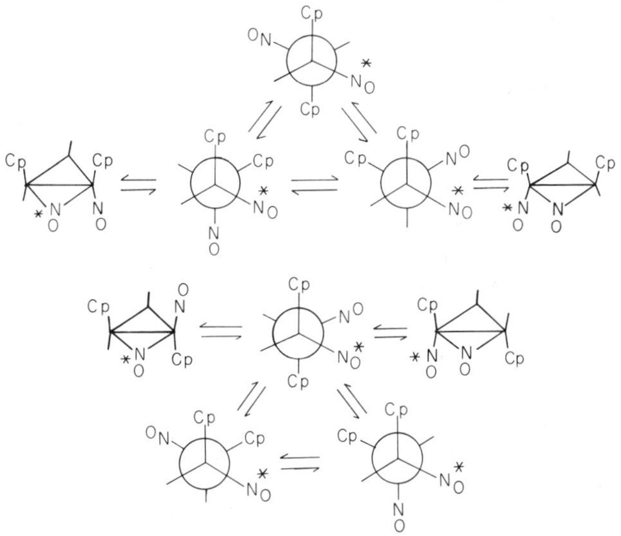

Fig. 14 Scheme showing possible rotamers of nonbridged intermediates obtainable from the observed *cis*- and *trans*-$(\eta^5\text{-}C_5H_5)_2Mn_2(CO)_2(NO)_2$ isomers.

This situation thus falls entirely within the framework of the previously discussed mechanism of Adams and Cotton for $(\eta^5\text{-}C_5H_5)_2Fe_2(CO)_4$ and its derivatives; as before the faster of the two processes should be the one entailing no internal rotation, which is in accord with observation.

It is to be noted that, as shown in Fig. 14, the only trans structures that could be derived from the cis isomer would be those with two NO or two CO bridges; since these are unobserved they are presumably thermodynamically unstable. Similarly, from the observed trans structure, only cis structures with two NO or two CO bridges can be reached, and again, these presumably are too unstable to be populated.

5. Dicobalt Octacarbonyl and $(\eta^5\text{-}C_5H_5)_2Ru_2(CO)_4$

As noted earlier, $Co_2(CO)_8$ exists in solution as a mixture of bridged and nonbridged tautomers (Noack, 1963; Bor, 1963). It would not be surprising, since the two tautomers are of approximately equal stability, if they were to interconvert relatively rapidly. Whether this is so remains uncertain. A ^{59}Co nmr study (Lucken et al., 1967) revealed only a broad line from $-10°$ to 30°C, but the significance of this observation is not clear. The intensity seemed to correspond to only about half the cobalt present; this could mean that only one of the tautomers was being observed, rather that the two were interconverting rapidly.

In the case of $(\eta^5\text{-}C_5H_5)_2Ru_2(CO)_4$ (Cotton and Yagupsky, 1967), the bridged and nonbridged tautomers are of similar stability; very probably cis and trans isomers of the bridged form are present though this has not been shown conclusively. It has been found (Bullitt et al., 1970) that whatever species are present interconvert rapidly even at $-100°C$, since the C_5H_5 proton resonance remains a sharp singlet down to this temperature.

6. $(\eta^5\text{-}C_5H_5)_2Mo_2(CO)_6$ and $(\eta^5\text{-}C_5H_5)_2Mo_2(CO)_5(CNCH_3)$

The hexacarbonyl appears to exist only in the nonbridged form under all conditions of observation, although the presence of traces of a bridged isomer below the level of instrumental sensitivity (roughly $\sim 1\%$) cannot, of course, be ruled out. In the crystal, the rotational conformation is trans; i.e., the molecule has a center of symmetry and belongs to the point group C_{2h}. This configuration is retained in solution in nonpolar solvents (Adams and Cotton, 1973a), but as solvent dielectric constant increases, an increasing proportion of the molecules rearrange to a gauche rotamer of C_2 symmetry. In acetone, there are comparable concentrations of both rotamers.

The rate of interconversion of gauche and trans rotamers of $(\eta^5\text{-}C_5H_5)_2\text{-}Mo_2(CO)_6$ has been determined at various temperatures by proton nmr measurements (see Fig. 15). The Arrhenius parameters are $E_a = 15.3 \pm 1.0$

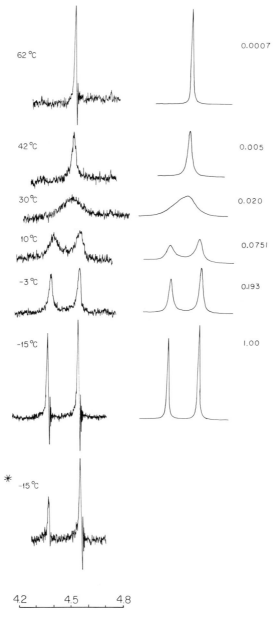

Fig. 15 Observed and computed pmr spectra (100 MHz) for $[(\eta^5\text{-}C_5H_5)Mo(CO)_3]_2$ in acetone. The spectrum marked with an asterisk was taken after addition of 20% toluene to the acetone. Numbers on computed spectra give residence times. (Reproduced by permission from Adams and Cotton, 1973a.)

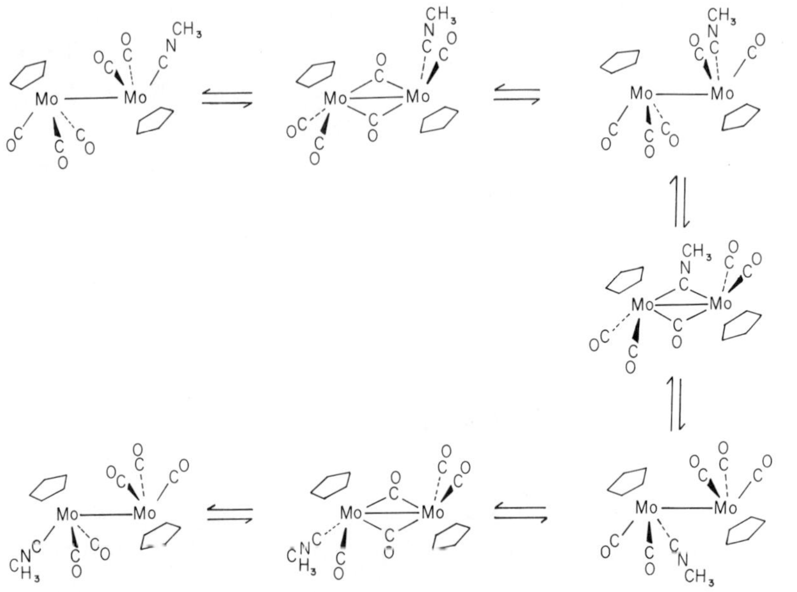

Fig. 16 A diagram showing how isomerization and scrambling of ligands can occur in $(\eta^5\text{-}C_5H_5)_2Mo_2(CO)_5(CNCH_3)$ by way of bridged intermediates. This scheme is simplified and omits a number of related pathways. A more complete scheme is given in Adams *et al.* (1973).

kcal/mole and $\log A = 13.0 \pm 1$. Thus, there is a substantial barrier to internal rotation; this result was important in drawing attention to the role that hindered rotation might play in many related cases, such as those of $(\eta^5\text{-}C_5H_5)_2Fe_2(CO)_4$ and its derivatives, which have been discussed earlier.

A study of the methyl isonitrile derivative of the hexacarbonyl, namely, $(\eta^5\text{-}C_5H_5)_2Mo_2(CO)_5(CNCH_3)$, has shown that even though the only directly observed structure of the molecule is one without bridges, there is rapid scrambling of the CH_3NC ligand among the six positions available to it (two trans to the Mo—Mo bond and four cis to that bond), and the only reasonable pathway for these scrambling processes involves both CO-bridged and CH_3NC-bridged intermediates (Adams *et al.*, 1973). Figure 16 illustrates the main rearrangement mechanisms.

This study may be thought of as supplying the second half of the complete picture of intramolecular rearrangements in binuclear metal carbonyls. The behavior of $(\eta^5\text{-}C_5H_5)_2Fe_2(CO)_4$ and its substitution products showed that a molecule with a stable bridged structure can isomerize and scramble ligands by opening and subsequently reestablishing bridges. The behavior of $(\eta^5\text{-}C_5H_5)_2Mo_2(CO)_5(CNCH_3)$ shows that, conversely, a molecule for which

12. Nonrigidity in Metal Carbonyl Compounds

the most stable structure is one without bridging ligands can undergo intramolecular isomerization and scrambling processes by way of bridged intermediates.

C. Polynuclear Systems

1. *The* $Rh_4(CO)_{12}$ *System*

After a protracted period of uncertainty, the structures of the three compounds, $M_4(CO)_{12}$, M = Co, Rh, Ir, were unequivocally established both in the crystal (Wei and Dahl, 1966; Wei *et al.*, 1967) and in solution (Haas and Sheline, 1967; Lucken *et al.*, 1967). The cobalt and rhodium compounds have structure **28**, while the iridium compound has structure **27**. During the period of uncertainty about the structures, a third structure, **29**, had been proposed (Smith, 1965) for the cobalt compound in solution.

In early 1966, before the appearance of the conclusive structural evidence, it was proposed (Cotton, 1966) that (a) all three of these structures would be likely to have similar energies, and (b) they could readily interconvert by concerted swinging of CO groups to and from bridging positions, with the nonbridged structure **27**, in which all CO groups are equivalent, serving as an intermediate between **28** and **29**. It is no longer necessary to postulate the existence of structure **29** (although it may well exist in small quantities) and thus the postulated interconversion of structures **27** and **28** is no longer needed for the original purpose of explaining how **29** might arise in solution, even if **28** is present in the crystal. However, the **28** → **27** → **28'** rearrangement now takes on importance in a different way. (**28'** is a structure of type **28** with some carbonyl ligands permuted.)

Although $Rh_4(CO)_{12}$, with structure **28**, has four different, equally populated sites for the CO groups, the ^{13}C nmr spectrum at room temperature has been found to consist of a single broad line (Cotton *et al.*, 1972). On cooling the solution, to about 7°C, this line collapses; on heating the solution to about 60°C, the line sharpens and becomes a 1:4:6:4:1 quintet.

The appearance of the quintet at high temperatures shows that all CO groups are being scrambled rapidly and nondissociatively over the four sites. The element rhodium consists entirely of ^{103}Rh with nuclear spin of $\frac{1}{2}$, so that only if each ^{13}C nucleus is coupled equally to all four ^{103}Rh nuclei can the observed quintet arise. Such equal coupling to all four ^{103}Rh nuclei requires movement of the CO groups from one type of site to another and from one rhodium atom to another, i.e., requires scrambling of the CO groups. The retention of $^{103}Rh-^{13}C$ coupling proves qualitatively that neither dissociation and recombination nor concerted intermolecular exchange can be responsible for the scrambling; it must be a unimolecular process.

It was proposed that the most reasonable way to account for the intramolecular scrambling is to assume that a given permutamer (of which there are 12!/3) of structure **28** can convert itself by concerted opening of its bridges to either of two permutamers (of which there are 12!/12) of structure **27**. Each permutamer of **27** can then generate with equal probability any of eight permutamers of **28** (including, of course, the one from which it arose) by reclosing a set of three bridges.

Further support for this proposal has been provided by low-temperature studies using 75% enriched material (J. Lewis and B. F. G. Johnson, private communication, 1973). In CD_2Cl_2, the slow-exchange limit was reached at $-65°C$ and showed separate signals for the four types of CO group, as expected for structure **28**. From the observed Rh–C couplings (Rh–Rh–C couplings were only about 1 Hz) an average quintet splitting of 16.9 Hz was calculated, which agrees satisfactorily with the value observed by Cotton *et al.* of 17.1 Hz. The significant observation was made that all four low-temperature signals broaden at the same rate. This is consistent with the proposed **28** → **27** → **28'** pathway. It is, of course, possible that structure **29** participates in the scrambling process, but there is no positive evidence of this, and so, by the principle of Occam's razor, it has not been explicitly considered.

2. $Co_4(CO)_{12}$

This molecule, in the crystal, also has structure **28**. It has been examined in solution by ^{59}Co nmr spectroscopy (Haas and Sheline, 1967; Lucken *et al.* 1967). Two resonances with relative intensities of 1:3 were observed. Thus, it appears that structure **28** persists in solution and that scrambling of CO groups proceeds relatively slowly. Since the cobalt atom is smaller than the rhodium atom, it would be expected that the nonbridged structure, **27**, would be less stable relative to **28** for $Co_4(CO)_{12}$ than for $Rh_4(CO)_{12}$. Thus, if scrambling of CO groups is to occur by the same pathway in the two cases, as seems likely, it should occur more slowly in the case of the cobalt compound.

3. $Fe_3(CO)_{12}$

This compound has a triangular structure of type **25**; there are two bridging CO groups across one edge, while all of the remaining CO groups are of the terminal type (Wei and Dahl, 1969). The molecule has C_{2v} symmetry and there are six nonequivalent types of CO group. It has long been suspected that this structure is altered in some way when the substance goes into solution, since the infrared spectrum is not easily reconciled with the complete retention of this structure. However, no definite proposal has been made regarding the species present in solution.

It has been reported (Gansow et al., 1972) that in the temperature range $-10°$ to $50°C$, only one sharp ^{13}C resonance is observed for $Fe_3(CO)_{12}$. It seems very likely that this result is evidence for stereochemically nonrigid character of the $Fe_3(CO)_{12}$ molecule.

Note Added in Proof

Perhaps no chapter in this book has suffered so greatly as this one because of rapid advances made since its completion. The manuscript was finished April 30, 1973, but in the ensuing year important new insights have been gained on a number of points. It was not feasible to attempt rewriting; a future review will be needed to treat the new developments. However, certain references have been updated, and a few guiding comments are given here to draw attention to the points where the foregoing discussion is most seriously obsolete.

1. *Grossly Unsymmetrical* CO *Bridges*

It is now considered that these fall into two distinct categories. In the first are pairs, or larger sets, where the distortions are compensatory; these alone are deemed relevant to the problem of CO scrambling. In the second are sets of one or two CO groups, which can be called "semibridging carbonyl groups" whose function is maintenance of charge balance in the molecule; these are not relevant, directly, to the majority of carbonyl scrambling processes. For brief discussion of the latter, see Cotton and Troup, 1974a. For a comprehensive treatment of both types, see Cotton, 1974).

2. CO *Scrambling in Mononuclear Carbonyls*

There is now some published (Kruczynski and Takats, 1974; Rigatti, *et al.* 1972; Kreiter, *et al.* 1974) and much unpublished work (Kruczynski and Takats; Cotton, Hunter and Lahuerta) on the scrambling of CO groups in species such as $(LL)Fe(CO)_3$, $C_7H_8Fe_2(CO)_6$, etc.

3. CO *Scrambling in Dinuclear Species*

Two new papers confirm the detailed mechanism proposed for $(\eta^5\text{-}C_5H_5)_2\text{-}Fe_2(CO)_4$ and its derivatives. These deal with $(\eta^5\text{-}C_5H_5)_2Fe_2(CO_2)(\mu\text{-}GeMe_2)\text{-}(\mu\text{-}CO)$ (Adams *et al.* 1974) and $(\eta^5\text{-}C_5H_5)_2Fe_2(\mu\text{-}CO)_2(CO)[P(OPh)_3]$ (Cotton *et al.*, 1974).

4. $Rh_4(CO)_{12}$

A ^{13}C study at low temperatures has been published (Evans, *et al.*, 1973). Since the slow-exchange limit spectrum was not published, and no further journal publication is intended, that spectrum is presented in Fig. 17.

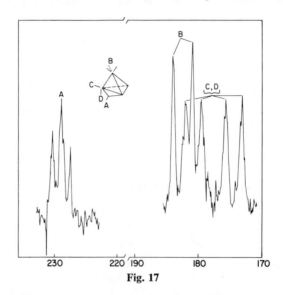

Fig. 17

5. $Fe_3(CO)_{12}$

Further X-ray study of this and related molecules (Cotton and Troup, 1974b) has shown that the bridges in $Fe_3(CO)_{12}$ are definitely unsymmetrical and that in this molecule and several derivatives, the energy variation in passing over the entire, continuous range of configurations, from a completely nonbridge structure to one with a pair of entirely symmetrical bridges, is probably very small (< 5 kcal/mol). It is therefore further suggested that (a) this molecule will scramble CO groups rapidly on the nmr time scale even at temperatures as low as the limit for observation, e.g., −150 to −170°C; (b) the puzzling infrared spectrum in solution is probably a convolution of spectra due to an entire range of structures, within the limits mentioned above, with those closer to the nonbridged limit more populated than those close to the symmetrically bridged limit.

REFERENCES

Adams, R. D. (1973). Ph.D. Thesis, Massachusetts Institute of Technology, Cambridge.
Adams, R. D., and Cotton, F. A. (1970). *J. Amer. Chem. Soc.* **92**, 5003.

12. Nonrigidity in Metal Carbonyl Compounds

Adams, R. D., and Cotton, F. A. (1973a). *Inorg. Chim. Acta* **7**, 153.
Adams, R. D., and Cotton, F. A. (1973b). *J. Amer. Chem. Soc.* **95**, 6589.
Adams, R. D., and Cotton, F. A. (1974). *Inorg. Chem.* **13**, 249.
Adams, R. D., Cotton, F. A., and Rusholme, G. A. (1971). *J. Coord. Chem.* **1**, 275.
Adams, R. D., Brice, M., and Cotton, F. A. (1973). *J. Amer. Chem. Soc.* **95**, 6594.
Adams, R. D., Brice, M. D., and Cotton, F. A. (1974). *Inorg. Chem.* **13**, 1080.
Akhtar, M., Ellis, P. D., MacDiarmid, A. G., and Odom, J. D. (1972). *Inorg. Chem.* **11**, 2917.
Bennett, M. A., Johnson, R. N., Robertson, G. B., Turney, T. W., and Whimp, P. O. (1972), *J. Amer. Chem. Soc.* **94**, 6540.
Berry, R. S. (1960). *J. Chem. Phys.* **32**, 933.
Bor, G. (1963), *Spectrochim. Acta* **19**, 2065.
Bramley, R., Figgis, B. N., and Nyholm, R. S. (1962). *Trans. Faraday Soc.* **58**, 1893.
Braterman, P. S. (1972). *Struct. Bonding (Berlin)* **10**, 57.
Brice, M. D., and Cotton, F. A. (1973). *J. Amer. Chem. Soc.* **95**, 4529.
Bryan, R. F., and Green, P. T. (1970). *J. Chem. Soc., A* p. 3064.
Bryan, R. F., Greene, P. T., Newlands, M. J., and Field, D. S. (1970). *J. Chem. Soc., A* p. 3068.
Bullitt, J. G., Cotton, F. A., and Marks, T. J. (1970). *J. Amer. Chem. Soc.* **92**, 2155.
Bullitt, J. G., Cotton, F. A., and Marks, T. J. (1972). *Inorg. Chem.* **11**, 671.
Cotton, F. A. (1966). *Inorg. Chem.* **5**, 1083.
Cotton, F. A. (1974). *Progr. Inorg. Chem.* In press.
Cotton, F. A., and Frenz, B. A. (1974) *Inorg. Chem.* **13**, 253.
Cotton, F. A., and Lukehart, C. M. (1972). *Progr. Inorg. Chem.* **16**, 487.
Cotton, F. A., and Troup, J. M. (1974a). *J. Amer. Chem. Soc.* **96**, 1233.
Cotton, F. A., and Troup, J. M. (1974b). *J. Amer. Chem. Soc.* **96**, 4155.
Cotton, F. A., and Wilkinson, G. (1972). "Advanced Inorganic Chemistry," 3rd ed., p. 684 *et seq*. Wiley (Interscience), New York.
Cotton, F. A., and Yagupsky, G. (1967). *Inorg. Chem.* **6**, 15.
Cotton, F. A., Danti, A., Waugh, J. S., and Fessenden, R. W. (1958). *J. Chem. Phys.* **29**, 1427.
Cotton, F. A., Dunne, T. G., and Wood, J. S. (1965). *Inorg. Chem.* **4**, 318.
Cotton, F. A., Kruczynski, L., Shapiro, B. L., and Johnson, L. F. (1972). *J. Amer. Chem. Soc.* **94**, 6191.
Cotton, F. A., Hardcastle, K. I., and Rusholme, G. A. (1973b). *J. Coord. Chem.* **2**, 217.
Cotton, F. A., Kruczynski, L., and White, A. J. (1974). *Inorg. Chem.* **13**, 1402.
Dahl, L. F., and Wei, C. H. (1963). *Inorg. Chem.* **2**, 328.
Dessy, R. E., Rheingold, A. L., and Howard, G. D. (1972). *J. Amer. Chem. Soc.* **94**, 746.
Evans, J., Johnson, B. F. G., Lewis, J., Norton, J. R., and Cotton, F. A. (1973), *J. Chem. Soc., Chem. Commun.* p. 807.
Faller, J. W., and Anderson, A. S. (1970). *J. Amer. Chem. Soc.* **92**, 5852.
Faller, J. W., Anderson, A. S., and Jakubowski, A. (1971). *J. Organometal. Chem.* **27**, C47.
Farrer, T. C., and Becker, E. D. (1971). "Pulse and Fourier Transform NMR." Academic Press, New York.
Gansow, O. A., Burke, A. R., and Vernon, W. D. (1972). *J. Amer. Chem. Soc.* **94**, 2550.
Grobe, J. (1968). *Z. Anorg. Allg. Chem.* **361**, 32.
Haas, H., and Sheline, R. K. (1967). *J. Inorg. Nucl. Chem.* **29**, 693.
Joshi, K. K., Mills, O. S., Pauson, P. L., Shaw, B. W., and Stubbs, W. H. (1965). *Chem. Commun.* p. 181.
King, R. B., and Saran, M. S. (1972). *Inorg. Chem.* **11**, 2112.

Kreiter, C. G., Stüber, S., and Wackerle, L. (1974). *J. Organometal. Chem.* **66**, C49.
Kruczynski, L., and Takats, J. (1974). *J. Amer. Chem. Soc.* **96**, 932.
Lucken, E. A. C., Noack, K., and Williams, D. F. (1967). *J. Chem. Soc., A* p. 148.
McArdle, P., and Manning, A. R. (1970). *J. Chem. Soc., A* p. 2119.
Manning, A. R. (1968). *J. Chem. Soc., A* p. 1319.
Marks, T. J. (1971). *J. Amer. Chem. Soc.* **93**, 7090.
Marks, T. J., and Kristoff, J. S. (1972). *J. Organometal. Chem.* **42**, 277.
Meakin, P., and Jesson, J. P. (1973). *J. Amer. Chem. Soc.* **95**, 1344 (and additional private communications of work to be published later in the same journal).
Mills, O. S., and Paulus, E. F. (1967). *J. Organometal. Chem.* **10**, 331.
Moss, J. R., and Graham, W. A. G. (1970). *Chem. Commun.* p. 835.
Muetterties, E. L. (1970). *Accounts Chem. Res.* **3**, 266.
Muetterties, E. L. (1973), *J. Chem. Soc., Chem. Commun.* p. 221.
Noack, K. (1963). *Spectrochim. Acta* **19**, 1925.
Noack, K. (1967). *J. Organometal. Chem.* **7**, 151.
Paulus, E. F., Fisher, E. O., Fritz, H. P., and Schuster-Woldan, H. (1967). *J. Organometal. Chem.* **10**, P3.
Rigatti, G., Baccalon, G., Ceccon, A., and Giacometti, G. (1972). *J. Chem. Soc., Chem. Commun.* p. 1165.
Sly, W. G. (1959). *J. Amer. Chem. Soc.* **81**, 18.
Smith, D. L. (1965). *J. Chem. Phys.* **42**, 1460.
Stammreich, H., Wilkinson, G., and Cotton, F. A. (1959). *J. Inorg. Nucl. Chem.* **9**, 3.
Sumner, G. G., Klug, H. P., and Alexander, L. E. (1964). *Acta Crystallogr.* **17**, 732.
Treichel, P. M., Douglas, W. M., and Dean, W. K. (1972). *Inorg. Chem.* **11**, 1615.
Udovich, C. A., Clark, R. J. (1969). *Inorg. Chem.* **8**, 938.
Udovich, C. A., Clark, R. J., and Haas, H. (1969). *Inorg. Chem.* **8**, 1067.
Wei, C. H., and Dahl, L. F. (1966). *J. Amer. Chem. Soc.* **88**, 1821.
Wei, C. H., and Dahl, L. F. (1969). *J. Amer. Chem. Soc.* **91**, 1351.
Wei, C. H., Wilkes, G. R., and Dahl, L. F. (1967). *J. Amer. Chem. Soc.* **89**, 4792.
Whitesides, G. M., and Mitchell, H. L. (1969). *J. Amer. Chem. Soc.* **91**, 5384.
Wright, G., and Mawby, R. J. (1971). *J. Organometal. Chem.* **29**, C29.

13

Dynamic NMR Studies of Carbonium Ion Rearrangements

L. A. TELKOWSKI and M. SAUNDERS

I. Introduction. 523
II. Generation of Stable Carbonium Ion Solutions. 524
III. Types of Carbonium Ion Rearrangements Studied 526
    A. Bond Rotations and Conformational Changes 526
    B. Hydride and Alkide Shifts 530
    C. Rearrangements via Protonated Cyclopropane Pathways 536
References. 539

I. Introduction

The bulk of the recent information available concerning stable carbonium ions in solution has been obtained by nmr spectroscopy. Evidence concerned with their ionic character, structure, energies, and rearrangement reactions has been obtained by proton and carbon magnetic resonance. Most of the questions that were under debate before these ions were prepared in stable solutions concerned the detailed mechanisms of reactions in which rearranged products are formed via carbonium ion intermediates. The rearrangement steps themselves have now been studied using the methods of dynamic nuclear resonance spectroscopy. A wealth of information—on the processes that occur readily and those which do not, on the relative energies of isomers, and the energies of transition states—is now available.

As in other studies of rates by methods of dynamic nmr spectroscopy, the full line-shape method has led to the most reliable results. It is possible to do the calculations required for virtually all carbonium ion rearrangements that have been studied through the use of readily available computer programs. Use of approximate formulas involving only coalescence temperature or some other single feature of the spectrum has led to erroneous results later demonstrated by study with the full line-shape method.

Log A of the Arrhenius equation has turned out to be around 13 (entropy of activation approximately zero) in all cases where careful experimental work has been done, so far, on carbonium ion rearrangements. This result implies that the transition states for these processes are solvated to the same degree as the starting ions. This conclusion is further confirmed by the absence of significant solvent effects in the examples of these reactions which have been studied in different solvents. Surprising though this conclusion might appear, there is precedent from other studies. Previous observations on the differences between free energies of substituted trityl cations in different solvents (water, acetonitrile) (Jensen and Taft, 1964; Taft and McKeever, 1965; Young et al., 1966) led to a similar conclusion about relative energies. The large size and the diffusion of the charge within the anionic counter-ions in these systems is the probable reason for this result. One should therefore regard reports of log A values much different from 13 (or substantial entropies of activation) in these systems with great suspicion.

It should also be noted that energy differences between carbonium ions measured in solution have been strikingly similar to differences between a number of appearance potentials for formation of carbonium ions in the mass spectrometer (Lossing and Semeluk, 1970; Yeo and Williams, 1970). Of course, solvation energies of ions in solution are certainly large, but these data again indicate that differences between the solvation energies of different carbonium ions are apparently small.

II. Generation of Stable Carbonium Ion Solutions

Most textbooks give the impression that evidence for the existence of carbonium ions was first obtained from studies of solvolysis reactions and acid-induced rearrangements where the ions are intermediates and that only recently have solutions containing permanent concentrations of these ions been obtained. But in fact, studies of stable solutions of carbonium ions considerably antedate the first suggestion of carbonium ion intermediates. In the period 1901–1910, studies (Norris, 1901a, b; Norris and Sanders, 1901; Kehrmann and Wentzel, 1901, 1902; Walden, 1902, 1903; Gomberg, 1902; von Baeyer, 1905; Hantzsch, 1907; Hoffmann and Kirmreuther, 1909; Gomberg and Cone, 1909, 1910) of triphenylmethyl chloride and alcohol in strong acids and in SO_2 indicated the formation of ions. Conductivity, freezing point depression in sulfuric acid, absorption spectroscopy, and the isolation of a crystalline perchlorate offered very strong evidence that organic cations were present. Baeyer originally used the name "*carboniumvalenz*" in order to describe these ions. The early work in this field has recently been reviewed (Nenitzescu, 1968).

13. Carbonium Ion Rearrangements

The extrapolation from the study of stable triaryl cations to considering ions as possible reaction intermediates occurred later (Meerwein and van Emster, 1922; Klipstein, 1926; Meerwein, 1927). Whitmore (1932) introduced an important unified perspective to organic chemistry by emphasizing that many reactions, regardless of their apparent complexity and diversity, may share the common mechanistic ground of carbonium ion intermediacy instead of being isolated anomalies.

Hughes and Ingold began their extensive kinetic analyses of solvolytic and other nucleophilic substitution reactions in the 1930s (Hughes and Ingold, 1933, 1935; Hughes et al., 1933, 1935; Hughes, 1935). They developed the nomenclature S_N1 and S_N2 to distinguish between reactions at saturated carbons which involve discrete carbonium ion intermediates and those in which synchronous bimolecular conversion from reactant to product occurs. The formidable body of solvolytic data accumulated in the period 1930–1950 provided the impetus for renewed efforts to isolate and study carbonium ions themselves. Solvolytic work continues to be a major source of knowledge of the nature of transition states for forming carbonium ions.

Progress in the realm of preparing solutions of carbonium ions was, prior to the early 1960s, restricted to cations that are highly stabilized by resonance. Tropylium ion (Doering and Knox, 1954) and heptamethyl benzenonium ion (Doering et al., 1958) are examples of such ions, which are even stable in aqueous acid. Actually, concentrated sulfuric acid had been, until this time, the medium predominantly used for the generation of ions. The potential of sulfuric acid for oxidation, coupled with its relatively high basicity, which encourages proton elimination from carbonium ions, strongly detract from its usefulness in generating these ions. Nevertheless, many studies of allylic cations (Deno, 1964b, 1970) were made using this medium, which supports ions in which both termini of the allylic system are alkylated. The simplest allylic ion isolated in sulfuric acid, 1,1,3,3-tetramethylallyl ion, was prepared (Deno et al., 1963) from the corresponding alcohol. However, many attempts to generate alkyl cations in sulfuric acid were unsuccessful. Addition of t-butyl alcohol or isobutylene to sulfuric acid (Deno, 1964a; Deno et al., 1962) yielded solutions of cyclic allylic ions, which are the products of polymerization, fragmentation, cyclization and rearrangement steps.

Further progress required media of much lower basicity. Brown and Pearsall (1952) showed that the combination $HCl–AlCl_3$ was capable of protonating aromatic compounds forming what he called sigma complexes (benzenonium ions). The idea of using Lewis acids alone to produce ions is found in Seel's report (1943) of the reaction of acetyl chloride with antimony pentachloride. Olah et al. (1962) extended this study to reactions of acetyl, propionyl, and benzoyl fluorides with SbF_5, AsF_5, and PF_5. He identified the structures of the products as oxocarbonium (acylium) ions and showed that the most stable solutions contained excess SbF_5. In 1963, Olah et al. (1963) prepared the t-butyl oxocarbonium ion from SbF_5 and trimethylacetyl fluoride. Decomposition occurred with the evolution of carbon monoxide and the appearance of a new peak in the nmr spectrum and one in the ir spectrum at 1835 cm^{-1}. The three singlets in the nmr spectrum were assigned to the acid chloride, t-butyl oxocarbonium ion, and t-butyl cation, respectively. The latter assignment was confirmed by dissolving t-butyl fluoride in excess SbF_5, yielding a single-line nmr spectrum of comparable chemical shift.

Following the successful generation of the t-butyl cation, Olah et al. (1964) treated various other alkyl halides with excess antimony pentafluoride. In each case, the thermodynamically most stable cation was the only product observed (e.g., 1-fluoropentane yielded t-amyl cation). Brouwer (Brouwer and Mackor, 1964) subsequently prepared stable solutions of alkyl cations using a variety of precursors: halides, alcohols, olefins, ethers, parafins, and alkyl-cyclopropanes dissolved in $HF-SbF_5$ mixtures. Antimony pentafluoride remains the most effective medium for generating stable carbonium ion solutions.

The development of vacuum line techniques has increased the number of carbonium ions that can be obtained in relatively pure solution. These methods were employed by Saunders and co-workers (1968) in preparing the secondary alkyl carbonium ion, sec-butyl. More recently, the highly unstable cyclopentadienyl cation (Saunders et al., 1973c) has been prepared by the use of an improved technique involving the formation of molecular beams of the reactants by flow through glass nozzles into a highly evacuated chamber and mixing on a surface cooled by liquid nitrogen (Saunders et al., 1973d).

Progress in the realm of carbonium ion chemistry has been a function of the ability to discover appropriate media and preparatory techniques for generating stable solutions of increasingly less stable ions to enable study through nmr spectroscopy.

III. Types of Carbonium Ion Rearrangements Studied

A. Bond Rotations and Conformational Changes

Olah (Olah and Bollinger, 1968; Olah et al., 1970c) and co-workers reported rotational barriers for several methyl substituted allyl cations. Unfortunately, complete line-shape analysis was not used, and only the coalescence temperatures were employed. In 1,1,2-trimethylallyl cation, peaks corresponding to the geminally related methyl groups coalesced at $-49°C$. A free energy of activation for the rotation process was reported as 11.7 ±

1 kcal/mole. In the 1,1,2,3-tetramethylallyl ion, substitution of an additional methyl in the 3 position was reported to raise the analogous barrier to about 16 kcal/mole ($T_c = +22°$; $\Delta G^\ddagger = 15.8 \pm 1$). Transition states for both processes should reasonably be tertiary ions unconjugated to the attached vinyl groups because of a 90° twist, and thus they should be of comparable energy.

However, the ground state of the tetramethyl species would be expected to be stabilized relative to the ground state of the trimethyl cation due to the C-3 methyl. The observed 4 kcal/mole excess in rotational barrier for the tetramethylallyl ion is consistent with this ground state effect. In the completely methylated species, 1,1,2,3,3-pentamethylallyl ion, consideration of 1,3 methyl–methyl steric repulsions in the ground state leads to a prediction of a lower barrier (Schleyer et al., 1966). The experimental value ($\Delta G^\ddagger = 13.8 \pm 1$ kcal/mole; $T_c = -11°C$) is consistent with this prediction.

For the isomeric 1,3-dimethylallyl ions, rotational barriers were measured on the chemical time scale by integration of the appropriate nmr peaks as they appeared. Saunders et al. (1969) reported an E_a of 17.5 ± 1.0 kcal/mole with log $A = 11.8 \pm 0.8$ for the interconversion cis,cis-dimethyl- to cis,trans-dimethylallyl ion and a barrier of 24.0 ± 1.0 kcal/mole with log $A = 14.0 \pm 1.0$ for the interconversion cis,trans to trans,trans isomer.

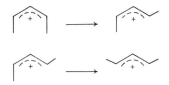

The 1,2,3-trimethylallyl systems have been studied (Olah et al., 1970c) by the same technique. The activation parameters $E_a = 16.6 \pm 1.0$ and log $A = 11.5 \pm 0.9$ for the reaction cis,cis → cis,trans and $E_a = 21.9 \pm 1.2$ with log $A = 11.6 \pm 1.0$ for cis,trans → trans,trans have been found. Within this group of rotations which involve secondary ion transition states, the similarity in barriers for the analogous dimethyl- and trimethylallyl interconversions indicates that 2-methyl substitution affects the ground and transition states to about the same extent. The higher barriers for the cis,trans → trans,trans processes (~ 23 kcal/mole) than for the cis,cis → cis,trans (~ 17 kcal/mole) are consistent with a ground state 1,3-CH_3,CH_3 interaction of 6 kcal/mole in the cis,cis isomers.

In light of the magnitudes of rotational barriers in allyl systems, it is interesting to note the experimental results in a cyclopropylcarbinyl system

(Namanworth and Kabakoff, 1970). The pmr spectrum of the dimethylcyclopropyl carbonium ion (Olah and Pittman, 1965a) shows two nonequivalent methyl groups with no temperature dependence in the range investigated. This aspect is duplicated in the ^{13}C spectrum (Olah et al., 1970a), in which the chemical shifts of the carbonium ion carbon and the cyclopropyl methylene carbons imply delocalization of the charge into the cyclopropane ring. A structure compatable with these features is a bisected one that allows overlap between the vacant p-orbital and the ring. Namanworth and Kabakoff (1970, Fig. 2b) were able to measure the rotational barrier that effects equilibration of the methyl peaks via the magnetization transfer technique. The transition state for this process is the tertiary ion conformation in which the vacant p orbital is "perpendicular" to the ring. The activation energy was found to

$$R = CH_4$$

be 13.7 ± 0.4 kcal/mole with a log A of 12.2 ± 0.4. This barrier is strikingly similar to the rotational barriers involving tertiary ion transition states in the allyl systems cited above. In all these systems, if one assumes that resonance stabilization of the carbonium ion is eliminated in a perpendicularly rotated transition state, the barriers are a direct measure of this stabilization.

The pmr spectrum of the 7-norbornadienyl cation was first reported in 1962 (Saunders and Story, 1962). The ion has an unsymmetrical structure, since the spectrum shows distinct peaks for the "bound" and "unbound" vinyl hydrogens, which remain sharp up to +45°C (Winstein et al., 1967a), the temperature at which decomposition occurred. In an effort to observe a scrambling process, Winstein and co-workers (1967b) prepared a precursor to the ion that contained a deuterium atom at all four vinyl positions. Upon ionization, a degenerate five-carbon scrambling process ($\Delta G^\ddagger$ = 16.7 kcal/ mole) redistributed the two deuterium atoms of the unbound vinyl group so that the resulting deuterium content at the unbound positions was 40% while the bound vinyls retained 100% D. By looking for protium incorporation into the bound vinyl positions, Winstein was able to detect a "bridge-flipping" process that exchanges the bound and unbound vinyls (see Winstein et al., 1967a, c). The free energy of activation is 19.6 kcal/mole ($k = 8 \times 10^{-4}$ sec^{-1} at -2.5°C) and represents the free energy difference between the unsymmetrical ion and the symmetrical transition state species.*

* Other possible mechanisms for this process are 1,2 shifts of the unbound vinyl carbon from C-1 to C-2 or C-4 to C-3 and ring contraction involving the unbound vinyl group.

13. Carbonium Ion Rearrangements

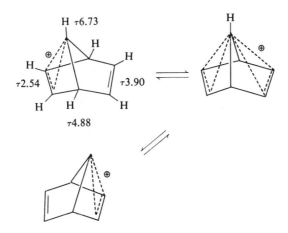

To confirm the bridge-flipping mechanism, Winstein (Winstein et al., 1967b) also prepared the 7-methylnorbornadienyl ion, whose nmr spectrum displayed temperature dependence. The bound and unbound vinyl peaks broadened as the temperature was raised above −45°C and eventually coalesced at −14°C. With the exception of sharpening of the bridgehead (C-1, C-4) proton signal, no other spectral changes were observed in this temperature range. Since only the vinyl protons were seen to exchange, the bridge-flip mechanism was confirmed. 7-Methyl substitution might have relatively little effect on the ground state ion, but it is expected to strongly stabilize the transition state, as more positive charge should reside at C-7 in the symmetrical species where olefinic electron involvement is reduced. The free energy barrier, 12.4 kcal/mole ($k = 189$ sec^{-1} at −14°C), is 7 kcal/mole lower than the barrier for the unmethylated ion and is consistent with this reasoning. Subsequently, Winstein (Winstein et al., 1967c) published the results of 7-phenyl- and 7-methoxynorbornadienyl ion syntheses, designed to reflect progressively greater stabilization of the bridge-flip transition state and concomitant lower barriers. The spectra of both ions exhibited a four-proton triplet in the vinyl region, which showed little or no broadening down to −100°C, indicative of the substantially lower ($\Delta G^\ddagger \leq 7.6$ kcal/mole) energy barrier to bridge flip in these species. A static symmetric structure was discounted for these ions. In the 7-phenylnorbornadienyl ion all phenyl protons absorb within a 9 Hz range, contrary to what would be predicted for the symmetric species with substantial charge residing at C-7.

A ring inversion barrier for the cis,cis-1,2,3,4,5,6-hexamethylcyclohexenyl cation has been reported by Hogeveen (Hogeveen and Volger, 1968; Hogeveen et al., 1969). From a line-shape analysis, the activation parameters E_a = 8.9 kcal/mole and log A = 10.8 were determined. This barrier seems surprisingly high in view of the very low inversion barriers in cyclohexyl rings containing one or two sp^2 hybridized carbons. In particular, Anet (Anet and Haq, 1965) has found a free energy of activation of 5.3 kcal/mole for cyclohexene, and $\Delta G^{\ddagger} \leq 5.1$ kcal/mole (Jensen and Beck, 1966) has been reported for cyclohexanone. Hogeveen has looked at several other cyclohexenyl cations, including 1,3,5,5-tetramethyl- (to $-100°C$), 1-hydroxy-3,5,5-trimethyl- (to $-130°C$), and 1,3-dihydroxy-5,5-dimethylcyclohexenyl (to $-130°C$), and was unable to observe any line broadening. In all cyclohexenyl ions, five of the ring carbons should be close to a plane in the ground state because of the three sp^2 carbons of the allyl system.

Rakshys et al. (1971) have examined trityl cations substituted with difluoromethyl groups that served as indicators of syn–anti isomerism and simultaneously of optical activity in the favored propellerlike conformation. The fluorine spectrum is expected to give separate A–B quartets for both syn and anti isomers, each being further split by the hydrogen on the same carbon into octets as observed. Over the temperature range $-40°$ to $+25°C$, the fluorine spectrum changed to a simple doublet. Line-shape analysis yielded rate constants for the syn–anti and racemization reactions. The ratio of these rate constants was compared with predictions generated from the several mechanisms that had to be considered. The "two ring flip" process, where a single ring remains conjugated with the carbonium carbon and the others pass through perpendicularity to the plane of the central carbon, was found to be most readily consistent with the experimental results after making certain assumptions.

B. Hydride and Alkide Shifts

Of the various hydride and alkide shifts found in alkyl carbonium ions (Brouwer and Hogeveen, 1972; Karabatsos and Fry, 1970), the degenerate 1,2 shifts have been found in general to be extremely facile. For example, an upper limit on the activation energy for the 3,2-hydride shift in the s-butyl cation has been set at 6 kcal/mole (Saunders et al., 1968). The 3,2-methide shift in the 2,2,3-trimethylpentyl ion has been reported (Namanworth, 1967) to have $\Delta G^{\ddagger} \leq 4.0$ kcal/mole. It should be noted that the proposed equilibrating open ion formulation of these species, as opposed to static, bridged structures, are favored by Olah (Olah and White, 1969c) as a result of the analysis of C-13 spectra of these ions.

A prominent exception to this trend is the norbornyl cation. Line-shape analysis (Saunders et al., 1964) revealed a relatively high barrier, $E_a = 10.8 \pm 0.6$ kcal/mole with $\log A = 12.3$, to the 3,2-hydride shift in this ion (from Brouwer and Hogeveen, 1972, Fig. 37a). Norbornyl cation has been assigned a protonated cyclopropane structure (Olah and White, 1969a, b) as a result of a study of its proton and carbon magnetic resonance spectra. If this species is stabilized relative to the open ion, it would also be stabilized relative to the transition state for the hydride shift, and the barrier would therefore be raised.

Three slow 1,2-alkide shifts have been reported in the literature. The first of these is the methide shift in the 9,9,10-trimethylphenanthrenonium ion

(Koptyug et al., 1968). The large activation energy ($E_a = 12.1 \pm 0.2$ kcal/mole; log $A = 13.5 \pm 0.2$) determined by a line-shape analysis may be attributed to resonance stabilization through delocalization of the charge in the ground state ion. The methide shift process passes through a bridged transition state close to a protonated cyclopropane. Such a structure would clearly not be stabilized to the same degree by resonance. Winstein and Poulter (1969) also observed what may be considered slow methyl interchange in the 1-butyl-1,3,3-trimethylallyl ion via line-shape changes (Winstein and Poulter, 1969, p. 3650).

Coalescence of the C-1 methyl and *t*-butyl peaks occurred at +75°C ($E_a = 14.3$ kcal/mole; log $A = 11.5$). The simplest suggested mechanism is the 1,2-methide shift to the open homoallyl ion. However, methyl exchange may equally well be envisioned to occur via direct isomerizaton to the related cyclopropylcarbinyl system, avoiding the homoallyl species as an intermediate. Finally, a slow 1,2-alkide shift formally accounts for the temperature dependence of the nmr spectrum of the 1,2,3,5,7-pentamethyladamantyl ion (Schleyer et al., 1974). In this case, coalescence of the C-2 methyl peak with the peak for the equivalent C-1, and C-3 methyls was seen, and line-shape fitting yielded $E_a = 12.1 \pm 0.4$ kcal/mole and log $A = 13.3 \pm$

0.3. The mechanism proposed is a series of three Wagner–Meerwein shifts involving two degenerate tertiary protoadamantyl intermediates:

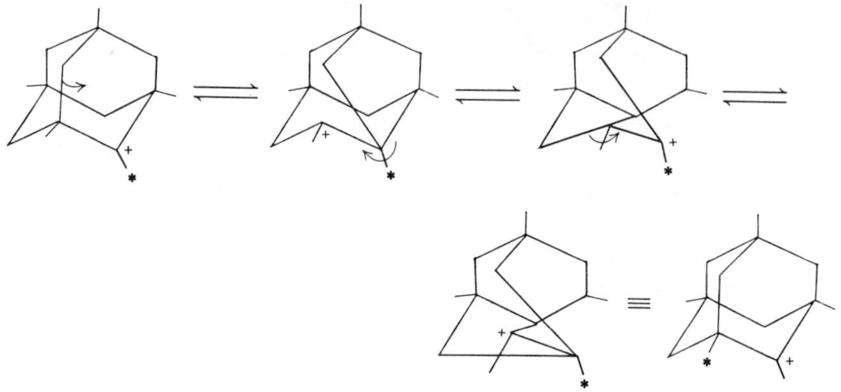

The experimental activation energy agrees well with a 9 kcal/mole energy difference between the tertiary adamantyl and protoadamantyl ions calculated (Schleyer et al., 1974) by molecular mechanics methods.

For systems in which 1,2-hydride shifts interconvert nondegenerate species, the energy barriers are considerably higher, reflecting chiefly the energy difference between the ions themselves. An example is the 1,2-shift in methylcyclopentyl ion, first reported by Brouwer (1968), which interchanges the α and β methylene hydrogens. Saunders and Rosenfeld (1969) determined the activation parameters, $E_a = 15.4 \pm 0.5$ kcal/mole and $\log A = 13.0 \pm 0.3$, for this process, which includes an initial hydride shift from a tertiary to a secondary ion. It is proposed (Brouwer and Hogeveen, 1972) that hydrogen migration alone, rather than alternating hydrogen and methyl migration, is the energetically more favorable mechanism, although in the analogous acyclic system (t-amyl cation) methide shifts are required (Brouwer and Hogeveen, 1972, Fig. 46b).

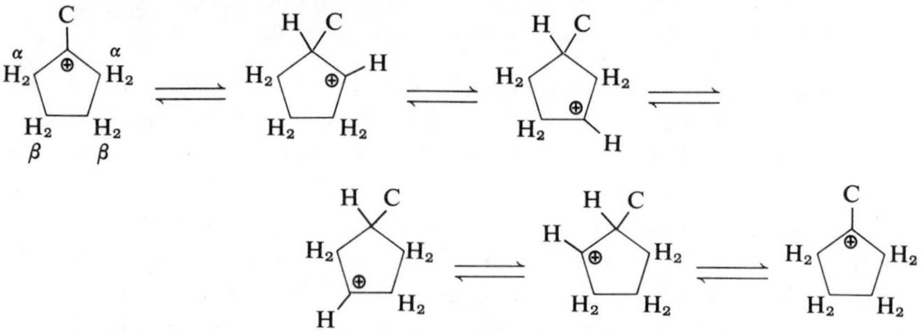

Several higher-order hydride shifts have been observed in acyclic ions. A rapid 1,3-hydride shift was shown (Brouwer and van Doorn, 1969) to occur

13. Carbonium Ion Rearrangements

in the 2,4-dimethyl-2-pentyl cation. Saunders and Stofko (1973, Fig. 2) have measured $E_a = 8.5 \pm 0.1$ kcal/mole and $\log A = 12.6 \pm 0.1$ for the shift from line-shape changes at low temperature. They have also found an

approximate barrier ($E_a = 12$–13 kcal/mole) for the 1,4-hydride shift in the 2,5-dimethyl-2-hexyl ion via the magnetization transfer technique. The next higher homolog, 2,6-dimethyl-2-heptyl cation, was also investigated. Only a single averaged methyl peak was observed down to $-90°C$, placing an upper limit of 6 or 7 kcal/mole on the activation energy for the 1,5-hydride shift.

Nondegenerate 1,2-shift rearrangements have been reported in cyclic alkenyl ions. Among the alkenyl species, temperature-dependent nmr spectra have been observed (Saunders and Berger, 1972) for the cyclopentenyl cation. A mechanism involving a hydride shift to an adjacent carbon to yield the unconjugated ion reversibly and also one involving a single step 1,4-hydride shift were considered. On the basis of the theoretically calculated curves, derived from the appropriate transition probability matrices, the unconjugation mechanism (Saunders and Berger, 1972) was found to account for the line broadening. The Arrhenius parameters $E_a = 18.0 \pm 0.9$ kcal/mole and $\log A = 12.2 \pm 0.6$ were obtained. Although no quantitative measurements

were made, a process that interchanges allylic and alkyl hydrogens in the cyclohexenyl cation (Saunders et al., 1973b) has been observed via deuterium scrambling at $-65°C$. A degenerate mixing process has also been reported for the 1,2,3,4,4,5-hexamethylcyclopentenyl cation (Brouwer and van Doorn, 1970, Eq. 2). Line-shape fit of the high temperature spectra yielded $\Delta G^{\ddagger} = 23.7$ kcal/mole, $\Delta H^{\ddagger} = 29.4$ kcal/mole, and $\Delta S^{\ddagger} = +13$ kcal/mole-deg at 150°C, which can be converted to $E_a = 30.2$ kcal/mole. The proposed mechanism is a 1,2-hydride shift (analogous to that in the unsubstituted cyclopentenyl ion) from C-5 to C-1, followed by a 1,2-methide shift from C-4 to C-5.

Many examples of degenerate shift rearrangements in cyclohexadienyl ions prepared by protonation of substituted benzenes have been published. The benzenonium ion itself (Olah et al., 1970b) displays temperature-dependent spectra below $-100°C$. Application of line-shape techniques gave an E_a of 10 kcal/mole with $\log A = 15.9 \pm 1.6$ for the 1,2-hydride shift process that interconverts degenerate protonated benzene species (Olah et al., 1970b). This barrier is reasonably comparable to others determined by line-

shape analysis for benzenonium ions undergoing rapid 1,2-hydride shifts. For example in the hexamethylbenzenonium ion (Brouwer et al., 1965) hydride migration around the ring has a barrier of 11.3 kcal/mole with $\log A$ of 13.6. Hydride exchange between two adjacent carbons in the prehnitenium (1,2,3,4-tetramethylbenzenonium) and 5,8-dimethyltetralinium ions (Brouwer et al., 1965) has activation parameters $E_a = 11.6$ kcal/mole with $\log A = 13.3$ and $E_a = 11.2$ kcal/mole with $\log A = 13.3$, respectively. No line broadening was seen in the spectrum of 2,3-dimethoxybenzenonium ion up to $-20°C$. The apparent slow hydride shift in this species suggested that the rate-determining factor for hydride migration in benzenonium ions is the magnitude of the partial positive charge at the adjacent carbon. The strong mesomeric effect of the methoxy groups tends to reduce this charge magnitude.

Migration barriers in benzenonium ions for groups other than hydrogen have also been measured. For example, line broadening and eventual coalescence of the methyl groups into two peaks was observed in spectra of the heptamethylbenzenonium ion (Saunders, 1967) above room temperature. This behavior could be explained in terms of discrete 1,2-methide shifts or a random shift mechanism involving a hexamethylbenzene–methyl ion π complex intermediate. The latter alternative was ruled out on the basis of

line-shape fitting using the appropriate probability matrices. Arrhenius parameters $E_a = 15.2$ kcal/mole and $\log A = 12.3$ were found, indicating a substantially higher barrier to methyl migration in benzenonium ions. An

13. Carbonium Ion Rearrangements

exchange process that scrambles all of the methyls in nitrohexamethylbenzenonium ion (Olah et al., 1972) has been reported. In this case, the line-shape changes were compatible with nitro group migration, and an activation energy of 16.8 ± 1.5 kcal/mole was calculated.

An interesting class of alkide migration is that composed of bridging group migrations. Three such migrations have been reported. The first is inferred (Winstein et al., 1967a, p. 6351) from a degenerate five-carbon scrambling process seen in the 7-norbornadienyl cation. The rate of this rearrangement was measured on the chemical time scale by means of deuterium labeling. The process was seen to sequentially scramble all carbons except those of the "bound" vinyl group. The mechanism proposed involved a ring contraction to a bicyclo[3.2.0]heptadienyl cation (via a 1,2-shift of one of the bound vinyl carbons) followed by a ring expansion (via a 1,2-shift of the other bound vinyl carbon) back to a norbornadienyl species. Repetition

of these steps results in migration of the bound vinyl group around the five-membered ring defined by the five other carbon atoms. The activation parameters $\Delta G^{\ddagger} = 16.7$ kcal/mole and $\Delta H^{\ddagger} = 14.5 \pm 2.0$ kcal/mole ($E_a = 15$ kcal/mole) were determined for this case.

Temperature dependence was reported (Winstein and Childs, 1968) in the nmr spectra of the heptamethylbicyclo[3.1.0]hexenyl cation, which was generated photochemically by ring closure of the heptamethylcyclohexenyl ion. At $-87°$C, coalescence of the five methyl groups on the cyclopentenyl ring was observed. At higher temperatures (up to $-9°$C), one sharp statistically averaged 15-proton singlet was seen, while the two methyl groups on the cyclopropane methylene carbon (C-6) remained distinct. The fivefold degenerate scrambling process apparently involves migration of C-6 around the cyclopentenyl ring (Winstein and Childs, 1968, Fig. V). According to the Woodward–Hoffmann formalism (Woodward and Hoffmann, 1969), it is a series of concerted sigmatropic 1,4 shifts that are constrained to be suprafacial by the present ring system. The thermally allowed shift is predicted to occur with inversion at the migrating carbon at each step. As a result of this criterion, the "inside" methyl on C-6 should remain "inside," and the "outside" methyl should remain "outside," exactly as experimentally observed. The free energy of activation for the scrambling process was calculated to be 9.0 kcal/mole. The fact that no line broadening of the C-6 methyls was seen up to $-9°C$ permits a $\Delta G^{\ddagger}$ of at least 14.7 kcal/mole to be calculated for the forbidden thermal 1,4 shift. Interestingly, the analogous fivefold degenerate rearrangements in the 1,2,3,4,5-pentamethylbicyclo[3.1.0]hexenyl and 1,2,3,4,5,6-hexamethylbicyclo[3.1.0]hexenyl ions could not be observed on the nmr time scale. Winstein speculated that this was probably a manifestation of the considerable positive charge borne by C-6 in the transition state. Evidence in this regard was presented in a study (Saunders *et al.*, 1971) of the unsubstituted bicyclo-[3.1.0]hexenyl cation. The free energy barrier for the fivefold degenerate rearrangement of this ion was found to be 15 ± 1 kcal/mole by measuring the rate of deuterium scrambling in the 2-deuterio cation.

C. Rearrangements via Protonated Cyclopropane Pathways

One species that has frequently been invoked as participating in rearrangements of carbonium ions is protonated cyclopropane (Saunders *et al.*, 1973a), the name assigned to any structure containing a three-membered ring with seven substituent atoms or groups. The exact role of this species, whether as transition state or intermediate on the reaction pathway, and the details of its structure (bond character, charge distribution, etc.) are subjects of continuing debate. However, the fact that such species are intermediates now rests on firm ground on the basis of observed ion rearrangements as well as kinetic and product studies of solvolytic reactions.

The simplest carbonium ion prepared thus far (Saunders and Hagen, 1968b) is the isopropyl cation. Temperature dependence of the pmr spectra in the range $0°-40°C$ was seen, indicating intramolecular exchange of the two types of hydrogen. The activation energy was found via line-shape analysis to be 16.4 ± 0.4 kcal/mole, with $\log A = 13.2 \pm 0.3$. The simplest mechanism that accounts for the observed exchange and is consistent with a 16 kcal/mole barrier is reversible isomerization to the primary propyl ion (Saunders *et al.*, 1973a, p. 55). Investigation of the [2-^{13}C]isopropyl cation revealed that

13. Carbonium Ion Rearrangements

carbon scrambling also occurs in this ion. This cannot be accounted for on the basis of the *n*-propyl cation mechanism and requires a mixing process involving a protonated cyclopropane.

The *s*-butyl cation is believed to isomerize via a protonated methylcyclopropane species (Saunders *et al.*, 1968). The rapid, degenerate 3,2-hydride shift in the *s*-butyl ion has already been mentioned. A second process is observed at high temperatures that scrambles all protons. The Arrhenius activation parameters $E_a = 7.5 \pm 0.1$ kcal/mole and $\log A = 12.3 \pm 0.1$ were obtained from a line-shape fitting of rates. The low magnitude of this barrier, compared to that for hydrogen mixing in the isopropyl cation, suggests that a different mechanism must be operative in this case. The most probable one involves closure to a corner-protonated methylcyclopropane intermediate, followed by degenerate corner-to-corner proton rearrangement and ring opening (Saunders *et al.*, 1973a, p. 56). Rearrangement of *s*-butyl ion

to *t*-butyl ion is also observed, with an activation energy calculated to be about 18 kcal/mole. The similarity of this barrier to that for proton interchange in isopropyl ion is consistent with the view that both rearrangements proceed through similar transition states, namely structures close to primary ions.

Two rearrangement processes were observed in the *t*-amyl cation, and their energy barriers were measured via line-shape techniques. The first is intramolecular interchange of the two types of methyl groups (Saunders and Hagen, 1968a), which occurs with an E_a of 15.3 ± 0.2 kcal/mole and $\log A = 13.2 \pm 0.2$. The simplest proposed mechanism involves isomerization to the secondary ion, followed by a typical Wagner–Meerwein 1,2-methyl shift, which by nature requires a protonated dimethylcyclopropane species, and 1,2-hydride shift back to the tertiary ion (Saunders *et al.*, 1973a, p. 57). Above 80°C, a slower process, which mixes the methylene protons with the methyl protons, was detected (Saunders and Rosenfeld, 1969). An activation energy of 18.8 ± 1.0 kcal/mole, with log A = 13.2 ± 0.5 was found. The matrix of probabilities used to obtain these values corresponds to a mechanism identical to that above, except that it involves corner-to-corner proton migration within the protonated dimethylcyclopropane. Isomerization via primary ions was considered unlikely in this instance, as their energies would be significantly higher than tertiary ion energies than the experimentally determined barriers warrant.

The behavior of the methylcyclopentyl cation is analogous to that of the t-amyl ion. Interchange of the α and β ring hydrogens has already been noted. At higher temperatures, a process that mixes all protons was seen (Saunders and Rosenfeld, 1969). An activation energy of 18.2 ± 0.1 kcal/mole with log A of 13.6 ± 0.1 was found for the interchange of ring and methyl hydrogens. To investigate further the nature of the rearrangement process, a mixture of methylcyclopentyl cations containing ^{13}C- and deuterium-labeled methyl groups was studied in order to simultaneously determine carbon and proton scrambling rates. The results were consistent with a mechanism in which the protonated bicyclo[3.1.0]hexane undergoes corner-to-corner proton migration to yield a species which then reversibly opens to cyclohexyl cation more rapidly than it returns to its predecessor (Saunders *et al.*, 1973a, p. 57).

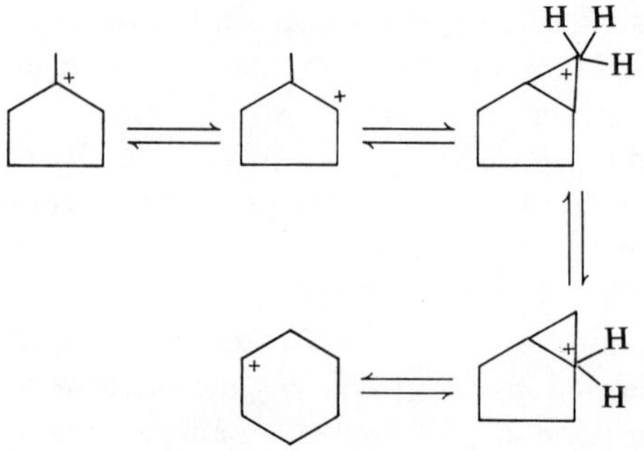

The methyl interchange rate in the dimethylisopropylcarbonium ion was measured (Saunders and Vogel, 1971) via temperature-dependent changes in the spectra of various deuterated versions of the ion. Line-shape analysis yielded an activation energy of 12.4–13.5 kcal/mole with log A of 12.5–13.5 for the overall process, which includes closure to a protonated cyclopropane, opening to a secondary ion, and closure to a different protonated cyclopropane.

Finally, the barrier to proton migration about a protonated cyclopropane ring system has been measured in the norbornyl cation (Olah and White, 1969a,b). Line-shape changes in spectra of the ion in the temperature range

−128° to −156°C were found to correspond to an E_a of 5.9 ± 0.2 kcal/mole with log A = 12.7 for the process.

REFERENCES

Anet, F. A. L., and Haq, M. A. (1965). *J. Amer. Chem. Soc.* **87**, 3147.
Brower, D. M., and Hogeveen, H. (1972). *Progr. Phys. Org. Chem.* **9**, 179.
Brower, D. M., and Mackor, E. (1964). *Proc. Chem. Soc., London* p. 147.
Brouwer, D. M., and Oelderik, J. M. (1968). *Rec. Trav. Chim. Pays-Bas* **87**, 721.
Brouwer, D. M., and van Doorn, J. A. (1969). *Rec. Trav. Chim. Pays-Bas* **88**, 573.
Brouwer, D. M., and van Doorn, J. A. (1970). *Rec. Trav. Chim. Pays-Bas* **89**, 333.
Brouwer, D. M., MacLean, C., and Mackor, E. L. (1965). *Discuss. Faraday Soc.* **39**, 121.
Brown, H. C., and Pearsall, H. W. (1952). *J. Amer. Chem. Soc.* **74**, 191.
Deno, N. C. (1964a). Quoted in Olah *et al.* (1964).
Deno, N. C. (1964b). *Progr. Phys. Org. Chem.* **2**, 129.
Deno, N. C. (1970). *In* "Carbonium Ions" (G. A. Olah and P. von R. Schleyer, eds.), Vol. 2, p. 783. Wiley (Interscience), New York.
Deno, N. C., Richey, H. G., Liu, J. S., Hodge, J. D., Houser, J. J., and Wisotsky, M. J. (1962). *J. Amer. Chem. Soc.* **84**, 2016.
Deno, N. C., Bollinger, J., Friedman, N., Hafer, K., Hodge, J. D., and Houser, J. J. (1963). *J. Amer. Chem. Soc.* **85**, 2998.
Doering, W. von E., and Knox, L. H. (1954). *J. Amer. Chem. Soc.* **76**, 3203.
Doering, W. von E., Saunders, M., Boynton, H. G., Earhart, H. W., Wadley, E. F., Edwards, W. R., and Laber, G. (1958). *Tetrahedron* **4**, 178.
Gomberg, M. (1902). *Ber. Deut. Chem. Ges.* **35**, 2397.
Gomberg, M., and Cone, L. (1909). *Justus Liebigs Ann. Chem.* **370**, 142 and 193.
Gomberg, M., and Cone, L. (1910). *Justus Liebigs Ann. Chem.* **376**, 183.
Hantzsch, A. (1907). *Z. Phys. Chem. (Leipzig)* **61**, 257.
Hoffmann, K., and Kirmreuther, H. (1909). *Ber. Deut. Chem. Ges.* **42**, 4856.
Hogeveen, H., and Volger, H. C. (1968). *Rec. Trav. Chim. Pays-Bas* **87**, 1047.
Hogeveen, H., Gaasbeek, C. J., and Volger, H. C. (1969). *Rec. Trav. Chim. Pays-Bas* **88**, 379.
Hughes, E. D. (1935). *J. Chem. Soc., London* p. 255.
Hughes, E. D., and Ingold, C. K. (1933). *J. Chem. Soc., London* p. 1517.
Hughes, E. D., and Ingold, C. K. (1935). *J. Chem. Soc., London* p. 244.
Hughes, E. D., Ingold, C. K., and Patel, C. S. (1933). *J. Chem. Soc., London* p. 526.
Hughes, E. D., Gleave, J. L., and Ingold, C. K. (1935). *J. Chem. Soc., London* p. 236.
Jensen, E. D., and Taft, R. W. (1964). *J. Amer. Chem. Soc.* **86**, 116.
Jensen, F. R., and Beck, B. H. (1966). *Tetrahedron Lett.* p. 4287.
Kehrmann, F., and Wentzel, F. (1901). *Ber. Deut. Chem. Ges.* **34**, 3815.
Kehrmann, F., and Wentzel, F. (1902). *Ber. Deut. Chem. Ges.* **35**, 622.
Klipstein, K. K. (1926). *Ind. Eng. Chem.* **18**, 1328.
Koptyug, V. A., Shubin, V. G., Korchagina, D. V., and Rezvukhin, A. I. (1968). *Dokl. Akad. Nauk SSSR* **179**, 119.
Lossing, F. P., and Semeluk, G. P. (1970). *Can. J. Chem.* **48**, 955.
Meerwein, H. (1927). *Justus Liebigs Ann. Chem.* **455**, 277.
Meerwein, H., and van Emster, K. (1922). *Ber. Deut. Chem. Ges.* **55**, 2500.
Namanworth, E. (1967). Quoted in Olah and Lukas (1967).

Namanworth, E., and Kabakoff, D. S. (1970). *J. Amer. Chem. Soc.* **92**, 3234.
Nenitzescu, C. D. (1968). *In* "Carbonium Ions" (G. A. Olah and P. von R. Schleyer, eds.), Vol. 1, p. 1. Wiley (Interscience), New York.
Norris, J. F. (1901a). *Amer. Chem. J.* **25**, 117.
Norris, J. F. (1901b). *Chem. Zentralbl.* **1**, 699.
Norris, J. F., and Sanders, W. W. (1901). *Amer. Chem. J.* **25**, 54.
Olah, G. A., and Lukas, J. (1967). *J. Amer. Chem. Soc.* **89**, 2227.
Olah, G. A., and Pittman, C. U., Jr. (1965). *J. Amer. Chem. Soc.* **87**, 2998(1) and 5123(b).
Olah, G. A., and White, A. M. (1969a). *J. Amer. Chem. Soc.* **91**, 3956.
Olah, G. A., and White, A. M. (1969b). *J. Amer. Chem. Soc.* **91**, 6883.
Olah, G. A., and White, A. M. (1969). *J. Amer. Chem. Soc.* **91**, 5801.
Olah, G. A., and Bollinger, J. M. (1968). *J. Amer. Chem. Soc.* **90**, 6082.
Olah, G. A., Kuhn, S., Tolgyesi, W., and Baker, E. (1962). *J. Amer. Chem. Soc.* **84**, 2733.
Olah, G. A., Tolgysei, W., Kuhn, S., Moffatt, M., Bastion, I., and Baker, E. (1963). *J. Amer. Chem. Soc.* **85**, 1328.
Olah, G. A., Baker, E., Evans, J., Tolgyesi, W., McIntyre, J., and Bastien, I. (1964). *J. Amer. Chem. Soc.* **86**, 1360.
Olah, G. A., Kelly, D. P., Jeuell, C. L., and Porter, R. D. (1970a). *J. Amer. Chem. Soc.* **92**, 2544.
Olah, G. A., Schlosberg, R. H., Kelly, D. P., and Mateescu, G. D. (1970b). *J. Amer. Chem. Soc.* **92**, 2546.
Olah, G. A., Bollinger, J. M., and Brinick, J. M. (1970c). *J. Amer. Chem. Soc.* **92**, 4025.
Olah, G. A., Lin, H. C., and Mo, Y. K. (1972). *J. Amer. Chem. Soc.* **94**, 3667.
Rakshys, J. W., Jr., McKinley, S. V., and Freedman, H. H. (1971). *J. Amer. Chem. Soc.* **93**, 6522.
Saunders, M. (1967). "Magnetic Resonance in Biological Systems." p. 85. Pergamon, Oxford.
Saunders, M., and Berger, R. (1972). *J. Amer. Chem. Soc.* **94**, 4049.
Saunders, M., and Hagen, E. L. (1968a). *J. Amer. Chem. Soc.* **90**, 2436.
Saunders, M., and Hagen, E. L. (1968b). *J. Amer. Chem. Soc.* **90**, 6881.
Saunders, M., and Rosenfeld, J. (1969). *J. Amer. Chem. Soc.* **91**, 7756.
Saunders, M., and Stofko, J. J. (1973). *J. Amer. Chem. Soc.* **95**, 252.
Saunders, M., and Story, P. R. (1962). *J. Amer. Chem. Soc.* **84**, 4876.
Saunders, M., and Vogel, P. (1971). *J. Amer. Chem. Soc.* **93**, 2561.
Saunders, M., Olah, G. A., and von R. Schleyer, P. (1964). *J. Amer. Chem. Soc.* **86**, 5680.
Saunders, M., Hagen, E. L., and Rosenfeld, J. (1968). *J. Amer. Chem. Soc.* **90**, 6882.
Saunders, M., Rosenfeld, J. C., von R. Schleyer, P., and Su, T. M. (1969). *J. Amer. Chem. Soc.* **91**, 5174.
Saunders, M., Vogel, P., Hasty, N. M., and Berson, J. A. (1971). *J. Amer. Chem. Soc.* **93**, 1551.
Saunders, M., Vogel, P., Hagen, E. L., and Rosenfeld, J. (1973a). *Accounts Chem. Res.* **6**, 53.
Saunders, M., Vogel, P., Seybold, G., and Wiberg, K. B. (1973b). *J. Amer. Chem. Soc.* **95**, 2045.
Saunders, M., Berger, R., O'Neil, J., Jaffe, A., McBride, J. M., Breslow, R., Hoffman, J. M., Perchonock, C., Wasserman, E., Hutton, R. S., and Kuck, V. J. (1973c). *J. Amer. Chem. Soc.* **95**, 3017.
Saunders, M., Cox, D., and Olmstead, W. (1973d). *J. Amer. Chem. Soc.* **95**, 3018.
Schleyer, P. von R., Van Dine, G. W., Schollkopf, U., and Paust, J. (1966). *J. Amer. Chem. Soc.* **88**, 2868.

Scheleyer, P. von R., Saunders, M., Telkowski, L., Vogel, P., Lenoir, D., and Mison, P. (1974). *J. Amer. Chem. Soc.* **96**, 2157.
Schleyer, P. von R., Saunders, M., Telkowski, L., et al. (1974). *J. Amer. Chem. Soc.* (submitted for publication).
Seel, F. (1943). *Z. Anorg. Allg. Chem.* **250**, 331.
Taft, R. W. and McKeever, L. D. (1965). *J. Amer. Chem. Soc.* **87**, 2489.
von Baeyer, A. (1905). *Ber. Deut. Chem. Ges.* **38**, 569.
Walden, P. (1902). *Ber. Deut. Chem. Ges.* **35**, 2018.
Walden, P. (1903). *Z. Phys. Chem.* **43**, 443.
Whitmore, F. C. (1932). *J. Amer. Chem. Soc.* **54**, 3274.
Winstein, S., and Childs, R. F. (1968). *J. Amer. Chem. Soc.* **90**, 7146.
Winstein, S., and Poulter, C. D. (1969). *J. Amer. Chem. Soc.* **91**, 3649.
Winstein, S., Lustgarten, R. K., and Brookhart, M. (1967a). *J. Amer. Chem. Soc.* **89**, 6350.
Winstein, S., Brookhart, M., and Lustgarten, R. K. (1967b). *J. Amer. Chem. Soc.* **89**, 6352.
Winstein, S., Brookhart, M., and Lustgarten, R. K. (1967c). *J. Amer. Chem. Soc.* **89**, 6354.
Woodward, R. B., and Hoffmann, R. (1969). "The Conservation of Orbital Symmetry," p. 128. Academic Press, New York.
Yeo, A. N. H., and Williams, D. H. (1970). *Chem. Commun.* p. 737.
Young, A. E., Sandel, V. R., and Freedman, H. H. (1966). *J. Amer. Chem. Soc.* **88**, 4532.

14

Conformational Processes in Rings

F. A. L. ANET and RAGINI ANET

I. Introduction.. 543
II. Definitions of Conformational Processes......................... 544
   A. General Considerations..................................... 544
   B. Ring Inversion... 546
   C. Local Ring Inversion....................................... 547
   D. Ring Pseudorotation.. 547
   E. Geometric Constraints in Rings............................. 550
   F. Conformational Depictions and Labeling Schemes............. 551
III. Conformational Energy Calculations............................. 554
   A. Force-Field (Strain-Energy) Calculations................... 554
   B. Applications of Force-Field Calculations to Conformational Processes... 558
   C. Quantum Mechanical Calculations............................ 572
IV. Experimental Data and General Discussion....................... 574
   A. Kinetic Parameters from DNMR.............................. 574
   B. Six-Membered Rings.. 578
   C. Seven-Membered Rings...................................... 592
   D. Eight-Membered Rings...................................... 597
   E. Nine-Membered and Larger Rings............................ 604
   References... 613

I. Introduction

Extensive experimental information on conformational processes in ring systems has been obtained during the last fifteen years, largely from applications of dynamic nuclear magnetic resonance (dnmr) spectroscopy. Other experimental techniques, including ultrasonic absorption, mechanical relaxation, nmr double resonance, and conventional low-temperature rate measurements have also been used, but only in isolated cases. Theoretical calculations of conformational processes have developed rapidly in recent years and have provided an important framework for interpreting the data obtained by experimental methods.

Certain processes, although taking place in rings, involve only minor conformational changes of the ring skeleton and will not be considered in this chapter. Some of these processes, however, are reviewed in Chapters 8 and 9. Other topics excluded from this chapter are atom inversion in rings (where this is the only process taking place as in aziridines) and rotation about single or partial double bonds abutting rings (again where this is the only process occurring, as in t-butylcyclohexane). Four- and five-membered rings have extremely low conformational barriers and cannot be investigated by dnmr. In these cases, an approach based on vibrational analysis is most useful (Laane, 1971) but will not be reviewed here. Thus, the main emphasis in this chapter will be on cycloalkanes and their derivatives and on heterocyclic analogs of the above compounds, where the conformational barriers are greater than about 3.5 kcal/mole. Because of the very large literature on applications of dnmr to conformational processes in rings, only illustrative examples will be discussed in this chapter. Additional data on conformational barriers are presented in tables, which, however, are not meant to be comprehensive in their coverage of compounds.

II. Definitions of Conformational Processes

A. General Considerations

Conformational processes can be defined as hindered rotations about single bonds (Eliel *et al.*, 1965; Hanack, 1965). The instantaneous rotational state about a particular bond in a cyclic molecule can be characterized by a torsional or dihedral angle. We shall use the sign convention shown in **1** whereby

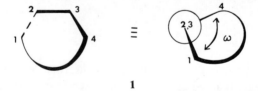

1

the dihedral angle made by the planes 123 and 234 is positive (Romers *et al.*, 1969). This angle will be referred to as the (ring) dihedral or torsional angle about the 2,3 bond, and will be given the symbol ω_{1234}, for simplicity ω_2. Any conformational process, in general, will involve changes in ring torsional angles and is best defined in terms of these angles. Variations in the internal

14. Conformational Processes in Rings

bond angles and bond lengths may of course also take place. Furthermore, changes in torsional and bond angles external to the ring have to be considered if substituents are present.

A formal permutational approach, as described in Chapter 2, is hardly feasible for analyzing processes in most rings. For example, the number of theoretically possible permutations for the protons in any one conformation of cyclodecane is 20!, a truly astronomical number ($\simeq 10^{18}$). Fortunately, the stereochemical rigidity of carbon and the intramolecularity of most processes under consideration allow only a few permutations to take place. In the cyclohexane chair, only one process is expected to be detected by dnmr, i.e., ring inversion, because the boat conformations occur in concentrations that are too small to be detected and the only exchange possible is one that does not break any bonds. In the nomenclature of Ugi *et al.* (1970), conformational processes are regular polytopal rearrangements; i.e., they do not involve breaking and reforming of bonds.

Conformational processes in rings that are fluxional structures, or which contain stereochemically nonrigid atoms (e.g., nitrogen in cyclic amines) can give overlapping dnmr effects and require careful analysis. Complications can also occur in cyclic amines and alcohols (because of intermolecular proton exchange) and in compounds such as the boron trifluoride complexes of cyclic ketones (because of dissociation and recombination).

Certain terms, such as ring inversion and pseudorotation, have been widely employed in discussions of conformational processes. Other expressions have found only restricted usage as, for example, the process called "*Version*" in the German literature (Kabuss *et al.*, 1966a). Binsch *et al.* (1971) have suggested that the exchange of identical ligands be called topomerization, and they have presented subclassifications of this phenomenon.

There are two aspects that need to be differentiated in a discussion of conformational processes. One aspect concerns the mechanistic details of transformations. This requires a knowledge of the energy as a function of arbitrary changes in the geometry of a molecule. Because of the number of atoms in molecules of interest, only approximate calculations can be carried out and it is generally necessary to severely limit the number of geometries investigated.

Experimental data, on the other hand, give direct information on the overall exchange process but provide only the enthalpy and entropy of activation at some unknown transition state along the reaction coordinates. Certain paths can of course be eliminated because they do not lead to the exchange required by the experimental data. The remaining paths can only be differentiated to some extent by qualitative energy considerations, but in general, calculations of the kind described previously are needed to make further progress.

B. Ring Inversion

The term "ring inversion" has been widely used to describe the change from one chair form to the alternate chair form in six-membered rings (Eliel *et al.*, 1965). It has also been used for analogous processes in other ring systems. Lambert and Oliver (1968) have suggested that the word "reversal" be used to describe ring inversion and that the word "inversion" be restricted to atom (e.g., nitrogen) inversion. However, these words have such general meanings that they are nearly always qualified in any case (e.g., ring inversion), and thus there seems no great advantage in the proposed nomenclature change, even when both ring and atom inversion occur in the same molecule.

Definitions of ring inversion have been discussed by Anet (1974). In a conformation like the chair form of cyclohexane, a strict definition of ring inversion is possible: The process is a ring inversion if all of the ring dihedral angles change their signs, but their absolute magnitudes remain unchanged. Bond lengths and internal angles do not change as a result of ring inversion. It should be noted that this definition says nothing about the path followed during the inversion process and therefore does not imply any particular mechanism. As mentioned in Section II,A, dnmr does not give direct informa-

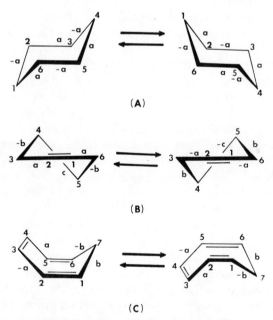

Fig. 1 Examples of ring inversion: (A) chair cyclohexane; (B) half-chair cyclohexene; (C) cycloheptatriene. Small letters indicate dihedral angles.

tion about mechanisms, and thus the term "ring inversion" is useful in describing appropriate exchange processes detected by dnmr.

While the definition given above is unambiguous, it does not apply to many situations. For example, ring inversion of axial methylcyclohexane to equatorial methylcyclohexane does not result in the ring torsional angles keeping accurately constant their absolute magnitudes, although the signs of these angles do change. Since the above usage of ring inversion is common and of value, Anet (1974) extended the strict definition of ring inversion as follows: Ring inversion causes all the signs of the ring dihedral angles to change (except that dihedral angles that are close to 0° or 180° may or may not change signs) and the absolute magnitudes of these angles are either unaffected or slightly changed. A further extension is possible and is useful in discussing nmr results: The dihedral angles can be averaged over rapidly interconverting conformations. Examples of ring inversions are shown in Fig. 1.

C. Local Ring Inversion

The change from the chair form to the boat form in a six-membered ring involves changes in signs of two dihedral angles, while two other dihedral angles remain virtually unchanged (Fig. 2). The remaining two dihedral angles have their values substantially altered, but the change is not simply a sign inversion. In this situation a "local ring inversion" can be said to occur in that part of the molecule where the torsional angles change their signs. A local ring inversion, as defined here, is precisely the same as the process which Friebolin and his co-workers call a "*Version*" in the German literature (Kabuss et al., 1966a; Schmid et al., 1969; von Bredow et al., 1970a, 1971).

The word "*Version*" is not convenient to describe this process in the English language, because of other meanings of this word. It is for this reason that we suggest acceptance of the expression "local ring inversion," which also has the merit of being self-explanatory. Other examples of local inversions are shown in Fig. 2 and include the changes from boat to chair forms of cycloheptene, and the changes from the chair–chair to the boat–chair and then to the boat–boat conformation in cyclooctane.

D. Ring Pseudorotation

The term pseudorotation was first applied to cyclopentane (Kilpatrick et al., 1947; Pitzer and Donath, 1959) to describe the dynamic conformational property of this molecule. Like inversion, it has an atomic analog in five-coordinate compounds, such as phosphorus pentafluoride (Berry, 1960). The expression "pseudorotation" means "false rotation," and it is therefore

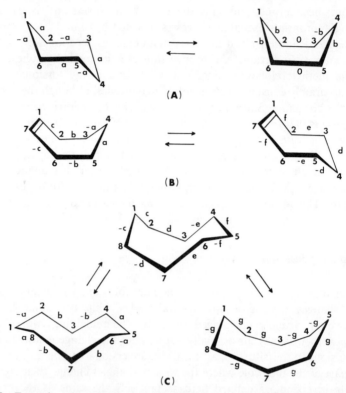

Fig. 2 Examples of local inversion: (A) chair to boat interconversion in cyclohexane; (B) chair to boat interconversion in cycloheptene; (C) chair-chair to boat-chair to boat-boat interconversion in cyclooctane. Small letters indicate dihedral angles.

appropriate for any conformational process that results in a conformation superposable on the original, but which may differ from the original in being rotated about one or more axes (Anet, 1974). Pseudorotation, in analogy with real molecular and internal rotations, can be free, as in cyclopentane, or more or less hindered, as in cycloheptane and higher cycloalkanes. In moderately to severely hindered pseudorotation, distinct stable conformations that are pseudorotation partners can be visualized, and these cases are often amenable to study by dnmr. If the barrier to pseudorotation is very low, or in the limit when pseudorotation is free, it is not really justified to consider separate conformations (e.g., the C_2 and C_s forms of cyclopentane), because there is actually only one stable conformation, and the pseudorotation is simply a molecular vibration.

Ring inversion (when strictly defined) of achiral conformations is actually a particular pseudorotation. Ring inversion in the cyclohexane chair by a

14. Conformational Processes in Rings

flattening mechanism that maintains the D_{3d} symmetry, for example, leaves the molecule apparently rotated by 60° along the C_3 axis. Nevertheless, in order to conform with common usage, we will exclude from the definition of pseudorotation any process that can be defined as ring inversion.

As with ring inversion, it is desirable to extend the (strict) definition of pseudorotation given above. In order to do this, it is convenient to introduce a suitable abstraction, which can be called the ring skeleton of a conformation (Anet, 1974). The ring skeleton is the geometric figure corresponding to the conformation, with ring bonds considered as straight lines, and ring atomic nuclei considered as the vertices of the figure. In the extended definition, pseudorotation, either free or hindered, is a process that results in the apparent rotation of the ring skeleton of the conformation, with minor changes in the skeleton being ignored. This definition can be extended still further to cover chiral conformations by allowing the comparison to be made with either the original skeleton or its mirror image. As before, it is desirable to exclude from this definition those processes that can be called ring inversion (in the extended definition).

The present definition of hindered pseudorotation, like the definition of ring inversion, is independent of the details of the interconversion path. Hendrickson (1964, 1967b) has used pseudorotation to indicate a particular mechanism of conformational interconversion, but we feel that this is a less desirable definition, particularly from the point of view of dnmr spectroscopy. On the whole, the (extended) definitions given here for both ring inversion and pseudorotation are in close agreement with the usage of these terms in the literature. Typical examples of pseudorotations are given in Fig. 3.

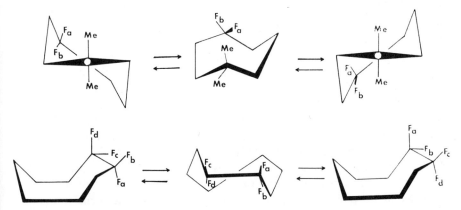

Fig. 3 Examples of ring pseudorotation: (*top*) interconversion of two enantiomeric twist-chair forms of 4,4-difluoro-1,1-dimethylcycloheptane, with a chair form as an intermediate; (*bottom*) interconversion of two superposable boat-chair forms of 1,1,2,2-tetrafluorocyclooctane, with a twist-boat form as an intermediate.

E. Geometric Constraints in Rings

Constraints of a mathematical (geometrical) character exist in rings. The most outstanding example is perhaps the contrast between the rigid chair and the flexible boats of cyclohexane, as first described by Sachse (1890). This difference can easily be seen on molecular (e.g., Dreiding) models, and is in agreement with the results of strain-energy calculations presented in Section III. The mechanical models, however, have no torsional constraints, unlike real molecules. Also, the models have fixed bond lengths and rather stiff bond angles, features that are not very close to reality.

Dunitz (1970) and Dunitz and Waser (1972) have inquired into the geometric factors that determine why certain rings are rigid while others are flexible. Their conclusion is that a nonplanar six-membered ring with fixed bond lengths and angles is flexible if it possesses a "nonintersecting" twofold axis of symmetry, otherwise it is rigid. A nonintersecting axis is one that does not pass through either atoms or bonds. Torsional effects are assumed to be absent, so that the geometric requirement for flexibility is correct for Dreiding models, but can be seriously in error for real molecules, as was carefully pointed out by the above authors. The nearly free pseudorotation in the boat form of cyclohexane is a result of a constant sum for the torsional and non-bonded strain, and, in a way, this is a mathematical accident, which cannot be expected to hold generally. The rigidity of the chair form is more interesting, since it is dependent only on having reasonably stiff internal bond angles. However, the absence of a nonintersecting twofold axis does not guarantee any real rigidity. For example, the boat forms of tetrahydropyran should be rigid according to the (strict) geometric criterion, but models show that the distortions required for pseudorotation to take place are insignificant. Thus, these boat forms are flexible in practice. The chair form of tetrahydropyran, on the other hand, is clearly rigid like that of cyclohexane.

Seven- and higher-membered rings with no torsional constraints have at least one degree of freedom and are thus flexible in the above geometric sense. In fact, with the exception of cycloheptane, conformational barriers in medium and large (up to C_{14}) cycloalkanes are in the range of 5 to 8 kcal/mole (Anet *et al.*, 1972), so that the geometric approach is not valuable in this series.

The introduction of an essentially rigid torsional constraint such as a double bond reduces the available degrees of freedom in a ring. Dunitz and Waser (1972) have discussed the situation in *cis,cis*-1,5-cyclooctadiene from their geometric point of view. The chair form (**2-C**) of the diene is rigid, whereas the twist-boat form (**2-TB**) is flexible and can pseudorotate to a boat (**2-B**) and a skew form (**2-S**), as can easily be seen on models. In the boat family, a twofold nonintersecting axis is maintained during the pseudorotation. Experimental nmr data (Anet and Kozerski, 1973; Montecalvo *et al.*,

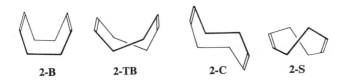

2-B   2-TB   2-C   2-S

1973) and strain-energy calculations (F. A. L. Anet and L. Kozerski, unpublished work, 1973) do show the chair to be more rigid than the twist-boat, but not by very much (see also Sections III and IV). In the case of *cis,cis*-1,4-cyclooctadiene, no conformation can have a twofold nonintersecting axis. Yet, examination of models shows that the 1,4-diene can exist in a rigid and an apparently flexible form. The skeleton of the latter form has nearly, but not quite, a twofold nonintersecting axis, thus explaining its virtual flexibility.

On the whole, the geometric approach along the lines described above is severely limited by the assumption of negligible nonbonded interactions and torsional barriers. It gives certain insights on the question of conformational rigidity and flexibility, but it is much more applicable to Dreiding molecular models than to real molecules.

F. Conformational Depictions and Labeling Schemes

Conformational names such as chair and boat are very useful in describing six-membered rings, and precisely defined extensions of these names to seven-, eight-, nine-, and ten-membered cycloalkanes have been suggested by Hendrickson (1964, 1967b). Examples of this naming scheme are given in Figs. 2 and 3. As in cyclohexane, a twist form has an intersecting C_2 axis, whereas a nontwist form always has at least one plane of symmetry. Hendrickson (1967b) has also proposed a shorthand notation to describe medium ring cycloalkane conformations that have symmetry. An axis or a plane of symmetry passing through the ring is represented by a horizontal line. The signs of the dihedral angles are then arranged above and below this line as shown in Fig. 4. A bond that lies on the reference element of symmetry has its torsional sign placed alongside the line.

Wiberg (1965) has used a representation which consists of a more or less regular polygon, with the corners labeled 0, +, or −, depending on whether the carbons are in, above, or below the general plane of the polygon, respectively. Although this representation is easy to visualize and can be written quickly, it is not applicable to many conformations, especially in the larger rings.

A very useful scheme, particularly for large cycloalkanes, has been proposed by Dale (1973a), who uses a wedge representation as shown in Fig. 5. These perspective drawings are views looking down on the general plane of the molecule and have some advantages over perspective drawings made with side views. Horizontal bonds are indicated by uniform lines and wedges are used to show inclined bonds, with the thick end of the wedge indicating the higher part of the bond. The addition of numbers representing torsional angles to these drawings (see Fig. 5) is very helpful. Conformations are considered to have "sides" and "corners." Corner atoms are situated between ring bonds whose torsional angles have the same sign (or where one torsional angle is zero). The "sides" of a conformation are terminated by "corner" atoms.

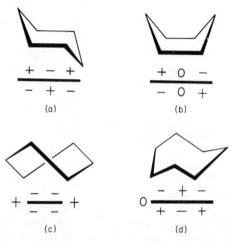

Fig. 4 Nomenclature for ring conformations, according to Hendrickson (1967b): (a) chair cyclohexane; (b) boat cyclohexane; (c) twist-boat cyclohexane; (d) chair cycloheptane.

Dale has introduced a shorthand notation based on the concepts described above. The numbers of bonds on consecutive sides of a conformation, starting with the shortest side, are concatenated and placed in square brackets, as shown in Fig. 5. Torsional angles of zero degrees are not found in medium- and large-ring cycloalkane conformations which are energy minima. A conformation having a single torsional angle of 0° is an energy maximum which is a potential transition state for some conformational process. In such a case, the "1" in the notation name is underlined to indicate this fact, as shown in Fig. 5.

In any general discussion of conformational interconversions, a need arises for a double-labelling scheme: one scheme is required for the various conformational sites, and a second scheme to label the atoms (or groups of atoms) in the molecule of interest. While formal double-labeling procedures are often not needed in very simple systems, a failure to use such schemes in more complex situations easily leads to confusion. A very general labeling

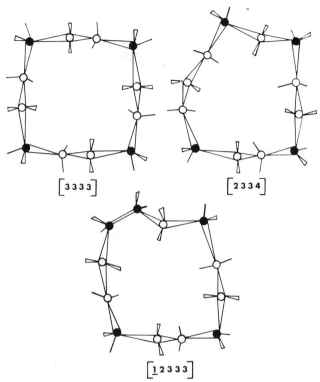

Fig. 5 Wedge representation of ring conformations for cyclododecane; [3333] and [2334] represent energy minima, [12333] is a transition state for the interconversion of the [3333] and [2334] forms (Dale, 1973a). Black circles indicate corner positions.

method has been used by Ugi *et al.* (1970) and Klemperer (Chapter 2). A permutational (e.g., conformational) isomer is assigned to a $2 \times n$ matrix

$$\binom{l}{s} = \binom{123 \ldots n}{ijk \ldots l},$$

where the top row lists the indices of *ligand* labels (ligands are not necessarily linked to a central atom in this context but can represent ring carbons and hydrogens) in an increasing sequence, and below each ligand index is placed the index of the *skeletal* (e.g., conformational) position which that ligand occupies. For example, 1,2,3 and 4 could be the indices of ring carbon atoms at sites indexed by i, j, k, and l respectively of a given conformation (the matrix may be superscripted to identify the conformation). A conformational process can then be defined mathematically by a permutation operation which rearranges the indices of the skeletal positions in the bottom row of the matrix.

The operation depicted as $(i)(jkl)$, for example, is defined according to standard permutation notation as follows: site i is unchanged, j is replaced by k, k is replaced by l, and l is replaced by j. Thus, carbons 1,2,3 and 4 are changed to occupy sites i, k, l, and j respectively as a result of the operation. In the mathematical model given above, both skeletal positions and ligands are indexed by numbers (i, j, k, etc. stand for numbers, the letters being used purely for convenient identification in the present discussion). The distinction between the indices for ligands and skeletal positions is clear because of their different positions in the $2 \times n$ matrix describing a particular isomer. In less general double-labeling schemes, ligands can be indexed with letters (Roman or Greek), while skeletal positions are indexed with numbers (or vice versa.) Sometimes the double-labeling scheme can be implicit, as in the interconversion of e-chlorocyclohexane and a-chlorocyclohexane, where e (equatorial) and a (axial) are skeletal labels in the chair form.

III. Conformational Energy Calculations

A. Force-Field (Strain-Energy) Calculations

Ideally, the properties of molecules should be calculated by quantum-mechanical methods. Because of the drastic assumptions that are required to make the computation time reasonable, these methods are of limited application and value (see Section III,C). Thus, by default, semiempirical calculations have found wide applicability and have developed a great deal since the early applications reviewed by Westheimer (1956). Hendrickson (1961) introduced the use of computers to reduce the time required for these calculations. He also critically examined the parameters that should be used for cycloalkane conformations. More recently, efficient iterative computer programs have been developed so that the computation cost of obtaining the strain-energy and geometry of a conformation of a medium-ring hydrocarbon amounts to only a few dollars. The reliability of strain-energy calculations, however, is not easy to evaluate for molecules that are structurally quite different from those molecules on which the calculations have been parametrized.

Despite the development of elaborate computer programs for strain-energy calculations, there still appears to be a place for very crude (but simple) calculations, which can even be done satisfactorily without a computer. The procedure used by Dale (1973a,b,c) for exploring the conformations and conformational processes in large cycloalkane rings, discussed later in this section, falls in this category. Even more drastic assumptions than Dale's have been made by Saunders (1967), who picked out the ring conformations belonging to the "diamond lattice." In this procedure, only those conformations with gauche ($\pm 60°$) and anti (180°) torsional angles are considered. Both Saunders' and Dale's approaches can be regarded as giving conformations that are appropriate starting points for more sophisticated computations.

A rather special kind of force-field calculation has been developed by Pickett and Strauss (1970), but has been applied so far only to a limited number of six-membered rings. It requires a knowledge of the normal modes of vibration, especially those (six in number for cyclohexane) connected with ring bending motions. A set of angle bending and torsional functions consistent with vibrational frequencies are then obtained. It should be noted that the torsional functions used here include nonbonded interactions and are not comparable to the torsional functions previously discussed in this chapter. With the assumption of constant bond lengths, and by treating methylene groups as single units, it is possible to describe in an explicit fashion all possible conformations in terms of only two parameters, which can be considered to be complex internal coordinates. This approximation allows the energy to be calculated for every conceivable geometry and to be plotted to give pretty contour maps in two dimensions. However, it must be remembered that the approximations made are quite significant, as shown by the fact that the calculated barrier for ring inversion in cyclohexane is too high by about 3 kcal/mole. It thus appears that this method is of limited generality, and is not particularly accurate even when it can be applied.

In semiempirical strain-energy calculations, the atoms in a molecule are considered to be constrained by a force field, which is defined in terms of internal molecular coordinates, e.g., bond lengths and bond angles. Some information about force fields can be obtained from vibrational spectroscopy and intermolecular interactions, but in the final analysis, the force field can be considered to be empirical and thus chosen to reproduce the geometries, energies, and vibrational frequencies for as large a number of molecules as possible. Some authors have concentrated on fitting energies and geometries well and more or less ignored vibrational frequencies. Others have concluded that all properties should be fitted equally well and have developed "consistent force fields." Only a brief mention of some of the force fields that have been used will be given here. More extensive reviews can be consulted for additional details (Williams *et al.*, 1968; Buckingham and Sargeson, 1971).

The force field typically consists of the following components, which are assumed to be independent of one another in the simplest treatment (i.e., cross-terms are assumed to be zero):

1. The energy contribution (i.e., excess strain) from bond lengths is given by a Hooke's law function $E(r) = \tfrac{1}{2}k_r(r - r_0)^2$, where r_0 and k_r are the unstrained bond length and the force constant, respectively.

2. The energy contribution for bond angle bending is also given by a

Hooke's law function $E(\theta) = \frac{1}{2}k_\theta(\theta - \theta_0)^2$, where θ_0 is the unstrained bond angle and k_θ is the force constant.

3. The torsional energy contribution for a three-fold barrier is given by $E(\phi) = \frac{1}{2}k_\phi[1 + \cos 3\phi]$, where k_ϕ is the barrier height corresponding to the torsional strain.

4. The energy contribution due to non-bonded interactions (excluding 1,3 interactions, which are included implicitly in the angle bending term) is given by

$$E_{nb}(r) = -A/r^6 + B\exp(-Cr) \quad \text{or by} \quad E_{nb}(r) = -A/r^6 + B/r^{12}.$$

The first term is attractive, but is overcome by the second and repulsive term at very short distances.

The force field described above can be elaborated to include anharmonic terms, dipole–dipole repulsions, and torsional barriers other than threefold. The main problem in the force-field approach is to determine values for the constants in the equations given above. The situation is not too difficult in saturated hydrocarbons, but many parameters are needed once more complex features, such as double bonds, heteroatoms and aromatic rings, are introduced. Some typical force-field parameters are given in Table I.

In the context of conformational processes, it is only necessary to have calculations of relative strain energies, since the chemical binding energy is constant. In other calculations, where different molecules are compared, or where quantities such as calculated heats of formation are required, it is necessary to include a term for the chemical binding energy, and this is generally done by an empirical additivity scheme.

As directly calculated, the strain energy refers to a fictitious motionless state at 0°K. In some calculations (Boyd, 1968; Lifson and Warshel, 1968), the vibrational frequencies are calculated, and hence the zero point energy and the vibrational enthalpy at temperatures above 0°K become available. In fact, all the thermodynamic parameters can, in principle, be obtained, provided that the calculated vibrational frequencies are accurate. In practice, these frequencies often have considerable errors, and some authors have chosen to include quantities such as the zero point energy in the empirical chemical binding energy, thus simplifying the calculations without introducing great errors.

Iterative programs can locate the position of an energy minimum (which may be a local minimum) and several mathematical approaches have been tried, the aim being to achieve rapid convergence. The method of steepest descent was used by Wiberg (1965) in the first application of the iterative approach to strain-energy calculations and has been modified and improved by Allinger *et al.* (1971). Others have used the Newton–Raphson procedure

TABLE I
Typical Force-Field Parameters for Saturated Hydrocarbons[a]

Strain-energy terms[b]	Unstrained values of angles or distances	Force constants		
Bond stretching[c]				
C—C	$r_0 = 1.53$	$k_r = 4.4$		
C—H	$r_0 = 1.09$	$k_r = 4.55$		
Bond angle bending[d]				
C—CH$_2$—C	$\theta_0 = 1.937$	$k_\theta = 0.8$		
C—CH—H	$\theta_0 = 1.91$	$k_\theta = 0.608$		
H—C—H	$\theta_0 = 1.88$	$k_\theta = 0.508$		
Torsional strain[e]				
>C—C<	$\phi = \pm\pi/3$ or π	$k_\phi = 0.0158$		
Nonbonded interactions[f]		*A*	*B*	*C*
C$\cdots$C	—	4.45	104	3.09
C$\cdots$H	—	0.96	30	3.415
H$\cdots$H	—	0.19	18.4	3.74

[a] From Wiberg and Boyd (1972).
[b] Energies in units of 10^{-11} erg/molecule; r in angstroms; θ, ϕ in radians.
[c] $E(r) = \frac{1}{2}k_r(r - r_0)^2$.
[d] $E(\theta) = \frac{1}{2}k_\theta(\theta - \theta_0)^2$.
[e] $E(\phi) = \frac{1}{2}k_\phi(1 + \cos 3\phi)$.
[f] $E^{bn}(r) = -A/r^6 + B\exp(-Cr)$.

or various pattern-search techniques (Williams *et al.*, 1968; Boyd 1968; Lifson and Warshel, 1968; Engler *et al.*, 1973).

The modified Newton–Raphson procedure used by Boyd (1968) and by Lifson and Warshel (1968) starts with a calculation of the strain energy for a given trial geometry and will be briefly described. The energy is written in terms of internal coordinates (bond distances, bond angles, etc.) and expanded in a Taylor series through quadratic terms about the trial geometry. It is then transformed to external (Cartesian) coordinates. Differentiation results in a set of linear algebraic equations for displacements from the trial coordinates. The displacements are obtained by solving the linear equations by matrix methods and are then used to calculate a new set of atomic coordinates, and the process is repeated until the displacements are nearly zero.

Wiberg and Boyd (1972) have recently modified Boyd's iterative program so that the energy minima can be sought for constrained conformations. The most useful constraint consists of fixing in advance the dihedral angle(s) of one bond or of several bonds. This is a great improvement for investigating conformational processes, compared to older procedures. For example,

Hendrickson (1967b), in his pioneering work on conformational interconversions in the medium-ring cycloalkanes, used a noniterative procedure that was satisfactory only provided that an element of symmetry was retained during the interconversion. Thus, asymmetric processes could not be investigated quantitatively by Hendrickson.

Finally, we include a brief description of Dale's exploratory calculations mentioned at the beginning of this section. Dale completely ignores bond length and bond angle strains. The nonbonded interactions are used as all-or-nothing effects, except for 1,4-butane interactions. That is, only conformations that have 1,5 and higher hydrogen interactions above 1.8 Å are considered for large rings; for medium rings, this requirement is somewhat relaxed. The strain energy is then calculated from the torsional angles by using a smooth (sine) function based on the observed torsional potential in butane. The strain energies for various butane conformations, relative to the anti form, used by Dale are as follows: gauche (torsional angle 65°), 0.8 kcal/mole; anti–gauche barrier (torsional angle 120°), 3.8 kcal/mole; eclipsed form (torsional angle 0°), 6.9 kcal/mole. The dihedral angles were measured directly from Fieser–Dreiding models to the nearest 5°. Because of the simplicity of the calculations, which merely require reading the energy from a graph for each torsional angle, Dale was able to examine a large number of conformations and to study conformational interconversions as described in the next section.

B. Applications of Force-Field Calculations to Conformational Processes

1. Noniterative Calculations

a. *Exploratory Calculations.* Dale (1973b) has used his exploratory method, which was described briefly in the previous section, to analyze interconversions in the cycloalkanes from C_9 to C_{16}. He concludes that the most favorable mechanism for conformational interconversions in this series involves an "elementary process" which causes the migration of "corner" positions (see Section II,F) as illustrated in Fig. 6 for the situation where the bent chain keeps its overall shape as much as possible. In larger rings, the bonds having torsional angles of 120° in Fig. 6 can relax from these eclipsed states. The transition state is assumed by Dale to have one bond with zero torsional angle. In medium and large rings, a succession of elementary processes can result in virtually arbitrary conformational interconversions. As a simple example, the conformational processes in cyclododecane will be discussed in some detail. The [3333] form, which is the lowest-energy conformation of this hydrocarbon (see Section II,F for this nomenclature), can

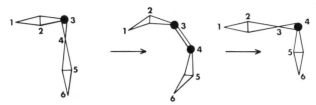

Fig. 6 Migration of a corner position (black circle) in an elementary process (Dale, 1973b).

undergo the changes shown in Fig. 7. This form can go over to the [2433] form, which lies 4.4 kcal/mole above the ground state conformation, by means of the transition state [12333], already depicted in Fig. 5. A succession of four very similar elementary processes returns the molecule to the [3333] form with all the corner carbons changed to noncorner carbons. The noncorner carbons, which incidentally are all located in similar sites in the [3333] form, do not all change during the above pseudorotation; one half remain noncorner and the other half become corner. There is also an equivalent process that causes the pseudorotation to take place in the opposite direction to that shown in Fig. 7. Both processes give rise to mirror images of the original chiral [3333] form. The changes in proton sites as a result of the pseudorotation shown in Fig. 7 can be described as follows: The [3333] form has eight isochronous isoclinal protons on corner carbons, and two sets of eight isochronous protons on noncorner carbons, one set being axial and the other equatorial. The pseudorotation causes one-half of the isoclinal protons to become axial and one-half to become equatorial. One-half of the axial (and equatorial) protons are unchanged, and the other half are changed to isoclinal protons. Only a limited amount of nmr data is available on cyclododecane (see Section IV,E),

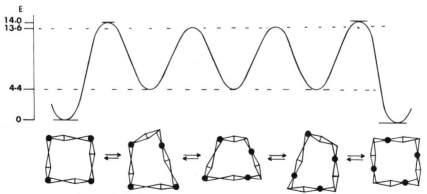

Fig. 7 Conformational processes in the [3333] or square conformation of cyclododecane (Dale, 1973b). Black circles represent carbon atoms and not conformational positions (*cf.* Fig. 5 and 6).

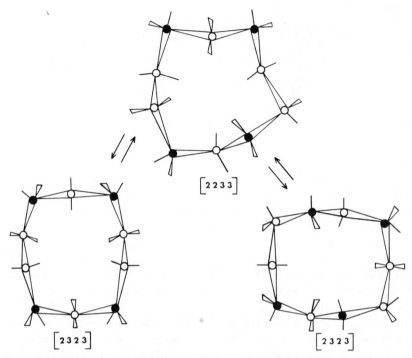

Fig. 8 Pseudorotation of the [2323] or boat–chair–boat conformation of cyclodecane via the [2233] conformation. Black circles indicate carbon atoms (see Fig. 7).

but what is known about the temperature dependence of the cmr and pmr spectra supports Dale's interconversion mechanism. However, the barrier calculated by Dale for cyclododecane (and other cycloalkanes) are all substantially too high because of the approximations made in the calculations.

Pseudorotation in the [2323] or boat–chair–boat conformation of cyclodecane can take place by two successive elementary processes operating on diametrically opposite corners. The intermediate [2233] conformation is calculated to be only 1.4 kcal/mole less stable than the [2323] form (Fig. 8). This particular mechanism has been discussed previously by Dunitz (1971). The barrier height calculated by Dale for this pseudorotation is 7.9 kcal/mole, which is somewhat higher than the experimentally determined barrier (Anet et al., 1972; Noe and Roberts, 1972), (see Section IV,E). By contrast, Hendrickson (1967b) was unable to find a low-energy path for this pseudorotation because of the symmetry restrictions of his approach.

Cyclotetradecane is an interesting hydrocarbon because it appears to be conformationally homogeneous and exists in the nearly unstrained diamond lattice [3434] conformation. Pseudorotation of this form can take place via

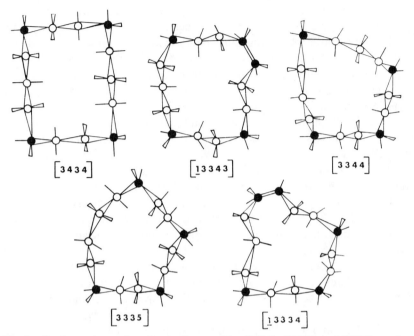

Fig. 9 Conformations of cyclotetradecane: The [3434], [3344], and [3335] are local energy minima, with the [3434] having the lowest energy; [13343] is a transition state for the interconversion of [3434] and [3344], while [13334] is a transition state for the interconversion of [3344] and [3335]. Black circles indicate corner positions.

the [3344] intermediate form, which is calculated to be 1.7 kcal/mole above the ground conformation. The [13343] barrier for this process is calculated to be 13.8 kcal/mole, again much higher than the observed value (about 7.0 kcal/mole). The actual interconversion scheme is possibly more complicated than this, however, since the [3344] conformation can itself pseudorotate via a relatively high-energy [3335] form over the [13334] barrier, which is calculated to be 13.0 kcal/mole. Wedge representations of these conformations are shown in Fig. 9.

F. A. L. Anet and A. K. Cheng (unpublished work, 1973) have developed a kinetic scheme that is applicable to the exchange of carbons in the [3434] conformation and takes into account all significant pathways involving the conformations and transition states given in Fig. 9. The kinetic scheme allows calculations of cmr line-shape changes to be made and it is found that the line shapes depend on the relative heights of the [13343] and [13334] barriers. Although the low-temperature cmr spectrum of cyclotetradecane contains four well-resolved lines in the ratio of 2:2:2:1 (F. A. L. Anet and A. K. Cheng, unpublished work, 1973; Cheng, 1973), spectra at intermediate rates of exchange have not as yet been obtained with a sufficient signal-to-noise ratio to make meaningful mechanistic conclusions. A further difficulty is that the low-temperature spectrum must be correctly assigned. In view of these uncertainties, we will not discuss the conformational processes in cyclotetradecane in further detail, but it is worth mentioning that this is one cycloalkane where nontrivial ^{13}C dnmr effects exist and are potentially capable of giving valuable information on interconversion paths.

Dale (1973b) has also investigated odd-membered cycloalkane conformations, but relatively little is known experimentally about these compounds except for cyclononane (Anet and Wagner, 1971). The principles outlined above for even-membered conformations also hold for the odd-membered rings and will not be further elaborated in this chapter.

The effect of placing *gem*-dimethyl groups on conformational energies and on barrier heights is extremely interesting (Dale, 1973c). In general, atoms or groups much larger than hydrogen strongly prefer to occupy corner sites, because these are always unhindered. Certain exchanges, however, require these groups to occupy hindered noncorner sites in the transition-state conformations, thus raising the barrier heights. Other exchanges do not have this requirement, and these barriers are relatively unaffected. The 1,1-dimethyl derivative of cyclodecane is an excellent example of this type (Dale, 1973c; F. A. L. Anet and J. J. Wagner, unpublished work, 1972). The two methyl groups are in nonequivalent sites when attached to a corner atom of the [2323] form. Pseudorotation of the kind described previously for the parent compound can now occur in two nonequivalent fashions. In one case, the methyls remain on the corner, but exchange sites. The barrier of this process should be, and actually is, close to that of the parent compound. The two geminal protons on C-2, however, do not exchange as a result of this pseudorotation. These protons only exchange as a result of further pseudorotations that force the methyls into noncorner positions (Fig. 10). The barrier (about 9 kcal/ mole) for this process is observed to be much higher than that in the parent compound (F. A. L. Anet and J. J. Wagner, unpublished work, 1972).

In 1,1-dimethylcyclononane (Anet and Wagner, 1971), the methyl groups are attached to a carbon which lies on a C_2 axis of symmetry in the [333] conformation and thus the cmr spectrum is temperature independent. The ring protons, however, do give a spectrum which is temperature dependent, because they are attached to carbons that do not lie on C_2 axes of symmetry. As in the ten-membered ring, the barrier (9 kcal/mole) for the exchange process visible in the ring proton spectrum, is substantially higher than barrier (6 kcal/mole) in cyclononane itself. The barrier in 1,1,4,4-tetramethylcyclononane, which also has the [333] conformation, is about 20 kcal/mole (Borgen *et al.*, 1972). In the tetramethyl compound, exchange requires either that the methyl groups take up positions with very large methyl–methyl

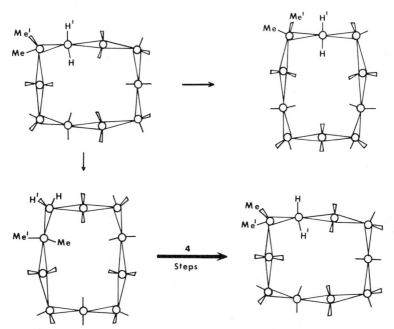

Fig. 10 Pseudorotations in 1,1-dimethylcyclodecane with the *gem*-dimethyl groups at a corner position of the [2323] conformation. One pseudorotation step (see Fig. 8) exchanges the methyl groups (top), but exchange of the protons on C-2 requires five successive pseudorotation steps, and forces the methyl groups into unfavorable noncorner positions (e.g. bottom left conformation).

interactions, or that an inherently very unfavorable interconversion path be followed. The barrier for the dimethyl compound involves only methyl–hydrogen interactions and it is thus understandable that the effects of *gem*-dimethyl groups on the barrier height is far from being additive.

b. *Vibrational Calculations.* Pickett and Strauss (1970) have applied their vibrational approach, which was described briefly in Section III,A, to cyclohexane, tetrahydropyran, 1,3- and 1,4-dioxane, and 1,3,5-trioxane. Table II shows calculated kinetic parameters for ring inversion in the above compounds, together with relevant experimental data. The calculations show that the transition state for ring inversion in cyclohexane should be considered as an almost freely pseudorotating structure, and not just in terms of the lowest-energy transition-state structure. The entropy of activation for ring inversion, however, is essentially the same for both models, being about 4 eu, and is also close to the best experimental value. The calculated entropies of activation for the other compounds listed in Table II vary from a high of

TABLE II

VIBRATIONAL CALCULATIONS OF THE CHAIR TO TWIST-BOAT CONVERSION IN SOME SIX-MEMBERED RINGS

Kinetic parameters[a]

Molecule	Calculated[b]			Experimental			References for experimental values
	$\Delta F^{\ddagger}$	$\Delta H^{\ddagger}$	$\Delta S^{\ddagger}$	$\Delta F^{\ddagger}$	$\Delta H^{\ddagger}$	$\Delta S^{\ddagger}$	
Cyclohexane	13.0	13.8	4.0	10.3	—	—	—[c]
Cyclohexane-d_{11}	13.1	13.8	3.7	10.25	10.71	2.2	—[c]
Tetrahydropyran	11.8	12.4	2.7	9.6 ± 0.4	10.1 ± 1.2	1 ± 6	Gatti et al. (1967)
1,3-Dioxane	11.0	11.3	1.3	9.6 ± 0.2	9.5 ± 0.5	−1 ± 3	Pedersen and Schaug (1968)
1,4-Dioxane	10.1	10.9	3.5	9.4	—	—	Anet and Sandström (1971)
				9.7	—	—	Jensen and Neese (1971)
1,3,5-Trioxane	13.9	14.7	4.2	10.6 ± 0.2	8.8 ± 1.2	−9 ± 6	Pedersen and Schaug (1968)

[a] $\Delta F^{\ddagger}$ and $\Delta H^{\ddagger}$ are in kcal/mole; $\Delta S^{\ddagger}$ in eu.
[b] Values quoted in Pickett and Strauss (1970).
[c] For references and additional experimental values see Table VII.

TABLE III
Calculated Energy Barriers for Ring Inversion in Cyclohexane

Energy barrier (kcal/mole) for C_2 and C_s transition states			
C_2	C_s	Method	Reference
12.7	14.1	Force field	Hendrickson (1961)
11.0	11.3	Force field	Hendrickson (1967b)
12.0		Force field	Allinger et al. (1967)
10.2	10.4	Force field	Bucourt and Hainaut (1967)
11.0		Force field	Allinger et al. (1968)
11.3	11.3	Force field	Schmid et al. (1969)
11.2		Ab initio	Hoyland (1969)
14.5	14.5	Vibrational	Pickett and Strauss (1970)
9.9	9.9	Extended Hückel	Hoffmann (1970)
6.0	13.9	MINDO/2	Dewar (1971)
6.4	6.3	MINDO/2	Komornicki and McIver (1973)
9.5	10.0	Force field	Wiberg and Boyd (1972)
10.9	11.7	Force field	Wiberg and Boyd (1972)
(10.7)		Experimental	See Table VII

4.2 eu for 1,3,5-trioxane to a low of 1.3 eu for 1,3-dioxane, but unfortunately the experimental data are not accurate enough for a meaningful comparison to be made. A more extensive discussion of ring inversion in cyclohexane is given in Section III,B,1c (see also Section III,C and IV,B,1).

c. *Full Force-Field Calculations.* The process of ring inversion in the chair conformation (**3**) of cyclohexane has been the subject of numerous force-field calculations (Hendrickson, 1961, 1967b; Schmid et al., 1969; Bucourt and Hainaut, 1967; Allinger et al., 1967, 1968, 1969, 1971). The twist-boat (**4**, C_2 symmetry) and its slightly higher-energy pseudorotation partner, the

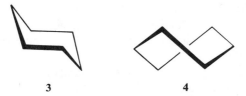

3   4

boat, (**5**, C_s symmetry) are found to be intermediates in the ring inversion of the chair. Thus, the significant transition state for ring inversion is that separating the boat family forms from the chair. The lowest energy path from the chair to the boat or twist-boat proceeds via a half-chair (**6**) as the maximum energy form. This half-chair, which has C_2 symmetry, does not have a zero-degree torsional angle.

5   6

Another form (7) that has often been considered as a possible transition state for ring inversion has C_s symmetry, with five carbons nearly but not quite in a plane. This form is only slightly higher in energy than the half-chair, as shown in Table III. Actually, as pointed out first by Hendrickson (1967b) and especially noted by Pickett and Strauss (1970) (see Section III,B,1b), as well as by later workers, the transition state for ring inversion in cyclohexane should be considered to be a more or less freely pseudorotating form, rather than the static structures 6 and 7. Quantum mechanical calculations, to be

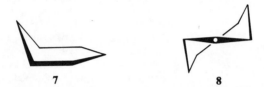

7   8

discussed in Section III,C, also suggest that the transition state for ring inversion is nearly freely pseudorotating. Table III includes results from the quantum mechanical calculations, and iterative force-field calculations (Section III,B,2), as well as the best experimental data (see Section IV,B,1). The agreement between experimental and calculated data for ring inversion in cyclohexane is quite good on the whole.

Schmid *et al.* (1969) have made force-field calculations of the barriers to ring inversion in 1,1-dimethyl-, 1,1,3,3-tetramethyl-, 1,1,4,4-tetramethyl-, and 1,1,3,3,5,5-hexamethylcyclohexane. The calculations agree moderately well with the experimental data (Section IV), and reproduce the experimental trends quite well; e.g., 1,1,3,3-tetramethylcyclohexane is calculated to have a lower barrier (10.7 kcal/mole) than that of the 1,1,4,4-isomer (13.8 kcal/mole). The experimental $\Delta G^{\ddagger}$ values are 9.6 and 11.7 kcal/mole respectively (Friebolin *et al.*, 1969b).

Barriers to ring inversion in cyclohexanone and cyclohexene have also been calculated by force-field methods. The transition state for ring inversion in the ground state half-chair form of cyclohexene is the boat (Bucourt and Hainaut, 1967; Allinger and Sprague, 1972), and thus the only energy minima in this system are the half-chairs. Ring inversion in cyclohexanone is similar to that in cyclohexane, but with a much lower barrier.

Hendrickson (1961, 1962, 1967b) finds that both the twist-chair (8) and twist-boat (9) forms of cycloheptane can pseudorotate very easily, the transi-

9   10   11

tion states being the chair (10) and the boat (11), respectively. There is, however, a relatively high barrier (8.1 kcal/mole) separating the boat and chair families. Because the twist-boat is 2.4 kcal/mole higher in energy than the twist-chair and the pseudorotation barrier in the latter conformation is only 1.4 kcal/mole, dnmr effects are not expected to be observed in cycloheptane. It is possible to decrease the rate of pseudorotation in the twist-chair by means of geminal dimethyl groups, for the reasons which have already been discussed qualitatively in Section III,B,1a. Hendrickson (1961, 1967a) has carried out extensive calculations on the conformational energies of methylcycloheptane, with the methyl group occupying all possible positions in the various cycloheptane conformations. Experimental data, however, is available only for cycloheptanes with two pairs of geminal substituents (Knorr et al., 1967; Borgen, 1972), and these systems have been examined theoretically only in a semi-quantitative fashion (Borgen and Dale, 1972). Further discussion of these and other saturated and unsaturated seven-membered rings is given in Section IV,C.

Conformational interconversions in the medium rings have been investigated by force-field calculations by Hendrickson (1967b), who used a procedure that is restricted to interconversion paths where an element of symmetry is maintained. This restriction is immaterial in the interconversion within the cyclooctane crown family conformations, for example, but becomes quite unsatisfactory for other cases, especially as the ring size increases, as already pointed out in Section III,B,1a. Iterative calculations based on the procedure of Wiberg and Boyd (1972) do not have this difficulty and calculations for cyclooctane using this procedure are presented in Section III,B,2 and are compared with Hendrickson's results. Force-field calculations of conformational processes in a variety of miscellaneous rings have been carried out and will be presented together with the relevant experimental data in Section IV.

2. Iterative Calculations

The iterative force-field procedure of Wiberg and Boyd (1972) was briefly described in Section III,A. It has been applied by these workers to the ring inversion process in cyclohexane, with results described below. Two different sets of force constants were used, and both sets were shown to reproduce satisfactorily the vibrational spectra of cyclohexane and cyclohexane-d_{12}. The change from the chair to a boat-family conformation can be achieved simply by "driving" one dihedral angle of the chair so that this angle changes sign. The transition state for this process is the half-chair (**6**) and is achieved when the dihedral angle is close to zero. Ultimately, a twist-boat is formed, as this is the lowest-energy conformation in the boat family. If two adjacent dihedral angles of the chair are driven in a symmetrical fashion towards zero, the transition state has five carbons in a plane, and the final conformation is the boat (the twist-boat cannot be formed because of the symmetry constraint) (see structures **3–7** and Table III).

Wiberg and Boyd (1972) prefer to consider the one-angle activated complex, i.e., the half-chair **6**, to have one low-frequency vibration (100 cm^{-1},) rather than to be a (freely) pseudorotating entity. They find that the entropy contribution from this low-frequency mode is essentially cancelled by the "missing" frequency of the activated complex, and thus the entropy of activation is mainly determined by the difference in symmetry numbers between the chair and the activated complex. The two-angle activated complex is found to be an energy maximum with respect to distortion to the half-chair and is thus not a true transition state. Additional discussion on the ring inversion process in cyclohexane is given in Sections III,B,1,c, III,B,2, and IV,B,1.

The iterative procedure of Wiberg and Boyd has been applied by Anet and Krane (1973) to cyclooctane and by F. A. L. Anet and L. Kozerski (unpublished work, 1973) to *cis,cis*-1,5-cyclooctadiene. Pseudorotation in the boat–chair conformation (**BC**) of cyclooctane (see Fig. 11) can take place via the twist-boat chair (**TBC**) as an intermediate, as originally envisaged by Anet and St. Jacques (1966). The transition state between the boat-chair and the twist-boat–chair has the unsymmetrical structure labeled (**BC** ⇌ **TBC**)‡. This pseudorotation can be studied theoretically by driving ω_{1234} or ω_{2345} of the boat-chair and has a barrier of only 3.3 kcal/mole. By driving other angles in the boat-chair various other conformational processes can be studied, for example, the boat-chair to twist-boat (**TB**) interconversion is produced by driving ω_{3456}. Certain processes are best investigated by starting from the twist-boat–chair, which, as mentioned above is separated from the boat–chair by quite a low barrier. For example, driving ω_{8123} toward zero in the twist-boat–chair leads smoothly to the twist-chair–chair (**TCC**) which is the lowest-energy conformation in the crown family. The transition state has C_2 symmetry, although the system is not in fact constrained to have any symmetry. Hendrickson studied the change from the twist-chair-chair to the twist-boat-chair, but constrained the transition state to have D_2 symmetry, resulting in a barrier which is slightly too high. The best mechanism for ring inversion in the boat-chair can be achieved by symmetrically driving ω_{8123} and ω_{4567} in the twist-boat-chair, the transition state being the chair (**C**). It is interesting that

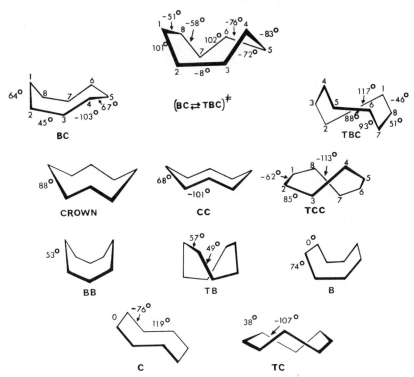

Fig. 11 Intermediate (twist-boat-chair, **TBC**), and transition state (**BC** ⇌ **TBC**)‡ for pseudorotation of the boat-chair (**BC**) in cyclooctane (*top*). Also shown are the crown, chair-chair (**CC**), twist-chair-chair (**TCC**), boat-boat (**BB**), twist-boat (**TB**), boat (**B**), chair (**C**), and twist-chair (**TC**) forms.

ring inversion can also be achieved by driving a single dihedral angle (namely, ω_{1234} or ω_{2345}) away from zero in the twist-boat-chair. The highest energy state along this path, which ultimately gives the twist-chair (**TC**), is an asymmetric form, and is of slightly higher energy than the chair, which appears to be the best transition state for ring inversion. Hendrickson (1967b) also found that the chair was a lower-energy intermediate than the twist-chair for ring inversion in the boat-chair. Pseudorotation of the twist-chair-chair can take place with the chair-chair (**CC**) as a transition state, and the barrier is very low, as was also found by Hendrickson (1967b). In fact, all three crown-family conformations, namely the symmetrical crown, the chair-chair, and the twist-chair are of closely similar energies and can interconvert very readily. These easy interconversions are to be expected, because torsional angles are not forced to become completely eclipsed (0°) at any stage. Tables IV and V contain the results of various force-field calculations and the available experi-

TABLE IV
CONFORMATIONAL ENERGIES FOR CYCLOOCTANE

Conformation	Symmetry group	Relative strain energies (kcal/mole)[a]					Observed
Family and name		Calculated					
		1	2	3	4	5	6
Boat-chair							
Boat-chair (BC)	C_s	0	0	0	0	0	
Twist-boat-chair (TBC)	C_2	1.7	2.0	—	—	—	1.7
Crown							
Twist-chair-chair (TCC)	D_2	0.8	1.7	1.89	−0.25	2.20	
Crown	D_{4d}	1.5	2.8	3.62	0.25	2.09	
Chair-chair (CC)	C_{2v}	1.8	1.9	—	—	2.25	
Boat							
Boat-boat (BB)	D_{2d}	2.8	1.4	—	4.44	—	⎫
Twist-boat (TB)	S_4	2.8	0.9	—	—	—	⎬ >2.0
Boat (B)	D_{2d}	11.2	10.8	—	8.96	—	⎭
Chair							
Chair (C)	C_{2h}	7.5	8.3	—	—	—	
Twist-chair (TC)	C_{2h}	7.7	8.7	—	6.09	—	

[a] References: (1) Anet and Krane (1973); (2) Hendrickson (1967a); (3) Bixon and Lifson (1967); (4) Wiberg (1965); (5) Allinger et al. (1968); (6) Anet and Basus (1973).

mental data for the conformations and conformational processes in cyclooctane. The agreement between theory and experiment is probably as good as can be expected, considering the approximations made in the calculations. Certain aspects of the experimental work on cyclooctane are discussed in detail in Section IV,D.

F. A. L. Anet and L. Kozerski (unpublished work, 1973) have investigated conformational processes in *cis,cis*-1,5-cyclooctadiene, which exists in the twist-boat conformation (**2-TB**). The lowest energy path is a pseudorotation via the skew form (**2-S**), which is very close in geometry and energy to the

TABLE V

MAXIMUM-ENERGY CONFORMATIONS FOR VARIOUS INTERCONVERSION PATHS IN CYCLOOCTANE

Conformation	Driven angle	Symmetry group	Energy[a]		
			1	2	Observed
(BC ⇌ TBC)‡	ω_2 or ω_3	C_1	3.3	—	<4[b]
(TBC ⇌ TC)‡	ω_2 or ω_3	C_1	7.9	(11.6)	—
(TBC ⇌ C)‡(≡C)	ω_1 and ω_5	C_{2h}	7.5	8.3	8.1[c]
(BC ⇌ TB)‡	ω_4	C_1	10.8	—	—
(BC ⇌ BB)‡	ω_4 and ω_5	C_s	10.7	≃20	—
(TBC ⇌ TCC)‡	ω_1	C_2	10.3	(11.4)	10.5[d]
(BB ⇌ B)‡(≡B)	$\omega_1, \omega_3, \omega_5,$ and ω_7	D_{2d}	11.2	—	—
(BB ⇌ TB)‡	ω_1 and ω_5	S_4	<0.2	—	—

[a] All energies are in kcal/mole above the energy of the boat-chair. (1) Data of Anet and Krane (1973); (2) data of Hendrickson (1967b). (Values given in parentheses are for paths with symmetries higher than those given in the table.)
[b] Anet (1974).
[c] Anet and Hartman (1963).
[d] Anet and Basus (1974).

transition state for this process. Next in energy is a pseudorotation with the boat (**2-B**) as the transition state. A local inversion of the boat can give the chair (**2-C**) in a process where a plane of symmetry is maintained, but this is of higher energy than a direct unsymmetrical path from the twist-boat to chair. These results (Table VI) are in satisfactory agreement with experiment (see also Section IV,D).

Conformational processes can be considered to occur on a multidimensional energy surface. The method of driving one dihedral angle successively by small amounts does not necessarily give the lowest barrier. This is a problem analogous to that of local minima in the determination of the lowest-energy conformation. It is conceivable that the lowest barrier for a given overall

TABLE VI
INTERCONVERSION PATHS IN cis-cis-1,5-CYCLOOCTADIENE

Path	Symmetry maintained	Calculated energy maximum[a] kcal/mole relative to twist-boat	Observed energy[b] (free energy)
Twist-boat to twist-boat via boat	C_2	6.6	4.9
Twist-boat to twist-boat via skew	C_2	4.2	4.2
Twist-boat to chair	C_1	7.2	—
Boat to chair	C_s	7.4	—

[a] F. A. L. Anet and L. Kozerski (unpublished work, 1973).
[b] Anet and Kozerski (1973).

interconversion might be reached in some cases only by driving several torsional angles simultaneously and unsymmetrically. The required changes may also vary as the process takes place. Fortunately, it appears that these problems are not too serious in practice.

C. Quantum Mechanical Calculations

1. Semiempirical Treatments

Quantum mechanical calculations (e.g., CNDO, INDO, MINDO) of the semiempirical kind can be applied to reasonably large molecules, and the computation times required are quite moderate (Dewar, 1969, 1971). Geometric optimization with these methods can be done by iterative procedures, such as the steepest descent technique (McIver and Komornicki, 1971). The computational simplicity of the semiempirical calculations, however, is offset by the approximate nature of the results, especially when small energy differences are required, as in conformational analysis. It is possible that further developments of the semiempirical methods, or possibly even improved parametrization, may make such calculations more accurate in the future. For the present, at any rate, CNDO, INDO, or MINDO calculations do not appear to be competitive with the empirical force-field calculations in conformational studies. It should perhaps be remarked that some proponents of semiempirical quantum mechanical methods appear to be very optimistic about the accuracy of their calculations.

Dewar (1971) has reported an application of the MINDO/2 method for the calculation of the transition state for ring inversion in cyclohexane. The calculated barrier, 6.0 kcal/mole, is very much less than the experimental enthalpy of activation, which is 10.8 kcal/mole (Anet and Bourn, 1967).

Dewar (1971) tries to contend that the experimental work may be in error, but in fact, cyclohexane is probably the one compound where not only accurate free energies of activation, but also quite accurate enthalpies and entropies of activation have been obtained (Section IV,A,1). Dewar quotes a value of 7.5 kcal/mole for the barrier to ring inversion in perfluorocyclohexane as justification for his suggestion. However, not only is perfluorocyclohexane a very poor model for cyclohexane, but Dewar neglects to mention later and more careful work (Gutowsky and Chen, 1965) on perfluorocyclohexane, where the barrier is found to be 9.9 kcal/mole. It is thus clear that the MINDO/2 calculation gives a barrier for ring inversion in cyclohexane that is substantially too low, a view that is also accepted by other workers who have applied the same method to this compound, as described below.

A very recent detailed study of ring inversion in cyclohexane also makes use of the MINDO/2 method (Komornicki and McIver, 1973). The ring inversion path that maintains C_s symmetry and has structure 7 (Section III,B,1c) as the transition state is found to have a barrier of 6.3 kcal/mole, very slightly lower than the barrier (6.4 kcal/mole) calculated for the path that keeps a C_2 axis of symmetry. Although the order of the two barriers is reversed from that given by all the force-field calculations, this is not particularly significant, since the differences in the two barriers are very small in both kinds of calculations. The absolute value of the barrier to ring inversion in cyclohexane, however, is much too low, as it was also in the other MINDO/2 calculation discussed earlier in this section. Komornicki and McIver have calculated that the entropies of activation for the C_s and C_2 transition states for ring inversion are 5.98 and 6.84 eu, respectively, which they find disturbingly higher than the experimental value of 2.8 eu (Anet and Bourn, 1967). Each transition state is calculated to have a vibration of extremely low frequency (15 cm^{-1} for the C_2 structure and 25 cm^{-1} for the C_s structure) corresponding to a pseudorotation mode that interconverts C_2 and C_s structure into one another. Although the two "transition states" are found to be energy minima for this pseudorotation itinerary, the barrier separating these transition states must be extremely small, and treating them separately is probably not really justified. In a proper calculation of the entropy, pseudorotation is considered explicitly, as has been done by Pickett and Strauss (1970) for the present system (Section III,B,1b), or by Kilpatrick *et al.* (1947) for cyclopentane, and is likely to lead to a lower entropy of activation than that calculated by Komornicki and McIver.

An extended Hückel calculation by Hoffmann (1970) gives a barrier of 9.9 kcal/mole for ring inversion in cyclohexane, substantially better than the MINDO/2 results discussed above. In general, however, the MINDO/2 calculations appear to be superior to the extended Hückel procedure (Dewar, 1971).

2. Ab Initio Treatments

Ab initio molecular orbital calculations require much more computer time than do the semiempirical methods, but involve no arbitrary parameters. However, the energies obtained are not correct in an absolute sense, because only a very limited number of basis orbitals can be employed for even small molecules if computation times are to be reasonable. Furthermore, corrections for electron correlation and relativistic effects are generally neglected (Richards and Horsley, 1970). Fortunately, the neglected terms largely cancel in comparisons of conformational energies, because isomeric molecules are being considered. *Ab initio* calculations have been used with considerable success for conformational calculations, particularly of barriers to hindered rotation and nitrogen inversion (Lehn, 1971). Ring inversion in cyclohexane has been studied by Hoyland (1969), who calculates a barrier of 11.2 kcal/mole in excellent agreement with experiment. A possible very useful application of the *ab initio* method is for the calculation of nonbonded repulsion and other parameters needed in force-field calculations.

IV. Experimental Data and General Discussion

A. Kinetic Parameters from DNMR

We will make some brief comments about the determination of kinetic parameters from dnmr measurements. Conformational processes are kinetically first-order reactions and are thus relatively simple. On the other hand, nmr spectra of cyclic compounds are often extremely complex. Cyclohexane at low temperatures (e.g., $-100°C$) has two proton chemical shifts because ring inversion is slow under these conditions. The spectrum, however, is so complicated that a complete analysis is virtually impossible, even at high magnetic fields when the chemical shift difference between axial and equatorial protons is much greater than any of the coupling constants. The complexity arises in part from the large number (twelve) of protons that are coupled together, and in part from the nonequivalence within each group of isochronous protons. The number of transitions for a twelve-spin system of the kind present in cyclohexane is enormous, and there is thus no possibility of obtaining a spectrum showing individual transitions even in a perfectly homogeneous magnetic field. The calculation of line shapes for such a system treated strictly correctly is completely out of the question, even with the most efficient programs and computers presently available, or even imaginable. Certain approximate methods can be applied, and quite accurate rate constants can easily be obtained in the present system in the fast exchange region, when the spectrum consists of a single broadened line which is essentially Lorentzian in shape.

In other systems, e.g., the cycloalkanones, the spectra are too complex for analysis even under very fast exchange conditions. In virtually all these cases, it is still possible to obtain rough rate constants at the points of maximum spectral change, and by the absolute rate theory, to calculate free energies of activation ($\Delta G^\ddagger$ values) that are reasonably accurate (± 0.5 kcal/mole). It is not possible, however, to obtain kinetic data accurate enough to calculate the enthalpy and entropy of activation (or the Arrhenius activation energy and preexponential factor) with any meaningful accuracy unless the spectra can be correctly analyzed. Systems that have only a few spins can usually be analyzed by line-shape methods by means of efficient computer programs based on the density matrix method (Binsch, 1969). Even then, carefully obtained data are required and the presence of systematic errors must be borne in mind. There are in fact very few systems where the enthalpies ($\Delta H^\ddagger$) and entropies ($\Delta S^\ddagger$) of activation have been obtained with a sufficient accuracy that discussions of these quantities are warranted.

Values of $\Delta S^\ddagger$ outside of the range of ± 10 eu have been reported many times in the recent literature. These large values of $\Delta S^\ddagger$ have often been rationalized by invoking changes in symmetry between the ground and transition states. However, simple calculations of the entropy contributions clearly show that symmetry effects amount to only a few entropy units. It is also highly likely that vibrational contributions to $\Delta S^\ddagger$ are quite small. Solvent contributions may be of importance when the polarities of the ground and transition states are very different, but this is seldom the case in conformational processes. In our opinion, there is not a single example where a large value of $\Delta S^\ddagger$ has been reliably demonstrated to occur in a conformational process. Thus, values of $\Delta H^\ddagger$ associated with large positive or negative values of $\Delta S^\ddagger$ are probably not meaningful. Similarly, reported values of the Arrhenius activation energy associated with preexponential factors very different from 10^{13} cannot be considered to be reliable. It should be noted that errors in $\Delta G^\ddagger$ are generally small, but that errors in $\Delta H^\ddagger$ and $\Delta S^\ddagger$ are large and correlated, since $\Delta G^\ddagger = \Delta H^\ddagger - T \Delta S^\ddagger$.

Complex proton spectra can always be simplified by means of deuteration, but only at the cost of more or less time-consuming syntheses. Furthermore, small isotope effects can occur so that the observed kinetic parameters are not necessarily the same as in the undeuterated compound. Although it has been claimed that complex spectra are advantageous in dnmr (Kleier, et al.

1970) we are not aware of any real experimental support for this point of view. As we have pointed out elsewhere (Anet and Anet, 1971), the best system for a dnmr investigation is one that is as simple as possible, and has the greatest possible frequency separation between lines that broaden and coalesce as the exchange rate increases. If the averaging process involves only chemical shifts, a single proton or other suitable nucleus is sufficient. In some cases the averaging involves mainly coupling constants, and two nuclei are then obviously required, and they are sufficient if they have different chemical shifts. If the two nuclei have the same or nearly the same chemical shifts, an additional nucleus (often ^{13}C) may be required to remove the degeneracy or near degeneracy in the system. For example, ring inversion in cyclohexane is best studied with cyclohexane-d_{11}, a compound that contains only a single proton. Other highly deuterated species ,e.g., d_{10}, d_9, or d_8, could be employed to give more or less complex spectra that can be analyzed, but these compounds would not offer any advantage over the d_{11} species in obtaining accurate kinetic parameters.

Further uses for deuteration and high magnetic fields occur in systems where the slow-exchange spectra are strongly coupled, as in cyclohexane and 1,4-dioxane. The low-temperature pmr spectra of cyclohexane at 40, 60, and 251 MHz are shown in Fig. 12, and it can be seen that the presence of only two chemical shifts is not obvious at 40 or 60 MHz. In 1,4-dioxane, the chemical shift between axial and equatorial protons is so small compared to the coupling constant between these protons that the spectrum at 100 MHz is

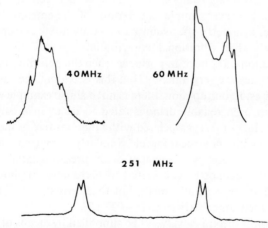

Fig. 12 Proton nmr spectra at $-100°$C of cyclohexane in carbon disulfide solution at 40 MHz (Moniz and Dixon, 1961), 60 MHz (Jensen *et al.*, 1962), and 251 MHz (F. A. L. Anet and V. J. Basus, unpublished work, 1973). The chemical shift difference between axial and equatorial protons is 0.462 ppm. The 40 and 60 MHz spectra are reprinted by permission of the American Chemical Society.

essentially independent of the rate of ring inversion. On the other hand, highly deuterated derivatives of 1,4-dioxane show very clear spectral changes with temperature and the barrier to ring inversion can easily be measured (Anet and Sandström, 1971; Jensen and Neese, 1971).

Carbon-13 nmr spectra, obtained in natural abundance with proton decoupling, are very simple because only chemical shifts occur. They therefore offer some advantages for dnmr work, including larger chemical shifts than in pmr spectroscopy, but this is to some extent counterbalanced by a generally low signal-to-noise ratio. Also, there is some difficulty in obtaining precise temperature measurements when a large proton radiofrequency power is used for decoupling, especially with dielectrically lossy samples, e.g., cations in strongly acidic solutions. Additionally, some processes, such as ring inversion in cyclohexane, cannot be studied by cmr spectroscopy.

Fluorine labeling, a valuable technique for the study of conformational processes, has been employed extensively by Roberts (1966). Most often, one CH_2 group is replaced by a CF_2 group. The ^{19}F spectrum, taken with proton decoupling if desired, is then simply an A_2 singlet when exchange is fast and an AB quartet (or several AB quartets) when exchange is slow. The generally large ^{19}F chemical shift is a distinct help in the application of dnmr. On the other hand, the large geminal F–F coupling constant is a nuisance if the chemical shifts are in fact fairly small. The use of CHF groups as a label overcomes this difficulty, but the loss in symmetry that usually occurs is a disadvantage. Simple fluorides are often difficult to prepare and are not as stable as compounds containing CF_2 groups. For example, 1,1-difluorocyclooctane has been studied by ^{19}F dnmr, but no preparation of cyclooctyl fluoride has been reported. However, both fluorocyclohexane and 1,1-difluorocyclohexane are well-known compounds and have been investigated in detail by ^{19}F dnmr, as will be described in Section IV,B,2. Fluorine labeling, of course, is a much more serious perturbation than is isotopic labeling, although the perturbations, mainly steric and inductive effects, can provide additional information not available from isotopic substitution.

Line broadening resulting from dipole–dipole relaxation becomes quite important for CH_2 resonances below about $-160°C$ in medium size molecules. In these cases, deuterated derivatives containing CHD groups give much narrower proton line widths at very low temperatures than do the parent undeuterated compounds (Anet, 1970; Anet et al., 1973b). Low-temperature measurements are also greatly helped by employing the highest possible magnetic field, at least when chemical shifts are the parameters being averaged (Anet et al., 1969, 1973b).

Although early work (Allerhand and Gutowsky, 1964; Allerhand et al., 1965) seemed to indicate that the spin-echo technique should be the most accurate method for obtaining activation parameters, it is now clear that there are systematic errors in this technique that tend to give low values of both $\Delta H^\ddagger$ and $\Delta S^\ddagger$, even though the free energy of activation is quite accurate (Inglefield et al., 1968). At present, it appears that quite accurate, and reliable measurements of $\Delta H^\ddagger$ and $\Delta S^\ddagger$ can be obtained if line-shape dnmr data is combined with either double-resonance saturation transfer data (Anet and Bourn, 1967) or with direct kinetic data (Bushweller et al., 1970), but very few such studies have been carried out.

In the tables of experimentally determined conformational energy barriers of this chapter, $\Delta G^\ddagger$ values will be shown as given by the original authors, or if not given, as calculated from rate data or coalescence temperatures. Solvents are not quoted in these tables, since conformational energy barriers in rings are virtually independent of solvent polarity. As mentioned above, $\Delta G^\ddagger$ values for conformational processes should be close to $\Delta H^\ddagger$ values and therefore only slightly temperature dependent. Temperatures at which $\Delta G^\ddagger$ values have been obtained are not explicitly given in the tables, but the dnmr method is such that similar values of $\Delta G^\ddagger$ are measured at similar temperatures. As a rough guide, barriers of 5, 10, 15, and 20 kcal/mole are measured by dnmr methods at about $-160°$, $-60°$, $+40°$, and $+130°C$, respectively. Barriers obtained by dynamic esr methods have a different time scale, and typically, a barrier of 5 kcal/mole will be measured at about $-60°C$. Racemization data obtained polarimetrically, or direct measurements of conformational interconversions, have still another time scale wherein barriers of 20 and 35 kcal/mole are obtained near room temperature and 150°C, respectively. Unless otherwise stated, $\Delta G^\ddagger$ values given in the tables have been obtained by dnmr methods. Although $\Delta G^\ddagger$ values are given in the tables for all compounds, values of $\Delta H^\ddagger$ are only occasionally quoted.

B. Six-Membered Rings

1. Cyclohexane

Ring inversion in cyclohexane and cyclohexane-d_{11} has been extensively studied by nmr methods. Experimental techniques have included the following: (a) line shape methods, (b) simple spin-echo T_2 measurements, (c) spin-echo T_2 measurements as a function of the pulse repetition rate, (d) $T_{1\rho}$ (T_1 in the rotating frame) measurements as a function of the H_1 radiofrequency magnetic field, and (e) double-resonance saturation transfer measurements. The experimental data are summarized in Table VII.

In general, recent dnmr measurements are more accurate than older data because of the continuing improvements in nmr instrumentation. In our view, the best values of $\Delta H^\ddagger$ and $\Delta S^\ddagger$ (Table VII) for the chair-to-boat process in cyclohexane-d_{11} are obtainable by combining line-shape and double-resonance

TABLE VII

EXPERIMENTALLY DETERMINED BARRIERS FOR CHAIR TO TWIST-BOAT CONVERSION IN CYCLOHEXANE AND CYCLOHEXANE-d_{11}

Temperature range (°K)	Method	$\Delta G^{\ddagger}$ (kcal/mole) ($-50°$ to $-60°$C)	$\Delta H^{\ddagger}$ (kcal/mole)	$\Delta S^{\ddagger}$ (eu)	Reference
		Cyclohexane			
—	Line-shape	10.1	11.5 ± 2	—	Jensen et al. (1960); Jensen et al. (1962)
203–253	Line-shape and wiggle decay	10.3	10.3 ± 0.2	−0.2 ± 1	Harris and Sheppard (1967)
213–248	Spin-echo	10.7	11.5	4.9	Meiboom (1962)
	Spin-echo	10.3	9.1 ± 0.5	−5.8 ± 2.4	Allerhand et al. (1965)
	Relaxation in rotating frame	≃10.2	—	—	Deverell et al. (1970)
223–300	Fourier-transform line-shape	10.25	10.4 ± 0.1	1.1 ± 0.5	F. A. L. Anet and V. J. Basus (unpublished work, 1973)
223–263	Spin-echo	10.3	—	−3.6	Inglefield et al. (1968)
		Cyclohexane-d_{11}			
179–241	Line-shape and wiggle decay	10.3	10.9 ± 0.6	2.9 ± 2.3	Anet et al. (1964a)
198–226	Line-shape	10.2	10.5 ± 0.5	1.4 ± 1.0	Bovey et al. (1964b)
175–298	Spin-echo	10.3	9.1 ± 0.1	−5.8 ± 0.4	Allerhand et al. (1965)
156–246	Line shape and saturation transfer	10.22	10.8 ± 0.1	2.8 ± 0.5	Anet and Bourn (1967)
223–300	Fourier transform line-shape	10.2	10.1 ± 0.1	0.0 ± 0.5	F. A. L. Anet and V. J. Basus (unpublished work, 1973)
156–300	Best overall value	10.25	10.71 ± 0.04	2.2 ± 0.2	Anet and Bourn (1967); F. A. L. Anet and V. J. Basus (unpublished work, 1973)

measurements made at 60 MHz (Anet and Bourn, 1967), with recent line-shape measurements obtained at 251 MHz (F. A. L. Anet and V. J. Basus, unpublished work, 1973). A Fourier transform pulse nmr technique, which takes into account the inhomogeneity of the magnetic field and any time-dependence of this inhomogeneity, was used in the latter work. As a result, accurate rate constants could be obtained at room temperature, where the broadening is much less than 1 Hz. The nmr measurements on cyclohexane-d_{11} span temperatures from $-117°$ to $25°C$ and the rate constants vary by a factor of over twenty million over this temperature range.

2. Carbocyclic Six-Membered Rings (Excluding Cyclohexane)

Barriers to ring inversion for a variety of cyclohexane derivatives are given in Table VIII. A single substituent on cyclohexane has very little effect on the barrier to ring inversion, but the presence of many substituents can give rise to a moderate increase in the barrier if the ground state is not too strained, as in the γ isomer of 1,2,3,4,5,6-hexachlorocyclohexane (**12**). In compounds

12

where the ground state is quite strained, as in 1,1,3,3-tetramethylcyclohexane, there can be a significant decrease in the barrier.

Wolfe and Campbell (1967a) have pointed out that a cis-1,2-disubstituted cyclohexane can undergo ring inversion in such a way that the substituents are not eclipsed in the half-chair transition state. Eclipsing of these substituents, which must take place at some stage in the ring inversion process, can occur during pseudorotation in the boat form, which is about 5 kcal/mole lower in energy than the half-chair. Thus, the barrier to ring inversion in a cis-1,2-disubstituted cyclohexane should be close to that in cyclohexane, unless the substituents are very large (Fig. 13). In the latter case, e.g., *cis*-1,2-di-*t*-butyl-cyclohexane, the eclipsing strain in the boat can be large enough to offset the energy difference between the half-chair and the boat, as well as the ground state strain in the chair.

The barrier to ring inversion in *cis*-decalin is distinctly higher than that of cyclohexane, because the mechanism presented above for a simple cis-1,2-disubstituted cyclohexane is not feasible. The dichair conformation can give the boat-chair by a local inversion, with a barrier which should be close to

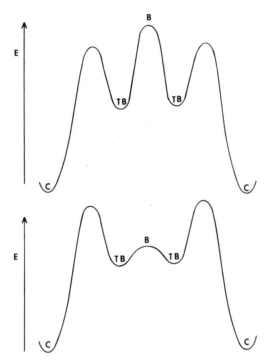

Fig. 13 Diagrammatic representation of the ring inversion path for a simple cyclohexane derivative (*bottom*), and for a cis-1,2-disubstituted derivative with large substituents (*top*). C = chair; TB = twist-boat; B = boat.

that in cyclohexane. The boat-chair form of *cis*-decalin, however, cannot pseudorotate easily because of the rigid chair moiety. It is necessary for the boat-chair to undergo a second local inversion so as to give the boat-boat conformation, before pseudorotation can occur (Fig. 14). The barrier for the second local inversion should be about the same as that for the first local inversion, but the starting point is the boat-chair, which should be about 5 kcal/mole above the chair-chair. Thus, the overall barrier should be some 5 kcal/mole higher than the barrier in cyclohexane. This crude treatment obviously overestimates the ring inversion barrier in *cis*-decalin, but the direction of the change is given correctly (Gerig and Roberts, 1966; Jensen and Beck, 1966).

Barriers to ring inversion in unsaturated carbocyclic six-membered rings (Table IX) tend to be much lower than those in cyclohexane and its derivatives. The low barrier to ring inversion in cyclohexene is understandable, since the ground state conformation is a half-chair and the change to the boat, which is the transition state (Section III,B,1c) involves torsional strain but

TABLE VIII

KINETIC PARAMETERS FOR RING INVERSION IN SOME SIX-MEMBERED SATURATED CARBOCYCLES

Compounds	$\Delta G^{\ddagger}_{CC}$ (kcal/mole)[a]	$\Delta H^{\ddagger}$ (kcal/mole)	References
Cyclohexane	10.5	10.8	—[b]
Fluorocyclohexane	10.14	9.87 ± 0.1	Bovey et al. (1964a)
Chlorocyclohexane	≃10.8	—	Reeves and Strømme (1960)
	11.7 ± 1	—	Neikam and Dailey (1963)
	10.9	11.6	Jensen and Bushweller (1969a)
Bromocyclohexane	≃10.9	—	Reeves and Strømme (1960)
	11.7 ± 1	—	Neikam and Dailey (1963)
Iodocyclohexane	11.7 ± 1	—	Neikam and Dailey (1963)
Cyclohexanol	11.7 ± 1	—	Neikam and Dailey (1963)
Trideuteriomethoxycyclohexane	9.4 ± 0.1	—	Jensen and Bushweller (1969a)
1,1-Difluorocyclohexane	9.71	10.43	Spassov et al. (1967)
	9.8	9.0	Jonáš et al. (1965)
trans-1,2-Dichlorocyclohexane	12	—	Reeves and Strømme (1961a)
trans-1,3-Dichlorocyclohexane	≃12.0	—	van Dort and Sekuur (1963)
trans-1,2-Dibromocyclohexane	≃11.9	—	Reeves and Strømme (1961a)
trans-1,3-Dibromocyclohexane	9.9	—	van Dort and Sekuur (1963)
1,1-Dimethylcyclohexane	10.5	11.3 ± 0.7	Dalling et al. (1971)
1,1-Dimethylcyclohexane-3,3,5,5-d_4	10.6	—	Friebolin et al. (1969b)
cis-1,2-Dimethylcyclohexane	10.6	9.3 ± 0.7	Dalling et al. (1971)
trans-1,3-Dimethylcyclohexane	10.1	11.1 ± 0.7	Dalling et al. (1971)
cis-1,4-Dimethylcyclohexane	10.1	11.0 ± 0.7	Dalling et al. (1971)
1,1-Dibenzylcyclohexane	10.7	—	Levin et al. (1972)
cis-1,2-Di-t-butylcyclohexane	16.3 ± 0.3	15.4 ± 1	Kessler et al. (1968)
cis-1,2-Bismethoxycarbonylcyclohexane-3,3,4,5,6,6-d_6	10.7–10.8	—	Wolfe and Campbell (1967a)
1,1-Dimethoxycyclohexane	10.8	—	Friebolin et al. (1969b)
1,1,2,2-Tetrafluorocyclohexane	10.87	7.51	Spassov et al. (1967)
1,1-Difluoro-4,4-dimethylcyclohexane	10.18	8.77	Spassov et al. (1967)
1,1,3,3-Tetramethylcyclohexane	9.6	—	Friebolin et al. (1969b)
1,1,4,4-Tetramethylcyclohexane	11.7	—	Friebolin et al. (1969b)

Compound	$\Delta G^{\ddagger}_{CC}$	$\Delta G^{\ddagger}$	Reference
1,1-Dibenzyl-4,4-dimethylcyclohexane	11.8	—	Farnham (1972)
(structure)			
1,1-Dimethyl-4,4-dimethoxycyclohexane	10.7	—	Lambert et al. (1970)
1,3-Cyclohexadione-bisethylene ketal	11.3	—	Friebolin et al. (1969b)
1,1,4,4-Tetramethoxycyclohexane	≃9.5	—	Friebolin et al. (1969b)
	10.7	—	Friebolin et al. (1969b)
	10.5 ± 0.4	—	Abraham and MacDonald (1966)
cis-anti-cis-1,2,4,5-Tetraacetoxycyclohexane-3,3,6,6-d_4	13.5	—	Wolfe and Campbell (1967b)
cis-syn-cis-1,2,4,5-Tetraacetoxycyclohexane-3,3,6,6-d_4	12.4	—	Wolfe and Campbell (1967b)
2,2,5,5-Tetramethylcyclohexanone ethylene ketal	12.0	—	Friebolin et al. (1969b)
5,5-Dimethyl-1,3-cyclohexadione bisethylene thioketal	9.4	—	Murray and Kaplan (1967)
	≃13	—	Harris and Sheppard (1964)
γ-Hexchlorocyclohexane (13)	>13.8	—	Wolfe and Campbell (1967b)
	11.9	9.9	Yamamoto et al. (1973)
cis-Inositol acetate	15.4[c]	—	Brownstein (1962)
allo-Inositol	12.6[c]	—	Brownstein (1962)
muco-Inositol acetate	10.5[c]	—	Brownstein (1962)
Perfluorocyclohexane	10.9	—	Tiers (1960)
	11.2	7.5 ± 0.3	Gutowsky and Chen (1965)
cis-Decalin	12.8	9.9 ± 0.2	Jensen and Beck (1966)
	13.0	—	Dalling et al. (1971)
2,2-Difluoro-cis-decalin	12.3	13.6 ± 0.7	Gerig and Roberts (1966)
9-Methyl-cis-decalin	13.0	14.5	Dalling et al. (1971)
cis-2,2,7,7-Tetramethoxydecalin	12.4	12.4 ± 0.7	Geens et al. (1967)
cis-syn-cis-2,2,6,6-Tetramethoxyperhydroanthracene	11.2	—	Pessemier et al. (1973)
3,3-Difluoro[4.4.4]propellane	15.7	15.1	Gilboa et al. (1969)

[a] $\Delta G^{\ddagger}_{CC}$ is for the overall chair-chair ring inversion process. If a single intermediate exists and two barriers of equal energies are crossed in the ring inversion process, $\Delta G^{\ddagger}$ (chair to intermediate) $= \Delta G^{\ddagger}_{CC} - RT \ln 2$. At $-50°C$, $RT \ln 2 = 0.3$ kcal/mole.

[b] See Table VII for references (note that $\Delta G^{\ddagger}$ values given in Table VII are for the conversion of the chair to the twist-boat in cyclohexane).

[c] Calculated from the spectral date given in the reference.

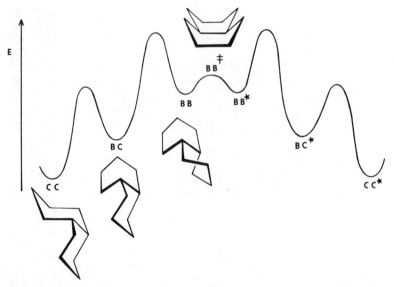

Fig. 14 Ring inversion path for *cis*-decalin. Mirror-image conformations are starred.

very little bond angle strain. The low barrier to ring inversion in cyclohexanone ($\Delta G^{\ddagger} = 4.0$ kcal/mole) is related to the very low barrier (0.8 kcal/mole) to internal rotation in acetone (Swalen and Costain, 1959). Since the barrier to internal rotation in isobutylene is 2.2 kcal/mole (Laurie, 1961), it is not surprising that the barrier to ring inversion in methylenecyclohexane ($\Delta G^{\ddagger} = 8.4$ kcal/mole) is substantially higher than the barrier in cyclohexanone (Table IX).

9,10-Dihydrophenanthrene is nonplanar, but the barrier to ring inversion of the (formal) 1,3-cyclohexadiene ring system has not yet been measured by dnmr. The presence of substituents at C-9 and C-10 or at C-4 and C-5, however, makes the rate of ring inversion easily observable by nmr methods (Table IX). These compounds can be considered to be bridged diphenyl derivatives (for other derivatives containing seven-membered rings, see Section IV,C).

3. *Heterocyclic Six-Membered Rings*

Barriers to ring inversion have been determined in a wide variety of heterocyclic six-membered rings (Table X). On the whole, the barriers are rather similar to those in analogous carbocyclic derivatives. Oxygen heterocycles tend to have lower barriers, and nitrogen heterocycles somewhat higher barriers than those of corresponding carbocycles. Many nitrogen heterocycles show nmr evidence for nitrogen inversion as well as for ring inversion. In

TABLE IX
KINETIC PARAMETERS FOR RING INVERSION IN SOME SIX-MEMBERED UNSATURATED CARBOCYCLES

Compounds	$\Delta G^{\ddagger a}$ (kcal/mole)	$\Delta H^{\ddagger}$ (kcal/mole)	References
Cyclohexanone-3,3,4,4,5,5-d_5	4.1 ± 0.1	—	Anet et al. (1973b)
2,2,5,5-Tetramethylcyclohexanone	8.1	—	Bernard and St. Jacques (1970)
Cyclohexanone oxime methyl ether	5.7 ± 0.5	—	Jensen and Beck (1968)
Methyleneecyclohexane	7.9 ± 0.5	—	Jensen and Beck (1968)
	8.4 ± 0.1	—	Gerig and Rimerman (1970)
2,2,5,5-Tetramethyl-3-methylenecyclohexanone	5.9	—	Bernard et al. (1972).
1,4-Dimethyleneecyclohexane	7.5 ± 0.5	—	St. Jacques and Bernard (1969)
2,2,5,5-Tetramethyl-1,3-dimethylenecyclohexane	7.6	—	Bernard et al. (1972)
1,4-Bis(difluoromethylene)cyclohexane	6.5 ± 0.5	—	St. Jacques and Bernard (1969)
Cyclohexene-d_6	5.4	—	Anet and Haq (1965)
Cyclohexene-d_8	5.37	5.5	Jensen and Bushweller (1969b)
4-Bromocyclohexene	6.2 ± 0.1	—	Jensen and Bushweller (1965)
Perfluorocyclohexene	6.83 ± 0.1	7.2 ± 0.2	Anderson and Roberts (1970)
1,2-Dichlorooctafluorocyclohexene	6.3 ± 0.1	6.3 ± 0.2	Anderson and Roberts (1970)
3,3-Dimethylcyclohexene	6.3	—	Bernard and St. Jacques (1973)
4,4-Dimethylcyclohexene	6.1	—	Bernard and St. Jacques (1973)
3,3,6,6-Tetramethylcyclohexene	8.4	—	Bernard and St. Jacques (1973)
Cyclohexanesemidione	3.8[b]	4.0	Russell et al. (1967)

(continued)

TABLE IX (continued)

Compounds	$\Delta G^{\ddagger a}$ (kcal/mole)	$\Delta H^{\ddagger}$ kcal/mole	References
3,3,5,5-Tetramethylcyclohexanesemidione	4.1[b]	2.6	Russell et al. (1967)
6,6-Difluoro-cis-decal-2-one	9.5	10.8	Lack et al. (1968)
10-Methyl-6,6,-difluoro-cis-decal-2-one	9.8	8.8	Lack et al. (1968)
2-Methylene-cis-decalin-1,1,3,3-d_4	12.6	11.2	Gerig and Ortiz (1970)
2,2-Difluoromethylene-cis-decalin	12.5	17.2	Gerig and Ortiz (1970)
9,10-Dihydrophenanthrene radical anion	6.9[b]	5.6 ± 0.1	Van der Kooij et al. (1971)
cis-9,10-Dimethyl-9,10-dihydrophenanthrene	10.8	—	Rabideau et al. (1969)
cis-9,10-Diphenyl-9,10-dihydrophenanthrene	9.7	—	Rabideau et al. (1969)
9,10-Dihydro-4,5-dimethylphenanthrene	24.1 ± 1.0	24.7 ± 1.0	Yamamoto and Nakanishi (1973)
9,10-Dihydro-4,5-dimethylphenanthrene	23.1[c]	—	Mislow and Hopps (1962)
1,2,3,6,7,8-Hexahydropyrene radical anion	7[b]	6.0 ± 0.6	Claridge et al. (1972); Pijpers et al. (1971)
	16.7	15.7	Gilboa et al. (1969)

[a] $\Delta G^{\ddagger}$ is for the overall process observed and literature values have been corrected where necessary (see Table VIII, footnote a).
[b] From line shape analysis of the esr spectrum.
[c] Racemization of optically active compound.

TABLE X

KINETIC PARAMETERS FOR RING INVERSION IN SOME HETEROCYCLIC SIX-MEMBERED RINGS

Compounds		Kinetic parameters[a]		References
		$\Delta G^{\ddagger}$	$\Delta H^{\ddagger}$	
 Tetrahydropyran	X = O	10.3	—	Lambert et al. (1973)
		10.3	—	Gatti et al. (1967)
Thiane	X = S	9.4	—	Lambert et al. (1973)
Selenane	X = Se	8.2	—	Lambert et al. (1973)
Tellurane	X = Te	7.3	—	Lambert et al. (1973)
Thiane 1-oxide	X = SO	10.1	—	Lambert and Keske (1966); Lambert et al. (1973)
Selenane 1-oxide	X = SeO	8.3	—	Lambert et al. (1973)
Thiane 1,1-dioxide	X = SO$_2$	10.3	—	Lambert et al. (1967); Lambert et al. (1973)
Selenane 1,1-dioxide	X = SeO$_2$	6.7	—	Lambert et al. (1973)
Piperidine	X = NH	10.4	14.1	Lambert et al. (1967)
N-Methylpiperidine	X = NMe	12.2	14.0	Lambert et al. (1971)
N-Chloropiperidine	X = NCl	13.5	16.4	Lambert et al. (1971)
Cyclopentamethylenedimethylsilane	X = Si(Me)$_2$	5.66 ± 0.25	—	Jensen and Bushweller (1968)
Cyclopentamethylenedimethylsilane	X = Si(Me)$_2$	5.5 ± 0.1	5.9 ± 0.3	Bushweller et al. (1971a)
β-D-Robopyranose tetraacetate	X = O	≃10.3	—	Bhacca and Horton (1967); Durrette et al. (1969)
β-D-Lyxopyranose tetraacetate	X = O	9.6 ± 0.4	—	Durrette and Horton (1969)
4,4-Difluoropiperidine	X = NH	10.0	—	Yousif and Roberts (1968)
3,3-Dimethylpiperidine	X = NH	10.5	11.7	Lambert et al. (1972)
1,3,3-Trimethylpiperidine	X = NMe	13.1	10.3	Lambert et al. (1972)
1,4,4-Trimethylpiperidine	X = NMe	12.7	10.6	Lambert et al. (1972)

(continued)

TABLE X (continued)

Compounds		Kinetic parameters[a]		References
		$\Delta G^{\ddagger}$	$\Delta H^{\ddagger}$	
$\Delta^{2,3}$-Dihydropyran	X = O	6.6 ± 0.3	—	Bushweller and O'Neill (1969)
3,3,6,6-Tetramethyl-1,2-dioxane	X = O; Y = O	14.3	15.7	Schmid et al. (1966)
		14.6	—	Claeson et al. (1961)
1,2-Dithiane-4,5-d_4	X = S; Y = S	11.6	—	Claeson et al. (1960)
3,3,6,6-Tetramethyl-1,2-dithiane	X = S; Y = S	13.8	—	Claeson et al. (1961)
3,3,5-Trimethyldiox-4-ene	X = O; Y = O	11.0	12.6 ± 1.8	Kaplan and Taylor (1973)
1,2-Dicarbomethoxy-3,6-diphenyl-1,2,3,6-tetra-hydropyridazine	X = NR; Y = NR	18.9	—	Breliere and Lehn (1965)
1,2-Dicarbethoxy-1,2,3,6-tetrahydropyridazine	X = NR; Y = NR	20.5	20.7 ± 1.4	Bittner and Gerig (1972)
1,2-Dicarbotrifluoroethoxy-1,2,3,6-tetra-hydropyridazine (13)	X = NR; Y = NR	20	—	Bittner and Gerig (1972)
1-Carbotrifluoroethoxy-1,2,3,6-tetrahydro-pyridazine (14)	X = NR; Y = NH	<10	—	Bittner and Gerig (1972)
1,3-Dioxane	X = O; Y = O	9.7 ± 0.2	—	Friebolin et al. (1962)
		9.9 ± 0.2	9.5 ± 0.5	Pedersen and Schaug (1968)
1,3-Dithiane	X = S; Y = S	10.3	—	Friebolin et al. (1969a)
1,3-Diselenane	X = Se; Y = Se	8.2 ± 0.1	—	Geens et al. (1969)

Compound	Substituents			Reference
5,5-Dimethyl-1,3-dioxane	X = O; Y = O	10.8	12.7	Schmid et al. (1966)
		10.5 ± 0.3	—	Anderson (1970)
cis-4,5,6-Tetramethyl-1,3-dioxane	X = O; Y = O	10.0	—	Tavernier et al. (1973)
5,5-Dimethylene-1,3-dioxane	X = O; Y = O	9.3 ± 0.3	—	Anderson (1970)
2,2-Pentamethylene-1,3-dioxane	X = O; Y = O	10.1	—	Tavernier et al. (1973)
5,5-Dimethyl-2,2-heptamethylene-1,3-dioxane	X = O; Y = O	8.0 ± 0.2	—	Anderson (1969)
2,2,5,5-Tetramethyl-3-thiadioxane	X = O; Y = S	11.5	12.1	Schmid et al. (1966)
N-Methyltetrahydro-1,3-oxazine	X = O; Y = NMe	9.8 ± 0.3	—	Lehn et al. (1967)
3,5,5-Trimethyltetrahydro-1,3-oxazine	X = O; Y = NMe	10.5 ± 0.3	10.4 ± 0.5	Lehn et al. (1967)
N,N'-Dimethylhexahydropyrimidine	X = NMe; Y = NMe	11.6 ± 0.4	—	Riddell (1967); Farmer and Hamer (1968)
		11.6 ± 0.1		
N,N'-Dimethylhexahydropyrimidine	X = NMe; Y = NMe	11.6 ± 0.4	—	Riddell (1967); Farmer and Hamer (1968)
		11.6 ± 0.1		

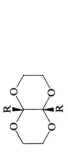

Compound	Substituents			Reference
1,4-Dioxane-d_6	X = O; Y = O	9.7	—	Anet and Sandström (1971)
1,4-Dioxane-d_7	X = O; Y = O	10.0	—	Jensen and Neese (1971)
Octafluoro-1,4-dithiane	X = S; Y = S	10.05 ± 0.1	9.74 ± 0.2	Anderson et al. (1970)
2-Keto-4-thioxane	X = O; Y = S	9.7	—	Jankowski and Coulombe (1971)
N-Methylmorpholine	X = NMe; Y = O	11.5	—	Harris and Spragg (1966)
N-N'-Dimethylpiperazine	X = NMe; Y = NMe	12.6	—	Harris and Spragg (1966)
		13.3 ± 0.3	—	Reeves and Strømme (1961b)
N,N,N',N'-Tetramethylpiperazinium chloride	X = $\overset{+}{N}Me_2$; Y = $\overset{+}{N}Me_2$	13.2 ± 0.3	9.3 ± 1.2	Abraham and MacDonald (1966)
1,4-Dioxene	X = O; Y = O	7.3 ± 0.2	7.3 ± 0.2	Larkin and Lord (1973)

Compound	Substituents			Reference
	R = H	≃ 9.8	—	Fraser and Reyes-Zamora (1967)
	R = CH$_2$Br	11.2	9.0 ± 0.9	Fuchs et al. (1972a)
		12.6	—	Fuchs et al. (1972b)

(continued)

TABLE X (continued)

Compounds		Kinetic parameters[a]		References
		$\Delta G^{\ddagger}$	$\Delta H^{\ddagger}$	
5,5-Dimethyl-1,3,2-dioxathiane	X = O; Y = S; Z = O	12.5 ± 0.2	—	Wood and Srivastava (1971)
1,2,3-Trithiacyclohexane	X = S; Y = S; Z = S	13.2	—	Kabuss et al. (1966b)
5,5-Dimethyl-1,2,3-trithiacyclohexane	X = S; Y = S; Z = S	14.7	—	Kabuss et al. (1966b)
5,5-Dimethyltrimethylene sulfate	X = O; Y = SO$_2$; Z = O	8.25 ± 0.15		Wood et al. (1970)
1,3,5-Trioxane		10.9 ± 0.2	8.8 ± 1.2	Pedersen and Schaug (1968)
Hexahydro-1,3,5-trimethyl-1,3,5-triazine		13.2 ± 0.2	15.2 ± 0.2	Gutowsky and Temussi (1967)
		12.8 ± 0.1	14.1 ± 0.2	Farmer and Hamer (1966)
		12.8	14.1	Riddell and Lehn (1966)
Acetone diperoxide (1,1,4,4-tetramethyl-2,3,5,6-tetraoxocyclohexane)		15.4	11.7	Murray et al. (1966)
1,1,4,4-Tetramethyl-2,3,5,6-tetrathiacyclohexane (Duplothioacetone)		16.0 ± 0.1[b]	16.5 ± 0.4	Bushweller et al. (1970)
		≃11[c]		Bushweller et al. (1971b)
		≃16[b]	—	

[a] $\Delta G^{\ddagger}$ (kcal/mole) is for the chair-chair process unless otherwise specified.
[b] Twist-boat to chair process.
[c] $\Delta G^{\ddagger}$ for ring inversion in the cyclohexane rings.

14. Conformational Processes in Rings

some cases, it is not easy to determine whether a particular dnmr effect results from nitrogen or ring inversion (Lambert *et al.*, 1971; Kessler and Leibfritz, 1970).

Compounds with two adjacent heteroatoms, such as cyclic disulfides, peroxides, and hydrazines have comparatively high barriers. These variations in the energy required for ring inversion undoubtedly reflect differences in torsional potentials about C—C, C—O, C—S, C—N, O—O, S—S, and N—N single bonds.

Duplodithioacetone (tetramethyl-*s*-tetrathiane) is particularly interesting, because it has been shown to exist in solution as a mixture of chair and twist-boat conformations, with the latter predominating (Bushweller, 1967, 1968, 1969). Bushweller *et al.* (1970) have investigated in detail the interconversions of these forms and find that the nmr spectrum at intermediate rates of exchange can only be fitted by line-shape calculations based on the scheme shown in Fig. 15. In contrast to cyclohexane, the twist-boats shown in Fig. 15 are not expected to interconvert easily by pseudorotation, since the

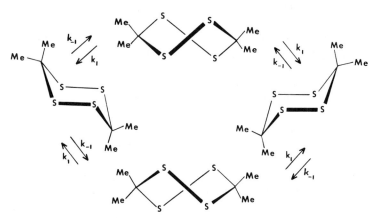

Fig. 15 Exchange scheme for duplodithioacetone. Rate constants between the two chair forms or between the two twist-boat forms are zero (Bushweller *et al.*, 1970).

intermediate symmetrical boat undoubtedly has a very high energy. The pseudorotation barrier, however, cannot be determined from the experimental data, but information on this point can be obtained by investigating related compounds as shown below. The nmr spectra, on the other hand, do show that the chairs are not interconverting directly, as fully expected. Particularly accurate and reliable activation parameters for the chair to twist-boat were obtained by combining dnmr data over the temperature range $-21°$ to 29°C with direct kinetic measurements of the interconversion at low temperatures ($-55°$ to $-69°$C).

Bushweller et al. (1971b) have shown that the barrier to direct interconversion of the two-mirror image twist-boats of 3,3:6,6-bis(pentamethylene)-s-tetrathiane is very probably larger than the barrier to ring inversion in the cyclohexane rings of this compound, and is thus at least 10 kcal/mole.

Barriers to ring inversion in tetrahydropyrazines are very much higher than those in simple cyclohexene derivatives, as was first shown by Breliere and Lehn (1965). Additional processes, such as restricted rotation, and nitrogen inversion, often occur in these systems and some controversy has arisen as to the precise assignments of all the dnmr effects which are observed. However, it is clear that the highest barrier observed is due to ring inversion. Bittner and Gerig (1972) have reviewed the evidence and have pointed out that high ring inversion barriers require substituents on both nitrogen atoms.

$$
\begin{array}{cc}
\text{N-CO}_2\text{CH}_2\text{CF}_3 & \text{N-CO}_2\text{CH}_2\text{CF}_3 \\
\text{N-CO}_2\text{CH}_2\text{CF}_3 & \text{N-H} \\
\mathbf{13} & \mathbf{14}
\end{array}
$$

Thus, whereas the free energy of activation for ring inversion in **13** is about 20 kcal/mole, that of **14** is less than 10 kcal/mole.

C. Seven-Membered Rings

Since an excellent review of the conformational mobility of seven-membered ring systems has been published recently (Tochtermann, 1970), this section will consist of only a brief review of some of the highlights of this topic. Energy barriers in seven-membered rings cover an extremely wide range. Pseudorotation in the twist-boat or twist-chair forms of cycloheptane have very low barriers according to strain energy calculations (Section III,B,1b). Thus, it is not surprising that neither cycloheptane nor 1,1,3,3-tetrafluorocycloheptane show any observable dnmr effect (Roberts, 1966). With substituents larger than fluorine or with the introduction of sulfur atoms, the barriers to pseudorotation become measurable by nmr methods (Table XI), as in 1,1-difluoro-4,4-dimethylcycloheptane and *trans*-4,5-dibromo-1,1-difluorocycloheptane (Knorr et al., 1967). Von Bredow et al. (1971) have found that 5,5-dimethyl-1,2-dithiacycloheptane exists in two different conformations. The major and minor forms are thought to belong to the chair and boat families, respectively.

The barriers to ring inversion of cycloheptene and 5,5-difluorocycloheptene are much smaller than those of benzocycloheptene and its 4,4- and 5,5-dimethyl derivatives (Table XI). Whereas all of the above compounds show

TABLE XI KINETIC PARAMETERS FOR CONFORMATIONAL PROCESSES IN SOME SEVEN-MEMBERED RINGS

Compounds	$\Delta G^{\ddagger a}$ (kcal/mole)	$\Delta H^{\ddagger}$ (kcal/mole)	References
1,1-Difluoro-4,4-dimethylcycloheptane	≃6	—	Roberts (1966)
4,5-*trans*-Dibromo-1,1-difluorocycloheptane	7.3	9.8 ± 0.3	Knorr et al. (1967)
5,5-Difluorocycloheptene	7.4	—	Knorr et al. (1967)
Cycloheptene	5.0	—	St. Jacques and Vaziri (1971)
Cycloheptene oxide	7.7	—	Servis et al. (1974)
1,2-Benzocycloheptene-4,4,6,6-d_4	10.9	—	Kabuss et al. (1965)
4,4-Dimethyl-1,2-benzocycloheptene	11.45	—	Kabuss et al. (1965)
5,5-Dimethyl-1,2-benzocycloheptene	11.8	—	Kabuss et al. (1965)
3,3,6,6-Tetramethylbenzocycloheptene	9.3 and 12.6	—	Grunwald and Price (1965)
3,3,5,5-Tetramethylcycloheptanone-6,6-d_2	6.3	—	St. Jacques and Vaziri (1972)
3,3,6,6-Tetramethylcycloheptanone	8.5	—	Borgen (1972)
2,3,5,6-Dibenzo-4-thiacycloheptanone	9.3 ± 0.1	—	Dürr (1967)
1,3,5-Cycloheptatriene	6.1	—	Anet (1964)
	5.7 ± 0.1	—	Jensen and Smith (1964)
1-Methyl-7-*t*-butylcycloheptatriene	15.2	—	Heyd and Cupas (1969)
1,2,5,6-Dibenzocycloheptatriene	9.0	—	Nógrádi et al. (1970)
1,2,3,4-Dibenzocycloheptatriene	18–19	—	Sutherland and Ramsay (1965)
7,7-Dimethoxy-1,2,5,6-dibenzocycloheptatriene	15	—	Tochtermann et al. (1964)
1,2,3,4,5,6-Tribenzocycloheptatriene	23.8	—	Nógrádi et al. (1970)
γγ-Difluoro-ε-caprolactone	10.0	—	Noe and Roberts (1971)
γγ-Difluoro-ε-caprolactam	10.4	—	Noe and Roberts (1971)
3,4,5,6-Tetrachloro-2,7-dihydrothiepin	8.0	8.8	Orrel et al. (1972)
3-Isopropyl-5-methylthiepin-1,1-dioxide	6.4	—	Anet et al. (1969)
5,5-Dimethyl-1,2-dithiacycloheptane-3,3,7,7-d_4	10.8 and 8.0	—	von Bredow et al. (1971)
4,6-Dithiacycloheptane	8.5	—	Kabuss et al. (1966b)
Benzo-5,5-dimethyl-4,6-dithiacycloheptene	12.4	14.6	Schmid et al. (1966)
1,2,3-Trithiacycloheptane	13.2	—	Kabuss et al. (1966b)
5,5-Dimethyl-1,2,3-trithiacycloheptane	14.6	—	Kabuss et al. (1966b)
4,5,6-Trithiacycloheptene	8.9	—	Kabuss et al. (1966b)
Benzo-4,5,6-trithiacycloheptene	17.4	—	Kabuss et al. (1966b)

(*continued*)

TABLE XI (continued)

Compounds	$\Delta G^{\ddagger a}$ (kcal/mole)	$\Delta H^{\ddagger}$ (kcal/mole)	References
R = H	17.9 and ≃10	—	von Bredow et al. (1970)
R = OMe	19.6 and 10.5 ± 0.3	—	von Bredow et al. (1970)
R = Me	19.75 and 11.2 ± 0.3	—	von Bredow et al. (1970)
R = Ph	21.2 and 11.4 ± 0.3	—	von Bredow et al. (1970)
2,3,5,6,7-pentathiacycloheptane	11.4	12.5	Moriarty et al. (1969)
2-Anilino-3-H-azepine	10.2 ± 0.4	—	Mannschreck et al. (1967)
2-Ethoxy-1-methyl-3,4-bismethoxycarbonyl-1-H-1-benzazepine	15.3 ± 0.6	—	Mannschreck et al. (1967)
X = S	15.5	—	Sutherland and Ramsay (1965)
X = SO$_2$	18.2	—	Sutherland and Ramsay (1965)
X = C(COOEt)$_2$	14.25	—	Sutherland and Ramsay (1965)
X = C=N—OH			
(syn)	13.7	—	Sutherland and Ramsay (1965)
(anti)	13.2	—	Sutherland and Ramsay (1965)

Structure	Value	Reference
(biphenyl lactone with H₃C, H₃C, C=O, O)	12.5	Sutherland and Ramsay (1965)
(dibenzo 7-membered ring with R and X); R = CMe₂OH; X = CH₂	17.5	Nógrádi et al. (1970)
R = CMe₂OH; X = CO	<9	Nógrádi et al. (1970)
R = CMe₂OH; X = S	17.7	Nógrádi et al. (1970)
R = CMe₂OH; X = O	10.3	Nógrádi et al. (1970)
(dibenzo ring with R and X); X = CO; R = Et	20.0	Nógrádi et al. (1970)
X = SiMe₂; R = H	>23	Corey and Corey (1972)
(benzodiazepine with R₁ and R₂)		

(continued)

TABLE XI (continued)

Compounds	$\Delta G^{\ddagger a}$ (kcal/mole)	$\Delta H^{\ddagger}$ (kcal/mole)	References
$R_1 = R_2 = $ Ph	12.6 ± 0.5	—	Mannschreck et al. (1967)
$R_1 = R_2 = $ Me	11.7 ± 0.4	—	Mannschreck et al. (1967)
$R_1 = $ Me; $R_2 = $ Ph	12.2 ± 0.4	—	Mannschreck et al. (1967)
R = H	12.2 ± 0.3	—	Linscheid and Lehn (1967)
R = Me	17.7	—	Linscheid and Lehn (1967)
7,12-Dihydropleiadene (**16**)	17.8	—	Sadée (1969)
8-Methyl-7,12-dihydropleiadene	13.6	—	Lansbury et al. (1966)
8,11-Dimethyl-7,12-dihydropleiadene	14.5	—	Lansbury et al. (1966)
7-Methylene-7,12-dihyropleiadene	15.6	—	Lansbury et al. (1966)
trans-7,12-Diacetoxy-7,12-dihydropleiadene	15.2	—	Lansbury et al. (1966)
trans-7,12-Dimethoxy-7,12-dihydropleiadene	14.3	—	Lansbury et al. (1966)
	15.2	—	Lansbury et al. (1966)

Structure: benzodiazepine with R on N, carbonyl (O), Ph substituent, and Cl on aromatic ring.

evidence for chair forms only, 3,3,6,6-tetramethylbenzocycloheptene and 5,5-dimethyl-3,7-dithiabenzocycloheptene exist in both chair and boat (or twist-boat) conformations. In the dithia compound, two distinct dnmr effects are observed; the higher energy process is the interconversion of the twist to the chair form. A lower energy process is observed in the twist form and is ascribed to hindered pseudorotation (von Bredow et al., 1970).

The increase in the barrier to ring inversion in cycloheptene that results from the addition of a fused benzene ring probably has its origin in two effects: (a) the transition state has internal bond angles which are larger than normal, and this distortion is resisted by the fused benzene ring, (b) nonbonded repulsions between the ortho hydrogens and the nearest hydrogens on the seven-membered ring are much greater in the nearly planar transition state than in the chair form.

Bridged diphenyls containing seven-membered rings exist in chiral twist forms, which in some cases can be isolated in optically active forms, and are closely related to the well-known isomerism in unbridged diphenyl derivatives. The barrier in the parent carbocyclic bridged biphenyl (**15**), however, is much too small for optically active forms to be isolated (Table IX).

Molecular models indicate that 7,12-dihydropleiadene (**16**) is highly nonplanar and requires relatively large bond-angle distortions for ring inversion to take place. Lansbury et al. (1966) have determined that the barrier to ring inversion in **16** is 13.6 kcal/mole and that the barrier is considerably

greater if a hydrogen on one of the methylene group is replaced by an alkyl group (Table XI).

The barrier to ring inversion in 1,3,5-cycloheptatriene is quite low (6 kcal/mole), but the presence of bulky substituents or fused benzene rings can greatly increase this barrier (Table XI) for reasons already mentioned in the case of cycloheptene and its derivatives.

D. Eight-Membered Rings

Conformational processes in eight-membered rings that have small torsional constraints, e.g., cyclooctane, cyclooctanone, and oxocane (but not cyclooctene), have recently been reviewed in detail (Anet, 1974), and the present discussion of this topic will therefore be very brief.

Conformational interconversions in cyclooctane itself have already been discussed in Section III,A,2. Substituted cyclooctanes have played a key role in elucidating the conformational picture in cyclooctane. For example, pseudorotation of the boat-chair via the twist-boat-chair has a barrier that is too small for this process to be studied by dmnr, but the introduction of a pair of geminal substituents raises the barrier sufficiently for observation (Table XII). The barrier to ring inversion in the boat-chair, by contrast, is not strongly affected by substituents. Another example concerns the occurrence of crown-family conformations, which are present to only a few percent in cyclooctane itself, but are important or even dominant conformations in 1,1,2,2-tetrafluorocyclooctane, *trans-syn-trans*-1,2,5,6-tetrabromocyclooctane and 1,3,5,7-tetraoxocane (Table XII). The barriers to ring inversion in the crown conformations are appreciably higher than those in boat-chair conformations.

cis-Cyclooctene exists in an unsymmetric distorted boat-chair and two processes can be clearly observed by dnmr (St. Jacques, 1967; Anet, 1970). The lower energy process (Table XII) introduces a C_2 axis of symmetry on a time-average basis. The suggestion (Allinger and Sprague, 1972) based on strain-energy calculations, that the lower energy process introduces a plane of symmetry (on a time-average basis) is rigorously excluded by the nmr evidence.

Barriers to ring inversion have been measured in *cis-cis*-1,3-cyclooctadiene, *cis-cis-cis*-1,3,5-cyclooctatriene, and in derivatives of *cis-cis-cis-cis*-1,3,5,7-cyclooctatetraene (Table XII). In *cis-cis*-1,5-cyclooctadiene, which exists in the twist-boat form, two processes are detectable by dnmr at low temperatures (Table XII) and possible interconversion paths have already been discussed in Section III,B,2.

Cyclooctatetraene and its derivatives can undergo two distinct processes: (a) bond shift and (b) ring inversion, with single and double bonds in unchanged positions (Anet *et al.*, 1964b; Oth, 1971). When both processes occur in the same compound, it is found that ring inversion has a lower barrier than has bond shift. Bond shift is not a conformational process, and it will not be discussed further in this chapter. Ring inversion, it should be noted, cannot be observed by dnmr methods in cyclooctatetraene itself, but the presence of diastereotopic nuclei on a substituent, e.g., a $C(CH_3)_2OH$ group, allows the inversion rate in the cyclooctatetraene derivative to be measured. As in the case of seven-membered rings discussed in the previous section, it is found that substituents or fused benzene rings lead to increased conformational barriers (Table XII). A number of calculations of the energy barrier for ring inversion in cyclooctatetraene and its derivatives have been

TABLE XII

KINETIC PARAMETERS FOR CONFORMATIONAL PROCESSES IN SOME EIGHT-MEMBERED RINGS

Compound	Process	$\Delta G^\ddagger$ (kcal/mole)	$\Delta H^\ddagger$ (kcal/mole)	Reference
Cyclooctane-d_{14}	Ring inversion	8.1	7.4	Anet and Hartman (1963)
1,1-Difluorocyclooctane	Ring inversion	7.5	—	Anderson et al. (1969)
	Pseudorotation	4.9	—	Anderson et al. (1969)
1,1,2,2-Tetrafluorocyclooctane	Ring inversion	10.3	—	Anderson et al. (1969);
	Ring inversion (boat-chair)	7.7	—	Anet (1974)
1,1,4,4-Tetrafluorocyclooctane	Pseudorotation	6.1	—	Anderson et al. (1969)
Perfluorocyclooctane	?	11	—	Anderson et al. (1969)
Cyclooctyl trimethylsilyl ether	Ring inversion	≃8	—	Schneider et al. (1972)
Cyclooctyl formate	Ring inversion	≃8	—	St. Jacques (1967)
Cyclooctyl acetate	Ring inversion	≃8	—	Schneider et al. (1972)
trans-1,2-Cyclooctanediol isopropylidene ketal	Ring inversion(?)	≃9.5	—	Anet and St. Jacques (1966)
1,5-cis-Diacetoxycyclooctane-d_8	Pseudorotation	6.5 ± 0.3	—	St. Jacques and Prud'homme (1972)
1,5-trans-Diacetoxycyclooctane-d_8	Ring inversion	8.0 ± 0.3	—	St. Jacques and Prud'homme (1972)
Cyclooctanone dithioethylene ketal	Pseudorotation	8.5	—	Anet (1971)
	Pseudorotation	6.6	—	Anet (1971)
Cyclooctanone ethylene ketal	Ring inversion	7.6	—	Anet (1971)
	Pseudorotation	5.3	—	Anet (1971)
Methylcyclooctane	Ring inversion	≃8	—	Anet and St. Jacques (1966)
t-Butylcyclooctane	Ring inversion	≃8	—	Anet and St. Jacques (1966)
1,1-Dimethylcyclooctane	Ring inversion and pseudorotation	8	—	Henrichs (1969); Anet (1974)
trans-syn-trans-1,2,5,6-Tetrabromocyclooctane	Ring inversion	11.1	—	Ferguson et al. (1968)
	Ring inversion (twist-chair-chair)	11.4	—	F. A. L. Anet and L. Kozerski (unpublished work, 1973)
	Ring inversion (boat-chair)	9.4	—	F. A. L. Anet and L. Kozerski (unpublished work, 1973)
	Pseudorotation (boat-chair)	7.7	—	F. A. L. Anet and L. Kozerski (unpublished work, 1973)

(continued)

TABLE XII (continued)

Compound	Process	$\Delta G^{\ddagger}$ (kcal/mole)	$\Delta H^{\ddagger}$ (kcal/mole)	Reference
cis-Cyclooctene-d_{13}	Ring inversion	8.2	—	St. Jacques (1967); Anet (1971)
1-Fluorocyclooctene	Pseudorotation	5.3	—	St. Jacques (1967); Anet (1971)
1-Fluorocyclooctene	Ring inversion	6.1	—	Weigert and Strobach (1970)
cis-Cyclooctene oxide-1,3,3-d_3	Ring inversion	8.0	—	Servis and Noe (1973)
Methylenecyclooctane	Ring inversion	8.1	—	Anet (1971)
4,4,7,7-Tetramethylcyclooctyne	Ring inversion	12.5 ± 0.5	—	Krebs (1968)
cis-cis-1,3-Cyclooctadiene	Ring inversion	$\simeq 8$	—	St. Jacques et al. (1966)
cis-cis-1,5-Cyclooctadiene	See Table VI			
Dibenzocycloocta-1,5-diene	Ring inversion	4.9 ± 0.1	—	Anet and Kozerski (1973)
Dibenzocycloocta-1,5-diene	Ring inversion	4.2 ± 0.2	—	Anet and Kozerski (1973)
Dibenzocycloocta-1,5-diene	Chair to boat	10.2 ± 1	—	Montecalvo et al. (1973)
Dibenzocycloocta-1,5-diene	Boat inversion	7.5 ± 1	—	Montecalvo et al. (1973)
4,4-Dimethyl-1,2:5,6-dibenzocyclooctadien-3-one	Ring inversion	18	—	Davies and Graham (1968)
3,4-Dicarbomethoxy-5-hydroxy-7,7-dimethyl-2,4-cyclooctadiene	Ring inversion	19.5	—	Cone et al. (1972)
cis-cis-cis-1,3,5-Cyclooctatriene-cis-7,8-d_2	Ring inversion	6.2	—	St. Jacques and Prud'homme (1970)
cis-1,2-Dibromo-cis-cis-1,3,5-cyclooctatriene	Ring inversion	13	—	Huisgen and Boche (1965)
Cycloocta-1,3,5-triene-7-one	Ring inversion	11.9 ± 0.5	—	Ganter et al. (1966)

Compound	Process	$\Delta G^\ddagger$	Reference
$X = Y = CH_2, R = Me, R^1 = H$	Ring inversion	>23.2	Lansbury and Klein (1968)
$X = SO, Y = CO, R = Me, R^1 = H$	Ring inversion	25.2	Johnson and Vegh (1969)
$X = Y = CO, R = H, R^1 = NO_2$	Ring inversion	23	Agosta (1967)
Cyclooctanone-d_{13}	Ring inversion	7.5	Anet et al. (1970)
	Pseudorotation	6.3	Anet et al. (1970)
5-t-Butylcyclooctanone	Pseudorotation	8.0	Anet (1971)
[structure with two D labels on ring fused to benzene]	Ring inversion	9.2 ± 0.2	F. A. L. Anet and G. W. Buchanan (unpublished work, 1970); Buchanan (1972)
Cyclooctatetraenyl-2,3,4,5,6,7-d_6-dimethylcarbinol	Ring inversion	14.7	Anet et al. (1964b)
Ethoxycyclooctatetraene	Ring inversion	12.47 ± 0.03	Oth (1971)
Isopropoxycyclooctatetraene	Ring inversion	12.47 ± 0.03	Oth (1971)
Isopropylcyclooctatetraene	Ring inversion	14.2 ± 0.2	Young (1971)
[structure with CH(CH_3)_2 group on fused bicyclic system]	Ring inversion	13.4	F. A. L. Anet and G. W. Buchanan (unpublished work, 1970); Buchanan (1972)
[structure with CF_2H group and COOH]	Ring inversion	12.3 ± 0.2	Senkler et al. (1972)
[structure with Br, COOH groups]	Ring inversion	≃31[a]	Mislow and Perlmutter (1962)

(continued)

TABLE XII (continued)

Compound	Process	$\Delta G^{\ddagger}$ (kcal/mole)	$\Delta H^{\ddagger}$ (kcal/mole)	Reference
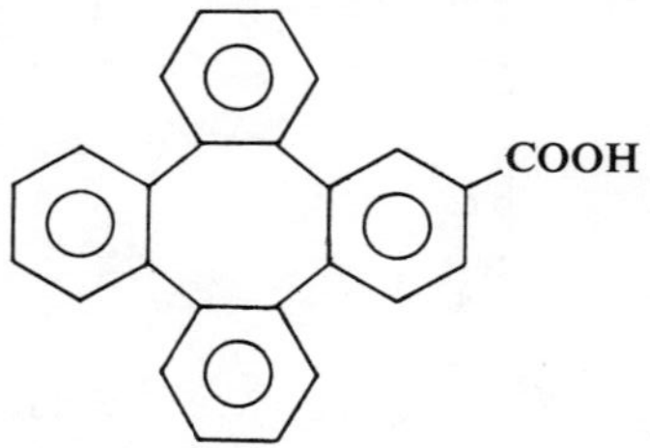 (structure with COOH)	Ring inversion	>45[a]	—	Gust et al. (1972)
Oxacyclooctane (oxocane)	Ring inversion	7.4	—	Anet and Degen (1972a)
1,3-Dioxocane	Ring inversion	7.3	—	Anet and Degen (1972a)
	Pseudorotation	5.7	—	Anet and Degen (1972a)
1,3,6-Trioxocane	Ring inversion	8.7	—	Anet and Degen (1972a)
	Pseudorotation	6.8	—	Anet and Degen (1972a)
1,3,5,7-Tetraoxocane	Ring inversion	12.0	—	Anet and Degen (1972a)
5-Oxocanone	Ring inversion	9.0 ± 0.2	—	Anet and Degen (1972b)
	Pseudorotation	7.8 ± 0.2	—	Anet and Degen (1972b)
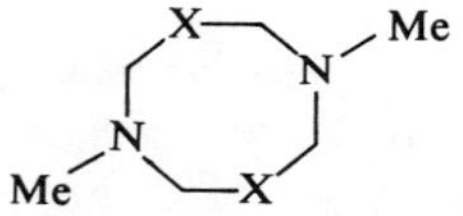 (N,N'-dimethyl diaza structure) X = S	Ring inversion	14.8	—	Lehn and Riddell (1966)
X = Se	Ring inversion	14.4	—	Lehn and Riddell (1966)

14. Conformational Processes in Rings

Structure	Process	ΔG	Reference
	X = CH Ring inversion	11 ± 2	Doyle et al. (1970)
	X = N Ring inversion	≈11	Coll et al. (1972)
	R_1 = Me; R_2 = isopropyl; X = O	17.7 ± 0.2	Ollis and Stoddart (1973)
	R_1 = isopropyl; R_2 = Me; X = O	18.4 ± 1	Ollis and Stoddart (1973)
	$R_1 = R_2$ = H; X = N—CH_2Ph	>27	Ollis and Stoddart (1973)
		21.4	Ollis and Stoddart (1973)
		14	Nelson and Gillespie (1973)

[a] Racemization of optically active compound.

made and the results are in satisfactory agreement with the experimental results (Allinger *et al.*, 1973; Dewar, 1971; Dewar *et al.*, 1969; Wipff *et al.*, 1971).

E. Nine-Membered and Larger Rings

Table XIII contains a selection of barriers in both carbocyclic and heterocyclic compounds with nine or more members in the ring. Conformational interconversions in the larger cycloalkanes have already been discussed in Section III,B,1a. The presence of benzene rings or of extensive unsaturation often simplifies the conformational picture because the rings become more rigid. Such systems, when highly nonplanar, often have large conformational barriers and dnmr effects can then be observed well above room temperature. For example, the barrier to ring inversion in cyclotriveratrylene (**17**) is so

[Structure of compound **17** showing cyclotriveratrylene with OCH$_3$ groups]

17

high that the nmr spectrum is still in the slow exchange region even at 200°C (Cookson *et al.*, 1968). By comparison, the barrier to ring inversion in 1,4,7-cyclononatriene is only 15 kcal/mole and dnmr effects are observable at room temperature (Untch and Kurland, 1963). Both of these compounds exist in crown conformations having C_3 symmetry. It has been shown, however, that the monoketone derived from cyclotriveratrylene has a flexible conformation (C_2 symmetry) that shows no observable dnmr effects at $-60°C$.

Salicyclic acid (*o*-hydroxybenzoic acid) and certain ring alkyl derivatives give cyclic twelve-membered lactones on dehydration. These compounds have interesting properties; e.g., tri-*o*-thymotide undergoes spontaneous resolution and occludes small organic molecules in the crystalline state (clathrates). Racemization, which is a conformational process, has been studied by optical rotation measurements, and more recently, dnmr methods have been applied (Downing *et al.*, 1970). While tri-*o*-thymotide (**18**) has a propeller conformation (C_3 symmetry) with all ester groups trans in the crystalline state, solutions of this compound show significant amounts of the helix conformation,

14. Conformational Processes in Rings

TABLE XIII KINETIC PARAMETERS FOR CONFORMATIONAL PROCESSES IN RINGS WITH MORE THAN EIGHT MEMBERS

Compounds	$\Delta G^\ddagger$ (kcal/mole)	$\Delta H^\ddagger$ (kcal/mole)	References
Cyclononane	$\simeq 6$	—	Anet and Wagner (1971)
1,1-Dimethylcyclononane	$\simeq 9$	—	Anet and Wagner (1971)
1,1,4,4-Tetramethylcyclononane	$\simeq 20$	—	Borgen and Dale (1970)
Cyclodecane	$\simeq 6$	—	Anet et al. (1972)
1,1-Dimethylcyclodecane	7.8 and 5.7	—	F. A. L. Anet and J. J. Wagner (unpublished work, 1972)
1,1-Difluorocyclodecane	5.7	—	Noe and Roberts (1972)
Cyclododecane	7	—	Anet et al. (1972)
Cyclotetradecane	$\simeq 7$	—	Anet et al. (1972)
Cyclononanone	5.0 ± 0.2 and 6.5 ± 0.2	—	Anet et al. (1973a)
1,1,4,4-Tetramethylcyclononanone	$\simeq 14$	—	Borgen and Dale (1970)
	15.2 ± 0.2	—	Dale (1971)
4,4,8,8-Tetramethyl-1,2-cyclononanedionebishydrazone	21	—	Blomquist and Miller (1968)
Cyclodecanone	6.5 ± 0.2 and 7.3 ± 0.2	—	Anet et al. (1973a)
4,4,8,8-Tetramethylcyclodecanone	$\simeq 11$	—	Dale (1971)
Cycloundecanone	6.0 ± 0.2 and 6.9 ± 0.2	—	Anet et al. (1973a)
Cyclododecanone	7.6 ± 0.4	—	Anet et al. (1973a)
5,5,9,9-Tetramethylcyclododecanone	$\simeq 11$	—	Dale (1971)
Cyclotridecanone	5.6 ± 0.4	—	Anet et al. (1973a)
Cyclotetradecanone	6.3 ± 0.4 and 6.7 ± 0.4	—	Anet et al. (1973a)
Cyclopentadecanone	5.0 ± 0.4	—	Anet et al. (1973a)
Cyclohexadecanone	5.6 ± 0.4	—	Anet et al. (1973a)
cis-cis-cis-1,4,7-Cyclononatriene	15.5	—	Untch and Kurland (1963)
	15.5	—	Radlick and Winstein (1963)
trans-Cyclodecene-1,2,4,4,9,9-d_6	10.7 ± 0.3	—	Binsch and Roberts (1965)
3,3-Difluoro-trans-cyclodecene	12.4	—	Noe et al. (1972)

(continued)

TABLE XIII (continued)

Compounds	$\Delta G^\ddagger$ (kcal/mole)	$\Delta H^\ddagger$ (kcal/mole)	References
Cyclodeca-1,6-dien-4,9-dione	7.9	—	Roberts et al. (1968)
Cyclodeca-1,6-dien-4,9-bisethylene ketal	18.8	—	Roberts et al. (1968); Dale et al. (1968)
4,9-Bismethylenecyclodeca-1,6-diene	12.1	—	Dale et al. (1968)
4,9-Dioxacyclodeca-1,6-diene	$\simeq 10 \pm 0.3$	—	Feigenbaum and Lehn 1(969)
4,9-Dithiacyclodeca-1,6-diene	12.3 ± 0.2	—	Feigenbaum and Lehn (1969)
1,4,7,10-Tetraoxocyclododecane	6.8 ± 0.3 and 5.5 ± 0.3	—	Anet et al. (1973c)
1,4,7,10-Tetraoxocyclododecanelithium thiocyanate	8.2 ± 0.3 and 5.6 ± 0.3	—	Anet et al. (1973c)
X = CH; Y = CH$_2$; R = H	20.2 ± 0.3	17.0 ± 0.5	Vögtle (1969d)
	19.7	—	Sherrod and Boekelheide (1973)
X = CD; Y = CH$_2$; R = H	19.6	—	Sherrod and Boekelheide (1973)
X = CH; Y = CH$_2$; R = CHO	17.7 ± 0.3	—	Hefelfinger and Cram (1970)
X = CH; Y = S; R = H	14.7 ± 0.4	—	Vögtle (1969d)
X = CF; Y = CH$_2$; R = H	>22.8	—	Vögtle (1969d)
[3.3]Paracyclophane	11.7 ± 0.5	—	Anet and Brown (1969)
1,10-Diketo[3.3]paracyclophane	8.8	—	Anet and Brown (1969)
[4.3]Paracyclophane-6-carboxylic acid	$\simeq 34^a$	—	Cram et al. (1958)
[4.4]Paracyclophane-6-carboxylic acid	$<20^a$	—	Cram et al. (1958)
	11.2	10.5 ± 0.3	Whitesides et al. (1968)

Structure	Value	Reference
(cyclophane-d structure)	13.5	12.8 ± 0.7	Whitesides et al. (1968)
Methyl 1,11-dioxa[11]paracyclophane carboxylate	31[a]	—	Lüttringhaus and Eyring (1957)
X = OMe; Y = NH	> 22.2	—	Sakamoto and Ōki (1973)
X = Me; Y = NH	> 10.8	—	Sakamoto and Ōki (1973)
X = OMe; Y = O	11.0	—	Sakamoto and Ōki (1973)
X = Me; Y = O	21.7	—	Sakamoto and Ōki (1973)
(cyclophane diene)	20	—	Reiner and Jenny (1969)
(dioxa bridged structure)	16.8	—	Gault et al. (1967)

(continued)

TABLE XIII (continued)

Compounds	$\Delta G^\ddagger$ (kcal/mole)	$\Delta H^\ddagger$ (kcal/mole)	References
X = N; Y = CH	≥ 20.5	—	Vögtle and Effler (1969)
X = CH; Y = N	20.5 ± 0.3	—	Vögtle and Effler (1969)
X = N; Y = N	<13.6	—	Vögtle and Effler (1969)
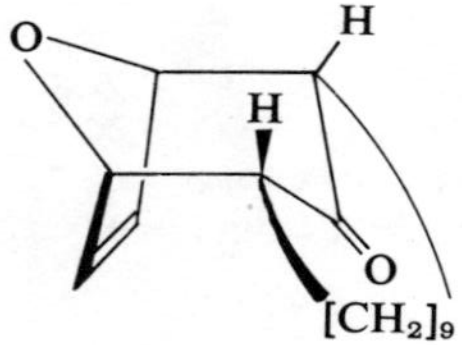	16	—	Vinter and Hoffmann (1973)
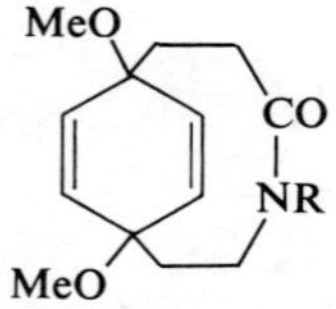 R = H	19.0	—	Hemmi et al. (1972)
R = Ac	12.2	—	Hemmi et al. (1972)

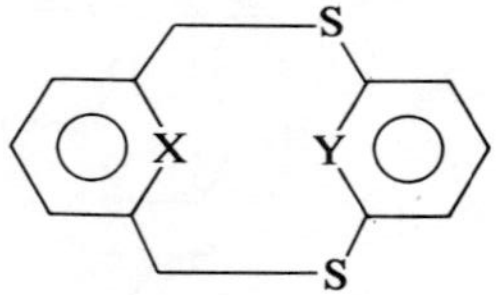

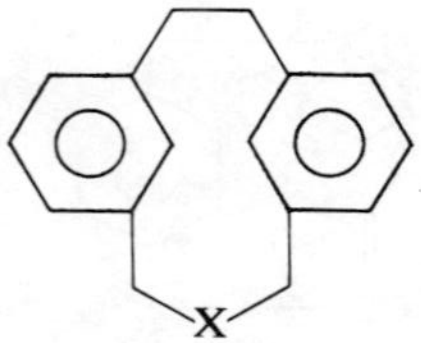

X = CHCOOH	16.8 ± 0.2	Griffin and Coburn (1964)
X = C(COOEt)$_2$	19.2	Griffin and Coburn (1967)
X = CHC(Me)$_2$OH	16.2 ± 0.3	Griffin and Coburn (1967)
X = CHC(Ph)$_2$OH	15.7 ± 0.3	Griffin and Coburn (1967)
X = S	8.32	Sato et al. (1968)
X = SO$_2$	13.7	Sato et al. (1968)
X = Y = CF	21.1 ± 0.5	Vögtle and Schunder (1969)
X = CF; Y = N	≃22.7	Vögtle and Schunder (1969)
X = Y = CCl	>23	Vögtle and Schunder (1969)
2,2″-Ethano-m-terphenyl	≃20 and ≃15	Vögtle (1969a)
X = CH; Y = S; n = 2	12.4	Vögtle (1969b)
X = CH; Y = CH$_2$; n = 3	≃11.5	Fujita et al. (1972)
X = CH; Y = S; n = 3	10.2 and 8.7	Mitchell and Boekelheide (1969)
X = CH; Y = S; n = 4	≃9.1	Vögtle (1969b)
X = CF; Y = S; n = 4	22.6 ± 0.5	Vögtle (1969c)
X = CH; Y = S; n = 5	<10.3	Vögtle (1969b)
X = CF; Y = S; n = 5	15.2	Vögtle (1969b)
X = CF; Y = N—Ts; n = 5	20.6 ± 0.6	Vögtle and Neumann (1970)
X = CF; Y = S; n = 6	10.5	Vögtle (1969b)
X = CF; Y = S; n = 7	<10.9	Vögtle (1969e)
X = C—Cl; Y = S; n = 7	23.5	Vögtle (1969e)
X = C—Br; Y = S; n = 7	<23.2	Vögtle (1969e)

(continued)

TABLE XIII (continued)

Compounds	$\Delta G^{\ddagger}$ (kcal/mole)	$\Delta H^{\ddagger}$ (kcal/mole)	References
X = C—CH$_3$; Y = S; n = 7	25–26	—	Vögtle (1969e)
X = C—Cl; Y = S; n = 8	15.4	—	Vögtle (1969e)
X = C—Br; Y = S; n = 8	16.6	—	Vögtle (1969e)
X = C—CH$_3$; Y = S; n = 8	22.5	—	Vögtle (1969e)
X = C—Cl; Y = S; n = 9	11.6	—	Vögtle (1969e)
X = C—Br; Y = S; n = 9	15.4	—	Vögtle (1969e)
X = C—CH$_3$; Y = S; n = 9	12.0	—	Vögtle (1969e)
X = C—Br; Y = S; n = 10	≃9.4	—	Vögtle (1969e)
1,4,7,10,13,16-Hexamethyl-5,6,11,12,17,18-hexahydrotribenzo[aei]-cyclododecaene	17.1 ± 0.3	—	Brickwood et al. (1973)
Cyclotetraveratrylene	13.2 ± 0.3	10.4 ± 0.2	White and Gesner (1968)
Tri-o-thymotide	20.5 ± 0.3	—	Downing et al. (1970)
Tri-o-carvacrotide	20.6 ± 0.2 and 17.6 ± 0.2	—	Downing et al. (1970)
Tri-3,6-dimethylsalicylide	18.0 ± 0.5 and 14.3 ± 0.5	—	Downing et al. (1970)
Protopine	≃13	—	Anet and Brown (1967)
Germacra-1(10),4,7(11)-triene-9-α-ol acetate	≃17	—	Horibe et al. (1973)

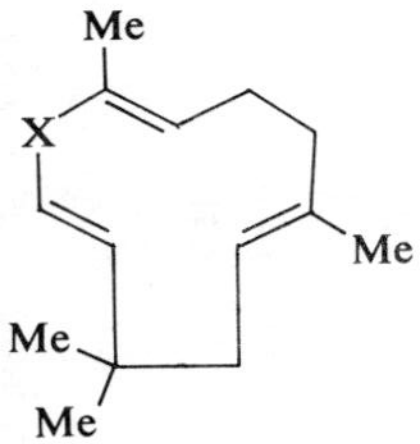			
X = C=O	15.8 ± 0.3	18.4 ± 0.8	Dev *et al.* (1968)
X = CH_2	10.6 ± 0.3	—	Dev *et al.* (1968)
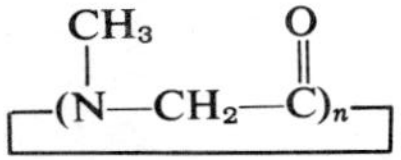			
n = 3	≃20	—	Dale (1971)
n = 4	≃22	—	Dale (1971)
n = 5	≃15	—	Dale (1971)
n = 8	≃15	—	Dale (1971)
d-Tubocurarine	≃16	—	Egan *et al.* (1973).

[a] Racemization of optically active compound.

                    18

which lacks symmetry. Both conformations are chiral, of course. Dnmr measurements show that the interconversion of the two mirror-image propeller conformations (ring inversion) proceeds by three successive local inversions as follows: propeller ⇌ helix ⇌ enantiomeric helix ⇌ enantiomeric propeller. Energy barriers for some of these systems are given in Table XIII.

A large number of cyclophanes and heterocyclophanes have been examined by dnmr methods, and a representative selection of compounds and their conformational free energy barriers are grouped together in Table XIII. Non-bonded interactions often dominate the barriers in this series, and therefore space-filling molecular models give good qualitative indications of the magnitudes of these barriers. The presence of large nonbonded interactions in the transition state for ring flipping in [2,2]metaparacyclophane is borne out by the large deuterium isotope effect ($\Delta\Delta G^{\ddagger} = 0.1$ kcal/mole) found by Sherrod and Boekelheide (1973) (see Table XIII). On the other hand, local inversion of the trimethylene chain in [3.3]paracyclophane involves mainly eclipsing and bond angle strain and is closely related to local ring inversion in unsaturated seven-membered rings. The barrier in 1,10-diketo[3.3]paracyclophane is about 3 kcal/mole less than in the parent hydrocarbon (Table XIII), presumably for the same reasons that cyclohexanone has a much lower barrier to ring inversion than does cyclohexane (Section IV,B,2).

The barriers to rotation of 1,4-bridged phenyl rings in cyclophanes depend strongly on the size of the bridge. Whereas [4.3]paracyclophane derivatives can be resolved and racemize only slowly at 150°, [4.4]paracyclophane derivatives have barriers which are too low for resolution (Cram *et al.*, 1958; Smith, 1964). The barrier in cyclophanes is strongly dependent on the nature of substituents present on the bridged phenyl residue. It is interesting that a methoxy substituent can be larger than a methyl group in some paracyclophanes, whereas the reverse is true when the substituent is hindering rotation in biphenyl derivatives (Sakamoto and Ōki, 1973).

14. Conformational Processes in Rings

Vögtle (see Table XIII for references) has determined conformational barriers in a wide range of dithia[n]metacyclophanes in which n varies from 6 to 13. The presence of a 2-substituent in these 1,3-bridged benzene derivatives has a strong effect on the barrier, which increases in the order H < F < Cl < CH_3 < Br, although the complete order cannot be determined with any one value of n.

REFERENCES

Abraham, R. J., and MacDonald, D. B. (1966). *Chem. Commun.* p. 188.
Agosta, W. C. (1967). *J. Amer. Chem. Soc.* **89**, 3926.
Allerhand, A., and Gutowsky, H. S. (1964). *J. Chem. Phys.* **41**, 2115.
Allerhand, A., Chen, F., and Gutowsky, H. S. (1965). *J. Chem. Phys.* **42**, 3040.
Allinger, N. L., and Sprague, J. T. (1972). *J. Amer. Chem. Soc.* **94**, 5734.
Allinger, N. L., Miller, M. A., Van Catledge, F. A., and Hirsch, J. A. (1967). *J. Amer. Chem. Soc.* **89**, 4345.
Allinger, N. L., Hirsch, J. A., Miller, M. A., Tyminski, I. J., and Van Catledge, F. A. (1968). *J. Amer. Chem. Soc.* **90**, 1199.
Allinger, N. L., Hirsch, J. A., Miller, M. A., and Tyminski, I. J. (1969). *J. Amer. Chem. Soc.* **91**, 337.
Allinger, N. L., Tribble, M. T., Miller, M. A., and Wertz, D. H. (1971). *J. Amer. Chem. Soc.* **93**, 1637.
Allinger, N. L., Sprague, J. T., and Finder, C. J. (1973). *Tetrahedron* **29**, 2519.
Anderson, J. E. (1969). *Chem. Commun.* p. 669.
Anderson, J. E. (1970). *Chem. Commun.* p. 417.
Anderson, J. E., and Roberts, J. D. (1970). *J. Amer. Chem. Soc.* **92**, 97.
Anderson, J. E., Glazer, E. D. Griffith, D. L., Knorr, R., and Roberts, J. D. (1969). *J. Amer. Chem. Soc.* **91**, 1386.
Anderson, J. E., Davis, D. R., and Roberts, J. D. (1970). *J. Org. Chem.* **35**, 1195.
Anet, F. A. L. (1964). *J. Amer. Chem. Soc.* **86**, 458.
Anet, F. A. L. (1971). *In* "Conformational Analysis" (G. Chiurdoglu, ed.), p. 15. Academic Press, New York.
Anet, F. A. L. (1974). *Fortschr. Chem. Forsch.* **45**, 169.
Anet, F. A. L., and Anet, R. (1971). *Determination Org. Struct. Phys. Methods* **3**, 344.
Anet, F. A. L., and Basus, V. J. (1973). *J. Amer. Chem. Soc.* **95**, 4424.
Anet, F. A. L., and Bourn, A. J. R. (1967). *J. Amer. Chem. Soc.* **89**, 760.
Anet, F. A. L., and Brown, M. A. (1967). *Tetrahedron Lett.* p. 4881.
Anet, F. A. L., and Brown, M. A. (1969). *J. Amer. Chem. Soc.* **91**, 2389.
Anet, F. A. L., and Degen, P. J. (1972a). *J. Amer. Chem. Soc.* **95**, 1390.
Anet, F. A. L., and Degen, P. J. (1972b). *Tetrahedron Lett.* p. 3613.
Anet, F. A. L., and Haq, M. Z. (1965). *J. Amer. Chem. Soc.* **87**, 3147.
Anet, F. A. L., and Hartman, J. S. (1963). *J. Amer. Chem. Soc.* **85**, 1204.
Anet, F. A. L., and Kozerski, L. (1973). *J. Amer. Chem. Soc.* **95**, 3407.
Anet, F. A. L., and Krane, J. (1973). *Tetrahedron Lett.* p. 5029.
Anet, F. A. L., and St. Jacques, M. (1966). *J. Amer. Chem. Soc.* **88**, 2586.
Anet, F. A. L., and Sandström, J. (1971). *Chem. Commun.* p. 1558.
Anet, F. A. L., and Wagner, J. J. (1971). *J. Amer. Chem. Soc.* **93**, 5266.
Anet, F. A. L., Ahmad, M., and Hall, L. D. (1964a). *Proc. Chem. Soc., London* p. 145.
Anet, F. A. L., Bourn, A. J. R., and Lin, Y. S. (1964b). *J. Amer. Chem. Soc.* **86**, 3576.

Anet, F. A. L., Bradley, C. H., Brown, M. A., Mock, W. L., and McCausland, J. H. (1969). *J. Amer. Chem. Soc.* **91**, 7782.
Anet, F. A. L., St. Jacques, M., and Henrichs, P. M. (1970). *Intra-Sci. Chem. Rep.* **4**, 251.
Anet, F. A. L., Cheng, A. K., and Wagner, J. J. (1972). *J. Amer. Chem. Soc.* **94**, 9250.
Anet, F. A. L., Cheng, A. K., and Krane, J. (1973a). *J. Amer. Chem. Soc.* **95**, 7877.
Anet, F. A. L., Chmurny, G. N., and Krane, J. (1973b). *J. Amer. Chem. Soc.* **95**, 4423.
Anet, F. A. L., Krane, J., Dale, J., Daavatn, K., and Kristiansen, P. O. (1973c). *Acta Chem. Scand.* **27**, 3395.
Bernard, M., and St. Jacques, M. (1970). *Chem. Commun.* p. 1097.
Bernard, M., and St. Jacques, M. (1973). *Tetrahedron* **29**, 2539.
Bernard, M., Sauriol, F., and St. Jacques, M. (1972). *J. Amer. Chem. Soc.* **94**, 8624.
Berry, R. S. (1960). *J. Chem. Phys.* **32**, 933.
Bhacca, N. S., and Horton, D. (1967). *J. Amer. Chem. Soc.* **89**, 5993.
Binsch, G. (1969). *J. Amer. Chem. Soc.* **91**, 1304.
Binsch, G., and Roberts, J. D. (1965). *J. Amer. Chem. Soc.* **87**, 5157.
Binsch, G., Eliel, E. L., and Kessler, H. (1971). *Angew. Chem., Int. Ed. Engl.* **10**, 570; *Angew. Chem.* **83**, 618 (1971).
Bittner, E. W., and Gerig, J. T. (1972). *J. Amer. Chem. Soc.* **94**, 913.
Bixon, M., and Lifson, S. (1967). *Tetrahedron* **23**, 769.
Blomquist, A. T., and Miller, R. D. (1968). *J. Amer. Chem. Soc.* **90**, 3233.
Borgen, G. (1972). *Acta Chem. Scand.* **26**, 1738.
Borgen, G., and Dale, J. (1970). *Chem. Commun.* p. 1105.
Borgen, G., and Dale, J. (1972). *Acta Chem. Scand.* **26**, 3593.
Borgen, G., Dale, J., and Schaug, J. (1972). *Acta Chem. Scand.* **26**, 1073.
Bovey, F. A., Anderson, E. W., Hood, F. P., and Kornegay, R. L. (1964a). *J. Chem. Phys.* **40**, 3099.
Bovey, F. A., Hood, F. P., Anderson, E. W., and Kornegay, R. L. (1964b). *Proc. Chem. Soc., London* p. 146.
Boyd, R. H. (1968). *J. Chem. Phys.* **49**, 2574.
Breliere, J. C., and Lehn, J. M. (1965). *Chem. Commun.* p. 426.
Brickwood, D. J., Ollis, W. D., and Stoddart, J. F. (1973). *Chem. Commun.* p. 638.
Brownstein, S. (1962). *Can. J. Chem.* **40**, 870.
Buchanan, G. W. (1972). *Tetrahedron Lett.* p. 665.
Buckingham, D. A., and Sargeson, A. M. (1971). *Top. Stereochem.* **6**, 219.
Bucourt, R., and Hainaut, D. (1967). *Bull. Soc. Chim. Fr.* [5] p. 4563.
Bushweller, C. H. (1967). *J. Amer. Chem. Soc.* **89**, 5978.
Bushweller, C. H. (1968). *J. Amer. Chem. Soc.* **90**, 2450.
Bushweller, C. H. (1969). *J. Amer. Chem. Soc.* **91**, 6019.
Bushweller, C. H., and O'Neill, J. W. (1969). *Tetrahedron Lett.* p. 4713.
Bushweller, C. H., Gollini, G., Rao, G. U., and O'Neil, J. W. (1970). *J. Amer. Chem. Soc.* **92**, 3055.
Bushweller, C. H., O'Neill, J. S., and Bilofsky, H. S. (1971a). *Tetrahedron* **27**, 3065.
Bushweller, C. H., Rao, G. U.. and Bissett, F. H. (1971b). *J. Amer. Chem. Soc.* **93**, 3058.
Cheng, A. K. (1973). Ph.D. Thesis, University of California, Los Angeles.
Claeson, G., Androes, G. M., and Calvin, M. (1960). *J. Amer. Chem. Soc.* **82**, 4428.
Claeson, G., Androes, G., and Calvin, M. (1961). *J. Amer. Chem. Soc.* **83**, 4357.
Claridge, R. F. C., Peake, B. M., and Golding, R. M. (1972). *J. Magn. Resonance* **6**, 29.
Coll, J. C., Crist, D. L., Barrio, M. del C. G., and Leonard, N. J. (1972). *J. Amer. Chem. Soc.* **94**, 7092.
Cone, E. J., Garner, R. H., and Hayes, A. W. (1972). *Tetrahedron Lett.* p. 57.

Cookson, R. C., Halton, B., and Stevens, I. D. R. (1968). *J. Chem. Soc.*, *B* p. 767.
Corey, J. Y., and Corey, E. R. (1972). *Tetrahedron Lett.* p. 4667.
Cram, D. J., Wechter, W. J., and Kierstead, R. W. (1958). *J. Amer. Chem. Soc.* **80**, 3126.
Dale, J. (1971). *Pure Appl. Chem.* **25**, 469.
Dale, J. (1973a). *Acta Chem. Scand.* **27**, 1115.
Dale, J. (1973b). *Acta Chem. Scand.* **27**, 1130.
Dale, J. (1973c). *Acta Chem. Scand.* **27**, 1149.
Dale, J., Ekeland, T., and Schaug, J. (1968). *Chem. Commun.* p. 1477.
Dalling, D. K., Grant, D. M., and Johnson, L. F. (1971). *J. Amer. Chem. Soc.* **93**, 3678.
Davies, J. M., and Graham, S. H. (1968). *Chem. Commun.* p. 542.
Dev, S., Anderson, J. E., Cormier, V., Damodaran, N. P., and Roberts, J. D. (1968). *J. Amer. Chem. Soc.* **90**, 1246.
Deverell, C., Morgan, R. E., and Strange, J. H. (1970). *Mol. Phys.* **18**, 555.
Dewar, M. J. S. (1969). "The Molecular Orbital Theory of Organic Chemistry." McGraw-Hill, New York.
Dewar, M. J. S. (1971). *Fortschr. Chem. Forsch.* **23**, 1.
Dewar, M. J. S., Harget, A., and Haselbach, E. (1969). *J. Amer. Chem. Soc.* **91**, 7521.
Downing, A. P., Ollis, W. D., and Sutherland, I. O. (1970). *J. Chem. Soc.*, *B* p. 24.
Doyle, M., Parker, W., Gunn, P. A., Martin, J., and MacNicol, D. D. (1970). *Tetrahedron Lett.* p. 3619.
Dunitz, J. D. (1970). *J. Chem. Educ.* **47**, 488.
Dunitz, J. D. (1971). *Pure Appl. Chem.* **25**, 495.
Dunitz, J. D., and Waser, J. (1972). *J. Amer. Chem. Soc.* **94**, 5645.
Durette, P. L., and Horton, D. (1969). *Chem. Commun.* p. 516.
Durette, P. L., Horton, D., and Bhacca, N. S. (1969). *Carbohydrate Res.* **10**, 565.
Dürr, H. (1967). *Z. Naturforsch. B* **22**, 786.
Egan, R. S., Stanaszek, R. S., and Williamson, D. E. (1973). *J. Chem. Soc., Perkin Trans.* II, p. 716.
Eliel, E. L., Allinger, N. L., Angyal, S. J., and Morrison, G. A. (1965). "Conformational Analysis." Wiley (Interscience), New York.
Engler, E. M., Andose, J. D., and Schleyer, P. von R. (1973). *J. Amer. Chem. Soc.* **95**, 8005.
Farmer, R. F., and Hamer, J. (1966). *Chem. Commun.* p. 866.
Farmer, R. F., and Hamer, J. (1968). *Tetrahedron* **24**, 829.
Farnham, W. B. (1972). *J. Amer. Chem. Soc.* **94**, 6857.
Feigenbaum, A., and Lehn, J. M. (1969). *Bull. Soc. Chim. Fr.* p. 3724.
Ferguson, M. N., MacNicol, D. D., Oberhausli, R., Raphael, R. A., and Zabkiewiez, R. (1968). *Chem. Commun.* p. 103.
Fraser, R. R., and Reyes-Zamora, C. (1967). *Can. J. Chem.* **45**, 1012.
Friebolin, H., Kabuss, S., Maier, W., and Lüttrringhaus, A. (1962). *Tetrahedron Lett.* p. 683.
Friebolin, H., Kabuss, S., Maier, W., and Lüttringhaus, A. (1969a). *Org. Magn. Resonance* **1**, 67.
Friebolin, H., Schmid, H. G., Kabuss, S., and Faisst, W. (1969b). *Org. Magn. Resonance* **1**, 147.
Fuchs, B., Goldberg, I., and Shmueli, U. (1972a). *J. Chem. Soc., Perkin Trans.* II, p. 357.
Fuchs, B., Auerbach, Y., and Sprecher, M. (1972b). *Tetrahedron Lett.* p. 2267.
Fujita, S., Hirano, S., and Nozaki, H. (1972). *Tetrahedron Lett.* p. 403.
Ganter, C., Pokras, S. M., and Roberts, J. D. (1966). *J. Amer. Chem. Soc.* **88**, 4235.
Gatti, G., Segré, A. L., and Morandi, C. (1967). *J. Chem. Soc.*, *B* p. 1203.

Gault, I., Price, B. J., and Sutherland, I. O. (1967). *Chem. Commun.* p. 540.
Geens, A., Tavernier, D., and Anteunis, M. (1967). *Chem. Commun.* p. 1088.
Geens, A., Swaelens, G., and Anteunis, M. (1969). *Chem. Commun.* p. 439.
Gerig, J. T., and Ortiz, C. E. (1970). *J. Amer. Chem. Soc.* **92**, 7121.
Gerig, J. T., and Rimerman, R. A. (1970). *J. Amer. Chem. Soc.* **92**, 1219.
Gerig, J. T., and Roberts, J. D. (1966). *J. Amer. Chem. Soc.* **88**, 2791.
Gilboa, H., Altman, J., and Loewenstein, A. (1969). *J. Amer. Chem. Soc.* **91**, 6062.
Griffin, R. W., Jr., and Coburn, R. A. (1964). *Tetrahedron Lett.* p. 2571.
Griffin, R. W., Jr., and Coburn, R. A. (1967). *J. Amer. Chem. Soc.* **89**, 4638.
Grunwald, E., and Price, E. (1965). *J. Amer. Chem. Soc.* **87**, 3139.
Gust, D., Senkler, G. H., and Mislow, K. (1972). *Chem. Commun.* p. 1345.
Gutowsky, H. S., and Chen, F. M. (1965). *J. Phys. Chem.* **69**, 3216.
Gutowsky, H. S., and Temussi, P. A. (1967). *J. Amer. Chem. Soc.* **89**, 4358.
Hanack, M. (1965). "Conformation Theory." Academic Press, New York.
Harris, R. K., and Sheppard, N. (1964). *Mol. Phys.* **7**, 595.
Harris, R. K., and Sheppard, N. (1967). *J. Mol. Spectrosc.* **23**, 231.
Harris, R. K., and Spragg, R. A. (1966). *Chem. Commun.* p. 314.
Hefelfinger, D. T., and Cram, D. J. (1970). *J. Amer. Chem. Soc.* **92**, 1073.
Hemmi, K., Nakai, H., Naruto, S., and Yonemitsu, O. (1972). *J. Chem. Soc., Perkin Trans.* II, p. 2252.
Hendrickson, J. B. (1961). *J. Amer. Chem. Soc.* **83**, 4537.
Hendrickson, J. B. (1962). *J. Amer. Chem. Soc.* **84**, 3355.
Hendrickson, J. B. (1964). *J. Amer. Chem. Soc.* **86**, 4854.
Hendrickson, J. B. (1967a). *J. Amer. Chem. Soc.* **89**, 7043.
Hendrickson, J. B. (1967b). *J. Amer. Chem. Soc.* **89**, 7047.
Henrichs, P. M. (1969). Ph.D. Thesis, University of California, Los Angeles.
Heyd, W. E., and Cupas, C. A. (1969). *J. Amer. Chem. Soc.* **91**, 1559.
Hoffmann, R. (1970). Quoted in Pickett and Strauss (1970).
Horibe, I., Tori, K., Takeda, K., and Ogino, T. (1973). *Tetrahedron Lett.* p. 735.
Hoyland, J. R. (1969). *J. Chem. Phys.* **50**, 2775.
Huisgen, R., and Boche, G. (1965). *Tetrahedron Lett.* p. 1769.
Inglefield, P. T., Krakower, E., Reeves, L. W., and Stewart, R. (1968). *Mol. Phys.* **15**, 65.
Jankowski, K., and Coulombe, R. (1971). *Tetrahedron Lett.* p. 991.
Jensen, F. R., and Beck, B. H. (1966). *Tetrahedron Lett.* p. 4523.
Jensen, F. R., and Beck, B. H. (1968). *J. Amer. Chem. Soc.* **90**, 1066.
Jensen, F. R., and Bushweller, C. H. (1965). *J. Amer. Chem. Soc.* **87**, 3285.
Jensen, F. R., and Bushweller, C. H. (1968). *Tetrahedron Lett.* p. 2825.
Jensen, F. R., and Bushweller, C. H. (1969a). *J. Amer. Chem. Soc.* **91**, 3223.
Jensen, F. R., and Bushweller, C. H. (1969b). *J. Amer. Chem. Soc.* **91**, 5774.
Jensen, F. R., and Neese, R. A. (1971). *J. Amer. Chem. Soc.* **93**, 6329.
Jensen, F. R., and Smith, L. A. (1964). *J. Amer. Chem. Soc.* **86**, 956.
Jensen, F. R., Noyce, D. S., Sederholm, C. H., and Berlin, A. J. (1960). *J. Amer. Chem. Soc.* **82**, 1256.
Jensen, F. R., Noyce, D. S., Sederholm, C. H., and Berlin, A. J. (1962). *J. Amer. Chem. Soc.* **84**, 386.
Johnson, C. R., and Vegh, D. C. (1969). *Chem. Commun.* p. 557.
Jonáš, J., Allerhand, A., and Gutowsky, H. S. (1965). *J. Chem. Phys.* **42**, 3396.
Kabuss, S., Friebolin, H., and Schmid, H. (1965). *Tetrahedron Lett.* p. 469.
Kabuss, S., Lüttringhaus, A., Friebolin, H., Schmid, H. G., and Mecke, R. (1966a). *Tetrahedron Lett.* p. 719.

14. Conformational Processes in Rings

Kabuss, S., Lüttringhaus, A., Friebolin, H., and Mecke, R. (1966b). *Z. Naturforsch. B* **21**, 320.
Kaplan, M. L., and Taylor, G. N. (1973). *Tetrahedron Lett.* p. 295.
Kessler, H., and Leibfritz, D. (1970). *Tetrahedron Lett.* p. 4297.
Kessler, H., Gusowski, V., and Hanack, M. (1968). *Tetrahedron Lett.* p. 4665.
Kilpatrick, J. E., Pitzer, K. S., and Spitzer, R. (1947). *J. Amer. Chem. Soc.* **69**, 2483.
Kleier, D. A., Binsch, G., Steigel, A., and Sauer, J. (1970). *J. Amer. Chem. Soc.* **92**, 3787.
Knorr, R., Ganter, C., and Roberts, J. D. (1967). *Angew. Chem., Int. Ed. Engl.* **6**, 556; *Angew. Chem.* **79**, 577 (1967).
Komornicki, A., and McIver, J. W., Jr. (1973). *J. Amer. Chem. Soc.* **95**, 4512.
Krebs, A. (1968). *Tetrahedron Lett.* p. 4511.
Laane, J. (1971). *Quart. Rev., Chem. Soc.* **25**, 533.
Lack, R. E., Ganter, C., and Roberts, J. D. (1968). *J. Amer. Chem. Soc.* **90**, 7001.
Lambert, J. B., and Keske, R. G. (1966). *J. Org. Chem.* **31**, 3429.
Lambert, J. B., and Oliver, W. L., Jr. (1968). *Tetrahedron Lett.* p. 6187.
Lambert, J. B., Keske, R. G., Cahart, R. E., and Jovanovitch, A. P. (1967). *J. Amer. Chem. Soc.* **89**, 3761.
Lambert, J. B., Gosnell, J. L., Jr., Bailey, D. S., and Greifenstein, L. G. (1970). *Chem. Commun.* p. 1004.
Lambert, J. B., Oliver, W. L., Jr., and Packard, B. S. (1971). *J. Amer. Chem. Soc.* **95**, 933.
Lambert, J. B., Bailey, D. S., and Michel, B. F. (1972). *J. Amer. Chem. Soc.* **94**, 3812.
Lambert, J. B., Mixan, C. E., and Johnson, D. H. (1973). *J. Amer. Chem. Soc.* **95**, 4634.
Lansbury, P. T., and Klein, M. (1968). *Tetrahedron Lett.* p. 1981.
Lansbury, P. T., Bieron, J. F., and Klein, M. (1966). *J. Amer. Chem. Soc.* **88**, 1477.
Larkin, R. H., and Lord, R. C. (1973). *J. Amer. Chem. Soc.* **95**, 5129.
Laurie, V. W. (1961). *J. Chem. Phys.* **34**, 1516.
Lehn, J. M. (1971). *In* "Conformational Analysis" (G. Chiurdoglu, ed.), p. 129. Academic Press, New York.
Lehn, J. M., and Riddell, F. G. (1966). *Chem. Commun.* p. 803.
Lehn, J. M., Linscheid, P., and Riddell, F. G. (1967). *Bull. Soc. Chim. Fr.* [5] p. 1172.
Levin, R. H., Roberts, J. D., Kwart, H., and Walls, F. (1972). *J. Amer. Chem. Soc.* **94**, 6856.
Lifson, S., and Warshel, A. (1968). *J. Chem. Phys.* **49**, 5116.
Linscheid, P., and Lehn, J. M. (1967). *Bull. Soc. Chim. Fr.* [5] p. 992.
Lüttringhaus, A., and Eyring, G. (1957). *Justus Liebigs Ann. Chem.* **604**, 111.
McIver, J. W., Jr., and Komornicki, A. (1971). *Chem. Phys. Lett.* **10**, 303.
Mannschreck, A., Rissmann, G., Vögtle, F., and Wild, D. (1967). *Chem. Ber.* **100**, 335.
Meiboom, S. (1962), *Pa. Amer. Chem. Soc. Symp.* (Boulder, Col.) p. 1.
Mislow, K., and Hopps, H. B. (1962). *J. Amer. Chem. Soc.* **84**, 3018.
Mislow, K., and Perlmutter, D., (1962). *J. Amer. Chem. Soc.* **84**, 3591.
Mitchell, R. H., and Boekelheide, V. (1969). *Tetrahedron Lett.* p. 2013.
Moniz, W. B., and Dixon, J. A. (1961). *J. Amer. Chem. Soc.* **83**, 1671.
Montecalvo, D., St. Jacques, M., and Wasylishen, R. (1973). *J. Amer. Chem. Soc.* **95**, 2023.
Moriarty, R. M., Ishibe, N., Kayser, M., Ramey, K. C., and Gisler, H. J., Jr. (1969). *Tetrahedron Lett.* p. 4883.
Murray, R. W., and Kaplan, M. L. (1967). *Tetrahedron* **23**, 1575.
Murray, R. W., Story, P. R., and Kaplan, M. L. (1966). *J. Amer. Chem. Soc.* **88**, 526.
Neikam, W. C., and Dailey, B. P. (1963). *J. Chem. Phys.* **38**, 445.

Nelsen, S. F., and Gillespie, J. P. (1973). *J. Amer. Chem. Soc.* **95**, 2940.
Noe, E. A., and Roberts, J. D. (1971). *J. Amer. Chem. Soc.* **93**, 7261.
Noe, E. A., and Roberts, J. D. (1972). *J. Amer. Chem. Soc.* **94**, 2020.
Noe, E. A., Wheland, R. C., Glazer, E. S., and Roberts, J. D. (1972). *J. Amer. Chem. Soc.* **94**, 3488.
Nógrádi, M., Ollis, W. D., and Sutherland, I. O. (1970). *Chem. Commun.* p. 158.
Ollis, W. D., and Stoddart, J. F. (1973). *Chem. Commun.* p. 571.
Orrel, K. G., Carruthers, W., and Pellat, M. G. (1972). *Spectrochim Acta A* **28**, 753.
Oth, J. M. (1971). *Pure Appl. Chem.* **25**, 573.
Pedersen, B., and Schaug, J. (1968). *Acta Chem. Scand.* **22**, 1705.
Pessemier, F. D., Tavernier, D., and Anteunis, M. (1973). *Chem. Commun.* p. 594.
Pickett, H. M., and Strauss, H. L. (1970). *J. Amer. Chem. Soc.* **92**, 7281.
Pijpers, F. W., Arick, M. R., Hendricks, B. M. P., and de Boer, A. E. (1971). *Mol. Phys.* **22**, 736.
Pitzer, K. S., and Donath, W. E. (1959). *J. Amer. Chem. Soc.* **81**, 3213.
Rabideau, P. W., Harvey, R. G., and Stothers, J. B. (1969). *Chem. Commun.* p. 1005.
Radlick, P., and Winstein, S. (1963). *J. Amer. Chem. Soc.* **85**, 344.
Reeves, L. W., and Strømme, K. O. (1960). *Can. J. Chem.* **38**, 1241.
Reeves, L. W., and Strømme, K. O. (1961a). *Trans. Faraday Soc.* **57**, Part 3, 390.
Reeves, L. W., and Strømme, K. O. (1961b). *J. Chem. Phys.* **34**, 1711.
Reiner, J., and Jenny, W. (1969). *Helv. Chim. Acta* **52**, 1624.
Richards, W. G., and Horsley, J. A. (1970). "Ab Initio Molecular Orbital Calculations for Chemists." Oxford Univ. Press (Clarendon), London and New York.
Riddell, F. G. (1967). *J. Chem. Soc., B* p. 560.
Riddell, F. G., and Lehn, J. M. (1966). *Chem. Commun.* p. 376.
Roberts, B. W., Vollmer, J. J., and Servis, K. L. (1968). *J. Amer. Chem. Soc.* **90**, 5264.
Roberts, J. D. (1966). *Chem. Brit.* **2**, 529.
Romers, C., Altona, C., Buys, H. R., and Havinga, E. (1969). *Top. Stereochem.* **4**, 39.
Russell, G. A., Underwood, G. R., and Lini, D. C. (1967). *J. Amer. Chem. Soc.* **89**, 6636.
Sachse, H. (1890). *Ber. Deut. Chem. Ges.* **23**, 1363.
Sadée, W. (1969). *Arch. Pharm. (Weinheim)* **302**, 769.
St. Jacques, M. (1967). Ph.D. Thesis, University of California, Los Angeles.
St. Jacques, M., and Bernard, M. (1969). *Can. J. Chem.* **47**, 2911.
St. Jacques, M., and Prud'homme, R. (1970). *Tetrahedron Lett.* p. 4833.
St. Jacques, M., and Prud'homme, R. (1972). *J. Amer. Chem. Soc.* **94**, 6479.
St. Jacques, M., and Vaziri, C. (1971). *Can. J. Chem.* **49**, 1256.
St. Jacques, M., and Vaziri, C. (1972). *Tetrahedron Lett.* p. 4331.
St. Jacques, M., Brown, M. A., and Anet, F. A. L. (1966). *Tetrahedron Lett.* p. 5947.
Sakamoto, K., and Ōki, M. (1973). *Tetrahedron Lett.* p. 3989.
Sato, T., Wakabayashi, M., and Kainosho, M. (1968). *Tetrahedron Lett.* p. 4185.
Saunders, M. (1967). *Tetrahedron* **23**, 2105.
Schmid, H. G., Friebolin, H., Kabuss, S., and Mecke, R. (1966). *Spectrochim. Acta* **22**, 623.
Schmid, H. G., Jaeschke, A., Friebolin, H., Kabuss, S., and Mecke, R. (1969). *Org. Magn. Resonance* **1**, 163.
Schneider, H. J., Keller, T., and Price, R. (1972). *Org. Magn. Resonance* **4**, 907.
Senkler, G. H., Jr., Gust, D., Riccobono, P. X., and Mislow, K. (1972). *J. Amer. Chem. Soc.* **94**, 8626.
Servis, K. L., and Noe, E. A. (1973). *J. Amer. Chem. Soc.* **95**, 171.
Servis, K. L., Noe, E. A., Easton, N. R., Jr., and Anet, F. A. L. (1974). *J. Amer. Chem. Soc.* **96**, 4185.

Sherrod, S. A., and Boekelheide, V. (1972). *J. Amer. Chem. Soc.* **94**, 5513.
Smith, B. H. (1964). "Bridged Aromatic Compounds," p. 333. Academic Press, New York.
Spassov, S. L., Griffith, D. L., Glazer, E. S., Nagarajan, K., and Roberts, J. D. (1967). *J. Amer. Chem. Soc.* **89**, 88.
Sutherland, I. O., and Ramsay, M. V. J. (1965). *Tetrahedron* **21**, 3401.
Swalen, J. D., and Costain, C. C. (1959). *J. Chem. Phys.* **31**, 1562.
Tavernier, D., Anteunis, M., and Hosten, N. (1973). *Tetrahedron Lett.* p. 75.
Tiers, G. V. D. (1960). *Proc. Chem. Soc., London* p. 389.
Tochtermann, W. (1970). *Fortschr. Chem. Forsch.* **15**, 378.
Tochtermann, W., Walter, U., and Mannschreck, A. (1964). *Tetrahedron Lett.* p. 2981.
Ugi, I., Marquarding, D., Klusacek, H., Gokel, G., and Gillespie, P. (1970). *Angew. Chem., Int. Ed. Engl.* **9**, 703.
Untch, K. G., and Kurland, R. J. (1963). *J. Amer. Chem. Soc.* **85**, 346.
Van der Kooij, J., Gooijer, C., Velthorst, N. H., and MacLean, C. (1971). *Rec. Trav. Chim.*, **90**, 732.
van Dort, H. M., and Sekuur, T. J. (1963). *Tetrahedron Lett.* p. 1301.
Vinter, J. G., and Hoffmann, H. M. R. (1973). *J. Amer. Chem. Soc.* **95**, 3051.
Vögtle, F. (1968). *Tetrahedron Lett.* 5221.
Vögtle, F. (1969a). *Chem. Ber.* **102**, 1449.
Vögtle, F. (1969b). *Tetrahedron* **25**, 3231.
Vögtle, F. (1969c). *Chem. Ber.* **102**, 3077.
Vögtle, F. (1969d). *Chem. Ber.* **102**, 1784.
Vögtle, F. (1969e). *Tetrahedron Lett.* p. 3193.
Vögtle, F., and Effler, A. H. (1969). *Chem. Ber.* **102**, 3071.
Vögtle, F., and Neumann, P. (1970). *Tetrahedron Lett.* p. 115.
Vögtle, F., and Schunder, L. (1969). *Chem. Ber.* **102**, 2677.
von Bredow, K., Jaeschke, A., Schmid, H. G., Friebolin, H., and Kabuss, S. (1970). *Org. Magn. Resonance* **2**, 543.
von Bredow, K., Friebolin, H., and Kabuss, S. (1971). *In* "Conformational Analysis" (G. Chiurdoglu, ed.), p. 51. Academic Press, New York.
Weigert, F. J., and Strobach, D. R. (1970). *Org. Magn. Resonance* **2**, 303.
Westheimer, F. H. (1956). *In* "Steric Effects in Organic Chemistry" (M. S. Newman, ed.), p. 523. Wiley, New York.
White, J. D., and Gesner, B. D. (1968). *Tetrahedron Lett.* p. 1591.
Whitesides, G. M., Pawson, B. A., and Cope, A. C. (1968). *J. Amer. Chem. Soc.*, **90**, 639.
Wiberg, K. B. (1965). *J. Amer. Chem. Soc.* **87**, 1070.
Wiberg, K. B., and Boyd, R. H. (1972). *J. Amer. Chem. Soc.* **94**, 8426.
Williams, J. E., Stang, P. J., and Schleyer, P. von R. (1968). *Annu. Rev. Phys. Chem.* **19**, 531.
Wipff, G., Wahlgren, U., Kochanski, E., and Lehn, J. M. (1971). *Chem. Phys. Lett.* **11**, 350.
Wolfe, S., and Campbell, J. R. (1967a). *Chem. Commun.* p. 874.
Wolfe, S., and Campbell, J. R. (1967b). *Chem. Commun.* p. 877.
Wood, G., and Srivastava, R. M. (1971). *Tetrahedron Lett.* p. 2937.
Wood, G., McIntosh, J. M., and Miskow, M. H. (1970). *Tetrahedron Lett.* p. 4895.
Yamamoto, O., and Nakanishi, H. (1973). *Tetrahedron* **29**, 781.
Yamamoto, O., Yanagisawa, M., Hayamizu, K., and Kotowycz, G. (1973). *J. Magn. Resonance* **9**, 216.
Young, G. J. (1971). Ph.D. Thesis, University of California, Los Angeles.
Yousif, G. A., and Roberts, J. D. (1968). *J. Amer. Chem. Soc.* **90**, 6428.

15

Proton Transfer Processes

ERNEST GRUNWALD and EARLE K. RALPH

    I. Introduction... 621
   II. Rate of Proton Exchange..................................... 622
 III. Kinetic Analysis... 624
       A. Kinetic Information from Different Subspectra................ 626
       B. Walden Inversion 627
       C. Number of Solvent Molecules 628
  IV. Comparison with Relaxation Methods 630
   V. Representative Results and Conclusions........................ 632
       A. Reactions of Hydronium Ion in Water....................... 632
       B. Reactions of Lyonium Ion in Nonaqueous Solvents 635
       C. Reactions of Hydroxide Ion in Water........................ 635
       D. Reactions of Lyate Ion in Nonaqueous Solvents............... 635
       E. Symmetrical Proton Exchange without Solvent Participation..... 636
       F. Symmetrical Proton Exchange with Solvent Participation....... 638
       G. Unsymmetrical Proton Exchange............................ 641
       H. Ion Pairs in *t*-Butyl Alcohol 641
       I. Ion Pairs in Acetic Acid 643
  VI. Concluding Remarks 646
       References.. 646

I. Introduction

The kinetic study of proton transfer reactions has been of interest to chemists throughout the history of chemistry. This is not only because of the practical importance of acids and bases as reactants and catalysts, but also because proton transfer reactions have been a gold mine of information concerning reaction mechanisms and solvation in liquids.

In the 1950s, with the development of relaxation spectrometry and nuclear magnetic resonance, it has become possible to extend these studies to very fast proton transfer reactions, in particular to those of oxygen and nitrogen acids and bases. In the relaxation methods, one displaces the system from equilibrium and then measures the rate of return to equilibrium (Eigen and DeMaeyer, 1961). In the nmr methods, one measures the rate of exchange of a given nucleus between magnetically distinguishable states while the chemical system remains at equilibrium. Since one is measuring a rate of exchange, there is a marked analogy between the nmr method, where the proton is labeled by virtue of its spin state, and isotopic exchange, where the hydrogen nucleus is labeled by virtue of its mass. For example, by suitable nuclear magnetic double labeling one can measure the rate of exchange of the proton among molecules belonging to the same chemical species.

The study of fast proton transfer reactions has proved to be so fruitful because one is not merely extending the classic studies to much higher rates, but qualitatively new phenomena appear. In particular, for many oxygen and nitrogen acids and bases, exoergic proton-transfer steps are ultrafast (Grunwald, 1965) and other processes, such as the rotation of a solvent molecule in, or its departure from, the solvation shell become rate-limiting for proton exchange. The rate constants of such processes can then be deduced by kinetic analysis. It also turns out that proton transfer from an acid to a base often proceeds with participation of one or more solvent molecules, and this in turn provides insight into solvation.

II. Rate of Proton Exchange

We shall assume the existence of dynamic equilibrium. As a result of reversible proton transfers, the chemical shift or spin-coupling of certain protons becomes time-dependent, so that such protons experience two or more discrete resonance frequencies. For example, there may be a change of chemical shift or of spin–spin interaction as a particular proton (underlined in the following examples) enters a new molecule.

$$CH_3O\underline{H} + HOH \rightleftharpoons CH_3OH + \underline{H}OH \qquad (1)$$

$$H^{16}O\underline{H} + H^{17}OH \rightleftharpoons H^{16}OH + H^{17}O\underline{H} \qquad (2)$$

For another example, the conversion of an acid to its conjugate base may change the chemical shift of other protons that are not being transferred, such as the CH_3 protons in Eq. (3).

$$\underset{\underset{CH_3}{|}}{\overset{\overset{CH_3}{|}}{C_6H_5P}}-H^+ + CH_3OH \rightleftharpoons \underset{\underset{CH_3}{|}}{\overset{\overset{CH_3}{|}}{C_6H_5P}} + CH_3OH_2^+ \qquad (3)$$

We may state as a general rule that whenever there is a change in resonance

15. Proton Transfer Processes

frequency, it is possible to extract information about the rate of proton transfer. The details of how this is done have been discussed in earlier chapters, and we shall not repeat them here. However, we do wish to define the characteristic time τ and rate R of proton exchange.

For definiteness, consider the underlined proton in Eq. (1). If we observe this proton over an extended period of time, it will undergo many cycles of exchange. The amount of time the proton spends at a given site (CH$_3$O$\underline{\text{H}}$ or HO$\underline{\text{H}}$) will vary in different cycles, following a Gaussian distribution. Let $\tau_{\text{CH}_3\text{OH}}$ and τ_{HOH} denote the mean time the proton spends at the respective site during one cycle of exchange. We then define τ according to (4).

$$\frac{1}{\tau} = \frac{1}{\tau_{\text{CH}_3\text{OH}}} + \frac{1}{\tau_{\text{HOH}}} \tag{4}$$

Note the analogy between $1/\tau$ and the rate constant for the approach to equilibrium of a reversible first-order reaction.

For the general case of proton exchange (5), $1/\tau$ is defined analogously in (6). From a chemist's point of view, $1/\tau_{\text{HA}}$ and $1/\tau_{\text{HB}}$ are pseudo-first-order

$$\text{HA} \rightleftharpoons \text{HB} \tag{5}$$

$$\frac{1}{\tau} = \frac{1}{\tau_{\text{HA}}} + \frac{1}{\tau_{\text{HB}}} \tag{6}$$

rate constants, as shown in (7), where R_+ denotes the rate (in moles per liter per second) at which protons move from HA sites to HB sites, R_- denotes the reverse rate, and [HA] and [HB] denote the moles per liter of HA sites and HB sites, respectively. Because the system is in dynamic equilibrium, $R_+ = R_-$.

$$\frac{1}{\tau_{\text{HA}}} = \frac{R_+}{[\text{HA}]}; \quad \frac{1}{\tau_{\text{HB}}} = \frac{R_-}{[\text{HB}]} \tag{7}$$

Let R denote the rate of chemical exchange, and let p be the probability that the chemical reaction will result in the transfer of the proton from an HA to to an HB site. [For an example in which $p \neq 1$, see Eq. (18)]. Hence (6) and (7) imply (8) and (9).

$$pR = \frac{[\text{HA}]}{\tau_{\text{HA}}} = \frac{[\text{HB}]}{\tau_{\text{HB}}} \tag{8}$$

$$R = \frac{[\text{HA}][\text{HB}]}{([\text{HA}] + [\text{HB}])p\tau} \tag{9}$$

In Eq. (9), $1/\tau$ is deduced from the nmr spectrum and p is a statistical factor whose value can in most cases be predicted *a priori*. [HA] and [HB] are obtained from the known composition of the solution. Thus, all quantities are at hand for the calculation of the rate of chemical exchange. Kinetic analysis of R by standard methods then leads to the rate law and to values for the rate constants.

III. Kinetic Analysis

The ability to measure the rates of very fast reactions is of course a great step forward. However, if we also wish to understand the molecular mechanisms of the fast reactions, the measurements must be accurate enough and extensive enough for kinetic analysis.

It is hardly necessary at this point to stress the importance of kinetic analysis. After all, some of the most important discoveries concerning reactivity are based on kinetic analysis. For example, the existence of radical chain reactions, general acid and base catalysis, carbonium ions, and enzyme–substrate complexes were all demonstrated in this way. Fortunately, for fast proton transfer reactions, the nmr technique is accurate enough to justify detailed kinetic analysis of the results. For definiteness, we shall give one example in detail.

When trimethylammonium ion is dissolved in water at pH < 5, it is possible to measure the rate of proton exchange between NH sites of $(CH_3)_3NH^+$ and OH sites of water. The following rate law was established (Loewenstein and Meiboom, 1957; Grunwald, 1963b).

$$R = \frac{k[(CH_3)_3NH^+]}{1 + k'[H^+]} + k_2[(CH_3)_3NH^+][(CH_3)_3N] \tag{10}$$

This rate law is relatively complex, suggesting that several reactions are significant. The second kinetic term indicates that $(CH_3)_3NH^+$ and $(CH_3)_3N$ are reactants, and the method of measurement requires that water be a direct participant. A plausible reaction mechanism is shown in Eq. (11). It has been

$$(CH_3)_3N\underline{H}^+ + \overset{H}{\underset{|}{O}}H + N(CH_3)_3 \longrightarrow (CH_3)_3N + \underline{H}\overset{H}{\underset{|}{O}} + HN^+(CH_3)_3 \tag{11}$$

found, by methods to be described later, that this reaction mechanism is correct in showing that one water molecule participates (Luz and Meiboom, 1963b).

The first kinetic term in Eq. (10) indicates a complex reaction mechanism. The kinetic form is typical of a reaction that involves a metastable intermediate. Before presenting the reaction mechanism that was proposed, we wish to point out that reaction will be incomplete if it proceeds only to the reaction intermediate. For full reaction, the proton must move all the way

15. Proton Transfer Processes

from the NH site of $(CH_3)_3NH^+$ to a water molecule in bulk water. The mechanism that was proposed is shown in Eqs. (12)–(14).

$$H_2O + \underset{1}{(CH_3)_3N\underline{H}^+ \cdot OH_2} \underset{k_{-a}}{\overset{k_a}{\rightleftharpoons}} \underset{2}{(CH_3)_3N \cdot \underline{H}OH + H_3O^+} \quad (12)$$

$$\underset{2}{(CH_3)_3N \cdot \underline{H}OH} + HOH(aq) \xrightarrow{k_H} \underset{3}{(CH_3)_3N \cdot HOH + \underline{H}OH(aq)} \quad (13)$$

$$\underset{3}{(CH_3)_3N \cdot HOH} + H_3O^+ \xrightarrow{k_{-a}} \underset{4}{(CH_3)_3NH^+ \cdot OH_2 + H_2O} \quad (14)$$

According to this mechanism, acid dissociation of trimethylammonium ion, shown as the hydrate (**1**), leads to trimethylamine hydrate (**2**) in which the original N$\underline{H}$ proton is still hydrogen bonded to nitrogen. **2** is a short-lived intermediate. Under the experimental conditions, the concentration of trimethylamine is less than 10^{-5} M, and its reaction with hydronium ion is diffusion controlled. If **2** reacts with hydronium ion, **1** is reformed and there is no proton exchange. In order to have proton exchange, the water molecule in **2** must be replaced (Eq. 13), forming **3** and moving the original N$\underline{H}$ proton to an OH site in bulk water. The new intermediate (**3**) then reacts with hydronium ion to give trimethylammonium ion (**4**) with a new proton, thus completing the cycle of exchange.

Because **3** is also a reaction intermediate at low concentration, its further reaction (15) of replacing the hydrogen-bonded water molecule would have no effect on the rate of proton exchange.

$$(CH_3)_3N \cdot HOH + HOH(aq) \xrightarrow{k_H} (CH_3)_3N \cdot HOH + HOH(aq) \quad (15)$$

On applying the steady-state approximation to this mechanism we obtain the rate law (16), which is of precisely the same mathematical form as the first term on the right in Eq. (10). Accordingly, we identify k with k_a and k' with k_H/k_{-a}.

$$R = \frac{k_a[(CH_3)_3NH^+]}{1 + k_{-a}[H^+]/k_H} \quad (16)$$

If we had to rely on nmr kinetic data alone, this would be as far as the problem can be carried. However, it is possible to go further. Because the solution is in dynamic equilibrium, k_a/k_{-a} must be equal to the acid dissociation constant, K_a, which can be measured independently. Since k_a is known from the kinetic measurements ($k_a = k$), this enables us to calculate k_{-a} and hence k_H. As a final check, the value of k_{-a} may be compared with results obtained by relaxation spectrometry (Grunwald and Ralph, 1971).

The validity of this mechanism will be presented later, but it should be noted that we are thus able to measure the rate constant k_H for exchange of a water molecule between the solvation shell of trimethylamine and bulk water.

In ordinary chemical kinetics, a reaction intermediate is a molecular species whose concentration is stoichiometrically insignificant. In dynamic nuclear magnetic resonance, a reaction intermediate is a species whose concentration is magnetically insignificant. That is to say, the presence of the intermediate has no significant effect on that portion of the nmr spectrum used to measure the rate of exchange. Apart from this subtle difference, the two concepts of reaction intermediate are the same. Thus, by adding suitable trapping reagents, the intermediate in an nmr experiment may be diverted to other products, and its lifetime may be changed.

A. Kinetic Information from Different Subspectra

A given reaction will affect each nmr subspectrum in a characteristic way so that different subspectra will give complementary information about the same reaction. For trimethylammonium ion in water, if we examine the NH–HOH proton system, we can deduce the rate of proton exchange between NH and HOH sites. However, there may be other processes of NH proton exchange that do not involve HOH sites; for example, a direct reaction between trimethylammonium ion and trimethylamine (Eq. 17). We cannot

$$(CH_3)_3NH^+ + (CH_3)_3\underline{N} \longrightarrow (CH_3)_3N + (CH_3)_3\underline{N}H^+ \qquad (17)$$

measure this rate separately, but we can measure the total rate of NH proton exchange of trimethylammonium ion (under conditions where trimethylamine is a reaction intermediate) and thus obtain the rate of reaction (17) by difference (Loewenstein and Meiboom, 1957). To do this, we examine the nmr of the CH_3 protons of trimethylammonium ion. Owing to spin–spin interaction with the NH proton, the shape of the CH_3 proton resonance can indicate the rate of NH-proton exchange, because each exchange is attended by a 50% chance of spin inversion of the coupled NH proton (Eq. 18). Consequently, the CH_3 proton resonance changes from a sharp doublet when $1/\tau = 0$ to a sharp singlet when $1/\tau$ is very large.

$$(CH_3)_3\overset{\oplus}{N}H\uparrow \xrightarrow{-\overset{\oplus}{H}\uparrow} (CH_3)_3N \begin{array}{c} \xrightarrow[50\%]{+\overset{\oplus}{H}\uparrow} (CH_3)_3\overset{\oplus}{N}H\uparrow \\ \xrightarrow[50\%]{+\overset{\oplus}{H}\downarrow} (CH_3)_3\overset{\oplus}{N}H\downarrow \end{array} \qquad (18)$$

Another method for measuring the total rate of interconversion of trimethylammonium ion and trimethylamine involves working under conditions

15. Proton Transfer Processes

where the concentrations of these species are comparable. The rate can then be measured by utilizing the difference in chemical shift of the CH_3 protons in the two species.* Comparison of this rate of interconversion with the rate of NH exchange of trimethylammonium could show processes that do not involve trimethylamine. A possible but unlikely mechanism is a concerted displacement reaction such as that shown in (19).

$$(CH_3)_3\overset{+}{N}\begin{matrix}H\\\diamond\\H\end{matrix}O-H \tag{19}$$

B. Walden Inversion

For substances of appropriate molecular structure, the nmr spectrum can be used to measure the rate of Walden inversion as well as the rate of proton exchange. A well studied example is dibenzylmethylammonium ion (**5**) in aqueous solution (Saunders and Yamada, 1963; Morgan and Leyden, 1970).

$$C_6H_5CH_2-\overset{\overset{H}{|}}{\underset{\underset{CH_3}{|}}{N}\oplus}-CH_2C_6H_5$$

5

The rate of NH proton exchange has been measured by examining the CH_3 proton resonance. Proton exchange could lead to Walden inversion about the nitrogen atom by the following mechanism (20). For **5**, the rate of

$$\begin{matrix}\diagdown\\ \diagup\end{matrix}\overset{\oplus}{N}-\underline{H} \longrightarrow \left[\begin{matrix}\diagdown\\ \diagup\end{matrix}N:\right] \longrightarrow \begin{matrix}\diagdown\\ \diagup\end{matrix}\overset{\oplus}{N}-H$$

$$\updownarrow \text{Walden inversion} \tag{20}$$

$$\left[:N\begin{matrix}\diagup\\ \diagdown\end{matrix}\right] \longrightarrow H-\overset{\oplus}{N}\begin{matrix}\diagup\\ \diagdown\end{matrix}$$

Walden inversion about nitrogen can be measured because this process causes the exchange of chemical shifts of the methylene protons in the benzyl groups. Figure 1 shows Newman projection formulas with one methylene carbon in front and the nitrogen atom behind. Formulas **6**, **7**, and **8** show the staggered

---

* The effect of NH spin coupling would of course have to be taken into account. In actual practice, the above method has not yet been used because in aqueous solution the rate of exchange has been too fast for measurement.

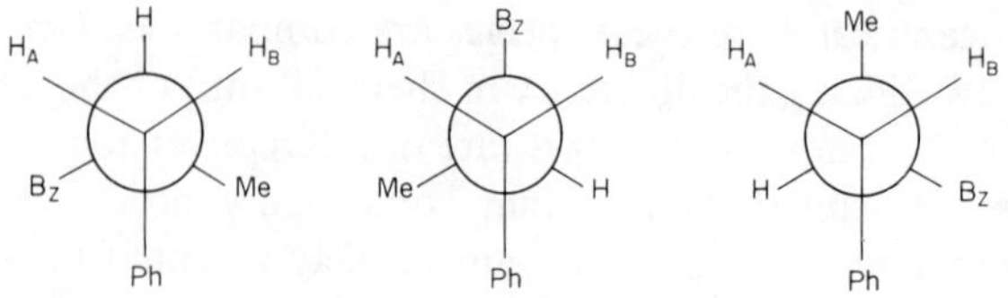

Fig. 1 Conformational isomers of dibenzylmethylammonium ion.

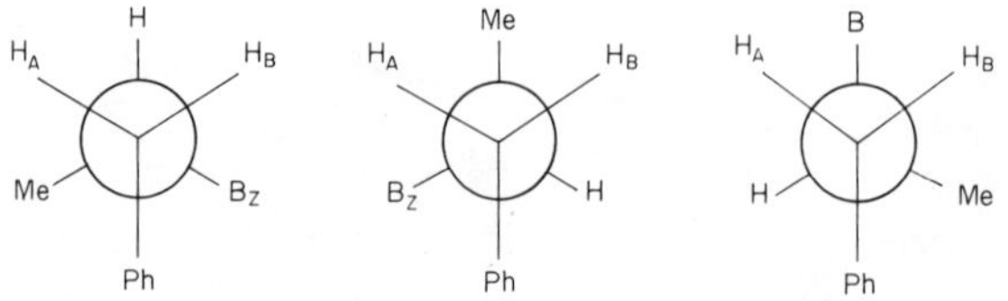

conformational isomers before Walden inversion, and **6,' 7'**, and **8'**, show the corresponding isomers after inversion. H_A and H_B have different chemical shifts because of the magnetic asymmetry of the molecules. It is obvious from Fig. 1 that the chemical shifts of H_A and H_B are interchanged upon inversion.

According to the mechanisms shown in (20), proton exchange involves the formation of the amine as a reaction intermediate. It is conceivable, of course, that the formation of the amine also proceeds with Walden inversion. How-

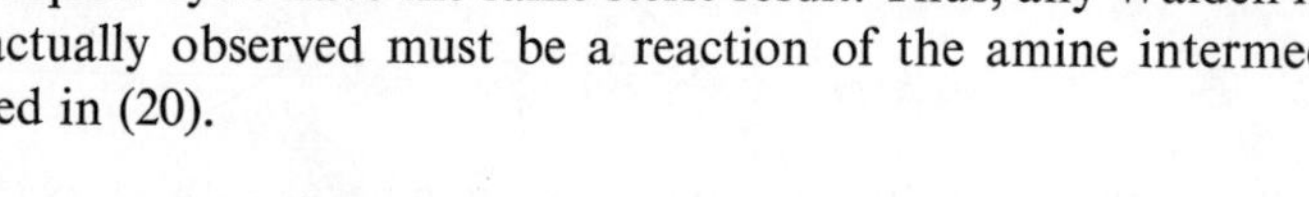

$$\diagdown\!\!\!\!\overset{+}{\underset{\diagup}{N}}\!\!-\!H \xrightarrow{\text{Walden inversion}} :N\!\!\!\overset{\diagup}{\underset{\diagdown}{}} \xrightarrow{\text{Walden inversion}} \diagdown\!\!\!\!\overset{+}{\underset{\diagup}{N}}\!\!-\!H \qquad (21)$$

ever, as shown in (21), proton exchange would then proceed without net Walden inversion, because microscopic reversibility requires that both steps in the complete cycle have the same steric result. Thus, any Walden inversion that is actually observed must be a reaction of the amine intermediate, as postulated in (20).

C. *Number of Solvent Molecules*

Consider the symbolic chemical equation, $X + nS \rightarrow$ Products. It then follows that $d[S]/dt = n\,d[X]/dt$. The same kinetic method of determining stoichiometry may be applied to proton exchange in order to find the number of solvent molecules that participate in a given proton-transfer process.

However, because the kinetics of proton exchange is often complex, it is necessary to make sure that $d[X]/dt$ and $d[S]/dt$ (for solute and solvent, respectively) refer to the same kinetic process. In practice, a kinetic analysis is made and corresponding rate constants are compared.

For example, by measuring the NH-proton exchange rate of purine (HPu) with water, one of the kinetic terms is found to be $R_{NH} = (1.0 \times 10^8)$-[HPu][Pu$^-$], where [Pu$^-$] denotes the concentration of the conjugate base, purinate ion. By measuring the OH–proton exchange rate among water molecules, the term with the corresponding concentration dependence is found to be $R_{OH} = (1.1 \times 10^8)$[HPu][Pu$^-$]. The ratio $R_{OH}/R_{NH} = 1.1/1.0 = 1.1 \pm 0.2$ (Marshall and Grunwald, 1969). Within the experimental error, this ratio could be precisely unity. Hence $n \sim 1$, and the reaction (or most of it) is termolecular, as in (22).

$$\text{Pu}\underline{\text{H}} + \overset{\text{H}}{\underset{|}{\text{O}}}\text{H} + \text{Pu}^- \longrightarrow \text{Pu}^- + \underline{\text{H}}\overset{\text{H}}{\underset{|}{\text{O}}} + \text{HPu} \qquad (22)$$

To give another example: By measuring the NH proton exchange rate of 2,4-lutidinium ion (LuH$^+$, 2,4-dimethylpyridinium ion) with water, one of the kinetic terms is found to be $R_{NH} = (1.05 \times 10^8)$[LuH$^+$][Lu]. By measuring the OH–proton exchange rate among water molecules, the term with the corresponding concentration dependence is found to be $R_{OH} = (1.82 \times 10^8)$[LuH$^+$][Lu]. The ratio $R_{OH}/R_{NH} = 1.82/1.05 = 1.74 \pm 0.3$ (Rosenthal and Grunwald, 1972). Within the experimental error, this ratio could be precisely 2. Hence $n \sim 2$, and the reaction (or most of it) is quatermolecular, as in (23).

$$\text{Lu}\overset{+}{\underline{\text{H}}} + \overset{\text{H}}{\underset{|}{\text{O}}}\text{H} + \overset{\text{H}}{\underset{|}{\text{O}}}\text{H} + \text{Lu} \longrightarrow \text{Lu} + \underline{\text{H}}\overset{\text{H}}{\underset{|}{\text{O}}} + \underline{\text{H}}\overset{\text{H}}{\underset{|}{\text{O}}} + \text{HL}\overset{+}{\text{u}} \qquad (23)$$

To measure the rate of proton exchange among water molecules, it is convenient to use ^{17}O-enriched water and measure the rate of exchange (Eq. 2) between H$_2^{17}$O and H$_2^{16}$O or H$_2^{18}$O (Meiboom, 1961; Luz and Meiboom, 1963b). Such measurement is possible because the nuclear spin of ^{17}O is $\frac{5}{2}$ while that of ^{16}O and ^{18}O is zero. The H$_2^{16}$O and H$_2^{18}$O proton resonance is of course a singlet, while that of H$_2^{17}$O is a spin–spin sextet. One usually assumes that the lifetime of a proton on H$_2^{17}$O is the same as that on H$_2^{16}$O or H$_2^{18}$O. The use of this method is limited to those reactions whose rate is at least comparable to the rate of proton exchange catalyzed by hydrogen ion and hydroxide ion. The latter rates are very high, being responsible for the anomalous conductance of the hydrogen ion and hydroxide ion (Meiboom, 1961). In practice, one usually works near neutral pH under conditions where the acidic and basic solutes (H$^+$, OH$^-$, HPu, LuH$^+$, etc.) may be treated as reaction intermediates.

In methanol and other hydroxylic solvents, limitations due to proton exchange with lyonium ion or lyate ion are less severe. In particular, methanol has been a useful solvent for measuring the kinetic order with respect to solvent in proton transfer reactions. A convenient experimental method is to examine the resonance of the CH_3 protons, which are spin-coupled to the exchanging OH proton (Grunwald et al., 1963).

IV. Comparison with Relaxation Methods

Examples that permit a direct comparison of nmr and relaxation methods are relatively scarce. However, where valid comparisons can be made, the agreement is good. Three examples, involving three different relaxation methods, are shown in Table I. In the first example, the acid dissociation of imidazolium ion, the agreement confirms the assumed reaction mechanism

TABLE I
Rate Constants by NMR and Relaxation Methods

(1) Imidazolium$^+$ + H_2O $\underset{13°C}{\overset{k_a}{\rightleftharpoons}}$ Imidazole + H_3O^+

NMR method[a]	$k_a = 1.0 \times 10^3$ sec^{-1}
Temperature jump[b]	$k_a = 1.1 \times 10^3$ sec^{-1}

(2) $(C_2H_5)_3NH^+ + OH^- \underset{30°C}{\overset{k_b}{\rightleftharpoons}} (C_2H_5)_3N + HOH$

NMR method[c]	$k_{-b} = 1.7 + 10^{10}$ sec^{-1} mole^{-1}
Ultrasonic relaxation[d]	$k_{-b} = 2.0 \times 10^{10}$ sec^{-1} mole^{-1}

(3) $Al(OH_2)_6^{3+} + H_2O \underset{25°C}{\overset{k_a}{\rightleftharpoons}} Al(OH_2)_5OH^{2+} + H_3O^+$

NMR method[e]	$k_a = 0.8 \times 10^5$ sec^{-1}
Electric-field jump[f]	$k_a = 1.1 \times 10^5$ sec^{-1}

[a] Grunwald and Ralph (1969).
[b] Eigen et al. (1960).
[c] Ralph and Grunwald (1967).
[d] Brundage and Kustin (1970).
[e] Fong and Grunwald (1969).
[f] Holmes et al. (1968).

15. Proton Transfer Processes

(which is analogous to that in Eqs. 12–14) by showing that the nmr-based rate constant k is indeed equal to k_a.

When rate constants obtained by nmr and relaxation methods are significantly different, this difference may be used to elucidate the reaction mechanism. For example, consider the acid dissociation of hexamminoplatinum-(IV) ion, Eq. (24). Rate constants, determined by the electric-field jump method

$$(H_3N)_5PtNH_3^{4+} + H_2O \underset{k_{-a}}{\overset{k_a}{\rightleftarrows}} (H_3N)_5PtNH_2^{3+} + H_3O^+ \qquad (24)$$

are $k_a \approx 140$ sec^{-1}, $k_{-a} \approx 2 \times 10^9$ sec^{-1} mole^{-1}. On the other hand, the nmr-based rate constant for first-order proton exchange is 2500 sec^{-1}, or about 18 times greater than k_a. It follows that acid dissociation leads to the exchange of all eighteen NH protons. Further nmr experiments show that the conjugate base undergoes rapid, bifunctional proton exchange with water, by the mechanism shown in (25) and (26) (Grunwald and Fong, 1972). To give

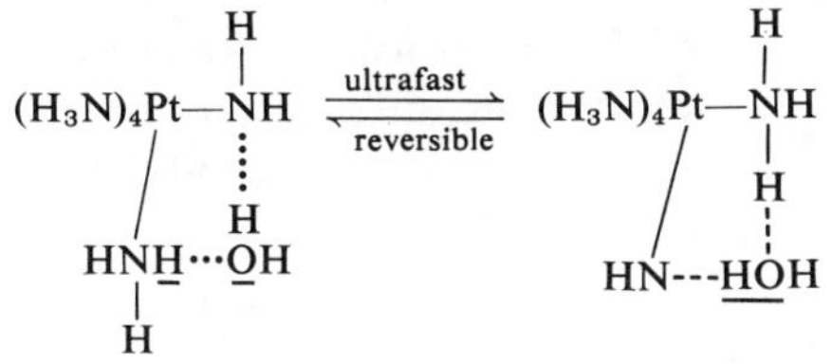

$$(H_3N)_4Pt\text{—}NH(H)(HNH\cdots OH)(H) \underset{\text{reversible}}{\overset{\text{ultrafast}}{\rightleftarrows}} (H_3N)_4Pt\text{—}NH(H)(HN\text{---}HOH)(H) \qquad (25)$$

$$H_2O(aq) + (H_3N)_4Pt\text{—}NH(H)(HN\text{---}HOH)(H) \xrightarrow{\text{rate determining}} HOH(aq) + (H_2N)_4Pt\text{—}NH(H)(HN\cdots HOH)(H) \qquad (26)$$

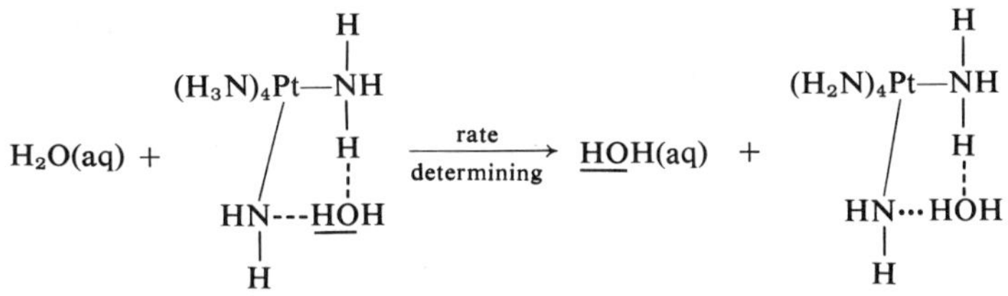

another example: For the acid dissociation of acetic acid in water, k_a as measured by electric-field jump in 7.8×10^5 sec^{-1}. By comparison, the rate constant for first-order proton exchange is 1.0×10^8 sec^{-1}, more than a hundred times greater. In this case, the nmr method detects the rate of a cyclic process suggested in Eq. (27). Since this process is symmetrical, it is undetected by the relaxation method (Luz and Meiboom, 1963a).

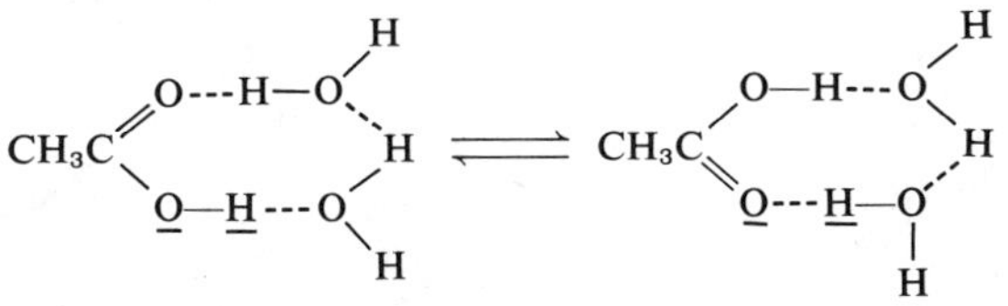

$$CH_3C\begin{pmatrix}O\text{---}H\text{—}O\\O\text{—}H\text{---}O\end{pmatrix}\begin{matrix}H\\H\end{matrix} \rightleftarrows CH_3C\begin{pmatrix}O\text{—}H\text{---}O\\O\text{---}H\text{—}O\end{pmatrix}\begin{matrix}H\\H\end{matrix} \qquad (27)$$

V. Representative Results and Conclusions

Representative results for proton exchange rates measured by the nmr method are listed in Tables II–X. Although the data in these tables are not a complete compilation of the literature, they illustrate most of the types of reactions that have been studied.

A. Reactions of Hydronium Ion in Water

Table II begins with the rate constant for proton transfer between H_3O^+ and H_2O. The rate constant for this process, 6×10^{11} sec^{-1}, is very high. This reaction is responsible for the anomalously high conductivity of the hydronium ion in aqueous solution (Meiboom, 1961).

Table II also lists rate constants for some exoergic reactions of hydronium ion with amines and with the NH_2 group of amino acids. In the latter case, the proton acceptor is the less stable uncharged form of the amino acid rather than the zwitterion. Most of the rate constants are of the correct magnitude for a diffusion-controlled reaction. Possible exceptions are trishydroxyethylamine and glycine, where there are hydrogen-bonded substituents near the reaction site, and trimethylphosphine.

The reaction of hydronium ion with phenol is probably the concerted reaction (28) rather than the two-step cycle (29 and 30), because the formation of the very unstable phenoxonium ion in Eq. (29) would be too slow to account for the observed rate (Grunwald and Puar, 1967a). On the other

$$H_2\overset{\oplus}{O}H + \underset{Ph}{O\underline{H}} + OH_2 \longrightarrow H_2O + \underset{Ph}{HO} + \underline{H}\overset{\oplus}{O}H_2 \qquad (28)$$

$$H_3O^+ + PhO\underline{H} \rightleftharpoons \underset{H}{Ph\overset{\oplus}{O}\underline{H}} + H_2O \qquad (29)$$

$$\underset{H}{Ph\overset{\oplus}{O}\underline{H}} + H_2O \longrightarrow PhOH + \underline{H}\overset{\oplus}{O}H_2 \qquad (30)$$

hand, the acid catalyzed NH–proton exchange of amides probably involves the two-step mechanism (31) (Berger et al., 1959). The proton exchange is

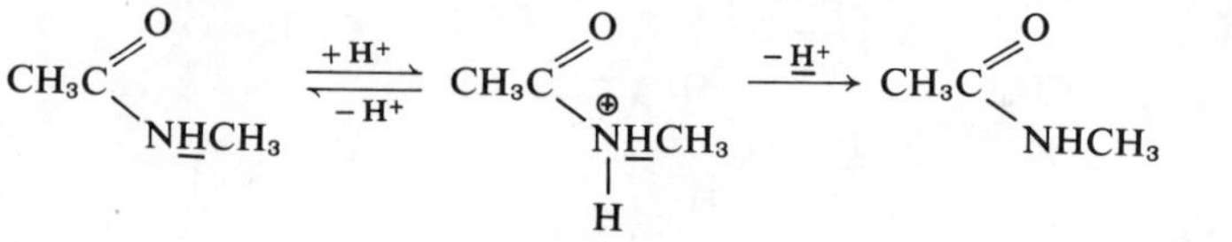

$$CH_3C\overset{O}{\underset{N\underline{H}CH_3}{\diagdown}} \underset{-H^+}{\overset{+H^+}{\rightleftharpoons}} CH_3C\overset{O}{\underset{\underset{H}{\overset{\oplus}{N}\underline{H}CH_3}}{\diagdown}} \overset{-\underline{H}^+}{\longrightarrow} CH_3C\overset{O}{\underset{NHCH_3}{\diagdown}} \qquad (31)$$

TABLE II
REACTIONS OF HYDRONIUM ION IN WATER

Reaction	k	Reference[a]
$H_3O^+ + H_2O \rightarrow H_2O + H_3O^+$	6×10^{11} sec^{-1} at 25°C	1
$H_3O^+ + NH_3 \rightarrow NH_4^+ + H_2O$	4.3×10^{10} sec^{-1} mole^{-1} at 25°C	2
$H_3O^+ + (CH_3)_3N \rightarrow (CH_3)_3NH^+ + H_2O$	3.0×10^{10} sec^{-1} mole^{-1} at 25°C	3
$H_3O^+ + \begin{array}{c}\text{imidazole}\\ N\!\!-\!\!H\end{array} \rightarrow \begin{array}{c}\text{imidazolium}\\ NH + H_2O\end{array}$	2.4×10^{10} sec^{-1} mole^{-1} at 25°C	4
$H_3O^+ + CH_3N(CH_2C_6H_5)_2 \rightarrow CH_3NH(CH_2C_6H_5)_2^+ + H_2O$	1.3×10^{10} sec^{-1} mol^{-1} at 30°C	5
$H_3O^+ + (HOCH_2CH_2)_3N \rightarrow (HOCH_2CH_2)_3NH^+ + H_2O$	8.5×10^{9} sec^{-1} mole^{-1} at 30°C	6
$HO_2CCH_2NH_2 + H_3O^+ \rightarrow HO_2CCH_2NH_3^+ + H_2O$	9×10^{9} sec^{-1} mole^{-1} at 23°C	7
$H_3O^+ + \text{(purine)} \rightarrow ?$	2.4×10^{10} sec^{-1} mole^{-1} at 20°C	8
$HO_2CCH_2NHCH_3 + H_3O^+ \rightarrow HO_2CCH_2NH_2^+CH_3 + H_2O$	2.3×10^{10} sec^{-1} mole^{-1} at 21°C	9
$H_3O^+ + C_6H_5O\underline{H} + OH_2^- \rightarrow H_2O + C_6H_5OH + \underline{H}OH_2^+$	1.5×10^{7} sec^{-1} mole^{-1} at 25°C	10
$H_3O^+ + (CH_3)_3P \rightarrow (CH_3)_3PH^+ + H_2O$	5×10^{9} sec^{-1} mole^{-1}	11
Reactions according to Eq. (31)		
$H_3O^+ + CH_3C(=O)NHCH_3$	3.8×10^{2} mole^{-1} sec^{-1} at 23°C	12
$H_3O^+ + CH_3C(=O)NHCH_2C(=O)N\underline{H}CH_3$	46 mole^{-1} sec^{-1} at 25°C	13

[a] References: (1) Loewenstein and Szöke (1962), Luz and Meiboom (1964), Meiboom (1961); (2) Emerson *et al.* (1960); (3) Grunwald (1963a, 1967); (4) Ralph and Grunwald (1969a); (5) Grunwald and Ralph (1967); (6) E. K. Ralph (unpublished work); (7) Sheinblatt and Gutowsky (1964); (8) Marshall and Grunwald (1969); (9) Sheinblatt (1962); (10) Grunwald and Puar (1967a); (11) Silver and Luz (1961); (12) Berger *et al.* (1959); (13) Molday and Kallen (1972).

TABLE III
REACTIONS OF LYONIUM ION IN NONAQUEOUS SOLVENTS[a]

Reaction	k	Reference[b]
$CH_3OH_2^+ + CH_3OH$	17.6×10^{10} sec^{-1} at 25°C	1
$NH_4^+ + NH_3$ in liquid ammonia	6.9×10^9 sec^{-1} at 25°C	2
$CH_3OH_2^+ + (CH_3)_3N$	6×10^9 sec^{-1} mole^{-1} at 25°C	3
$CH_3OH_2^+ + $ p-CH$_3$-C$_6$H$_4$-NH$_2$	1.0×10^{10} sec^{-1} mole^{-1} at 25°C	4
$CH_3OH_2^+ + (CH_3)_2PC_6H_5$	1.3×10^8 sec^{-1} mole^{-1} at 25°C	4
$CH_3OH_2^+ + C_6H_5CO_2H + CH_3OH$	2.5×10^8 sec^{-1} mole^{-1} at 25°C	5
$CH_3OH_2^+ + C_6H_5OH + CH_3OH$	6.4×10^7 sec^{-1} mole^{-1} at 25°C	6
$CH_3OH_2^+ + $ 3,5-(CH$_3$)$_2$-C$_6$H$_3$-N(C$_2$H$_5$)$_2$	1.3×10^9 sec^{-1} mole^{-1} at 30°C	7
Lyonium$^+ + $ 3,5-(CH$_3$)$_2$-C$_6$H$_3$-N(C$_2$H$_5$)$_2$	3.9×10^9 sec^{-1} mole^{-1} at 25°C (t-BuOH–H$_2$O, 11.47 mole % t-BuOH)	7

[a] Viscosity: methanol, 0.00545 poise; t-BuOH–HOH, 0.0343 poise; t-BuOH, 0.0089 poise; water, 0.0089 poise.
[b] References: (1) Grunwald et al. (1962); (2) Clutter and Swift (1968); (3) Grunwald (1963a, 1967); (4) Cocivera et al. (1964); (5) Grunwald and Jumper (1963); (6) Puar and Grunwald (1968); (7) Grunwald et al. (1969).

subject to an interesting kinetic salt effect. Cations may coordinate with the carbonyl oxygen atom and thus form a less reactive species (Schleich et al., 1968).

In contrast to the amides, purine reacts with hydronium ion with a rate constant that is of the correct magnitude for diffusion-controlled reaction (Marshall and Grunwald, 1969). In view of the formal analogy between the HN—C=O group of amides and the HN—C=N group of purine, this fact is puzzling.

B. Reactions of Lyonium Ion in Nonaqueous Solvents

The reactions shown in Table III are very similar to those of the hydronium in water. For example, the reactions of methoxoniun ion with methanol and of ammonium ion with ammonia are analogous to that of hydronium ion with water. Note, however, that the rate constants are smaller.

In methanol, the exoergic reactions of lyonium ion with amines mostly have rate constants of the correct magnitude for diffusion-controlled reaction. The values are somewhat smaller than in water, largely because of the lower mobility of the methoxonium ion. A possible exception is the reaction with N,N-diethyl-m-toluidine; a probable exception is the reaction with phenyldimethylphosphine.

C. Reactions of Hydroxide Ion in Water

In Table IV, the rate constant for proton exchange between hydroxide ion and water is very high and accounts for the anomalously high conductivity of hydroxide ion (Meiboom, 1961). Exoergic reactions of hydroxide ion with alkylammonium ions seem to be diffusion controlled. However, the corresponding reaction with trimethylphosphonium ion is distinctly slower.

The reaction of hydroxide ion with amides is endoergic and appears to proceed by a mechanism involving proton abstraction. The rate constant for proton exchange is approximately proportional to the acid strength of the NH proton (Sheinblatt, 1970).

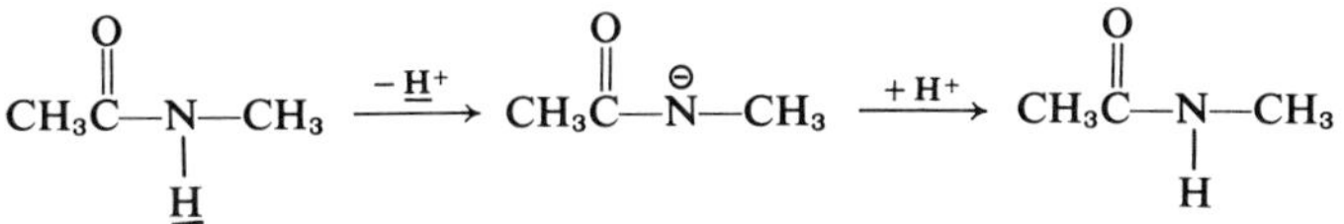

$$CH_3\overset{O}{\overset{\|}{C}}-\underset{\underline{H}}{N}-CH_3 \xrightarrow{-H^+} CH_3\overset{O}{\overset{\|}{C}}-\overset{\ominus}{N}-CH_3 \xrightarrow{+H^+} CH_3\overset{O}{\overset{\|}{C}}-\underset{H}{N}-CH_3 \quad (32)$$

D. Reactions of Lyate Ion in Nonaqueous Solvents

Table V shows rate constants for the reaction of methoxide ion in methanol and amide ion in liquid ammonia. These processes are formally analogous to

TABLE IV
REACTIONS OF HYDROXIDE ION IN WATER

Reaction	k (sec^{-1} mole^{-1})	Reference[a]
$OH^- + H_2O$	2.1×10^{11} at 25°C	1
$OH^- + (C_2H_5)_3NH^+$	1.7×10^{10} at 30°C	2
$OH^- + (CH_3)_3PH^+$	4.6×10^7 at 22°C	3
$OH^- + CH_3\overset{O}{\overset{\|}{C}}N\underline{H}CH_3 \rightarrow$ $CH_3\overset{O}{\overset{\|}{C}}N^-CH_3 + HOH$	5.2×10^6 at 21°C	4
$OH^- + CH_3\overset{O}{\overset{\|}{C}}NHCH_2\overset{O}{\overset{\|}{C}}N\underline{H}CH_3$	1.7×10^7 at 25°C	5
$OH^- + NH_3^{\oplus}CH_2\overset{O}{\overset{\|}{C}}N\underline{H}CH_2CO_2^{\ominus}$	8×10^8 at room temperature	6

[a] References: (1) Meiboom (1961); (2) Ralph and Grunwald (1967); (3) Silver and Luz (1961); (4) Berger et al. (1959); (5) Molday and Kallen (1972); (6) Sheinblatt (1970).

proton exchange between hydroxide ion and water. However, the rate constants are much lower. The table also shows the rate constant for the exoergic reaction of methoxide ion with *p*-bromophenol in a termolecular reaction. The rate constant, 1.5×10^9 sec^{-1} mole^{-1} at -81.6°C, is that expected for a diffusion-controlled reaction.

E. Symmetrical Proton Exchange without Solvent Participation

Rate constants for the few reactions in which the proton exchange without solvent participation has been resolved are listed in Table VI. Note the striking decrease in rate constant with increasing methyl substitution. It is reasonable to describe these reactions as taking place in three kinetically distinct reversible steps, shown in Eqs. (33)–(35). The rate-determining step

$$\underline{B}H^+(aq) + B(aq) \rightleftharpoons \underline{B}H^+ \cdot \overset{H}{O}H \cdot B \quad (33)$$

$$\underline{B}H^+ \cdot \overset{H}{O}H \cdot B \underset{\text{determining}}{\overset{\text{rate}}{\rightleftharpoons}} \underline{B}H^+ \cdot B + H_2O \quad (34)$$

$$\underline{B}H^+ \cdot B \overset{\text{fast}}{\rightleftharpoons} B \cdot H\underline{B}^+ \quad (35)$$

TABLE V
REACTIONS OF LYATE ION IN NONAQUEOUS SOLVENTS

Reaction	k	Reference[a]
NH_2^- + NH_3 in liquid ammonia	0.8×10^9 sec^{-1} at 25°C	1
CH_3O^- + CH_3OH in liquid methanol	$1.85 + 10^{10}$ sec^{-1} at 25°C	2
CH_3O^- + OH + HO–C$_6$H$_3$(CH$_3$)–Br in liquid methanol	1.5×10^9 sec^{-1} mole^{-1} at −81.6°C	3

[a] References: (1) Clutter and Swift (1968); (2) Grunwald et al. (1962); (3) Grunwald et al. (1967).

TABLE VI
Rate Constants for Symmetrical Proton Transfer Reactions without Solvent Participation in Aqueous Solution at 25°C[a]

Reaction	k (sec^{-1} mole^{-1})
$NH_4^+ + NH_3$	11.7×10^8
$CH_3NH_3^+ + CH_3NH_2$	4.0×10^8
$(CH_3)_2NH_2^+ + (CH_3)_2NH$	0.5×10^8
$(CH_3)_3NH^+ + (CH_3)_3N$	0.0×10^8

[a] Grunwald and Cocivera (1965).

appears to be (34), the dehydration of the solvent-separated complex $BH^+ \cdot OH_2 \cdot B$. Theory and other evidence both indicate that the proton-transfer step (35) is ultrafast. It is worth noting that in this mechanism, the formation of the reactive encounter complex $BH^+ \cdot B$, proceeds with an overall rate constant that is much smaller than the value of about 10^{10} sec^{-1} mole^{-1} usually assumed for diffusion-controlled reactions in water (Grunwald and Ku, 1968).

F. Symmetrical Proton Exchange with Solvent Participation

One of the remarkable facts about proton transfer reactions is that so many of them proceed with participation by solvent molecules. In fact, there are examples where two or more solvent molecules participate. For example, the reaction of 2,4-lutidinium ion with 2,4-lutidine has been shown to be quatermolecular (Eq. 23). Similarly, proton exchange of benzoic acid in methanol involves two methanol molecules (Eq. 36). On the other hand, proton exchange of neopentyl alcohol in acetic acid involves only one molecule of alcohol (Eq. 37) (Grunwald et al., 1963; Cocivera and Grunwald, 1965).

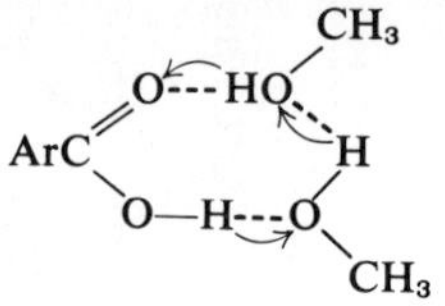

$$ArC\begin{matrix} O{\cdots}HO\text{―}CH_3 \\ \\ O\text{―}H{\cdots}O\text{―}CH_3 \end{matrix}H \qquad (36)$$

$$CH_3C\begin{matrix} O{\cdots}H \\ \\ O\text{―}H \end{matrix}O\text{―}CH_2C(CH_3)_3 \qquad (37)$$

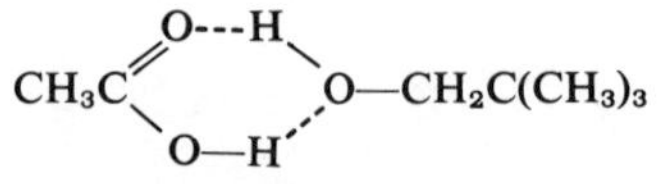

Representative data for symmetrical proton exchange with solvent participation are listed in Table VII. Although the free-energy change is uniformly

TABLE VII
Rate Constants for Symmetrical Proton Transfer with Solvent Participation

Reaction	k	Reference[a]
A. Reactions in Water[b]	(sec^{-1} mole^{-1})	
$NH_4^+ + H_2O + NH_3$	0.50×10^8 at 30°C	1
$CH_3NH_3^+ + H_2O + CH_3NH_2$	6.1×10^8 at 25°C	2
$(CH_3)_2NH_2^+ + H_2O + (CH_3)_2NH$	9.0×10^8 at 25°C	2
$(CH_3)_3NH^+ + (1.0 \pm 0.2)H_2O + (CH_3)_3N$	3.7×10^8 at 25°C	2
$(C_2H_5)_3NH^+ + H_2O + (C_2H_5)_3N$	1.8×10^8 at 30°C	3
$(C_6H_5CH_2)_2\overset{H^+}{\overset{|}{N}}\!-\!CH_3 + H_2O + (C_6H_5CH_2)_2NCH_3$	1.5×10^7 at 30°C	4
Purine + $(1.1 \pm 0.2)H_2O$ + purinate$^-$	1.0×10^8 at 20°C	5
2,4-lutidinium$^+$ + $(1.74 \pm 0.3)H_2O$ + 2,4-lutidine	1.05×10^8 at 25°C	6
2,4-lutidine H$^+$ + $(1.8 \pm 0.3)D_2O$ + 2,4-lutidine	0.26×10^8 at 25°C	6
imidazolium$^+$ + $(1.4 \pm 0.2)H_2O$ + imidazole	1.07×10^8 at 25°C	7
$C_6H_5OH + (1.0 \pm 0.2)H_2O + C_6H_5O^-$	5.7×10^8 at 25°C	8
$CH_3NH_2CH_2CO_2^- + H_2O + CH_3NHCH_2CO_2^-$	1.5×10^8 at 21°C	9
$(HOCH_2CH_2)_3NH^+ + H_2O + N(CH_2CH_2OH)_3$	3.6×10^6 at 30°C	—
B. Reactions in 11.47 mole % t-BuOH–88.53 mole % HOH[b]	(sec^{-1} mole^{-1})	
imidazolium$^+$ + $(1.0 \pm 0.2)H_2O$ + imidazole	2.0×10^8 at 25°C	10
imidazolium$^+$ + $(CH_3)_3COH$ + imidazole	1.6×10^7 at 25°C	10

(*continued*)

TABLE VII (continued)

Reaction	k	Reference[a]
$m\text{-}O_2NC_6H_4OH + H_2O + m\text{-}O_2NC_6H_4O^-$	5.4×10^8 at 15.7°C	11
$m\text{-}O_2NC_6H_4OH + (CH_3)_3COH + m\text{-}O_2NC_6H_4O^-$	0.17×10^8 at 15.7°C	11
$o\text{-}BrC_6H_4OH + H_2O + o\text{-}BrC_6H_4O^-$	1.25×10^8 at 15.7°C	11
$o\text{-}BrC_6H_4OH + (CH_3)_3COH + o\text{-}BrC_6H_4O^-$	0.091×10^8 at 15.7°C	11
$\text{Ar}\overset{\oplus}{N}H(C_2H_5)_2 + \text{SolvOH} + \text{ArN}(C_2H_5)_2$	6.0×10^5 at 25°C	12

C. Reactions in Methanol[b]

Reaction	k	Reference[a]
$p\text{-}CH_3C_6H_4NH_3^+ + (1.0 \pm 0.2)CH_3OH + p\text{-}CH_3C_6H_4NH_2$	0.81×10^8 at 25°C	13
$C_6H_5CO_2H + (1.0 \pm 0.2)CH_3OH + C_6H_5CO_2^-$	1.2×10^8 at 24.8°C	14
$m\text{-}O_2NC_6H_4CO_2H + CH_3OH + m\text{-}O_2NC_6H_4CO_2^-$	2.5×10^8 at 24.8°C	15
$p\text{-}O_2NC_6H_4CO_2H + CH_3OH + p\text{-}O_2NC_6H_4CO_2^-$	2.7×10^8 at 24.8°C	15
$o\text{-}O_2NC_6H_4CO_2H + CH_3OH + o\text{-}O_2NC_6H_4CO_2^-$	1.8×10^8 at 24.8°C	15
$\text{Ar}\overset{\oplus}{N}H(C_2H_5)_2 + (1.0 \pm 0.2)CH_3OH + \text{ArN}(C_2H_5)_2$	4.2×10^5 at 30°C	12

D. Reactions according to Eq. (36)[c]

Reaction	k (sec^{-1} at 25°C)	Reference[a]
$C_6H_5CO_2H + 2CH_3OH$	1.3×10^5	14
$m\text{-}O_2NC_6H_4CO_2H + \text{methanol}$	3.6×10^5	15
$p\text{-}O_2NC_6H_4CO_2H + \text{methanol}$	4.5×10^5	15
$o\text{-}O_2NC_6H_4CO_2H + \text{methanol}$	1.9×10^6	15

[a] References: (1) Grunwald and Ku (1969); (2) Grunwald and Cocivera (1965); (3) Ralph and Grunwald (1967); (4) Grunwald and Ralph (1967); (5) Marshall and Grunwald (1969); (6) Rosenthal and Grunwald (1972); (7) Ralph and Grunwald (1969a); (8) Grunwald and Puar (1967a); (9) Sheinblatt (1963); (10) Ralph and Grunwald (1969b); (11) Grunwald et al. (1971); (12) Grunwald et al. (1968); (13) Cocivera et al. (1964); (14) Grunwald et al. (1963); (15) Grunwald and Meiboom (1963).

[b] These rate constants are pseudo second-order rate constants based on proton exchange in the solute acid. If the solvent is shown without a numerical coefficient, we know only that at least one solvent molecule participates, but the precise number of solvent molecules has not been determined.

[c] These rate constants are pseudo first-order rate constants.

zero (except for one experiment performed in HOD–D$_2$O), the rate constants cover rather a wide range, from 10^9 to 10^5 sec^{-1} mole^{-1}. We believe this is because the acid and base centers must be well aligned with the intervening solvent molecules in order for reaction to occur. Subtle changes in the steric features of hydrogen-bonding, therefore, have a marked effect on the reaction rate.

G. Unsymmetrical Proton Exchange

Although a great deal is known about unsymmetrical proton exchange from relaxation spectrometry (Eigen, 1963), the information obtained by the nmr method is relatively scanty. Results are listed in Table VIII. The measurement of unsymmetrical proton-exchange rates by the nmr method is kinetically complicated by parallel processes of symmetrical proton exchange, and this no doubt accounts for the paucity of data. Nevertheless, considerable further effort is justified, because much could be learned by comparing symmetrical with unsymmetrical processes (Marcus, 1968) and by comparing nmr with relaxation results.

H. Ion Pairs in t-Butyl Alcohol

Proton exchange has been used to study the structure of ion pairs in solvents of low dielectric constant. For example, Table IX shows rate constants for proton transfer of methyl- and trimethylammonium salts with the corresponding amines in *t*-butyl alcohol. These salts exist largely in the form of ion pairs in this solvent, and it is clear from the kinetics of the reaction that the proton donor is the ion pair rather than the free ammonium ion.

Table IX shows a striking difference between the behavior of methylammonium and trimethylammonium salts. For the former, the rate constants are high and almost independent of the nature of the anion. For the latter, the rate constants are substantially lower and highly dependent on the nature of the anion. A plausible explanation is that the ion pairs are intimate (rather than solvent-separated) with the alkylammonium ion hydrogen bonded to the anion, and that the NH proton thus tied to the anion is relatively unreactive. This interaction leaves two relatively reactive NH protons in the methylammonium ion pairs **9** but none in the trimethylammonium ion pairs **10**.

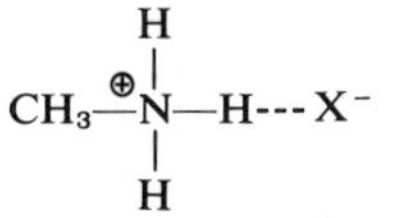

$$CH_3-\overset{\oplus}{\underset{\underset{H}{|}}{\overset{\overset{H}{|}}{N}}}-H---X^- \qquad\qquad (CH_3)_3\overset{\oplus}{N}H---X^-$$

$$\textbf{9} \qquad\qquad\qquad\qquad \textbf{10}$$

TABLE VIII
Unsymmetrical Proton Exchange in Aqueous Solution

Reaction	$k_f(\text{sec}^{-1}\text{ mole}^{-1})$	$k_r(\text{sec}^{-1}\text{ mole}^{-1})$	Reference[a]
$(CH_3)_3NH^+ + NH_3 \rightleftharpoons (CH_3)_3N + NH_4^+$ (30°C)	$(0.1 \pm 0.1) \times 10^8$	$(0.4 \pm 0.5) \times 10^8$	1
$(CH_3)_3NH^+ + OH_3 + NH_3$ (30°C)	$(1.65 \pm 0.1) \times 10^8$	$(7.6 \pm 0.4) \times 10^8$	1
$H\overset{\oplus}{N}H_2CH_2CO_2^- + NH_2CH_2CO_2H \xrightarrow{+HOH(?)} NH_2CH_2CO_2^- + H\overset{\oplus}{N}H_2CH_2CO_2H$ (23°C)	3.8×10^8	—	2
$NH_3^+CH_2CO_2^- \xrightarrow{+HOH(?)} H_2NCH_2CO_2H$ (23°C)	175 (sec^{-1})	7.1×10^7 (sec^{-1})	2

[a] References: (1) Grunwald and Ku (1968); (2) Sheinblatt and Gutowsky (1964).

TABLE IX
Reactions of Ion Pairs in t-Butyl Alcohol at 35°C[a]
$$\underline{BH^+X^-} + t\text{-BuOH} + B \rightarrow \underline{B} + \underline{H}OBu\text{-}t + BH^+X^-$$

BH^+	X^-	k (sec^{-1} mole^{-1})
$CH_3NH_3^+$	Cl^-	2.6×10^7
	OTs^-	2.4×10^7
$(CH_3)_3NH^+$	Cl^-	1.1×10^5
	$F_3CCO_2^-$	1.8×10^5
	Br^-	5.3×10^5
	OTs^-	7.0×10^5

[a] Cocivera and Grunwald (1964).

I. Ion Pairs in Acetic Acid

Table X lists kinetic data for several reactions of ion pairs in acetic acid. The first reaction, proton exchange of ammonium acetate salts with solvent, proceeds by the two-step mechanism shown in the table. The data strongly suggest that the proton transfer step is fast and that the breaking of the N·HAc hydrogen bond is rate determining; that is, $k_H \ll k_2$. Consequently when the equilibrium constant (k_1/k_2) is known, it becomes possible to deduce k_H from the kinetic data. Some values obtained in this way are included in Table X.

The second reaction in Table X, that of p-toluenesulfonic acid with anilinium acetate ion pairs, is exoergic and appears to be diffusion controlled.

One of the pay-offs of kinetic analysis is that it gives precise values of the rate constants for certain processes of ion-pair exchange (Table X, reactions 3 and 4). These rate constants show considerable specificity. The large difference between the values for $C_6H_5NH(C_2H_5)_2^+$ salts and $CH_3C_6H_4NH_3^+$ salts can be rationalized in terms of the double ion pairs (**11** and **12**) as reaction intermediates. The symmetrical hydrogen-bonded structure of **11** would

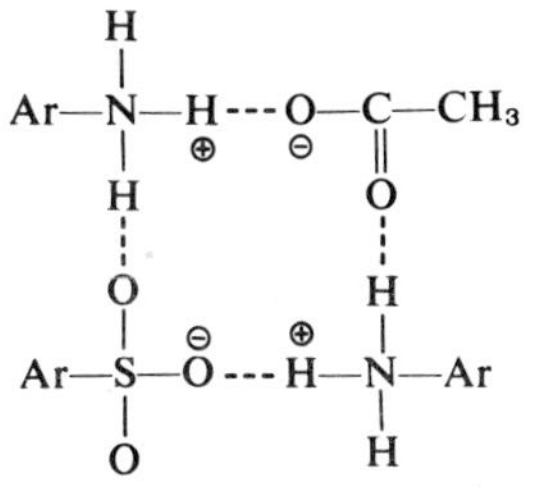

11 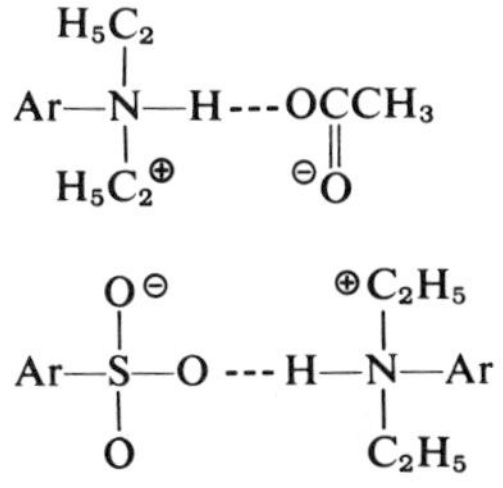 **12**

TABLE X
REACTIONS OF ION PAIRS IN ACETIC ACID

(1) $\underline{BH}^+\underline{Ac}^- \underset{k_2}{\overset{k_1}{\rightleftharpoons}} B \cdot \underline{HAc}$

$B \cdot \underline{HAc} + HAc \text{ (solv)} \xrightarrow{k_H} B \cdot HAc + \underline{HAc} \text{ (solv)}$

BH^+	$k_1 k_H/(k_2 + k_H)$ (sec^{-1})	$10^{-8} k_H$ (sec^{-1})	pK_A in H$_2$O (25°C)	Reference[a]
NH_4^+	6050 (25°)	≈2.2	9.25	1
$CH_3NH_3^+$	230 (25°)	≈1.9	10.62	1
$(CH_3)_3NH^+$	945 (25°)	≈1.2	9.80	1
$p\text{-}CH_3C_6H_4NH_3^+$	1.0×10^8 (30°)	20	4.99	2
$C_6H_5NH(C_2H_5)_2^+$	5.5×10^6 (26°)	6.0	6.56	3

(2) $BH^+Ac^- + HT_s \underset{k}{\overset{k'}{\rightleftharpoons}} BH^+T_s^- + HAc \text{ (solv)}$

BH^+	k' (sec^{-1} mole^{-1})	k'/k	pK_A in H$_2$O (25°C)	Reference[a]
$p\text{-}CH_3C_6H_4NH_3^+$	7.5×10^9 (30°C)	0.4×10^7	4.99	2
$C_6H_5NH(CH_3)_2^+$	5.0×10^9 (26°C)	0.6×10^7	5.15	3
$C_6H_5NH(C_2H_5)_2^+$	$11. \times 10^9$ (26°C)	1.1×10^7	6.56	3
$C_6H_5NH(C_3H_7)_2^+$	8.5×10^9 (26°C)	1.1×10^7	5.59	3

(3) $\underline{BH^+T_s^-} + BH^+Ac^- \xrightarrow{k} \underline{BH^+Ac^-} + BH^+T_s^-$

BH^+	k (sec^{-1} mole^{-1})	Reference[a]
$p\text{-}CH_3C_6H_4NH_3^+$	8.2×10^8 (30°C)	2
$C_6H_5NH(C_2H_5)_2^+$	6×10^7 (26°C)	3

(4) $p\text{-}CH_3C_6H_4NH_3^+T_s^- + M^+Ac^- \underset{k_r}{\overset{k_f}{\rightleftharpoons}} M^+T_s^- + p\text{-}CH_3C_6H_4NH_3^+Ac^-$ (Reference 2)

M^+	k_f (sec^{-1} mole^{-1})	k_f/k_r (at 30°C)
Li$^+$	3.0×10^8	0.58
Tl$^+$	7.5×10^8	0.80
BH$^+$	8.2×10^8	1.00
K$^+$	8.8×10^8	1.60
Cs$^+$	13.2×10^8	2.20
$(n\text{-}Bu)_4N^+$	21×10^8	2.75

[a] References: (1) Grunwald and Price (1964); (2) Crampton and Grunwald (1971); (3) Grunwald and Puar (1967b).

permit dissociation with or without exchange of the original ionic partners with equal probability. On the other hand, the unsymmetrical structure of **12** would require rearrangement of hydrogen bonds as a necessary condition for dissociation with ion-pair exchange, and this would be relatively improbable. For similar reasons, in unsymmetrical ion-pair exchange (Table X, reaction 4), cations producing the most stable or tightest ion pairs with acetate ion are the least reactive.

VI. Concluding Remarks

It is clear from the preceding examples that the nmr method can elaborate the elementary steps in fast proton transfer reactions with the same accuracy that classic methods reveal the mechanisms of slower reactions. In particular, the nmr method is of sufficient accuracy to establish complex rate laws, and it can be used to measure participation by solvent molecules and Walden inversion. Because the proton-transfer steps are often ultrafast, processes that involve the transport of solvent molecules into, out of, or within the solvation shell often become rate-determining, and we are thus able to glance at least a glimmer of dynamic processes involving the solvation shell. We know of no other single method that can give such a variety of directly measured information.

REFERENCES

Berger, A., Loewenstein, A., and Meiboom, S. (1959). *J. Amer. Chem. Soc.* **81**, 62.
Brundage, R. S., and Kustin, K. (1970). *J. Phys. Chem.* **74**, 672.
Clutter, D. R., and Swift, T. J. (1968). *J. Amer. Chem. Soc.* **90**, 601.
Cocivera, M., and Grunwald, E. (1964). *J. Amer. Chem. Soc.* **87**, 2070.
Cocivera, M., and Grunwald, E. (1965). *J. Amer. Chem. Soc.* **87**, 2551.
Cocivera, M., Grunwald, E., and Jumper, C. F. (1964). *J. Phys. Chem.* **68**, 3234.
Crampton, M. R., and Grunwald, E. (1971). *J. Amer. Chem. Soc.* **93**, 2987 and 2990.
Eigen, M. (1963). *Angew. Chem.* **75**, 489.
Eigen, M., and DeMaeyer, L. (1961). In "Technique of Organic Chemistry" (S. L. Friess, E. S. Lewis, and A. Weissberger, eds.), 2nd ed., vol. 8, chap. 18. Wiley (Interscience), New York.
Eigen, M., Hammes, G. H., and Kustin, K. (1960). *J. Amer. Chem. Soc.* **82**, 3482 (recalculated by Prof. G. H. Hammes, Aug. 29, 1967).
Emerson, M. T., Grunwald, E., and Kromhout, R. A. (1960) *J. Chem. Phys.* **33**, 547.
Fong, D.-W., and Grunwald, E. (1969). *J. Amer Chem. Soc.* **91**, 2413.
Grunwald, E. (1963a). *J. Phys. Chem.* **67**, 2208.
Grunwald, E. (1963b). *J. Phys. Chem.* **67**, 2211.
Grunwald, E. (1965). *Progr. Phys. Org. Chem.* **3**, 317.
Grunwald, E. (1967). *J. Phys. Chem.* **71**, 1846.

Grunwald, E., and Cocivera, M. (1965). *Discuss. Faraday Soc.* **39**, 105 (and references cited therein).
Grunwald, E., and Fong, D.-W. (1972). *J. Amer. Chem. Soc.* **94**, 7371.
Grunwald, E., and Jumper, C. F. (1963). *J. Amer. Chem. Soc.* **85**, 2051.
Grunwald, E., and Ku, A. Y. (1968). *J. Amer. Chem. Soc.* **90**, 29.
Grunwald, E., and Meiboom, S. (1963). *J. Amer. Chem. Soc.* **85**, 2047.
Grunwald, E., and Price, E. (1964). *J. Amer. Chem. Soc.* **86**, 2970.
Grunwald, E., and Puar, M. S. (1967a). *J. Phys. Chem.* **71**, 1842.
Grunwald, E., and Puar, M. S. (1967b). *J. Amer. Chem. Soc.* **89**, 6842.
Grunwald, E., and Ralph, E. K. (1967). *J. Amer. Chem. Soc.* **89**, 4405.
Grunwald, E., and Ralph, E. K. (1969). *J. Amer. Chem. Soc.* **91**, 2422.
Grunwald, E., and Ralph, E. K. (1971). *Accounts Chem. Res.* **4**, 107.
Grunwald, E., Jumper, C. F., and Meiboom, S. (1962). *J. Amer. Chem. Soc.* **84**, 4664.
Grunwald, E., Jumper, C. F., and Meiboom, S. (1963). *J. Amer. Chem. Soc.* **85**, 522.
Grunwald, E., Jumper, C. F., and Puar, M. S. (1967). *J. Phys. Chem.* **71**, 492.
Grunwald, E., Lipnick, R. L., and Ralph, E. K. (1969). *J. Amer. Chem. Soc.* **91**, 4333.
Grunwald, E., Fong, D.-W., and Ralph, E. K. (1971). *Bull. Isr. Chem. Soc.* **9**, 287.
Holmes, L. P., Cole, D. L., and Eyring, E. M. (1968). *J. Phys. Chem.* **72**, 301.
Loewenstein, A., and Meiboom, S. (1957). *J. Chem. Phys.* **27**, 1067.
Loewenstein, A., and Szöke, A. (1962). *J. Amer. Chem. Soc.* **84**, 1151.
Luz, Z., and Meiboom, S. (1963a). *J. Amer. Chem. Soc.* **85**, 3923.
Luz, Z., and Meiboom, S. (1963b). *J. Chem. Phys.* **39**, 366.
Luz, Z., and Meiboom, S. (1964). *J. Amer. Chem. Soc.* **86**, 4768.
Marcus, R. A. (1968). *J. Phys. Chem.* **72**, 891.
Marshall, T. H., and Grunwald, E. (1969). *J. Amer. Chem. Soc.* **91**, 4541.
Meiboom, S. (1961). *J. Chem. Phys.* **34**, 375.
Molday, R. S., and Kallen, R. G. (1972). *J. Amer. Chem. Soc.* **94**, 6739.
Morgan, W. R., and Leyden, D. E. (1970). *J. Amer. Chem. Soc.* **92**, 4527.
Puar, M. S., and Grunwald, E. (1968). *Tetrahedron* **24**, 2603.
Ralph, E. K., and Grunwald, E. (1967). *J. Amer. Chem. Soc.* **89**, 2963.
Ralph, E. K., and Grunwald, E. (1969a). *J. Amer. Chem. Soc.* **91**, 2422.
Ralph, E. K., and Grunwald, E. (1969b). *J. Amer. Chem. Soc.* **91**, 2429.
Rosenthal, D., and Grunwald, E. (1972). *J. Amer. Chem. Soc.* **94**, 5956.
Saunders, M., and Yamada, F. (1963). *J. Amer. Chem. Soc.* **85**, 1882.
Schleich, T., Gentzler, R., and von Hippel, P. H. (1968). *J. Amer. Chem. Soc.* **90**, 5954.
Sheinblatt, M. (1962). *J. Chem. Phys.* **36**, 3103.
Sheinblatt, M. (1963). *J. Chem. Phys.* **39**, 2005.
Sheinblatt, M. (1970). *J. Amer. Chem. Soc.* **92**, 2505.
Sheinblatt, M., and Gutowsky, H. S. (1964). *J. Amer. Chem. Soc.* **86**, 4814.
Silver, B., and Luz, Z. (1961). *J. Amer. Chem. Soc.* **83**, 786.

Subject Index

A

Absolute-value display, 142
Acetic acid
  proton exchange, 631
  solvent, 643
N-Acetylpyrrole, rotational barrier in, 206
Achiral skeletal frameworks, 27
Acids, proton exchange in, 11
Activation energy, determination of, 4, 75
Activation parameters
  calculation of, 75
  errors in, 76
Acylides, rotational barriers in, 235
Alkide shifts in carbonium ions, 530
Alkylbenzenes, 173
Allenemetal compounds, 427
Allyl cations, 256
Allyl complexes (of various elements, by element)
  $B(\eta^1-C_3H_5)_3$, 456, 481
  $Cd(\eta^1-C_3H_5)_2$, 447, 448, 482
  $Co(\eta^3-C_3H_5)(CN)_4^{2-}$, 447
  $Co(\eta^1-C_3H_5)(CN)_5^{3-}$, 447
  $Co(\eta^3-C_3H_5)(CO)_3$, 446
  $Fe(\eta^3-C_3H_5)(CO)_3I$, 459
  $Fe(\eta^3-C_3H_5)_2(CO)_2$, 459, 464
  $Ge(\eta^1-C_3H_5)_4$, 447, 481
  $Hf(\eta^3-C_3H_5)_4$, 462, 463
  $Hg(\eta^1-C_3H_5)Cl$, 457
  $Hg(\eta^1-C_3H_5)_2$, 447, 457, 481
  $Ir(\eta^3-C_3H_5)_3$, 460
  $Ir(\eta^3-CH_3 \cdot C_3H_4)Cl_2L_2$, 464
  KC_3H_5, 442, 449
  LiC_3H_5, 442, 448, 457, 482
  $Mg[1,1-(CH_3)_2C_3H_3]Br$, 482
  MgC_3H_5Br, 442
  $Mg(C_3H_5)_2$, 447, 448, 457, 463, 482
  $Mn(\eta^1-C_3H_5)(CO)_5$, 447
  $Mn(\eta^3-C_3H_5)(CO)_4$, 447
  $Mo(\eta^3-C_3H_5)(CO)_2(\eta^5-C_5H_5)$, 452, 458, 460, 464
  $Mo(\eta^3-C_3H_4R)(CO)_2(\eta^5-C_5H_5)$ $(R=CH_3)$, 458
  $Mo(1,2,7-\eta^3-3,5[(CH_3)_2CH]_2C_6H_2CH_2)$ $(\eta^5-C_5H_5)(CO)_2$, 463
  $Mo(1,2,7-\eta^3-4CH_3 \cdot C_6H_4CH_2)(\eta^5-C_5H_5)(CO)_2$, 463
  NaC_3H_5, 442, 448, 482
  $Ni(\eta^3-C_3H_5)_2$, 445, 461, 465, 466
  $[Ni(\eta3-C_3H_5)Cl]_2$, 445
  $Ni(\eta^3-C_5H_5)Cl(PPh_3)$, 445, 466
  $Ni(\eta^3-2-CH_3 \cdot C_3H_4)_2$, 461
  $Pd(\eta^3-C_3H_5)Cl(\alpha$-phenylethylamine), 470, 471, 479
  $Pd(\eta^1-C_3H_5)(O_2CC_5H_4N)(PPhMe_2)$, 456, 481
  $Pd(\eta^3-C_3H_5)$ (oxinate), 472
  $Pd(\eta^3-C_3H_5)$ (picolinate), 472, 473
  $[Pd(\eta^3-C_4H_7)(\acute{S}CN)]_2$, 466
  $Pd(\eta^3-C_3H_5)(SnCl_3)(PPh_3)$, 445
  $Pd(\eta^1-C_3H_5)(S_2CNMe_2)(PPhMe_2)$, 456
  $Pd(\eta^3-C_3H_5)_2$, 445, 461, 466, 467
  $[Pd(\eta^3-C_3H_5)Cl]_2$, 443, 446, 466
  $Pd(\eta^3-1-CH_3 \cdot C_3H_4)X(2$-picoline), 471, 475, 478
  $[Pd(\eta^3-1-CH_3 \cdot C_3H_4)Cl]_2$, 475
  $Pd(\eta^3-2-CH_3 \cdot C_3H_4)Cl(AsPh_3)$, 469, 472, 473
  $Pd(\eta^3-2-CH_3 \cdot C_3H_4)Cl(PPh_3)$, 445, 451, 466, 467, 470, 472, 473, 474, 476
  $Pd(\eta^1-2-CH_3 \cdot C_3H_4)(S_2CNMe_2)(PPhMe_2)$, 481
  $Pd(\eta^3-2-CH_3 \cdot C_3H_4)Cl(PPhMe_2)$, 454
  $[Pd(\eta^3-2-Cl \cdot C_3H_4)Cl]_2$, 477
  $Pd(\eta^3-C_4H_7)(CN)L$, 466
  $Pd(\eta^3-C_4H_7)(OAc)(PPh_3)$, 466, 473, 477
  $Pd(\eta^3-C_4H_7)Cl(\alpha$-phenylethylamine), 475
  $[Pd(\eta^3-C_4H_7)(PR_3)_2]^+Cl^-$, 472

649

Pd(η^3-C_4H_7)(SDBM), 451
Pd(η^3-C_4H_7)(xanthate)PPhMe$_2$, 477
[Pd(η^3-C_4H_7)Cl]$_2$, 466, 467, 468, 469, 470, 472, 473, 474
Pd(η^3-1-acetyl-2-C_3H_4)Cl(S-α-phenylethylamine), 478, 479
Pd(η^3-1-acetyl-2-C_3H_4)Cl(pyridine), 479
Pd(η^3-1-acetyl-2,3-dimethylallyl)Cl(S-α-phenylethylamine), 480
[Pd(η^3-2-CH_3 · C_3H_4)(OAc)]$_2$, 461, 462
[Pd(η^3-syn-1-i-C_3H_7 · C_3H_4)Cl]$_2$, 456
Pd(η^3-2-i-C_3H_7 · C_3H_4)Cl(PPh$_3$), 455
Pt(η^3-C_3H_5)C1L, 466
Pt(η^3-C_3H_5)$_2$, 445, 461, 466, 467
[Pt(η^3-C_3H_5)Cl]$_2$, 445
[Pt(η_3-C_4H_7)Cl]$_2$, 466
Pt(η^3-2-CH_3 · C_3H_5)$_2$, 461
[Rh(η^3-C_3H_5)Cl]$_2$, 460, 464
Rh(η^3-C_3H_5)$_3$, 460, 464
Rh(η^3-2-CH_3 · C_3H_4)Cl$_2$L$_2$, 464, 465
Ru(η^3-C_3H_5)$_2$(CO)$_2$, 453, 459, 460, 464
Si(η^1-C_3H_5)$_4$, 447, 481
Sn(η^1-C_3H_5)$_4$, 447
Th(η^3-C_3H_5)$_4$, 461, 463
Ti(η^1-C_3H_5)[(CH$_3$)$_2$N]$_3$, 457, 481
Ti(η^1-2-CH_3 · C_3H_4)[(CH$_3$)$_2$N]$_3$, 481
Ti(η^1-2-CH_3 · C_3H_4)[(C$_2$H$_5$)$_2$N]$_3$, 481
Ti(η^1-2-CH_3 · C_3H_4)$_2$[(C$_2$H$_5$)$_2$N]$_2$, 481
W(η^3-C_3H_5)(CO)$_4$X, 458
W(η^3-C_3H_5)(η^5-C_5H_5)(CO)$_2$, 458, 460, 464
W(η^3-C_3H_4R)(η^5-C_5H_5)(CO)$_2$ (R=CH$_3$), 458
W(1,2,7-η^3-4CH$_3$ · C$_6$H$_4$CH$_2$)(η^5-C_5H_5)(CO)$_2$, 463
Zn(η^1-C_3H_5)$_2$, 447, 448, 457, 463, 482
Zn(η^1-2-CH_3 · C_3H_4)$_2$, 482
Zr(η^3-C_3H_5)$_4$, 462, 463
Allyl-metal bonding, 442
  π-bonding, 442
  σ-bonding, 442, 447
  ionic bonding, 442, 448
Aluminum(III) β-diketonate complexes, 339, 348-351, 355-358, 363
  kinetic data, 351, 356, 357
Aluminum(III) tropolonate complexes, 350-358
  kinetic data, 351, 356

Amides, 12, 13, 182, 189
  acid catalysis and rotational barriers in, 214
  effect of solvent on rotational barriers in, 212
  effect of substitution on rotational barriers in, 210
Amidines, rotational barriers in, 215
Amines, 186
Aminoboranes, rotational barriers in, 237
Angle
  dihedral, sign convention, 544
  torsional, sign convention, 544
Anti-Berry mechanism, 269
Arylaldehydes, 179, 181
Arylalkyl ketones, 179, 181

B

Band-shape analysis, practical, 9-13, 17-18, 70
  computer programs for, 72
  dynamic parameters, 74
  experimental procedures, 70
  static parameters, 73
Band-shape formulas, classical
  approximate, 50
  many-site exchange, 51
  quadrupole broadened, 63
  single line, 48
  two-site, 49
Band-shape formulas, quantum-mechanical
  approximate, 63
  double resonance, 64
  effects of finite sweep rates on, 69
  for quadrupole broadened spectra, 63
  saturated, 67
  unsaturated steady-state, 60, 61, 62
Band shape
  steady-state absorption, 48
  steady-state dispersion, 48
  unsaturated steady-state, 48
Band-shape theory
  classical, 47
  quantum-mechanical, 53
Band width
  approximate, 50
  effective, 48

Subject Index

natural, 48, 49
saturated, 48, 49
Barriers to rotation, see Rotational barriers
Benzamides, N,N-dimethyl, effect of substituents on the rotational barriers in, 210
Benzyne complexes, 434
Berry mechanism, 276ff., 491
Berry rearrangement, 269
Biological cells, measurement of restricted diffusion, 96, 97
Biphenyl derivatives, 179, 180
Bisbidentate complexes, nonrigidity, 320-328, 333-338
Bisfluorenylidenes, rotational barriers in, 243
Bloch equations, 9, 10, 47
  modified, 49, 51, 52
  with diffusion, 84, 90
Bond rupture mechanisms, 341-343, 347-350, 355, 358
Boron compounds
  closoboranes, 294
  intramolecular rearrangements, 294
  metalloboranes, 296
  nidoboranes, 295
Bridge-terminal rearrangements in metal carbonyls, 500
3-Bromothiophene-2-aldehyde, T_2 measurements in, 152
Bullvalene, degenerate Cope rearrangement of, 121, 122
1,3-Butadienes, 185

C

Cadmium(II) β-aminothionate complexes, 325-328
  kinetic data, 326
Carbanions, 191
Carbocycles, kinetic parameters
  6-membered saturated, 582
  6-membered, unsaturated, 585
Carbon-13 nmr,
  determination of rotational barriers in amides, 205
Carbon-13 shifts in Ni(II) complexes, 322
Carbon-13 spectra, 153

Carbonium ions, 191
  alkide shifts in, 530
  conformational changes in, 526
  generation of, 524
  hydride shifts in, 530
  rotations in, 526
1-Carbotrifluoroethoxy-1,2,3,6-tetrahydropyridazine, ring inversion, 592
Carr-Purcell method, effect of field drift and oscillation, 125
Carr-Purcell sequence, 133
  Meiboom-Gill modification, 138, 139, 141
  phase alternation, 138, 141
Cerium(IV) complexes, 369
Chemical exchange, 132, 149
  A matrix, 118, 119
  by Carr-Purcell sequence, two sites uncoupled, 113
  comparison of steady state and pulse methods, 122
  decay constants for two coupled sites, 120
  fast exchange limit, 109
  first echo height, 108
  free induction decay, 107
  inter- and intramolecular cases, 119, 120
  matrix operators for Carr-Purcell sequence, 113
  measurement in the rotating frame, 125
  modulation of Larmor frequency, 106
  modulation of relaxation rates, 106
  N,N-dimethylnitrosamine, 114
  numerical solution in Bloch equations, 109
  probability theory in phase angles, 110
  pulse experiments, density matrix treatment, 115, 116
  pulse methods, errors, 120, 121, 122
  pulse methods, random errors, 110
  spin-echo studies, 83, 105
  symmetric case, approximate results, 114
  systematic errors, 109
Chemical exchange effects, 8-14
Chemical exchange rates
  N,N-dimethyl carbonyl chloride, 110
  N,N-dimethylformamide, 123
  N,N-dimethyltrichloroacetamide, 110
  hindered rotation in 1,1-difluoro-1,2,-dibromodichloroethane, 123
  inversion 1,1-difluorocyclohexane, 121

inversion of perfluorocyclohexane, 121
Chiral skeletal frameworks, 27
Chromium(III) complexes, 348, 357
Coalescence formulas, 50, 63
Cobalt(III) β-diketonate complexes, 319, 347, 348, 354-358
Cobalt(III) dithiocarbamate complexes, 319, 358, 362, 363
Cobalt(III) tropolonate complexes, 319, 350-358
  kinetic data, 351, 356
Cobalt(III) xanthate complexes, 319, 363
Coherent time-dependent fields, 15, 16
Configurational symmetry groups, 26
Configurations, 24
  intermediate, 29
Conformational energy calculations, 554, 558
  applications to six-membered rings, 564
  force-field, 554
  force-field parameters for saturated hydrocarbons, 557
  quantum-mechanical, *ab initio*, 574
  quantum-mechanical, semi-empirical, 572
  vibrational, 563
Conformational interconversions
  in carbonium ions, 526
  labeling scheme, 553
Conformational labeling, notation in cycloalkanes, 551
Contact shifts
  definition, 321
  in planar-tetrahedral mixture, 321
Copper(I) complex, 296
Correlation function, 57
Correlation time, 50, 57, 58
Cumulenes, rotational barriers in, 243
Cyclic amines, pseudorotation, 263
Cyclododecane
  conformational notation, 553
  conformational processes, 558, 559, 560
Cycloheptane
  conformational energies, 567
  conformational notation, 552
1,3,5-Cycloheptatriene, ring inversion, 546, 597
Cycloheptatrienylmetal compounds, 400
  monohapto, 400
  pentahapto, 403
  trihapto, 401, 417

Cycloheptene
  local ring inversion, 548
  ring inversion, 593
Cyclohexadienyl complexes, fluxional, 431
Cyclohexane
  calculated energy barriers for inversion, 565, 573, 574
  conformational notation, 552
  experimental barrier for inversion, 579
  local ring inversion, 548
  nmr spectra, 576
  ring inversion, 546
Cyclohexane-d_{11}, ring inversion barrier, 576, 578, 579
Cyclohexanes, *cis*-1,2-disubstituted, ring-inversion path, 580, 581
Cyclohexanone, ring inversion, 566
Cyclohexatrienemetal compounds, 416, 425
Cyclohexene, ring inversion, 546, 566, 581
Cyclononane, barrier to pseudorotation, 562
1,4,7-Cyclononatriene, ring inversion, 604
cis-cis-1,3-Cyclooctadiene, barriers to ring inversion, 598
cis-cis-1,5-Cyclooctadiene
  force-field calculation, 568
  interconversion paths, 572
1,5-Cyclooctadiene complexes
  [Ir(1,5-C_8H_{12})Cl]$_2$, 473
  Ir(1,5-C_8H_{12})Cl(AsPh$_3$), 451, 473
  Ir(1,5-C_8H_{12})Cl(PPh$_3$), 472
  Ir(1,5-C_8H_{12})(PPhMe$_2$)$_2$(CH$_3$), 451, 473
  [Rh(1,5-C_8H_{12})Cl]$_2$, 467, 470, 474
  Rh(1,5-C_8H_{12})Cl(AsPh$_3$), 467, 470, 248
  Rh(1,5-C_8H_{12})Cl(PPh$_3$), 467, 470, 474
  Rh(1,5-C_8H_{12})(PPhMe$_2$)$_2$(CH$_3$), 451
Cyclooctane
  conformational energies, 570
  conformations, 569
  force-field calculation, 568
  local ring inversion, 548
  maximum-energy conformations for interconversion paths, 571
Cyclooctatetraene, ring inversion, 598
Cyclooctatetraenemetal compounds, 403
  of chromium, 406
  of iron, 403, 418
  of molybdenum, 406
  of osmium, 403
  of ruthenium, 403, 412

Subject Index

of tungsten, 406
cis-cis-cis-1,3,5-Cyclooctatriene, barriers to
  ring inversion, 598
Cyclooctatrienemetal compounds, 413
cis-Cyclooctene, conformational barriers,
  598
Cyclopentadienylmetal compounds, 378, 419
  of aluminum, 398
  of chromium, 385
  of copper, 385, 388
  of germanium, 392
  of gold, 385, 389
  of iron, 378
  of mercury, 387, 389
  of molybdenum, 382, 391, 417, 420
  of phosphorus, 399
  of platinum, 391
  of silicon, 392
  of thallium, 399
  of tin, 392
Cyclopentenyl cation, 532
Cyclopropyl carbonium ions, 528
Cyclotetradecane
  conformations, 560, 561
  conformational processes, 561
Cyclotriveratrylene, ring inversion, 604

D

cis-Decalin, ring inversion path, 580, 583,
  584
Density matrix, 53
Density operator, 53
Density vector, 54, 56
Diastereoisomers, NI (II) complexes,
  322-324
Diastereomeric reactions, 28
Diastereotopic averaging in metal complex
  rearrangements, 325-327, 339, 345, 349,
  353, 354, 359-361, 364-365, 368
Diastereotopic group, 261
Diazoketones, rotational barriers in, 184,
  236
1,2-Dicarbotrifluoroethoxy-1,2,3,6-tetra-
  hydropyridazine, ring inversion, 592
o-Dichlorobenzene, T_2 measurements in,
  159
Dichloromethyl group, rotation of, 173, 174
Dicobalt octacarbonyl, 500, 514, 518

Diffusion
  advantages of spin-echo methods, 92
  anisotropic, 93, 94
  benzene liquid, 99
  both components of binary solution, 99
  comparison of standard and unkown
    liquid, 100
  disparate values for water, 92
  echo amplitude change, 92
  experimental errors of measurement, 99,
    101, 102, 103
  Ficks Law, 87
  and flow, 84
  gaseous helium-3, 98
  lamellar system, 95, 96
  linear siloxanes, 98
  liquid lithium metal, 99
  liquid methane, 99
  liquid NH_3, ND_3, 99
  liquid SF_6, 99
  low-molecular-weight polyisobutylene, 99
  measurement in rotating frame experiments,
    98
  methane, propane, high pressure gas, 99
  multispin-echo experiments, 89
  normal paraffins, 98
  phospholipid/water system, 97
  plastic crystals, solids, 97, 98
  pulsed field gradient, advantages, 92
  random walk approach, 86
  restricted, 92, 93, 95
  restricted, steady field gradient, 96, 97
  stimultaneous measurements of several
    components, 93
  smectic phases, 97
  tensor, 93, 94
  tetramethylsilane, 98
Diffusion constants
  measurement by varied gradient, 104
  measurement by varied pulse interval,
    104, 105
  ranges of values measureable by MR
    methods, 98
Diffusion rates, 14
4,4-Difluoro-1,1-dimethylcycloheptane,
  ring pseudorotation, 549
Digonal twist, 266
Dihydridotetrakis (diethoxyphenylphos-
  phine) ruthenium (II), 39

9,10-Dihydrophenanthrene, conformational barriers in derivatives, 584
7,12-Dihydropleiadene, ring inversion, 597
Dihydropyridines, 4-methylene derivatives, rotational barriers in, 241
1,10-Diketo [3,3] paracyclophane, conformational barrier, 612
5,5-Dimethyl-3,7-dithiabenzocycloheptene, conformational barriers, 597
N,N-Dimethylacetamide
  effect of cations on rotational barrier, 215
  rotational barrier in, 206
N,N-Dimethylacetamide-2-d_3, rotational barrier in, 206
N,N-Dimethylamides, effect of substitution on rotational barriers in, 209
1,1-Dimethylcyclodecane, pseudorotation, 562, 563
1,1-Dimethylcyclohexane, force-field calculation, 566
1,1-Dimethylcyclononane, conformational barriers, 562
N,N-Dimethylformamide, rotational barrier, 204-206
N,N-Dimethylselenourea, rotational barrier in, 217
1,4-Dioxane
  chemical shifts, 577
  vibrational calculations, 564
Diphenoquinones, rotational barriers in, 242
1,2-bis(diphenylphosphino) ethane (2-methylallyl) palladium (II) cation, 38
Dipolar broadening, 5-7
Dipole-dipole interactions, 3-8
Discovery of nmr, 2
Dithia [n] methacyclophanes, conformational barriers, 613
Double resonance, 64
  in the linear approximation, 64
  nonlinear effects in, 67
Double resonance effects, 15, 16
Duplothiacetone, exchange scheme, 591
Dynamic stereochemistry, 24

E

Echo modulation, 135, 145
  amplitude modulation, 137
  phase modulation, 136
Effective nmr symmetry groups, 30

Eight-coordinate molecules
  dodecahedral structure, 291
  intramolecular rearrangement, 291
Eight-membered rings
  kinetic parameters, 599
Enamines, 190
  effect of substituents on rotational barrier in, 225, 226, 227
  rotational barriers in, 218
Enantiomeric reactions, 28
Enol ethers, rotational barriers in, 237
Enolate ions, rotational barriers in, 233
Enumeration of nmr differentiable reactions, 34
Ethanol, 11, 13
Exchange
  matrix, 51
  mechanism, 52, 59, 75
  operator, 59
  rate constant, 49, 50, 51, 75
  superoperator, 58, 64
Exchange, classical
  between many sites, 51
  between two sites, 49
Exchange, quantum-mechanical
  intermolecular, 59
  intramolecular mutual, 58
  intramolecular nonmutual, 59
Extreme narrowing, 6, 50, 51, 58

F

Fast exchange limit, definition, 11, 254
Field gradient by shaped pole caps, 100
Field gradient, sinusoidal variation, 91
Field gradient coils
  calibration, 101, 103
  design, 100
Field gradient vector, 90
Field-inhomogeneity broadening, 47, 50, 60, 64, 66, 70
Five-coordinate molecules, intramolecular rearrangements, 266 ff.
Flip angle, 132
Flow, measurement by nmr, 97
Fluorene derivatives, 170
Formamide, rotational barrier in ^{15}N-labeled, 206
Four-coordinate molecules, intramolecular rearrangements, 265ff.

Fourier transform, 136
Fourier transform nmr, 259
Fourier transformation, 135
Free induction decay, 259
  Bessel function with linear field gradient, 101
  with diffusion, 85
Free induction signal, 135
Frequency dependence studies, 19
Full configurational symmetry groups, 27
Fulvene, rotational barriers in 6-N,N-dimethylamino derivatives, 231

G

Gallium (III) complexes, 348-357
  kinetic data, 351, 356, 357
Geometric constraints in rings, 550
Germanium (IV) complexes, 333, 334, 336, 350, 356
  kinetic data, 337
Group of allowed permutations, 26
Guanidines, rotational barriers in, 217

H

$H_2Fe[P(OC_2H_5)_3]_4$, 279 ff.
HML_4 species
  exchange barriers, 274
  intramolecular rearrangement, 273 ff.
  tetrahedral jump model, 275
H_2ML_4 species
  basic permutational sets, 283
  cis isomers, 278
  intramolecular rearrangements, 278 ff.
  mechanistic information, 280 ff.
  nonconcerted process, 286
  tetrahedral jump, 284
  trans isomers, 288
  x-ray structure, 284
H_4ML_4, intramolecular rearrangement, 291
$H_4Mo[P(C_6H_5)_2CH_3]_4$, 291 ff.
  x-ray structure, 291
$H_2{}^{17}O$, use of, 629
$HRh(PF_3)_4$, 273
$HTa(CO)_2[(CH_3)_2PCH_2CH_2P(CH_3)_2]_2$, 291
Hafnium (IV) complexes, 336, 338, 369
  cyclopentadienyl derivatives, 369-371
  kinetic data, 371

Hahn conditions for a pulse experiment, 84, 85
Halogenated alkanes, 164
Halogens, bulk influence on rotational barrier, 164, 173
Heterocycles, 6-membered, kinetic parameters, 587
Heteronuclear lock, 259
γ-1,2,3,4,5,6-Hexachlorocyclohexane, ring inversion, 580
1,1,2,2,5,5-Hexamethylcyclohexane, force-field calculation, 566
Hexamminoplatinum (IV), 631
Holmes, Sherlock, 435
Homonuclear lock, 259
Hydride shifts in carbonium ions, 530
Hydrogen fluoride, 8-9, 14
Hydronium ion
  reactions, of, 632
  table of, 633
Hydroxide ion
  reactions of, 635
  table of, 636
3-Hydroxy-2,4-dimethylcyclobutanone, 35

I

Imines
  calculated rates of inversion of, 249
  mechanism of inversion of, 244-249
  rates of inversion of, 245, 246, 247
Idenylmetal compounds, 382
  of iron, 382
  of silicon, 396
  of tin, 394
Indium (III) complexes, 348, 355, 356
Inhomogeneity of magnet, 91
Intermediate configurations, 29
Intermolecular exchange, 299 ff.
  calculational aspects, 301
  electron exchange effects, 300
  general, 311
  paramagnetic effects, 300
  quadrupole effects, 300
Intermolecular processes, definition, 254, 257, 258
Internal rotation, *see* Rotation
Intramolecular rearrangements, definition, 254, 255, 256, 257

Inversion
  local, 547
  ring, 546
Inversion of configuration
  bischelate complexes, 325-328
  sexadentate complexes, 368-369
  trischelate complexes, 349-368
Ion pairs in acetic acid
  reactions of, 643
  table of, 644
Ion pairs in t-butyl alcohol
  proton exchange, 641
  table of, 643
Ionic metal-allyl compounds
  bonding, 448
  fluxional movements, 457, 482
Iron (II) complexes, 368, 369
Iron (III) complexes, 348, 355, 356, 358
Iron (IV) dithiocarbamate complexes, 358-362
Iron dithiocarbamate dithiolene complexes, 363-368
Isochromats, 133
Isocyanide complexes, 507

J

J-modulation, 137
J-spectra, 149

K

$K(3\text{-Ph-}C_3H_4)$, 448, 482
Kinetic analysis, 624

L

Larmor frequency, 48
Ligand association, 306
  in L_2PtX_2 species, 311
  in $trans$-RhCl(CO)[P(CH$_3$)$_2$(C$_6$H$_5$)]$_2$, 308 ff.
Ligand dissociation, 303 ff.
  in $H_2RhCl[P(C_6H_5)_3]_3$, 304 ff.
  in $Rh[P(OCH_3)_3]_5$, 303
Ligand exchange, 50
Ligand kinetic series, 355-357
Line averaging computer, 258
Line shape, see Band shape
Line width, see Band width
Linear field gradient, 85, 86
Linear response, 60, 64, 65

Liouville representation, 53
  composite, 55
  interaction, 56, 59
  primative, 55
  special, 55, 65
Liouville space, 55
Liouville superoperator, 55
Liouville-von Neumann equation, 54
Li$(1,3\text{-Ph}_2C_3H_3)$, 448, 457, 482
Li$(3\text{-Ph-}C_3H_4)$, 448, 457, 482
Low temperature solvent, CHClF$_2$, 261
2,4-Lutidine, 629
Lyonium ions
  reactions of, 635
  table of, 636

M

Magnetic equivalence factorization, 62
Magnetization
  complex transverse, 48
  macroscopic, 47
ν-Magnetization with diffusion, 88
Manganese (III) complexes, 348, 350, 352, 356
Memory effects, 29
Metal-π-allyl compounds
  bonding, 442
  cis-trans isomerization, 450, 472, 473, 474, 478
  fluxional movements, 449
  kinetics, 467, 468, 469, 470
  retention of configuration, 473
  syn-syn, anti-anti exchange, 450, 457, 466
  syn-anti exchange, 454, 462, 476
Metal-σ-allyl compounds
  bonding, 447
  fluxional movements, 456, 480, 481
  infrared spectroscopy, 442, 253
  photoelectron spectroscopy, 443
  Raman spectroscopy, 442
  ^{1}H and ^{13}C nmr spectroscopy, 442
  thermochemistry, 443
  x-ray spectroscopy, 443
Metal carbonyls, 489
  bridge-terminal rearrangements, 500
  double bridged, 490
  polytopal rearrangements of, 491, 519
  terminal, bonding of, 489
  triply bridged, 490
  unsymmetrically bridged, 490, 519

Subject Index

Metal kinetic series, 336, 355-357, 362
Metallocyclic complexes, fluxional, 433
[2,2]Metaparacyclophane, isotope effect on conformational barrier, 612
Methylcycloheptane, conformational energies, 567
Methylcyclohexane, ring inversion, 547
Methylenecyclohexane, conformational barriers, 584
Methylenecyclopropene and derivatives, rotational barriers in, 239
Methyl group, rotation of, 167
Methyl iodide, 155
Microscopic reversibility, 29
ML_5 species
  basic permutational sets, 268, 269
  exchange barriers, 267, 268
  intramolecular rearrangement, 273
  mechanistic information, 268
  steric effects, 267
  trigonal bipyramidal stereochemistry, 273
Modulation sidebands, 16
Molecular reorientation, 3-6
Molybdenum (IV) complexes, 333, 335
  kinetic data, 337
Monobidentate complexes, nonrigidity, 331, 332
Motional narrowing, 5-7
Multiple resonance, 67
Mutual exchange, definition, 255

N

Na (3-Ph-C_3H_4), 448, 482
Neopentyl derivatives, 170, 173
NH_3, 261
Nickel (II) aminotroponeiminate complexes, 321, 322
Nickel (II) β-ketoaminate complexes, 321, 322
Nickel (II) phosphine complexes, 328-330
  kinetic data, 330
Nickel (II) salicylaldiminate complexes, 321-324
Nine- and higher coordinate molecules, intramolecular rearrangement, 294
Nine-membered and larger rings, kinetic parameters, 605
Nitroso derivatives, 190

NMR differentiable reactions, 33
  complete set of, 33
NRM nondifferentiable reactions, 32
Nonlinear response, 66
  in multiple resonance, 67
  in single resonance, 66
Nonmutual exchange, 255
Nonsecular effects, 58, 63
Norbornadiene complexes
  [Rh(C_7H_8)Cl]$_2$, 467, 473
  Rh(C_7H_8)Cl(AsPH$_3$), 473
  Rh(C_7H_8)Cl(PPh$_3$), 474
  Rh(C_7H_8)(SDBM), 451
Norbornadienyl cation, 528, 535
Nuclear configuration, 56
Nuclear Overhauser effect, determination of rotational barriers in amides, 206
Nuclear Overhauser enhancement, 153, 157

O

Olefin complexes, 428
Operator
  annihilation, 68
  creation, 68
  exchange, 59
  projection, 54, 65, 69
Organometallic compounds, 377, 441
Overhauser effect, 16

P

[3.3]Paracyclophane, conformational barrier, 612
3,3,6,6-bis(pentamethylene)-s-tetrathiane, ring inversion, 592
Perfluorcyclohexane, ring inversion barrier, 573
Permutation of indices (PI) method, 60
Permutational analysis, 286
Phase angles
  Gaussian distribution, 87
  mean square deviation with diffusion, 87
Phase distribution with diffusion, 85, 88
Phase locking, 259
Planar-tetrahedral equilibrium (d^8), 265
Planar-tetrahedral interconversion, 321-324, 238-330

Polytopal rearrangements
  Berry mechanism for, 491
  of cobalt (I) compounds, 492
  of cyclopentadienylmolybdenum compounds, 494
  of dimethylgermyl-bridged compounds, 496
  of iron pentacarbonyl, 491
  of metal carbonyls, 491
  of rhodium (I) compounds, 492
Population vector, 51, 52
Proper configurational symmetry groups, 27
Proton exchange rate, definitions, 622
Proton transfer, kinetics, 624
Protonated cyclopropanes in carbonium ions, 536
Pseudorotation, 545
Pulsed field gradient, 89, 90
Pulse sequence
  Carr-Purcell, 87, 133
  Freeman-Wittekock, 138
  Meiboom-Gill, 138
Pulses
  nonselective, 134
  selective, 134
Purines, 629
  rotational barriers in 1-N,N-dimethylamino derivatives, 231
P-values, 166
Pyramidal inversion, 261 ff.
Pyramidal molecules
  electronic effects, 263
  intramolecular rearrangements, 261 ff.
  inversion barriers, 265
  nomograph (Baehler), 265
  quasi four-coordinate, 261
  ring strain effects, 263
  steric effects, 263 ff.
Pyrimidine, rotational barrier in 4-N,N-dimethylamino, 231

Q

Quadrupolar effects, 63, 71
Quinonemethides, rotational barriers in, 242

R

Radio frequency pulses, 132
  nonselective, 135
  selective, 134, 144

Reaction sequences, 29
Reactions, 24
  diastereomeric, 28
  differentiable in a chiral environment, 28
  enantiomeric, 28
  nmr differentiable, 32
  nmr nondifferentiable, 32
  nondifferentiable in a chiral environment, 27
  symmetry equivalent, 27
Reference configurations, 25
Relaxation
  broadening, transfer of, 50
  interaction tensor, 51, 57
  matrix, 51, 64
  quadrupole, 63
  superoperator, 61
  time, effective, 75
  time, longitudinal, 47
  time, transverse, 47
  trivial, 60
Relaxation spectrometry, 630
Reverse reactions, 26
$Rh[P(OCH_3)_3]_5^+$, 269
Rhodium (III) complexes, 348, 350, 352, 355, 356
Rotary echoes, 134
Rotating coordinate system, 47, 56, 64
Rotating reference frame, 132
  double resonance in, 155
Rotation
  in carbonium ions, 526
  about C-C bonds
    sp^2-sp^2, 178
    sp^3-sp^2, 170
    sp^3-sp^3, 164
  about C-N bonds, 185
  about C-O bonds, 192
  about C-P bonds, 193
  about C-S Bonds, 193
  about Mo-Mo bonds, 514
  about N-N bonds, 194
  about N-O bonds, 196
  about N-X bonds, 195, 196
  about partial double bonds, 163
  about S-S bonds, 193
Rotational barriers, *see* individual compounds or classes of compounds
Ruthenium (III) complexes, 319, 348, 350, 352, 355-358

Subject Index

Ruthenium (III) dithiocarbamate complexes, 319, 362, 363

S

Saturation, 48, 66, 70
Scandium (III) complexes, 348, 355
Second quantization, 67, 68, 69
Secondary echoes, 89
Selfdiffusion, 144, 145, 151
Seven-coordinate molecules, intramolecular rearrangements, 289 ff.
Seven-membered rings, kinetic parameters, 593
Sexadentate complexes, 368, 369
Shift reagents, use in determination of rotational barriers in amides, 205
Signal to noise, 259
Silicon (IV) complexes, 336, 338, 348, 352, 356
Six-coordinate molecules
   chelate rearrangements, 277 ff.
   intramolecular rearrangements, 277 ff.
Skeletal frameworks
   achiral, 27
   chiral, 27
Slow exchange limit, definition, 254
Solvent effects on rotational barriers in amides, 214
Solvent participation, 628
Spectral density, 57, 63
Spin decoupling, 153
   effect on spin echoes, 153
Spin echoes, 83, 133
   effect of heteronuclear decoupling, 153
   effect of pulse imperfections, 138, 139, 140
   effect of sample spinning, 140
   exchange studies, 11, 17
   J spectra, 149, 151
   spin echo spectra, 150
Spin Hamiltonian
   for multiple resonance, 65
   for single resonance, 56
Spin-lattice relaxation (T_2), 3-5
Spin-locking, 146, 147, 156
Spinning sidebands, 16
Spin-spin coupling
   modulation of echo amplitudes, 115
Spin-spin relaxation time, 132

Statistical matrix, 52, 63
Stereochemistry, dynamic, 24
Steric courses, 29
Stimulated echo, 89
Stochastic model
   for exchange, 49
   for quadrupolar relaxation, 63
Subspectra, information from, 626
Sulfur tetrafluoride, 40, 312
Superconducting nmr systems, 260
Superoperator
   derivation, 55
   exchange, 58, 59, 60
   Liouville, 55
   relaxation, 56, 57
Symmetrical proton exchange
   with solvent participation, 638
      table, 639
   without solvent participation, 636
      table, 638
Symmetry equivalent reactions, 27
Symmetry groups
   configurational, 26
   effective nmr, 30
   full configurational, 27
   proper configurational, 27

T

$T_1 \rho$, 148
T_2, 132
T_2^*, 133
Temperature control in nmr, 20, 260
Temperature determination, 207, 260
Tertiaryl butyl group, rotation of, 166, 168
trans-syn-trans-1,2,5,6-Tetrabromocyclo-octane, conformation, 598
Tetracobalt dodecacarbonyl, 502
1,1,2,2-Tetrafluorocyclooctane
   conformations, 598
   ring pseudorotation, 549
Tetrahedral jump, 284
3,3,6,6-Tetramethylbenzocycloheptene, conformations, 597
1,1,3,3-Tetramethylcyclohexane, 582
   force-field calculation, 566
1,1,4,4-Tetramethylcyclohexane, 582
   force-field calculation, 566
1,1,4-4-Tetramethylcyclonane, conformational barriers, 562

Tetramethylcyclooctatetraenemetal compounds, 407
1,3,5,7-Tetraoxocane, conformations, 598
Tetrarhodium dodecacarbonyl, 517, 520
Thioamides, rotational barriers in, 217
Thioenol ethers, rotational barriers in, 234
Thorium (IV) complexes, 369, 370
Tin (IV) complexes, 336-338, 355
  kinetic data, 337
Titanium (IV) complexes, 331-337
  kinetic data, 337
Transient effects, 69, 70
Transient nutation, 71
Transverse magnetization, 132
  with chemical exchange, 111
Transverse relaxation, 4-6, 84
1,1,2-Trichloroethane, T_2 measurements in, 136
2,4,5-Trichloronitrobenzene, T_2 measurements in, 137
Trifluoromethyl derivatives, 164
Trigonal twist, 283
Triiron dodecacarbonyl, 518, 520
Trimethylammonium ion, 624
Tri-o-thymotide, conformational barriers, 604
Triruthenium dodecacarbonyl, 501
Trisbidentate complexes, nonrigidity, 338-368
Tris (β-diketonate) complexes
  kinetic classification, 351, 356, 357
Tropolonate complexes, rearrangement kinetics and mechanism, 347, 350-358
Turnstile mechanism, 269
Twist angle, definition, 338
  in rearrangements, 358, 363, 367, 369, 373
Twist mechanism, 341, 344-361, 365-368
  permutational analysis, 344-347, 367
Two-site exchange, 9-13

U

Uranium (IV) complexes, 370

Unsymmetrical proton exchange, 641
  table of, 642
Ureas, barriers to rotation in, 210

V

Vanadium (III) complexes, 348, 350, 352, 354, 356
Version, 545
Virtual coupling, 308

W

Walden inversion of amines, 627
WBR (Wangsness-Block-Redfield) theory, 55
  master equation, 57

X

XML_4 species
  basic permutational sets, 271
  diastereotopic groups, 275
  double bond character, 272
  intramolecular rearrangement, 270 ff.
X_2ML_3 species
  intramolecular rearrangements, 276
  polarity rules, 277

Y

Yttrium (IV) complexes, 355, 369

Z

Zinc (II) β-aminothionate complexes, 325-328
  kinetic data, 326
Zinc (II) β-ketoaminate complexes, 325, 326
Zirconium (IV) complexes, 336, 338, 369
  cyclopentadienyl derivatives, 369-372
  kinetic data, 371